SIXTH EDITION 6

Technical Communication

Rebecca E. Burnett

IOWA STATE UNIVERSITY

>

THOMSON WADSWORTH

Australia • Canada • Mexico • Singapore
Spain • United Kingdom • United States

Technical Communucation, Sixth Edition
Rebecca E. Burnett

Publisher: *Michael Rosenberg*
Acquisitions Editor: *Dickson Musslewhite*
Development Editor: *Michell Phifer*
Production Project Manager: *Lianne Ames*
Marketing Manager: *Katrina Byrd*
Print Buyer: *Mary Beth Hennebury*
Compositor: *Graphic World*

Project Manager: *Beth Callaway*
Photography Manager: *Sheri Blaney*
Permissions Manager: *Karyn Morrison*
Photo Researcher: *Christina Micek*
Illustrator: *Graphic World*
Cover Designer/Text Designer: *Linda Beaupre*
Printer: *QuebecorWorld*

Cover Image: © Owaki - Kulla/CORBIS

Printed in the United States of America.
2 3 4 5 6 — 07 06 05

For more information contact Thomson Wadsworth, 25 Thomson Place, Boston, Massachusetts 02210 USA, or you can visit our Internet site at http://www.thomson.com

Credits appear on pages 825–826, which constitutes a continuation of the copyright page.

Library of Congress Control Number: 2004107037

ISBN: 1-4130-0189-0
(Student Edition)
ISBN: 1-4130-0660-4
(Instructor's Edition)

CONTENTS IN BRIEF

CONTENTS

15. Creating Technical Descriptions 547

16. Creating Process Explanations 579

PART V Preparing Professional Communication 605

17. Engaging in Oral Communication 607

18. Preparing Correspondence 645

PREFACE

Technical Communication, Sixth Edition, has a clear goal: to help students and workplace professionals communicate technical information — written, oral, and visual — to audiences in a variety of complex workplace contexts.

Technical Communication, Sixth Edition, presents straightforward explanations and guidelines based on current theory, research, and practice. The elegant, usable design of this edition enables students to use many of the text's pages, figures, and tables as models. The discussions and the classroom-tested exercises and assignments, both individual and collaborative, help users become better communicators. This highly readable and teachable edition stresses the integrated, recursive nature of producing effective print and electronic documents, encouraging students to think of invention and revision as ongoing processes, to think of visuals as ways to present information, and to think of language as having the power to shape and influence users' perceptions.

What's the approach? This edition emphasizes a rhetorical, problem-solving approach. Users of this text will learn to make decisions about rhetorical elements such as context, content, purpose, audience, organization, visuals, and design as they engage in the process of communicating technical information. In addition, they will learn the *reasons* behind their communication decisions.

Users of this text — both traditional and nontraditional students as well as workplace professionals — will learn that effective technical communication contains both creativity and craft, a point I make by the continued use of Leonardo da Vinci's technical drawings on all the chapters openers throughout the book. Like Leonardo da Vinci's technical drawings, technical communication should be precise, detailed, functional, and focused. That's not all it should be; it's the least it should be.

Creativity and craft play out in another way with this edition's new cover featuring a photograph of wind turbines. The image is not only aesthetically pleasing, but it also represents the leading edge of a centuries' old technology that captures many of the challenges that technical professionals and technical communicators face. Traditions exist, but to endure they must adapt to a dramatically changing world.

Throughout the text, the traditional concerns of technical communication — techniques such as definitions, descriptions, and processes; and forms such as correspondence, instructions, proposals, and reports — are always related to rhetorical elements. Beyond these concerns, the text continues to include detailed information about collaboration, ethics, visuals, and design and has significantly expanded its discussion of international communication, usability testing, and technology.

What's new in this edition? This sixth edition of *Technical Communication* maintains the strengths of the previous editions while incorporating important new information. The changes in this edition are drawn from current theory and research, leading-edge practices from the workplace, and helpful tools and strategies from practitioners. Twelve of the changes are particularly important:

- *Evaluation criteria.* Chapter 1 introduces easy-to-use evaluation criteria — *accessibility, comprehensibility,* and *usability* — which are used as a heuristic throughout the book to analyze and assess written, oral, and visual communication.
- *New examples.* New examples come from a variety of disciplines and professions. Many relate to global themes — scientific/technological innovation, medical advances/disease control, food production, and ecological/environmental balance — which demonstrate that scientific and technological concerns are global.
- *International emphasis.* The international workplace is emphasized with textual and visual examples from many countries and in many languages.
- *Special features.* Two-page spreads highlight interesting situations where technical documents, oral presentations, and visuals are created and used.
- *New chapter about culture.* Chapter 2, "Understanding Culture and the Workplace," addresses international and organizational cultures as well as characteristics that influence individuals in those cultures.
- *New chapter about usability.* Chapter 9, "Ensuring Usability," characterizes usability and usability testing, introduces guidelines for conducting various types of tests, and differentiates usability from accessibility.
- *New chapter about technology.* Throughout the new edition, you will find increased information about technology, including Chapter 13, "Designing Electronic Communication," which highlights critical features of new media, ranging from e-mail and PowerPoint® to animated Web sites.
- *More annotations.* More extensive marginal annotations for the sample documents include questions and comments to encourage critical evaluation.
- *New Ethics Sidebars.* Each chapter includes an Ethics Sidebar that focuses on a complex ethical issue considered by workplace professionals. Several of these sidebars are new to this edition.
- *WEBLINKS.* Numerous WEBLINKS in each chapter are accessible through the book's Companion Web site, which also has a wide array of relevant examples and articles for discussion.

- *Photographs throughout.* New photographs throughout the text illustrate workplaces where technical documents, oral presentations, and visuals are created and used.
- *Improved design.* The sophisticated, usable design features full-color pages that demonstrate ways in which color can be functional and contribute to interpreting and using technical information.

What critical concerns are addressed? While accuracy is arguably the most critical aspect of any technical document, oral presentation, or visual, technical accuracy is not enough. This text balances theory and research, pedagogy and practice, and classroom and workplace needs. The balance is demonstrated in ten critical concerns:

- *Rhetorical base.* Most problems communicating technical information occur in complex workplace environments. In addressing this challenge, *Technical Communication,* Sixth Edition, uses rhetorical elements — for example, the context of the communication; the constraints of the situation; needs of the audience(s); purposes of the writers, speakers, and designers; conventions of the genres; strategies for organizing and designing information — which influence the process of creating documents, oral presentations, and visuals.
- *Communicating with audiences, including international audiences.* The text provides early and ongoing in-depth coverage of audience analysis, focusing on practical suggestions for evaluating audiences and developing strategies for adjusting material to different audiences, including those in a variety of international and organizational cultures.
- *Visuals and document design.* The text emphasizes the rhetorical nature of visuals and establishes parallels between visual and verbal information. The text also explores the role of color and emphasizes the impact of information design, both in print and electronic documents. The design of this text models ways to make information accessible, comprehensible, and usable.
- *Collaboration.* The text not only presents a chapter about collaboration, but the end of each chapter includes collaborative assignments. Throughout the text, students are reminded that the workplace is a collaborative environment and that written, oral, and visual information is seldom produced or used in isolation. Students not only learn the importance of collaboration and teamwork, but they also learn how to be productive collaborators.
- *Testing.* Creating effective documents, oral presentations, and visuals should include text-based, expert-based, and user-based testing. The results of testing can be used as the basis for revising and editing.
- *Process and product.* This edition shows how technical professionals are involved in a complex process to create as well as interpret technical

documents, presentations, and visuals. The text discusses ways to approach communication problems; explains options available to writers, speakers, and designers; offers suggestions about logical organization; and illustrates appropriate language use.

- *Technology.* Because technical communication takes place in a rapidly changing electronic environment, this text discusses the impact of technology on creating and interpreting documents, oral presentations, and visuals. Users learn ways to make informed decisions about technology as well as ways that enable them to take advantage of the power of new media.
- *Examples.* Throughout the text, annotated examples from students and workplace professionals around the world illustrate the key points and serve as models.
- *Style. Technical Communication* is a reader-based text; it directly addresses students and workplace professionals in a straightforward manner that is both appealing and accessible.
- *Apparatus and computer support.* The extensive marginal annotations prompt critical thinking and offer opportunities to discuss ideas and apply the practices presented in *Technical Communication.* The Individual and Collaborative Assignments at the end of each chapter encourage planning and developing a wide array of documents, oral presentations, and visuals.

What support materials are available? This new edition of *Technical Communication* has a greatly expanded *Instructor's Manual* and a Book Companion Web Site.

- The *Instructor's Manual (IM)* is a valuable resource both for highly experienced and for new instructors. It offers strategies for teaching each chapter successfully, including the use of a new convenient heuristic: RADAR (Read, Act, Discuss, Assess, and Reflect). The *IM* also includes suggestions for planning and preparing course material, additional activities and assignments, strategies for managing different classroom formats (traditional, computer lab, and online), PowerPoint® presentations for introducing or reviewing concepts, and tools and strategies for assessing student work (written, oral, and visual; individual and collaborative; print and electronic). In addition, the *IM* includes essays by nationally recognized researchers and educators; written specifically to accompany *Technical Communication,* Sixth Edition, these essays discuss innovative approaches and practices being used in technical communication classrooms. Many of the articles are new to this edition's manual. The *IM* is available free to all adopters of this new edition of *Technical Communication.* (You may request a copy of the *IM* from your Thomson representative or download the PDF version from the instructor side of the Book Companion Web Site, available at www.english.wadsworth.com/burnett6e.)

- *Student side of the Book Companion Web Site.* The student side of the Web site includes for each chapter an overview, tools and tips, "Workplace Realities" (videotapes with workplace professionals), annotated interactive examples, a variety of classroom and Internet activities, key terms, Web links to chapter-related content, and a tutorial quiz. Additionally, the student side of the Web site includes many cases, new assignments, a complete handbook, and career resources. To access the Book Companion Web Site, go to **www.english.wadsworth.com/burnett6e**.
- *Instructor side of the Book Companion Web Site.* The instructor side of the Web site includes for each chapter an instructor overview, chapter teaching tips, PowerPoint® lecture slides, and answers to chapter quizzes as well as suggestions for integrating the additional examples and assignments into the course. The instructor side of the Web site also includes course management tools, sample syllabi, rubrics for assessing student work, and suggestions for using the Web site's cases and career resources. To access the Book Companion Web Site, go to **www.english.wadsworth.com/burnett6e**.

I have made changes in each edition based on the recommendations of colleagues from colleges, universities, agencies, businesses, and companies around the world. If you have suggestions about changes I should consider, please contact me. If you have examples that would work well for the next edition, please contact me as well. I value your feedback.

Rebecca E. Burnett

Rebecca E. Burnett, PhD
University Professor of Rhetoric & Professional Communication
Iowa State University
c/o Thomson Higher Education
25 Thomson Place
Boston, MA 02210

Acknowledgments

Technical Communication would not exist without the personal and professional support of family, friends, and colleagues.

Researchers. In preparing this edition, I have been thankful for the skillful and thorough researchers from Iowa State University who assisted me in preparing this revision: Brian Hentz, Katherine Miles, and Rebecca Pope-Ruark. They posed alternative approaches, updated information, located examples, suggested two-page spreads, provided valuable editing, and drafted new activities, glossary lists, PowerPoint presentations, and quizzes. Their assistance has been essential.

Co-author. Donna Kain, at East Carolina University, has been my trusted and expert co-author for Chapters 9 and 13, my co-author for the expanded and updated *Instructor's Manual,* and the creative and insightful developer of the content on the Book Companion Web Site. She has been invaluable in this revision.

Personal thanks. As always, Dorothea Burnett, Margaret Burnett, Christopher Burnett, and Paula Thompson provide me with the confidence and support to make the revision possible. William Jeffries continues in his unwavering support and serves as a voice of reason.

Friends and colleagues. I appreciate the support and contributions provided by friends and colleagues around the world: Philippa Benson, Mort Boyd, Shelly Boyd, Linda Driskill, Marcia Greenman Lebeau, Muriel McGrann, Susie Poague, Bev Sauer, Don Stanford, Judith Stanford, Anette van der Mescht, and Ben Xu. They have helped to shape my thinking as I went about this revision.

I want to thank my friends and colleagues at Iowa State University who see wisdom in balancing theory, research, workplace practice, and pedagogy. Their curiosity, dedication, and insight are an inspiration to me. My undergraduate and graduate students at Iowa State University and technical professionals in university seminars have also been important to this edition. I want to acknowledge the valuable discussions and contributions of various kinds provided by ISU friends and colleagues, most especially Gloria Betcher, Daniel Coffey, Dan Douglas, Laura Hannasch, Carl Herndl, Lee Honeycutt, Denny Howe, Robert Martin, Linta Meetz, Michael Mendelson, Neil Nakadate, Elizabeth Orcutt-Kroeger, Lorrie Pellack, Lee Poague, Diane Prince-Herndl, David Roberts, Caskey Russell, Geoff Sauer, Sarah Stambaugh, Lana Voga, and Loren Zachary. And in this revision, I also want to thank my friends and colleagues at Rice University who provided many opportunities that have enabled me to select useful new examples.

Contributors. I especially appreciate the colleagues and workplace professionals who contributed examples for this edition: Paul Boyd from the US Army Corps of Engineers; Maria Cochran, Marisa Corzanego, Mark Gleason, Oskana Hlyva, Ken Jolls, and Arvid Osterberg — all from Iowa State University; Sarah Helland from Pioneer; Zachary Lavicky from Emporia State University; Melissa Poague from University of Washington; Lee Tesdell from Minnesota State University–Mankato; and Tara Barrett Tarnowski from University of Georgia.

Previous editions. The contributions by the following people to the fifth edition have been substantially retained: Jill Bigley, Arricka Brouwer, Christopher Burnett, Larry Chan, David Clark, Irene Faass, Patty Harms, Elizabeth Herman, William Jeffries, Ken Jolls, Kari Krumpel, Muriel McGrann, Walden Miller, Kate Molitor, Matt Turner, Peggy Pollock, Janet Renze, Daryl Seay, Doug Schaapveld, Clay Spinuzzi, Don Stanford, Judith Stanford, Melissa Waltman, and Julie Zeleznik as well as friends, colleagues, and students at Iowa State University.

The contributions by the following people to the fourth edition have been substantially retained: Susan Booker, Kaelin Chappelle, David Clark, Andrea Breemer Frantz, Woody Hart, William Jeffries, Lee-Ann Kastman, Elenor Long, Muriel McGrann, Ron Myers, Tom Myers, Mike Peery, Clay Spinuzzi, Don Stanford, Judith Stanford, Gary Tarcy, Lee Tesdell, Christianna White, Dorothy Winchester, Mark Zachry, and Stephanie Zeluck as well as friends, colleagues, and students at Iowa State University.

The contributions by the following people to the third edition have been substantially retained: Reva Daniel, Michael Hassett, William Jeffries, Muriel McGrann, Cindy Myers, and Christianna White as well as colleagues and students at Iowa State University.

The contributions by the following people to the second edition have been substantially retained: Philippa Benson, William Jeffries, and Barbara Sitko as well as my friends, colleagues, and students at Carnegie Mellon University.

The contributions by the following people to the first edition have been substantially retained: Geraldine Branca, Christopher Burnett, Robert Carosso, Bernard DiNatale, Arline Dupras, Elizabeth Foster, Nancy Irish, Elizabeth Carros Keroack, Marcia Greenman Lebeau, Muriel McGrann, Stephen Meidell, Leon Sommers, and Judith Dupras Stanford as well as students at Northern Essex Community College, Merrimack College, and the University of Massachusetts at Lowell.

Reviewers. I appreciate the helpful feedback from colleagues around the country who suggested revisions, sometimes in a quick e-mail message, sometimes in a short conversation at a conference. The official reviewers' practical and often insightful suggestions were, of course, instrumental in revisions for this edition. Detailed reviews were provided by

Scott Chadwick	*Creighton University*
Dave Clark	*University of Wisconsin–Milwaukee*
Lee Ann Kastman Breuch	*University of Minnesota*
Cezar Ornatowski	*San Diego State University*
Penny Sansbury	*Florence-Darlington Technical College*
Geoffrey Sauer	*Iowa State University*
Karen Schnackenberg	*Carnegie Mellon University*
Stuart Selber	*Pennsylvania State University*
Clay Spinuzzi	*University of Texas–Austin*
Cindy Raisor	*Texas A&M University*

The Thomson Wadsworth team for *Technical Communication* has been extraordinary. Dickson Musslewhite, my trusted and supportive Acquisitions Editor, provided excellent editorial direction. Michell Phifer, Senior Development Editor, kept me on track in this complex revision with intelligence, patience, expertise, and humor — for which I am immensely grateful. Linda Beaupre redesigned the book, creating a design that is elegant and usable (a remarkable and wonderful combination); her design reflects the spirit of the text. Joe Gallagher and Cara Douglas-Graff supported the development of the Web site for the book. Janet McCartney copyedited the manuscript and the endnotes, saving me from embarrassing errors. Sally Cogliano, Senior Production Editor, began the production process with efficiency and good humor. Lianne Ames, Senior Production Editor, saw the book through the production process with remarkable energy, insight, flexibility, patience, and grace. Christina Micek was invaluable as the photo researcher for this edition. Karyn Morrison handled the complex task of permissions with thoroughness. Joy Westberg wrote wonderful marketing materials. Beth Callaway from Graphic World supervised and coordinated the composition process with care and deliberation. For all, I am immensely thankful.

PART I

Communicating in the Workplace

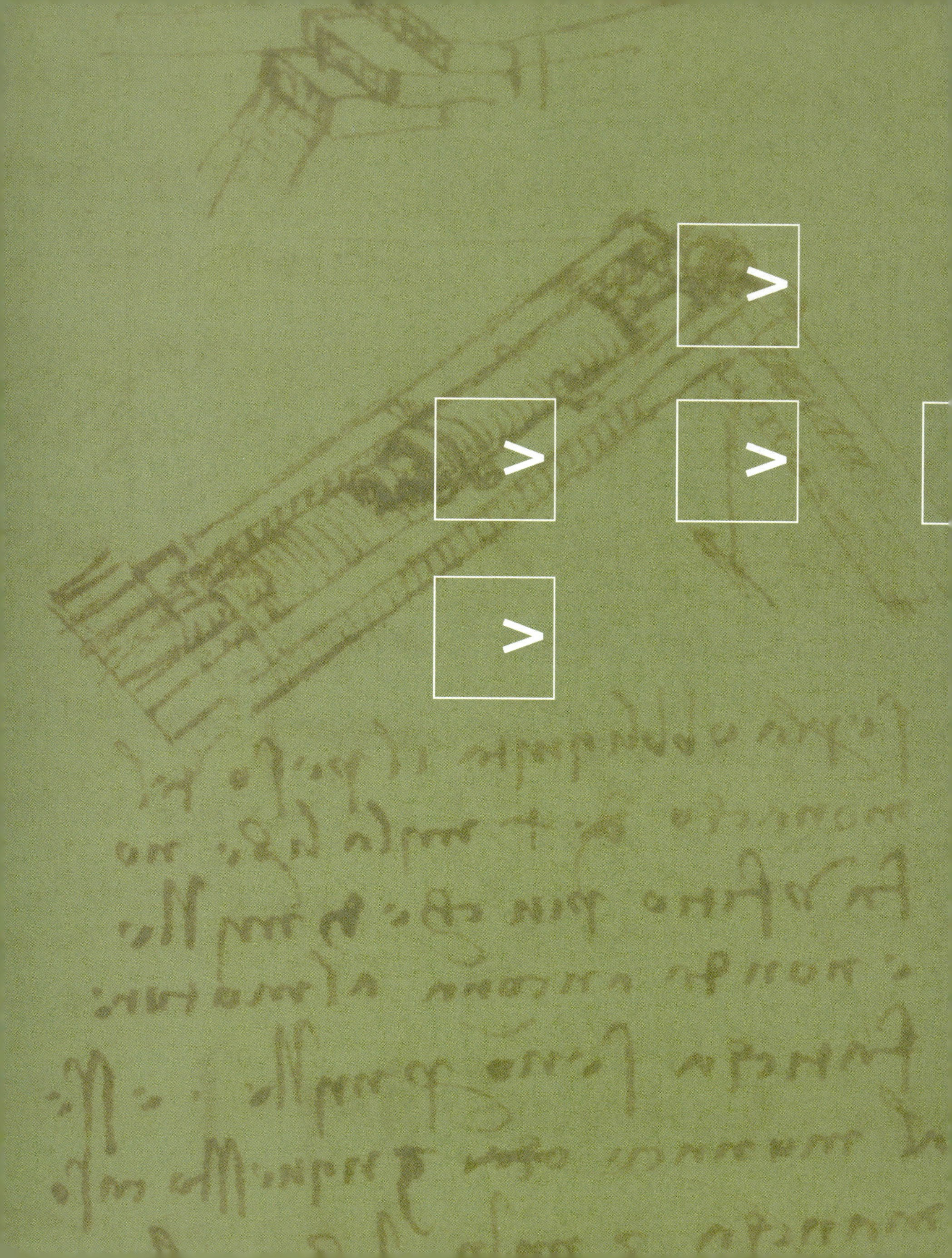

CHAPTER 1

Characterizing Workplace Communication

Objectives and Outcomes

This chapter will help you accomplish these outcomes:

- Define technical communication, its criticality in the workplace, and its relation to job success
- Understand the role of genre, technology, and ethics in technical communication
- Describe rhetorical elements that experienced communicators consider, including context, purpose, audience, organization, and document design
- Identify factors that contribute to accessibility, comprehensibility, and usability
- Identify constraints that affect workplace communication

What do astrophysicists, obstetricians, electrical engineers, ecologists, farmers, musicians, and veterinarians have in common? All create and interpret technical documents, oral presentations, and technical visuals. What do the moons of Jupiter, in vitro fertilization, silicon chips, wetlands conservation, soybean crops, flutes, and foals have in common? All are subjects of technical communication, a broad field that touches nearly every profession because it connects ideas, people, and their activities. Technical communication defines, describes, and directs activities in business and industry, government and research institutions, hospitals and farms.

Although certain professions such as engineering have traditionally been associated with technical communication, virtually all disciplines and professions have technical documents, oral presentations, and visuals. For example, detailed information about sound formation is important for speech pathologists, computer engineers, and human factors experts designing voice-activated computer systems. Knowledge about muscle conditioning is equally relevant to physical therapists, ballet dancers, and veterinarians. Data about weather changes are crucial to meteorologists, airline pilots, and commercial fishers.

Importance of Effective Communication

Two broad categories of professionals are responsible for virtually all the technical documents, oral presentations, and visuals in any organization: technical communicators and technical experts. Technical communicators have evolved as specialists whose primary responsibility is to design, develop, and produce a wide range of documents, oral presentations, and visuals. As essential members of project teams, technical communicators support the products and services of an organization. Equally important are technical experts who communicate as a regular part of their job. These scientists, engineers, technicians, and managers regularly plan and prepare a wide range of technical communication.

> *How much do you anticipate communicating when you're in the workplace? Examine your profession's journals, trade magazines, and Web sites to see if you can determine the role of reading, writing, and speaking in your field.*

Professionals who communicate effectively, whether technical communicators or technical experts, usually achieve more career success and have greater job satisfaction than those without the skills to communicate their technical knowledge. Successful professionals see reading and writing, listening and speaking, and viewing and designing visuals as integral parts of their job, not as something extra that they have to fit in. They assume responsibility for their own communication — from organizing ideas to final proofreading of documents.

For more than 20 years, surveys have touted the importance of communication skills in the workplace: More than 90 percent of technical professionals have reported that speaking and writing skills are important to their success. They have also reported that the amount of time they spend writing increases as their responsibilities increase: Nearly half spend up to 40 percent of their time writing and more than one-quarter spend between 40 percent and 100 percent of their time writing.[1]

Results from a variety of recent surveys, four of which are summarized here, support these long-accepted views and illustrate the importance that successful professionals place on effective communication:

- A survey by Robert Half International of 1,400 chief financial officers identified interpersonal skills such as communication and listening as critical for professional success.[2]
- A survey by the American Management Association of nearly 300 U.S. administrators identified written and oral communication skills as the highest ranked performance skills for professional success.[3]
- A survey by American Express of nearly 800 small business owners reported that 86 percent believe oral communication skills are very important, 77 percent believe that interpersonal skills are very important, and 60 percent believe that written skills are very important.[4]
- A survey of more than 200 Canadian technical professionals who are members of L'association canadienne de l'informatique (Canadian Information Processing Society) and a survey of more than 100 technology-based Canadian organizations identified critical job skills. Technical professionals identified top skills as the ability to conduct analyses and make recommendations; use project management tools; and write, speak, and listen well. Employers identified top skills as project management, communication skills, teamwork, and leadership.[5]

These representative surveys show that you will gain tremendous professional advantages if you write and speak well, work on teams and manage projects effectively, and listen carefully. Your communication needs to be technically complete and accurate, logically organized for the audience, visually appealing, and interesting; it also needs to be mechanically and grammatically conventional, and it must say something worthwhile.

As customer satisfaction with products and services becomes more important, companies increase the attention they give to paper and electronic documents, which are now seen as essential parts of the product by most companies. Since operations increasingly depend on rapid, accurate communication, companies expect all employees to be good communicators. In fact, a poll by the National Association of Manufacturers stated that poor reading or writing skills prevent 32 percent of entry-level applicants from being hired; poor oral skills prevent 18 percent from being hired.[6]

Defining Technical Communication

Although technical communication has existed as long as people have recorded information, technical communication as a profession evolved exponentially during the second half of the twentieth century. Recent definitions of technical communication consider it rhetorical.

Why the term *rhetorical? Rhetoric,* the art and craft of communicating appropriately and persuasively, is concerned with the ways in which written, oral, and visual information is planned, conveyed, and interpreted in particular contexts, for particular audiences, for particular purposes. Technical communication is rhetorical because it is the art and craft of communicating technical information appropriately and persuasively to intended audiences, in complex contexts, for particular purposes. The rhetorical elements identified in Figure 1.1, considered collectively, provide a broad and useful way to characterize technical communication. These elements apply to all modes of technical communication, whether paper or electronic texts, oral presentations, or visuals.

Can you find an example of a technical document that fulfills most of the characteristics in Figure 1.1?

FIGURE 1.1 **Rhetorical Elements Characterizing Technical Communication**

Rhetorical Elements	Effective Technical Documents, Oral Presentations, and Visuals
Content	■ Present accurate, appropriate technical information adjusted to the audience(s) ■ Provide appropriate source citations and documentation as necessary
Context	■ Respond to the organizational situation ■ Fulfill the identified task ■ Are revised frequently for currency
Purpose(s)	■ Inform and persuade the intended audience(s) ■ Identify the position being taken
Audience(s)	■ Address identified audience(s) — readers, listeners, or viewers — who often have different needs and constraints ■ Recognize that multiple interpretations of documents, oral presentations, and visuals occur
Organization	■ Organize information so that it is logical, accessible, and retrievable, and so that it is easy to comprehend, navigate, and recall
Visuals	■ Convey content through various kinds of visuals that aid audience understanding and decision making
Document design	■ Design information so that it is accessible, comprehensible, and usable
Usability	■ Provide functional and usable information for the audience(s)
Language conventions	■ Provide straightforward information that differentiates opinions from verifiable information ■ Use clear and direct language without unnecessary complexity; often use short- to medium-length sentences and subject-verb-object word order; provide simple but stylistically varied information

These rhetorical elements also are important because they are the very factors that experienced communicators typically consider when planning, drafting, and revising documents, oral presentations, and visuals. For example, experienced communicators consider far more than the content. They ask themselves about the context in which they're working and in which the information will be used. They identify the purpose and audience and then determine ways to adapt, organize, and support the information appropriately. They design effective visuals and create an appealing, usable design. And they conform to the conventions of their organization and their profession. However, they can take an effective shortcut and use these evaluation criteria: Is the information *accessible, comprehensible,* and *usable*? The following scenario about Jon Baliene, an experienced manufacturing supervisor and skillful communicator, illustrates these rhetorical elements and evaluation criteria in a typical workplace situation.

As a reader, listener, or viewer, what criteria do you use to determine if a document, oral presentation, or visual you use is effective? As a writer, speaker, or designer, what criteria do you typically use to determine if a document, oral presentation, or visual you prepare is effective?

Division manager Sandy Schaeffer asks her manufacturing supervisor, Jon Baliene, to recommend solutions to production problems in his department. During his initial planning, Jon considers the immediate situation as well as expectations and constraints that could influence what he might say and how Sandy might interpret it.

Context
Jon thinks about his company — the corporate culture as well as the particular circumstance that he's responding to.

Context
He understands the specific task he is asked to do and thinks about what's involved in completing it.

As Jon plans his recommendation report, he identifies his purposes: provide and analyze verifiable information; recommend credible solutions that persuade Sandy to accept his analysis of the situation and, thus, his recommendations; and submit a report by the deadline.

Purposes
1. *Provide information*
2. *Offer persuasive recommendations*
3. *Meet deadline*

Because Jon selects his information in response to Sandy's request, he knows he cannot use a boilerplate report format and simply fill in the blanks; the dynamic nature of the situation requires that he adapt the form and content to meet his manager's specific needs and expectations. And he knows he'll submit the report as an electronic attachment that will probably be read on-screen (though it could also be printed), so he chooses a font appropriate for on-screen reading and plans to hyperlink the section headings in the table of contents to the headings in the report.

Context
Jon selects information appropriate to the task.

Audience
Jon adapts material to Sandy, the decision maker.

Document Degisn
He adapts the design for the place the report will most likely be read — on a computer screen.

Accessibility
He ensures that Sandy will be able to easily access the information in an electronic format.

Because the company has a strong corporate identity program for internal and external communication, Jon knows he needs to use the report

Document Design
Jon uses a standard report format appropriate for the workplace culture of his company.

conventions detailed in the company's style guide, easily available on the company's intranet.

Jon organizes an audience-based report — that is, one geared specifically to Sandy's needs, experience, and capabilities. Because her background is in business, Jon adjusts the technical material by adding explanations of highly specialized information. In addition, he includes an appendix with calculations and specifications appropriate for secondary readers who have more technical experience than his manager.

Beyond identifying and analyzing the problems, Jon includes recommendations, which, he emphasizes, are based on current projections and costs; in six months, he would have to write a different report. He hopes the background information along with his carefully selected support and persuasive arguments convince Sandy to accept his recommendations. However, even though Jon has worked hard to prepare an effective report, he can't guarantee the way that Sandy Schaeffer will interpret the information and his argument. Jon checks the text as well as the visuals to make sure nothing seems ambiguous or confusing.

Because Jon knows that Sandy likes well-designed tables and graphs to support a narrative, he organizes numerical data into a table and creates a graph that shows the trends he has identified from the tabular data. He places these visuals in the text immediately following his discussion of the key points.

Despite Jon's enthusiasm for his recommendations, he explicitly explains the problems and proposes his recommended solution. He checks that all the headings signal the following text correctly and double-checks that the hyperlinks in the table of contents take readers to the appropriate sections.

Jon realizes that his primary reader (Sandy Schaeffer) and secondary readers (others in management and manufacturing) are intelligent, well educated, and well informed; however, they have little time to read an unnecessarily complex report. He checks that his text and visuals are understandable and that his recommendations are easy to find and well supported in a clear timeline for implementation.

Comprehensibility
He ensures that Sandy will be able to understand information needed for decisions.

Audience
Jon consider more than the primary reader.

Organization
Jon uses a conventional sequence of information to present and argue for his recommendations.

Audience
Jon recognizes the report is open to interpretation.

Visuals
Jon conveys data in a table and graph that are integrated into the text.

Language Conventions
He uses clear, direct, and accessible style.

Content
Jon presents accurate technical content adjusted to the decision maker.

Content
He provides current information that will become rapidly outdated.

Usability
He ensures that readers will be able to locate information needed for decision making.

Document Design
Jon provides headings and subheadings to signal main sections and ease reading.

Accessibility, Comprehensibility, and Usability
He ensures that all intended readers will be able to access, understand, and use the information.

Meaning does not reside in any of the documents, presentations, or visuals that you create or use. Instead, meaning is constructed from your interpretation of the information. Each individual's construction of meaning helps account for intelligent, well-informed readers, listeners, and viewers sometimes having different interpretations of the same information. Attitudes, education, experience, and contexts shape interpretations. Understanding some of the factors that affect people's interpretations and uses of technical information will help you become a more skillful communicator.

Genre in Technical Communication

As the situation with Jon shows, technical information does not occur in isolation. A document, oral presentation, or visual *and* the rhetorical situation in which it is created and used combine to constitute the *genre.* Researchers Carol Berkenkotter and Thomas Huckin explain that genres (information and situations) are *dynamic:* they change synergistically in response to particular circumstances — that is, each affects the other. They are also situated in a particular community's workplace tasks or activities.[7]

In these workplace communities, technical professionals use and reproduce genres as part of their regular work; that is, they regularly prepare information and modify it for particular situations and purposes. This book as a whole is about genres in technical communication — the ways documents, oral presentations, and visuals are created and used for particular purposes, in particular situations and cultures, at particular points in time.

Like Jon, you can have the goal of creating audience-based documents, oral presentations, and visuals that consider audiences' needs and reactions and that organize information for audiences' understanding. Creating audience-based materials requires awareness of the rhetorical situation in which they are generated and used. Audiences are, of course, influenced by the actual words or images, but these words and images may carry different meanings for each member of the audience. And, whether written, spoken or visually presented, they are only part of what contribute to understanding. Audience comprehension of and response to information are influenced by a range of factors, including attitudes, cultural perspectives, values, education, job function, political position, personality, and experiences.

What other factors can you think of that could affect the way members of an audience respond to documents, oral presentations, and visuals? What might affect their comprehension? What might affect their recall of information?

Because meaning is constructed by your audiences, you need to be especially careful to select words and visuals and then organize information so the meaning your audience constructs will be similar to the one you intend. No document, oral presentation, or visual fully conveys your intended meaning, nor does any reader, listener, or viewer ever understand everything you intend. As a communicator, you both add and omit things that were part of the original rhetorical situation. As a reader, listener, or viewer, your interpretation is affected by what linguists call *exuberance* (that is, your interpretation always adds something not intended) and *deficiency* (that is, your interpretation always ignores things you could have noticed).

Communities

As the brief discussion about genre shows, one of the important characteristics of workplace professionals is that they belong to communities that influence their understanding and their activities. Although people belong to multiple communities simultaneously, two types of community are particularly important to think about in a work context.

Workplace professionals belong to various *discourse communities,* identifiable groups with a common, often specialized, language. They can be members of discourse communities without working together or even seeing each other. For example, optical engineers, organic chemists, horticulturalists, and athletic trainers each have a specialized and shared vocabulary, whether or not they know each other personally. Thus, an optical engineer in Tuscaloosa can understand an article written by an optical engineer in Toronto, even if they are strangers. Their discipline and profession share concepts, practices, and vocabulary.

> *Do you think communities of practice are formed as permanent groups, or do you think that they go through stages of development — coalescing into a community, actively working together, and eventually dispersing?*

Workplace professionals also belong to various *communities of practice.* Etienne Wenger, internationally recognized researcher and consultant, defines a community of practice as a group of people who have a joint enterprise, mutual engagement, and a shared repertoire of resources.[8] With his collaborator, Jean Lave, Wenger explains that people in a community of practice establish a relationship that over time enables them to work together doing things that matter to them. They not only share concepts and language, but they also share commitment and resources. They develop a history — the way they do things in their community of practice. Their common experiences bind them together and enable them to undertake even more challenging activities.[9]

As a workplace professional, you will belong to several discourse communities and communities of practice. You will also regularly communicate with people outside your own discourse communities and communities of practice, so you need to consider what will make information accessible, comprehensible, and usable to those in your own communities as well as those outside your communities.

WEBLINK

To learn more about communities of practice in the workplace, go to www.english.wadsworth.com/burnett6e for a link to a more detailed article.
CLICK ON WEBLINK
CLICK on Chapter 1/communities of practice

Technology in Technical Communication

Interpretation is not only influenced by genres and communities, it is also influenced by the technology you use. What is broadly called *computer-mediated communication* (CMC) is a process of human communication via computers. CMC provides a way for people to communicate, interact with, retrieve, and interpret information in a variety of contexts and to shape communication for a variety of purposes. Media influences the shape of the information and its interpretation. For example, print and oral information are necessarily linear — that is, one sentence follows the next sentence, one paragraph follows the next paragraph. In contrast, electronic and visual information are not necessarily linear; for example, hypertext allows users to sequence information in multiple ways.

How does technology affect people's interaction with and interpretation of information? In general, it is based on their familiarity and comfort with the technology. More specifically, technology influences several factors:

- Sustaining reading of lengthy text or reading for extended periods
- Keeping track of your place in the text
- Managing and maintaining multiple, active on-screen windows
- Locating and reviewing difficult or confusing information
- Taking notes, highlighting relevant text, adding personal comments and questions
- Checking other places in the text and returning to your original place

Technology further influences what you say and how you say it, affecting privacy, immediacy, and permanency.

- *Privacy.* Virtually all electronic communication in the workplace can be monitored. It is simply not private — ever. Knowing that someone other than the intended audience might read your e-mail should promote some caution in what you choose to write and send. For example, e-mail messages and online chats are not necessarily protected by privacy laws. As much as you may object, some legal experts argue that "the writer of an e-mail message is implicitly consenting to its recording."[10]

To learn more about the lack of privacy in electronic communication — especially e-mail and online chats — go to **www.english.wadsworth.com/burnett6e** for a link to an article about cyberlaw.

CLICK ON WEBLINK

CLICK on Chapter 1/cyberlaw

WEBLINK

- *Immediacy.* You can decide when to communicate by participating in synchronous or asynchronous communication. *Synchronous communication* is concurrent or simultaneous (real-time) digital communication such as computer conferencing, chat rooms, or white board environments. Such environments often promote comments that are not carefully thought through. *Asynchronous communication* is digital communication that takes place independently in time; accessing and responding to communication exchanges such as voice mail, e-mail, or online newsgroups can be delayed by minutes, hours, or days. While asynchronous communication allows more time for reflection, the informal nature often promotes casual, hurried responses.

> *What factors besides privacy, immediacy, and permanency are influenced by computer-mediated communication?*

- *Permanency.* Information on a computer — especially on corporate and institutional servers — is frequently backed up and archived. You can record and then delete information on your computer; however, that information can often still be retrieved. Just because you can no longer see or access a file does not mean that no one else can access it. Furthermore, information you think of as yours (such as e-mail) resides on a server and can be accessed, even if you have deleted that information on your own computer. In fact, some companies' primary business is retrieving supposedly inaccessible information.

Ethics in Technical Communication

Preparing documents, oral presentations, and visuals that are accessible, comprehensible, and usable is not enough for workplace professionals. Certainly, to meet the needs of your audience, you must initially decide about your purposes to inform and persuade. Beyond these considerations, you must also respond to the context and culture, define and focus the content, analyze the task and audience, organize the information, and design the specific document. But you still need more.

To be an effective communicator, you must also consider factors that influence you and your audience as you and they construct meaning. What's one of the most important factors that influences everyone involved in writing, reading, and responding to technical documents? Ethics. In this textbook, ethics are a focus of every chapter, presented prominently in sidebars.

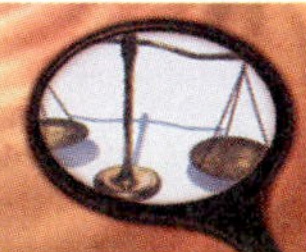

ETHICS SIDEBAR

Public vs. Private: Ethics and the Technical Professional

Imagine the following workplace situation:

You are working on your first major report, a technical description of the company's latest product. You know your technical description should highlight the product's state-of-the-art features. Those features are a key selling point. However, some field tests indicate those features are unreliable under adverse weather conditions—conditions common in the intended market for this product. Other tests, though, are planned. You are faced with a decision: Do you include the current test results in your technical description? You know what your company would prefer, but you feel an obligation to let customers know about the possibility of problems. The description is due by the end of the day. What do you do?

How can you decide what information to include in a document and what information to exclude?

Writing situations like this one require technical professionals to make ethical decisions. Ethics determine what we are willing to communicate and how we are willing to communicate. Ethics come from many different perspectives: personal beliefs, professional guidelines, organizational practices, cultural expectations, and legal or judicial requirements.

Communication researcher Cezar Ornatowski[11] believes ethical dilemmas for technical professionals occur because of two "incompatible" goals: serving the interests of employers while attempting to write technical documents that are "objective, plain, factual." Due to these conflicting pressures, technical professionals are at increased risk for ethical violations — either producing a document that does not meet the employer's goals or producing a document that may not be verifiable.

Dealing with these ethical dilemmas is not always easy: Sometimes, technical professionals must make difficult decisions. However, many ethical dilemmas can be addressed by knowing that these dilemmas exist and understanding various ways to resolve them. Even if you don't find ethical violations that challenge your personal beliefs, you should be aware of ways they might affect your standing among your colleagues and in your profession. Ethical violations affect a document's credibility, which, in turn, affects how you will be professionally perceived.

Reread the scenario at the beginning of this sidebar. Upon consideration, how would you deal with this conflict? Would you include the questionable test results? Would excluding them be an ethical violation?

Accessibility, Comprehensibility, and Usability

Regardless of the genre, communities, technology, or ethics involved, a document, oral presentation, or visual must meet three essential criteria, ones that are central to this book. Each document, presentation, or visual must meet these criteria:

1. Be physically accessible, so a reader, listener, or viewer can see or hear it
2. Be comprehensible, so a reader, listener, or viewer can understand it
3. Be usable, so a reader, listener, or viewer can use it easily and productively

If you take one thing from this introductory chapter, remember that effective communicators need to make information accessible, understandable, and usable. Figure 1.2 shows how these three important criteria are affected by rhetorical elements that experienced communicators consider. Each cell in Figure 1.2 identifies

FIGURE 1.2 **Factors Affecting Accessibility, Comprehensibility, and Usability**

Although the table identifies the factors that affect accessibility, comprehensibility, and usability of information, it does not explicitly convey the interactive nature of writing/reading, speaking/listening, designing/viewing. These factors all presume that effective communication — whether written, oral, or visual — is accurate, purposeful, appropriate to the context, adapted to the audience, organized, well designed, and responsive to the conventions of its medium.

		ACCESSIBILITY How easily can the audience see, hear, and/or touch the information?	COMPREHENSIBILITY How easily can the audience understand the information?	USABILITY How easily can the audience use the information?
Audience		**What physical factors of the audience affect their ability to access information?** Visual acuity? Aural acuity? Tactility? Coordination? Cognitive ability?	**What social and cultural factors affect the audience's understanding of information?** Language? Prior knowledge? Cultural presumptions? Reaction to gestures? Genre knowledge?	**What pragmatic factors affect the audience's use of the information?** User representation of task? User role? User attitude and experience?
Context		**What physical factors in the environment affect the audience's ability to access information?** Technology? Lighting? Temperature? Noise? Distractions? Interruptions?	**What situational factors affect the audience's understanding of information?** Stated or implied purpose of information? Perceived value of information?	**What situational factors affect the audience's use of the information?** Benefits to user? Sufficiency of time and/or support? Division of labor? Organizational structure?
Information	Text	**What physical characteristics of text affect the accessibility of information?** Headings? Typography? White space? Resolution?	**How does the text affect the audience's understanding of information?** Terminology? Organization? Coherence?	**How does the text affect the audience's use of the information?** Navigability? Credibility of sources?
	Oral	**What physical characteristics of an oral presentation affect the accessibility of information?** Volume? Clarity? Articulation? Pace? Gesticulation? Speaker movement?	**How does delivery of an oral presentation affect the audience's understanding of information?** Arrangement? Coherence? Dialect/accent? Modulation?	**How does the delivery of the oral presentation affect the usability of the information?** Repetition and emphasis? Forecasting? Signposting?
	Visuals	**What physical characteristics of visuals affect the accessibility of information?** Image quality? White space? Paper type/quality? Color? Screen size/resolution? Figure-ground contrast?	**How do the visuals affect the audience's understanding of information?** Recognizable images? Scale? Coherence? Nature of argument?	**How do the visuals affect the audience's use of the information?** Sequencing of images? Coordination with text?

a key question that you can ask yourself and minimal factors you need to consider when planning or evaluating technical communication. Throughout the book, you'll learn more about these factors and others that you need to consider.

What additional factors could you add to any of the cells?

In applying the factors identified in Figure 1.2, you can draw on the work of Herbert Paul Grice, a philosopher of language who taught at Oxford University and at the University of California, Berkeley.[12] Although Grice's work dealt specifically with conversation, it also applies to written communication.[13] Grice's overarching principle suggests that your communication should be timely and purposeful, which can happen if you follow four maxims:

- ***Quality.*** What you say should be accurate and verifiable.
- ***Quantity.*** What you say should be as informative as necessary — not too much information or too little information.
- ***Relevance.*** What you say should be relevant.
- ***Manner.*** What you say should be "perspicuous"; thus, you need to avoid obscurity and ambiguity and also be brief and well organized.

Can you think of situations in which H. P. Grice's cooperative principle and/or the four maxims are contradictory if you try to apply them at the same time?

Communication in the Workplace

If you identified every workplace situation that requires some type of technical document, oral presentation, or visual, the list would extend for pages, both because so many types exist and because the terminology is not standardized. Several easy ways exist to differentiate technical documents, presentations, and visuals: by the workplace situation necessitating them, by genre, by the audience, or by the purpose.

What are the most common types of technical documents, presentations, and visuals in your career area? When are they used? What are their purposes? Who are their audiences? How much does it matter that you be able to create and use them?

In the next section, you will read excerpts from five common types of workplace documents. They are similar to the ones you are likely to prepare and use: a technical brochure, a PowerPoint presentation, a technical report, a Web site, and a set of instructions. The examples are annotated to identify ways in which each is accessible, comprehensible, and usable. As you read the examples and their annotations, note that they incorporate elements that you can use in your own communication:

- Preview what's to come.
- Define critical terms.
- Use headings to call attention to key points.
- Select details appropriate for the audience's level of understanding.
- Use a design that contributes to accessibility, comprehensibility, and usability.
- Use an accessible font appropriate for print or on-screen use.
- Select typographic devices (such as bullets, italics, and boldfacing) to call attention to information.
- Use visuals (such as a table) to reinforce, illustrate, or explain the text.

One common function of technical documents, oral presentations, and visuals is providing information, sometimes for background, sometimes for decision making. What genres provide information? Sales, marketing, and promotional materials sent to customers. Product and process specifications sent to purchasing agents. Brochures for patients in medical and dental offices.

Figure 1.3 presents an excerpt from ***Allergy Relief Guide,*** a booklet for people suffering from respiratory problems. The booklet has been prepared by the American Academy of Allergy and Immunology, a group of physicians "who specialize in the diagnosis and management of asthma and allergy disorders of adults and children." The introduction explains that information in the booklet helps readers know more about their own health and, thus, take better care of themselves and their families.

This excerpt is from a section that begins with a simple definition of asthma (an allergy that occurs in the lungs) and its triggers: allergens (such as pollen, mold), viral infections, irritants (such as tobacco smoke, air pollutants), distance running, sensitivity to drugs, and emotional anxiety. Thus, the information in this excerpt is presented after readers have some general background.

What other genre (information and situations) can help with background and decision making?

Following this excerpt, the booklet includes a two-column list of "industrial materials known to cause occupational asthma." For example, flour can cause occupational asthma for bakers as well as for farmers and grain handlers. Similarly, phenylglycine acid chloride and sulphone chloramides can cause occupational asthma for workers in the pharmaceutical industry.

Accessibility

- *Sufficiently large serif font is easy to read.*
- *Related information is chunked into short paragraphs.*
- *Individual topics are separated by white space.*
- *Bulleted items draw attention and make reading easier.*

Comprehensibility

- *Questions as headings make key points easy to understand.*
- *Sequence of information is logical: what before how.*
- *Limited use of technical information; necessary terms are defined and explained.*

Usability

- *Bold headings and italics make text easy to scan to locate information.*
- *The question-and-answer format makes information usable because it is easy to locate.*
- *Each cause is discussed in a separate paragraph.*
- *Examples have sufficient details for readers to relate to their own experiences.*
- *Cause-effect explanations help readers understand symptoms.*

OCCUPATIONAL ASTHMA

What is Occupational Asthma?

Occupational asthma is generally defined as a respiratory disorder directly related to inhaling fumes, gases, or dust while on the job. Symptoms include wheezing, chest tightness, coughing, and may also include runny nose, nasal congestion, and eye irritation. The cause may be allergic or nonallergic in nature. The disease may persist for lengthy periods in some workers even if they are no longer exposed. Many workers with symptoms have been incorrectly diagnosed as having bronchitis.

It's important to remember that persons living in residential areas near factories are also exposed to fumes and may suffer symptoms as well.

AP Photo/John Moore

Covered in flour, an employee working for a flour wholesaler unloads a shipment in Haiti.

In many cases, a previous family history of allergy will make a person more likely to suffer from occupational asthma. Yet many individuals who have no such history still develop this disease. Studies show that the length of exposure varies and can range from 4 to 36 months before symptoms occur.

How Prevalent is Occupational Asthma?

The exact prevalence of occupational asthma is not known. Researchers have found, however, that at least 15% of all male cases of asthma in Japan result from exposure to industrial vapors, dust, gases, or fumes. In the U.S., 5–10% of all cases of asthma are thought to have job-related origins.

What are the Causes of Occupational Asthma?

Occupational asthma may be caused by one of three mechanisms: irritants, allergic, or environmental factors:

- Examples of ***irritants*** that provoke cases of occupational asthma include exposure to hydrochloric acid, sulfur dioxide, or ammonia found in the petroleum or chemical industries. People who may already have asthma or some other respiratory disorder are particularly affected when exposure occurs. In instances where irritants are responsible for causing the asthma, allergic mechanisms are not actively involved. (An allergic mechanism refers to the body's immune system responding adversely to an offending substance.)
- ***Allergic factors*** do play a role, however, in the following instances: exposures to the enzymes of Bacillus subtilis in the washing powder industry, exposure to castor and green coffee beans, and contact with papain in the food processing industry. Other examples include complex salts of platinum in metal refining and other agents such as ethylene diamine, phthalic anhydride, toluene diisocyanate (TDI), and trimellitic anhydride (TMA) in the plastics, rubber, and resin industries.
- ***Environmental factors*** include the inhalation of dust or liquid extracts of dust from cotton, flax, or hemp. In these cases, tissue cells in the lung directly release chemicals, such as histamine, which can cause bronchial smooth muscle to contract and, thus, block airways. Yet this reaction can occur in subjects who are not allergic to any of these substances. The exact role that these factors play in producing this drug-like effect is not yet completely known.

FIGURE 1.3 **Excerpt from an Extended Definition from a Medical Booklet for Nonexperts**[14]

Technical communicators use a variety of strategies to identify problems before they propose feasible solutions. The following example shows five slides from a PowerPoint presentation in which a student team argued that dependency on external energy sources is environmentally and economically unsound for their state. They argued instead that their state has the resources to become largely energy self-sufficient using wind turbines. As part of a class project, they formed a fictitious company that sought funding for an alternative energy project. The five-slide excerpt in Figure 1.4 is from their PowerPoint presentation to persuade an agency to provide funds for research about and expansion of the wind turbines in their state. The light tan bubbles present excerpts from their oral comments. During your classes as well as during your career, you'll need to create effective PowerPoint presentations for a variety of purposes and audiences.

ISU Wind Energy Project

Windsock Incorporated

Clean Energy for Our Environment and Economy

Jane Smith
Eugene Hollaner
Kerry Johansson

FIGURE 1.4A **Title Slide**[15]

Accessibility

- *The sans serif font, Trebuchet MS, is used on all the slides. It is a font especially designed for easy screen readability.*
- *The font of the title is large enough — 80 point — to visually signal it as the title.*
- *The header that appears on all the slides provides a visually coherent reminder of the topic.*

Comprehensibility

- *The slide identifies the topic, title, goal, and presenters*

"To explore the benefits of wind turbine production as an alternative energy source, Windsock Incorporated looks forward to showing you our proposal for creating viable, cost-efficient turbines over the next five to seven years. Our presentation explains the impact this clean energy can have on our environment and economy."

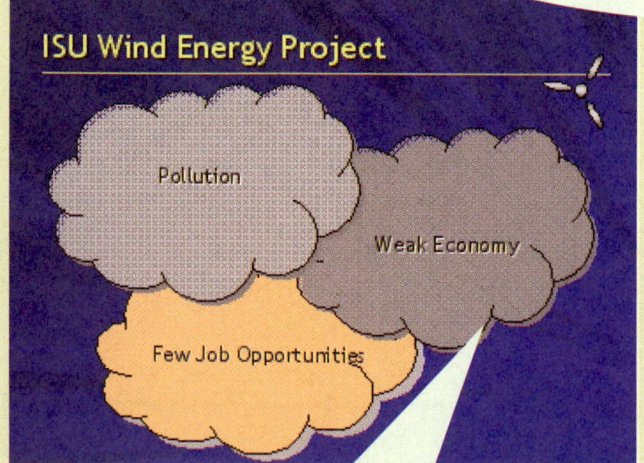

FIGURE 1.4B **Overview Slide**

Accessibility

- *The main text is easy to see because of the figure-ground contrast: dark text on the light background.*

Comprehensibility

- *The slide introduces the three key topics in the presentation.*
- *The windmill icon is repeated on each slide for visual coherence — and whimsy.*

Usability

- *The pollution clouds help the audience connect the critical concepts to a visual image that is used throughout the presentation.*

Many factors currently contribute to the state's financial and energy crisis. Not only are pollution rates alarmingly high, but fewer job opportunities and a stagnant economy have dissuaded many struggling citizens from pursuing postsecondary education.

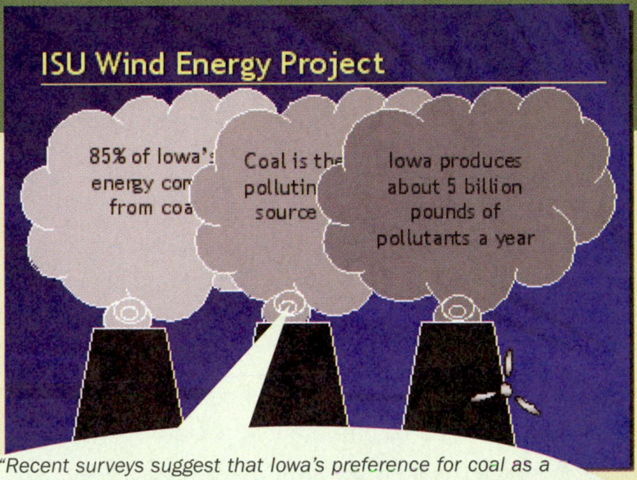

ISU Wind Energy Project

Iowa produces about 5 billion pounds of pollutants a year

"Recent surveys suggest that Iowa's preference for coal as a primary energy source remains strong. In fact, coal continues to be the most popular energy source in the state. Specifically, state officials estimate that Iowans produce about five billion pounds of coal-based pollutants every year. . . ."

FIGURE 1.4C **Slide Introducing Pollution Problem**

Accessibility

- *The text is easy to read because the 24-point sans serif font is large enough to see when projected on a screen.*

Comprehensibility

- *Each key point is easy to distinguish (in its own pollution cloud).*
- *The sentences are short and easy to understand.*

Usability

- *When the presentation is animated, each point is projected individually.*
- *The slide presents only three main points, easy for the audience to hold in short-term memory.*

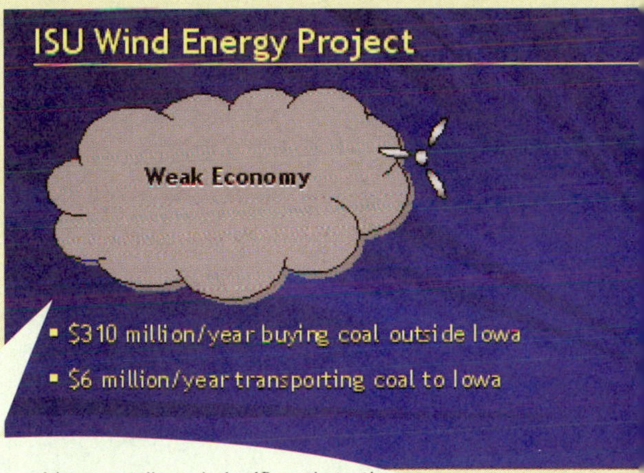

ISU Wind Energy Project

Weak Economy

- $310 million/year buying coal outside Iowa
- $6 million/year transporting coal to Iowa

"However, Iowans' dependence on coal has contributed significantly to the weakened economy. More than $300 million was spent last year purchasing and transporting coal from neighboring states to regional energy centers throughout Iowa. While our proposed wind turbine system would require significant start-up costs, the annual expenditures for coal would eventually be phased out of the state budget."

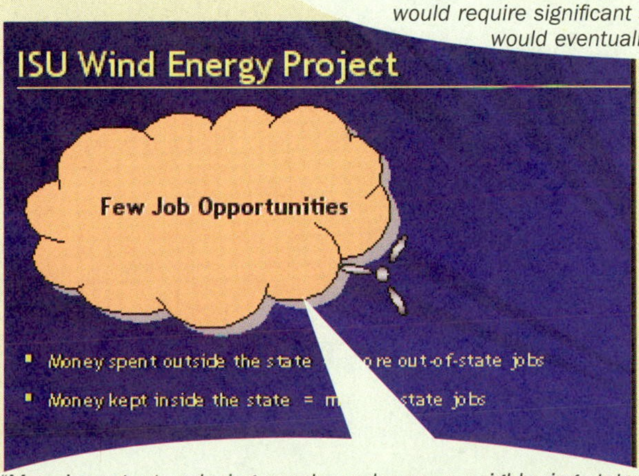

FIGURE 1.4D **Slide Introducing Economic Problem**

Accessibility

- *The text under the cloud is easy to see because of the high figure-ground contrast: light text on a dark background.*
- *Bullets distinguish the key points.*

Comprehensibility

- *Two key points are in easy-to-read phrases, with the most important information first.*

Usability

- *For fast recognition, the main topic is in the same visual form as when it was initially introduced — the same size and color of pollution cloud.*

ISU Wind Energy Project

Few Job Opportunities

"More important, reducing our dependence on neighboring states for energy sources will generate jobs for Iowans. Our proposal addresses a fundamental aspect of the state's energy budget — that we spend an inordinate amount of our annual energy budget on out-of-state expenses — and seeks a turbine system to create jobs and ultimately give Iowa's economy a much-needed boost."

FIGURE 1.4E **Slide Introducing Job Opportunity Problem**

Accessibility

- *A shadow font in the cloud distinguishes it from the primary text.*

Comprehensibility

- *An equation with the = sign rather than words is faster to understand than the word "equals."*

Usability

- *Slides with the same general purpose use the same general design — such as this slide and the preceding one.*
- *The color of the heading identifying the project and the color of the key text are identical, creating a topical connection.*

REPORTING PROGRESS

Much of the technical communication you do will be reporting progress on a project. Sometimes the report will be a brief, informal, oral presentation at a small meeting; at other times, it will be a lengthy formal report with a large distribution.

The example in Figure 1.5 is an excerpt from a long internal company report presenting an assessment of the capabilities of one of several battery manufacturers being considered as a supplier for the company. The readers are company employees — engineers, industrial buyers, managers — who are members of a committee to review the capabilities of ten major battery manufacturers. (The names of all the companies have been changed.) This committee will recommend which manufacturers the company should select as primary and secondary suppliers. A major portion of the committee's decision rests on the capability of the battery manufacturer and the quality of the batteries. And that quality depends on the production processes — that is, the way the batteries are manufactured. Based on information obtained from site visits to each manufacturing facility, one of the engineers on the committee produced a separate, detailed, and lengthy report for each manufacturer being considered.

The engineer/writer decided that each of the ten reports — completed over a six-month period, coinciding with the site visits — would include the same eight sections so that comparisons among the various manufacturers could be made easily. The excerpt in Figure 1.5, from the report about the site visit to Dalton Battery Company, is a subsection of the third main division of the report about chemical processes.

Paul A. Souders/CORBIS

The quality of paste production is critical in the manufacture of aluminum.

Accessibility

- *Sufficiently large font is easy to read.*
- *A boldface section heading matches the table of contents (and has a parallel heading in all ten reports assessing the other battery companies).*
- *Related information is chunked into short paragraphs.*
- *Each parameter is discussed in a separate paragraph so that it is easy to identify.*

Comprehensibility

- *Important information is effectively signaled and organized. For example, two kinds of paste are clearly distinguished.*
- *Definitions and explanations enable the less technical members of the committee to follow the discussion but do not annoy technical readers who don't need as much explanation.*
- *Critical definitions and explanations help members of the committee who do not have an engineering background.*
- *Technical details are important to the engineers on the committee.*
- *Words such as* only *and* instead *help readers interpret the information.*

Usability

- *Information in the table reinforces and illustrates the text.*
- *Foregrounding of critical parameters prepares readers for the next three paragraphs.*
- *Effective use of standard comparisons helps readers assess the information made over a six-month period.*

PASTE PRODUCTION

Paste production involves mixing lead oxide with water, sulfuric acid, organic fibers, and red lead to produce a dough-like substance called *paste*. In this section, I discuss Dalton's general paste formula and control parameters but not the exact formula. No comparison with other vendors can be made because these formulas are closely guarded secrets. However, I do provide a comparison of Dalton's control parameters both to industry standards and to optimum conditions for known parameters.

Formulations for the positive and negative paste differ primarily because red lead is added to the positive paste and not to the negative, while expander is added to the negative paste and not to the positive. In the battery industry, red lead is usually 10–30% of the total lead oxide: the higher the percentage of red lead the better. Dalton uses only 10% red lead. The expander, a mixture of lignin, carbon black, and barium sulfate, is usually added at about the 1% level. The best expander is Anzon-Kx. Instead, Dalton uses HL-120, from Harrington Lead Company. Harrington has the reputation for occasionally producing erratic quality expanders. Dalton's paste formulation — rounded off to 1% — is given in the table below.

	Positive Formula	Negative Formula
Oxide	72%	82%
Red Lead	8%	None
Water	11%	10%
Acid	9%	7%
Expander	None	2%

The quality of the paste is determined by three parameters: peak temperature, dump temperature, and cube weight. The temperatures are monitored continuously but are controlled by the rate of reaction between the lead oxide and sulfuric acid. The other two parameters are manually measured with each batch of paste prior to dumping into the pasting machine. Usually only one measurement is taken per batch.

Peak temperature is probably the most important of the control parameters because it sets the amount of nuclei of tetrabasic lead sulfate, the phase necessary for long life in the battery. The hotter the peak temperature the better. The industry standard for this critical parameter is 160°F; in contrast, Dalton's control point is 140°F ± 10°F. (Note: I would prefer to see peak temperatures of 190°F.)

Unlike peak temperatures, ***dump temperatures*** should be kept as low as possible to prevent premature curing of the paste. The industry standard is 110–115°F, which is considerably cooler than Dalton's 125°F. Perhaps Dalton Battery is partially compensating for the low peak temperatures with high dump temperatures to increase their tetrabasic lead sulfate nucleation count.

Cube weights, which measure the density of the paste, are best if kept between 72–75 g/in^3 for the positive in order to achieve long battery life. The negative is less critical but traditionally has cube weights of 68–78 g/in^3. In comparison, Dalton uses cube weights of 66–68 g/in^3 and 72–74 g/in^3 for the positive and negative, respectively.

FIGURE 1.5 **Excerpt from a Technical Report**[16]

How big is the Internet? Worldwide in 2002, the host count (a *host* has an active and permanent IP address that is directly connected to the Internet) was nearly 175 million. While the United States has more Internet hosts that any other country, with an annual growth rate of about 10 percent, the growth rate of Internet hosts in many other countries is much higher. For example, Belgium, Denmark, Estonia, and Turkey have annual growth rates exceeding 50 percent, while China, the Czech Republic, Italy, and Ukraine exceed 65 percent.[17]

What does this mean in practical terms for individuals and businesses in the United States? The Stanford Institute for the Quantitative Study of Society (SIQSS) reports that 65 percent of American households have at least one computer, and 43 percent are connected to the Internet.[18] On the business side, the Internet is a $200 billion industry in the United States,[19] with at least two-thirds of small U.S. businesses using the Internet.[20]

Figure 1.6 contains three excerpts from Science@NASA, identified as best science site on the Internet and recognized for excellence in science communication and "the spread of science culture" using the Internet.[21] NASA representatives summarize the goal of the site: "The communication of newly acquired knowledge and understanding [are placed] on an equal footing with the generation of that knowledge and understanding."

The annotations identify some of the more than 100 critical features for Web site accessibility, comprehensibility, and usability that Jakob Nielson and Marie Tahir discuss and illustrate in their book, *Homepage Usability.*[22]

The first excerpt, from the top of Science@NASA's home page, provides quick descriptions for the general public of a range of discoveries, with links to more detailed information. The information focuses on detecting earthquakes before they strike.

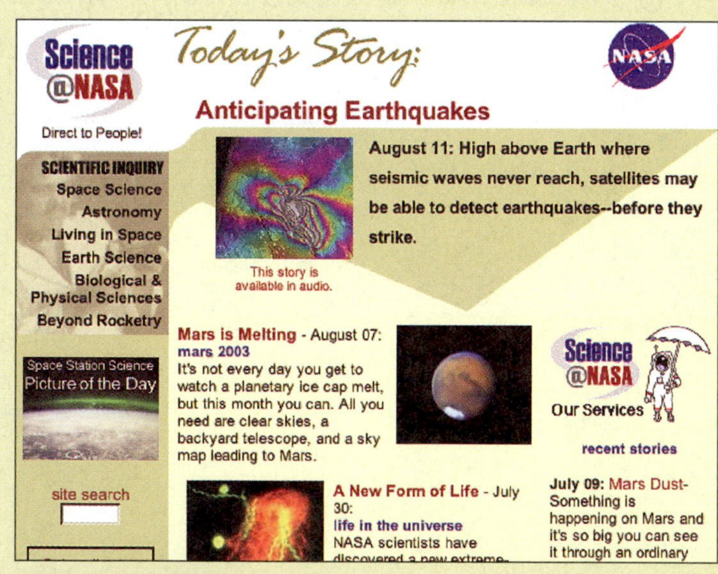

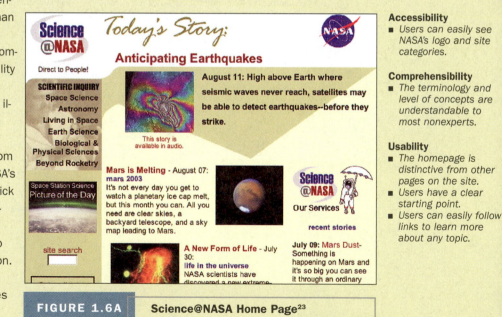

Accessibility
- *Users can easily see NASA's logo and site categories.*

Comprehensibility
- *The terminology and level of concepts are understandable to most nonexperts.*

Usability
- *The homepage is distinctive from other pages on the site.*
- *Users have a clear starting point.*
- *Users can easily follow links to learn more about any topic.*

FIGURE 1.6A Science@NASA Home Page[23]

The second excerpt drills down one level into the Web site to an article intended for knowledgeable, interested nonexperts.

Accessibility

- *This page gives users the option of accessing streaming audio and animated photos.*

Comprehensibility

- *The text is written at about the level of the weekly science page in the* New York Times.

Usability

- *Critical identifying information is available on the initial screen*
- *The bottom of this page has direct links to more technical information.*

Science @NASA
Direct to People!
science.nasa.gov

Anticipating Earthquakes

High above Earth where seismic waves never reach, satellites may be able to detect earthquakes--before they strike.

Listen to this story via streaming audio, a downloadable file, or get help.

"... temperatures will be in the high 40s to low 50s, fair and breezy -- another chilly day here in the San Francisco Bay Area," the T.V. weatherman says. *"The satellite earthquake forecast shows low to normal risk, with no critical crustal stresses or infrared signatures around the San Andreas fault ..."*

August 11, 2003: For many people, earthquakes are synonymous with unpredictability. They strike suddenly on otherwise normal days, and despite all the achievements of seismology, scientists still can't provide warning of an impending quake in the way that weathermen warn of approaching storms.

Although earthquakes seem to strike out of the blue, the furious energy that a quake releases builds up for months and years beforehand in the form of stresses within Earth's crust. At the moment, forecasters have no direct way of seeing these stresses or detecting when they reach critically high levels.

Above: A 3-D visualization of seismic energy during the 1994 Northridge earthquake in California. Credit: Kim Olsen, University of California, Santa Barbara. [more]

FIGURE 1.6B **Excerpt of Article for Interested Nonexperts[24]**

Figure 2.1 Evolution of Coulomb stresses prior to an earthquake. Each figure shows the progression of the surface Coulomb stress due to earthquakes and deep fault creep on a fault segment that will experience a future earthquake. Warm colors indicate that the change in stress favors a future earthquake. Thus, in addition to the steady-state tectonic loading of the future earthquake segment, the positive Coulomb stress caused by the surrounding fault segments increases the likelihood of an event on the future earthquake segment. (Sammis and Ivins, 2002)

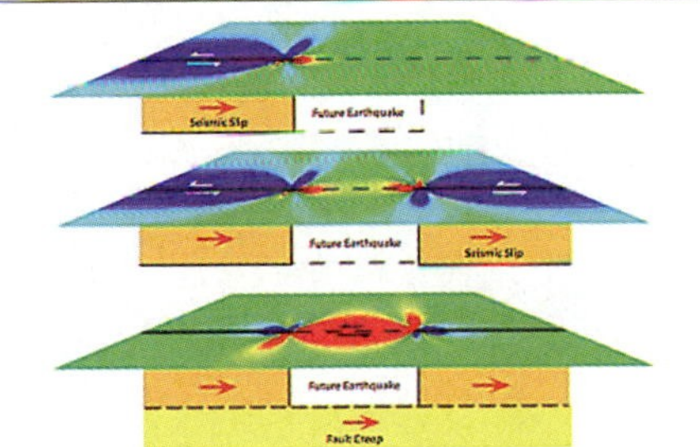

Significant improvement in observation of earthquake crustal deformation provided by GPS and InSAR during the past decade placed critical constraints on some existing models and forced significant revision of others. Perhaps the most significant inference we can draw from these advances is that the feedback loop between data and models is critical, and that future advances will require better data, particularly InSAR data.

As stated in Chapter 1, we solicited studies to define requirements for an observational system that could address specific outstanding questions in earthquake science. The results of the studies are discussed here. In the following section, we have renumbered the original six study questions slightly, combining questions 3 and 4 to emphasize the relationship between complex and triggered earthquakes, and postseismic processes.

1. ***How does the crust deform during the interseismic period between earthquakes and what are its temporal characteristics (if any) before major earthquakes?***

Detecting signals precursory to large earthquakes has been one of the most sought after and debated aspects of earthquake physics. Observations of precursory signals have been sporadic and often without a clear link to the subsequent earthquake. In the cases where the connection is clear, the measurements have generally been point location measurements, sometimes requiring measurement sensitivities that are not possible with satellite systems.

At the core of this debate is whether or not earthquakes are fundamentally predictable. Some have argued that the crust is continuously in a state of self-organized criticality (SOC) with the probability of earthquake size and location remaining steady. Sammis and

The third excerpt drills down another level to a technical article intended for experts.

Accessibility

- *The sans serif font makes long reports easy to read.*
- *Two column text is easy to read.*

Comprehensibility

- *The final GESS report is a 104-page technical discussion that is understandable by a broad scientific audience*
- *This report incorporates an executive summary as well as a large number of visuals.*
- *Appropriate use of color makes the figures easier to understand and interpret.*

Usability

- *The captions are placed next to the relevant figure and provide clear, complete explanations and citations.*

FIGURE 1.6C **Excerpt of Article for Experts[25]**

DIRECTING ACTION

Instructions can address everything from the operation of manufacturing equipment to procedures for personnel practices. Every product and process — whether for the commercial or consumer market — can be accompanied by paper or online manuals to direct assembly, guide operation, recommend maintenance, specify steps, caution about safety, and troubleshoot common problems.

Virtually every airline provides its passengers with an in-flight magazine. One regular feature in some of these magazines is instructions about ways to reduce the occurrence of deep venous thrombosis (DVT), which has become an increasing health concern due in part to cramped seating and extended trips. The example in Figure 1.7, originally published in *Royal Wings,* the in-flight magazine for Royal Jordanian Airline, presents six illustrated steps to reduce the likelihood of DVT.

What causes DVT? Seated immobility for extended periods is a big factor. Who's at risk? People at risk include those who smoke, are obese, have a history of malignancy, or have had recent surgery. Women are more at risk more than men, especially those taking oral contraceptives or hormone replacements. People over 40 are usually more at risk; however, young people with hereditary incidents are at risk as well.[26] Many airlines actively educate passengers in order to reduce the likelihood of DVT.

Accessibility

- *Both the English and Arabic texts are large enough for most people to see easily.*
- *The size and quality of the images are sufficient.*

Comprehensibility

- *Many passengers on Royal Jordanian Airline read Arabic as their first language; many also read English.*
- *The introduction defines DVT but doesn't explain the criticality—that is, that a blood clot in one of the large veins can block the flow of blood back to the heart. The most serious risk of DVT is a pulmonary embolus, which can cause death. More typically, DVT causes chronic pain and swelling.*
- *Except for the name of each exercise, the terminology is adjusted to nonexperts.*
- *The scale in the visuals realistically shows the cramped space in which these exercises are done.*
- *Most of the visuals are sufficiently accurate and detailed to enable people to do some of the exercises even if they can't read Arabic or English.*

Usability

- *The numbered steps follow the English text by moving from left to right. If the sequence followed the Arabic text, it would move from right to left.*
- *The instructions do not make clear that seated immobility for extended periods is a major cause of DVT and, therefore, that doing the exercises is very important.*
- *The visuals include clearly marked directions for movement.*
- *The visuals are coordinated with the relevant text.*

ECONOMY CLASS SYNDROME: FICTION OR FACT?

Clotting of blood, usually occurring in the lower legs is known as deep venous thrombosis 'DVT'. This has become known in the press as 'Economy Class Syndrome', but the term is misleading. Individuals seated in cars, buses and trains may all be at risk, and cases of DVT occurring in flight have been reported in premium cabins as well as economy. The term 'traveller's thrombosis' is much more accurate.

Certain people are at increased risk of developing DVT, such as those who have previously been affected by this condition in the past, cancer patients, smokers, elderly people and those who are obese. Women who are taking oral contraceptive pills may also have a higher risk.

RECOMMENDATIONS

1. Wear loose-fitting comfortable clothes
2. Avoid knee-length stockings
3. Avoid alcohol and caffeine
4. Drink plenty of water
5. While seated, avoid crossing legs, and do simple physical exercises for feet and knees for two to three minutes every 30 minutes as shown in the pictures below.

ظاهرة الدرجة السياحية حقيقة أم وهم

يحدث تخثر الدم عادة في الساقين و هو ما يدعى تخثر الأوردة العميق و هو ما اصبح يسمى بالصحافة ظاهرة الدرجة السياحية لكن هذا المصطلح فيه بعض التضليل.

إن الأشخاص الذين يستقلون السيارات الصغيرة ، الحافلات و القطارات يتعرضون لحدوث هذا الاضطراب ، و هذه الحالة تحدث في الطيران أيضا لدى مسافري الدرجة الأولى كما هو الأمر لمسافري الدرجة السياحية ، و ربما كان مصطلح تخثر الأوردة لدى المسافرين اكثر دقة. يحدث هذا الأمر عند أشخاص لديهم التاهب و القابلية لذلك كالذين أصيبوا به سابقا، المدخنين، كبار السن، و المصابون بالسمنة . إن النساء اللواتي يتناولن حبوب منع الحمل لديهن قابلية وخطورة اكبر.

التوصيات :

١- ارتداء الملابس الفضفاضة

٢- تجنب ارتداء الجرابات الطويلة التي تصل إلى الركبتين

٣- تجنب شرب الكحول و السوائل التي تحتوي مادة الكافئين

٤- اكثر من تناول الماء و العصائر

٥- بينما أنت جالس على مقعدك تجنب وضعية تقاطع الساقين و قم بإجراء التمارين البسيطة للقدمين و الركبتين لمدة (٢-٣) دقائق كل حوالي ٣٠ دقيقة كما هو موضح بالصورة .

1 (Knee flex): Lift knee towards chest, decreasing the amount of joint space at the back of the knee. Repeat with other leg.

١ **عطف الركبة :** قم برفع الركبة نحو الصدر، مقلصا مساحة المفصل من الناحية الخلفية للركبة.- كرر ذلك للساق الأخرى.

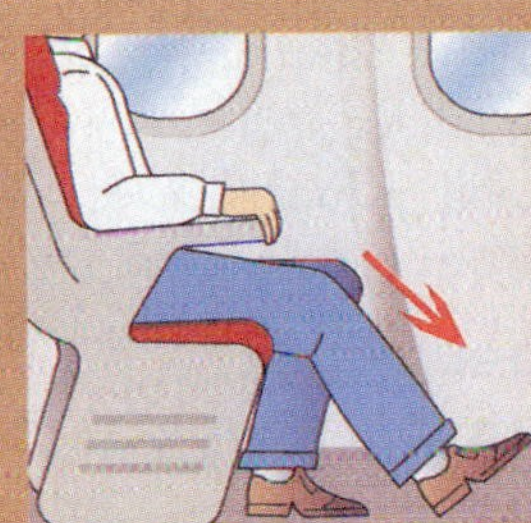

2 (Knee Extension): Straighten knee, increasing the amount of joint space at the back of the knee to its full range. Repeat with other leg.

٢ **بسط الركبة :** قم بمد الركبة بحيث تزيد مساحة المفصل من الناحية الخلفية للركبة.

- كرر ذلك للساق الأخرى.

3 (Dorsiflexion): With heel on floor, point toes upwards, decreasing the angle between the foot and front of the leg. Repeat with other foot.

٣ **عطف ظهر القدم :** بينما العقب على الأرض، ارفع أصابع القدم للأعلى مقلصا الزاوية بين القدم و مقدمة الساق.

- كرر ذلك للقدم الأخرى

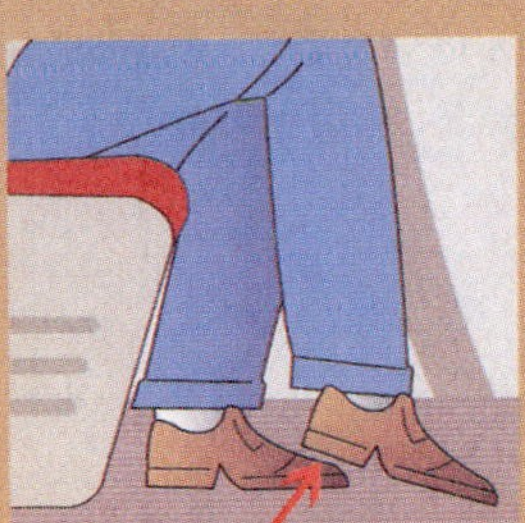

4 (Plantar flex): Stretch the foot and toes down and back, increasing the angle between top of the foot and front of the leg. Repeat with other foot.

٤ **عطف أخمص القدم :** ابسط القدم بينما أصابع القدم للأسفل و الخلف بحيث تزيد الزاوية بين مقدمة القدم و مقدمة الساق.

- كرر ذلك للقدم الأخرى

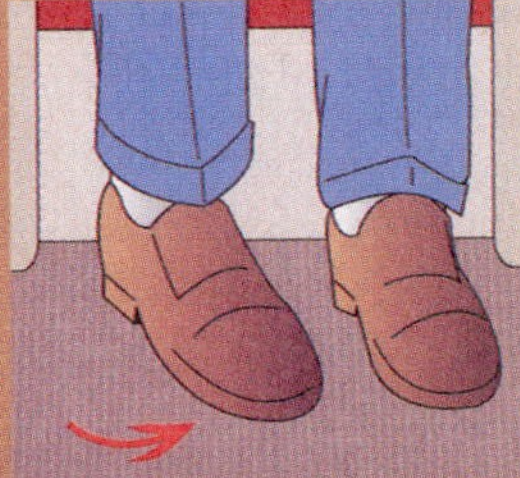

5 (Inversion): With foot on floor, gently roll the sole of the foot inward. Repeat with other foot.

٥ **دوران داخلي :** بينما القدم على الأرض، اسحب بلطف أخمص القدم للداخل.

- كرر ذلك للقدم الأخرى.

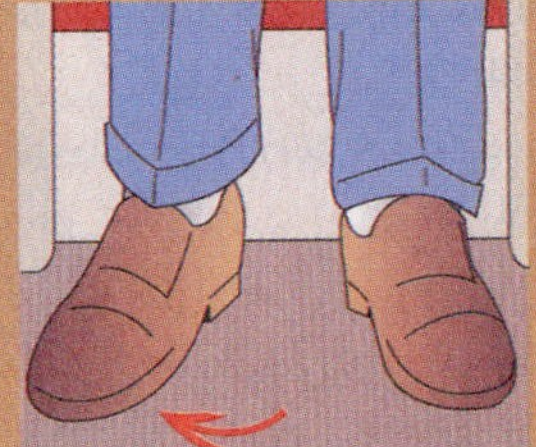

6 (Enversion): With foot on floor, gently roll the sole of the foot outward. Repeat with other foot.

٦ **دوران خارجي :** بينما القدم على الأرض، اسحب بلطف أخمص القدم للخارج.

- كرر ذلك للقدم الأخرى

FIGURE 1.7 Instructions to Reduce Occurrence of Deep Venous Thrombosis[27]

Constraints that Communicators Encounter

What kinds of constraints have you already faced in the workplace? How did you manage these constraints so they did not cause major problems?

Regardless of the kinds of technical documents, oral presentations, and visuals you create and use (such as those in Figures 1.3–1.7), you'll need to manage constraints that affect the entire process of creating and interpreting technical information.

Time Constraints

You'll often work within time limits that seem unreasonable; for example, the schedule may have been set by someone unaware of the demands of documenting technical projects. In practice, time considerations and deadlines limit every stage of preparing technical information. One of your responsibilities is deciding how much time you can devote to each project.

Subject and Format Constraints

The subject for a technical document, oral presentation, or visual is usually predetermined. However, you'll face the challenge of focusing on details appropriate for the intended audience and presenting them in an interesting and useful manner. You will seldom be given a specific approach. Rather, you'll need to be able to narrow your subject, select and organize content, and then determine the limitations. To do this, you'll need to assess the audience's needs and identify its purposes.

Audience Constraints

Sometimes you'll prepare technical information for an audience in which everyone has similar education, experience, and expectations. Frequently, however, the members of the audience will have varied backgrounds as well as different reasons for reading a document or listening to a presentation. For example, a business manager might have an advanced degree in management but little expertise in the company's technical field. As a result, you'll need to make decisions about the level of technical complexity, organization, diction, and design to respond to the needs of a variety of readers or listeners. (See Chapter 4 for a more detailed discussion about audience.)

Collaboration as a Constraint

Sometimes you'll work independently, sometimes collaboratively. Your actual role in preparing any given document or presentation will probably fall somewhere between two extremes. When you work independently, you'll assume responsibility for exploring, planning, drafting, and perhaps even editing and publishing a document. When you work collaboratively, you'll share the responsibility for one or more stages of the process. (See Chapter 5 for a more detailed discussion about community and collaboration.)

Constraints in Data Collection

The information you need won't always be easily or quickly accessible. To collect information, you must confidently approach new equipment and unfamiliar processes. As you learn to operate the equipment or complete the process, you'll often discover what your audience needs to know. Don't be surprised, however, when you are hampered by limited access to the necessary equipment or when the available information seems skimpy. Sometimes the deadlines will be too tight to learn everything you need to know. In these situations, you must depend on others for the information you need. Develop persuasive skills so that people who have busy schedules agree to make time to talk with you. Once you've located the people who have the information and have made an appointment with them, you need strong interviewing skills to collect the data. (Chapters 7 and 19 and the Companion Web site [www.english.wadsworth.com/burnett6e] provide guidelines for conducting or participating in various kinds of interviews.)

WEBLINK

Constraints in Technology

Technology is part of communication in every workplace, from agribusiness to nanoscale physics laboratories, from endangered species breeding programs to software engineering departments. You face several potential constraints. First, you will need access to appropriate technology as well as the capability and training to use it. You will be at a tremendous disadvantage if you aren't familiar with and don't have the training to use a wide range of appropriate electronic tools. Other constraints you might encounter include limited access to electronic tools, lack of technical support for the hardware and software, and users' unfamiliarity with the technology if they need to access information electronically.

Constraints Caused by Noise

Interference, or what is sometimes called *noise,* can come from the context or environment, from any of the communicators, or from the document, oral presentation, or visual itself.

What constraints do you anticipate you might have in the future when preparing a technical document or presentation? How do you plan to deal with these constraints?

- *Noise from the context or environment.* One kind of environmental interference occurs because the physical reception is garbled or masked. Physical interference may be as simple as a noisy air conditioner in a meeting room or static on a phone line. Another kind of environmental interference occurs because the social or political environment influences the way in which people interpret information. Social or political interference may be caused by organizational policies and practices. For example, if safety updates are typically given only cursory attention in a particular company, a writer has a more difficult time than if these updates are carefully read and implemented. No environment is neutral; some form of interference always exists.

- *Noise from any of the communicators.* Writers, speakers, and designers as well as readers, listeners, and viewers can generate noise. The experiences and attitudes of the writers, speakers, or designers (factors such as vested interests, emotional biases, perceptions) affect their selection and slant of information. This interference may also come from problems that the writers, speakers, or designers have in formulating the information (for example, ambiguity or failure to consider audience needs). The experiences and attitudes of readers, listeners, and viewers (factors such as boredom, doubt, disagreement, anger, or indifference) affect their interpretation of information. No communicator is neutral; interference of some type is always part of those who generate the information as well as those who interpret it.
- *Noise embedded in information and in its delivery.* Information can be incomplete, inaccurate, illogical, unsupported, or inflammatory, thus causing interference in understanding. Beyond that, interference can occur because of confusing design, low-quality reproduction, or problems with electronic transmission. Interference can occur in oral presentations due to inadequate amplification or inappropriate visuals for the room size. No delivery is neutral; some form of interference always affects the information and its reception.

Being aware of the constraints that communicators face — time, subject and format, audience, collaboration, data collection, technology, and various kinds of noise — should help you anticipate and manage some of the problems that these constraints generate.

Individual and Collaborative Assignments

1. **Define technical communication.** Create a definition of technical communication for colleagues in your profession that accurately reflects the complexity of creating and interpreting technical information so that it is accessible, comprehensible, and usable. Use information in this chapter, information from two organizations, the Society for Technical Communication in the U.S. (www.stc.org) and the Institute of Scientific and Technical Communicators in the UK (www.istc.org.uk; www.istc.org.uk/site/news.asp), and information from the Web sites of the organizations in your own profession. Compare your definition to those created by others in the class.
2. **Identify general audiences.** Read the following list of subjects. Identify at least two realistic technical audiences for any five of these subjects. Explain what technical information each audience might need to learn or use.

   Example: Renaissance paint might be of interest to any of these technical professionals: a museum curator for authenticating a painting; an art historian for characterizing materials in a particular period; a chemist for conducting a spectrographic analysis.

arrows	ethanol	magnifying glass	running shoes
ballet	fax machine	modem	soybeans
butter	frisbees	motor oil	steroids
CDs	furniture stripper	peanuts	styrofoam
cell phone	golf balls	perfume	tires
coffee	hot dogs	plastic film	tricycles
condors	jogging	poetry	trout
eggs	legumes	rainbows	ZIP code

3. **Identify accessibility, comprehensibility, and usability.** Work in a small group to visit any three of the following Web home pages, which represent a range of national and multinational corporations as well as nonprofit organizations. Consider these questions:

   - **Who's the sponsor?** Can users easily identify a name, logo, and tag line on the home page describing what the organization does?
   - ***Who are the intended audiences?*** Who are the probable users? What are their purposes for using the site?
   - ***How accessible is the site?*** How is the site accessible? To users reading other languages? To users with visual disabilities? To users with limited or old computer resources? Does the home page have distracting or competing elements?
   - ***How easy is the site to understand?*** Can users easily determine what's most important on the home page? What prior knowledge is assumed? What conventions does the home page use? What is the technical level of definitions and explanations?
   - ***How easy is the site to use?*** How easily can users determine the site's organization from the home page? Can users find a clear place to start? How are links to other parts of the site indicated? How easy is the site to navigate?

www.admworld.com	www.fedex.com	www.dupont.com
www.bayerus.com	www.nestle.com	www.oracle.com
www.birdseye.com	www.lucent.com	www.ups.com
www.cocacola.com	www.merck.com	www.whymilk.com

4. **Assess a short memo.** The following memo was submitted as a weekly progress report about project activities. It is typical in several ways — amount of information, tone, length, and attention to detail — of the memos and reports produced by Wayne Jackson, an associate engineer. Because Mr. Jackson has always been extremely careful to double-check his content for accuracy, he is surprised that his semiannual review identifies his writing as a block to rapid advancement. Examine the memo and use the factors for assessing documents presented in this chapter to help explain

whether you agree with the assessment by Mr. Jackson's supervisor. What advice would you give Mr. Jackson to help him become a more effective communicator? Revise the memo.

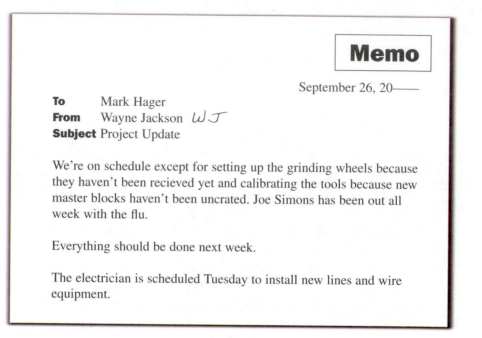

Memo

September 26, 20——

To Mark Hager
From Wayne Jackson *WJ*
Subject Project Update

We're on schedule except for setting up the grinding wheels because they haven't been recieved yet and calibrating the tools because new master blocks haven't been uncrated. Joe Simons has been out all week with the flu.

Everything should be done next week.

The electrician is scheduled Tuesday to install new lines and wire equipment.

5. **Assess a technical explanation.** Read the following text, "Understanding and Avoiding the Leading Cause of Death among Pilots," written by an experienced pilot for beginning student pilots. In the partially completed rubric on page 32, read the abbreviated criteria for *excellent* for each feature listed. While managers and supervisors (and teachers) use rubrics to help assess various kinds of communication, rubrics can provide the most help to writers, presenters, and designers as they revise their work to meet the criteria.

Using the information in this chapter and the generic rubrics on this textbook's Companion Web site, **www.english.wadsworth.com/burnett6e**, expand each criteria to provide more detailed help to writers as to what will make the document to pilots excellent. Then complete the rubric's cells for *unacceptable, weak,* and *acceptable* in relation to the document for student pilots. The *weak* and *unacceptable* columns will help you identify elements to avoid. Based on your completed rubric, explain whether the writer has created an effective technical document.

Note: Throughout this text, you will be asked to create and use rubrics as tools to help you assess documents, oral presentations, and visuals. This textbook's Companion Web site, **www.english.wadsworth.com/burnett6e**, presents generic rubrics that you can modify as needed for specific print documents, oral presentations, and visuals.

Understanding and Avoiding the Leading Cause of Death among Pilots[28]

More pilots are killed in ***stall-spin accidents*** than in any other type of incident. The next cause of fatalities is inadvertent flight into bad weather by pilots not trained in instrument flight, which usually leads to a stall-spin scenario. At the end of this discussion, you should be able to answer these questions:

- What holds the airplane up?
- What is a stall?
- What is a spin?
- How can you recover from a stall?

The engine doesn't hold the airplane up in the air, the wing doesn't hold it up in the air, and (although we're reluctant to admit it) the pilot's skill doesn't hold it up in the air. Airplanes are held aloft by a physical force called lift. ***Lift*** is an upward force generated by the smooth flow of air over the wing's surfaces.

The illustration below uses tiny balloons to represent paired air molecules passing above and below a wing. In 1738, a Swiss physicist named Daniel Bernoulli suggested that the air mass remains undisturbed by the wing and that each pair of balloons would join again at the back of the wing. Wings have a certain shape: the top is convex; the bottom is flat. Slats at the leading edge of the wing and flaps at the trailing edge of the wing exaggerate the difference. As the illustration shows, the curved line traveled by the balloons above the wing is a greater distance than the straight line traveled by balloons below the wing.

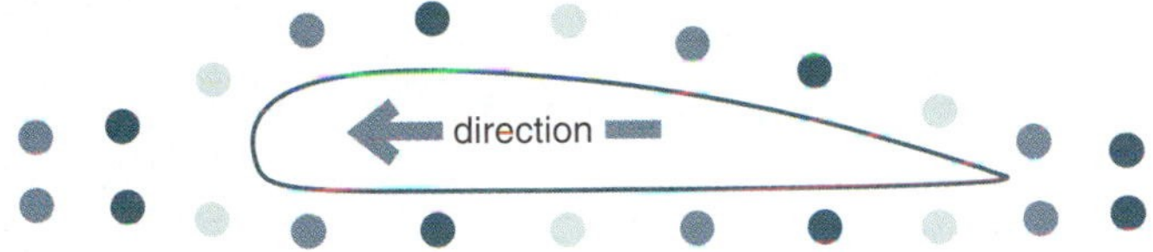

In order for the balloons to rejoin, the speed and spacing of the upper balloons must be greater than the speed and spacing of the lower balloons; the difference in speed and distance causes a relatively lower *air pressure* along the top of the wing. This relative suction translates into a force called lift. Lift increases as the flow of air increases over the wings. When the speed of the airflow over the wings generates a lifting force stronger than gravity, the plane flies. The engine does not thrust the plane into the sky; the engine moves the plane forward at a speed permitting the wings to generate sufficient lift.

What happens when this goes wrong? When the airflow is disturbed, the subtle relative suction disappears, and gravity re-asserts itself over the mass of the aircraft. The airflow can be disturbed by ice or by an extremely nose-high attitude. The loss of lift is called a ***stall,*** a phrase that has nothing to do with the way a jalopy quits on the highway. One warning: Never attempt to takeoff with ice or frost on the wings. Keep the wings clean of dirt and bugs. Lift requires clean wings.

When one wing stalls before the other, the aircraft becomes asymmetrical, resulting in a disorienting rotation dive called a ***spin.*** Low-time pilots can be overwhelmed by the physical forces of a sustained spin and may be unable to regain controlled flight.

Recovery from the stall is straightforward: Level the wings to avoid a spin, drop the nose to increase airflow over the wings and restore lift, increase to full engine power to climb to safety, and reflect on the audacity of flight.

Your experienced flight instructor will show you how to intentionally stall and spin your aircraft to practice recovering from these situations. When you start these maneuvers at a safe altitude and with your instructor's guidance, you will learn how to avoid, recognize, and resolve the situations that put novice pilots at risk. Remember: Level the wings, drop the nose, and apply full power.

Rubric for Assessing the Explanation about Spin and Stall

Document Features	Unacceptable	Weak	Acceptable	Excellent
Accurate information	Includes many errors	Includes several errors	Includes few errors	Includes accurate and sufficient information
Accessible information				Makes information completely accessible to the intended audience
Response to context				Responds to major concern of beginning pilots
Understandable information				Defines/explains information at a level appropriate for beginning pilots
Clear purpose				Announces purpose/goal up front
Adapted to audience				Uses appropriate vocabulary and level of explanation
Effective organization				Identifies and develops key topics in order
Appropriate visuals				Uses diagram to clearly illustrate textual discussion
Effective design				Uses coherent paragraphs, bulleted list, boldface terms, and an embedded diagram appropriate for nonexperts
Usable information				Enables beginning pilots to answer key questions
Conventional language				Uses conventional language with no errors

Chapter 1 Endnotes

1 Barnum, C., & Fischer, R. (1984). Engineering technologists as writers: Results of a survey. *Technical Communication, 31*(2), 9–11.

2 "The [2003] survey was developed by Robert Half Finance & Accounting. . . . Conducted by an independent research firm, the survey includes responses from 1,400 CFOs from a stratified random sample of U.S. companies with more than 20 employees." Various Web sites about the survey are listed below:

- Retrieved August 16, 2003, from http://www.roberthalfmr.com/PressRoom;jessionid=AEBifnjywmB1ecw5VvgtYj2KokMNGTx2C0Yo3G1fyKq1BvYLCX!-221591521!NONE
- Retrieved August 16, 2003, from http://www.roberthalffinance.com/PressRoom?LOBName=RH&releaseid=357
- Retrieved August 16, 2003, from http://www.ivc.ca/studies/us.html

3 *AMA 2003 Views on Administrative Professionals' Tasks and Competencies.* Retrieved January 13, 2004, from http://www.amanet.org/research/pdf.htm

4 The National Institute for Literacy (NIFL) was created by a bipartisan congressional coalition and currently authorized under the Workforce Investment Act of 1998. NIFL, which supports the development of high quality regional, state, and national literacy services, is administered by the Secretaries of Education, Labor, and Health and Human Services with the help of a ten-member advisory board. Data are from the survey of Workforce Skill Requirements. Retrieved August 16, 2003, from http://www.nifl.gov/nifl/facts/workforce.html

5 The 2000 survey of 204 members of L'association canadienne de l'informatique were compared with results from the Information & Advanced Technology Survey 2000, which surveyed over 100 technology-based organizations in Canada. Retrieved August 16, 2003, from http://www.wynfordgroup.com/cips/CIPS_2000_Results.htm

6 Data are from the survey of Corporate Concerns. Retrieved August 16, 2003, from http://www.nifl.gov/nifl/facts/workforce.html

7 Berkenkotter, C., & Huckin, T. N. (1995). *Genre knowledge in disciplinary communication: Cognition/culture/power.* Hillsdale, NJ: Lawrence Erlbaum. (See especially pages 1–25.)

8 Wenger, E. (1998). *Communities of practice: Learning as a social system.* Retrieved January 13, 2004, from http://www.co-i-l.com/coil/knowledge-garden/cop/lss.shtml

9 Smith, M. K. (2003). Communities of practice. *The encyclopedia of informal education.* Retrieved September 7, 2003, from http://www.infed.org/biblio/communities_of_practice.htm

10 Kaplan, C. S. (2000, 14 January). Judge says recording of electronic chats is legal. *New York Times.* Retrieved October 1, 2003, from http://www.nytimes.comibrary/tech/00/01/cyber/cyberlaw/14law.html

11 Ornatowski, C. (1992). Between efficiency and politics: Rhetoric and ethics in technical writing. *Technical Communication Quarterly, 1*(1), 91–103.

12 For biographical details about H. P. Grice, check this Web site: http://www.artsci.wustl.edu/-philos/MindDict/grice.html

13 For interesting discussions about Grice's cooperative principle and related maxims, refer to these two articles:

- Davies, B. (2000). Grice's cooperative principle: Getting the meaning across [Electronic version]. In D. Nelson & P. Foulkes (Eds.), *Leeds Working Papers in Linguistics, 8,* 1–26.
- White, R. (2000). Adapting Grice's maxims in the teaching of writing [Electronic version]. *ELT Journal, 55*(1), 62–69.

14 American Academy of Allergy and Immunology. *Allergy relief guide.* [Brochure]. Reprinted by permission of the Academy.

15 The slides are reproduced here with the permission of the students who prepared them in a technical communication class at Iowa State University, Summer, 2003.

16 Gary Tarcy, "Site Visit to Dalton Battery." Reprinted by permission of the author.

17 "Data on internet activity worldwide." Updated March 16, 2003. Retrieved August 18, 2003 from http://www.gandalf.it/data/data1.htm

18 Nie, N. H., & Erbring, L. (February 2000). *Internet and society: A preliminary report.* Retrieved August 17, 2003, from Stanford University, Stanford Institute for the Quantitative Study of Society Web site.

19 In Stat MDR. (2001, January 16). *Internet infrastructure continues to grow: U.S. businesses to spend nearly $200 billion in 2004.* Retrieved August 18, 2003, from http://www.instat.com/pr/2001/ec0008ms_pr.htm

20 Intellisites. (2002, March 29). *Two thirds of small businesses using the Internet.* Retrieved August 18, 2003, from http://www.intellisitedesign.com/articles/art-Two-Thirds-of-Small-Businesses-Using-the-Internet.htm

21 See press releases:

- NASA Space Science News. (1999, January 19). *The Oscars of the Internet.* Retrieved August 23, 2003, from http://www.southpole.com/newhome/headlines/sc19jan99_1.htm
- Science@NASA. (2001, April 4). *Tireless Science Communication Pays Off for Science@NASA.* Retrieved August 23, 2003, from http://science.nasa.gov/headlines/y2001/ast04apr_1.htm

22 Nielson, J., & Tahir, M. (2002). *Homepage usability: 50 Websites deconstructed.* Indianapolis, IN: New Riders Publishing.

23 Retrieved August 17, 2003, from http://science.nasa.gov/

24 Retrieved August 17, 2003, from http://science.nasa.gov/headlines/y2003/11aug_earthquakes.htm

25 *Global Earthquake Satellite System: A 20-year plan to enable earthquake prediction.* Displayed is an excerpt from page 18. Retrieved August 17, 2003, from http://solidearth.jpl.nasa.gov/gess2.html

26 Information compiled from these sources:

- Cleveland, M. (n.d.). Deep venous thrombosis. In *Diseases and conditions encyclopedia.* Retrieved August 17, 2003, from http://health.discovery.com/diseasesandcond/encyclopedia/590.html
- *Deep venous thrombosis (DVT) and air travel.* (n.d.). Retrieved August 17, 2003, from http://www.caa.co.za/resource%20center/av.%20medicine/docs/thrombosis.html
- National Air and Space Administration, Occupational Health. (2001, July 1). *Flight-related deep vein thrombosis (DVT) — "Economy class syndrome."* Retrieved August 17, 2003, from http://ohp.ksc.nasa.gov/alerts/dvt.html

27 "Economy class syndrome: Fiction or fact?" (2003). *Royal Wings,* (July-August), 51.

28 Quigley, E. (nd.) Understanding and avoiding the leading cause of death. School of Communication and Information Systems. Robert Morris College. Pittsburgh, PA. Reprinted by permission of the author.

CHAPTER 2

Understanding Culture and the Workplace

Objectives and Outcomes

This chapter will help you accomplish these outcomes:

- Define culture and recognize the critical role that it plays in the workplace
- Understand factors that contribute to various cultures
- Analyze and understand the ways in which culture affects workplace communication
- Work more productively and respectfully with colleagues from various cultures

> *What are negative words or phrases that people sometimes use to describe attitudes, actions, tools, and artifacts that are unfamiliar? Weird. Funny. Stupid. What else? Why do people label the unfamiliar negatively?*

Professionals who understand workplace cultures have a remarkable advantage. This chapter encourages you to think about culture — both your own culture (which may be strange to others) and the cultures of others (which may be strange to you). A culture seems strange when it is unfamiliar, uncomfortable, and sometimes even inaccessible. Learning how to make sense of and function productively with unfamiliar cultures in the workplace increases your value as an employee and improves your performance as a professional.

This chapter encourages you to think about culture in a new way: not just as nationality and ethnicity but also as workplace and community organizations, as a factor that influences our way of living, and as a factor that comprises our demographic identity and personal way of life. While we can discuss these aspects of culture separately, in practice they are synergistically integrated.

Noticing Culture in the Workplace

A recent survey of more than 1,000 CEOs around the world showed these findings:[1]

	Companies with customers in six or more countries	**Companies with suppliers in six or more countries**	**Companies with employees in six or more countries**
Europe	83%	76%	67%
Australia/New Zealand	75%	75%	64%
Asia	68%	69%	56%
North America	62%	55%	43%
Latin America	57%	50%	33%

Given findings such as these, your chances are better than average of working for a company that has international customers, suppliers, and/or employees.

You don't even have to leave the United States or work for a multinational corporation to find cultural differences. The U.S. workplace itself is multicultural and multiracial. For example, the U.S. Census for 2000 shows the following demographics:[2]

- Nearly 18% of people in the United States speak a language other than English at home.
- Racial minorities own nearly 15% of U.S. companies.
- Slightly less than 70% of people counted by the U.S. census are white. Slightly more than 12% are black. Another 12% are of Hispanic or Latino origin. Nearly 4% are Asian. About 1% are American Indian or Alaska Native.
- Slightly more than 11% of people counted by the U.S. census were born in another country.

And this multicultural and multiracial workforce exports goods and services to more than 200 countries around the world[3] and imports goods and services from nearly 250 other countries.[4]

How do the gender and racial descriptions of the workforce in your community and state differ from the national percentages? Go to www.english.wadsworth.com/burnett6e for links to search for this information.
CLICK ON WEBLINK
CLICK on Chapter 2/census

WEBLINK

Your ability to communicate successfully in the workplace includes your ability to recognize and positively respond to culture-specific attitudes, actions, tools (processes and procedures as well as physical tools), and artifacts such as documents, oral presentations, and visuals. You can start by thinking about cultures as being societies that have structures, purposes, and group-specific meanings.

Culture can be associated with nationality, race, ethnicity, or religion. For example, South Africa, a country with black, Indian, multiracial, and white citizens who speak many different languages, has both a national culture striving for post-apartheid unity and a role in the world economic community as well as many subcultures defined by race, religion, and tribal heritage. Similarly, the United States has a national culture that includes Native Americans as well as people from many other countries, who in turn form subcultures defined by race, ethnicity, religion, and many other factors such as economics, sexual orientation, gender, and education.

Culture, however, is not just about these factors. You can also accurately talk about a culture of disability — that is, people whose physical or mental disabilities, whether visible or not, shape their lives. Or you can talk about the culture of specific types of companies. While Apple, Dell, Gateway, IBM, and Microsoft each has a distinctive corporate culture, collectively these computer companies differ from the corporate culture of Ford, General Motors, Nissan, Saturn, and Toyota.

Teams, which are the norm in many workplaces, typically include people from different cultures. Taking advantage of multiple perspectives usually increases the quality of the work the team does.

What's Normal?

What other factors besides clothing, speech, and meeting conventions signal differences that affect workplace attitudes, interactions, and performance?

Whether you can understand a culture sometimes depends on its visibility to you. Parts of other cultures tend to be visible to us; our own culture tends to be invisible because we see it as "normal." We notice, for example, when others wear clothing at work that attracts or distracts us; we tend not to notice our own clothing because we see it as "normal." We notice when people at work speak with a different accent, volume, inflection, or pace than we're accustomed to; we tend not to notice our own speech because we hear it as "normal." We notice when people at meetings change the order or focus of business, such as taking time at the beginning of a meeting to discuss personal concerns or share stories; we tend not to notice our own meeting conventions because we consider them "normal." Part of becoming culturally aware is looking at what we do through someone else's eyes, through someone else's experiences. When we encounter something outside our own experience, we can ask ourselves if it is, perhaps, normal or familiar to someone else.

Cultural Blindness

This notion of "normal" can also make us blind to certain aspects of other cultures. Just as we don't give a second thought to what we believe is normal, we also don't usually realize that a comparable event, circumstance, or accommodation may exist in another culture.

Try this quick question: "What do July 5, August 24, December 6, and December 12 mean to you?" The question would perhaps be easier if the question were changed slightly: "What do July 4, July 5, August 24, December 6, and December 12 mean to you?" Then you would have one clue that they are all national holidays celebrating independence. When you are working with people from other cultures, their holidays are likely to be meaningful to them the same way that your holidays are meaningful to you.

What other cultural differences besides holidays and building accommodations surround us but are often invisible? How can these differences affect workplace attitudes, interactions, and performance?

Maybe you're thinking to yourself that knowing these dates is unrealistic (and unfair), so let's test your cultural sensitivity closer to home. Think of five buildings that are regularly used by the majority of people at your college (e.g., library, medical center, computer facilities, theater/auditorium, police department, post office, offices of institutional officials, gymnasium/sports facilities). Try this quick question: "Where, exactly, is the wheelchair access to each building, where is the elevator once you're inside, and where are the wheelchair-accessible restrooms?" If you have a disability that requires a wheelchair (or have a close friend who does), you probably know the answers. If you are not using a wheelchair, you may have to think about, or not even know, the locations.

Defining Culture

What is culture? A number of researchers have spent a great deal of time investigating culture and trying to create a workable definition. Most see culture as a system of learned beliefs and values that influences attitudes and actions. In completing these actions, people in a particular group use common tools to create artifacts. In technical communication, these artifacts range from documents, oral presentations, and visuals to meeting behaviors, team interactions, and organization hierarchies. According to anthropologist John Bodley, "Culture involves . . . what people think, what they do [along with the tools they use], and the material products they produce."[5]

When a multinational corporation opens offices in another country, should the corporation adhere to the workday practices of the corporation's home country or adapt to those of the new country? For example, should IBM offices in Mexico have a siesta like many other Mexican businesses, or follow U.S. business practices and have no siesta?

These learned attitudes, actions, tools, and the resulting artifacts comprise a culture — whether defined by nation or race or disability or organization or any number of other factors. Clearly then, culture is sometimes more easily identified in other people than in ourselves. Raymond Williams provides an explanation for this: "Culture is ordinary."[6] Simply put, what we do in our communities and our corporations seems ordinary to us, so we don't identify it as "culture"; it's just the way things are. For example, U.S. culture includes predictable workdays — typically Monday through Friday from 8:00 a.m. to 5:00 p.m. But your vendors in Israel may want to reach you on a Sunday, which is a normal workday for them. And some of your vendors in Mexico may not be in their offices Monday through Friday from 1:00 to 3:00 p.m. Each of these cultures has a different sense of what constitutes a "normal" workday and workweek.

Understanding the Importance of Culture

More and more companies are paying attention to culture as part of making their organization a good place to work. Many organizations are addressing cultural differences by encouraging diversity and including forthright statements about culture, diversity, and human rights for its employees, contractors, and suppliers. Figure 2.1 presents an example from GlaxoSmithKline, a major pharmaceutical company headquartered in the UK with operations in the United States. This part of GlaxoSmithKline's Web site demonstrates the ways in which cultural differences are valued and respected in the corporation and the ways its human rights position are enforced.

What are the pros and cons of making an organization's diversity statement explicit? How does naming or omitting particular groups affect the inclusiveness of the statement?

Every year *Fortune* magazine publishes a list of the 100 best companies to work for in America. In what ways does workplace culture contribute to a company being named as a "best company"? Go to www.english.wadsworth.com/burnett6e for links to the "best companies" lists published by the Great Place to Work Institute. You can check the Web site to view the companies by year, by location, or by type of organization.

CLICK ON WEBLINK

CLICK on Chapter 2/best companies

WEBLINK

FIGURE 2.1 GlaxoSmithKline Web Site Statement about Corporate Culture, Diversity, and Human Rights[7]

gsk GlaxoSmithKline
The Impact of Medicines
Corporate and Social Responsibility Report 2002
REPORTING FRONT PAGE
DOWNLOADS
CSR 2002 home
A global challenge
The scope of our business
Our contribution to society
Medicines for the developing world
Community investment
Research and development
Valuing people
Case study
Environment, health and safety
Business ethics and integrity
Management of CSR
Web references
Corporate Social Responsibility Committee
People are the greatest single source of competitive advantage for any company.
At GlaxoSmithKline, we believe that attracting, retaining and motivating the very best people is the foundation for our future success.
Valuing people
GSK employs over 100,000 people and operates in more than 150 countries.
PREVIOUS
NEXT
We are committed to providing the opportunity for our employees to do meaningful and challenging work in pursuit of our goal to improve the quality of human life by enabling people to do more, feel better and live longer.
Last year we reported on the principles that underpin our approach to people management, and a range of programmes which were designed to deliver our global human resources strategy. We are now pleased to report the progress we have made and set out, where we are able, how we intend in future to measure our performance on issues such as diversity.
THE GSK CULTURE
At GSK, great emphasis is placed not only on what we must achieve as a company, but also on how we deliver our achievements. Our culture is summed up in the GSK Spirit which defines the qualities we expect all our employees to embrace:
• performance with integrity
• entrepreneurial spirit
• focus on innovation
• a sense of urgency
• passion for achievement

Globalization and Localization

Why should this concern with culture matter to you? You are likely to work in situations that put you in contact with other cultures. When that occurs, you'll be more successful if you understand and respect other cultures, whether those cultural differences are based on nationality, race, ethnicity, religion, economics, sexual orientation, gender, education, or disability.

FIGURE 2.1 **GlaxoSmithKline Web Site Statement about Corporate Culture, Diversity, and Human Rights[7], continued**

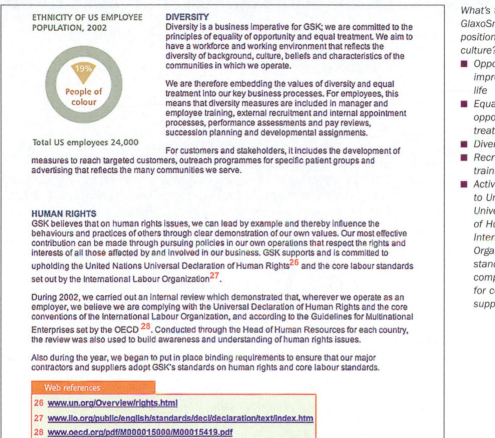

ETHNICITY OF US EMPLOYEE POPULATION, 2002

Total US employees 24,000

DIVERSITY

Diversity is a business imperative for GSK; we are committed to the principles of equality of opportunity and equal treatment. We aim to have a workforce and working environment that reflects the diversity of background, culture, beliefs and characteristics of the communities in which we operate.

We are therefore embedding the values of diversity and equal treatment into our key business processes. For employees, this means that diversity measures are included in manager and employee training, external recruitment and internal appointment processes, performance assessments and pay reviews, succession planning and developmental assignments.

For customers and stakeholders, it includes the development of measures to reach targeted customers, outreach programmes for specific patient groups and advertising that reflects the many communities we serve.

HUMAN RIGHTS

GSK believes that on human rights issues, we can lead by example and thereby influence the behaviours and practices of others through clear demonstration of our own values. Our most effective contribution can be made through pursuing policies in our own operations that respect the rights and interests of all those affected by and involved in our business. GSK supports and is committed to upholding the United Nations Universal Declaration of Human Rights[26] and the core labour standards set out by the International Labour Organization[27].

During 2002, we carried out an internal review which demonstrated that, wherever we operate as an employer, we believe we are complying with the Universal Declaration of Human Rights and the core conventions of the International Labour Organization, and according to the Guidelines for Multinational Enterprises set by the OECD [28]. Conducted through the Head of Human Resources for each country, the review was also used to build awareness and understanding of human rights issues.

Also during the year, we began to put in place binding requirements to ensure that our major contractors and suppliers adopt GSK's standards on human rights and core labour standards.

Web references

26 www.un.org/Overview/rights.html

27 www.ilo.org/public/english/standards/decl/declaration/text/index.htm

28 www.oecd.org/pdf/M000015000/M00015419.pdf

What's the focus in GlaxoSmithKline's position on workplace culture?

- *Opportunity for improved quality of life*
- *Equality of opportunity and treatment*
- *Diverse workforce*
- *Recruitment and training*
- *Active commitment to United Nations' Universal Declaration of Human Rights and International Labour Organization standards for their company as well as for contractors and suppliers*

Let's look at the statistics for some organizations where you might work, even if you never leave your own community. The nearly 40,000 organizations that are often called "global" or "multinational" account for approximately four-fifths of world trade.[8] Many of the organizations are for-profit companies, but others are organizations such as the United Nations or non-government organizations (NGOs) such as the Red Cross.

According to the Financial Times–Stock Exchange (FTSE) and Standard & Poor's (S&P) World Index, a company is a "multinational" if 30 percent or more of its sales are outside its own region of the world. For example, a German company is considered multinational if 30 percent of its sales are outside Europe. Of the 2,218 companies listed on the FTSE/S&P World Index, 506 are multinationals.[9]

ETHICS SIDEBAR

Minimizing Cultural Bias in International Web Sites[10]

Although you can find Web sites in Chinese, Japanese, Portugese, and Ukrainian (and hundreds of other languages), English dominates the Web in international business, industry, government, and science, which necessitates new ways of working across national boundaries and presents cross-cultural challenges in creating bilingual and multilingual Web sites. These challenges mean that Web developers must accommodate various technical and linguistic differences, which can be difficult, not only from a technical standpoint, but also from an ethical one. Current practice suggests three possible strategies, each of which has ethical implications.

Why may bias that privileges a dominant culture be unethical? Why may cultural insensitivity be unethical? What are some other advantages and disadvantages of globalization, localization, and collaboration beyond those discussed in the sidebar?

One communication strategy that many companies have implemented is *globalization,* which often includes standardization of writing styles in several ways:

- terminology guidelines — using consistent nomenclature and avoiding words with multiple common meanings
- cultural guidelines — avoiding humor and culture-specific examples in text, photos, and symbols
- readability guidelines — using short sentences and keeping subjects and verbs close together to make writing unambiguous and predictable

Unfortunately, standardization often ignores culture differences because the standards may resemble the "dominant" culture and marginalize or even ignore the local culture.

A second strategy is *localization.* This attends to cultural differences by incorporating verbal and visual colloquialisms as well as the history and culture of the audience into the Web site. While this strategy is increasingly viable, few companies have adopted it because it is time consuming and expensive. Localization can also backfire because not everyone is receptive to outsiders' attempts to blend into the local culture. If a translation is naive, incomplete, or haphazard, the result reflects insensitivity to the local culture — the opposite of the intended effect.

A third strategy is collaboration by people working on *multicultural/multilinguistic teams.* This strategy shows the most promise for companies looking for creative ways to address cultural differences. Technical communicator Steve Chu compares the collaborative process to the dilemma of which utensils to offer at a Chinese restaurant outside a Chinese-speaking country:

> The last time you dined at a Chinese restaurant . . . were forks, a pair of chopsticks, or both on your table? Did it seem odd that you had to request a pair of chopsticks because, after all, you were in a Chinese restaurant? Or did it seem odd that you had to request a fork because, after all, you were in a Western European country? (206)

According to Chu, one solution — the combined availability of chopsticks and a fork — is an appropriate metaphor for a bilingual Web site. The combination is a cultural hybrid that neither ignores cultural differences nor naively attempts to localize. In practical terms, this means Web sites that have features such as these:

- Parallel hypertext, or versions of the same page in different languages
- Similar layout and color schemes of the pages to create cohesion
- Links for visitors to switch from one language version to another
- Simple design that avoids cultural icons to minimize misinterpretation by international users

As markets continue to expand on a global scale, technical professionals will face decisions about creating Web sites that efficiently, accurately, and fairly communicate across many cultures and languages.

As a Web developer or a manager for an organization, which of the three communication strategies would you advocate? How might you defend your chosen strategy as a sound workplace practice, as a matter concerning both economics and ethics?

Globalization. Defining a global organization is more difficult because the definitions of "global" vary. Social theorist Anthony Giddens explains that globalization links local and distant cultures in ways that affect both.[11] In practice, *globalization* is the unrestricted movement of ideas and people, services and systems, goods and money across national borders around the world, made possible because of cooperative global economics, politics, and technologies.

One example of global impact of text and visuals is the series of warnings on cigarette packages in Canada. These new warnings provide dramatic, full-color images that in Canada must cover the top 50 percent of the front and back of the cigarette package along with explicit text warnings about the dangers of smoking. Do the warnings work? A study of more than 2,000 Canadians by the Société canadienne du cancer reports that 58 percent of the smokers interviewed said that the full-color pictures made them think more about the health effects of smoking. The new warnings increased the motivation of 44 percent of the smokers to quit smoking.[12] Figure 2.2 shows three of the 16 images and warnings that cigarette manufacturers are legally obligated to put on their Canadian packaging.

To see all 16 of the images and warnings in the Canadian series and the Société canadienne du cancer report about the impact of the warnings, go to www.english.wadsworth.com/burnett6e.

CLICK ON WEBLINK

CLICK on Chapter 2/warning images

CLICK on Chapter 2/Société canadienne du cancer evaluation

WEBLINK

FIGURE 2.2 Visuals and Text of Three of the Sixteen Warnings on Canadian Cigarette Packs[13]

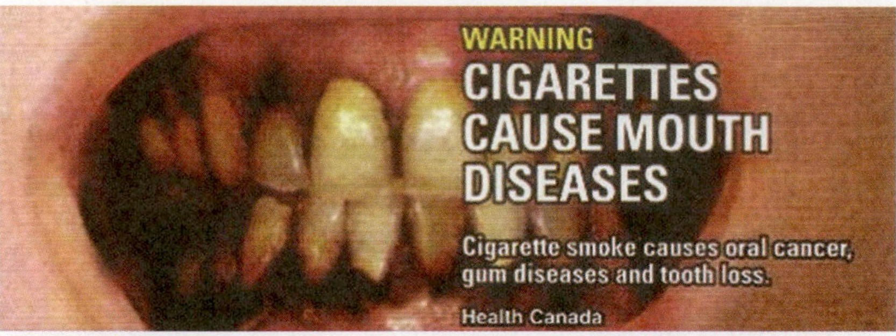

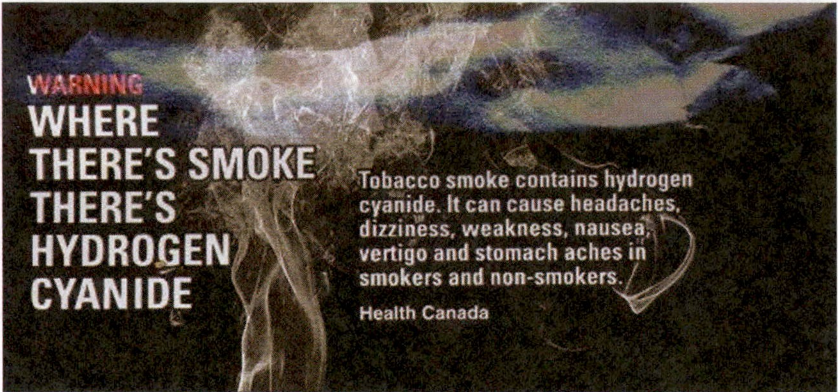

The effectiveness of these warnings has been noticed beyond Canada's borders, sparking global interest in similar graphic, explicit warnings. Others interested in this approach include the World Health Organization (WHO) and the United Nations, a number of countries in Africa and the European Union, Brazil, New Zealand, Thailand, and the United States.[14] A national warning may become a global warning.

While the warnings used on Canadian cigarette packs are a highly visible side of global interest in public health, such warnings are supported by a less visible and more technical side: the ongoing research reported in technical reports. Figure 2.3 shows the first page of a WHO press release summarizing a major technical report about global cancer rates.

FIGURE 2.3 **Page 1 of Press Release about World Health Organization Technical Report**[15]

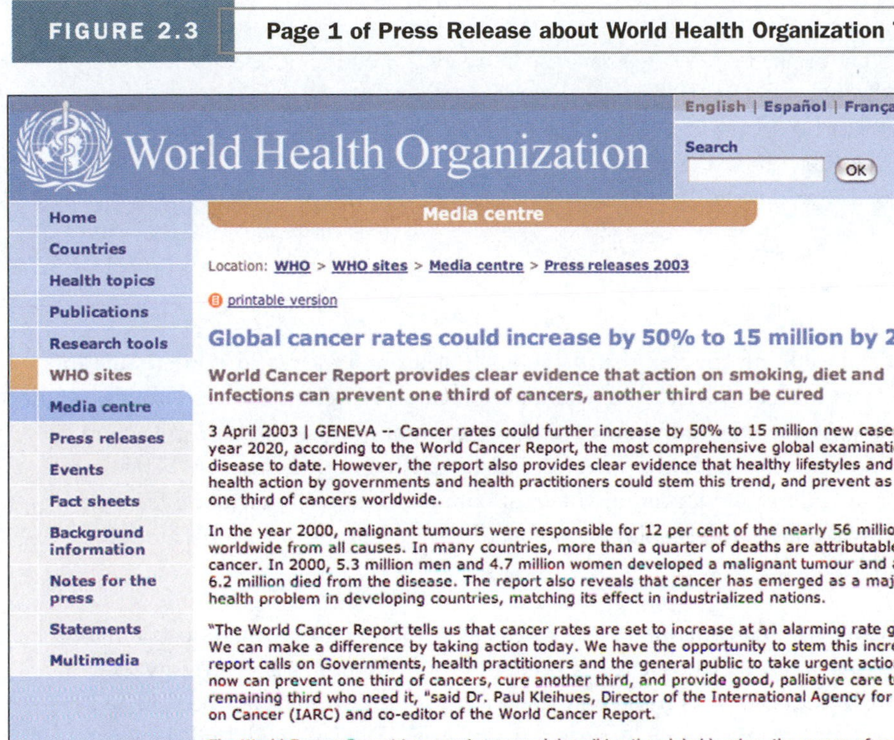

Media centre

Location: WHO > WHO sites > Media centre > Press releases 2003

printable version

Global cancer rates could increase by 50% to 15 million by 2020

World Cancer Report provides clear evidence that action on smoking, diet and infections can prevent one third of cancers, another third can be cured

3 April 2003 | GENEVA -- Cancer rates could further increase by 50% to 15 million new cases in the year 2020, according to the World Cancer Report, the most comprehensive global examination of the disease to date. However, the report also provides clear evidence that healthy lifestyles and public health action by governments and health practitioners could stem this trend, and prevent as many as one third of cancers worldwide.

In the year 2000, malignant tumours were responsible for 12 per cent of the nearly 56 million deaths worldwide from all causes. In many countries, more than a quarter of deaths are attributable to cancer. In 2000, 5.3 million men and 4.7 million women developed a malignant tumour and altogether 6.2 million died from the disease. The report also reveals that cancer has emerged as a major public health problem in developing countries, matching its effect in industrialized nations.

"The World Cancer Report tells us that cancer rates are set to increase at an alarming rate globally. We can make a difference by taking action today. We have the opportunity to stem this increase. This report calls on Governments, health practitioners and the general public to take urgent action. Action now can prevent one third of cancers, cure another third, and provide good, palliative care to the remaining third who need it, "said Dr. Paul Kleihues, Director of the International Agency for Research on Cancer (IARC) and co-editor of the World Cancer Report.

The World Cancer Report is a concise manual describing the global burden, the causes of cancer, major types of malignancies, early detection and treatment. The 351-page global report is issued by IARC, which is part of the World Health Organization (WHO).

Dr Gro Harlem Brundtland, Director-General of WHO, states: "The report provides a basis for public health action and assists us in our goal to reduce the morbidity and mortality from cancer, and to improve the quality of life of cancer patients and their families, everywhere in the world,"

Examples of areas where action can make a difference to stemming the increase of cancer rates and preventing a third of cases are:

- Reduction of tobacco consumption. It remains the most important avoidable cancer risk. In the 20th century, approximately 100 million people died world-wide from tobacco-associated diseases
- A healthy lifestyle and diet can help. Frequent consumption of fruit and vegetables and physical activity can make a difference.
- Early detection through screening, particularly for cervical and breast cancers, allow for prevention and successful cure.

The predicted sharp increase in new cases - from 10 million new cases globally in 2000, to 15 million in 2020 - will mainly be due to steadily ageing populations in both developed and developing countries and also to current trends in smoking prevalence and the growing adoption of unhealthy lifestyles.

"Governments, physicians, and health educators at all levels could do much more to help people change their behaviour to avoid preventable cancers," says Bernard W. Stewart, Ph.D., co-editor of the report, Director of Cancer Services, and Professor, Faculty of Medicine, University of New South Wales, Australia. "If the knowledge, technology and control strategies outlined in the World Cancer Report were applied globally, we would make major advances in preventing and treating cancers over the next twenty years and beyond."

From a global perspective, there is strong justification for focusing cancer prevention activities particularly on two main cancer-causing factors - tobacco and diet. We also need to continue efforts to curb infections which cause cancers," said Dr Rafael Bengoa, Director, Management of Non-communicable disease at WHO. "These factors were responsible for 43 per cent of all cancer deaths in 2000, that is 2.7 million fatalities, and 40 per cent of all new cases, that is four million new cancer cases."

ACCESSIBILITY

- *The document can be read in English, Spanish, or French.*
- *The document can be read online in a sans serif font, which is easy for reading on a monitor.*
- *Users can select a printed version, which automatically converts to a serif font, which is easier for reading on paper.*

COMPREHENSIBILITY

- *This is for a nonexpert audience, so the vocabulary is accessible and the mortality statistics are simplified.*
- *A clear problem-solution is established in the opening paragraph.*

USABILITY

- *The most important information is put in the headline and repeated in the opening paragraph.*
- *Possible actions are identified in the secondary headline.*
- *Information is presented in descending order of importance.*
- *The scope of the investigation is defined.*

WEBLINK

To see the complete WHO press release in Figure 2.3 and to access the executive summary and the report, go to www.english.wadsworth.com/burnett6e.

CLICK ON WEBLINK

CLICK on Chapter 2/WHO press release

CLICK on Chapter 2/WHO summary and report

The health information that WHO collects around the world is interesting to an array of global readers, all of whom have common interests in world health, regardless of national boundaries. Standardization of processes of both collecting and reporting information is one of the characteristics of globalization. While such standardization reduces, modifies, or eliminates idiosyncratic, local features, it also makes more information accessible to a larger number of people around the world.

Localization. The attention given to globalization has given rise to its counterpart: *localization.* Richard Harris, an economist at Simon Fraser University, explains that localization is based on preferences for familiar goods, processes, and services that have recognizable characteristics, even if global goods and services may have a higher quality or a lower price.[16] Increasing globalization has resulted in the reduction of goods and services that are localized — that is, ones that are distinctive because they reflect local culture.

A European-based organization, Localisation Industry Standards Association (LISA) works with its members to encourage global responsibility balanced with local sensitivity and usability. Companies around the world have a responsibility to respect the cultures with which they do business. The values advocated by LISA — responsibility, entrepreneurship, leadership, and cooperation — can be used by any organization that is localizing information in order to improve communication.[17]

Sometimes you'll hear the terms *globalization* and *localization* used as if they're mutually exclusive opposites, but they don't have to be.[18] In fact, both globalization and localization have compatible advantages that technical professionals need to consider. Many products that are used all over the world (for example, agricultural seed and equipment; computer hardware and software; medical equipment and supplies) require design and development for an international market. This part of design and development involves considering what capabilities users will need, regardless of their culture. For example, word processing software needs to be able to cut, copy, and paste highlighted text, regardless of the language. Similarly, computer hardware needs to be installed in the same way, regardless of the culture; however, making the information available in multiple languages increases the likelihood that the installation will be done correctly. Figure 2.4a shows the languages available for the installation of an HP Storage-Works SDLT 220 GB internal tape drive. Figure 2.4b shows one of the steps in English and then partial images of the same step in six other languages.

Products and services for people all over the world often need to be localized so they are appropriate for specific users or small groups of users, taking into account needs that are unique to their country, customers, and/or language. Sometime this means idiomatic translation of the text and redesign of visuals, including icons. For example, word-processing software often uses an icon of a trash can as a place for users to deposit their deleted text; however, in many parts of the world, trash cans don't look like those in the United States, so the icon — selected by US designers to relate the visual to the function — is an arbitrary image for much of the world.

FIGURE 2.4A Languages for Tape Drive Installation Instructions[19]

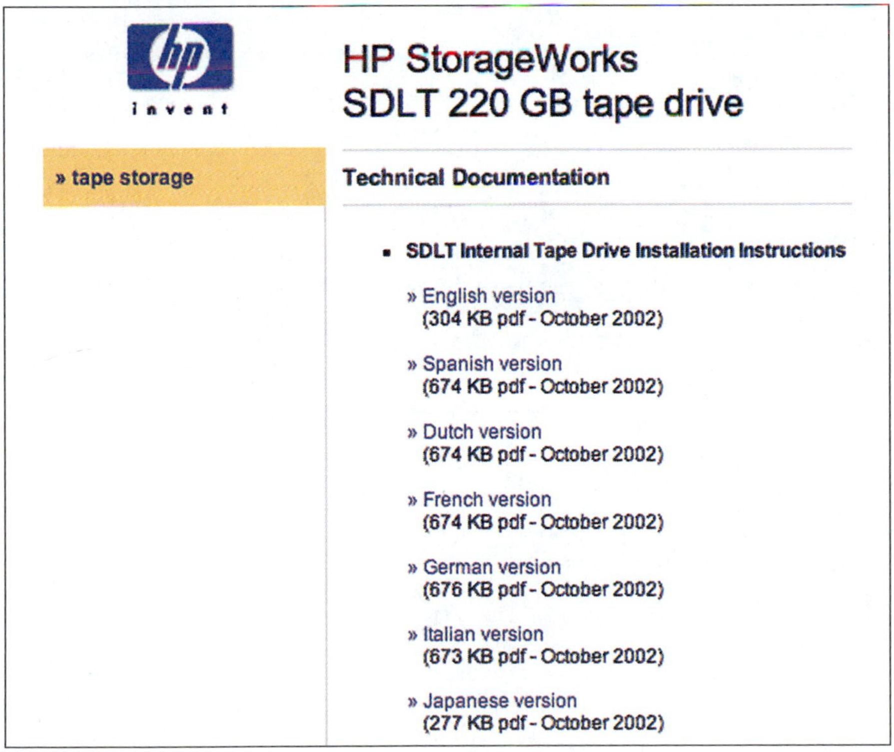

HP StorageWorks
SDLT 220 GB tape drive

» tape storage

Technical Documentation

- **SDLT Internal Tape Drive Installation Instructions**
  - » English version (304 KB pdf - October 2002)
  - » Spanish version (674 KB pdf - October 2002)
  - » Dutch version (674 KB pdf - October 2002)
  - » French version (674 KB pdf - October 2002)
  - » German version (676 KB pdf - October 2002)
  - » Italian version (673 KB pdf - October 2002)
  - » Japanese version (277 KB pdf - October 2002)

FIGURE 2.4B Languages for Tape Drive Installation Instructions[19]

ACCESSIBILITY

- *Steps are clearly labeled with a number and a task.*
- *The accompanying figure is labeled and placed next to the relevant step.*

COMPREHENSIBILITY

- *The imperative verbs (e.g., "install," "slide," "remove") are unambiguous.*
- *The introductory conditional phrases ("If . . .") are short enough to not interfere with the main task.*
- *Languages: English, Spanish, Dutch, French, German, Italian, Japanese*

USABILITY

- *The two situations (with and without rails) are clearly differentiated.*
- *Illustrations are easy to use because the diagrams are simplified and important parts are shaded.*

Step 3, Set the Terminator Power (Figure A, insert right)

The drive ships from the factory with terminator power (TERM PWR) enabled. This allows the drive, in addition to the controller, to provide the termination power.

To enable TERM PWR, a jumper must be placed on 7 (see Figure A, insert right) of the termination block.

Step 4, Install the Drive (Figure C)

With Drive Rails

If your computer requires drive rails, slide the tape drive into two available removable media bays and secure with screws provided.

Without Drive Rails

If your computer does not require drive rails, remove and discard them. Slide the tape drive into two available removable media bays and secure with the screws that were removed from the rails.

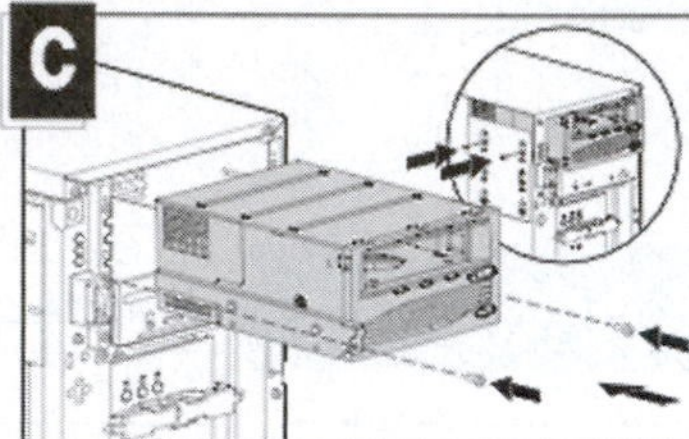

Step 5, Connect the Cables (Figure D)

Connect an available power cable and the SCSI signal cable (provided in the kit) to the tape drive. If you are not using the SCSI signal cable in the kit, make sure the existing cable is properly terminated.

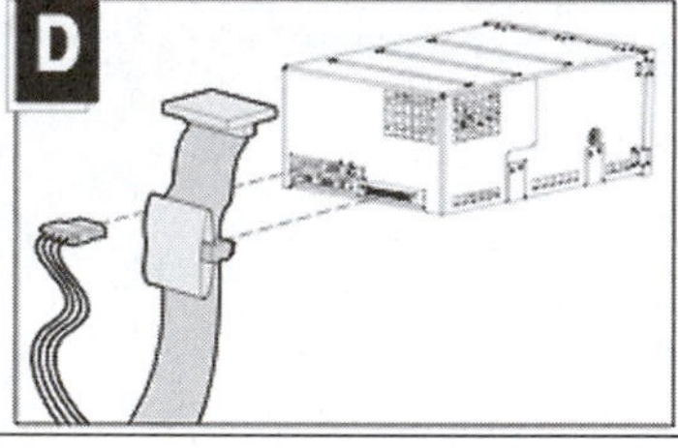

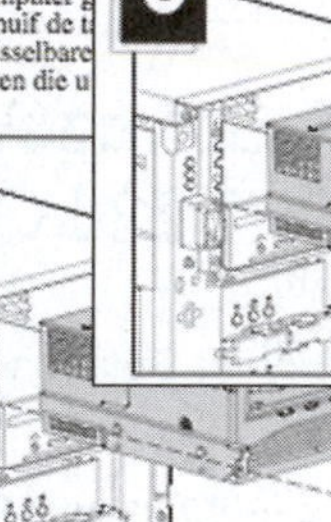

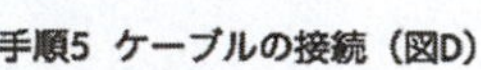

手順5　ケーブルの接続（図D）

テープ　ドライブに使用可能な電源ケーブルとSCSI信号ケーブル（キットに同梱）を接続します。キットに同梱されているSCSI信号ケーブルを使用しない場合は、既存のケーブルが適切に終端されていることを確認してください。

By definition, cultures are necessarily about collectives, whether characterized by nationality, race, disability, or some other defining feature, not individuals. However, no culture — even those that are remarkably consistent — is monolithic. No culture is comprised of identical individuals, and no culture is the same in different contexts and circumstances.

WEBLINK

The very term *globalization* provokes strong reactions. Go to **www.english.wadsworth.com/burnett6e** for links to different definitions, most of which reinforce territoriality and only one of which argues for a fundamental change in social geography. Then check the implications of delocalization.

CLICK ON WEBLINK

CLICK on Chapter 2/globalization

CLICK on Chapter 2/delocalization

Cultural Values

Characterizing cultures provides information that technical professionals can use. However, technical professionals don't communicate to an entire culture; instead, they direct documents, presentations, and visuals to specific audiences. The issue is how much to localize documents — that is, how much to adapt communication to local culture in terms of language, practices, values, and so on.

While cultural awareness often involves a global perspective, in many multicultural countries (for example, Brazil, Canada, Russia, South Africa, the United States), technical professionals need such awareness to be skillful and productive *within* their own national borders. Let's consider why all of us should learn about other cultures. Understanding and respecting other cultures increases our effectiveness, so we are less likely to misinterpret attitudes, actions, and artifacts that reflect a different cultural perspective. Increased knowledge about and sensitivity to other cultures can improve performance in several ways:

- ***Response.*** Documents, presentations, and visuals that are adapted to people from other cultures are likely to get more attention and response.
- ***Productivity.*** Meetings and other interactions with people who have different cultural perspectives are often more productive and pleasant.
- ***Explanations.*** Alternative points of view can often explain attitudes, actions, tools, and artifacts that aren't immediately or obviously compatible with our own perspectives.

- ***Compliance.*** Attention to cultural perspectives is likely to increase compliance with organizational policies and procedures and thus increase consistency, safety, productivity, and compatibility within an organization.

WEBLINK

The Society of Technical Communication has a special interest group (SIG) that focuses on international technical communication. Go to www.english.wadsworth.com/burnett6e for a link to more information about technical communication outside the United States.

CLICK ON WEBLINK

CLICK on Chapter 2/international

Analyzing Culture

This section focuses on ways to analyze national culture, organizational culture, and individuals in cultures. While virtually all researchers who study culture explain that culture is about groups, they also understand that culture shapes and is shaped by individuals. Individuals typically work for and with various kinds of organizations that have distinct cultures. And both individuals and organizations are located in nations with geographic, political, and cultural boundaries.

National Cultures

Research indicates that having a long-term perspective and a focus on the individual rather than the group are both correlated with national economic growth.[21] Why do you think this may be an accurate correlation?

Cultural differences are interesting, but do they really matter? Pricewaterhouse Coopers reports that "67% of international managers blame cultural differences for the difficulties they are confronted with working internationally."[20] That's enough to get some attention — especially if understanding cultural differences can help improve productivity and cooperation.

Sometimes even understanding that differences exist doesn't prevent a cultural gap. For example, several years ago I was in Tokyo, Japan, meeting with senior managers at Fuju-Xerox. I was interested in learning how their document design and development teams functioned and how their technical manuals were produced. After lengthy and formal greetings, introductions, and refreshments, we turned to the business of the meeting: the preliminary stages of my investigation of their collaborative processes and document production. Time and again during the ensuing discussion, I had to remind myself that their smiling and nodding simply signaled that they had heard me, not that they necessarily agreed with me. In the United States, their actions would have signaled agreement with what I was saying. In Japan, their actions signaled acknowledgement and politeness, not agreement.

What's the point? Like I was in Japan, many of us are strongly influenced by our own cultural frameworks. Common sense tells us that cultures in various countries are different, but we have don't always have sufficient knowledge to function comfortably in other cultures. Figure 2.5 provides a series of binaries and questions that have been identified by anthropologists and psychologists as useful ways to distinguish national cultures. These binaries provide a way to begin discussions about culture. However, binaries tend to treat cultures as monolithic, that is, assuming that all Brazilian citizens are alike or that all Russian citizens are alike, which can lead to stereotyping extremely diverse national and organizational cultures. Despite the limitations, however, the binaries and questions in Figure 2.5 are a useful starting place.

The binaries in Figure 2.5 are not the only distinctions that characterize nations. Factors such as language, proxemics, and time, usually learned at an early age, are a few additional factors.

Languages. Language is one of the most visible aspects of national culture. English is spoken as a first or a second language on five continents by people who make up about 49 percent of the world's population.[22] This means that many of the technical documents and presentations you prepare can be in English, regardless of the geographic location of the audience. However, despite the dominant role of English in education, research, government, and business, it is often not the first language of a country; in fact, the use of English is sometimes resented and aggressively challenged. What's a reasonable professional response? Acknowledging that language is an inherent part of culture, so its use is always part of a social context and is always political.[23]

Since language is a political hot button, demanding the use of a particular language or making joking or demeaning remarks about a country's first language is a certain way to alienate colleagues. When you're trying to negotiate the complex relationships between globalization and localization, be respectful of local languages.

Unquestionably, English is the common denominator in the international workplace — but whose English? The term *world Englishes* refers to the fact that English isn't the same from country to country (or even from region to region within countries). So, even when everyone is speaking English, speakers from, for example, Australia, Canada, Hong Kong, India, Singapore, South Africa, the United Kingdom, and the United States may have remarkable distinctions. For example, some words started in a particular area but are now widely used: *bungalow,* a term from South Asia, is now used wherever English is spoken. Or a single artifact can have a number of different names — for example, *gym shoes, plimsolls, sneakers,* and *tackies* are different names for the same kind of shoes.[24]

Many people resent English being used as the primary language of the international workplace. One cause of such resentment is the belief that the use of English diminishes respect for the native language. What else might cause such resentment? How can people from English-speaking countries defuse such resentment?

A common misconception is that everyone in the United States speaks English. In fact, according to Census 2000, 28 million U.S. residents age five and over speak Spanish at home; more than half of these also report speaking

FIGURE 2.5 **A Summary of Factors Differentiating National Cultures**[25]

Individualism	***Individual - - - - - - Group*** How much focus is on the individual, and how much focus is on the collective — that is, the group or team?
Task focus	***Tasks - - - - - - Relationship*** How much focus is on the tasks being completed, and how much is on the relationship of the people doing the tasks?
Forthrightness	***Indirectness - - - - - - Directness*** Do people tend to learn about each other indirectly, through a third party? Or do they tend to learn about each other when they speak directly to each other about their backgrounds, roles, and responsibilities?
Power distance	***Hierarchy - - - - - - Equality*** How is power concentrated — at the top of a hierarchy or distributed equitably? What's the distance between people who make decisions and people who do the work resulting from those decisions? Are opportunities equally available to everyone?
Interaction	***Nonverbal interaction - - - - - - Verbal interaction*** Do people tend to be cautious, using clichés and assuming that body language and tone of voice convey hidden messages, or do they speak openly, expressing opinions explicitly?
Source of credibility	***Formal status - - - - - - Achievement*** How much attention is paid to status: age, rank, title, longevity in the organization, education, and how much attention is paid to a person's actual accomplishments relevant to the task?
Perspective	***Short-term perspective - - - - - - Long-term perspective*** Do people have a short-term perspective, concerned with fulfilling social obligations and protecting their reputation (saving "face")? Or do they have a long-term perspective, forward thinking, valuing perseverance and thrift (work hard now and get rewarded later)?
Materiality	***Material success - - - - - - Quality of life*** Are the people more concerned with material success or with quality of life?
Uncertainly avoidance	***Preference for certainty - - - - - - Tolerance for ambiguity*** Does the culture generally prefer certainty (and thus focus on laws, rules, regulations, and controls to reduce the uncertainty)? Or does the culture have a tolerance for ambiguity (and thus more readily accepts change and takes more and greater risks)?

fluent English.[26] Few people are surprised that Arizona, California, Florida, New Mexico, and Texas have a large number of native speakers of Spanish. But English isn't necessarily the first language of workers elsewhere in the United States. For example, in Iowa, many meat processing and packing companies employ a large number of Hispanic employees. Communication in these workplaces is often bilingual.

Recently I was in Durban, South Africa, conducting a seminar for about 30 professors in various areas of science and technology. We were all speaking English, but I was only one of four other people in the room for whom English was a native language. South Africa is home to 31 languages,[27] 11 of which are the official languages of the country: Sepedi, Sesotho, Setswana, siSwati, Tshivenda, Xitsonga, Afrikaans, English, isiNdebele, isiXhosa, and isiZulu.[28] That meant that more than 90 percent of the people in the seminar were speaking and writing in their second or third (and, in some cases, their fourth) language. And while everyone conceded that English was necessary, not everyone was happy about the concession, because of ethnic loyalties.

Cultural differences are often conveyed with specific vocabulary. For example, what is called the *trunk* of a car in the United States is called the *boot* in countries including Australia, the UK, and South Africa. In the United States, a colleague on a business trip who wants to make sure you awake for an early-morning meeting might say, "Would you like me to give you a wake-up call?" while a colleague from the UK might say, "Would you like me to knock you up?"

Once you're on the highway driving to a meeting in Durban, South Africa, you might see a warning vehicle like the one in this photo: "Abnormal Load Ahead." And just ahead is a large truck (also called a *lorry*) with a large sign, "Abnormal," mounted on the front and rear, like the one in the next photo. In the United States, the signs would probably read, "Wide Load Ahead" on the warning vehicle and "Wide Load" on the truck itself.

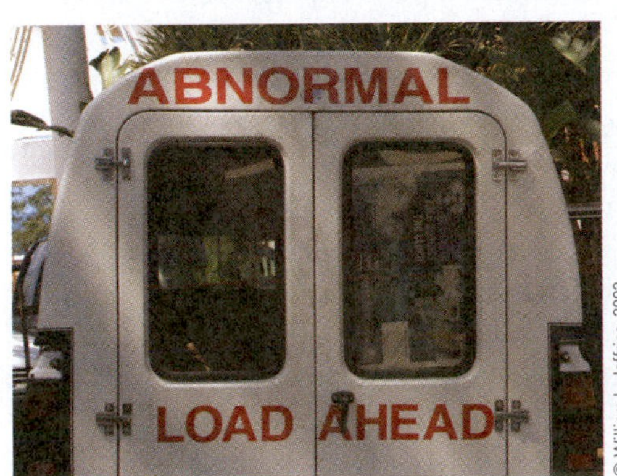

A warning vehicle in South Africa signals an extra-wide load ahead

A truck in South Africa carries a sign signaling an extra-wide load

Simply put, people who immediately understand "Abnormal Load Ahead" or "Abnormal" on a truck belong to a different *discourse community* than people who immediately understand "Wide Load Ahead" and "Wide Load." Thus, a discourse community is comprised of people who share a number of common cultural characteristics. For example, they may belong to the same gender, race, nationality, or profession. Or they may belong to the same political party, religion, athletic team, residential community, or corporation. However, their most important shared characteristic is their common use of a specific language to accomplish something and to get work done.

> *What discourse communities do you belong to? What are beliefs and behaviors from different discourse communities that might conflict with each other? How do you decide which beliefs and behaviors to follow in a given situation?*

Clearly, everyone belongs to multiple discourse communities and usually shifts seamlessly between them. Discourse communities can be layered; you can be a member of a discourse community comprised of people who work for the same organization, while also being a member of various discourse communities within the organization. For example, you might work in research and development, accounting, manufacturing, QC/QA, training, or marketing. And within each department, you may belong to the group of people who use PCs (or Macs), participate in flextime (or telecommuting), prefer e-mail (or face-to-face) conversations, or coach the organization's softball team (or run 5k charity events).

Proxemics. The study of the physical distance between people, called *proxemics,* shows that distance means different things in different cultures. We have layers of an invisible bubble around us. We allow family members, lovers, and very close friends into our *intimate space,* which in the United States is up to about 18 inches. We allow friends and close colleagues into our *personal space,* which for most people in the United States ranges from about 18 inches to about four feet. We allow colleagues and business acquaintances into our *social space,* which in the United States is typically 4 feet to about 10–12 feet. And our *public space,* in the United States beyond 10–12 feet, is for public events: business meetings, oral presentations, lectures, performances. Different cultures vary in what is considered acceptable for intimate, personal, social, and public distance.

Generally, people from North America expect larger intimate and personal space than people from many other countries, which sometimes leads to problems. For example, North American workers expect other workers to stay outside their intimate space, that is, to stay more than 18 inches away. They can misunderstand the closeness of coworkers who regularly move inside the 18-inch invisible boundary. The following photos illustrate differences in cultural practice.

Time. Time means different things in different cultures. How late can someone be and still be "on time"? What is a "long" meeting? What does "right away" mean? What does "multitasking" mean?

© Reza Estakhrian/Getty Images

© VCL/Chris Ryan/Getty Images

Professionals from different parts of the world have different perspectives about the acceptable distance between people in a discussion. What role does gender play in acceptable distances for workplace discussions?

Some of the early work about people's sense of time was done more than 50 years ago by Edward and Mildred Hall. In a recent book, they extend their early work, which explains that cultures tend to treat time as *monochronic* or *polychronic,* as these two lists show.[29]

Monochronic people

- Tend to do one thing at a time
- Concentrate on the specific job
- Take commitments seriously; set and comply with deadlines
- Are committed to the job
- Adhere to established plans
- Are concerned about not disturbing others; follow rules of privacy and consideration
- Emphasize promptness
- Build short-term relationships

Polychronic people

- Tend to do many things at once
- Respond to interruptions
- Consider time commitments as desirable, not absolute
- Are committed to human relationships
- Change plans often and easily
- Are more concerned with family, friends, close business associates than with privacy
- Base promptness on the relationship
- Build lifetime relationships

Of course, general lists like this don't apply to specific individuals but, instead, characterize many people. For example, monochronic people tend to see time as inflexible and tangible, so work is often done in isolation, and it may take priority over people. In contrast, polychronic people tend to see time as flexible and fluid, so, for example, appointments are shifted around at the last minute and changes are made to accommodate more important people. Because work time is not clearly separable from personal time, relationships may be more important than tasks.

WEBLINK

If you're interested in learning a lot more about concepts and practices time, go to www.english.wadsworth.com/burnett6e for a link to a detailed site devoted to the study of time.

CLICK ON WEBLINK

CLICK on Chapter 2/time

Organizational Cultures

When you're working in a multinational or global organization (whether a large corporation, small business, government agency, NGO, or nonprofit organization), you need to be concerned both with national culture and with organizational culture. When the national cultures are remarkably different, the organizational culture can provide a common bond that crosses national boundaries.

The shared beliefs and behaviors of people in an organization comprise the organizational culture. One way to describe organizational culture is to analyze dimensions that can be expressed as binaries — that is, contrasting practices that characterize activities in the organization, as shown in Figure 2.6. These are similar to the binaries for national cultures, but they apply specifically to organizations. As with nations, no single factor is sufficient to determine whether a particular practice is productive or effective, as shown in the photos on page 62. While the binaries in Figure 2.6 do not reflect all the factors that characterize an organization, they can be compiled to create a remarkably full portrait.

Sometimes organizational culture is conveyed in formal ways: policy statements, organizational charts, performance criteria. These formal procedures and policies need to be consistent with the actual policies. And sometimes organizational culture is conveyed in informal ways: common practices, lore, gossip, e-mail.

FIGURE 2.6 Factors and Questions about Organizational Cultures[30]

Factor	Continuum and Questions
Focus on people	***Performance oriented*** ---------- ***People oriented*** Does the organization seem more interested in tasks or in the employees completing those tasks?
Approach to processes	***Results oriented*** ---------- ***Process oriented*** Are employees and managers more concerned with getting the job done or with problem-solving processes that might be generalized to other tasks?
Attitude toward cooperation	***Competitive*** ---------- ***Cooperative*** Is the relationship between managers and employees competitive or cooperative? Are employees competitive or cooperative with each other?
Adherence to standards	***Parochial*** ---------- ***Professional*** Does the organization follow idiosyncratic local practices or professional standards?
Receptivity to change	***Closed to change*** ---------- ***Open to change*** Does the organization have a closed system that follows existing rules carefully and is resistant to change, regardless of the benefits or an open system that is receptive to new ideas and activities?
Access to resources	***Restricted*** ---------- ***Available*** Do employees have ready access to sufficient supplies, tools, and equipment including computer resources necessary for doing their jobs well?
Motivation to work	***Required work*** ---------- ***Pride in work*** Do employees see their work as a necessary but not integral or enjoyable part of life? Or do employees see their work as interesting, important, and fulfilling?
Relation to life outside the organization	***Insular*** ---------- ***Engaged*** Do managers and employees see work as part of a balanced life along with family and community commitments? Does the organization insulate itself from the activities or the community and the lives of its employees? Does the organization support people's lives outside the organization?
Respect toward employees	***Cogs in the machinery*** ---------- ***Valued members*** Are employees valued for their contributions? Are their opinions valued and taken seriously?
Management style	***Tightly controlled*** ---------- ***Loosely controlled*** Is the management tight and controlling, typical of hierarchical organizations, or loose and flexible? Do employees have assigned projects, set work hours, and dress codes?
Communication channels	***One-way*** ---------- ***Two-way*** Is communication always top-down? Or are employees encouraged to communicate with their managers, participate in quality-control projects, and voice their opinions?
Rules and regulations	***Normative*** ---------- ***Pragmatic*** Are the rules and regulations prescriptive and applied without regard to individual circumstances, or are they adapted and modified to fit specific situations?

GANSU PROVINCE PROJECT IN THE PEOPLE'S REPUBLIC OF CHINA

Iowa has a unique connection to Gansu Province. The Loess Hills in western Iowa contain the same type of soil as that found in Gansu. Deposited during the Glacier Age, these highly erosive, wind-blown soils are not very stable for building roadways. US experience in developing roadways on this soil meant that we could offer assistance to the Chinese in developing effective roadways.

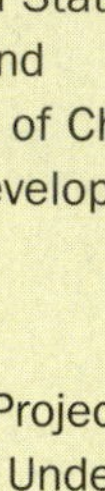

Duane Smith, a civil engineer and trainer for Iowa State University's Center for Transportation Research and Education (CTRE), traveled to the Gansu province of China to negotiate an agreement to improve highway development in this struggling area.

For the Gansu Province Project, the American and Chinese engineers developed a Memorandum of Understanding (MOU) that identified specific individual and group tasks. Because of cultural expectations in the People's Republic of China, Duane took special care to include particular officials and groups. An excerpt from the 4-page MOU is shown below.

> The purpose of the Coordinating Council will be to facilitate research, education, extension and training services for the Government agencies of the Gansu Province, for government agencies of other Provinces of China as well as the Central Government of China and for domestic and foreign private firms that are participating in the development of improved transportation systems within the Gansu Province and throughout China.

> 府机构、为中国其它省份及中央的政府机构，以及参与发展、改善甘肃及全国公路系统状况的中国及外国公司提供研究、教育和延伸服务。

Duane commented on the process: "As we discussed different aspects of the second document, the Training Agreement, I needed to have visuals and graphics to help me express my points. When we were all drawing on the board, we really started to communicate. When we could create an engineering drawing and put numbers on it, things clicked."

To make this cross-cultural experience successful, both parties needed individual preparations, specified in a Training Agreement. Party A (Gansu) agreed to provide additional English training to its engineers as well as obtain necessary visas to travel to the United States. Party B (the US hosts) agreed to provide a detailed training program that would allow the engineers to use a series of technical skills upon returning to China. An excerpt from the 4-page Training Agreement is shown below.

Parties Obligations and Responsibilities:

Party A:

1. Select and send trainees to ISU/CTRE.
2. Provide English training prior to the departure to ISU/CTRE.
3. Pay the ISU/CTRE billings within 30 days after receipt.
4. Prepare all the necessary trainee travel documents to travel to ISU/CTRE.

Party B:

1. Develop a detailed training program.
2. Provide the coordination and instruction for trainees to implement and practice the techniques provided in the training program.
3. Complete trainee evaluations for the English training, techniques training, field practice, and the training summary report.
4. Provide the necessary documentation to assist Party A in preparing all necessary trainee travel documents to travel to ISU/CTRE.

四．双方的责任和义务

甲方：

1．负责选派培训对象；
2．负责选派对象出国前的英语培训；
3．在接到培训费用通知的 30 天内，支付 ISU/CTRE 培训费用；
4．负责办理培训人员的出国手续；
5．负责审查培训计划并在 15 日内将审查意见反馈给乙方。

乙方：

1．负责制订详细的培训计划，该计划应于 2002 年 12 月 31 日前提交给甘肃省交通规划勘察设计院；
2．对理论培训组织具体的实施；
3．对每位学员作出切合实际的评价；
4．负责办理培训人员赴美的签证所需文件。
5．负责将培训费用账单寄给甘肃省交通规划勘察设计院。

五．培训时间和人数

Duane commented: "The Chinese are very serious about this project. This is one of their national initiatives. They see this project as a model that can be used for the rest of China. We plan to create a technology transfer center in Gansu, so that this initial effect will lead to future projects and ongoing training." The signatures on the Training Agreement below reflect mutual agreement.

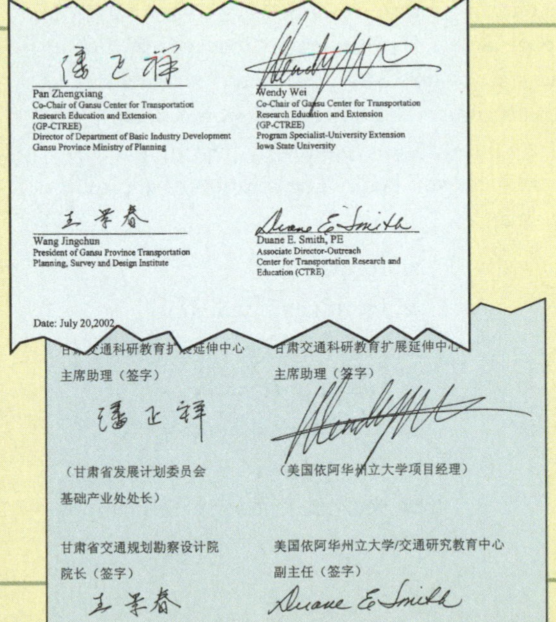

Pan Zhengxiang
Co-Chair of Gansu Center for Transportation Research Education and Extension (GP-CTREE)
Director of Department of Basic Industry Development
Gansu Province Ministry of Planning

Wendy Wei
Co-Chair of Gansu Center for Transportation Research Education and Extension (GP-CTREE)
Program Specialist-University Extension
Iowa State University

Wang Jingchun
President of Gansu Province Transportation Planning, Survey and Design Institute

Duane E. Smith, PE
Associate Director-Outreach
Center for Transportation Research and Education (CTRE)

Date: July 20,2002

甘肃交通科研教育扩展延伸中心
主席助理（签字）
（甘肃省发展计划委员会
基础产业处处长）

甘肃交通科研教育扩展延伸中心
主席助理（签字）
（美国依阿华州立大学项目经理）

甘肃省交通规划勘察设计院
院长（签字）

美国依阿华州立大学/交通研究教育中心
副主任（签字）

© Alex Tossi/Alamy

© Patrik Giardino/CORBIS

Is formality or informality a factor in productivity? Can you tell which employees work harder, are more innovative, are more productive, or enjoy their jobs more?

Organizational culture is also conveyed by the manner of communication. For example, people who go to the Dell Computer home page at **www.dell.com** can identify their country and, within seconds, can read the entire Dell site in their native language. If Mexico is their country (as in Figure 2.7a), the site will be in Spanish (Figure 2.7b). This concern for users and respect for their own language contributes to Dell being identified as one of the top five most admired companies in the world.[31]

Individuals and Their Culture

While all workplace professionals are part of a national culture as well as an organizational culture, they all have personal factors that influence the ways in which they participate in, contribute to, and react to these cultures. These personal factors also make us members of several subcultures that affect our professional opinions and, in many cases, complicate our decision making by providing contradictory information. This section helps you begin a description of yourself and the subcultures to which you belong:

How do you characterize yourself for each of these factors? Are any factors mutually exclusive? Which of them affect others? Do you find any of the boundaries of these factors fuzzy or ambiguous?

- Current and ancestral nationalit(ies)
- Races and ethnicities
- Age
- Gender
- Residences: Rural, suburban, urban
- Job status: white, pink, or blue collar
- Education
- Income
- Health and disabilities
- Marital status
- Sexual orientation
- Role(s) in community
- Political views and affiliations
- Religious views and affiliations

FIGURE 2.7A **Choosing the Country and Language for Dell Computer Company Web Site from Dell's U.S. Home Page**[32]

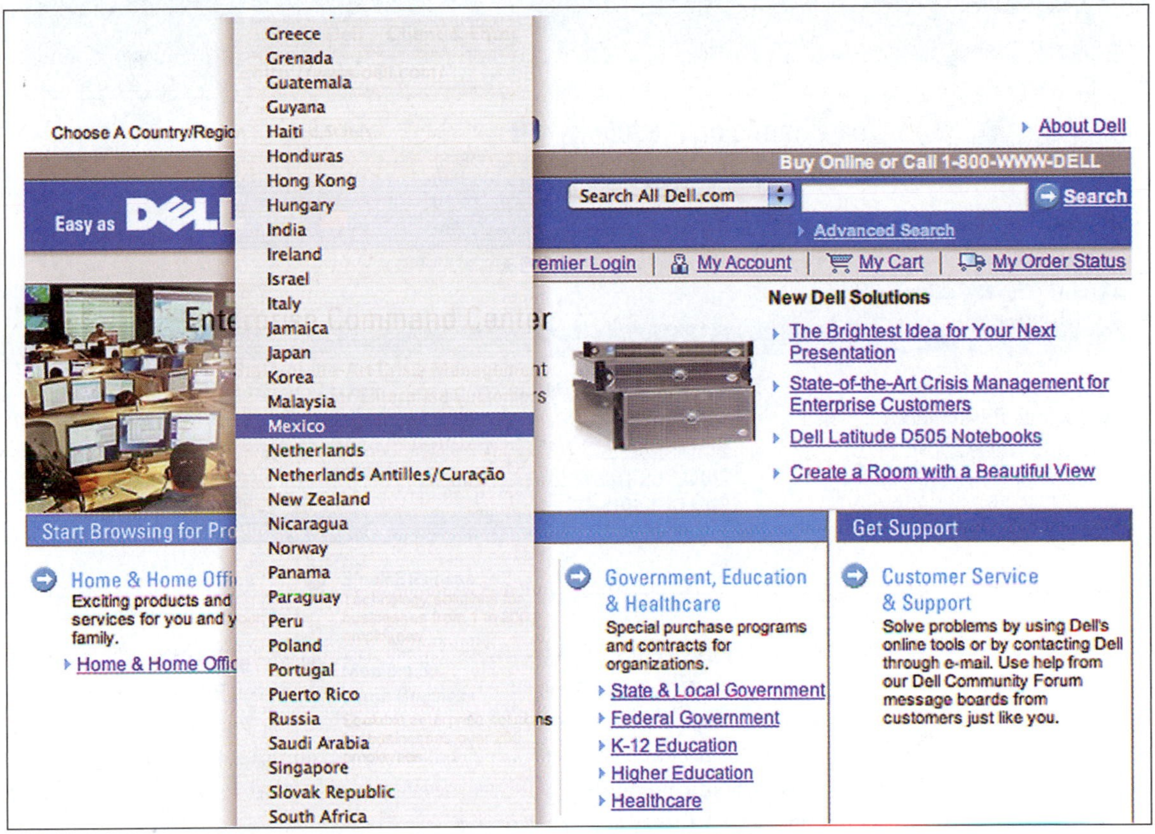

Greece
Grenada
Guatemala
Guyana
Haiti
Honduras
Hong Kong
Hungary
India
Ireland
Israel
Italy
Jamaica
Japan
Korea
Malaysia
Mexico
Netherlands
Netherlands Antilles/Curaçao
New Zealand
Nicaragua
Norway
Panama
Paraguay
Peru
Poland
Portugal
Puerto Rico
Russia
Saudi Arabia
Singapore
Slovak Republic
South Africa

Choose A Country/Regio
About Dell
Buy Online or Call 1-800-WWW-DELL
Easy as DELL
Search All Dell.com
Search
Advanced Search
remier Login | My Account | My Cart | My Order Status
New Dell Solutions
The Brightest Idea for Your Next Presentation
State-of-the-Art Crisis Management for Enterprise Customers
Dell Latitude D505 Notebooks
Create a Room with a Beautiful View
Start Browsing for Pro
Home & Home Offi
Exciting products and services for you and y family.
Home & Home Offic
Government, Education & Healthcare
Special purchase programs and contracts for organizations.
State & Local Government
Federal Government
K-12 Education
Higher Education
Healthcare
Get Support
Customer Service & Support
Solve problems by using Dell's online tools or by contacting Dell through e-mail. Use help from our Dell Community Forum message boards from customers just like you.

Do you want to better understand your personal cultural preferences? Go to www.english.wadsworth.com/burnett6e for a link to take a self-test that may help you see one version of ways you might react in certain organizations around the world.

CLICK ON WEBLINK

CLICK on Chapter 2/self-test

WEBLINK

Any professional must be qualified to fulfill the essential functions of a job. Sometimes personal factors prevent this, and no reasonable accommodation is

FIGURE 2.7B Dell Computer Company's Latin American Home Page[33]

available or possible. For example, public safety and medical professionals who are required to move (lift or carry or turn) people as a significant portion of their job must be physically able to do so. Personal factors affect job performance in other ways as well, so someone with certain kinds of severe allergies might not be able to work as a chemical engineer, forester, or pharmacist. And someone who is blind is unlikely to be a firefighter or a surgeon.

DELL
México
Sobre Dell (en inglés)
Soporte
Contacto
Premier Login
Estado de Mi Pedido
Dell Dimension™ 8300
Ideal para los Entusiastas de Aplicaciones de Multimedia
Bienvenido a Dell
Usuarios Domésticos y Oficinas Particulares
Excelentes compras de Multimedia para familias, estudiantes, y consultores independientes.
Empresas Micro, Pequeñas y Medianas
Sistemas hechos a la medida para empresas con menos de 400 empleados
Empresas Grandes y Sector Público
Desde sistemas expandibles hasta soluciones avanzadas para compañías privadas y del sector público con más de 400 empleados.

WEBLINK

The American with Disabilities Act (ADA), passed by the U.S. Congress in 1990, prohibits discrimination on the basis of disability. Go to www.english.wadsworth.com/burnett6e for a link to the text of the ADA.
CLICK ON WEBLINK
CLICK on Chapter 2/ADA

Some personal factors can indirectly affect professional performance. For example, a person's economic circumstances often affect access to education and computer technology and, thus, to workplace opportunities. And, similarly, a person's education affects job qualifications as well as job performance.

However, except for circumstances similar to those just mentioned, personal factors seldom influence professional competence. A person's success as, for exam-

© Walter Hodges/Getty Images

Workplaces benefit from the experience and expertise of professionals with a variety of abilities.

ple, an environmental engineer, physician's assistant, nanoscale physicist, or software designer typically does not depend on personal factors. For example, while race and ethnicity are powerful contributors to culture, they do not affect professional competence; our reactions (and prejudices) about race and ethnicity are learned rather than inherent. Similarly, nationality does not affect professional competence; it is often a circumstance of birth: on which side of a border a child is born or what nationality the parents claim. And a person's age, gender, sexual orientation, marital status, politics, and religion don't play any role in job performance. As the photo illustrates, doing a job well depends more on a person's professional competence than on physical abilities.

What factors affect people's access to workplace opportunities and to technology?

What are examples of personal factors that have been mistakenly correlated with job access or performance?

Increase Cultural Awareness

How can you work more productively and respectfully with colleagues from various cultures? This final section of the chapter offers some suggestions that may make your cross-cultural experiences more successful whether you're working on a cross-cultural team in your own organization, moving to a new job in a different organization, or traveling halfway around the world to work in another country.

Make some initial decisions about your own behavior and attitude. How you choose to approach a new culture may be a strong influence on your success.

- Be a keen observer. Some things are often more important than words: seating arrangements, order of speaking, body language, pauses and silences, the way people dress, interactions among people. Watch what's going on. Think about what it might mean, both in order to learn and to avoid making unintended gaffes.
- Demonstrate respect for people and practices. How? Defer judgement on things that are unfamiliar. Ask questions. Presume that input from everyone involved will inprove the output. Consider all possibilities equally.
- Demonstrate genuine interest in other cultures by your actions and attitudes. Try new foods. Adopt new ways of doing things that enhance your regular practices. Be aware of local politics, events, and controversies.
- Listen carefully, which is difficult when you don't know what you might hear. Listening for new information is challenging especially when you don't have criteria for effectively differentiating very useful but unexpected information from less useful and unexpected information. Also, tune your listening to better understand unfamiliar accents and vocabulary.
- Consider alternative interpretations; things may not be the way they seem. Does a smile mean friendliness or embarassment? Does nodding mean "I agree" or "Continue your explanation"?

WEBLINK

The Society of Technical Communication publishes a bibliography on international technical communication with an extensive list of resources. Go to www.english.wadsworth.com/burnett6e for a link to the bibliography.

CLICK ON WEBLINK

CLICK on Chapter 2/bibliography

Before you enter a new culture, whether a new organization or a new country, do your homework. Learn some basics so that you'll be less surprised when unfamiliar things happen and more prepared to act appropriately.

- Understand the hierarchy (whether an organizational chart or the government's structure).
- Make a list of people with whom you'll be working — names, positions, areas of responsibility, history with the organization, relation to others with whom you'll work. And get photos if possible so you'll recognize the people.

- Learn some common cultural customs such as giving business cards and gifts, greetings and handshakes, behaving conventionally at meetings and meals. Learn what's typically considered rude and polite, especially in relation to time, proxemics, and social behavior.
- Understand what contributes to credibility. For example, in some cultures summarizing accomplishments relevant to the current project is seen as boasting. In some cultures skipping the personal conversation and jumping directly into the business agenda is a personal affront.
- Learn some basic language ("Hello" "Thank you" "Excuse me" "How do you say . . . ?" "Where is . . . ?" "How do I get to . . . ?" "What is the cost of . . . ?").
- Know the basic layout; try to get a site plan or geographic map before you leave.

Once you arrive, welcome new experiences. To have a productive experience, you need to be receptive and considerate, even when things don't go the way you expected. The situation may be as unfamiliar to those you're working with as it is to you.

- Expect that things you never anticipated will happen, sometimes challenging your best intentions.
- Be patient about how long things may take to accomplish and how things may be done.
- Identify absolutes that affect your daily life (e.g., dietary restrictions, severe allergies, physical limitations, need for a refrigerator for medication, and so on).
- Be flexible. You may need to do some things differently. For example, you may need to offer indirect suggestions rather than give straightforward directions, revise contractual items, or acknowledge that your plans are seen as impossible — or at least intrusive and inflexible (regardless of your intentions).
- Make every effort to meet individuals and engage in conversations. Even when you're awkward and uncomfortable using unfamiliar language, keep trying.

WEBLINK

Global corporate trainer Sivasailam Thiagarajan, from Chennai, India, offers a list of guidelines and brief explanations that make intercultural communication easier. Go to www.english.wadsworth.com/burnett6e for links to his site.
CLICK ON WEBLINK
CLICK on Chapter 2/global training

Individual and Collaborative Assignments

1. Assess global citizenship. HP's Web site positions the company as a responsible global citizen:

> At a minimum, "global citizenship" is a commitment made by companies striving to do business in a manner that upholds local and international standards and values everywhere the business invests and operates, in areas including the environment, ethics, labor and human rights.[34]

Does HP meet its own high standard? According to this definition, how do the corporate cultures of Lucent Technologies and Kodak stack up? Search their Web sites and general news articles for more information to help you form an opinion. Be prepared to make a two-minute oral argument in class supporting your position.

Lucent Technologies' Culture and Diversity Statements[35]

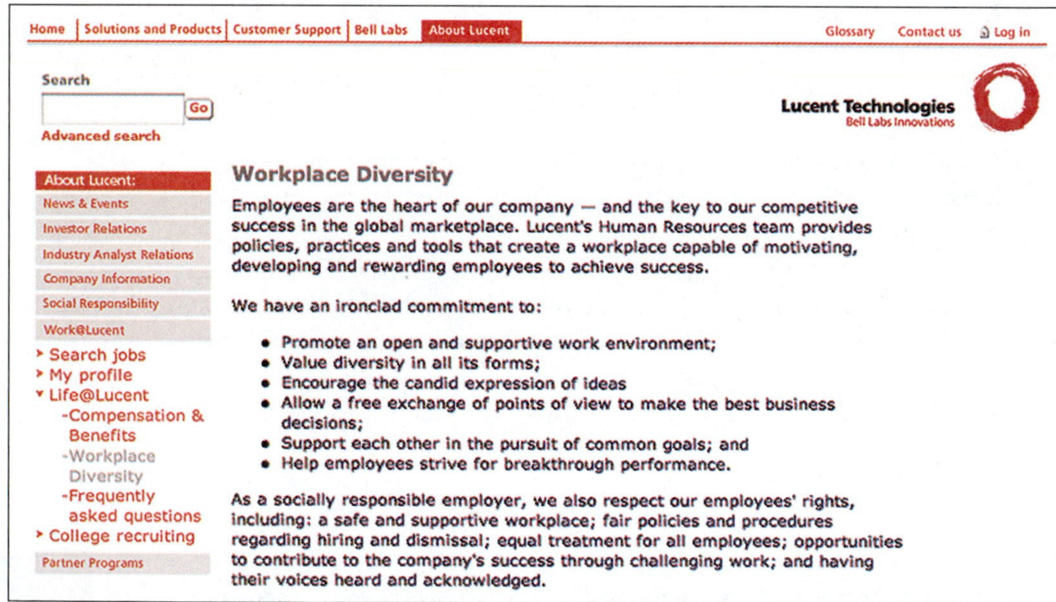

Home
Solutions and Products
Customer Support
Bell Labs
About Lucent
Glossary
Contact us
Log in
Search
Go
Advanced search
Lucent Technologies
Bell Labs Innovations
About Lucent:
News & Events
Investor Relations
Industry Analyst Relations
Company Information
Social Responsibility
Work@Lucent
Search jobs
My profile
Life@Lucent
-Compensation & Benefits
-Workplace Diversity
-Frequently asked questions
College recruiting
Partner Programs
Workplace Diversity
Employees are the heart of our company — and the key to our competitive success in the global marketplace. Lucent's Human Resources team provides policies, practices and tools that create a workplace capable of motivating, developing and rewarding employees to achieve success.
We have an ironclad commitment to:
• Promote an open and supportive work environment;
• Value diversity in all its forms;
• Encourage the candid expression of ideas
• Allow a free exchange of points of view to make the best business decisions;
• Support each other in the pursuit of common goals; and
• Help employees strive for breakthrough performance.
As a socially responsible employer, we also respect our employees' rights, including: a safe and supportive workplace; fair policies and procedures regarding hiring and dismissal; equal treatment for all employees; opportunities to contribute to the company's success through challenging work; and having their voices heard and acknowledged.

Kodak's Diversity Statement[36]

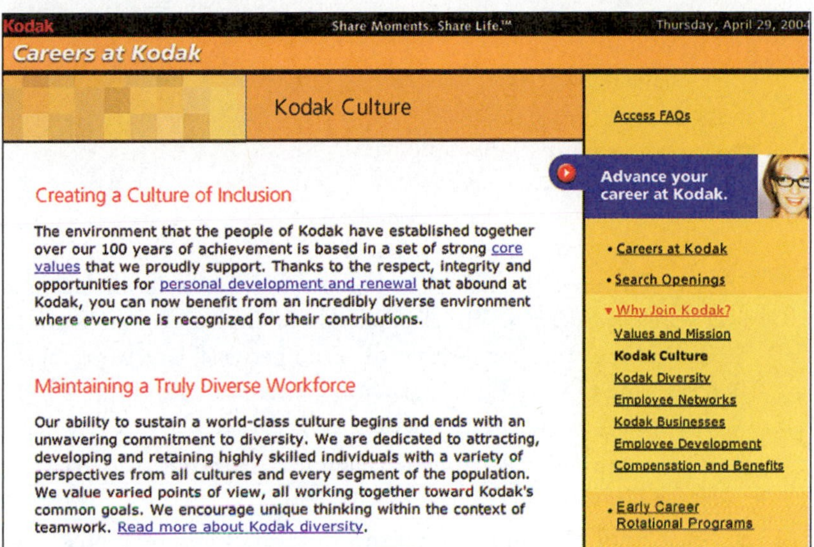

Kodak
Share Moments. Share Life.™
Thursday, April 29, 2004
Careers at Kodak
Kodak Culture
Access FAQs
Advance your career at Kodak.
Creating a Culture of Inclusion
The environment that the people of Kodak have established together over our 100 years of achievement is based in a set of strong core values that we proudly support. Thanks to the respect, integrity and opportunities for personal development and renewal that abound at Kodak, you can now benefit from an incredibly diverse environment where everyone is recognized for their contributions.
Maintaining a Truly Diverse Workforce
Our ability to sustain a world-class culture begins and ends with an unwavering commitment to diversity. We are dedicated to attracting, developing and retaining highly skilled individuals with a variety of perspectives from all cultures and every segment of the population. We value varied points of view, all working together toward Kodak's common goals. We encourage unique thinking within the context of teamwork. Read more about Kodak diversity.
• Careers at Kodak
• Search Openings
▾ Why Join Kodak?
Values and Mission
Kodak Culture
Kodak Diversity
Employee Networks
Kodak Businesses
Employee Development
Compensation and Benefits
• Early Career Rotational Programs

2. **Tell a story.** Study the following sketches and jot down the story of what you think is happening in each sketch. Then get together in a small group to see if your interpretation agrees with or differs from the interpretations of others in your group. Compare the interpretations in your group with those of the entire class. Discuss aspects of culture that might contribute to differences in interpretations.[37]

WEBLINK

Check out the culture, diversity, and human rights statements for HP, Lucent, and Kodak. Go to www.english.wadsworth.com/burnett6e for links to these sites.

CLICK ON WEBLINK

CLICK on Chapter 2/HP

CLICK on Chapter 2/Lucent

CLICK on Chapter 2/Kodak

3. **Identify cultures.** Examine this photograph of black South African dancers and porters working at Shakaland, a cultural heritage site in South Africa. Shakaland preserves and displays the Zulu heritage, largely for tourists (similar to Living History Farm, Plymouth Plantation, Sturbridge Village, or Williamsburg in the United States).
   - What organizational and national cultures are explicitly depicted or implied in this photo?
   - What kinds of technical infrastructures (e.g., transportation systems, sanitation systems) need to be in place for a cultural heritage site to function?
   - Investigate the current Zulu culture. How much does this photo reflect the lives of contemporary Zulus in South Africa?

South African dancers double as porters at Shakaland, a cultural heritage site in South Africa.

4. **Argue an opinion.** Read the three-article series "What is Global Language?" by Barbara Wallraff. These article about English as a global language challenge some common misperceptions about the nature of languages in international business, science, and technology. Select a country you're interested in learning more about and imagine that you are managing a project in that country. Investigate the country's culture (for example, check www.cia.gov for general background about the country). Write a memo to your manager in which you argue that conducting all business related to your project in English in this particular country is (or is not) a good idea.

Select a culture outside the United States and do some library or online investigation to identify the intimate, personal, social, and public distance in that culture.

5. **Characterize an organization.** Work with a partner to select an organization (government agency, nonprofit organization, small business, or large corporation) that interests both of you.
   - With your partner, create a list of questions based on the binaries presented in this chapter that you both can ask during a face-to-face or electronic interview with an employee in order to help you characterize the organization, at least from this one employee's perspective.
   - Encourage the employee to provide a general description as well as specific examples.

Access the three-article series by going to www.english.wadsworth.com/burnett6e for links to the site.
CLICK ON WEBLINK
CLICK on Chapter 2/ Wallraff Part 1
CLICK on Chapter 2/ Wallraff Part 2
CLICK on Chapter 2/ Wallraff Part 3

WWW
WEBLINK

   - Arrange a 20- to 30-minute interview. Take careful notes. Tape record it if you have permission.
   - Use the information from your interview notes to write a memo to potential employees that describes the culture of the organization.

   Your goal is to help potential employees decide if the organization is a good match for them. Your informant (the employee you interviewed) and your audience (potential employees) do not necessarily know the technical terms used to describe culture, so you need to explain these terms.

6. **Recommend an effective diversity statement.** Reread the three diversity statements in this chapter (GlaxoSmithKline, Lucent Technologies, Kodak) to review the key components. Do a Web search to select diversity statements from at least two other organizations. Analyze these statements to determine the focus, identify the kinds of examples provided, and assess whether the ramifications of noncompliance are identified. Select an

organization that needs to add or update statements regarding culture, diversity, and human rights. In your memo describe the range of options in composing such statements (content, tone, level of detail, compliance, and so on) and offer your recommendations for this specific organization.

7. **Identify your own cultural communication style.** Is your communication style linear, moving straight to your point? Do you circle around your subject until you get to your point? Different styles reflect different cultures, but spiral communicators might perceive linear communicators as abrupt or rude, and linear communicators might perceive spiral communicators as deceptive or indecisive. Write a one-paragraph description of yourself as a workplace communicator. Include examples to support your description, which can serve as preparation for a face-to-face job interview when you are asked questions about how you interact with others and how effective you are as a communicator.
   - **Courtesy:** How do you greet colleagues? How tactful are you in e-mail, discussions, and meetings? How do you respond to criticism? Are you an active listener?
   - **Directness and simplicity:** Do your explanations tend to be simple and direct or long and complex? Do you tend to use a linear or spiral approach?
   - **Assertiveness:** How assertive or reticent are you? How likely are you to raise issues you see as problems? Do you avoid ever presenting an opinion?
   - **Timing:** How do you decide the appropriate time to discuss a problem? How do you bring up problems or disagreements?
   - **Argument:** What's your style of argument? Confrontational? Emotional? Logical and rational? In making an argument, do you focus on immediate goals or long-term goals? How do you manage conflicts and disagreements?
8. **Identify cultural similarities.** One of the best ways to gain some understanding of another culture is to look for things that are common both in your own culture and the other culture. Identify a cultural group on your own campus that you know nothing about. It might be a group defined by ethnicity, nationality, lifestyle, athletics, or hobby. Attend at least one meeting of the group. Conduct 10-minute interviews with at least two members of the group. Prepare a three-minute oral presentation in which you identify the group and then explain three things you have in common with people from that culture.
9. **Describe the culture of your best company.** Access www.fortune.com/fortune/bestcompanies, one of the WEBLINKS presented earlier in the chapter. Select one of the companies that interests you. Using both print and online resources, identify the organizational culture of that company, drawing on Figure 2.9 as a starting place. Write a one-paragraph explanation that can serve as preparation for a face-to-face job interview when you are asked questions about why you think the particular organization is a good match for you.

Chapter 2 Endnotes

1 Rosen, R., Digh, P., Singer, M., & Phillips, C. (2000). *Global literacies: Lessons on business leadership and national cultures* (p. 21). New York: Simon & Schuster.

2 U.S. Census Bureau. (2003, July 15). *State and county quick facts.* Retrieved August 23, 2003, from http://quickfacts.census.gov/qfd/states/00000.html

3 U.S. Census Bureau. (2003, July 11). *U.S. exports to all countries from 1998 to 2002 by 5-digit end-use code.* Retrieved August 23, 2003, from http://www.census.gov/foreign-trade/statistics/product/enduse/exports/index.html

4 U.S. Census Bureau. (2003, July 11). *U.S. imports from all countries from 1998 to 2002 by 5-digit end-use code.* Retrieved August 23, 2003, from http://www.census.gov/foreign-trade/statistics/product/enduse/imports/index.html

5 Bodley, J. H. (1994). *An anthropological perspective.* Retrieved August 23, 2003, from www.wsu.edu:8001/vcwsu/commons/topics/culture/culture-definitions/bodley-text.html

6 Williams, R. (1958). *Moving from high culture to ordinary culture.* Retrieved August 23, 2003, from www.wsu.edu:8001/vcwsu/commons/topics/culture/culture-definitions/raymond-williams.html

7 GlaxoSmithKline. (n.d.). *The impact of medicines: Corporate and social responsibility report 2002.* Retrieved August 23, 2003, from http://www.gsk.com/financial/reps02/CSR02/GSKcsr-12.htm

8 Pitchon, P. (1997). Globalization versus localization. *Share International,* (April). Retrieved June 9, 2003, from http://www.shareintl.org/archives/economics/ec_ppglobal.htm

9 Simmons, P. (1999). *Isda forms new committee for equity deratives.* "Equity Risk Special Report." Retrieved August 23, 2003, from http://www.financewise.com/public/edit/riskm/equity/equity-news-p.htm
F.T.S.E. acronym. (n.d.). Retrieved August 23, 2003, from http://havenworks.com/acronyms/a-z/f/ftse/

10 The following sources were consulted for this ethics sidebar:

- Arnold, M. (1998). Building a truly World Wide Web: A review of the essentials of international communication. *Technical Communication,* (May), 197–207.
- Chu, S. W. (1999). Using chopsticks and a fork together: Challenges and strategies of developing a Chinese/English bilingual Web site. *Technical Communication,* (May), 206–219.
- Hoft, N. (1999). Global issues, local concerns. *Technical Communication,* (May), 145–148.
- Kohl, John R. (1999). Improving translatability and readability with syntactic cues. *Technical Communication,* (May), 149–166.

11 Smith, M. K. (2002). Globalization. In *The encyclopedia of informal education.* Retrieved August 23, 2003, from http://www.infed.org/biblio/globalization.htm

12 Canadian Cancer Society. (2003, August 29). *Evaluation of new warnings on cigarette packages—Highlights.* Retrieved August 30, 2003, from http://www.cancer.ca/ccs/internet/standard/0,2939,3172_334419_436465_langId-en,00.html

13 *Canadian health warnings.* (n.d.). Retrieved August 30, 2003, from http://home.online.no/~smpeders/ind-can6.htm

14 Summarized from these sites:

- Schneider, K. L. (2002, July 22). *Gross pics intended to help Canadian smokers.* Retrieved August 30, 2003, from http://healthfactsandfears.com/featured_articles/jul2002/pics072202.html
- *Grisly new smoking warnings.* (2002, March 6). Retrieved August 30, 2003, from http://health.iafrica.com/healthnews/905729.htm
- Canadian Cancer Society. (2002, April 1). *Canadians overwhelmingly support graphic cigarette warnings.* Retrieved August 30, 2003, from http://www.cancer.ca/ccs/internet/mediareleaselist/0,,3172_15232_356973_langId-en.html

15 World Health Organization. (n.d.). *Global cancer rates could increase by 50% to 15 million by 2020.* Retrieved August 30, 2003, from http://www.who.int/mediacentre/releases/2003/pr27/en/

16 Harris, R. G. (1998). Economic approaches to language and bilingualism. In A. Breton (Ed.), *New Canadian Perspectives.* Retrieved on 9 June 2003, from http://www.pch.gc.ca/progs/lo-ol/perspectives/english/economic/ch2_02.html

17 This information is from the LISA Web site. Retrieved September 2, 2003, from http://www.lisa.org/info/about.html

18 For a more detailed explanation that differentiates these terms, see Esselink, B. (2001). *A practical guide to localization.* Amsterdam/Philadelphia: John Benjamins. Chapter 1 of this book, which differentiates critical terms, is available at http://www.lionbridge.com/kc/ec.asp?kb=content_g11n&content=g11n_terms

19 Retrieved September 2, 2003, from http://h18000.www1.hp.com/products/storageworks/sdlt110220/documentation.html

20 ITIM Culture and Management Specialists. (n.d.). *A short introduction.* Retrieved September 7, 2003, from http://www.itim.org/3.html

21 Retrieved September 2, 2003, from http://www.itim.org/3.html /kubnw5.kub.nl/web/iric/hofstede/page3.htm

22 Oxford University Press. (n.d.). *More about world English.* Retrieved August 23, 2003, from http://www.askoxford.com/globalenglish/worldenglish/

23 Pennycook, A. (2001). *Critical applied linguistics: A critical introduction.* (pp. 59, 112) Mahwah, NJ: Lawrence Erlbaum Associates.

24 Oxford University Press. (n.d.). *More about world English: Summary.* Retrieved August 23, 2003, from http://www.askoxford.com/globalenglish/worldenglish/summary/

25 Adapted from these sources:

- Gundling, E. (2003). *Working globalsmart: 12 people skills for doing business across borders.* Palo Alto, CA: Davier-Black.
- Hall, E. T. & Hall, M. R. (1990). *Understanding cultural differences.* Yarmouth, ME: Intercultural Press.
- Hofstede, G. (1997). *Cultures and organizations: Software of the mind.* New York: McGraw-Hill.
- Hofstede, G. (2001). *Cultural consequences: Comparing values, behaviors, institutions, and organizations across nations.* Thousand Oaks, CA: Sage.

26 U.S. Census Bureau. (2002, September 3). *Hispanic heritage month 2002: Sept. 15–Oct. 15.* Retrieved August 24 2003, from http://www.census.gov/Press-Release/www/2002/cb02ff15.html

27 SIL International. (n.d.). *Languages of South Africa.* Retrieved September 7, 2003, from http://www.ethnologue.com/show_country.asp?name=South+Africa

28 Constitutional Court of South Africa. (n.d.). *Chapter 1 Founding Provisions: Languages.* Retrieved September 7, 2003, from http://www.concourt.gov.za/constitution/const01.html#6

29 Adapted from Hall, E. T. & Hall, M. R. (1990). *Understanding cultural differences.* Yarmouth, ME: Intercultural Press.

30 Retrieved September 2, 2003, from kubnw5.kub.nl/web/iric/hofstede/page4.htm
Team Dynamics. (n.d.). (n.t.). Retrieved September 2, 2003, from http://teamdynamics.org/workplac.htm
ABS Consulting. (n.d.). *Changing the workplace culture to reap the benefits of good reliability programs.* Retrieved September 2, 2003, from http://www.jbfa.com/changing-workplace-culture.html

31 *2003 global most admired companies.* (n.d.). Retrieved August 31, 2003, from http://www.fortune.com/fortune/globaladmired

32 Retrieved September 2, 2003, from http://www.dell.com

33 Retrieved September 2, 2003, from http://www.dell.com/la/mx/es/gen/default.htm

34 Hewlett-Packard Company. (n.d.). *Our commitment to global citizenship.* Retrieved August 23, 2003, from http://www.hp.com/hpinfo/globalcitizenship/commitment.html

35 Retrieved September 1, 2003, from http://www.lucent.com/work/culture.html

36 Kodak Culture. Retrieved on April 29, 2004, from http://www.kodak.com/US/en/corp/careers/why/culture.jhtml

37 Drawings done by Professor Donna Kain, Clarkson University, Potsdam, NY.

CHAPTER 3

Reading Technical Information

Objectives and Outcomes

This chapter will help you accomplish these outcomes:

- Understand that workplace professionals read documents, listen to conversations and presentations, and view visuals for a variety of purposes: assessing and making decisions, learning background, learning to do a task, and actually doing a task.
- Recognize that reading and writing are synergistically linked activities — each affects the other.
- Use the strategies of experienced communicators:
  - Skim, scan, and predict.
  - Identify structure/hierarchy: document features, visual displays, and organization.
  - Determine the main points.
  - Draw inferences: tacit assumptions, implications, ethics, and impact of implications.
  - Generate questions and examples.
  - Monitor and adapt strategies before, during, and after reading.

>

What is reading? Some people think of reading as simply looking at a page or screen with words and decoding them — that is, figuring out their pronunciation and definitions. But the process of reading is not so simple.

Experts in reading consider it a complex activity and believe that making meaning involves interpreting rather than just decoding ideas. Meaning comes from more than simply the words on a page or screen because their interpretation is strongly influenced by an individual's prior experiences and knowledge as well as by the context in which the document is written and the purpose for which it's read. As a result, the same words can be interpreted in different ways by different readers. For example, the sentence "The cell has been examined" doesn't mean much, even when you decode the words. You need to know whether the sentence is about a cell in a spreadsheet, a cartoon animation, a Braille transcript, a biological organism, a terrorist organization, a political party, a monastery, a prison, a geographic area covered by a mobile phone, or a battery. Simply put, you need to do more than decode the words in order to make meaning from that document.

Throughout your professional life, you'll read and respond to a variety of documents. Some of these documents will require very little effort or attention; others will be exceedingly difficult to understand. This chapter suggests ways to make all of your reading easier and more productive. Your workplace literacy, your ability to both read and create paper and electronic documents, is highly correlated with your professional success.

The chapter begins by summarizing reasons that workplace professionals might have for reading and then discusses the strong relationship between reading and writing. Most of the chapter is devoted to discussing six critical strategies used by effective readers.

Identifying Purposes

As a workplace professional, you read for a variety of purposes. Virtually all of these workplace purposes can be categorized into one of four often overlapping categories, as shown in Figure 3.1[1] — reading to assess, reading to learn, reading to learn to do, and reading to do.

> *What purposes for reading are you likely to have in your professional work?*

Regardless of the purposes that workplace professionals have for reading, they expect information in documents to be accessible, accurate, comprehensible, and usable. These expectations can usually be met if a document has a clear purpose, well-organized and appropriately developed points, and an effective balance of verbal and visual information.

Depending on the reader's needs in a particular context, a document such as a technical report will probably be read for a number of purposes. For example, an engineer might read an interim project report for any of the four purposes described in Figure 3.1. She might skim to *assess* whether she should read the report thoroughly later; she might read to *learn* how well a modification she designed for

FIGURE 3.1 Purposes for Reading

Purposes	Examples
Reading to assess. This reading, which is often skimming, enables you to decide whether a document will be useful for you or someone else, usually at a later time.	A forester might skim summaries of legislation about support for reforestation that he or she might need to study.
Reading to learn. This reading enables you to learn information for problem solving, decision making, and background knowledge.	A small-animal veterinarian might read an article in a professional journal to learn about new medication for feline colitis.
Reading to learn to do. This reading enables you to learn how to complete tasks.	A dental assistant might read a continuing education booklet about ways to minimize pain and anxiety for pediatric patients.
Reading to do. This reading, which serves largely as an external prompt, enables you to complete tasks.	A furniture refinisher might read the product label and information sheet to determine safety requirements for using a new paint remover.

the production line is working; she might read to *learn to do* the calibration of the new furnace controls; she might read to *do* a procedure for mechanical inspection that the report outlines.

Reading-Writing Relationships

Your reading and writing are closely related, whether you're working on a hard copy or electronic document. Consider these typical examples:

- A shipping manager writes instructions for sending products overseas; she rereads these instructions as she revises them following user testing. Later, her revised instructions are read by shipping clerks preparing the paperwork to accompany international orders.
- A greenhouse worker reads the package insert that came with a pesticide, specifically studying the section about application. Later he accesses the greenhouse's online database to record changes in greenhouse plants treated with this pesticide.
- A heating contractor reads the troubleshooting section of the manual for a new computer-controlled thermostat that isn't working properly. Later she completes a form she downloaded from the company's Web site to return the defective thermostat to the manufacturer.

What kinds of online documents are you likely to read? What complicating factors of online reading are likely to be the most problematic for readers?

- A hospice nurse makes notes about a patient's weekly medication; he rereads these notes before he writes a recommendation to the supervising physician for modifying the patient's care. Later, the patient's physician reads the recommendation in order to make her decision.

As a writer, imagine yourself as a member of your intended audience and then try to read the document you're preparing from that reader's perspective. For example, if you are working on a food safety Web site that discusses food irradiation, you should try to anticipate reader questions. One of the symbols that would probably appear on this site is the *radura,* an international symbol identifying foods that have been irradiated. You realize that some readers will not recognize the radura, so it will need to be defined. Other readers will recognize the symbol but will not necessarily know its purpose. The marginal questions next to Figure 3.2 show some of those questions readers might have about the radura before reading the information on this Web site.

What are your strongest strategies as a reader? Which features of electronic text do you find most influential in reducing your comprehension and/or speed in reading?

As a reader, you need to be conscious of a text's features. On a Web site, for example, this means recognizing the links that you can follow to obtain necessary information or knowing that you can conduct a search to locate additional information. As an active reader, you ask yourself questions that you hope the text answers. Of course, the questions you have are not necessarily the ones that other readers will have. Readers using a Web site like the example in Figure 3.2 will

FIGURE 3.2 Predict Readers' Questions[2]

Before reading, readers might ask these questions:

- *What is a radura?*
- *What are the legal requirements of using a radura?*
- *What is the history of using the radura?*
- *What protection does the use of the radura provide consumers?*
- *What links provide further information about the radura?*

As a writer, you might ask: What sequence of links could readers follow to create various understandings?

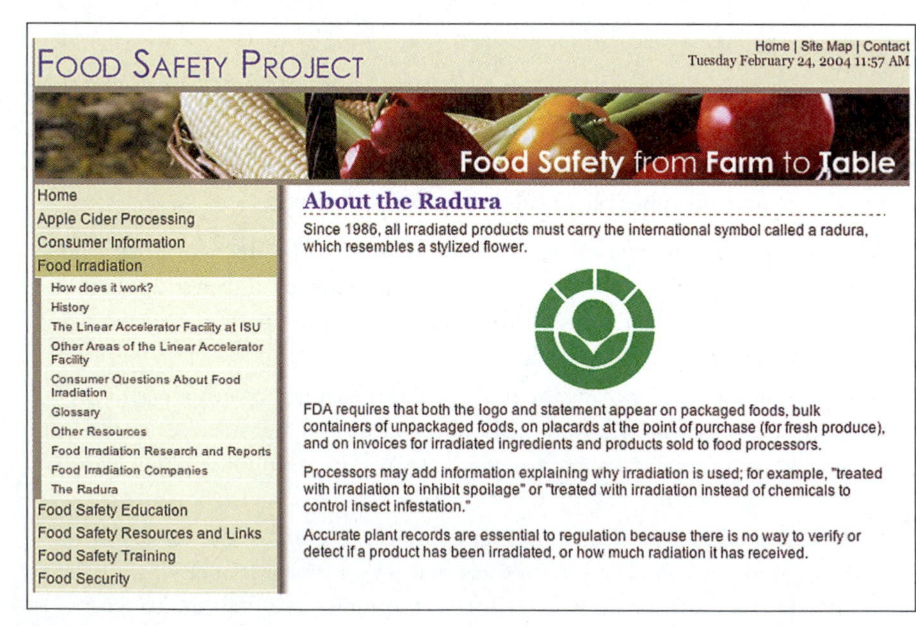

construct different meanings depending on which links they follow or what follow-up searches they do; thus, the same Web site — or print document — can be interpreted in different ways by different readers.

Readers have strong reactions to symbols as well as to words. Go to www.english.wadsworth.com/burnett6e for a link to review the FDA's description of *irradiation* and then two other links to review some reactions.
CLICK ON WEBLINK
CLICK on Chapter 3/radura

WEBLINK

Strategies for Effective Reading

How do people read technical documents in the workplace? You may have started this chapter thinking of reading as decoding words. Now you know that researchers see reading as remarkably complex, having less to do with decoding than with the way readers interpret meanings in particular situations. How well people read has a great deal to do with both their prior knowledge and their reading strategies.

Experienced readers tend to see technical documents and the situation they're part of as a whole; in other words, documents cannot be separated from the situations in which they're created and used. (See the discussion of genre in Chapter 1.) Both the documents and the situations have recurring, evolving patterns that are familiar to experienced readers. These familiar patterns help experienced readers select and adapt appropriate strategies to read both print and electronic documents. This chapter is largely about these strategies.

Figure 3.3 identifies six strategies for critical reading. The first column identifies ways in which skillful writers use these strategies to make documents more accessible, understandable, and usable. If you conscientiously use these strategies as you plan, draft, and revise documents, your documents are likely to improve. The second column identifies strategies used by skillful readers to increase their comprehension and use of information. If you conscientiously use these reading strategies, you are likely to find that your reading is more productive because you'll understand, remember, and be able to use more of what you read.

Skim, Scan, and Predict

When you need to read either a print or electronic document, you need to do three things before carefully reading the sentences and paragraphs, tables and graphs: Skim, scan, and predict.

FIGURE 3.3 **Strategies Used by Writers and Readers**[3]

Strategies	Ways for Writers to Create More Readable Documents	Ways for Readers to Increase Comprehension and Recall
Skim, scan, predict	Writers make a document easy to *skim* and *scan* so readers have an overview and can locate key information. They also know skilled readers *predict* what's coming and often ask themselves questions before reading.	Readers save time and increase their comprehension by *skimming* a document, *scanning* for key points, and then *predicting* what's coming before reading carefully.
Identify structure	Writers use familiar *document features* to help readers recognize the text's purposes, arrange *visual cues* to depict relationships, and reinforce the structure by *previewing the organization.*	Readers who relate *document features, visual cues,* and *previewed information* to a document's purpose usually remember more information than readers who don't.
Distinguish hierarchy	Writers help readers understand the *hierarchy of ideas* by using visual cues (e.g., headings), coherence devices (e.g., transitions), and textual cues (e.g., indicating relationships) that distinguish main from subordinate ideas.	Readers who understand the *hierarchy of ideas* usually recall more important information. Skillful readers often summarize the main points, which they connect to their prior knowledge.
Draw inferences	Writers anticipate readers' *purposes* for reading, which influences the information they include and the ways they imagine readers will use that information. Writers also anticipate possible *inferences* readers might make. They know readers *interpret* text.	Readers with specific *purposes* for reading are more focused. Readers who draw *inferences* usually increase understanding and recall of information. Skillful readers recognize that their understanding is an *interpretation;* they often extend ideas by applying them to new situations.
Generate questions and examples	Writers expect readers to ask *questions* and think of *examples* to increase their comprehension and recall of information.	Readers who ask themselves *questions* increase their comprehension. Skillful readers often think of examples to confirm or counter information in the document and their own prior knowledge.
Monitor and adapt reading strategies	Writers use *visual cues* (e.g., boldfacing, italics, and marginal annotations), *text support* (e.g., glossaries and checklists), and *reader appeals* (e.g., narratives, examples, and details) to increase reader engagement and comprehension.	Readers who monitor and reflect on their own reading process are aware of the ways in which *visual cues, text, support,* and *appeals* affect comprehension, and they make adjustments when their comprehension is low.

Skim. Begin reading a document by skimming — go through the entire document but do it very quickly. The idea is to get a quick, overall sense of what's in the document and where it's located. These specific strategies work well with most workplace documents:

- Determine the overall structure and organization of the document by looking through the table of contents to identify the chapter or section topics and sequence.
- Get an overview of the content and a sense of the scope and approach by examining the table of contents, the preface, and the index.
- Identify the main ideas by skimming the title, abstract or summary, headings, topic sentences, and first and final paragraphs.
- Check the bibliography or works cited to determine the authors, publishers, and currency of the references, so you can better situate the document.

Which of these skimming strategies do you currently use when reading complex documents? Which strategies would you be willing to start using?

The small amount of extra time you take to skim a document will be made up in increased comprehension when you read the document more carefully. You'll have a good sense of what you need to concentrate on carefully and what you can read fairly quickly. Skimming a print document can be easier than skimming an electronic document because readers moving through an electronic document sometimes lose their sense of place in the text.

WEBLINK

To learn more about the difference between paper and on-screen reading, go to www.english.wadsworth.com/burnett6e for a link to a study comparing the two.
CLICK ON WEBLINK
CLICK on Chapter 3/comparison

Scan. Before reading a document carefully, you should also scan it — run your eyes down a page or screen to look for specific information. Scanning is the kind of reading you do to locate a word in a dictionary. You scan a document to accomplish these tasks:

- Identifying key terms that are new to you
- Identifying the location of particular text features such as figures, examples, and sidebars
- Locating critical information that you must remember exactly, such as definitions, steps in a process, principles, laws, and formulas

Scanning is often aided by document design features that signal the presence of specific information. For example, italic or boldface type often signals key terms, and numbers often signal steps in a process.

WEBLINK

To learn more about reading strategies, go to www.english.wadsworth.com/burnett6e for a link to a Web site with more information about reading and additional links with practical suggestions.

CLICK ON WEBLINK

CLICK on Chapter 3/reading

Predict. While you're skimming and scanning a document before you begin careful reading, you can anticipate what it will contain and can formulate questions to be answered by more careful reading. You can get ideas about useful questions by thinking of key terms you'd like to know more about in relation to the document. For example, a document detailing a new product release could trigger thoughts about availability, so you might ask questions like these: "When will this new product be available?" "Is the availability of this product superior to the availability of the product I'm already using?" "What will be the availability of trained service technicians?" "What will be the availability of replacement parts?" You can predict whether the document will address your questions. The following words are a few that could trigger questions during your own reading:

- accessibility
- accuracy
- adaptability
- availability
- cost
- credibility
- development
- disadvantages
- efficiency
- long-range outlook
- maintenance
- materials
- specifications
- technical merit
- testing

Formulating questions that focus on what you want to get from your reading is a productive way to spend a few minutes before you jump into a document.

Identify Structure and Hierarchy

Many technical documents have an identifiable structure and hierarchy. They use conventional document features such as headings, visual cues, and previews of what's coming up in the document.

Document Features. Standard features of a document can help readers know what to anticipate. For example, the features of abstracts, which are often described as the most important part of a technical report, are influenced by their purpose: to provide an abbreviated overview of a report.

Well-written abstracts maintain the tone and focus of the original document, presenting key points. The *Journal of the American Medical Association* (*JAMA*) recommends that articles about a clinical investigation be accompanied by an abstract with these features:

> the objective(s) or purpose, the design (e.g., randomized, double-blind, placebo-controlled, multicenter trial), the setting (e.g., university clinic, hospital), the patients or participants, the intervention(s), the measurements and main results, . . . the conclusion . . . ; [and] important outcome measures or endpoints.

JAMA recommends that review articles intended to "identify, assess, and synthesize" information be accompanied by an abstract that contains these features:

> purpose, data identification (a summary of data sources), study selection (how many studies were chosen and how they were selected), data extraction (guidelines for abstracting articles), results of data synthesis, and conclusions (which should include potential applications and research needs).[4]

You are likely to read abstracts more quickly and with greater comprehension if you anticipate their features. This is also true of other kinds of documents you read. The example in Figure 3.4 annotates the four kinds of information that most often appears in abstracts:

- objective, purpose, or rationale
- methodology
- results
- conclusions

If you look for these features every time you read an abstract, your reading time may decrease and your comprehension of key points may increase.

The example in Figure 3.4 is an abstract from the *Journal of Food Production,* a professional journal for experts interested in research about food science and technology, including food safety and quality. It is published by the International Association for Food Protection (IAFP), whose mission, according to their Web site, is "[t]o provide food safety professionals worldwide with a forum to exchange information on protecting the food supply." The journal, recognized as the leading international publication in food microbiology, is read by more than 11,000 scientists from 69 countries.[5] The abstract in Figure 3.4 accompanied an article written by researchers who were interested in finding a simple way to eliminate bacteria from alfalfa and mung bean seeds before the seeds sprouted. Most readers, even those who are not food scientists, would have an easy time identifying

FIGURE 3.4 **Online Abstract from the *Journal of Food Production* with Features That Readers Should Anticipate**[6]

AN = *accession number in the library*

AU = *authors*

PY = *year of publication*

NU = *ISSN or ISBN*

LA = *language of text*

AB = *abstract*

1. ***Purpose of study:*** *The study investigated whether ammonia fumigation is effective against selected pathogens when seeds are used for sprouting.*

2. ***Methodology:*** *Alfalfa and mung bean seeds were sealed in glass jars with ammonia. Samples were collected at regular intervals.*

DE = descriptors (all of which are links in the online version of this abstract, as are the authors and journal itself above)

AN: 2002-04-J0805
TI: Reduction of Escherichia coli O157:H7 and Salmonella Typhimurium in artificially contaminated alfalfa seeds and mung beans by fumigation with ammonia.
AU: Sakchai-Himathongkham; Suphachai-Nuanualsuwan; Riemann-H; Cliver-DO
AD: Correspondence (Reprint) address, D. O. Cliver, Dep. of Population Health & Reproduction, Sch. of Vet. Med., Univ. of California, Davis, CA 95616, USA. Tel. 530-754-9120. Fax 530-752-5845. E-mail docliver@ucdavis.edu
PY: 2001
SO: Journal-of-Food-Protection; 64 (11) 1817-1819, 17 ref.
NU: 0362-028X
DT: Journal-Article
LA: En (**English**)
SC: J Fruits-vegetables-and-nuts
AB: Sprouts eaten raw are increasingly perceived as hazardous foods because they have been vehicles in outbreaks of food-borne disease, often involving Escherichia coli O157:H7 and Salmonella Typhimurium. Although the source of these pathogens has not been established, it is known that the seeds usually are already contaminated at the time sprouting begins. Earlier studies had shown that ammonia was lethal to these same pathogens in manure, therefore this study sought to determine whether ammonia fumigation is effective against them when associated with seeds to be used for sprouting. Experimentally contaminated (10-8-10-9 cfu/g **E. coli** O157:H7 and Salmonella Typhimurium) and dried alfalfa and mung bean seeds, intended for sprouting, were sealed in glass jars in which 180 or 300 mg ammonia/l of air space was generated by the action of ammonium sulphate and sodium hydroxide. Samples were taken after intervals up to 22 h at 20 degrees C. Destruction of approx. 2–3 logs was observed with both bacteria associated with alfalfa seeds, vs. 5–6 logs with mung beans. Greater kills were apparently associated with lower initial bacterial loads. Germination of these seeds was unaffected by the treatment. It appeared that this simple treatment could contribute significantly to the safety of sprout production of alfalfa seeds and mung beans.
DE: AMMONIA-; BACTERIA-; ESCHERICHIA-; FOOD-SAFETY-PLANT-FOODS; FUMIGATION-; INHIBITION-; LEGUMES-; MUNG-BEANS; SALMONELLA-; VEGETABLES-SPECIFIC; ALFALFA-SPROUTS; ALFALFA-; ANTIBACTERIAL-ACTIVITY; BEAN-SPROUTS; ESCHERICHIA-COLI; SALMONELLA-TYPHIMURIUM

TI = *title of the article*

AD = *information for correspondence (mail, phone, fax, e-mail)*

SO = *source for bibliographic citation*

DT = *type of document*

SC = *subject code*

3. ***Results:*** *Bacteria that had contaminated the seeds were killed at various rates; seed germination was unaffected by treatment.*

4. ***Conclusions:*** *Treatment could increase safety of sprout production of alfalfa seeds and mung beans.*

what this article is about if they look for purpose, methodology, results, and conclusion in the abstract.

Whatever you are reading, you can make good use of your time by knowing the features of that particular type of document. Just as abstracts have predictable features, so do the other documents you'll read. Recognizing and then anticipating the features and their organization will help you read efficiently.

Visual Cues. Various kind of visual cues help readers distinguish the hierarchy of ideas in a document. Typically, readers have an easier time understanding and using documents that have the following features:

1. Information is chunked so that relationships are clear.
   - Related information is topically chunked.
   - Headings and subheadings signal and separate topics.
   - Leading (spacing) between lines and sections separates related information.
2. Information is arranged so that the sequence is clear.
   - The sequence of information is logical, and explicit transitions indicate the relationship between the points.
   - Numbers or letters indicate the sequence or hierarchy of the items in a list; bullets indicate equivalency of the items.
3. Information is emphasized so that important elements are signaled.
   - Placement on a page or screen can indicate hierarchy. Main ideas can be signaled with centered and left-justified text. Subordinate ideas can be signaled by indentation.
   - Type size can indicate the hierarchy of headings. A level-1 heading might be in larger type than a subordinate level-2 heading.
   - Changes in typeface can differentiate headings and text. Headings might be in a sans serif font such as **Arial Black** while the body of the text might be in a serif font such as Times New Roman.
   - Type-style variations — for example, SMALL CAPS, **bold,** or *italics* — draw readers' attention to important terms or ideas.
   - Icons can signal important categories.

Readers use visual cues to help them decide what's important in a document, but careful writers use such cues judiciously. They don't overdo their use, because when too many cues are used, they lose their effectiveness.

A well-designed table of contents demonstrates how these visual cues can be used. (See Chapters 10 and 12 for additional discussion about designing information.) The table of contents in Figure 3.5, from *NASA Tech Briefs*, gives readers multiple visual cues about chunking, arrangement, and emphasis. The annotations identify cues you might use in documents that you design.

FIGURE 3.5 **Table of Contents Showing Structure**[7]

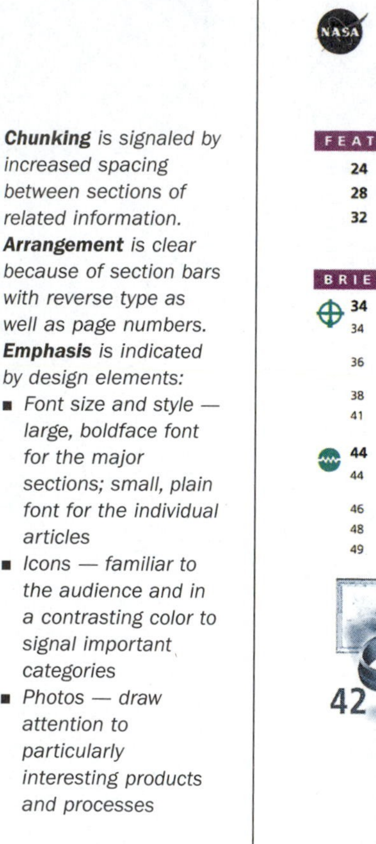

Chunking *is signaled by increased spacing between sections of related information.* ***Arrangement*** *is clear because of section bars with reverse type as well as page numbers.* ***Emphasis*** *is indicated by design elements:*

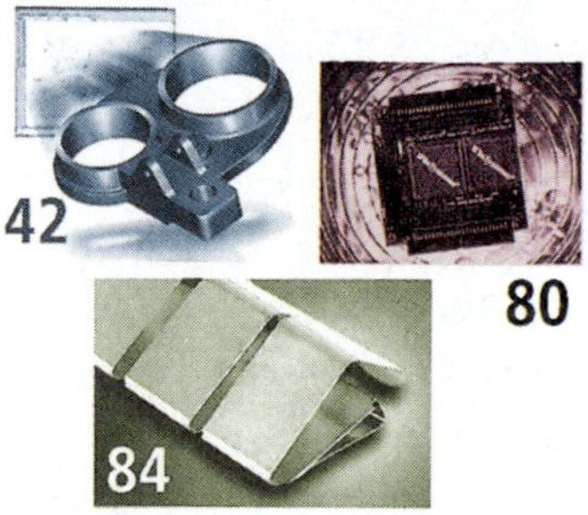

- *Font size and style — large, boldface font for the major sections; small, plain font for the individual articles*
- *Icons — familiar to the audience and in a contrasting color to signal important categories*
- *Photos — draw attention to particularly interesting products and processes*

NASA TECH BRIEFS

ENGINEERING SOLUTIONS FOR DESIGN & MANUFACTURING

March 2004 • Vol. 24 No. 3

6 www.nasatech.com NASA Tech Briefs, March 2004

You can read a remarkably informative interview with one of the country's leading experts on workplace literacy. Go to www.english.wadsworth.com/burnett6e for a link this interview.

CLICK ON WEBLINK

CLICK on Chapter 3/NCAL publications

WWW WEBLINK

Previews and Reviews. Previewing and reviewing give you a chance to identify and then reinforce the structure or organization of a document. Helping readers identify the structure of a document increases the likelihood that they will understand and recall the information. Visual and verbal cues, such as numbered or bulleted lists, parallel terms, design and typography, can effectively signal the relative relationship among terms.

Figure 3.6 shows the first half of a page from a manual for farmers, *How to Grow Shiitake Mushrooms Outdoors on Natural Logs.* In this preview, readers are signaled about the purpose of the section: "Let's quickly preview the factors to consider. Then, we'll take a close look at each of them." Readers are introduced to five critical factors that are distinguished by a series of visual cues:

- **bold,** SMALL-CAP section heading (acceptable for these very short phrases)
- numbered list with hanging indents
- boldface terms
- a colon separating each term from the comment

The wording of the first item in the preview ("Tree Selection") anticipates the wording of the first subheading ("What Tree Species Should be Selected?"). Similar repetition is used throughout the section to reduce the chance of misinterpretation; the wording in the preview matches the wording in the subheadings and in the text itself. Knowing what to anticipate helps readers call on their relevant prior knowledge and begin to organize what they'll be reading.

Determine the Main Points

Distinguishing between the main points and the subordinate points helps readers understand the relationships among ideas in a document, the hierarchy of information. Sometimes the main points are easy to distinguish because the writer uses typographic cues. For example, if you look again at Figure 3.6, you'll see that in the first paragraph the writer has italicized the terms *noncommercial* and *commercial,* so readers immediately know that this distinction matters.

FIGURE 3.6 Previewed Information[8]

Readers are clearly signaled about the purpose of this section.

The five critical factors are previewed.

These visual cues help readers:
- *numbered list with hanging indents*
- *boldface terms*
- *factors set off by colon from comment about each factor*

The language of the first item in the preview anticipates the language of the first heading.

PART 2:
Bedlogs: Selecting, Cutting, and Growing

If you want to grow a few shiitake *noncommercially,* all you really need to know about choosing bedlogs is to get fresh, healthy logs of the proper size and species that were cut in late winter. However, if you have a *commercial* interest, a lot more is worth learning about bedlog selection.

Let's quickly preview the factors to consider. Then, we'll take a close look at each of them.

1. **Tree Selection:** You must decide which tree species to use and how to select the best individual trees to fell.
2. **Bedlog Selection:** You must know how to cut suitable logs from the trees felled.
3. **Season of Felling:** The trees must be felled at the right time of year.
4. **Bedlog Handling:** Harvested bedlogs must be carefully handled and properly stored until they are inoculated.
5. **Bedlog Silviculture:** If you plan to have an ongoing operation, you probably need to carry out some woodland management practices to ensure a future supply of bedlogs. Or, if you plan to buy logs, you have to find a logger who is willing to meet your specifications.

What Tree Species Should be Selected?

Sometimes, though, the main points are more difficult to identify, and you must be able to generate a series of questions to help you identify them. The example in Figure 3.7 is an abstract for a technical article about management architecture for wireless sensor networks. Experienced readers might begin by asking — and expecting their reading to answer — questions such as these:

- What is the main issue or problem?
- Who and what are involved in the issue or problem?
- What is the approach to the solution to the problem?
- How will the approach be addressed or the solution be implemented?

Answers to these questions will help you identify the main points of virtually any technical document.

FIGURE 3.7 **Abstract with Main Points Identified by Reader Questions**[9]

"MANNA: A management architecture for wireless sensor networks," Linnyer Beatrys Ruiz, José Marcos Nogueira and Antonio A. F. Loureiro. IEEE Communications Magazine, vol. 41, no. 2, Feb. 2003, p. 116–125.

Wireless sensor networks (WSNs) are becoming an increasingly important technology that will be used in a variety of applications such as environmental monitoring, infrastructure management, public safety, medical, home and office security, transportation, and military. WSNs will also play a key role in pervasive computing where computing devices and people are connected to the Internet. Until now, WSNs and their applications have been developed without considering a management solution. This is a critical problem since networks comprised of tens of thousands of nodes are expected to be used in some of the applications above.

This article proposes the MANNA management architecture for WSNs. In particular, it presents the functional, information, and physical management architectures that take into account specific characteristics of this type of network. Some of them are restricted physical resources such as energy and computing power, frequent reconfiguration and adaptation, and faults caused by unavailable nodes. The MANNA architecture considers three management dimensions: functional areas, management levels, and WSN functionalities. These dimensions are specified to the management of a WSN and are the basis for a list of management functions. The article also proposes WSN models to guide the management activities and the use of correlation in the WSN management. This is a first step into a largely unexplored research area.

What is the issue or problem?
The main problem is signaled by the wording: "Until now, WSNs and their applications have been developed without considering a management solution."

Who and what are involved in the issue or problem?
The breadth of the problem is clearly indicated: "This is a critical problem since networks comprised of tens of thousands of nodes are expected to be used in some of the applications above."

What is the approach to the solution to the problem?
The approach to the solution is straightforward: "This article proposes the MANNA management architecture for WSNs."

How will the approach be addressed or solution be implemented?
The details of the solution are outlined: "The MANNA architecture considers three management dimensions: functional areas, management levels, and WSN functionalities. These dimensions are specified to the management of a WSN and are the basis for a list of management functions. The article also proposes WSN models to guide the management activities and the use of correlation in the WSN management."

Draw Inferences

Not everything that readers learn from a text is explicitly stated. Sometimes readers draw inferences: They make connections and draw conclusions beyond the words and visuals that are presented.

Most experienced readers draw inferences as they read, forming and reforming their opinions as they move through a document. Three specific strategies help you draw inferences:

- Identify the tacit assumptions on which you believe the document is based — that is, what's presumed but not articulated.

- Extend the ideas to pose reasonable but unstated implications — what's implied but not articulated.
- Speculate on the impact of the implications — what's possible but not articulated.

Drawing inferences involves identifying tacit assumptions, extending ideas, and speculating on the possible impact of what you've read in relation to what you already know. Figure 3.8 provides an example of a reader drawing inferences while reading. The excerpt is the first part of an online article about the Xanadu system, a concept for a worldwide electronic repository of documents that has received little attention. The comments are the questions and inferences drawn by Dave Clark, a cautious, perhaps even skeptical, reader who draws on his expertise to frame his questions. As you read the excerpt and Clark's comments you can consider how he identified tacit assumptions, extended the article's ideas, and speculated on the possible impact of the ideas in the article.

Figure 3.8 shows one way in which readers can be engaged with a text by articulating their own reactions as they read. Readers' reactions are shaped, in part, by prior experiences and knowledge as well as by the context in which the document is written and read.

The ethics sidebar on page 94 explains the importance of context in reading, showing how it affects our inferences and interpretations.

Generate Questions and Examples

Readers can ask questions to help themselves understand a document. Sometimes readers benefit from using a traditional taxonomy (a formal method of classification) that increases their chances of understanding important information. One widely used taxonomy uses six levels of questions:[10]

- **Knowledge questions** emphasize the recall of specifics. Such questions require recall of specialized terminology and symbols, quantifiable facts, conventions of organizing information, awareness of trends, knowledge of classification systems, evaluation criteria, methodology, principles, and theories.
- **Comprehension questions** require responses that incorporate knowledge as well as understanding. Responses to comprehension questions can involve translation, interpretation, or extrapolation.
- **Application questions** require specific applications of principles or theories.
- **Analysis questions** emphasize the separation of objects, mechanisms, systems, organisms, operations, or ideas into constituent parts, clearly establishing the relationship between these parts.
- **Synthesis questions** expect the reader to focus on organizing or structuring the parts to form a unique whole. The response may either serve as an overall plan or explain a particular phenomenon.

FIGURE 3.8 **Drawing Inferences while Reading**[11]

THE XANADU IDEAL

Theodor Holm Nelson

WHAT IS XANADU PUBLISHING?

The term universal electronic library has been suggested. Perhaps universal bookstore is more like it. World Publishing Repository™ is perhaps the most appropriate term.

What problems in understanding a document might occur if the writer and reader have different tacit assumptions about the subject?

The Xanadu system has been designed from the literary point of view, the computer point of view, the business point of view and the legal point of view.

The Xanadu publishing system will be a licensed method of online electronic publication provided by vendors throughout the world. "Publication" consists of placing a digital document somewhere in the repository network. A document may include text, pictures, audio, movies and any other form of digital information. Readers, or users, are of course at screens. Any user in the world may send for any document, or any part of a document. The publisher pays only for the storage; the user pays for delivery, including a royalty to the publisher. **Clark's Comment:** Nelson assumes that users can afford these costs and will want to pay for documents. Isn't this exclusionary, making it possible only for people with money to have access to this "World Publishing Repository"? Even if the charge is only a few cents per document, it's still more expensive than local libraries.

Clark questions one of Nelson's tacit assumptions. Clark sees economic barriers as serious and suggests that Nelson doesn't sufficiently consider cost.

The user obtains a digital copy of everything he or she sends for — to keep or discard. The user may point and click to travel among documents, obtaining only the small part needed to keep going. **Clark's Comment:** Nelson's assumption is that pointing and clicking are better, or more efficient, than shelf browsing at the library. How will users know which documents they want? Will they be paying for documents whether they "keep or discard" them? If so, every click would be an investment, and this doesn't seem like an improvement over shelf browsing.

Clark again questions one of Nelson's assumptions and speculates that Nelson's plan isn't an improvement over current practices.

Staying within the Xanadu online world, anyone may publish a connection to a document — a comment, illustration, disagreement, or link of any other type; and anyone may quote from a published Xanadu document, since the quotation is bought from the original publisher at the time of delivery. The publisher agrees to be legally responsible for the contents and agrees to interconnection by anyone. **Clark's Comment:** Again, we see "anyone"; do "anyone's" documents cost the same as famous authors' documents? This would be an interesting democratic move, but how would readers find credible and useful information if they must slog through piles of information by people they've never heard of, whose work may well not be useful or interesting?

Clark extends his concern about Nelson's tacit assumptions to include credibility of materials as a potential problem.

ETHICS SIDEBAR

"It's only a report": Ethics and Context

How can you make sure you don't lose sight of the context for a document you are writing?

The following is an excerpt from a technical report:

> The van's normal load is usually nine [pieces] per square yard. In Saurer vehicles, which are very spacious, maximum use of space is impossible, not because of any possible overload, but because loading to full capacity would affect the vehicle's stability. So, reduction of the load space seems necessary. It must absolutely be reduced by a yard, instead of trying to solve the problem, as hitherto, by reducing the number of pieces loaded.

Researcher Steven Katz[12] points out that this report meets many of the formal criteria necessary for an effective document. The report addresses a concern shared by both the readers and the writer and provides answers relevant to their concerns. However, Katz also points out that we must look at this document in the context in which it was written.

This report comes from Nazi Germany in 1942. The "pieces loaded" are Jews, Poles, gypsies, and other humans who are being transported to their deaths in concentration camps.

Katz examines this report to suggest that not only the content of a document can make it unethical. How readers interpret it and use it also matters. Technical professionals must recognize the purposes behind technical documents and the effects those documents might have. Many involved in indirectly supporting the Holocaust argued that they should not be held responsible for their actions. They were "only doing their jobs." Most scholars believe that we cannot accept this argument. When individuals are not held accountable for the contexts in which they perform their jobs, when they are not held responsible for interpreting and implementing ideas, then atrocities like the Holocaust continue.

The context in which technical professionals work is not as horrific as the Holocaust of World War II. Even so, technical documents affect readers. We must be aware of those effects and be willing to examine their ethical consequences. As you write, ask yourself: Do I know who is affected by this document? Am I willing to be responsible for those effects? Perhaps, if the writer of the 1942 report had asked himself this question, fewer such documents would exist.

- **Evaluation questions** require readers to judge something's qualitative and quantitative value. Such questions examine internal elements for logic and consistency as well as examining external comparisons to establish the relationship of the subject with accepted principles, theories, and works of recognized excellence.

Figure 3.9 presents "Slowing the Progress of Aortic Regurgitation" from the *Harvard Heart Letter,* a monthly newsletter written to interpret complex medical information for general readers interested in heart-related medical issues. In addition, Figure 3.9 presents questions based on the above taxonomy that a reader asked when reading this article. You can see how the reader used the taxonomy to develop questions that could help check comprehension. Whenever you read, you should ask yourself questions about the text and the accompanying visuals. You can jot these questions in the margins of the document you're reading (if it belongs to you), or you can keep a separate paper or electronic notepad for jotting notes and recording questions.

Monitor and Adapt Reading Strategies

Effective readers are aware of what they're doing when they move through a document. They're actively engaged in their reading; they're aware of their comprehension of concepts and terms. When their comprehension or speed decreases, they adjust their strategies to meet the needs of the situation.

Your goal is to develop reading strategies that work well for you. Your strategies will vary depending on your purpose. Most likely, you haven't had any instruction about how to be a better reader since elementary school. And you do just fine until the stakes are raised when the documents are more complex, the information is more important to understand and recall, and you have less time. Figure 3.10 gives you practical strategies to understand the visual and verbal information in technical documents. Effective readers usually engage in certain strategies before, during, and after reading very difficult or complex documents.

Monitoring your reading process may feel uncomfortable at first but, like reflecting on your writing processes as you plan, draft, and revise documents, monitoring your reading processes using the questions in Figure 3.10 will soon become second nature to you. Like other skillful readers, you'll learn what works well for you so that your reading is productive. Just like experienced writers, experienced readers are distinguished by their ability to adjust strategies to match the situation they're in.

To demonstrate the differences that may occur with readers of the same text, you'll have a chance to "listen in" on two different professionals as they read the opening three paragraphs from an article in *Scientific American,* "Smart Rooms."[13] The article is about the potential benefits of a project at MIT to create computer systems that recognize human faces, expressions, and gestures. Two expert readers were asked to read these paragraphs and to "interrogate" the article — to record the questions and comments they had while reading.

- Figure 3.11 presents the introductory paragraphs that embed the questions and comments made by Clay, who has a computer science degree. Clay read the article as preliminary background for work he was doing about the computer-human interface.

FIGURE 3.9 **Applying Taxonomy of Questions to Reading**[14]

SLOWING THE PROGRESS OF AORTIC REGURGITATION

A leaky aortic valve allows blood that has just been pumped out of the heart and into the aorta—the large blood vessel that distributes blood to the body—to flow backwards, or regurgitate, into the left ventricle. This problem, which is also known as aortic regurgitation or aortic insufficiency, forces the left ventricle to do extra work, since this blood must be pumped out a second time. Only by doing this additional work can the heart sustain a normal flow of blood to the body.

Over time, this added burden can take a toll on the heart. The left ventricle enlarges to accommodate the blood flowing back into it from the aorta and works less efficiently when its architecture becomes distorted. Eventually, some people with aortic regurgitation develop congestive heart failure, which may not be reversible even if the valve is replaced surgically. In heart failure, the heart is unable to pump enough blood to maintain normal circulation; the result is a buildup of fluid in the body.

Comprehension question: *How does the extra work the heart must do because of aortic regurgitation lead to an irreversible problem?*

Until recently, aortic-valve surgery was the only way to stop this process. Although the surgery is usually successful, the patient must live thereafter with the problems associated with an artificial valve. However, scientists have been searching for a medication that might decrease the damage from aortic regurgitation and thereby delay or eliminate the need for surgery. A recent randomized trial reports the first long-term positive results of such drug therapy.

New Role for Nifedipine

Knowledge question: *What side effects might be caused by taking nifedipine?*

The drug used in this study was nifedipine—a medication that is used most commonly to treat high blood pressure and angina (the chest pain or discomfort that occurs when the heart muscle is not getting enough blood from the coronary arteries). Nifedipine (Procardia, Adalat) is a calcium-channel-blocking agent, which causes muscle cells to relax by blocking the movement of calcium across cell membranes. This medication is believed to help people with coronary artery disease, in part by dilating the coronary arteries and preventing spasm of these blood vessels. It also lowers blood pressure by dilating arteries in the rest of the body.

Application question: *How does a vasodilator such as nifedipine delay the need for surgery?*

These vasodilator effects of nifedipine are believed to be the beneficial mechanism for patients with aortic regurgitation. The dilated arteries and the lowered blood pressure decrease the force that "pushes" blood back through the leaky aortic valve. Since less blood flows the wrong way, the strain on the left ventricle decreases.

Analysis question: *How does nifedipine treat the symptoms of high blood pressure and angina?*

Two relatively short-term studies previously showed that nifedipine and another vasodilator, hydralazine, might have a beneficial effect on the heart's function, but whether these drugs could actually prevent the need for surgery remained uncertain. Therefore, a team of researchers from Padua, Italy, designed a long-term study to test this idea (New England Journal of Medicine, 494, pp. 689–694).

Before Symptoms Begin

All 143 people who were enrolled in this study had severe aortic regurgitation, which was documented by echocardiography, an examination that uses sound waves to produce a video of the beating heart (see *Harvard Heart Letter*). None of the patients started with symptoms or evidence of damage to the left ventricle. About half were assigned to receive daily digoxin (0.25 mg) to increase the strength of the heart's contractions; the rest took nifedipine (20 mg) twice a day.

Over the next six years, the patients who received nifedipine were about half as likely to need an aortic-valve replacement as those who took digoxin. During the first two years, there were no valve replacements in the nifedipine group versus about 10% of those who took digoxin. At the end of six years, 34% of the patients who took digoxin had undergone valve replacement compared with only 15% of those assigned to nifedipine. The decision to perform surgery was made according to specific criteria, so it is unlikely that knowledge of which drug the patient was taking influenced the rate of valve replacements.

Synthesis question: *How does the use of a drug such as nifedipine contribute to the treatment of cardiac patients?*

Nifedipine's benefits were not without a price. At least one new mild symptom was reported during the first three months by 42% of the patients who took nifedipine versus 12% of those who used digoxin. The most common side effects were a sensation of a rapid heartbeat and headache among those who used nifedipine, and fatigue among those who used digoxin. At six years of follow-up, some side effects were still reported by 5% of those who used nifedipine, but no patient in this study discontinued medication because of complications.

Encouraging Data

This encouraging study suggests that the course of aortic regurgitation can be altered and that nifedipine can delay the need for surgery if given early enough in the course of this disease. Whether other vasodilators can reduce the need for surgery is unknown; in theory, other drugs could be as effective, but none have been studied as comprehensively as nifedipine.

Evaluation question: *In what ways does research into the use of vasodilators appear to be promising?*

Vasodilator therapy is not a replacement for surgery. If medical therapy with a vasodilator is unsuccessful, patients and their physicians should not continue to rely on that treatment. If the left ventricle is substantially dilated or damaged, patients should have surgery before the problem becomes so severe that even replacing the valve cannot restore normal function. Nevertheless, this study represents very positive news to many of those with aortic regurgitation. People who have this condition should consult with their physician to determine whether the findings from this investigation are relevant to them.

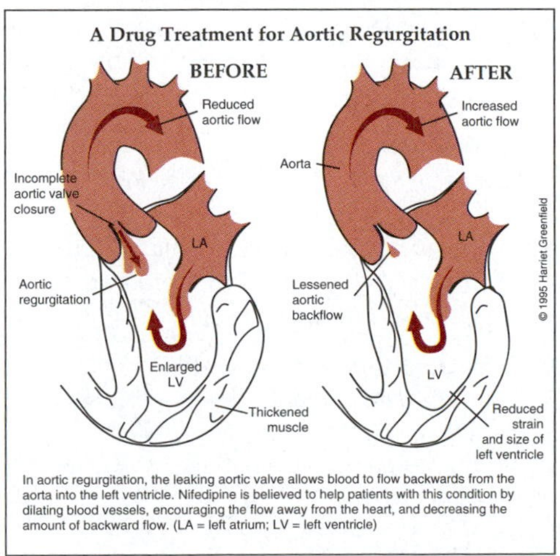

In aortic regurgitation, the leaking aortic valve allows blood to flow backwards from the aorta into the left ventricle. Nifedipine is believed to help patients with this condition by dilating blood vessels, encouraging the flow away from the heart, and decreasing the amount of backward flow. (LA = left atrium; LV = left ventricle)

FIGURE 3.10 Practical Approaches for Reading Complex Technical Documents

Before Reading

What pre-reading strategies are helpful?

- Skim the title, abstract, headings, opening and concluding paragraphs of major sections and visuals as well as their titles and captions,
- Mentally review what you already know about the subject. Identify what's likely to be difficult about the text and visuals. Complexity? Familiarity? Time constraints? Decide how you will manage these difficulties.
- Note and check meanings of unfamiliar terms or visual displays that catch your eye.
- Check the works cited or bibliography of the document to review the sources the writer used.
- Decide what's potentially relevant about the document.

What's your purpose(s) for reading a particular document?

- Determine your purpose for reading a particular document. *Reading to assess* (to decide whether to read thoroughly)? *Reading to learn* (to acquire background information)? *Reading to learn to do* (to understand a process you're supervising)? *Reading to do* (to complete a task or to make a decision)?

What's likely to be difficult about a particular document?

- Anticipate possible problems. Unfamiliar genre? New topic? Limited time to read? Difficult concepts? No supporting examples? Unfamiliar types of visuals? Complex visuals? Limited or inadequate visuals? Poor document design?

During Reading

How will you reduce (or prevent) distractions or interruptions during your reading?

- Given the difficulty and importance of the document, estimate the amount of time you'll need, and set other work aside during that time.
- Work in an isolated area, wear headphones, or post a DO NOT DISTURB sign.

How will you annotate your reading?

- Determine the approach you'll use for annotating important information: Take marginal notes? Highlight important phrases? Use self-stick notes? Make an outline? Create diagrams or charts?

How will you ask and record your questions about the reading?

- Determine the kinds of questions you should ask about the text and visuals: knowledge, comprehension, application, analysis, synthesis, and/or evaluation questions.
- Decide how you'll record these questions. On a note pad? In the margins? In a computer file?

After Reading

How will you make meaning from your reading?

- Determine the possible interpretations of the text and visuals. Identify possible alternative interpretations.
- Identify apparent omissions, gaps, distortions, and contradictions in the text and visuals.
- Decide the most effective way to increase your recall of the information in the text and the visuals. Summarize the key ideas? Write your reactions?

What is the most productive way to respond to the document?

- Identify the writer's key point(s) in the text and visuals. Identify the support for these points. Decide whether you agree with the writer.
- Consider how information adds to your knowledge or influences your position about the topic.

- Figure 3.12 presents the introductory paragraphs that embed the questions and comments made by Susan, who has a degree in journalism. Susan read the article to learn more about the interaction that takes place in smart rooms for a feature article she was writing for a local newspaper.

Can identical paragraphs mean different things to different readers? The questions posed by Clay and Susan show that from the very beginning they had different reactions. While their general purpose was the same (to learn), their specific purposes were quite different. Clay was interested in the technical interface and Susan in the human interaction and application. The meaning of a document is shaped by the reading process and interpretation brought to the document by the readers. As Figures 3.11 and 3.12 show, the opening paragraphs of this *Scientific American* article sometimes provoked similar comments and questions from Clay and Susan. At other times, though, their interpretations were startlingly different. The following list highlights some of their similarities and differences:

- Both remark that the term "smart room" sounds familiar. Both make connections to their prior knowledge, Clay mentioning a journal he reads and Susan drawing on her prior knowledge.
- Both are curious about who wrote the article. Susan immediately searches through the article until she finds the author's bio at the end. The information in this bio shapes Susan's reading, since she wonders whether the listed sponsors might influence the nature of the article.
- Clay and Susan both start eagerly, but they read the article in different sequences. Clay starts at the beginning and reads straight through the first three paragraphs, referring to photographs as he needs them. Susan reads the title, teaser, and author's name; then she searches for biographical information. Before moving back to the beginning of the article, she reads one of the article's sidebars, looks at a couple of the color photos in the article, and reads their captions.
- Clay and Susan both draw frequently on their prior knowledge, which shapes their reading. But their prior knowledge is remarkably different. Clay talks about UNIX-based systems, Silicon Graphic terminals, and limitations of VR goggles. Susan talks about the value of imperative mood, balanced examples, and personal anecdotes as well as ways to make information relevant to general readers.
- Clay and Susan form different views of the author. When Clay reads about software developed by the author and another programmer, he talks about the author's credibility going up: "This isn't a journalist trying to interpret an expert's words. . . ." Susan talks about the author's credibility having "taken a large dive" when she anticipates that he may be promoting his own work.

FIGURE 3.11 Computer Scientist "Interrogating" a Text (Reader Comments in Blue)

SMART ROOMS

Smart rooms. I heard about these in a *Communications of the ACM* a few years ago. The picture looks interesting. [Clay reads the caption.] **Okay, it's a giant screen and the user appears on it, sort of like a shadow projected on the wall, only more colorful. It's not 3D. How is it better?**

[article's teaser] In creating computer systems that can identify people and interpret their actions, researchers have come one step closer to building helpful home and work environments.

Not sure if I buy this. What do they mean by "interpret actions"? Physical actions, maybe — that's what the picture looks like. It reminds me of the haunted house ride in Disneyland; there's one point at which you pass a mirror and see ghosts sitting beside you in the reflection. You turn around and they're not there. Same with this screen.

by Alex P. Pentland

Who is this guy? Guess I'll find out.

Imagine a house that always knows where your kids are and tells you if they are getting into trouble. **Doesn't look like it'll be purely you facing the screen, then. I wonder how they're managing that? Cameras? How are they rendering the people they monitor — C++ objects? Shades of Big Brother here.** Or an office that sees when you are having an important meeting and shields you from interruptions. **Artificial intelligence — this isn't strictly graphical interface.** Or a car that senses when you are tired and warns you to pull over. **How?** Scientists have long tried to design computer systems that could accomplish such feats. **Examples? What research was this built from?** Despite their efforts, modern machines are still no match for babysitters or secretaries. But they could be.

The problem, in my opinion, is that our current computers are both deaf and blind; they experience the world only by way of a keyboard and a mouse. **And that's the way some of us like them.** Even multimedia machines, those that handle audiovisual signals as well as text, simply transport strings of data. **So this probably won't be referring to a UNIX-based system, since UNIX relies on data strings. Probably an object-oriented operating system. I see that the diagram has what looks like a Macintosh in it. Is this technology Mac-based? No, the next page shows someone rendering his face on a Silicon Graphics terminal, thank God.** [Clay skims the article looking for a mention of hardware or software.] **Here's a piece of software called Pfinder — apparently the writer and another programmer developed it. Well, the writer's credibility just went up: this isn't a journalist trying to interpret an expert's words, but someone who knows what he's talking about! But he doesn't want to tell us what OS he's using. Perhaps Pfinder IS the OS.**

They do not understand the meanings behind the characters, sounds, and pictures they convey. I believe computers must be able to see and hear what we do before they can prove truly helpful. **So is he talking about artificial intelligence? The experts say that true AI is a long ways away. Or is he speaking metaphorically?** What's more, they must be able to recognize who we are and, as much as another person or even a dog would, make sense of what we are thinking. **Yep, that's AI all right, or an impoverished imitation of it.**

To that end, my group at the Media Laboratory at the Massachusetts Institute of Technology has recently developed a family of computer systems for recognizing faces, expressions, and gestures. **That might be interesting. But maybe irritating. I mean, voice-activated technology at present involves having the user train herself to speak the way the computer wants him to — voice-pilots tend to sound pretty stilted when talking to the computer. Are we going to have to exaggerate our expressions to operate the software now? I don't like the idea of taking mime lessons just to get to my files. And different cultures use different expressions to express their emotions; would we have to have a library of expressions for each culture?** The technology has enabled us to build environments that behave somewhat like the house, office, and car described above. These areas, which we call smart rooms, are furnished with cameras and microphones that relay their recordings to a nearby network of computers. **Sounds expensive. High-end users, office security, etc. Could lessen need for security guards. Military installations? Corporate headquarters?** The computers assess what people in the smart room are saying and doing. Thanks to this connection, visitors can use their actions, voices, and expressions — instead of keyboards, sensors, or goggles — to control computer programs, browse multimedia information, or venture into realms of virtual reality. . . . **Arrgh. VR. Well, I have heard that VR goggles are too limiting for real work and collaboration — where was that, *Wired*? Makes sense. Maybe this would be useful, but I'd like to see more examples. This sounds too specialized and too expensive for the sort of tasks desktop users would perform — writing code, word processing, etc.**

FIGURE 3.12 Journalist "Interrogating" a Text (Reader Comments in Blue)

SMART ROOMS

OK, this is a term I have heard before, maybe in relation to missiles and other military applications. Hmmm. Wonder what smart rooms are like. Is this a virtual reality thing? I bet the subheads will help clarify.

[article's teaser] In creating computer systems that can identify people and interpret their actions, researchers have come one step closer to building helpful home and work environments.

Oh, my. This sounds incredible. So, how will the writer "sell" this to readers? Assume that they are sympathetic, excited about this technology? Approach the subject with soft sell? Do these work environments and home environments interact? Are they separate smart rooms?

by Alex P. Pentland

Who wrote this article? [Susan searches through the article to find a bio of the author.] **It's not at the bottom of the first page, which would have been more convenient; instead, it's at the end of the article next to the references. Geez. He has a Ph.D. from MIT — so, he's a technical person who writes rather than a reporter who's covering a technology beat. Awards, awards, awards. Honors, honors, honors. Oh, and he's a researcher on smart rooms. The sponsors of his research are listed; I'll be reading the article with an eye toward whether he mentions these sponsors explicitly. If so, I have to wonder about the piece.**

[Susan doesn't immediately go back to the beginning of the article. Instead, she looks at the PERSON FINDER sidebar and glances from it to the four-color photo to its right.]

So, now I know how this person finder application works. Why would I want to track this? A general reader of a feature article would be asking for this kind of answer, too.

[Susan's attention is attracted to the SURPRISED caption.] **The information is intriguing, but why? This looks a lot like the artists' renditions of suspects in crimes. Sort of the computerized drawings (a la Unabomber!) but these images are a comparison of "smart-room users" and the visual images of animated models. And, again, who are these smart room users?**

Imagine a house that always knows where your kids are and tells you if they are getting into trouble. **All right! The connection to the general reader of my feature! It's imperative mood. That's good. Those blasted question leads are deadly.** Or an office that sees when you are having an important meeting and shields you from interruptions. Or a car that senses when you are tired and warns you to pull over. **The three examples — that nice, odd number — offering general readers different, but recognizable examples.** Scientists have long tried to design computer systems that could accomplish such feats. **"Feats" makes me think that this technology is a good thing.** Despite their efforts, modern machines are still no match for babysitters or secretaries. **Pentland is reaching for the connection with the general reader with "babysitters and secretaries."** But they could be.

The problem, in my opinion, is that our current computers are both deaf and blind; they experience the world only by way of a keyboard and a mouse. **This is the same sentiment of the guy who wrote *Being Digital* — that computers are great, but they can be so much greater. So much more interactive.** Even multimedia machines, those that handle audiovisual signals as well as text, **Good. I needed a simple definition of what "multimedia machines" are.** simply transport strings of data. They do not understand the meanings behind the characters, sounds, and pictures they convey. I believe computers must be able to see and hear what we do before they can prove truly helpful. **Here's the opinion, an odd addition for the reporter to put himself into the story — even for a magazine. I guess I would have expected a first-person approach to be made explicitly from the start. Beginning with a personal anecdote.** What is more, they must be able to recognize who we are and, as much as another person or even a dog would, make sense of what we are thinking. **But why?**

To that end, my group at the Media Laboratory at the Massachusetts Institute of Technology **Uh, huh, here it comes. The ad for this guy's own work. His credibility has just taken a large dive.** has recently developed a family of computer systems for recognizing faces, expressions, and gestures. The technology has enabled us to build environments that behave somewhat like the house, office, and car described above. These areas, which we call smart rooms, **At last, a definition of smart room. I wonder if this wouldn't have been more effectively introduced earlier. I wonder if other readers remember the smart bomb rhetoric from the Gulf War and might have more easily connected with that. Probably endangers the negative rather than the positive for this guy's research.** are furnished with cameras and microphones that relay their recordings to a nearby network of computers. **Nearby, where?** The computers assess what people in the smart room are saying and doing. Thanks to this connection, visitors can use their actions, voices and expressions — instead of keyboards, sensors or goggles — to control computer programs, browse multimedia information or venture into realms of virtual reality. **The why? is missing . . . or at least not made explicit and real and desirable for general readers. Why do we want smart rooms?**

Explain why it's inaccurate to say that if a document is written clearly enough, each reader will gain the same meaning from it.

Regardless of the differences in the details of their reading, Clay and Susan both interrogate the article at about the same level of detail and for about the same amount of time, inserting occasional wry comments and regularly reminding themselves about their own purpose for reading. They both read critically; the meaning they create is shaped by their prior knowledge and their use of various reading strategies.

Individual and Collaborative Assignments

1. **Identify the purpose, methodology, result(s), and conclusion.** The abstract that follows is from the Web site of the journal *Medicine and Science in Sports and Exercise.* The headings have been removed from the abstract. Read the abstract and identify the purpose, methods, results, and conclusion. After you have identified them, check the Web site at http://www.ms-se.com/ to see if your categories agree with the published version.[15]

   Abstract. Previous studies have found conflicting relationships between type of playing surface and injury in American football but have not taken into account possible variations in the surface conditions of outdoor stadiums due to changing weather. A total of 5910 National Football League team games between 1989 and 1998 inclusive were studied to determine associations between knee and ankle sprains, playing surface, and the weather conditions on the day of the game. There was reduced risk of significant ankle sprains (at least 7-d time loss) for games in natural grass stadiums compared with domes (indoor stadiums using AstroTurf) (RR 0.69, 95% CI 0.58–0.83). There was also reduced risk of significant knee sprains on grass compared with domes (RR 0.77, 95% CI 0.66–0.91), although most of this reduction was related to cold and wet weather on grass (RR 0.66, 95% CI 0.47–0.93 compared with hot and dry weather on grass). In open (outdoor) AstroTurf stadiums, cold weather was associated with a lower risk of significant ankle sprains (RR 0.68, 95% CI 0.51–0.91), significant knee sprains (RR 0.60, 95% CI 0.47–0.77) and ACL injuries (RR 0.50, 95% CI 0.31–0.81) compared with hot weather in the same stadiums. Weather did not have any significant effects on the injury risk in domes. The ACL incidence rate was lower during the later (cooler) months of the season in open stadiums (both AstroTurf and natural grass) but not in domes. Cold weather is associated with lower knee and ankle injury risk in outdoor stadiums (both natural grass and AstroTurf), probably because of reduced shoe-surface traction.

2. **Identify the purpose, methodology, result(s), and conclusion.** The example that follows is an abstract from the journal *Science.* Carefully read the abstract and identify the purpose, methodology, result(s), and conclusion.

Pollution Monitoring of Puget Sound with Honey Bees[16]

Abstract. To show that honey bees are effective biological monitors of environmental contaminants over large geographic areas, beekeepers of Puget Sound, Washington, collected pollen and bees for chemical analysis. From these data, kriging maps of arsenic, cadmium, and fluoride were generated. Results, based on actual concentrations of contaminants in bee tissues, show that the greatest concentrations of contaminants occur close to Commencement Bay and that honey bees are effective as large-scale monitors.

3. **Identify the main points.** The following abstract is from an internal technical report, *Correcting the Field Failure of the XC-2000.*[17] Read it carefully and then answer the following questions:
   - What is the main issue or problem?
   - Who and what are involved in the issue or problem?
   - What is the approach to the solution to the problem?
   - How will the approach be addressed or the solution be implemented?

Abstract. The XC-2000 display has a serious problem — a 20% field failure rate since July — that has resulted in a shipping delay. The failure is caused when the display's 4305 driver components short internally, producing smoke that concerned customers.

Failure evaluations of both the 4305 driver and the 4503 receiver have been conducted by Design Component Engineering. The evaluations included pin-to-pin electrical testing, external and internal visual inspection, and failure-mode duplication testing in and out of the application. The cause of failure is a 10,000-volt transient on the 12-volt supply (pin 8) of the 4305, and a similar transient on the input (pin 5) of the 4503 receiver.

Investigation confirms that the damage occurs in transit, between final system test and delivery to the customer site. Because of the difficulty encountered in locating the exact cause of the transient damage, efforts have been channeled toward protecting the parts rather than removing the cause.

XC-2000 display driver/receiver boards were tested with several bypass protection techniques. The most effective and economical protection is the placement of two 1N7211 zener diodes across selected pins on the J1 and J4 connectors. The 300 reworked units have been shipped through normal methods and have arrived in working order at the customer sites. Therefore, a corrective action has been identified and should be used on all future production of the XC-2000.

4. **Determine kinds of reading.** Read the following brief scenarios and determine the primary kind(s) of reading each workplace professional is doing: reading to assess, reading to learn, reading to learn to do, or reading to do.
   - Engineer reading a company newsletter to learn about what other research projects in the division have been funded

- Agronomy technician reading product label and instruction sheet to select proportions for mixing an herbicide to apply on an experimental plot
- Division manager reading a report about market trends in order to propose a new product line
- Chemical technician reading a product abstract to see if the report will be useful for herself and the rest of her research group to read
- Engineer reading an operator's manual to learn to calibrate a piece of equipment
- Mechanical engineer reading product descriptions to decide which spectrometer to order for the lab
- Hospital dietician reading product labels to see if any of the goods in the supply room meet the restricted needs of a new patient
- Office supervisor reading an operator's manual about ways to change the toner cartridge in the photocopying machine
- Lab technician reading an article from published conference proceedings to learn general background information about a process that a competing firm is using
- Equipment supervisor reading product safety sheets to learn to do the most appropriate first-aid treatment in the case of an emergency
- Maintenance team leader reading the online manual that identifies step-by-step troubleshooting procedures in order to repair a piece of equipment
- Microbiologist skimming specifications sheets to decide which air filtration systems to consider closely for a new clean room

5. **Compare differences in reading purposes.** (a) Work with a small group to obtain additional interrogations of the following excerpt — three paragraphs exactly as they appeared in the *Scientific American* article (or use another text of similar interest and length). Locate two people who would be interested in reading the document but have different purposes for reading it and who would bring different backgrounds to their reading. Ask the two to read and interrogate the text.

   You can access an online version of these paragraphs by going to www.english.wadsworth.com/burnett6e for a link to this excerpt. If you can, give the participants the online version, so they can write their comments directly into the online text, as Clay and Susan did. If they are reading a paper version, they can record their comments on a tape recorder or jot them in the margins. (b) With your group, compare the differences and speculate about the possible reasons for the differences. (c) Write a memo to the class in which you report what your group identifies as the key similarities and differences between the two readers' interpretations.

Rhetorical Elements Characterizing Technical Communication[18]

Smart Rooms

In creating computer systems that can identify people and interpret their actions, researchers have come one step closer to building helpful home and work environments.

by Alex P. Pentland

Imagine a house that always knows where your kids are and tells you if they are getting into trouble. Or an office that sees when you are having an important meeting and shields you from interruptions. Or a car that senses when you are tired and warns you to pull over. Scientists have long tried to design computer systems that could accomplish such feats. Despite their efforts, modern machines are still no match for babysitters or secretaries. But they could be.

The problem, in my opinion, is that our current computers are both deaf and blind; they experience the world only by way of a keyboard and a mouse. Even multimedia machines, those that handle audiovisual signals as well as text, simply transport strings of data. They do not understand the meanings behind the characters, sounds, and pictures they convey. I believe computers must be able to see and hear what we do before they can prove truly helpful. What is more, they must be able to recognize who we are and, as much as another person or even a dog would, make sense of what we are thinking.

To that end, my group at the Media Laboratory at the Massachusetts Institute of Technology has recently developed a family of computer systems for recognizing faces, expressions, and gestures. The technology has enabled us to build environments that behave somewhat like the house, office, and car described above. These areas, which we call smart rooms, are furnished with cameras and microphones that relay their recordings to a nearby network of computers. The computers assess what people in the smart room are saying and doing. Thanks to this connection, visitors can use their actions, voices and expressions instead of keyboards, sensors, or goggles to control computer programs, browse multimedia information or venture into realms of virtual reality. . . .

6. **Interrogate a document.** Use the World Wide Web to locate a document that addresses some controversial aspect of a subject in your professional discipline. As Dave Clark did with the online document in Figure 3.8, interrogate the document:
   - Identify the tacit assumptions on which you believe the document is based — what's presumed but not articulated?

- Extend the ideas to pose reasonable but unstated implications — what's implied but not articulated?
- Speculate on the impact of the implication — what's possible but not articulated?

Chapter 3 Endnotes

1 Some of the discussion in this section is based on types of reading identified in two articles that have become standards: Diehl, W., & Mikulecky, L. (1988). Making written information fit workers purposes. *IEEE Transactions on Professional Communication, 24,* 5–9; and Redish, J. C. (1988). Reading to learn to do. *The Technical Writing Teacher, 15*(3), 223–233.

2 Retrieved September 12, 2003, from www.extension.iastate.edu/foodsafety/rad/radura.html

3 The discussion related to Figure 3.3 is based on a synthesis of information from several sources.

- Dole, J. A., Duffy, G. G., Roehler, L. R., & Pearson, P. D. (1991). Moving from the old to the new: Research on reading comprehension instruction. *Review of Educational Research, 61*(2) 239–264.
- Duin, A. H. (1988). How people read: Implications for writers. *The Technical Writing Teacher, 15*(3), 185–193.
- Flower, L. (1990). Negotiating academic discourse. In Flower, L., Stein, V., Ackerman, J., Kantz, M. J., McCormick, K., & Peck, W. C. *Reading to write: Exploring a cognitive and social process* (pp. 221–252). New York: Oxford University Press.
- King, K. (n.d.). *Reading strategies.* Retrieved September 12, 2003, from http://www.isu.edu/~kingkath/readstrt.html
- Knuth, R. A., & Jones, B. F. (1991). *What does research say about reading?* Retrieved September 12, 2003, from http://www.ncrel.org/sdrs/areas/stw_esys/str_read.htm
- Mulcahy, P. (1988). Writing reader-based instructions: Strategies to build coherence. *The Technical Writing Teacher, 15*(3), 234–243.
- Slater, W. H. (1988). Current theory and research on what constitutes readable expository text. *The Technical Writing Teacher, 15*(3), 195–206.

4 *Journal of the American Medical Association (JAMA).* (1991). Structuring abstracts to make them more informative. *JAMA, 266,* 116–117.

5 Retrieved September 13, 2003, from http://www.foodprotection.org/

6 Sakchai, H., Suphachai, N., Riemann, H., & Cliver, D.O. (2001). Reduction of Escherichia coli O157:H7 and Salmonella Typhimurium in artificially contaminated alfalfa seeds and mung beans by fumigation with ammonia. *Journal of Food Protection, 64*(11), 1817–1819. Abstract retrieved, September 13, 2003, from Food Science & Technology Abstracts.

7 *NASA Tech Briefs,* Vol. 24, No. 3, March 2004.

8 Burnett, C. D. (1993). *How to grow shiitake mushrooms outdoors on natural logs.* (Marquette, MI: Big Creek Farm. Reprinted with permission.

9 This abstract is an amalgam of four slightly different versions of the published abstracts for this article. One is available on the Web site of Dr. Linnyer Beatrys Ruiz; retrieved September 13, 2003, from http://www.dcc.ufmg.br/~linnyer/linnyer.html; the second is on the IEEE Web site at http://www.comsoc.org/ci1/Public/2003/feb/current.html; the third is with the PDF version of the article; the fourth is available online through the Iowa State University Park's Library, connected to IEEE Xplore.

10 Bloom, B. S. (Ed.). (1956). *Taxonomy of educational objectives.* New York: David McKay.

11 Adaptation from Nelson, T. H. (1993) *The Xanadu Ideal.* Retrieved September 13, 2003, from http://xanadu.com.au/general/ideal.html. Originally retrieved June 1996, from http://www.xanadu.com/au/xanadu/idea.html/

12 Katz, S. (1992). "The Ethic of Expediency: Classical Rhetoric, Technology, and the Holocaust." *College English, 43,* 255–275.

13 © 1996 Scientific American, Inc. All rights reserved. Reprinted with permission.

14 Slowing the course of aortic regulation. (1995). *Harvard Health Letter,* (February), 3–4. Illustration © 1995 Harriet Greenfield.

15 Orchard, J. W., & Powell, J. W. (2003). Risk of knee and ankle sprains under various weather conditions in American football. *Medicine & Science in Sports & Exercise, 35*(7), 1118–1123. Retrieved September 27, 2003, from http://www.ms-se.com/

16 Bromenshenk, J. J., et al. (1985). Pollution monitoring of Puget Sound with honey bees. *Science 227* (February 8), 632–634. Copyright 1985 by AAS. Reprinted with permission.

17 Megnin, C. (n.d.). *Correcting the field failure of the XC-2000.* (Technical and Scientific Writing 42.225) University of Lowell. Reprinted with permission.

18 © 1996 Scientific American, Inc. All rights reserved. Reprinted with permission.

CHAPTER 4

Addressing Audiences

Objectives and Outcomes

This chapter will help you accomplish these outcomes:

- Understand that professionals who plan, prepare, and present technical documents, presentations, and visuals usually have two broad purposes: to convey verifiable information and to persuade the audience to attend to this information
- Identify initial, primary, secondary, and external audiences, and determine whether these audiences are experts, professional nonexperts, technicians, equipment operators, students, generalists, or children
- Collect and analyze information about your audiences — the context in which they work, their attitudes and motivations, education, professional experiences, reading level, and organizational role
- Adjust material for different audiences in two broad ways: (1) address audiences with different levels of expertise by adjusting the complexity of the material; (2) address audiences with different organizational roles by shifting the focus of the discussion and the choice of details, including an acknowledgement of audiences' ethical stance or point of view

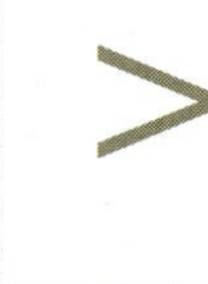

Nearly anything can be the subject of a technical document, oral presentation, or visual, but the treatment of that subject changes for different audiences. For example, the same polymer resin used to make baby bottles is also used for media storage disks. Many audiences could be interested in the same subject but with different purposes: Parents and pediatricians might want to know about polymer safety; media librarians and managers might focus on material stability; chemists might be interested in ways to modify material composition; production supervisors might want to know how to best extrude the resin.

The way you define and visualize your audiences influences your choices as you prepare documents, oral presentations, and visuals. This chapter discusses the importance of identifying your purpose, categorizing types of audiences, and analyzing factors that influence them.

Identifying Purposes

Professionals who prepare technical documents, presentations, and visuals usually have two broad purposes: to concisely and accurately convey verifiable information and to persuade audiences to attend to this information. Figure 4.1 shows questions you need to ask yourself about your audiences during planning.

- *Do you think technical documents convey neutral or value-free information?*
- *In what ways are technical documents persuasive? How could you convince a colleague who thinks that technical documents are only to convey information that they also have a persuasive function?*

FIGURE 4.1 Questions Stimulated by Purpose

Purpose	Questions
To convey information	▪ What information do I want my audience to learn? Why do I want them to learn this information? ▪ What decisions does the audience need to make? What information does the audience need in order to make a decision? ▪ What background information do I need to provide? What questions do I want to answer?
To argue/To persuade	▪ What ideas or actions of the audience do I want to influence? ▪ What information and approaches will persuade the audience? ▪ What constraints will affect the persuasiveness of my argument? ▪ What objections might the audience have? What logical argument would enable me to overcome those objections?

People may have both primary and secondary purposes for reading. As you read in Chapter 3, typical reasons include reading *to assess, to learn, to learn to do,* or *to do.* For example, a primary task might be to *do:* to make a decision that approves or rejects a proposal. A secondary task might be to *learn:* to gather information about important areas for future research and/or implementation.

Identifying Audiences

Before you analyze an audience, you must identify it. For example, every technical document has an *intended audience,* a specific individual (Elizabeth Jones, research and development [R&D] director) or a category of users (owners of Dell computers), with identifiable needs. Audiences of technical documents often want information about specific rather than general issues.

Complicating matters, the same information is often read by a number of different audiences in several different ways. For example, a proposal for new product development could be read by people on various levels in several areas in a company — finance, marketing, engineering, manufacturing. As a result, you often are expected to create material that simultaneously meets the needs of several

These workplace professionals are all reading information about the same new product, but the situations are different and the form in which the information is accessed is different.

categories of audiences. You can accomplish this most easily by directing different audiences to particular sections of a proposal. For example, managers or executives would usually be most interested in the executive summary and the major recommendations, whereas engineers would usually be most interested in the application of these recommendations.

Audience roles can be separated into four general categories based on factors such as people's position in the organization, their connection to the particular issues, and their part in decision making. The same individual may be a primary audience in one situation and a secondary audience in another situation.

- The *initial audience* is usually the person to whom you submit a document, though not necessarily the ultimate decision maker. This initial audience directs your document to the appropriate primary audience.
- The *primary audience* is the person for whom your document is actually intended, the one who will actually use the information, the decision maker.
- *Secondary audiences* receive and read your document; they have an interest because they are affected by the information or by decisions based on it.
- Many documents also have *external audiences* who are outside the immediate organization but are affected by the information or by decisions based on it.

> *What might be the ramifications of misjudging who is the primary audience of a document?*

For example, a general contractor could be the initial audience of a request for quotation (RFQ) on wiring a condominium's new community center. The primary audience could be the electrical contractor. The secondary audiences might include the condominium's board of directors. An external reader might be a local building inspector.

Audiences also are distinguished by their level of expertise and form a continuum of knowledge about the subject, from those with expertise and interest to those with little knowledge and some interest. Audiences typically fit on more than one place on a continuum, especially if they have multiple interests and responsibilities.

> *Which audience(s) do you think you will address most frequently in your professional work?*

This continuum of audiences is not absolute; an individual reader often fits into more than one category. For example, a polymer engineer who is comfortable reading reports about composite materials may be a general reader or a professional nonexpert when reading about plant pathology, traffic patterns, or kidney failure. Similarly, a medical technician who is comfortable reading biopsy reports may be a general or a professional nonexpert when reading about astronomy, learning disabilities, or woodworking. A company vice president is an expert in business matters but may be a student reader in an aviation class and a general reader regarding agronomy or microwaves. Figure 4.2 identifies the level of familiarity these audiences typically have with various kinds of subject matter, the kinds of expectations they often hold, and the education many of them have. Such generalizations, of course, don't fit the profile of every reader, but may help you make preliminary decisions about ways to approach a particular audience.

> *What are specific examples of technical documents that target children as audiences?*

FIGURE 4.2 **Audience Familiarity with Subject, Expectations, and Education**

Readers	Familiarity with Subject	Expectations	Prior Knowledge
Experts	Most know theories, practical applications, jargon, and technical information in their own field	Most want an explicit purpose; prefer direct explanations; use information to assess, learn, learn to do, and do	Usually have undergraduate or graduate degrees or equivalent experience in specialized fields
	For example, engineers, ecologists, and economists typically read a wide range of print and electronic material pertaining to their field, from practical applications (specifications) to theoretical concepts (professional journals).		
Professional nonexperts	Most know general concepts of the field in which they're working	Most prefer definitions and explanations of concepts and procedures; prefer information for decision making	Usually have undergraduate or graduate degrees or equivalent experience, often in areas peripheral to those for which they are responsible
	For example, a department manager in a computer firm may have a degree in business but little specialized knowledge of computers. A manager of an engineering department may have earned an engineering degree twenty years ago or have a degree in industrial management, not engineering.		
Technicians	Most know the specialized area in which they're working; may have theoretical understanding	Most prefer straightforward definitions and explanations; prefer information that focuses on learning to do and doing	Usually have a degree from a two-year or four-year college or on-the-job training
	For example, chemical technicians helping R & D engineers develop applications for carbon fibers are able to contribute to the project because of their theoretical background in polymer chemistry.		
Equipment operators	Most know tasks they're assigned to do; may improve performance if they understand the way their job relates to broader processes	Most prefer explanations that let them focus on *doing*	Usually have on-the-job training (often oral rather than written) instead of formal education related to the job
	For example, machine operators do a better job if they know how their work is impacted by and impacts larger processes. Most equipment operators have no theoretical knowledge of the field.		
Students	Often know generalizations in a field; typically need technical details as well as implications for broader processes	Most prefer information that helps them assess, learn, learn to do, and do; interested in theory as well as practice	May have specialized training from summer or part-time jobs, internships, or co-op programs
	For example, students are interested in learning disciplinary knowledge and forming opinions in order to become professionals in a specialized field.		
Generalists	Often know generalizations in a field; want information that explains the how and why	Most prefer information that helps them assess, learn, learn to do, and do; varied interests	May be highly educated but not in what they're doing; may have specialized knowledge
	For example, general readers want to read about science and technology, but their interests usually have little or nothing to do with their job.		
Children	Often want information that explains how and why things happen; often have limited concepts and vocabulary	Most prefer information that helps them to learn, learn to do, and do; widely varied interests	Usually have completed some school; may have specialized knowledge from hobbies and activities
	For example, many children read science books, have their own subscriptions to children's science magazines, and regularly surf Web sites. They also read the technical documents that come with their DVD player and video games.		

WEBLINK

Identifying, analyzing, and adapting to your audiences is a major concern in government projects and presentations. Go to www.english.wadsworth.com/burnett6e for links to sites that include specific discussions about audience as part of preparing major projects and presentations. Access these links and examine the ways that audience is discussed. You might be surprised at the detailed attention that is given to audience.

CLICK ON WEBLINK

CLICK on Chapter 4/audience government

Analyzing Audiences

Once you have identified the purposes for a document, oral presentation, or visual and the general category of the audience(s) (expert, technician, and so on), you need to learn more about the target audience(s). Several strategies exist for analyzing audiences. While no approach is foolproof, one of the most frequently used involves considering these characteristics as you plan to meet audiences' needs and expectations:

- Context in which a document is interpreted and used
- Purpose and motivation of the audiences
- Prior knowledge of the audiences, including education and professional experiences
- Reading level of the audiences
- Organizational role of the audiences

One of the most effective ways to assess your intended audiences is to talk with them and the people who work with them. Imagine that you are responsible for writing the user's manual and online help for a new computer system. Whether the manual and online help are usable for the intended audience is critical. So you need to find out about the actual users — what they're likely to have trouble with, what they'll probably find helpful, what they're likely to know and not know. To gain a sense of what should and shouldn't be included, you could talk with people in these areas:

- *Design and development* to learn what features the system will have and what kind of technical background a user would need
- *Marketing* to learn how the target market has been characterized

- *Sales* to learn how customers have reacted to similar manuals and online help systems
- *Customer service* to find out what causes the most difficulties for users of similar products

Even more useful would be talking with representative users — the actual people who will use the computer system and the manual and online help you plan to write. You can interview them about their needs, their expectations, and the problems they've had with other systems and documentation. You also should arrange for some users to test the manual at various stages during its drafting.

To help you collect information about the users of a document you're preparing, refer to the questions on the audience analysis worksheet in Figure 4.3. This worksheet will help you focus your audience analysis by answering specific questions. You will initially find it helpful to complete the entire worksheet. The more frequently you use it, however, the more familiar you will become with the questions. Eventually, you will automatically ask yourself the questions whenever you plan a document, oral presentation, or visual.

To use an electronic version of the Audience Analysis Worksheet, go to **www.english.wadsworth.com/burnett6e**.
CLICK ON WEBLINK
CLICK on Chapter 4/audience analysis

WEBLINK

Context

An often ignored but vital consideration for technical communicators is the physical and political context and the general working conditions in which audiences will interpret and use documents, oral presentations, and visuals. Regardless of a person's cognitive ability or knowledge of the subject, accessing difficult material — that is, seeing or hearing it — may be nearly impossible if the physical context is distracting or noisy. In preparing a technical document, a writer might adjust elements such as paragraph division, headings, page design, type and size of illustrations, and binding to make it more accessible. For example, a repair manual used by technicians while they are troubleshooting manufacturing equipment needs a sturdy cover, pages that lie flat, a detailed and usable table of contents and index, and headings and visuals that can be scanned easily. Similarly, a technical report read by a busy executive needs an abstract, clear headings, an initial statement of conclusions and recommendations, and brief explanations and justifications.

Just as the physical context is important, so, too, is the political context. For example, the attitude people within an organization have toward particular documents affects how these documents are read. Is a new employee told, "The manager's

FIGURE 4.3 **Audience Analysis Worksheet**

Context: Where will the audiences access and use the material? What distractions will they face in the physical and political context? What are the audiences' working conditions (e.g., access to e-mail, available time)? ______________________________

____

____

Audience: Who are the intended audiences? What are their purposes? Do you need to consider differences among the initial, primary, secondary, and external audiences? ______________________________

____

____

Purposes, attitudes and motivations: What are the attitudes of the audiences? What factors might contribute to their resistance or receptivity? How might these factors influence the way they read a document, listen to an oral presentation, or view a visual? How can you decrease resistance and increase receptivity? ______________________________

____

____

Prior knowledge: What level (high school, 2- or 4-year college, graduate school) and kind (theoretical or practical focus) of education do you anticipate that the audiences have? How does this affect your plans for the document, oral presentation, or visual? ______________________________

____

____

How much on-the-job experience do the audiences have? How does this affect your plans? ______________________________

____

____

Reading level: What is your best estimate about the level of material the audiences can handle without difficulty? How can you make the document more accessible to audiences? How have you used visuals and document design to aid reading? ______________________________

____

____

Organizational role: What are the audiences' positions in the organization? What professional experiences and organizational roles do the audiences have? Job title? Areas of responsibility? Years of experience? Familiarity with the subject? How do these factors affect what they need to read? What constraints are at work? ______________________________

____

____

memos are a waste of time; don't even bother reading them" or told, in contrast, "The manager's memos are important; make sure to keep them for reference"?

The general working conditions of an organization, including how information is presented to audiences, affect not only the way the documents and presentations are interpreted but whether they're attended to at all. For example, if important memos are simply posted on hallway or lunchroom bulletin boards, many audiences will skip them, thinking the memos are unimportant. If important internal documents are sent as attachments to e-mail, but some employees cannot access attachments, those readers may presume that they are unimportant, that the attachments are unimportant, or both. In these cases, audiences are making assumptions about information based on how it is disseminated, not on the content.

Another aspect of working conditions is the available time the audience has to read or listen to the material. Because time is valuable, documents, oral presentations, and visuals should be designed so that audiences can understand and use the information as quickly and easily as possible. Technical communicators use organizational or graphic devices that ease the audiences' task. The following strategies are particularly helpful:

- Initial abstracts or summaries
- Headings and subheadings
- Use of descending order so that the most important information comes first
- Definition of terms, if necessary
- Transitions that show how sections of the document relate to each other
- Visuals that make information easily accessible
- Page layout that is not crowded or cluttered

Purpose and Motivation

Analyzing an audience's purpose and motivation is relatively easy if you know the intended audience. Professional responsibilities can dictate a need to access and use particular information. For example, a computer technician might need to read an operations manual, whereas a manager might need to read theory about design or system architecture. A patient might need the instructions and side effects printed on a medication's package insert but probably would not care much about the chemical elements used to synthesize the medication, which might be of great interest to a pharmacist.

Knowing an audience's purpose and motivation helps you adjust the organization of information so that you increase audience receptivity and decrease resistance.

- *Receptive audiences:* You can present recommendations initially and then support them in subsequent sections.
- *Resistant audiences:* You can present the problem, discuss the alternatives, and then lead to the most appropriate and feasible solution, hoping audiences are persuaded by your interpretation.

Explain whether you think professional colleagues might ignore verifiable information in a report because they didn't agree with the overall position of the writer.

As the Baby Blues cartoon[1] shows, assessing and adjusting to your audience may make all the difference in whether the audience is receptive or resistant.

Receptivity is influenced by an individual's motivation, but motivation is an abstract quality that's difficult to quantify. So, how do you assess receptivity and motivation? Imagine an actual member of your intended audience. How will this person feel about the document — pleased, neutral, negative? How much does the person need the information? For example, a repair technician may be highly motivated in referring to the repair manual when equipment breaks down but may be unmotivated to check repair procedures if there's no problem. A manager may feel neutral about the proposal he has to read and completely unmotivated to begin reading it.

Prior Knowledge

If you can estimate audiences' prior knowledge, you will be able to determine the appropriate vocabulary and content. Both education and workplace experience influence prior knowledge.

The level, type, and duration of a person's education strongly influence prior knowledge, affecting a person's comprehension of concepts and their application.

- *Vocational-technical training* focuses on providing a practical or applied knowledge.
- *Professional or academic training* focuses on providing a theoretical understanding as well as a practical experience.

People's education directly affects the kinds of documents they use. For example, a physician or nurse practitioner might consult the *Physician's Desk Reference* to review the side effects of a new medication. A patient would read the much less technical package insert accompanying the medication. A child might just see the Mr. Yuk poison symbol on the bottle and know enough to leave it alone.

"Mr. Yuk," developed in 1971 in Pittsburgh, has had a successful history of warning children to stay away from hazardous substances such as medication and household chemicals. However, occasional studies over a 20-year period suggest that Mr. Yuk is so familiar that children are sometimes attracted to rather than warned away from containers with a Mr. Yuk warning sticker. You can begin learning more about the this issue by going to www.english.wadsworth.com/burnett6e.

CLICK ON WEBLINK

CLICK on Chapter 4/Mr. Yuk

WEBLINK

People's organizational and disciplinary experience influences their reactions to a document. For example, experienced professionals are more likely than entry-level employees to be aware of corporate policies, politics, and personalities that may influence a document's recommendations. An equally important part of professional experience involves people's expertise in the subject. This includes the audiences' level of responsibility, years of professional experience, and familiarity with the field.

Although estimating audiences' prior knowledge is difficult, you can generalize about the amount of prior knowledge a person needs to hold a job; then you can make a reasonable guess about the kind of information that a person in such a position would need and respond to — appropriate vocabulary, technical complexity, and format.

Why is equating people's ability or intelligence with their amount of formal education a risky thing to do?

Reading Level

Reading level refers to the degree of difficulty of material that audiences are able to comprehend. Writing for an audience's level is important; if audiences cannot understand and act on the written information, it is useless. Technical accuracy, completeness, and logical organization are all irrelevant if the intended audience cannot comprehend the material.

Knowing an audience's reading ability helps you adjust content and approach. Writers should not automatically assume, however, that the smarter the audiences, the more difficult the material should be. A very intelligent person may not have a high reading level; another person may be able to read complex material in one specialized area but not in another. Someone capable of reading nearly any material might be constrained by lack of time or interest and thus prefer short, easy-to-read information. Your writing should be as easy as possible to read without oversimplifying or distorting the content.

Readability Formulas Your computer's word-processing program very likely comes with a function that supposedly calculates the readability of any text you write. The problem is that readability calculations simply use one of several formulas to determine a ratio between word length and sentence length. Despite its name, no readability formula tells you how easy or difficult a text is for audiences.

Readability formulas operate on the premise that shorter words and sentences are always easier to read. This is simply not so. The three-syllable word *elephant* is easier for virtually everyone to understand than the three-letter word *erg.* Such an oversimplified view of reading ignores many factors, including the complexity of the concepts; inaccuracies or biases in the presentation; the nature, development, and support of the argument; the use of visuals; as well as the document's overall design.

Using the Flesch-Kincaid formula to calculate readability for the paragraphs in this section of the text determines this section is appropriate for grade 11.9. However, with only a few changes to the text, the formulas can be manipulated to show a higher or lower level of difficulty. For example, making the text into many more short, simple sentences reduces the grade level. And other changes (such as making the entire section one long, dense, undifferentiated paragraph) result in no change at all to the readability results. Trusting readability formulas to assess the appropriateness of your documents for audiences is risky, at best.

> *If readability formulas don't really measure how easy or difficult a text is to read, why are they so widely used?*

Although word and sentence length certainly do influence the accessibility of information and ideas in a document, readability formulas do not take into account many of the factors that affect reading. Figure 4.4 identifies some factors simply not considered by readability formulas that affect the ease or difficulty of a text. Of course, all the factors in Figure 4.4 have to be considered comparatively and in context. For example, for a food scientist, understanding how beans grow is probably as easy as understanding why beans cause flatulence; to a botany student, understanding how beans grow is probably much easier; and to a first-grade student, both topics are probably beyond reach.

Figure 4.4 provides questions that offer an alternative way to help you create documents, oral presentations, and visuals that are accessible, comprehensible, and usable. (See Figure 1.2 in Chapter 1 to review these critical concepts.)

Despite the value of understanding the factors in Figure 4.4, by themselves they do not make a text easy or difficult to read. In fact, how easy a document is to read depends largely on the interaction between the reader and the document, on the meaning that the reader creates during reading. As a writer, you can try to make a document accurate, accessible, appropriate, and appealing, but the ultimate readability test is what the reader gets out of the document. (Usability testing, discussed in Chapter 9, is a common method for making a document more accessible, comprehensible, and usable.)

FIGURE 4.4 **Factors That Affect the Ease or Difficulty of a Text**

Factor	Questions about Ease or Difficulty of Text	✔ Yes
Content	■ Is the content concrete?	❑
	■ Are abstractions supported with explanations and/or examples?	❑
	■ Do audiences have the necessary prior knowledge to understand the content?	❑
Context	■ Does the document, oral presentation, or visual sufficiently acknowledge and explain the context?	❑
	■ Is the type of document, oral presentation, or visual familiar to the audiences?	❑
Purpose	■ Is the purpose clearly stated?	❑
	■ Are audiences likely to agree with the stated purpose?	❑
Audience	■ Is the document, oral presentation, or visual adapted to the audience's level of understanding, needs and expectations, and organizational role?	❑
Organization	■ Is the content logically organized? Are audiences given cues about the organization?	❑
	■ Is the information coherent (for example, by using topic sentences and transitions)?	❑
	■ Is the argument clearly presented and well supported?	❑
Visuals	■ Is the verbal and visual information appropriately integrated?	❑
	■ Are the kinds of visuals appropriate and appealing?	❑
	■ Does the document, oral presentation, or visual conform to conventions?	❑
Design	■ Is information appropriately chunked and labeled? Is the hierarchy clear?	❑
	■ Are the choices for font, type style, type size, white space, and line length appropriate for the type of document, oral presentation, or visual as well as the audiences?	❑
	■ Is the overall design accessible and appealing?	❑
Usability	■ Is the information usable for the intended context, purpose, and audience?	❑
Language conventions	■ Does the document, oral presentation, or visual conform to conventions of language, both in grammar and mechanics?	❑

The use of readability formulas has been highly controversial for at least three decades. For "everything you ever wanted to know about readability tests but were afraid to ask," go to **www.english.wadsworth.com/burnett6e** for links that provide more information about this issue.

CLICK ON WEBLINK

CLICK on Chapter 4/readability

Limited Literacy When people talk about "limited literacy," they can mean two different things: (1) people who are skillful readers in some circumstances (for example, reading engineering reports) but not necessarily skillful in other circumstances (for example, reading legal briefs); or (2) people who don't get much information from any written documents. This section discusses this second kind of limited literacy.

Workers with limited literary may see documents as difficult, intimidating, confusing, or even incomprehensible; thus, they may ignore the documents entirely. These workers can often read single words and short phrases, but they have great difficulty reading sentences, paragraphs, and entire documents. They also may not understand document conventions, which may result in literacy problems such as these:

- Not knowing how to use a table of contents or an index
- Not using headings to preview upcoming text
- Not treating boxed, shaded, boldfaced, or italicized text as more important than plain text
- Skipping parts of the text they don't understand
- Misreading parts of the text but believing they do understand it
- Expecting that all relevant information will be in the visuals

The problem of limited literacy is widespread in the workplace. The National Adult Literacy Survey reports the results of interviews with more than 26,000 randomly selected adults in the United States. Between 47 and 51 percent of the adult population (the equivalent of 90 to 94 million adults) could not complete very simple literacy tasks. These simple tasks included locating a single piece of information in a short article, entering a signature on a form, and locating information about eligibility for employee benefits in a table.

Surprisingly, the majority of people who performed at this level described themselves as being able to read and write English "well" or "very well," so their limited skills may "allow them to meet some or most of their personal and occupational literacy needs." In the same National Adult Literacy Survey, only 18 to

21 percent of the adult population (the equivalent of 34 to 40 million adults) performed at the highest literacy levels. More complex literacy tasks included interpreting information from an article, identifying patterns from information in a table, and summarizing information in a table.[2]

A more concrete example, reported by the Canadian Association of Speech-Language Pathologists and Audiologists, reinforces the seriousness of limited literacy. Specifically, adults with limited literacy skills often cannot read the labels and instructions on both prescription and over-the-counter medication; as a result, they sometimes take medication incorrectly. They also have difficulty following prescribed treatment plans from their physicians, such as complying with directions for treating diabetes or following preventative care, self-care, and follow-up care after an illness or injury. They also have difficulty understanding appointment slips, informed consent forms, discharge information and oral instructions.[3]

What compensatory skills might a person develop in order to make up for limited literacy skills?

As a person responsible for creating technical documents, you need to be especially careful if your audience has limited literacy skills. If time and budget allow, consider alternative forms of presentation such as one-on-one conferences, face-to-face small group meetings, training programs, videocassettes, videodisks, or multimedia computer programs.

If these options are out of the question, create documents that are largely visual. However, keep in mind that sequence of information on a page and ways of interpreting color are not universal; U.S. conventions are not worldwide conventions. Consider that moving through a document from left to right and from top to bottom is only a convention for reading in certain languages, including English. If a person doesn't read, this left-to-right, top-to-bottom movement may not be a familiar convention.

In many cases, design elements such as color and icons can help signal critical information. For example, arrows in addition to numbers may be useful for signaling sequence. Similarly, high risks can be signaled in words such as *Caution* or *Danger,* with yellow (caution) or red (danger) to reinforce the words. Color use should comply with standards advised by the American National Standards Institute (ANSI). (For more information on using color, refer to Chapter 12, Using Visual Forms.)

WEBLINK

You may need to prepare information for limited literacy audiences. If you want to know more about how to effectively address such audiences, go to **www.english.wadsworth.com/burnett6e** for links to sites that provide helpful suggestions and identify resources.

CLICK ON WEBLINK

CLICK on Chapter 4/limited literacy

Organizational Role

Organizations are generally categorized as hierarchical or nonhierarchical. In practice, most organizations incorporate both hierarchical and nonhierarchical characteristics.

> *Explain where you would prefer to work — a hierarchical organization or a nonhierarchical organization. Why?*

- *A hierarchical organization* has bosses at the top, managers in the middle, and workers at the bottom. Hierarchical organizations generally assume that people work best when directed.
- *A nonhierarchical organization* has everyone contributing equally to the productivity of the organization. Nonhierarchical organizations regard people as most productive when they participate in decision making.

A hierarchical organization can be analyzed by examining where a person fits into the organizational structure of the company — who reports to whom, who is a decision maker, who holds ultimate responsibility for a process or a project. Curiously, e-mail is flattening hierarchies in some organizations because all employees are easily accessible simply by sending an e-mail message. But the ease of sending such a message doesn't mean it's necessarily appropriate. An assessment of hierarchy in an organization requires sensitivity to and awareness of any subtle differences between the formal structure and the actual working structure. For example, you cannot automatically assume that the most senior person in an organization is the most influential decision maker. A general manager or a CEO might collect all the relevant information and make decisions herself, or she might make decisions based largely on the recommendations of middle-level managers.

So far, the discussion has focused on internal audiences, those within the writer's organization. Frequently, however, audiences are external, belonging to another organization. Such audiences might be customers or a funding agency. They would not know many of the things that are apparent to internal audiences, so the writer needs to provide more background information than may be customary for an internal reader. This background might include elements such as definitions, a brief history, and the relation of the subject to the overall operation.

The tone used for external audiences is often more formal than the tone used for internal audiences. However, *formal* does not mean pompous; rather, the document just does not include casual language common in internal memos. For example, an internal memo might say, "I'll get back to you by Monday about the revised schedule," whereas a document for an external reader would be more likely to say, "I will send you information about the revised schedule by Monday, October 22" or "Information about the revised schedule will reach you by Monday, October 22."

Adjusting to Audiences

After you have identified and analyzed your audiences, you can prepare documents or oral presentations that respond to their need for information as well as to factors such as context, level of expertise, organizational roles, and prior knowledge. This section focuses on adjusting content complexity for different audiences in three broad ways:

- Address audiences with *different levels of expertise* by adjusting the complexity of the material.
- Address audiences with *different organizational roles* by shifting the focus of the discussion and the choice of details.
- Address audiences by designing Web sites that enable audiences to *construct unique sequences of information* to meet their own needs and interests.

Differences in Expertise

Audiences vary in their level of technical knowledge. When you adjust material for audiences with various levels of technical competence, you need to change the complexity of the concepts, language, details, and examples. The following examples in Figures 4.5A, B, and C show how workplace professionals can adapt information to meet the needs of different audiences. The director of medical technology for an urban hospital produced three paragraphs about coagulation, each intended for a different audience.[4]

FIGURE 4.5A **Paragraph for High School Students**

What is coagulation? If a blood vessel ruptures, the blood thickens and forms a gel called a clot, which slows the flow of blood from the wound. This process is called *coagulation,* a mechanism to prevent blood loss when a blood vessel is ruptured. This process is initiated both by the damaged blood vessel tissue and by substances released from the damaged tissue. The substances activate proteins in the blood called *procoagulants.* The activated procoagulants act as enzymes in a series of chemical reactions that culminate in the conversion of a molecule of fibrinogen to a smaller molecule, fibrin. The fibrin molecules link together into strands. These strands form a tight mesh that is known as a *fibrin clot.*

COMPREHENSIBILITY
- *presumes a minimal level of prior knowledge about coagulation*
- *minimizes technical vocabulary*
- *defines new terms*
- *reviews common knowledge before presenting causal relationship*

USABILITY
- *signals key point with run-in heading*
- *italicizes new terms*

FIGURE 4.5B Paragraph for Medical Technology Majors in College

The Process of Coagulation. A major function of the hemostatic mechanism of the cardiovascular system is the coagulation of blood. Coagulation plays a vital role in preventing blood loss in episodes of vessel injury, allowing the body to maintain blood volume and retain blood products. The process of coagulation is achieved through the chain of chemical reactions of procoagulants (a group of plasma proteins) and tissue cell constituents. The procoagulants circulate through the bloodstream as inert enzymes, activated by damaged tissue, phospholipid from cell membranes, and calcium ions. After a series of enzymatic reactions, the procoagulants cleave a protein peptide from fibrinogen, a glycoprotein, to form a fibrin monomer. The fibrin monomers polymerize into strands that form an insoluble mesh known as a fibrin clot.

COMPREHENSIBILITY

- *presumes a modest level of prior knowledge about coagulation*
- *introduces and defines new technical vocabulary*
- *offers an explicit causal explanation of the process*

USABILITY

- *signals key point with run-in heading*

FIGURE 4.5C Paragraph for Professional Medical Technologists in a Hospital Review Course

Pathways in coagulation. Three pathways are involved in the coagulation of the hemostatic mechanism: intrinsic, extrinsic, and common pathways. Each pathway consists of a cascade of proteolytic enzyme reactions in which procoagulants, a series of inert circulating proteolytic enzymes called coagulation factors, activate one another. The *intrinsic pathway* is activated by the presence of damaged endothelium tissue and a high molecular weight activator, kallikren. The *extrinsic pathway* is activated by tissue factor, a phospholipid of endothelium, and calcium ions. The *common pathway* is activated by the resultant coagulation factors of the intrinsic and extrinsic pathways. The result of the common pathway is the conversion of fibrinogen to fibrin monomers, which polymerize into strands. The strands mesh through covalent bonding to form an insoluble fibrin clot.

COMPREHENSIBILITY

- *presumes a high level of prior knowledge about coagulation*
- *uses technical vocabulary consistently*
- *uses a parts/whole approach (types of pathways)*
- *provides a causal explanation of each type of pathway*

USABILITY

- *signals key point with run-in heading*
- *uses a topic sentence to state the main point*
- *italicizes key terms*

The first paragraph (Figure 4.5A) is for high school biology students learning new concepts. These students need to learn the vocabulary as well as the process of coagulation. The writer uses the run-in heading to pose a question to focus the students' attention and let them know what to anticipate. The writer then reviews common knowledge about clots before presenting and defining new terms. He organizes details using a familiar cause-and-effect organization.

The second paragraph (Figure 4.5B) is for college students majoring in a four-year medical technology degree program. These students can handle technical vocabulary and already understand the basic causal relationships. What they need to learn, though, are concepts that explain coagulation. The writer signals this focus with the run-in heading. The paragraph itself explains the process of coagulation, which includes introducing and defining new terms (such as procoagulants and fibrinogen) and providing specific details (for instance, procoagulants are activated by damaged tissue, phospholipids, and calcium ions).

The third paragraph (Figure 4.5C) is for professional medical technologists who are enrolled in a review course given in their hospital. These medical professionals understand the basic process of coagulation, but they need a review. After using a run-in heading to signal the topic, the writer uses a topic sentence to preview the content and the organization of the paragraph (intrinsic, extrinsic, and common pathways). Italicizing each main term provides a typographic cue that helps audiences keep track of where they are in relation to the whole.

Differences in Roles and Stances

You should maintain a similar level of complexity for audiences with parallel roles in an organization — for example, people who are all managers but in different areas such as manufacturing, marketing, research and development, or quality control. One way to adjust to such audiences is emphasizing aspects of the subject that are relevant to their organization role. A second way is acknowledging their particular point of view.

The three brief e-mail messages in Figures 4.6A, B, and C, all written by the vice president of plant operations, are about the same subject: the purchase of a mill that is intended to increase product uniformity, reduce waste, and increase product availability. Each e-mail message is directed to a separate manager, each of whom has specialized interests and responsibilities. All three have approximately the same level of complexity, yet the focus shifts in each one to emphasize the content that is relevant to each reader's area of workplace responsibility.

In the first e-mail (Figure 4.6A) the writer addresses the production supervisor of the mill department. Because the recipient is interested primarily in production, the writer identifies a critical problem that will be eliminated with the new machine and identifies a series of benefits, including product uniformity and a reduction in rejections and downtime, that should increase productivity.

FIGURE 4.6A **E-mail from the VP of Plant Operations to the Production Supervisor**

To	Peter Smith, Prod. Supervisor, Mill Dept.
From	Thomas White, VP Plant Operations
Date	8 September 200-
Subject	Expected increase in product availability

The recently purchased mill is scheduled to be on line by October 15. This machine will solve the problems of the thermal stabilization times you have been plagued with, while increasing product uniformity. When the department has less product rejection and machine downtime, I look forward to seeing increased rates per shift. You will be notified of the new rate production schedule on October 15.

The writer focuses on the reader's interest — product availability. This topic is signaled in the subject line of the e-mail and is the primary topic of the paragraph.

The e-mail provides a critical date and identifies anticipated production benefits.

FIGURE 4.6B **Memo from the VP of Plant Operations to the Controller**

To	Elizabeth Daley, Controller
From	Thomas White, VP Plant Operations
Date	8 September 200-
Subject	New Wolverine Cereal Flaking Mill

The addition of the new mill to our production line is an effective solution to the production waste problems we have been experiencing. With the minimal waste expected combined with the machine's operating efficiency, the payback period should be within the next fiscal year. Preliminary information indicates that purchasing the mill has been a smart investment.

In writing to the controller, the writer focuses on financial matters.

Because the controller does not regularly work with production line equipment, she needs to be reminded (tactfully, in the subject line) about the specific machine.

The e-mail identifies ways the mill will save money and reviews the length of the payback period.

FIGURE 4.6C **Memo from the VP of Plant Operations to the Director of Marketing and Sales**

```
To       David Parker, Director of Mkting & Sales
From     Thomas White, VP Plant Operations
Date     8 September 200-
Subject  Expected increase in product availability

Due to the addition of capital equipment in the
production department, product availability will
increase by approximately 50%.

A short-term plan in marketing strategy should
be initiated to increase sales and expand to
new territories. Product stockpiling begins
October 15.
```

The writer wants the director of marketing and sales to concentrate on a marketing plan.

Mentioning the increase in availability and the product stockpiling should reinforce the need for new territories.

What are the advantages and disadvantages of a writer preparing separate e-mail messages (or other correspondence such as memos) for audiences who need similar information?

The second e-mail (Figure 4.6B) is written to the controller, the person responsible for the company's financial operations. A controller is primarily concerned about money matters, so she would be interested in cost savings that come about because of waste reduction and increased equipment efficiency. The controller would also be interested in the payback period. The difficulty of the content, sentences, and vocabulary in this e-mail is much the same as that in Figure 4.6A.

In the third e-mail, to the director of marketing and sales (Figure 4.6C), the writer concentrates on product availability — amount, date, implications. The e-mail includes no mention of production details or costs, which are only peripherally relevant to this audience. The emphasis is on the need to develop a new short-term marketing plan.

Should you write three separate e-mail messages? Yes, if you have the time and believe that the special attention will increase reader comprehension and support. However, you may not have the time to write separate e-mail messages, and sometimes political realities dictate that all audiences receive an identical e-mail. In these circumstances, you need to write one e-mail for multiple audiences who differ in their interests, familiarity with the project, and levels of expertise. To make this one as effective as possible, consider these suggestions:

- *Identify and write for the primary audiences:* Organize the information for the people who need the information for activities or decision making, putting first what is most important to them. In Figures 4.6A, B, and C, all three recipients have the vested interest of a primary audience in the purchase of the mill. But their interests are different — manufacturing operations, finance, and marketing.

- *Identify and consider the secondary audiences:* Include information of interest to secondary audiences based on its value to the primary audiences and the influence these secondary audiences have on the primary audiences. In Figures 4.6A, B, and C, possible secondary audiences might include line supervisors, accountants, and sales reps. Their opinions about the new mill might well affect the perceptions of their managers.
- *Use design elements to make information accessible in both paper and electronic messages:* Use headings, typographic elements (such as boldface and italic), and typographic devices (such as bullets and boxes) to signal important topics.

Explain which is easier for you to do — write for audiences with different levels of expertise or for audiences with different organizational roles.

So far, you've read about adjusting communication to meet needs established largely by audiences' organizational roles and responsibilities. Sometimes, however, concentrating solely on the verifiable information relevant to particular roles is not sufficient. For example, communication that ignores audiences' stance — their points of view, attitudes, values, and beliefs — is likely to be ineffective. In fact, audiences may dismiss documents and presentations that ignore affective concerns, feeling that such documents or presentations simply do not adequately consider them or the situation, despite being filled with verifiable information. The sidebar below addresses this important issue.

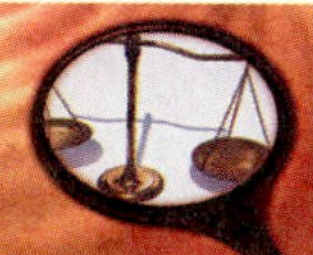

ETHICS SIDEBAR

"Their approaches are culturally insensitive": Ethics and Public Policies

Effective technical documents typically present clear, precise, verifiable information. Focusing on such details reflects the belief that the more audiences know, the better decisions they will make. However, technical professionals who rely too much on precise information sometimes ignore other, equally important aspects of a document. Focusing on precision can sometimes lead a writer to provide too much information or to present information in confusing or unfamiliar terms. Often, the result is a document that contains all the relevant information but does not sufficiently consider the ethics of the context, content, or purpose as it affects the audience.

Researchers Patricia Hynds and Wanda Martin, reviewing the public outcry over a city water project, outline how an unethical approach can derail a much-needed project.[5] Hynds and Martin describe a debate that arose when the City of Albuquerque, New Mexico, began the process of drilling a well in one of the city's suburbs, the South Valley. Albuquerque relies heavily on wells to provide water for its communities; city planners assumed that South Valley's residents would readily recognize the value of the new well. When South Valley residents voiced concerns about the effects of drilling the new well, Albuquerque's mayor asked the public works department to meet with South Valley residents to work out a new location, although resident approval for a new well's location was not legally necessary. Three community meetings later, communication between the city and residents had degenerated, and lawsuits were filed. The well has yet to be drilled.

How can you determine when your audience has concerns beyond the technical data?

Why did such a necessary project reach this impasse? Hynds and Martin believe the problem occurred because the city's representatives failed to adequately recognize the cultural and social viewpoints of the residents. The residents voiced concerns about the negative impact of the well; they did not want their small community to become too industrial or overcrowded with new construction. The city's representatives, however, viewed these concerns as simply a lack of understanding by the residents. In each meeting, the city's representatives continually emphasized the technical aspects of the well to support their position: they believed "that when the technical facts are clearly communicated, all reasonable hearers will arrive at similar conclusions." The residents, however, felt the technical explanations were patronizing and did not feel that their concerns were respected. For the residents, the issue was about more than facts; it was about respecting the traditions and heritage of the area. Eventually, the animosity the residents felt towards the city reached a point where the residents boycotted the last meeting and filed lawsuits to stop the drilling.

Hynds and Martin believe that despite the city's good intentions, the city planners failed to treat the audience ethically. Instead of talking to the residents about their concerns, they continually tried to make the residents see the situation from the city's point of view. They relied too much on precise facts and too little on the concerns of their audience. The unfortunate result was a breakdown in communication. If the city planners had used an ethical approach to communication, they would have considered more than their verifiable information about water needs and well locations.

How would you convince the city planners that listening to their audience matters? What would you do if your audience rejected carefully presented and verifiable information in favor of emotional and unverifiable opinions?

Individual and Collaborative Assignments

1. **Determine the audiences.** Read the following sets of sentences intended for different audiences. Determine the audience for each sentence and then answer the following questions:
   - How has the writer adjusted for the different audiences? Which set is adjusted for differences in audience expertise? Which set is adjusted for differences in organizational role?

- What information did the writer need to make these adjustments?
- Why are the changes necessary if the statements in each set convey essentially the same information?

Set 1 The sun damages many human-made materials.

The sun causes degeneration and weakening of many synthetic fibers.

Ultraviolet radiation causes polymer degeneration by attacking links in the polymer chain and reducing the molecular weight of the material.

Set 2 The shipment was delayed because of manufacturing equipment malfunction and misjudgment about the resulting delays.

Because your shipment has again been unavoidably delayed, we will credit your rental fee for fourteen days — the anticipated length of this second delay.

Occasional brief delays in shipping have a minimal impact on attaining projected third-quarter goals.

2. **Determine the audiences.** Visit the following Web sites and ask yourself these questions:
   - Who are the various audiences for the Web site?
   - How have the writer and designer adjusted for the different audiences? How have the writer and designer acknowledged and responded to the needs and interests of specific segments of the audience, such as people of color and international audiences?
   - What information did the writer and designer need to make these adjustments?

www.discovery.com	www.science.doe.gov/bes/NNI.htm
www.ibm.com	www.sch-plough.com
www.dairynetwork.com	www.campbellsoup.com
www.ecotourism.org/	www.nsf.com
www.wipo.org/	www.aciic.org.au/
www.futureharvest.org/	www.scientact.gr/

3. For each of these three paragraphs about neonatal herpes (a venereal disease that is transmittable to newborns), identify a likely intended audience and explain your choice by using specific textual examples as support.

   (a) Neonatal herpes simplex virus infections can result in serious morbidity and mortality. Many of the infections result from asymptomatic cervical shedding of virus after a primary episode of genital HSV in the third trimester. Antibodies to HSV-2 have been detected in approximately

20 percent of pregnant women, but only 5 percent report a history of symptomatic infection. All primary episodes of HSV and secondary episodes near term or at the time of delivery should be treated with antiviral therapy. If active HSV infection is present at the time of delivery, cesarean section should be performed. Symptomatic and asymptomatic primary genital HSV infections are associated with preterm labor and low-birth-weight infants. The diagnosis of neonatal HSV can be difficult, but it should be suspected in any newborn with irritability, lethargy, fever or poor feeding at one week of age. Diagnosis is made by culturing the blood, cerebrospinal fluid, urine and fluid from eyes, nose and mucous membranes. All newborns suspected to have or who are diagnosed with HSV infection should be treated with parenteral acyclovir.[6]

(b) About 25 percent of adults in the United States are infected with genital herpes. The virus is not curable. Babies born to mothers who have an active genital herpes infection at or near the time of delivery can become infected. This can be serious and sometimes fatal for newborns. If you have had genital herpes and are considering pregnancy or are pregnant, be sure to tell your doctor. He or she may give you antiviral medicines so you will be less likely to have an outbreak of herpes at or near the time you have your baby. If you do have an outbreak of genital herpes at the time of delivery, your doctor will want to deliver your baby by cesarean section so your baby will be less likely to get herpes infection.[7]

(c) Genital herpes is the most common cause of genital ulceration in developed countries, yet four in five people with HSV-2 go undiagnosed, either because they are truly asymptomatic or because they present with atypical symptoms which go undiagnosed.[8] It is now believed that what were previously regarded as atypical manifestations are, in fact, common presentations. Now it is becoming more important to support a clinical diagnosis of genital herpes with a diagnostic test.[9]

4. **Rewrite a paragraph.** Read the following paragraph and rewrite it to lower the difficulty.

A new concept that greatly improves the economics of outdoor storage of grain and other free-flowing bulk materials includes a self-erecting cover and new methods of aeration. The cover is laced together from triangular sections of vinyl-coated nylon fabric and raised to the top of a central filling tower before the filling operation is begun. During filling, it is lifted and spread by the growing grain pile and provides a form-fitting cover that completely encases the pile during all stages of fill. This form-fitting cover is highly resistant to lifting effects of high winds and protects the pile against rainfall. This protection allows the filling operation to be continued during the heaviest downpours. It also completely contains dust particles that would otherwise be carried into the atmosphere to settle on the surrounding countryside. When full, the covering, tightly fitted to the finished pile, is held in place without need of the usual tires or cable tie-downs and provides long-term protection against crusting and other surface spoilage effects caused by the weather. This protection is obtained at a very low cost, as the cost of the cover can be completely paid off by the spoilage eliminated by a single year of the cover's use.

5. **Analyze audiences for different paragraphs.**

   (a) Read and analyze the audiences and some of the related factors for the following two paragraphs about the same subject — why stars twinkle.

   Paragraph A. The stars twinkle because the earth's air scatters the light that comes from them. So during windy weather, stars appear to twinkle more than when the weather is calm. Also, stars near the horizon twinkle more than stars higher in the sky because the light has a thicker layer of air to go through. Observed from the moon, stars do not appear to twinkle at all because the moon does not have any air. Planets do not appear to twinkle as much as most stars because planets usually appear much brighter than most stars, making it harder for their light to be scattered.

   Paragraph B. Stars twinkle because of atmospheric diffraction, the scattering of light caused by the earth's atmosphere. Stars appear to twinkle much more when the atmosphere is extremely turbulent, as it is just before a weather front passes through. Stars near the horizon appear to twinkle much more than those near the zenith because light has to pass through a thicker layer of atmosphere and has more of a chance to be diffracted. The moon's lack of an atmosphere causes a very small amount of diffraction. The light from planets is not diffracted as much as that from stars because planetary disks subtend a much larger area than do stellar disks (40 arc seconds as compared to 0.001 arc second). This larger area causes the light to be more intense and results in less diffraction.

   (b) Complete a matrix based on your analysis of the paragraphs. Describe, explain, and illustrate each factor.

WEBLINK

Go to www.english.wadsworth.com/burnett6e for an electronic version of the matrix that follows.

CLICK ON WEBLINK

CLICK on Chapter 4/matrix

	Paragraph A	*Paragraph B*
Audience		
Context		
Content		
Purpose		
Vocabulary		

6. Contrast articles written for different audiences.

(a) Select an article from *Science Digest, Science News,* the *New York Times* science section (published on Tuesdays), or a local newspaper's science section, and then locate the original article from which it was abridged or on which it was based. Identify and comment on the changes.

(b) Create a table to present your analysis — one column to identify the features you examined and the other two columns (one for each publication) to present examples of the differences. Provide complete citation information for each article.

(c) Find a Web site that deals with the same information. Identify and comment on differences in the presentation of the information on the Web site. Revise your table (additional column and rows) to present the information from your Web site analysis.

7. Write memos for two audiences.

(a) Read the following two memos written by a production supervisor about the purchase of hand cutters for assembly workers on a production line. Note the places that make each memo appropriate for the intended audience.

- The first memo is to an engineer in manufacturing whose projects are affected by the quality of the work the assembly workers do. He wasn't involved in the testing, so he needs a summary of test information. He is particularly interested in the function and capability of the cutters.
- The second memo is to a finance representative. Her primary concern is cost — initially, total expenditure, then cost comparison. Least important to her is the reason for the choice.

(b) After reading the two memos, write two additional memos about the same subject. Direct one of the memos to the assembly workers themselves. Direct the other memo to the manager of the manufacturing division, who is the immediate supervisor of the production supervisor who is writing the memos.

(c) In a small group, choose people to role-play the responses of the recipients to the memos. Discuss your decisions as a group.

Memo from Production Supervisor to Engineer in Manufacturing Engineering

Memo

Marlborough Science, Inc.

October 11, 200 —

To Charles Faria
PCB Manufacturing Engineering
From Marty Holmann *mh*
Subject EREM Hand Cutters

From September 23, 200-, to October 10, 200-, the SPS assembly personnel evaluated three styles of EREM ergonomic hand cutters. The assemblers preferred cutter style 711E-BH to cutter styles 511E-BH or 512E-BH.

Nine line assemblers as well as six post solder operators had the opportunity to use the three styles of cutters. They all agreed that the 711E-BH cutter was the most comfortable, durable, and versatile. The padded handles and quick-action internal spring reduce the fatigue factor experienced by the assemblers with their current cutters. The cutters were used to cut leads varying from 0.019 inch to 0.050 inch and showed little sign of wear, even after being used to cut soldered leads.

In summary, the assemblers strongly preferred the EREM hand cutters to the Lindstrom or Utica cutters. They are still willing to evaluate any new cutters that you may consider equivalent or applicable to their needs.

Memo from Production Supervisor to Finance Department Representative

Memo

Marlborough Science, Inc.

October 11, 200 —

To Diane Martin
Finance Representative
From Marty Holmann *mh*
Subject EREM Hand Cutters

PCB Manufacturing Engineering and PCB Management are requesting approval from the Finance Department to purchase 400 EREM hand cutters for the assembly-line workers for a one-time total purchase price of $9,200.

From September 23, 200-, to October 10, 200-, three styles of EREM hand cutters were evaluated by the assembly-line workers and the manufacturing engineer of the department. The three styles are similar in size and shape and contrast only in material and price.

Style #	Unit Price	Discount Price (over 100)
711E-BH	**$30.00**	**$23.00**
511E-BH	$31.00	$25.00
512E-BH	$32.00	$25.00

The assemblers on the line preferred cutter style 711E-BH for its comfort, versatility, and durability. They also found that the cutter reduced their operator fatigue, which is a factor in our processes.

8. **Write for three audiences.** Write three separate paragraphs about the same subject for three different audiences.

Example: If you are writing about treating runners' knee injuries, you could write for any of these audiences. Each reader would be interested in some aspect of treating knee injuries, but the focus, as well as the amount and kind of detail, would differ.

EMT (emergency medical technician)
coach
athlete
athletic trainer
sports physician
physical therapist
therapy equipment manufacturer
running shoe designer

Be sure to make appropriate adjustments in content focus and complexity, clarity, diction, sentence structure, and organization. Label each paragraph to indicate the intended audience. Consider one of the following suggested topics, or select your own:

architecture for Web site
artificial insemination
cable TV operation
capability of computer software
conformation of show animal
construction of road bed
development of photographic film
effects of aerobic exercise
effects of cigarette smoking
fetal alcohol syndrome
gene splicing
hybrid seed development
laser optics
manufacture of plastic film
new treatment for cavities
operation of microwave oven
propagation of house plants
safety standards in paint shop
setting a fractured bone
shifting sand barriers
treatment for hangovers
weather tracking

Chapter 4 Endnotes

1 © 1996, reprinted with special permission King Feature Syndicate.

2 Kirsch, I. S., Jungeblut, A., Jenkins, L., & Kolstad A. (1993). *Adult literacy in America, Educational Testing Services and the National Center for Educational Statistics.* Washington, DC: Government Printing Office.

3 Myers, C. (n.d.). *Fact Sheets: Literacy, ESL, and Health Care.* Retrieved September 28, 2003, from http://www.caslpa.ca/english/resources/literacy.asp

4 Used with permission of James R. Carey, Lowell General, Lowell, Massachusetts.

5 Hynds, P., & Martin, W. (1995). Atrisco Well #5: A case study of failure in professional communication, *IEEE Transactions on Professional Communication,* 38.3, 139–145.

6 Rudnick, C. M., & Hoekzema, G. S. (2002). Neonatal herpes simplex virus infections. *American Family Physician, 65*(6), 1138.

7 American Academy of Family Physicians. (April 2003). *Herpes During Pregnancy: What You Should Know.* Family Doctor.org. Retrieved January 14, 2004, from http://www.familydoctor.org/x2425.xml

8 Herpes Web. (n.d.). *Issues in Genital Herpes.* Retrieved July 16, 2003, from http://www.herpesweb.net/profess/issues/issue.htm

9 Corey, L. (1994). The current trend in genital herpes: progress in prevention. *Sexually Transmitted Diseases, 21*(2 Suppl), S38-S44.

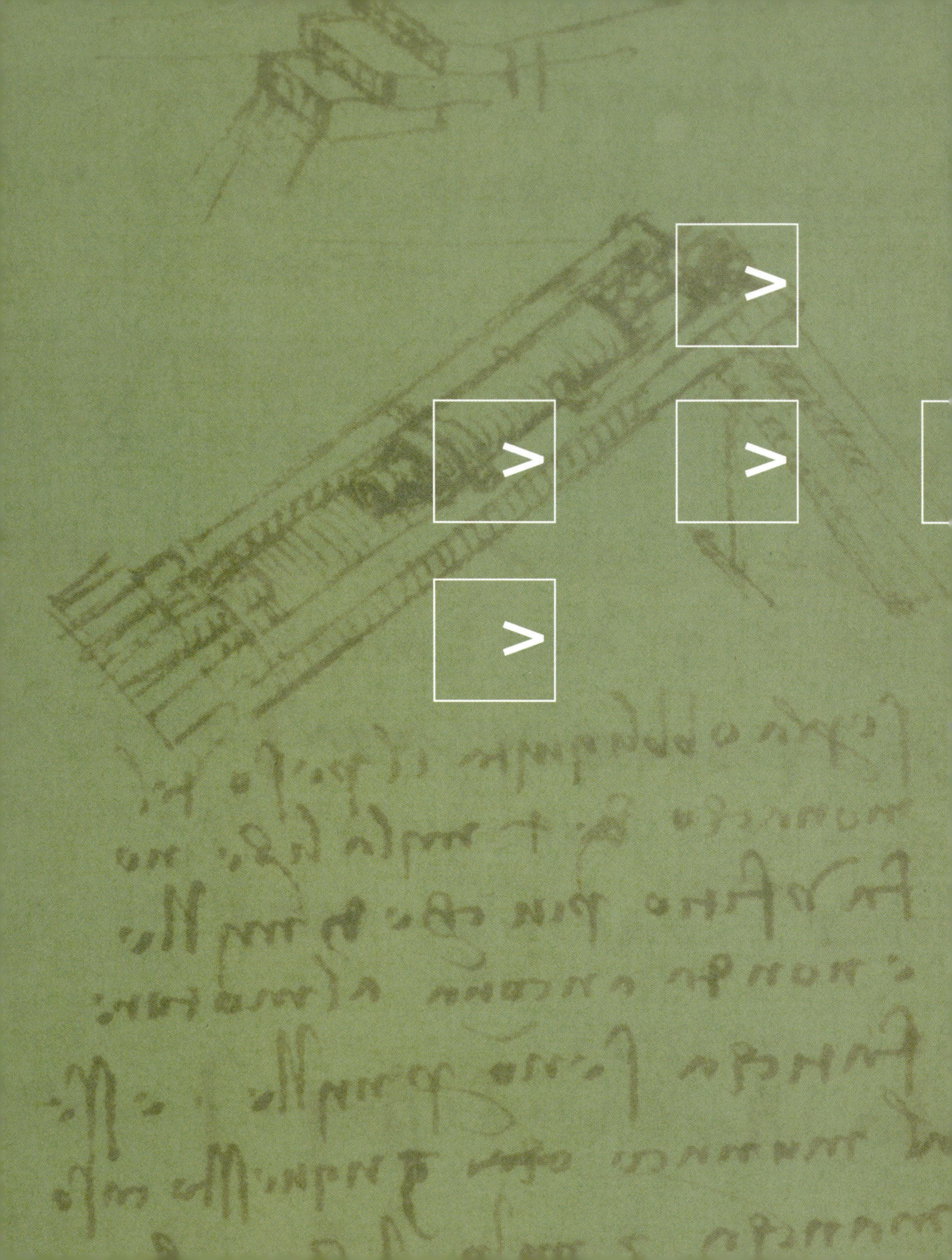

CHAPTER 5

Collaborating in the Workplace

Objectives and Outcomes

This chapter will help you accomplish these outcomes:

- Understand that much of the writing you will do in the workplace will involve collaboration
- Recognize situations that make collaboration appropriate — subject, process, product, and benefits — as well as situations that cause problems
- Develop skills to participate in different types of collaboration: coauthoring, consulting with colleagues, and contributing to team projects
- Develop behaviors typical of skillful, engaged, and cooperative collaborators: listen, ask questions, share, use technology effectively, and reflect
- Discourage *affective* (interpersonal) conflicts; negotiate potential *procedural* conflicts; and encourage *substantive* conflicts. Recognize that conflicts are sometimes the result of differences in cultural attitudes, beliefs, and practices

"I look for a team player. . . . I want someone who understands that people are human and doesn't pick at the behavior of others until it becomes disruptive. I want someone who is not afraid to throw a bad idea out, in the hopes that it may spark a good idea." So says Sharon Burton-Hardin, the Chief Operations Officer at Anthrobytes Consulting in Riverside, California, and President of the Inland Empire chapter of the Society for Technical Communication (www.iestc.org). Burton-Hardin posted her comment on TECHWR-L, an electronic mailing list for professional technical communicators. "Too many times all the employees are doing what they want, regardless of what they have been told the joint effort is. Bickering, deception, destruction."[1] Her opinions give some indication of things that can go wrong with collaboration: disruption, bickering, deception, destruction. And she hints at another problem — the reluctance to be wrong.

Because 75 to 85 percent of the writing you will do in the workplace may involve collaborating with another person, knowing how to be effective is critical. What do you need to know to be an effective collaborator? What pitfalls do you need to be aware of in collaboration? In what ways can collaboration be more productive or effective than writing individually? This chapter helps you answer these questions. You'll learn about the reasons professionals give for collaboration, various types of collaboration, techniques for effective collaboration, and ways to deal with conflicts and resolve problems in collaboration.

Collaborative relationships generally have some common elements: two or more people cooperatively completing some agreed-on task and a specific goal or product to be accomplished over time. Physical proximity is not a necessity; in fact, computer-mediated communication makes long-distance collaboration commonplace.

While collaboration can take place with as few as two people, some collaborative efforts involve teams or networks with dozens or even hundreds of people. One profession that has large teams is high-energy physics (also known as elementary particle physics). Sometimes the teams are so large that the articles appear under the names of institutions rather than the individual authors. One big international high-energy physics collaborative, called H1, is conducting an experiment at HERA, a German national laboratory in Hamburg. The H1 collaborative has built and operates the H1 detector for collecting data with colliding positron-proton beams.

WEBLINK

Large-scale collaborations are often international. Go to **www.english.wadsworth.com/burnett6e** for links to a series of sites that describe large-scale collaborative efforts.

CLICK ON WEBLINK

CLICK on Chapter 5/teams and networks

The people in front of the H1 detector are more than half of the members of the international H1 collaborative. This collaborative is comprised of about 400 people from 12 countries and about 40 institutes. Complex projects in particle physics require international teams to design, construct, and operate the detectors. The research focus of the H1 collaborative is "to measure the structure of the proton, to study the fundamental interactions between particles," and to "search . . . beyond . . . elementary particles."[2]

Reasons to Collaborate . . . or Not

Regardless of the number of people involved, professionals can enter collaborative relationships by individual choice, by circumstance, and even by management directive. Your reasons will generally fall into one or more of these four categories:

- Subject of the project
- Process used in the project
- Product that collaborators create
- Benefits of collaboration

WEBLINK

Collaboration is widespread in the workplace. Go to **www.english.wadsworth.com/burnett6e** for links to several workplace publications tht provide additional reasons for collaboration.
CLICK ON WEBLINK
CLICK on Chapter 5/reasons to collaborate

Despite all the reasons to collaborate, sometimes collaboration is inappropriate, inconvenient, or unsuccessful. After you review the reasons for collaboration, you'll have a chance to see some of the problems.

Subject

The *subject* of some projects makes collaboration essential. Some tasks require expertise from more than one person. For example, a research project to help paraplegics walk might require professionals with backgrounds in biomedical engineering, electronics, neurophysiology, and physical therapy. An article resulting from such a collaboration could include input from all the individuals involved; thus, each could be acknowledged as a coauthor. Examining articles in professional journals for various branches of science shows that many are coauthored. Traditionally, collaboration in technical disciplines has been determined by calculating the percentage of published articles that are co-authored; for example, as long as 30 years ago, 83 percent of the journal articles in chemistry were co-authored.[3] More recently, however, collaborative has become so widespread that the National Science Board now determines collaboration by calculating the percentage of published articles that reflect cross-institutional authorship. Between 1981 and 1985, 35 percent of the published articles in science and technology worldwide reflected cross-institutional authorship; between 1991 and 1995, that increased to 47 percent.[4]

> *What subjects in your profession are well suited to collaboration?*

More similar to what you'll probably do is an example from Kaufer Computing, Inc. The general manager heads a team that is preparing a proposal to install a new computer-aided design/computer-aided manufacturing (CAD/CAM) system in a lab for a local technical college. His team includes professionals with experience in drafting and design, education, electronics, computer systems and software, and, of course, technical communication — expertise unlikely to be found in a single individual. When you're working on a project that requires experience and expertise beyond your own, you should feel comfortable in forming or being a member of a team that works collaboratively to reach a common goal.

Process

Collaboration sometimes enriches the *process* of preparing a document in ways that are impossible for an individual working alone to experience. One major advantage of collaboration is considering a variety of alternative viewpoints. Imagine that an industrial engineer is exploring ways to improve the production capability for Northern Pacific Cabinet Company, Inc., an original equipment manufacturer (OEM) that provides handcrafted wood cabinets for high-quality electronic entertainment systems. The engineer realizes that this investigation will be much more productive if it involves materials buyers, cabinetmakers, schedulers, and shippers, all of whom can offer unique perspectives about ways to increase production.

Another way that collaboration can help you prepare documents is by providing feedback that is seldom available if you're working independently. Collaborators often reflect about the content and product as well as about the process. For example, the industrial engineer for Northern Pacific Cabinet Company not only will receive a variety of alternatives for improving production but also will receive feedback about all the proposed suggestions, her own as well as everyone else's. This type of feedback will enrich any work you do by providing you with reactions that you're not likely to think of alone. Your collaborators' ideas and reactions usually stimulate your own ideas and reactions, which not only makes the process more enjoyable but also improves the document.

How do you react when a collaborator poses an alternative that counters what you've already suggested?

Some companies build their reputations on specialized collaborative teams. For example, Omnica Product Development in Irvine, California, is a small company (25 employees) that promotes its industrial design team and its mechanical engineering team as the anchors of its design and engineering group. Their industrial design team creates concepts and designs, including photo-realistic renderings and working prototypes. Their mechanical engineering team makes sure that the most innovative designs can actually be manufactured.[5]

Omnica Corporation — Irvine, CA

Omnica's industrial design team says, "Much of the value of a product may come from the ability of the user to easily understand it."

Product

Sometimes the type of *product* signals the need for collaboration, simply because some products, such as Web sites, newsletters, user manuals, and proposals, are usually better if they're created collaboratively. Imagine that your task is creating a Web site for families of people with cystic fibrosis. The Web site might be more valuable and appealing if it contains topics written by a variety of health care professionals, by patients and their relatives, and by medical researchers. Written by you alone, this Web site is not likely to have the range of perspectives necessary to sustain the interest of such a variety of readers. Figure 5.1 presents an excerpt from a cystic fibrosis Web site published in English and Spanish by the Methodist Health Care System in Houston.

FIGURE 5.1 **Methodist Health Care System Cystic Fibrosis Web Site in English and Spanish**[6]

Accessibility

- *Since approximately one-third of the metro Houston area is Hispanic,[7] publishing the newsletter in both Spanish and English increases the number of readers.*
- *Media options for video, slides, and a model are clearly marked on the English-language site.*

Comprehensibility

- *Basic information is presented up front: definition, demographics, genetic explanations.*
- *Related information is logically chunked.*

Usability

- *The links are visible for the search engine, patient registration, scheduling, and appointments as well as a range of related services and basic medical background (anatomy, respiratory distress), equipment, and conditions.*

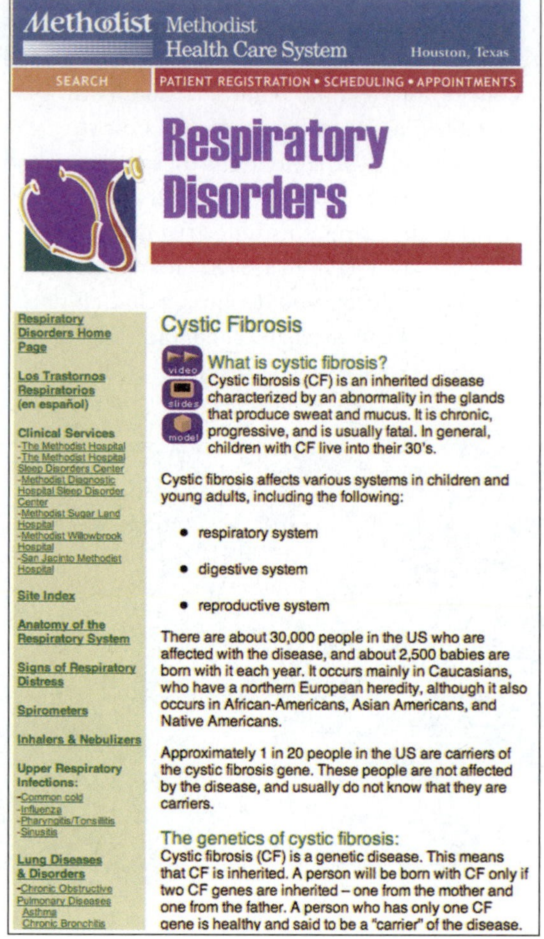

Methodist Methodist Health Care System Houston, Texas
SEARCH
PATIENT REGISTRATION • SCHEDULING • APPOINTMENTS
Respiratory Disorders
Respiratory Disorders Home Page
Los Trastornos Respiratorios (en español)
Clinical Services
-The Methodist Hospital
-The Methodist Hospital Sleep Disorders Center
-Methodist Diagnostic Hospital Sleep Disorder Center
-Methodist Sugar Land Hospital
-Methodist Willowbrook Hospital
-San Jacinto Methodist Hospital
Site Index
Anatomy of the Respiratory System
Signs of Respiratory Distress
Spirometers
Inhalers & Nebulizers
Upper Respiratory Infections:
-Common cold
-Influenza
-Pharyngitis/Tonsillitis
-Sinusitis
Lung Diseases & Disorders
-Chronic Obstructive Pulmonary Diseases
Asthma
Chronic Bronchitis
Cystic Fibrosis
video
slides
model
What is cystic fibrosis?
Cystic fibrosis (CF) is an inherited disease characterized by an abnormality in the glands that produce sweat and mucus. It is chronic, progressive, and is usually fatal. In general, children with CF live into their 30's.
Cystic fibrosis affects various systems in children and young adults, including the following:
• respiratory system
• digestive system
• reproductive system
There are about 30,000 people in the US who are affected with the disease, and about 2,500 babies are born with it each year. It occurs mainly in Caucasians, who have a northern European heredity, although it also occurs in African-Americans, Asian Americans, and Native Americans.
Approximately 1 in 20 people in the US are carriers of the cystic fibrosis gene. These people are not affected by the disease, and usually do not know that they are carriers.
The genetics of cystic fibrosis:
Cystic fibrosis (CF) is a genetic disease. This means that CF is inherited. A person will be born with CF only if two CF genes are inherited – one from the mother and one from the father. A person who has only one CF gene is healthy and said to be a "carrier" of the disease.

WEBLINK

The Methodist Health Care System Cystic Fibrosis Web site provides options for video, slides, and a model. To view these, go to www.english.wadsworth.com/burnett6e for links to the Web site in English and Spanish.
CLICK ON WEBLINK
CLICK on Chapter 5/CF

Typically, long documents such as user manuals or major proposals demand a collaborative effort because of the complexity of their content. For example, Ecological Recycling Associates wants to submit a proposal to the town of Sheffield for dealing with recyclable trash. One of the company's environmental engineers is responsi-

FIGURE 5.1 **Methodist Health Care System Cystic Fibrosis Web Site in English and Spanish[6] (continued)**

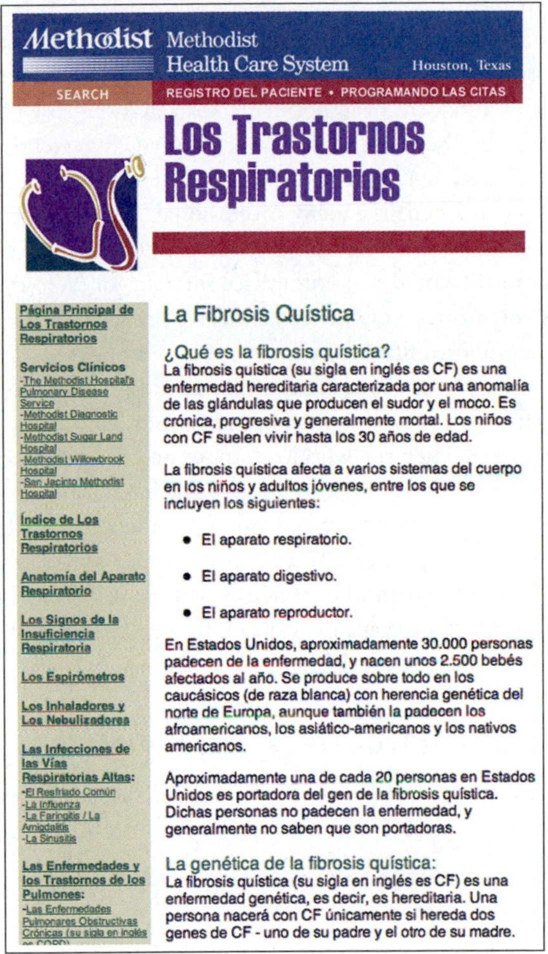

ble for coordinating a five-member team to prepare the proposal. The team meets to plan the proposal, but various sections of the proposal are prepared individually by different members of the team. The comprehensive nature of the proposal and the amount of technical detail require expertise impossible to be found in one person, especially given the deadlines for submission. Having a project coordinator contributes to the collaborative proposal's consistency in style, tone, and format.

When you're working on a team in which each person is responsible for a separate section, how could you ensure overall consistency in style, tone, and format?

Benefits

Many professionals advocate collaboration because of the *benefits,* both workplace and personal. Since the terrorist attacks of September 11, 2001, many workplace professionals are interested in alternatives to travel, including collaborative technologies. According to responses from approximately 700 business travelers to a poll conducted by Wainhouse Research, Web conferencing has increased 61.5 percent, video conferencing has increased 24.5 percent, and voice conferencing has increased 11.7 percent.[8]

While the initial increase in use of collaborative technologies was due to a concern for safety, since then workplace professionals have discovered additional benefits. "More than 60 percent of respondents indicated that use of conferencing technologies would allow them to get more work done (78 percent), make faster decisions (66 percent) and be more competitive (64 percent)."[9]

Beyond the pragmatic benefits, many professionals say the collaborative process is often more enjoyable than working independently. They can work with colleagues with whom they might not otherwise interact. In fact, some professionals admit that they sometimes establish collaborative projects, rather than individual ones, just for the opportunity to work with particular people.

> *How can you increase the interpersonal benefits of belonging to a team?*

Working collaboratively also has organizational and personal benefits. Many people like being part of a team. As a member of a team, you have a group identity that provides support when the workload and schedule are especially difficult. While companies like having employees who are satisfied with their work environment, a smoothly running team also can be more productive than the same number of individuals working separately. And finally, of those polled by Wainhouse Research, 72 percent identified an unexpected benefit of collaborative technologies: The increased productivity gave them more time to spend with family and friends.

Reasons Collaboration is a Problem

Even though collaboration is widespread in the workplace, it is not necessarily easy, and it's not always enjoyable. These are ten typical problems that can occur:

- ***Time.*** Collaborative work is often more time consuming than individual work.
- ***Discomfort.*** When people who don't know each other are assigned to the same team, interaction may initially be awkward and frustrating.
- ***Control.*** Managers may be unwilling to let your team make its own decisions.
- ***Credit.*** Some individuals are unwilling to share credit, whether in the form of coauthorship, patents, royalties, or public acclaim (though they may be willing to let others take the blame).
- ***Conflict.*** Unproductive conflict can become overwhelming. Productive conflict may not be recognized and may be mistaken for unproductive conflict. (See "Negotiating Conflicts," beginning on page 164.)
- ***Criticism.*** Some people do not gracefully accept criticism or modification of their ideas.

- ***Ethics.*** You may observe attitudes, behaviors, or decisions of other team members that you believe are unethical.
- ***Style.*** People may have very different ways of approaching tasks and problems, so they may seem abrasive, unfocused, uncooperative, unavailable, shy, directive, or critical.
- ***Responsibility.*** If individuals are accustomed to being in charge, they may find working on a team to be uncomfortable because it requires sharing power and responsibility.
- ***Technology.*** The organization may not have the technology necessary to carry out efficient collaboration.

What can you do if faced with each of these problems? Can you pose at least two ways of dealing with each problem?

Despite these problems, collaboration has so many benefits that it's likely to play a large part in your professional life. Since you can anticipate that these problems will occur in some of your collaborations, you need to figure out ways to prevent or counter problems so that your collaborative efforts are not derailed or undermined.

Types of Collaboration

Collaboration in the workplace often takes many different forms. As long as the collaborators agree on their interpersonal roles and their tasks, numerous collaborative relationships are possible. For example, collaborators might work together during only one phase of a project, such as the planning stage, but then one person might draft and revise the entire document or oral presentation. Or, after the initial collaborative planning, each person might complete a different portion of a major document or presentation. Another example occurs when collaborators work together on only one portion of a project. An engineer might collaborate with a colleague in finance when preparing the budget section for a proposal and with another colleague in quality control when preparing the testing and evaluation section of the same proposal. Regardless of the nature of the collaboration, when collaborators assume unequal responsibility for different aspects of a project, they should expect to receive unequal recognition and compensation.

What types of collaboration are you likely to do as an entry-level professional? As a more senior professional?

One useful way to examine collaboration is to analyze the role of the individual in relation to the other collaborators. For example, consider the work of Paula Thompson, a manager at Dover Systems, a company that designs hardware and software for commercial architects and contractors. Three of Paula's current projects involve different types of collaboration: coauthoring, consulting with colleagues, and contributing to team projects.

Coauthoring

To many people, collaboration means that the contributions of the collaborators are equivalent and the exchange of information is reciprocal. In a coequal collaboration, each collaborator contributes ideas but is willing to abandon or modify them. Coequal collaborators both analyze the strengths and weaknesses of their own ideas and also evaluate the ideas of the other collaborator(s). Any arguments that arise are explored in light of mutually agreed-on purposes, tasks, and products. Because each collaborator assumes responsibility for creating and criticizing, the team is able to build new ideas by taking the best from each collaborator. Each collaborator takes equal responsibility and receives equal credit as coauthor.

One project Paula Thompson is responsible for involves working with a colleague, Chris McCaffrey, to prepare an article for publication in the company's technical report series. As coauthors, Paula and Chris share responsibilities for planning, drafting, and revising their technical report.

Paula and Chris have agreed to an initial meeting to discuss content, purpose, audience, organization, and evidence, as well as to plan their schedule. They agree to rough out ideas separately and then meet again. Although they each take responsibility for writing separate sections of the report, they read each other's drafts and make suggestions that influence the other's work. Because they both write on a networked computer system, they can share their drafts and exchange e-mail about them. Even though Paula and Chris communicate regularly by e-mail, they also have face-to-face meetings to resolve problems. Once the draft of their coauthored report is completed, they establish another series of collaborative relationships as the document moves through the review and editing cycles.

Consulting with Colleagues

Another of Paula's projects is preparing a packet of information sheets for a new product line. Paula initially creates thumbnail sketches for information sheets that should accomplish these five goals: draw attention to a new product or product line that solves a problem, unify the product conceptually with the design of the information sheet, create a sense of corporate competence, arouse customers' interest enough for them to seek additional information, and summarize the marketable features of the product.

Paula also knows that information sheets should not become merely a list of technical features, so she works with colleagues in a number of departments to complete her project:

- Engineering — to get accurate product specifications
- Marketing — to get information about the target market
- Graphic design — to work with the artist to design the information sheets and pocket folder they'll go in
- Communication — to work with the editor assigned to the project
- Publication — to get corporate publishing specifications and printing costs
- Quality control — to arrange for technical review and user testing

Although Paula is responsible for the development of the information sheet packet, she consults with a variety of colleagues who have expertise that she needs.

This type of collaboration — consulting with colleagues — is common and expected. You should willingly seek information from colleagues and be prepared to reciprocate when they need information from you. Naturally, Paula lets her colleagues (and their managers) know how helpful they have been.

Contributing to Team Projects

A third project Paula is responsible for is coordinating the efforts of a team of hardware engineers, software designers, systems analysts, information designers, and budget planners. She must organize their individual efforts into a proposal for a state-of-the-art laboratory. The team has had numerous planning meetings, but each person is responsible for an individual section of the document. Part of Paula's job is to facilitate the planning meetings and to coordinate the separate sections that people have prepared into a coherent, cohesive, persuasive proposal.

However, even on a team, collaboration does not necessarily mean coequal contributions. When one person assumes a greater responsibility, that person might initially identify the purpose of the collaboration, define the task, and perhaps even specify the product. Individual contributions do not have to be equivalent. However, if the collaboration is to be successful, all the individuals involved need to have some say in what responsibilities each person should assume and suggest the collaborative approach that should be used.

Being a Good Collaborator

Allan Ackerson, lead technical writer working for the Colorado Springs division of Logicon Applied Technology, Inc., commented on the TECHWR-L electronic mailing list that collaborators — especially if they're team leaders — should follow three guidelines:

1. Treat people like you'd like to be treated.
2. Identify strengths and weaknesses of team members and use them accordingly.
3. Insist on excellence.[10]

Ackerson's advice is valuable because it is a reminder that being an effective collaborator and team leader requires common sense and courtesy. Following Ackerson's advice is a good start. Whether you are preparing a document as a coauthor, as a consultant with colleagues, or as a member of a project team, you can do several things that will help you be more effective:

- Self-assess
- Be engaged and cooperative
- Listen
- Conform to conversation conventions
- Ask questions
- Share
- Use technology effectively
- Reflect

Self-Assess

Before you consider the self-assessment in Figure 5.2, identify your strengths and weaknesses as a collaborator. What areas would you like to improve?

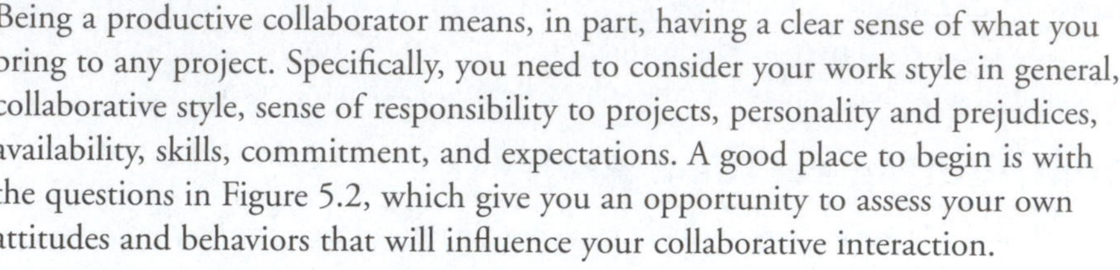

Being a productive collaborator means, in part, having a clear sense of what you bring to any project. Specifically, you need to consider your work style in general, collaborative style, sense of responsibility to projects, personality and prejudices, availability, skills, commitment, and expectations. A good place to begin is with the questions in Figure 5.2, which give you an opportunity to assess your own attitudes and behaviors that will influence your collaborative interaction.

WEBLINK

For an electronic version of Figure 5.2, go to www.english.wadsworth.com/burnett6e for a link to the questions.

CLICK ON WEBLINK

CLICK on Chapter 5/self-assessment

WEBLINK

Deciding what to tell your collaborators about yourself is sometimes difficult. The International Personality Item Pool (IPIP) is a questionnaire (50-item version or 100-item version) that makes no pretense of being scientific; it doesn't have norms or validity. However, if each member of your team completes the questionnaire, the responses will enable you to learn something about each other. Go to www.english.wadsworth.com/burnett6e for a link to the IPIP site.

CLICK ON WEBLINK

CLICK on Chapter 5/discussion inventory

After you have answered the questions in Figure 5.2 (on page 156–157), you can decide whether your attitudes and behaviors will probably have a positive or negative effect on your own collaboration, the collaboration of others on your team, and on the completion of the task itself. At that point, you and your collaborators might choose to discuss some or all of each other's self-assessments together.

Be Engaged and Cooperative

Several things increase the likelihood that your collaborations can be enjoyable and productive. You can begin by coming to the collaborative session with an open mind. Leave at the door any prejudices or negative preconceptions about the people you're working with or the project you're working on. Be receptive to the likelihood that the collaboration will improve whatever ideas you already have. Once you have an open mind and a receptive attitude, you will find the following general guidelines useful:

- Always *come prepared* for any collaborative meeting, which means gathering necessary information and having something to contribute to the collaboration.

- Be able to *articulate the purpose* of your collaborative work. What's the goal? What's the task? Do others on the team have similar ideas? What's the process you'll use to complete the project? How can you get all the collaborators in your group to represent or interpret the goals, task, and process in the same way?
- *Be articulate* in expressing your views. Don't assume body language or silence will sufficiently (or accurately) convey your views.
- *Be cooperative* and *supportive* rather than competitive and antagonistic.
- Be direct in stating your own opinions, but *don't trample* on the ideas of other collaborators.

Listen

You can learn a great deal if you are an *active listener* instead of a passive listener. An active listener is engaged — attentive, involved, interested. Being an active listener means paying attention to what your collaborators say, and what they don't say. Be aware, too, of their manner of speaking. Tone of voice, pacing, and inflection all can reflect attitudes about the content. Sometimes knowing those attitudes is as important as knowing the content. (Refer to Chapter 2 for a discussion of cultural differences in listening behaviors.)

Being involved and interested while listening to your collaborators means asking questions when you're confused and periodically summarizing what you hear the speakers saying. Getting immediate clarification lessens the chance that you'll misinterpret what you hear. Paraphrasing what speakers say is a good way to verify that your understanding matches their intent.

Active listeners often have a number of ways to respond to and encourage ongoing conversation. Try some of the beginnings listed on page 158,[11] followed by the open-ended questions suggested in Figure 5.3 on page 160.

Despite background distractions, an active listener pays close attention, periodically reflects back what is being said, and follows up with open-ended questions.

Lou Jones/Index Stock Imagery

FIGURE 5.2 Questions for Self-Assessment about Collaboration

Critical Factors	Self-assessment	Importance/Impact on Self, Team, Task?	Maintain or Change?
Your work style/work space: ■ How much do you procrastinate when starting or completing tasks? ■ What physical or psychological conditions do you have that could affect the work? ***Additional questions*** ■ Do you complete one task at a time (single focus) or work on several at once (multitask)? ■ When during the work process do you carefully check your work? ■ Do you work quickly? Slowly? ■ Are you fully engaged in tasks or do you lose concentration? ■ How do you prefer your workspace (sound, heat, light, interruptions)?			
Your collaborative style: How well do you listen and share? ■ How important is sharing the workload? How much do you expect collaborators to share the workload? How much of the workload will you share? ■ How do you prefer that collaborators inform you when you're not fairly sharing the workload? ■ How do you seek team input/agreement about team processes and procedures? ***Additional questions*** ■ How often do you challenge others' ideas for the purpose of considering alternatives and, thus, improving the outcome? ■ How do you prefer to resolve unproductive conflicts? ■ Do you prefer to share leadership or have a leader? How much do you like to be in charge? ■ What are your preferred strategies for communicating with team members? How often do you communicate about your progress? ■ How do you deal with team members who don't pull their own weight or assume their fair share of responsibilities?			
Your project responsibilities: How much of the work do you plan to do? ■ How active are you in establishing criteria to determine project quality? ■ What kind of project planning to you prefer to do? ■ What kind of written or oral feedback do you provide to others on the team? What kind do you expect them to provide you? ***Additional questions*** ■ How do you deal with team members who don't have the professional/technical skills to do their part of the project? ■ How do you deal with people who are uncooperative, belligerent, or unengaged?			
Your interpersonal/affective nature: How well do you get along? ■ What biases or prejudices of yours might influence team function or productivity (your reactions to race, gender, nationality, ethnicity, age, dialect, physical condition, sexual orientation, political preference, religion, etc.)?			

FIGURE 5.2 **Questions for Self-Assessment about Collaboration (continued)**

- How do you respond to criticism of your work?
- How do you respond to arguments or disagreements among team members?

Additional questions

- How important is having a friendly relationship with team members to your performance/productivity?
- How shy or outgoing are you? How likely are you to engage others in conversation? How reticent or voluble are you in conversation?
- How abrupt or welcoming are you when people contact you?
- How important is humor to you?

Your sense of time/availability: How much time will you put into the project?

- When are you available to work with the team?
- How much time each day/week are you planning to devote to this project?
- How easy are you to contact? How often do you check and respond to your voice mail? E-mail?

Additional questions

- What is your most productive time of day (or night) to work?
- Do you like to work for long or short periods? Do you take frequent breaks?
- What personal commitments might interrupt your work? Which cannot be changed?
- What do you think makes the most productive use of face-to-face meeting time?

Your skills or knowledge: What skills will you contribute to the team?

- Leadership skills?
- Project management skills?
- Problem-solving skills?
- Specialized technical knowledge?
- Writing and editing skills?
- Computer skills? Software? Hardware?
- Networking (people) knowledge?
- Knowledge of the organization(s)?
- Graphic design and document design skills?
- Translation skills?
- Usability testing skills?

Your commitment/expectations: How committed are you to the project?

- How important is doing excellent work? Is this a "done is good" project? How important is public acknowledgment of the project?
- How important is the implied or articulated commitment that comes with being on this team?
- How do you deal with free riders (team members who are lazy, careless, irresponsible, unproductive, or nonparticipating)?

Additional questions

- How willing are you to do the work of team members who don't do their work or don't do it well?
- What do you expect of other team members? How do you articulate these expectations?
- How will you find out what others expect of you?

- It sounds like you're saying . . .
- It seems that . . .
- I wonder if . . .
- Am I right that your main argument is . . . ?
- Am I right in thinking that your main points are . . . ?
- Tell me if I'm understanding that . . .
- I have the impression that . . .
- Am I right to think that . . . ?
- Do I understand that . . . ?
- You seem to be saying that . . .
- It seems to me that . . .
- Can you tell me if what's most important here is that . . . ?

An additional benefit is that if you listen actively to your collaborators, they're more likely to listen to you. As a result, you'll all have a better understanding of one another's views, and the project will have an increased chance of being successful.

Conform to Conversation Conventions

Collaboration requires conversation, either face-to-face or at a distance. Certainly having something to say matters, but, beyond that, following a few conventions makes face-to-face, telephone, and electronic conversations move more smoothly and productively.

All conversations

- Select an appropriate location with minimal background distractions.
- Look, sound, and act interested.
- State your points clearly.
- Provide explanations and examples as needed.
- Share the turn taking.
- Respond to the other person's ideas.

Face-to-face conversations

- Make direct eye contact (be aware of gaze, part of a person's cultural comfort zone).
- Respect personal space (be aware of proxemics, a person's spatial comfort zone).

Telephone conversations

- Don't eat, drink, or chew gum while you're on the phone.
- Don't put people on hold for more than a few seconds.

Electronic conversations (See Chapter 13 for netiquette about e-mail.)

- Keep in mind that all electronic messages can be forwarded, printed, or permanently stored by any recipient, so be sensible — even cautious — about what you send electronically.

- Copy only enough of the message you're responding to in order to provide a context or a reminder; do not copy the entire message unless you need a legal record.

What are your best habits — and your worst — in each kind of conversation?

Ask Questions

To be an effective collaborator, you need to ask questions to get the information you need. These questions can be powerful tools for helping you to assess what you already know and what you need to know to prepare an effective document. Two guidelines are particularly useful:

- Ask open-ended questions that require comments or discussions rather than questions that ask for *yes* or *no* responses.
- Ask questions that focus attention on a range of rhetorical elements important to the project: content, context, purpose, key points, audience, conventions of organization and support, and conventions of document design.

When you have primary responsibility for preparing a document, you can get a great deal of help from a supporter who asks you questions such as those in Figure 5.3. A good supporter encourages you to extend your ideas, questions you about points that seem inconsistent or contradictory, and offers suggestions about ways to improve the document. Of course, you can work with a supporter at any stage in planning, drafting, and revising a document. This kind of collaboration can help you focus your ideas, consider alternatives that may not have occurred to you if working alone, and give you feedback for improving the document. You can also reciprocate by acting as a supporter for colleagues.

The questions in Figure 5.3 are designed to help collaborators, whether they are acting as supporters for another writer, working as coauthors, or participating as members of a team. Although these questions are listed in separate categories, many concerns are relevant to more than one area. For example, the following question integrates several concerns: "How will the *organization* and *design* convey our *purpose* to the *audience*?" Periodically, you may want to consolidate two or more concerns into one question.

Share

Not only do you need to get information from your colleagues by listening and asking questions, but you also must be willing to provide them with detailed and accurate information. Unless prior agreements among all parties restrict sharing information (usually for proprietary reasons), you should make every effort to keep your collaborators informed about the work you're doing. You should also provide feedback to your collaborators about their work. This exchange of information should be done regularly.

The most effective way to ensure that information is exchanged regularly is to establish a schedule and a method when the project begins. For example, you can

FIGURE 5.3 **Questions to Help Collaborators Deal with Rhetorical Elements**[12]

Rhetorical Elements	Questions
Content	■ What critical information needs to be in this document or oral presentation? ■ What additional information might you/we include? ■ Have you considered including _______? ■ What content can be omitted?
Context	■ What is the context or situation in which this document or oral presentation will be used? ■ How will the situation be changed as a result of this document or oral presentation? ■ What aspects of the context or situation will influence the interpretation of the document or oral presentation?
Purpose and key point(s)	■ What do you see as your/our main purpose? ■ What main point(s) do you/we want to make? ■ How will the audience react to the purpose and points? ■ I see a conflict between _______ and _______. How can you/we resolve it?
Audience	■ Who is your/our intended audience? ■ What does the audience expect to learn? ■ How do you think the audience will react to _______? ■ What problems, conflicts, inconsistencies, or gaps might the audience see?
Organization and support	■ How can you/we organize the content to achieve the purpose? ■ What evidence can you/we use to support the purpose and appeal to the audience? ■ What examples (anecdotal, statistical, visual) should you/we use? ■ How are you/we going to connect _______ and _______?
Document design elements	■ How can design features be used to convey the main point(s)? ■ What design features will the audience expect? What will the audience respond to? ■ How can verbal and visual information be balanced? ■ How can the design be used to reflect the organization of the content?

> *What questions would you like to add to Figure 5.3? Why are these useful questions to add?*

all agree to provide weekly update memos. Or you can provide read-only privileges for selected electronic files. Or you can have weekly meetings to discuss the status of the project. Or you can circulate updated drafts of the document, either on paper or electronically. The method is not nearly as important as the fact that regular communication occurs among collaborators.

How widespread are global teams? Here are just two examples. Intel helps top Malaysian engineers launch their own businesses and then contracts with these new companies to work alongside Intel's U.S. employees. Some Hewlett-Packard software teams rarely have face-to-face meetings because the engineers are from places as far apart as Colorado, Australia, Germany, India and Japan.[13]

Use Technology Effectively

Most teams, especially global teams, depend on technology, specifically on groupware. This increased use of groupware draws attention to two important issues: awareness of collaborators and privacy. While these issues are important in any collaboration, they have the potential to be especially problematic in collaborative interactions that involve many people working asynchronously over long distances.

WEBLINK

Technologies that enable collaboration have moved from the experimental laboratories to test sites to large-scale implementation and integration in worldwide business and industry. Go to www.english.wadsworth.com/burnett6e for a link to a site that gives you access to a series of white papers about collaboration and technology from industry research groups and leading-edge companies. These papers are definitely about application and practice.

CLICK ON WEBLINK

CLICK on Chapter 5/white papers

Groupware *Groupware,* software designed to facilitate group interaction, usually changes the way collaborators plan, share documents, give and receive feedback, and make decisions. It enables the formation of groups for special projects. Two important factors characterize groupware:

1. *Time:* Are you and your collaborators working together at the same time (synchronous) or at different times (asynchronous)?
2. *Location:* Are you and your collaborators working in the same place (colocated; face-to-face) or in different places (noncolocated; distance)?

Whether you need synchronous or asynchronous communication, whether you and your collaborators are in the same room or separated by thousands of miles, groupware can bring together multiple perspectives and expertise. Figure 5.4 categorizes types of groupware that aid collaboration and teamwork.

Groupware designed for team interaction generally provides tools for a variety of complex activities: managing the information the team collects; managing factors that affect interaction, such as the context, agenda, and workspace; deciding access to information (who, when, and where); enabling team mobility so

FIGURE 5.4 Common Types of Groupware[14]

LOCATION	TIME: *Synchronous interaction (same time)*	TIME: *Asynchronous interaction (different time)*
Colocated interaction (same location)	■ **Presentation support systems** aid speakers in formal or informal presentations with computer slides, often accompanied by audio and animation.	■ **Shared computers** with word processing software enable collaborators to track changes and annotate each other's work.
Colocated interaction (same location) / *Distance interaction (different locations)*	■ **Decision support systems** (DSS) facilitate colocated and distance group decision making by helping collaborators with brainstorming, critiquing, weighing probabilities and alternatives, and voting. DSS may facilitate meetings by encouraging equal participation through anonymity or enforcing turn taking.	
Distance interaction (different locations)	■ **Shared whiteboards** permit two or more people to view and draw on a shared drawing from different locations. Color-coded telepointers identify each person. ■ **Video communications systems** allow two-way or multiway calling with live video. Video is advantageous when visual information is being discussed. ■ **Chat systems** permit multiple people in real time to write messages in a shared space. Many chat rooms control access or permit moderators for discussions. Issues in real-time groupware include anonymity, following the stream of conversation, scalability with number of users, and abusive users.	■ **E-mail** can send and forward messages to individuals or groups; filter, sort, and file messages; and attach files to messages. ■ **Newsgroups and electronic mailing lists,** which deliver information to groups, are similar; however, newsgroups only show waiting messages when requested, while mailing lists automatically deliver messages as they're sent. ■ **Hypertext** links related documents. Any authors can create links to new sites. Tools allow identifying visitors to a site and tracking the number of visitors on a site. ■ **Group calendars** let people coordinate their schedules for projects, meetings, equipment and facilities use, out-of-town conferences and trips, and so on. ■ **Workflow systems** help documents move through organizations electronically by a relatively fixed process. For example, employees can submit an expense report by routing it to a manager who electronically approves it for payment. The report is simultaneously debited from the group's account and then forwarded to the accounting department for payment.
Distance interaction (different locations)	■ **Collaborative writing systems** provide both synchronous and asynchronous support. Collaborators can use various tools to plan and coordinate as well as link separately authored documents, track changes by various collaborators, and annotate their work in progress. Synchronous support usually provides an additional communication channel to the authors as they work, such as videophones or chat. Asynchronous support usually uses shared computers (see above).	

members aren't restricted to a particular location; and encouraging team innovation and invention. Groupware also structures interactions, so aggressive people are less likely to dominate conversations.

Awareness Responding to what's happening is easier when team members are working face to face than when the team is separated by physical distance and by technology. Figure 5.5 describes a scenario in which two members of a team are using a shared, scrollable whiteboard. Sometimes awareness can be improved

FIGURE 5.5 **Increasing Awareness by Increasing Information**[15]

A *shared whiteboard* system allows two people to work on the same image at the same time. The changes that one person makes are copied to the other person's computer so that both people's workspaces contain the same object. If both people are able to scroll up, down, and across, . . . each person . . . [might have] a different view of the object . . . [with] only a small degree of overlap between the two views [as shown below]. In this situation, one could hardly say that either person is truly aware of the other's activity.

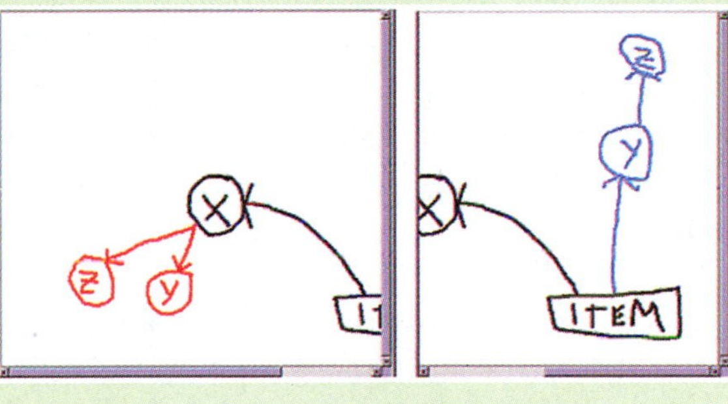

In such a case, even the smallest piece of additional information regarding the other person's actions would constitute an improvement in the collaboration between the two. [The figure below] illustrates how a mutual understanding of the overall context can be facilitated if each . . . is aware of the other's activities, and can adjust his or her own activities accordingly.

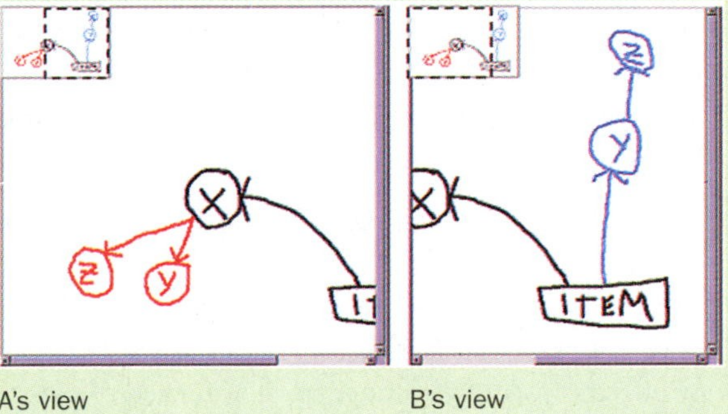

when the collaborators simply pay more attention to the context, asking about things that would be understood in a face-to-face meeting.

Team awareness includes being tuned in to what's happening in the interaction, in the workspace, and with the information.[16] The team members, both individually and collectively, should consider responses to the questions in Figure 5.6. These questions are representative rather than comprehensive, so you might be able to add others to the list. Why so many questions? Worldwide practice indicates that *distributed teams* — that is, teams distributed across time and space — have many problems that don't affect face-to-face teams to the same extent. One way to reduce the problems is for team members to be aware of potential problems. A second way is for team members to develop strategies for either preventing those problems or managing them if they do occur.

> *What objections regarding privacy and accuracy might a person have about using a department-wide electronic calendar that helps schedule meetings, reserve rooms and equipment, manage projects, and coordinate activities among people?*

Privacy Despite the convenience of groupware, users worry about privacy. Why? Security, control, safety, and accuracy. Electronically sharing information raises concerns about security: proprietary information needs to stay secure. Sharing information also raises concerns about control — that is, who has access to information and who decides what and how much to share. Sharing information raises concerns about safety; people worry that sharing relevant, project-specific information may open a path to private information that they want to keep private. And, finally, sharing information raises concerns about accuracy; for example, if a person doesn't keep a careful record of all meetings and appointments on a department calendar, people may have a distorted view of the person's availability.[17]

Reflect

Take time to reflect. Mull over the ideas you've heard. A good collaborator carefully considers ideas and opinions from other people. Planning a specific time for reflection is difficult when time is scarce — when deadlines slip, when your day is filled with back-to-back responsibilities, when you barely have time to eat lunch — but it is nonetheless an important part of the collaborative process. You need to sift through the information you've collected and assess its value, decide what information to use and what to discard, and pinpoint areas of confusion that should be clarified and areas of disagreement that should be resolved. Your reflection may also show that you need to go back to your colleagues or team for additional information or clarification; more decisions may be needed. In short, without reflection, you won't be able to effectively use the information you gain through collaboration.

Negotiating Conflicts

As you learn to be a more effective collaborator by self-assessing, being engaged and cooperative, listening, adhering to conventions, asking questions, sharing, using technology effectively, and reflecting, you'll also need to manage problems and negotiate the various kinds of conflicts — affective, procedural, and substantive — that arise.

FIGURE 5.6 Team Awareness

Areas for Team Awareness	Topics to Consider
Interaction	■ **Who** — Who's involved on the team? Who's coordinating and facilitating team interaction? ■ **What** — What might cause distortion or distraction in the interaction? ■ **When** — When do people interact? How are differences in time zones and work schedules managed? How often do people interact? ■ **Where** — Where is the interaction occurring? In virtual spaces? ■ **Why** — Why might conflicts occur? How will conflicts be resolved? ■ **How** — What will be the nature of the team interaction? How is the project history maintained so that team members can review who has done (or is doing) what? How are various versions of work being tracked or labeled? How will team members be able to see each other's planning sketches? Drafts?
Workspace	■ **Who** — Who's in the workspace? Who's doing what? What's the order of authorship? ■ **What** — What goals have been agreed on? What activities are necessary to achieve the goals? What activities are going on? ■ **When** — When is the beginning and end of the project? What's the timeline? When are the various milestones? When does synchronous communication occur? How long is acceptable between asynchronous responses? ■ **Where** — Where are team members located? How does the location affect participation and performance? ■ **Why** — Why is each team member involved? What's the motivation? Purpose? Expectation? ■ **How** — How will the team produce the deliverables?
Information	■ **Who** — Who on the team already knows relevant information? ■ **What** — What information is being gathered? By whom? What has been done so far? What does the entire picture look like (rather than a small segment)? ■ **When** — When are the interim and final deadlines? When is the product and document testing? ■ **Where** — Where are the most useful, reliable sources of information? What are people looking at? Using? Working with? ■ **Why** — Why is the process being used to categorize, archive, disseminate the information the most appropriate? ■ **How** — How is information being interpreted? How will the information be transformed into the deliverables?

Is too big a deal made about workplace conflict? Go to www.english.wadsworth.com/burnett6e for a link to an article that summarizes more than ten years of workplace surveys and incidents about workplace conflict.
CLICK ON WEBLINK
CLICK on Chapter 5/conflict

Affective Conflicts

Affect involves your attitudes and biases, your personality and values. These factors shape your interpersonal communication and influence the way you approach collaboration. With effective interpersonal communication, your collaborations are more likely to be successful.

Interpersonal relationships are as important to collaboration as the task itself. A project is more likely to succeed if you respect your collaborators and get along with them, if you're comfortable working with them and feel free to share ideas and voice concerns. You and your collaborators need to trust one another; you can help build and maintain this trust by frequent contact and sharing of information.

Affective conflict, defined simply as interpersonal disagreement, can seriously impede any collaborative interaction, perhaps even signaling the end of the collaborative relationship. Simply put, it's unproductive. Affective conflicts might arise because of strongly held values and beliefs about religion, politics, business ethics, or gender, racial, or ethnic prejudices. For example, a collaborator who says, "I'll never work with a ____" (fill in the blank with some personal bias) has let affective conflict interfere with a working relationship.

Affective conflict may also occur in unexpected ways. For example, if your team is using a lot of technology to communicate and to complete your work, you may find that people have habits or behaviors that cause affective conflict, such as downloading software or music illegally, accessing porn sites, sending inappropriate jokes, or engaging in sexual harassment.[18]

One way you can avoid affective conflict is to first acknowledge your biases and prejudices and then make a special effort to not be negatively influenced by them during your collaboration. Think about the reasons that particular people or behaviors annoy or distress you and decide whether the reasons are relevant in a particular collaborative interaction. Sometimes people don't know their attitudes or behavior is causing a problem; in such situations, talking to your colleague privately may help. But sometimes it doesn't. And sometimes the problem is yours, not the other person's. In either case, you have a choice: Change your attitude or figure out professionally acceptable work-arounds so that prejudices don't derail the project.

Another way you can avoid affective conflict is by paying attention to differences and changes in footing during collaboration. *Footing* is a term used by cultural anthropologists to describe the underlying assumptions people make about a particular situation; those assumptions govern the way people act. Of course, your assumptions change as you learn new information about the people you're working with — their knowledge, their attitudes, their behavior — and the task you're working on. Even though you have established a relationship with your collaborators, you should recognize that changes are inevitable. In fact, as you react to new information and recognize subtle shifts in footing, your views about your collaborators and the task will gradually change. Being aware of this evolution will make it easier to handle.

What affective problems have you had when collaborating? What might you have done to decrease or eliminate those problems?

WEBLINK

For a quick read about ways to influence uncooperative people, go to www.english.wadsworth.com/burnett6e for a link to a short article with practical suggestions.

CLICK ON WEBLINK

CLICK on Chapter 5/conflict management

Procedural Conflicts

For a collaborative group to function smoothly, you and your collaborators need to discuss how the group sessions will run — that is, the procedures that will govern the group's operation. Many experienced collaborators find it efficient and productive to begin their work together by explicitly discussing and agreeing on several critical factors that affect procedures:

1. ***Meeting details***
   - Settle details of meetings: time, place, duration.
   - Agree on what preparation should be done for meetings.
   - Discuss the collaborative approach the group will use.
2. ***Team roles and responsibilities***
   - Identify the responsibilities each individual will assume.
   - Determine how to monitor the group's progress.
   - Decide on order of authorship based on some mutually acceptable criteria.
3. ***Productive management of conflict***
   - Agree on ways to minimize affective and procedural conflict.
   - Agree on ways to encourage substantive conflict.
   - Decide how to negotiate among alternatives and resolve disagreements.

Deciding such concerns up front can reduce your chances of having a group derailed by procedural conflict, which is usually unproductive. Open discussions about procedures can strengthen the group cohesiveness, both the feeling of group identity and the group's commitment to the task.

Anticipating procedural conflict can be critical to a group's success. Don't wait until your third 4:30 meeting to discover that one member always has to leave early because of family responsibilities or a second job. Don't wait until people are shouting at each other to agree that the team should withhold criticism until all the options have been presented. Don't wait until someone fails to meet a deadline to discover that he or she doesn't have the time or skills to do that part of the project effectively. Don't wait until the first progress report is due to discover that no one is keeping track of what's already completed, what is being done, and what needs to be done. Sometimes just asking questions about procedural concerns makes people aware of potential problems that can be easily managed by some preliminary discussions and agreements.

A group of workplace professionals in a course to study collaboration compiled a list of surefire ways to sabotage any collaborative effort. If you want to ensure failure, follow the suggestions in Figure 5.7.

FIGURE 5.7 **Ways to Sabotage Excellence in Collaboration**[19]

What would you add to this list of ways to sabotage collaboration?

Miss meetings.	Expect that many hands make quick work.
Show up late.	Keep meetings going for more than two hours.
Don't bother to talk about what the group goal is.	Compete for individual recognition.
Don't ever allow discussion of anything not on the agenda.	Elevate personal success above group success.
Don't establish an agenda.	Don't bother to keep track of what is discussed.
Ignore established agendas and procedures.	Don't establish ways of communicating outside of scheduled meetings.
Disagree with everything just because you don't want to be there.	Rely solely on guesses to make decisions.
Agree with everything just because you don't want to be there.	Don't establish any way to assess individual progress.
Keep quiet, even when you have another idea.	Attempt to include everything anyone ever said.
Hoard your information.	Expect the worst from your group members.
Give up on your idea without even explaining it.	Don't let the group get to know you.
Make yourself responsible for everything.	Strive for quick consensus on every issue.
Don't take responsibility for anything.	Believe that everything runs smoothly in good collaboration.
Plan to get everything done in one big work session.	Never laugh.

Other procedural concerns have to do with the costs and materials of a project — procedures regarding the allocation of resources. Because collaborative projects often take a long time to complete, they can be expensive. And because the projects are often complex, they require purchase of or access to a variety of equipment, tools, supplies, and services, which also can be costly. A limited budget can affect the nature of a collaborative project, restricting employee time and access to necessary resources. However, realistic long-range planning often allows an organization to budget for collaborative projects.

The extraordinary problems created by inattention to or short circuiting of procedural concerns are discussed in the ethics sidebar below. Initial discussions rather than assumptions about who would receive credit for important work would have reduced the chances that people would have been passed over. And as the situation discussed in the sidebar illustrates, procedural conflicts quite predictably spilled over into affective conflicts as well.

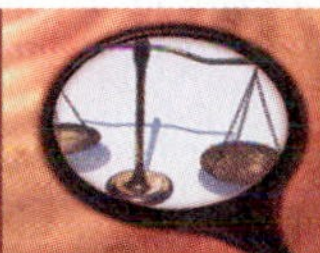

ETHICS SIDEBAR

"They were there to grasp all the credit for themselves": Ethics and the Culture of Credit

Picture the story of Susan Berget. Berget was a postdoctoral student researching human genes in a lab at the Massachusetts Institute of Technology (MIT). Berget's research revealed a puzzling connection between different strands of human chromosomes. Berget, working with Phillip Sharp, who headed the lab, later uncovered the concept of gene splicing, which was a major development in the field of molecular biology. The concept was so important that the 1993 Nobel Prize for medicine was awarded to its discovers: Phillip Sharp of MIT and Richard Roberts of New York's Cold Spring Harbor Laboratory (CSHL), a lab that had discovered gene splicing at about the same time. Notably missing from the list of prize winners was Susan Berget's name. Berget says that she and Sharp have settled their differences about the lack of acknowledgment for her role in the discovery, but "she acknowledges that the distribution of credit was not a perfectly smooth process."[20]

Reporters Jon Cohen and Gary Taubles investigated gene splice research and found that Berget was not the only researcher left out. As results from separate steps in the research were publicly released, contributions from many researchers went unacknowledged. John Hassell, a member of the CSHL group, believes that "[n]inety percent of the people at Cold Spring Harbor didn't get the credit they deserved." Cohen and Taubles attribute the winnowing of contributors to the "drive for individual credit — stoked by prize committees and the media." Focusing on individual credit "overshadows the collaborative nature of the scientific process, distorting the record and leaving researchers feeling bruised."[21]

How would you prevent the need for individual recognition from damaging a collaborative project?

How can technical professionals avoid these conflicts in giving credit? Researchers Mark Haskins, Jeanne Liedtka, and John Rosenblum believe a possible solution is to develop an ethic of collaboration that is based on relationships, not tasks. They identify four important elements in a collaboration that focuses on relationships:

- a caring attitude about the well-being of colleagues and about the work
- a recognition of the importance of the work for themselves and others
- a personal passion for the work
- a creative energy that fuels risk and innovation[22]

Focusing on relationships, Haskins and his colleagues believe, induces an ethic of collaboration that helps members see the advantages of group accomplishments over individual credit. If an ethic of collaboration had existed in Susan Berget's case, all the different researchers who played a role in the discovery would have been acknowledged for their role in the group's achievements.

Do you approach teamwork with an ethic of collaboration? Would you be willing to forgo individual credit for the sake of the group's success? Think of Susan Berget's story. How would you feel if your contributions to a project were not acknowledged?

Substantive Conflicts

Substantive concerns deal with the substance of a document or presentation and include decisions about issues such as content, purpose, audience, conventions of organization and support, and conventions of design. Just as you may make preliminary agreements about procedural concerns, you should also adopt the practice of experienced collaborators who often address substantive concerns early in their discussions:

- Agree on the purpose of the collaboration.
- Agree on project objectives and outcomes.

Collaborators who do not agree about these substantive concerns run the risk of producing unfocused, poorly designed documents that don't respond to the needs or interests of any particular audience. Making decisions about these concerns doesn't lock you in; it gives you a direction. Replanning is always possible. You always have the flexibility to change any of your initial decisions, realizing that each one influences the others. For example, changing the purpose influences the organization and design of a document. Changing the design can change its impact on the audience.

Collaborators need to reach consensus, but not too quickly. One effective and productive way of deferring consensus is to engage purposely in substantive conflict. Unlike affective conflict (which deals with interpersonal tensions) or procedural conflict (which deals with disagreements about things such as meeting times), substantive conflict focuses on the issues and ideas of your collaborative work. Substantive conflict, which is usually productive, includes two specific types of interaction:

- ***Voicing explicit disagreements:*** For example, when one collaborator suggests, "Let's do Z," the other collaborator might respond, "No," or "I disagree," or "I think that's wrong."
- ***Considering alternatives:*** For example, when one collaborator suggests, "Let's do X," the other collaborator might respond, "Yes, X is a possibility, but let's consider Y as another way to solve the problem."

Deferring consensus can give you more time to identify and consider opposing points, more time to generate and critically examine issues and ideas. Resolution of substantive conflict can lead to increased commitment to your team effort and to a potentially better product. Because engaging in substantive conflict should be cooperative rather than competitive (after all, you're all going after the same goal: high-quality decisions and high-quality products), the resolution can be described as win-win. Here are some suggestions for engaging in substantive conflict during a cooperative collaborative session:

1. ***Ask provocative questions.***
   - Ask questions that focus on potential problems between various elements: "How can we use the design to help show our purpose?" "How can we explain these examples so the readers will be able to understand them?"
   - Ask your collaborators for elaborations, clarifications, and explanations of statements, and be prepared to offer clarifications and explanations of your own statements.
   - Ask for reasons to support arguments and work on developing and supporting well-formed arguments of your own.
2. ***Take a productive and critical perspective.***
   - Try never to settle on one solution or decision without first having considered a couple of alternatives.
   - Assume the role of devil's advocate.
   - When you disagree with something, say so; support your disagreement with a reason and, if possible, an alternative that will work.
   - If other collaborators don't generate substantive conflict by raising alternatives and voicing disagreements about your ideas, bring up alternatives and possible objections yourself.
3. ***Separate ideas and personality.***
   - Don't mistake an objection to your ideas as an attack on your character, personality, or intellect.

Substantive conflict should be viewed as a way to strengthen a project rather than weaken it. Such conflict enables collaborators to rethink their own positions, delineate the perimeters of their views, and make decisions about key issues. Collaborators who establish interpersonal relationships know that disagreements aren't personal attacks; instead, they're ways of reexamining issues critical to the project. Of course, as the Dilbert cartoon[23] shows, successful collaboration depends on an ability to think and communicate clearly.

Explain whether you believe the attitudes and behaviors in the Dilbert cartoon accurately reflect the workplace.

Successful collaboration depends on an ability to think and communicate clearly, but it also depends on all the participants being willing to interact and being receptive to the ideas they hear — even when they don't necessarily agree. Summarily dismissing collaborators when they disagree is not productive.

Where can agreement begin? Collaborators should agree about their project's goals, the conceptual core that reflects their ideas and ideals. Collaborators should be able to define their concepts and explain why they're important. Some documents are confusing because they seem to present inconsistent, sometimes even conflicting, views.

During planning, collaborators need to explore the concepts from a variety of angles: accuracy, logic, underlying theory, related issues, and implicit values. If the ideas aren't clear to the collaborators, they certainly won't be clear to the readers. Such investigating helps collaborators examine alternative views, bolster arguments against attack, refine explanations, delete weak positions, and clarify vague or misleading statements. Considering these alternatives makes the eventual agreement much stronger.

The International Online Training Program on Intractable Conflict offers very useful information about conflict management and constructive confrontation. For a link, go to www.english.wadsworth.com/burnett6e. Once you've read how to use the site and taken the quick tour, you'll probably find the problem-solving and self-study section the most useful place to start.

CLICK ON WEBLINK

CLICK on Chapter 5/constructive confrontation

WEBLINK

Cultural Differences and Expectations

When a conflict arises, you need to ask yourself not only if it is affective, procedural, or substantive but also if it is generated by cultural differences. Drawing on what you learned about national and organizational cultures in Chapter 2, you can see that certain conflicts might occur during collaboration because attitudes and behaviors are unfamiliar.

Collaboration is a complex process of decision making and compromise that is influenced by cultural beliefs and practices. You can't assume that people from different cultural groups, backgrounds, or geographic regions will react the same way. Although all individuals have unique attitudes and behaviors that influence their collaborative style, within cultural groups you can often find common patterns of behavior.

Collaborators may inadvertently misjudge their colleagues' behaviors, not realizing that these behaviors are perfectly normal in another culture. For example, a colleague may agree with the clearly conflicting positions of two collaborators to avoid rejecting the ideas of either one. This behavior, regarded by some as indecisive or even wimpy, may simply be a reflection of the person's cultural practice of not insulting another by publicly rejecting his or her ideas.

In some cultures colleagues seldom directly express objections to problematic plans that are on the table for discussion. This reluctance to voice explicit disagreements may stem from a cultural practice that avoids criticizing others (which could embarrass them) and perceives disagreements as criticism or from a cultural perspective that values group harmony above potential benefits that substantive conflict might generate.[24] All sorts of behaviors — a reluctance to engage in substantive conflict, a hesitancy to spell out the procedural expectations, or an unwillingness to volunteer — may be the result of cultural practices different from yours.

You can help guard against inadvertent cultural bias by encouraging people you work with to exchange information about their collaborative patterns and practices, both their individual preferences as well as their cultural behaviors.

When your group is establishing the procedures to follow, collaborators can ask each other questions in categories such as these:

1. ***Moving the collaboration forward***
   - Are you willing to suggest an idea or plan that differs from the one accepted by the majority of the group?
   - When you don't understand something, do you ask for explanations and clarifications?
2. ***Reacting to disagreement***
   - When you don't agree with something, are you willing to express objections? Criticisms? How do you express your disagreement?
   - How do you think disagreements should be resolved?
3. ***Reacting to personal comments***
   - How do you react to being put on the spot?
   - How do you respond to criticism?
   - How do you respond to praise?
4. ***Considering team roles***
   - Do you think the group should have a leader? What are the responsibilities of the leader?
   - What are the responsibilities of the other collaborators?

Talking about these aspects of individual behavior may help collaborators understand that people respond in different ways. Understanding and perhaps even anticipating behaviors different from your own will help you be a more responsive and effective collaborator.

WEBLINK

If you want to learn more about collaboration, go to www.english.wadsworth.com/burnett6e for a link to tc.eserver.org, a portal that lists more than 60 workplace publications about collaboration.

CLICK ON WEBLINK

CLICK on Chapter 5/tc.eserver.org

Individual and Collaborative Assignments

1. **Address a complex problem.** Before coming to class or meeting with your collaborator, carefully read the case situation for Sundance Systems, Inc., and the Sundance information sheet. Before you meet with your collabora-

tor, prepare notes about a way to approach the technical memo and the information sheet redesign.

(a) **Write a memo.** Work with a collaborator to prepare a technical memo in response to the Sundance Systems case situation. The memo needs to include an analysis of the problems and propose a plan to address them. Take into account the task, the content, and concerns such as purpose, audience, organization and support, document design, and consolidation of these concerns. As part of your planning, discuss your approach to the issues and your collaborator's approach before you both decide how to prepare the memo.

Case Situation: Sundance Systems, Inc.

Examine the accompanying information sheet. In response to mail-in and telephone inquiries from potential customers (small businesses and homeowners), Sundance Systems, Inc., has sent out approximately 5,500 information sheets during the past 12 months. During this same period, the company has sold and installed 470 systems, at costs ranging from $5,000 to $25,000.

Unfortunately, a recent survey of customers indicates that although 47 percent of customers who purchased a Sundance System said they had read the information sheet, only 17 percent of customers who read it said it positively influenced their decision. In contrast, 22 percent of those who read it said it was confusing, and the remaining 61 percent said it didn't affect them one way or the other.

The information sheet was prepared by Lou Battle, founder of the company, who is considered an industry leader. Lou likes the idea of responding to inquiries with factual information. The current information sheet has been used for the past two years. In last week's staff meeting, Lou mentioned that he's heard some people want to update it, but he thinks it's pretty good as it is — it's short and accurate.

You've just been hired as an information designer in Sundance Systems' expanding corporate communication group. This group has consolidated the management of all published company documents, documents that until recently were handled by individual departments as needed. You were the top choice of Judith Dupras, the company's general manager. Part of your job is to work with another information designer to decide what to do about the survey results. What's the problem? What should be done? Present your collaborative assessment. (*Note:* You can create appropriate information to complete the assessment as long as it doesn't conflict with the minimal information provided.)

(b) **Revise the information sheet.** Continue working with the same collaborator. Your task now is to revise the Sundance information sheet. Take into account the content as well as purpose, audience, organization and support, visuals, and document design. Discuss your and your collaborator's approach to the issues as you decide how to revise the information sheet. Create a completely revised document.

SUNDANCE INFORMATION SHEET

Although most active solar systems are fairly simple in design and operation, the air-cooled system is the simplest. Unlike a liquid system, an air system needs needs no considerations for freeze protection, expansion, air bleeding, or vacuum relief. Since air-cooled systems run at lower operating temperatures than liquid-cooled systems, heat loss throughout the system is kept at a minimum while efficiency is kept at a maximum. The simplicity of the air-cooled solar system, its high efficiency, and its low cost should convince you to install a SUNDANCE SYSTEM in your home or business.

The heart of the air-cooled system is the solar collectors. The SUNDANCE solar collectors collect and convert the solar energy into usefull heat energy. The SUNDANCE collector is basically an air-tight insulated box. The inside of this box contains a black aluminum absorber plate, and the box is covered with a special transparent cover that allows light to pass through, but doesn't allow heat to escape. Light from the sun enters the box through the transparent cover, strikes the black aluminum absorber plate, and is converted to heat. Cool air enters the bottom of the solar collector, passes over the absorber plate, and exits from the top of the collector as heated air. Typically, this cool air entering the SUNDANCE collector is about 60F, and the heated air leaving the collector is about 140F. This heated air is brought via the ductwork into the building for water or space heating.

The ductwork is used to convey the cool air from the building (the return line) to the SUNDANCE collectors, and the heated air from the SUNDANCE colectors (the supply line) to the building. Ductwork is also used to connect the individual SUNDANCE collectors together to form an array and to vent this array during a period of nonuse such as a vacation. It is imperative that the ductwork system be thoroughly sealed from leakage. Leakage adversely affects the efficiency of the SUNDANCE SYSTEM by drawing cold outside air into the return line or by leaking heated air from the supply line. All ductwork done by SUNDANCE SYSTEMS is heavily insulated to keep heat loss at a minimum. Dampers are used to direct air into the house for space heating in the winter or to divert this heated air through a special water heater for summer operation.

A one-third horsepower two-speed blower provides air movement through the SUNDANCE SYSTEM. The fast speed is more suitable for winter operation since the system maintains a maximum output temperature of about 140F at this speed. This 140F air is usefull for space heating since it is about 70F warmer than the air already in the house. The difference between the temperature of the air leaving the house and the temperature of the air returning to the house is called the temperature differential. During summer operation however the lower operation speed is used since a higher operating temperature is necessary. The lower operating speed provides a collector output temperature of about 180F. This higher operating temperature is necessary for heating water since as the water becomes increasingly hotter the temperature differential and therefor the efficiency of the system are decrease. Therefore the SUNDANCE SYSTEM has a two-speed blower to maximize output year-round.

The SUNDANCE water heater is a three-shell design. The stainless steel inner shell contains the pressurized water. The middle shell provides the space for the heated air to circulate around the inner water tank. The outer shell contains the fiberglass insullation that is wrapped around the middle shell. Heated air enters the top of the water heater and circulates around the inner water tank, giving up its heat to the tank. Cooler air leaves the bottom of the water heater and is returned to the SUNDANCE collectors for reheating and the cycle continues. The heated water from the SUNDANCE solar water tank is piped into the existing electric hot water heater. By providing preheated water to the electric water heater, the amount of time this water water would be required to run is substantially reduced. This arrangement results in a continuous flow of hot water and an uncomplicated system

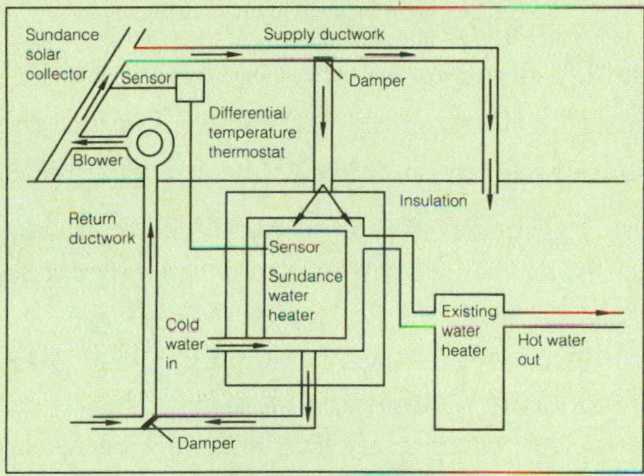

The differential temperature thermostat provides control over the entire SUNDANCE SYSTEM by controling the operation of the blower. In the winter operational mode, the unit turns the blower on when the collectors reach a preset temperature and off when they reach a preset minimum temperature. There is also a thermostat that will override the differential temperature thermostat and turn the system off if the building becomes too hot. In the summer mode, the operation of the temperature differential thermostat is a little more complex. The thermostat has two sensors connected to it. One from the SUNDANCE collector, and the other from the SUNDANCE hot water tank. The thermostat continuously monitors the temperatures at these two locations. The thermostat turns on the SUNDANCE SYSTEM when the temperature differential between the water tank and the collectors is greater than 40F, and off when the temperature differential is less than 20F. By maintaining a proper temperature difference between the collectors and the water tank and the collectors and the building, the SUNDANCE SYSTEM maintains maximum output and effeciency.

(c) **Write an analysis of the revised information sheet.** With your collaborator, prepare a memo to accompany the revised information sheet. This memo should identify the major changes you made and provide a rationale for them. Including "before" and "after" examples could illustrate your points. The memo should specifically address how the purpose, audience, organization and support, visuals, and document design affected your decisions.

2. **Evaluate your collaboration.** While conducting assignment #1, tape-record a planning session with your collaborator. After the assignment is completed, arrange a time when both you and your collaborator can listen to the tape together and then discuss the following questions. What interesting behaviors and patterns did you notice? Did anything surprise either of you when you listened to your collaboration? What does the tape reveal about the following issues:

   - Who was in charge? Was authority shared between the collaborators?
   - Was the time productively used?
   - Where did the ideas come from?
   - What did each person contribute to the collaborative effort?
   - What kinds of problems occurred? How were the problems resolved?

   Prepare a one-page memo to your instructor characterizing the the interaction and suggesting changes in your own collaborative style that you will work toward.

3. **Identify appropriate technology.** Identify technology that would enable you to be a productive team member in these situations: team members working in the same or different places and working synchronously or asynchronously.

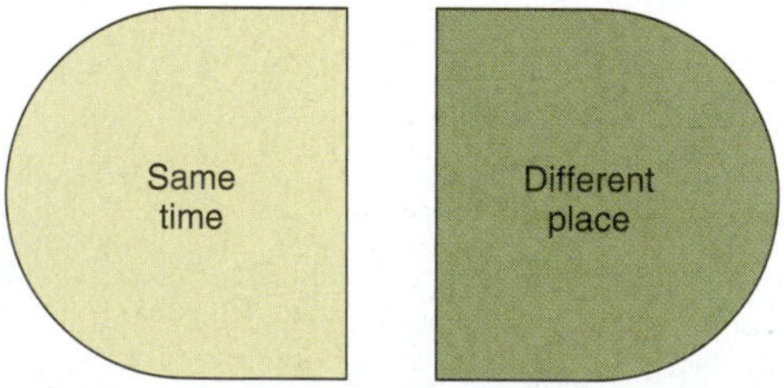

Same time | Different place

a. ______________________________

____

____

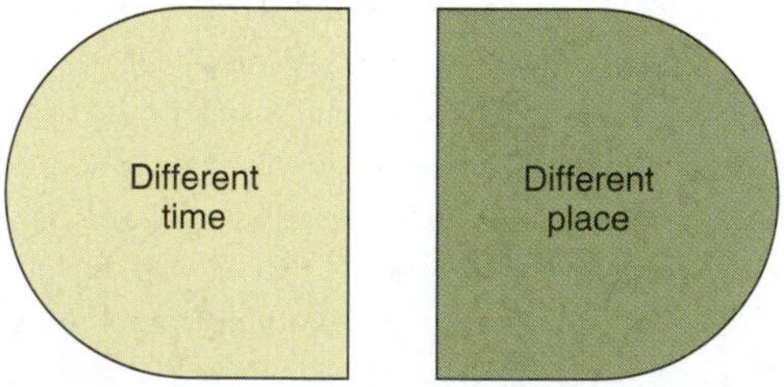

Different time | Different place

b. ______________________________

____

____

Same time

Same place

Different time

Same place

c. ______________________

d. ______________________

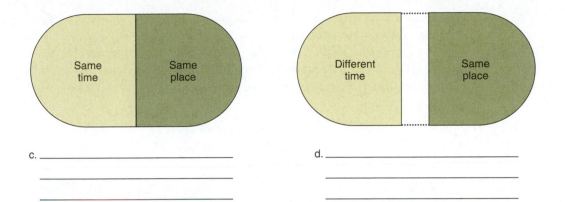

To learn more about workspace awareness on teams, go to www.english.wadsworth.com/burnett6e for a link to an article that summarizes a number of issues about computer-supported cooperative work.
CLICK ON WEBLINK
CLICK on Chapter 5/team awareness

WEBLINK

4. **Examine collaboration in your discipline.**
   (a) Survey a key professional journal in a specific discipline (your own field or another that interests you). Select three years — say, 1950, 1970, and 1990; 1960, 1980, and 2000; or 1983, 1993, 2003 — to determine changes in the percentage of coauthored articles. Present your findings in a table or a graph, and prepare a brief oral explanation of the results.
   (b) Interview a professional in your discipline (or one that interests you) to learn about the kinds of collaboration in which you'll be expected to participate. Prepare a series of questions based on the information presented in this chapter. Ask for specific examples of collaboration. In a paper for your classmates, discuss the information you learn. Be prepared to give an oral summary of your findings.
5. **Identify parallels in writing and design.** The processes that Omnica's industrial design team (highlighted earlier in the chapter) follows in designing products[25] are similar to those that good communication teams follow when designing effective documents and presentations. Identify at least six ways in which product design and development teams are similar to document design and development teams in processes and outcomes. Prepare a list with examples to share during class discussion.

6. **Draft a team agreement.** Successful collaborations often survive because the collaborators understand their roles and responsibilities. New collaborative relationships work better when people know what's expected of them. Draft a memo of understanding that details the expectations of all the people involved on your team. Here are some of the things that your agreement should address:

   - Understanding of the goals and outcomes the team will complete
   - Tasks each person is responsible for completing
   - Dates drafts and completed work are due
   - Ways collaborators will communicate: phone, e-mail, face-to-face meetings
   - Frequency of collaborators' communication with each other
   - Process for resolving disagreements
   - Order of authorship
   - Process of peer review
   - Quality of work expected

7. **Evaluate computer software.** Establish criteria for the effectiveness of a software program that enables some form of collaboration. Then use your criteria to determine whether the software is effective. Based on your own testing and review, write a short review suitable for a trade publication.

8. **What kind of tools that can aid collaboration have you used?**

   - ☐ E-mail
   - ☐ Group e-list messaging
   - ☐ Whiteboard
   - ☐ Electronic calendars
   - ☐ Threaded discussions
   - ☐ Shared databases
   - ☐ Track Changes tool for response and editing
   - ☐ Document archive and management
   - ☐ Electronic meeting rooms
   - ☐ Web conferences
   - ☐ Video conferences
   - ☐ Project and task management
   - ☐ Real-time chats
   - ☐ Group-decision support
   - ☐ ____________________
   - ☐ ____________________

9. **Track your own collaborative behavior.** One way to learn more about your collaborative processes is to track yourself as you prepare a technical document or oral presentation. The tracking process has three basic steps.

   (a) Keep a log (in a notebook or online) in which you record every time you work on the document or oral presentation (even just thinking about it counts). Indicate whether you are working individually or collaboratively. Record the date and the time you start and stop. Describe what happens during the time — what gets accomplished, what goes wrong, what you expected to get done but didn't (and why), and so on. Focus particularly on the interaction. Record the facts, but also jot down your attitudes and feelings.

(b) Review and analyze the log, looking for interesting incidents and patterns of behavior that you believe characterize your collaboration.

(c) Use your analysis to describe the nature of your own collaborative behavior. Prepare a detailed memo that focuses on no more than four specific aspects of your own collaborative behavior — for example, the way you approach tasks, the way you handle disagreements, and the way you approach leadership. Include some discussion about the way the collaboration affected the document or oral presentation you prepared.

Chapter 5 Endnotes

1 Courtesy of Sharon Burton-Hardin.

2 H1. (n.d.). *The H1 Experiment at HERA.* Retrieved October 3, 2003, from http://www-h1.desy.de/

3 Meadows, A.J. (1974). Chapter 7: Scientific Collaboration and Status. *Communication in Science.* London: Butterworth. p. 197.

4 National Science Board. (1998). *Science and Engineering Indicators 1998.* Chapter 5: Academic Research and Development: Financial and Personnel Resources, Integration With Graduate Education, and Outputs. Retrieved January 14, 2004, from http://www.nsf.gov/sbe/srs/seind98/c5/c5s4.htm

5 Omnica Product Development. (n.d.). *Industrial design.* Retrieved October 4, 2003, from http://www.omnica.com/industrial_design.htm

6 Methodist Health Care System Cystic Fibrosis Web site in Spanish and English. Retrieved October 4, 2003, from http://www.methodisthealth.com/spanish/respiratory/cystic.htm and http://www.methodisthealth.com/pulmonary/cystic.htm

7 Klineberg, S. (2002, September). *The Houston area survey 1982–present: Demographic transformations.* Retrieved October 5, 2003, from http://cohesion.rice.edu/centersandinst/has/hasreport.cfm?doc_id=1238

8 Walker, B. *Is corporate America kicking the travel habit? Poll finds collaboration technologies taking hold as predicted.* Retrieved October 4, 2003, from http://www.phoneplusmag.com/articles/2b1collaborate2.html.

Wainhouse Research (2002, September 4). *Wainhouse survey of business travelers tracks use of collaborative technologues.* Retrieved October 4, 2003, from http://www.wainhouse.com/prtravhab0902.html

9 Walker, B. *Is corporate America kicking the travel habit? Poll finds collaboration technologies taking hold as predicted.* Retrieved October 4, 2003, from http://www.phoneplusmag.com/articles/2b1collaborate2.html.

Wainhouse Research (2002, September 4). *Wainhouse survey of business travelers tracks use of collaborative technologues.* Retrieved October 4, 2003, from http://www.wainhouse.com/prtravhab0902.html

10 Courtesy of Allan Ackerson.

11 Berdondini, L., & Fantacci, F. (n.d.). *Training on listening and peer support.* Retrieved October 5, 2003, from http://www.comune.torino.it/novasres/_private/ascolto.pdf

12 A version of these questions was originally developed as part of the author's work for the Making Thinking Visible Project for the Center for the Study of Writing, Berkeley and Carnegie Mellon, funded by the Heinz Endowment of the Pittsburgh Foundation.

13 © 1998 Massachusetts Institute of Technology Alumni Association.

14 Modified from information at usabilityfirst.com: (1) Brinck, T. (1998). *Groupware: Introduction.* Retrieved October 4, 2003, from http://www.usabilityfirst.com/groupware/intro.txl; (2) Brinck, T. (1998). *Groupware: Applications.* Retrieved October 4, 2003, from http://www.usabilityfirst.com/groupware/applications.txl.

15 This scenario is from Haeussler, A. (n.d.). *Supporting groupware with awareness.* Retrieved October 4, 2003, from http://www.sapdesignguild.org/editions/edition5/awareness.asp.

16 Team awareness is a widely considered computer-supported cooperative work (CSCW) topic. Information from these sources contributed to the discussion in this chapter: (1) Ferscha, A. (n.d.). *Workplace awareness in mobile virtual teams.* Retrieved October 4, 2003, from http://www.soft.uni-linz.ac.at/Research/Publications/_Documents/WETICE2000.pdf; (2) Jang, C. Y., Steinfield, C., & Pfaff B. (December 2, 2000). *Supporting awareness among virtual teams in a web-based collaborative system: The TeamSCOPE system.* Retrieved October 4, 2003, from http://216.239.37.104/search?q=cache:xajSfQOCiQsJ:cscw.msu.edu/reports/cscw2000.pdf+%22Supporting+awareness+among+virtual+teams+in+a+web-based+collaborative+system:+The+TeamSCOPE=system%22&hl=en&ie=UTF-8

17 Modified from information at (1) Brinck, T. (1998). *Groupware: Design issues.* Retrieved October 4, 2003, from http://www.usabilityfirst.com/groupware/design-issues.txl; (2) Haeussler, A. (n.d.). *Supporting groupware with awareness.* Retrieved October 4, 2003, from http://www.sapdesignguild.org/editions/edition5/awareness.asp

18 Ford, J. (2003, May). *Integrating the internet into conflict management systems.* Retrieved October 4, 2003, from http://www.mediate.com/articles/ford10.cfm

19 Adapted from Gillette, B., Johnson, R., Polashek, E., Thornburg, J., & White, C. *The art of working together: A guide to collaboration* (p. 5), Ames, IA: Iowa State University, Department of English.

20 Cohen, J., & Taubles, G. (1995). The culture of credit. *Science, 268*(5218), 1707.

21 Cohen, J., & Taubles, G. (1995). The culture of credit. *Science, 268*(5218), 1706.

22 Haskins, M., Liedtka, J., & Rosenblum, J. (1988). Beyond teams: Toward an ethic of collaboration. *Organizational Dynamics, 26*(4), 34–51.

23 Dilbert reprinted by permission of United Features Syndicate, Inc.

24 See Bosley, D. S. (1993). Whose culture is it anyway? *Technical Communication Quarterly,* (Winter), 186–200.

25 Omnica Product Development. (n.d.). *Ergonomics is more than a buzzword.* Retrieved October 4, 2003, from http://www.omnica.com/omniview_industrial_design2.htm

PART II

Managing Critical Processes

CHAPTER 6

Locating and Using Information

Objectives and Outcomes

This chapter will help you accomplish these outcomes:

- Identify and use electronic database resources relevant to your research
- Execute a specific, targeted electronic search using advanced search techniques
- Gather research data using several nonelectronic techniques
- Integrate research successfully into your text
- Avoid plagiarism

This chapter is about the beginning steps of *knowledge management,* which, broadly defined, is a system for finding, selecting, organizing, synthesizing, and using information in order to improve your audience's understanding of a particular topic.[1] Your success as a communicator depends in large part on your ability to master and use the information in this chapter, to turn it into knowledge for yourself. More specifically, the quality of your communication depends on your care and thoroughness in effectively using appropriate information. The credibility and completeness of this information affect the audience's acceptance of your points. The following diagram shows one way to think of the knowledge management process, which may be a new name for a process you already use.[2] Although this process is displayed as linear, it is in fact recursive, with the possibility of returning to any earlier step at anytime during the process.

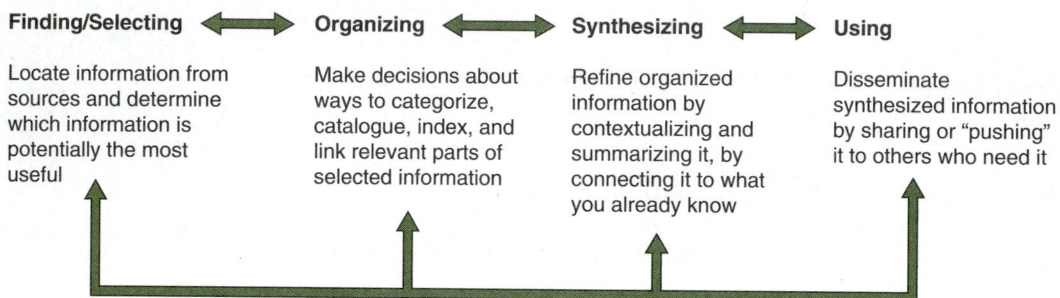

Finding/Selecting
Locate information from sources and determine which information is potentially the most useful
Organizing
Make decisions about ways to categorize, catalogue, index, and link relevant parts of selected information
Synthesizing
Refine organized information by contextualizing and summarizing it, by connecting it to what you already know
Using
Disseminate synthesized information by sharing or "pushing" it to others who need it

What's the difference between information and knowledge? The two terms are often used interchangeably, but many workplace professionals prefer not to use them as synonyms. *Information* is what we use as the building blocks for knowledge. We construct information from what we read, observe, hear, and experience in many other ways. *Knowledge* is what we *know.* It involves thinking about information, learning information, interacting with others as we shape and are shaped by information, and, probably most important, connecting information to other aspects of our prior knowledge. We construct both our tacit, unarticulated knowledge and our explicit, articulated knowledge. We express our knowledge in what we say, write, and design as well as in how we act — gestures, body language, and actions, including the attitudes we express and the processes we use.[3]

This chapter is about one of the first steps in knowledge management: finding and selecting information from a variety of sources. The following examples are typical of situations in which workplace professionals might collect information for technical documents, presentations, or visuals:

- A biomechanical engineer calculates the expected reduction in noise pollution in a call center office from using cubicles made primarily from recycled rubber and sends the results on to the marketing team.

- A polymer chemist spends several hours searching a computerized database to locate information for use in an internal proposal about new techniques to strengthen carbon fibers.
- An industrial engineer designs a survey for technicians as part of a study to assess the effectiveness of industrial degreasers, which will contribute to decisions about whether to change the current formula.
- A technical communicator has been corresponding with his senior application designer about changes made to the new version of their storage server management application, which he is documenting.
- A financial analyst examines online government documents to check details of recent changes in the tax code, which affects recommendations to the writers and graphic designers working on the organization's annual report.
- A hydrologist has spent 30 minutes reading abstracts in *Pollution Abstracts and Water Resources Abstracts,* an index that includes research on contamination of ground water, in order to decide what articles to read in their entirety.

How does each situation require knowledge management as it's defined on the preceding page?

What situations in your discipline require professionals to collect information?

Careful researchers collect information from multiple sources so that they can *triangulate* their data — that is, they can determine if they obtain the same or similar results from different sources. Nancy's team collected data from a range of sources including interviews, laboratory experiments, observations, and product analyses.

© Double Exposure/Getty Images

© Spencer Grant/PhotoEdit

© David Young-Wolff/PhotoEdit

© AP Photo/Michael Stravato

© Comstock Images/Alamy

For major projects in your own discipline, you'll need to collect information in the workplace as well as in the library. For example, Nancy Irish, a nutritionist who has worked for government agencies as well as private groups, is investigating the ecological and nutritional benefits of soy protein and its potential for greater use in U.S. diets. Her preliminary work involves academic research, using library resources to review available literature about soy proteins and tofu. Then she conducts interviews and surveys to determine people's attitudes toward tofu and obtains tofu samples from local stores and food cooperatives to determine availability and quality. Some of her research techniques are included as examples in this chapter.

Locating Primary and Secondary Sources

Technical professionals can collect information and data from a variety of primary and secondary sources. Primary source information is first-hand information reported by people directly involved with an action or event, not interpreted by a second or third party. Secondary source information does not come directly from people involved in the action or event. Instead, the information is interpreted and reported by a second or third person. Most information has the potential to be both a primary and a secondary source. For example, if an engineer conducts an experiment and then reports her findings, her technical report is a primary source. However, if that same engineer is interviewed about her experiment by a technical writer who writes an article about the experiment, that article is a secondary source.

Whether you decide to use primary or secondary sources, check the credentials of the source, the sponsoring agency, and the date the information was obtained because both primary and secondary sources can potentially be slanted or biased. For example, two firefighters who worked to put out a chemical fire in a warehouse could give first-hand, primary reports about fighting chemical fires. But if one of the firefighters had been injured during the chemical fire, that firefighter might give a different account of the specific fire and have a different opinion about safety in firefighting. On the other hand, a local politician supported financially by the company where the fire occurred might try to use the primary sources to downplay the company's accountability for the fire.

Another important point to consider when planning a technical document, presentation, or visual is whether it calls for quantitative or qualitative data, or both, as evidence to support your positions. *Quantitative data* describe measurable elements (quantities) in mathematical or statistical terms. *Qualitative data* describe observed or reported information (qualities). However, after qualitative data are recorded, the information can sometimes be treated quantitatively. For example, a field biologist might record observations of a cane toad population over a period of months (qualitative data). The number of observations of a particular phenomenon could then be counted (quantitative data).

The following list identifies some of the ways of managing the information you collect that this chapter discusses:

- *Print and electronic resources.* You can search, read, and analyze information from print resources (college, corporate, and community library resources and government publications) and large digital databases (library and government databases, as well as Web sites).
- *Internal records.* You can analyze an organization's proprietary records, both print and electronic, in order to gather information about practices and activities.
- *Personal observations.* You record and analyze information that you see, smell, taste, or feel in order to document a process or experience first-hand.
- *Interviews and letters of inquiry.* You can use face-to-face meetings or correspondence with knowledgeable people to gather information not readily available.
- *Surveys and polls.* Using a list of targeted questions, you can gather information from a large representative sample of people.

- *Empirical research.* You can collect information by conducting empirical research, which is an observational study or an experiment to answer a question, address a problem, or test a hypothesis or observable pattern. During empirical research, you collect various kinds of data, including physical samples and specimens that you categorize and analyze in order to identify features, and use standard formulas to help analyze, interpret, and assess your data. You record and categorize information, analyze and interpret data in ways that make sense to others, and then develop various narratives to explain your interpretation.

Empirical and *experimental* are not synonyms: Empirical research is the superordinate category; experimental research is one of the subcategories. What is an example of a research project in your professional area that is empirical but not experimental? What is an example of a research project that is empirical and also experimental?

All of these common ways to collect information are not mutually exclusive. For example, you could conduct an empirical investigation that involves an analysis of an organization's internal records. During a face-to-face interview, you could make specific reference to that person's Web site and online publications. The primary value of this list is in providing you with a range of options for collecting technical information when you have a problem to solve. The next section of the chapter focuses on electronic resources.

WEBLINK

Keeping a list of useful portals will make your work life much easier. Go to www.english.wadsworth.com/burnett6e for links to useful portals.
CLICK ON WEBLINK
CLICK on Chapter 6/portals

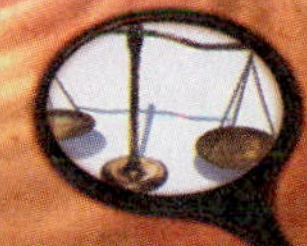

ETHICS SIDEBAR

On-the-Job Pressures: Ethics and Technology in the Workplace

Consider the following on-the-job actions:

- Sabotaging the systems or data of a current or former coworker or employer
- Accessing private computer files without permission
- Listening to a private cellular phone conversation
- Accessing pornographic Web sites using office equipment
- Using office equipment to search for another job
- Copying the company's software for home use
- Wrongly blaming an error on a technological glitch

Have you engaged in any of the unethical actions listed on the previous page?

Did you consider them unethical at the time?

What would be the most effective deterrent to keep you from engaging in them in the future?

- Making multiple copies of software for office use
- Using office equipment to shop on the Internet for personal needs

Which of these actions would you find unethical? All of them are unethical, according to the Ethics Officer Association (EOA), the professional association for managers of corporate ethics and compliance programs. In 1998, the EOA, along with the American Society of Chartered Life Underwriters and Chartered Financial Consultants, conducted a nationwide survey of American workers "to measure whether or not the presence of new technologies in the workplace increases the risk of unethical and illegal business practices."[4] Nearly half (45%) of the respondents to the survey indicated they had engaged in at least one unethical act.

The survey found that business and technical professionals feel pressured by new workplace technologies. The pressures caused by workplace technologies include increased productivity expectations; the continual need to change or update the technology; increased frustration with coworkers who are not up to date on the technology; less tolerance for errors; inadequate manuals and/or training; and the fear of losing data. Female respondents felt more pressure (70%) than male respondents (51%), largely due to a "lack of understanding terminology or lingo." "New technologies have changed the way we do our jobs and the way we work with one another," said EOA Executive Director Ed Petry when the survey results were announced. "We're expected to do more, to do it faster, and we're also working more independently."

What are some solutions to these ethical problems? Survey respondents offered several, including creating guidelines for the personal use of company technology resources, installing Internet-blocking software, and encouraging employees to police themselves. For technical professionals, perhaps the best individual advice is to be aware of how these actions can cause unnecessary workplace conflicts. Avoiding these conflicts will help make workplace technology more of an asset and less of a liability. (See discussion of unproductive conflicts in Chapter 5.)

Finding Information Using Electronic Resources

Odds are that whenever you begin research for a new paper or project, you probably start on the World Wide Web or in your university's online catalog. But less than 15 years ago, electronic resources were available only for a fee and with the assistance of a specially trained librarian. Now, electronic resources are widely available — not only in public, college, and corporate library databases, but also through the Internet. Knowing where to look for information can be a daunting task in the face of the thousands of available research databases and millions of public Web sites. The following section describes several types of electronic databases and catalogs where you can begin your research; the section after that offers specific pointers to help you search these electronic resources effectively.

Scientific and technical databases are remarkably important when you're searching for information. Go to www.english.wadsworth.com/burnett6e for a link to a detailed discussion of databases.

CLICK ON WEBLINK

CLICK on Chapter 6/databases

WEBLINK

Finding Electronic Resources

Your college, company, and community libraries, as well as your link to the Internet, give you access to a vast array of computerized databases and indices, which are searchable digital repositories of published material. Electronic databases have largely replaced print card catalogs and reference books in most well-funded libraries because they can be far more easily and cheaply updated on a regular basis. Because available information is increasing at a rate faster than anyone can manage to read it, professionals in most fields learn to use computerized databases that give them an overview of current information and research in their fields. Database searches can be helpful to you in several different ways:

- To prepare a literature review
- To stay abreast of current information and research
- To locate answers to specific questions
- To locate resources in the form of citations, abstracts, or full-text articles

When using electronic resources, you receive specific types of information for each returned "hit." You receive the name of the book or article, its author(s), and the date of publication. Other bibliographic information depends on the database. For example, library catalogs often include the publication's call number, ISBN or ISSN, and location in the library, as well as the category of a particular book. External databases generally do not provide call numbers but do identify where the article was originally published, so you can locate it in the library.

Most electronic databases return three types of hits to your topical search: citations, abstracts, and full-text articles. A *citation* provides only bibliographic information such as title, author(s), and journal or publisher; to use the article or book you must locate it in the original form. An *abstract* lists bibliographic information about the article and a short summary, either provided by the author(s) or written by a database indexer. Abstracts can be used to eliminate articles that are not related or are only tangentially related to your research, which can save a great deal of time when you are conducting a broad search. And finally, a *full-text article* is usually an HTML version of the entire article as it was published in the original setting, including the full bibliographic citation. Full-text articles are easy to print and use immediately but can pose problems when you attempt to appropriately cite page numbers from the article in your final document. Whether your search returns a citation, abstract, or full-text article depends on the database, the library holdings, and that company's agreement with the original source.

When researching a new topic, consider searching electronic sources in the following order:

- Library's online catalog
- Online reference materials
- General reference databases
- Discipline-specific databases
- Government databases
- The Internet and World Wide Web

What do you think is the rationale for this order of investigtion?

What specific electronic resources are available in your own college or university library?

Online Catalog. The online catalog in a specific library is a database that lists all of the library's holdings: books, bound periodicals (a year's issues of a particular journal bound into a single volume), audiovisual (AV) holdings (films, records, audio- and videotapes), and electronic resources (e-books, e-journals).

You can use the online catalog to determine if the library has books, bound periodicals, or AV materials relevant to your subject.

Each book or other item is assigned an individual number based on either the Dewey decimal system or the Library of Congress classification system. Public libraries generally use the Dewey decimal system, whereas college and university libraries use the Library of Congress system. The Dewey decimal system divides books into ten categories, each identified by a range of numbers. The Library of Congress system divides books into twenty categories, each identified by a letter followed by specific numbers. The broad letter categories represent major subject areas.

In both the Dewey decimal and the Library of Congress systems, the broad categories are further subdivided until each book has an individual number — the call number — that helps you locate specific volumes in library stacks. Being familiar with call numbers enables you to go into the stacks and browse though the books in your specific field of interest. Browsing does not constitute efficient use of time, but you often find volumes of interest that you might have ignored had you only read the online information about a book.

© Frank Chmura/Index Stock Imagery

Nancy Irish began her search using the online catalog at Iowa State University in order to locate current resources about nutritional benefits of soy protein. Following the on-screen directions, she had several choices: She could search for information by author, title, subject, or keyword. The keyword option is the one Nancy decided to use. Figure 6.1 shows selected screens that appeared when Nancy was conducting her search.

FIGURE 6.1 Online Catalog Screen[5]

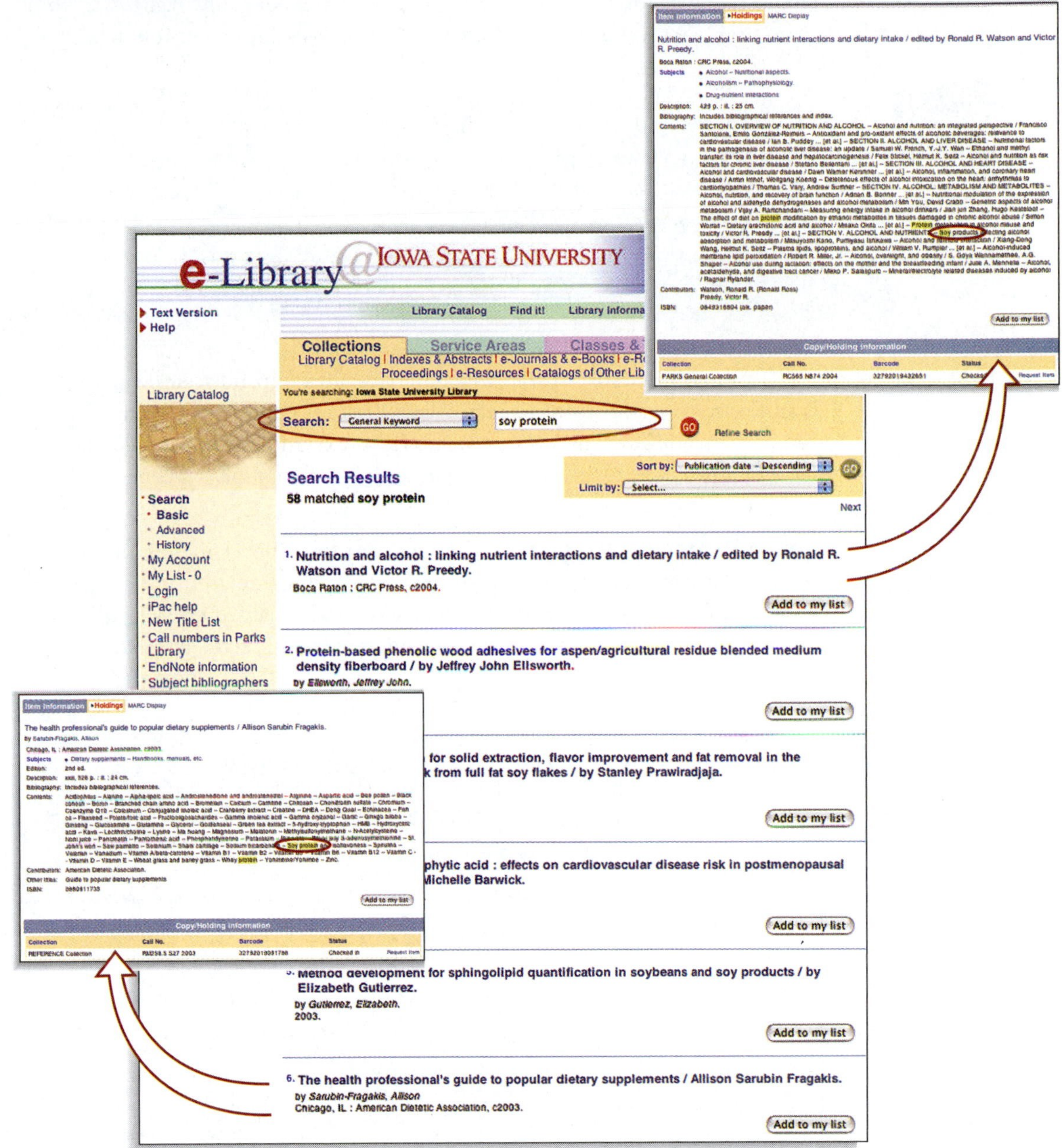

The *Library of Congress Guide to Subject Headings* is also valuable to use. After one or two attempts at identifying the topic(s) under which a subject is cataloged, check this guide for alternative labels and related areas of study. Figure 6.2 reproduces a section of the guide used by Nancy Irish during her research.

WEBLINK

To learn more about the Library of Congress subjects headings, go to **www.english.wadsworth.com/burnett6e.** This page defines and explains the relationships that are common in the subject headings, which will help you use the entries more productively.

CLICK WEBLINK

CLICK Chapter 6/Library of Congress

If you are interested in searching library catalogs beyond your own, WorldCat is an electronic service that links the catalogs of over 23,000 libraries in 63 countries. WorldCat allows you to search for any text found in the holdings of these libraries and provides bibliographic and interlibrary loan information for all listed texts.

Online Reference Materials. The majority of reference indices and materials used by reference libraries are online and available through electronic library resources. Two valuable references, handbooks and specialized encyclopedias, give a general overview of a topic before you continue more specific research.

FIGURE 6.2 **Excerpt from *Library of Congress Guide to Subject Headings*[6]**

"May Subd Geog" → "May Be Subdivided Geographically"
UF → "Used For"
BT → "Broader Topic"
RT → "Related Topic"
SA → "See Also"
NT → "Narrower Topic"
Topics can be tagged on to the subject headings to narrow the subject search, signaled by dashed lines (- -).

Soy
  USE Soy sauce
Soy-bean
  USE Soybean
Soy-bean as feed
  USE Soybean as feed
Soy-bean as food
  USE Soyfoods
Soy-bean flour
  USE Soybean as flour
Soy-bean meal
  USE Soybean meal
Soy-bean meal as feed
  USE Soybean meal as feed
Soy-bean milk
  USE Soymilk
Soy-bean mosaic virus
  USE Soybean mosaic virus
Soy-bean oil
  USE Soy oil
Soy... oil mills

  UF Soybean oil industry
    [Former heading]
  BT Oil industries
    Soybean industry
Soy proteins *(may Subd Geog)*
  UF Proteins, Soy
  BT Plant proteins
    Soybean products
  NT Textured soy proteins
Soy pulp
  USE Okara
Soy sauce *(may Subd Geog)*
  [TP438.56 (Manufacture)]
  [TX407.569 (Nutrition)]
  UF Sauce, Soy
    Soy
    Soya sauce
  BT Fermented soyfoods
  RT Cookery (Soy sauce)
Soy sauce industry *(may Subd Geog)*

Soybean curd
  USE Tofu
Soybean cyst nematode *(May Subd Geog)*
  [QL391.N4 (Zoology)]
  [SB9098.S68 (Pest)]
  UF Heterodera glycines
    [Former Heading]
    Soybean nematode
  BT Heterodera
Soybean flour
  USE Soy flour
Soybean glue
  BT Glue
    Soybean products
Soybean hulls
  USE Soy bran
Soybean industry *(May Subd Geog)*
  [HD9235.S6-HD9235.S62]
  BT Vegetable trade
  NT Soy oil industry

Handbooks usually present compact, carefully organized information in tables, charts, diagrams, graphs, and glossaries. Handbooks in both print and electronic format are available in astronomy, biological science, biomedical science, chemistry, computer science, engineering, environmental science, geoscience, mathematics, physics, and virtually every other major field of technical and scientific inquiry. Electronic handbooks include *CHEMNetBASE* for chemists and *ENGNetBASE* for engineers; however, using the print form may be easier since you can flip to the section very quickly.

Specialized encyclopedias and a variety of business and industry guides are invaluable for initial inquiries because the entries provide an overview, summarizing and discussing essential facts and theories about a subject. The entries also often present statistical data in various visual forms, provide historical background, and list bibliographic references. Encyclopedias for a wide range of technical topics are published in both print and electronic versions. For example, the *Encyclopedia of Life Science* and the *Encyclopedia of Virology* are available in both print and electronic forms.

General Reference Databases. Once you have completed a general topic search in the library catalog and reference books, you can do a topic search to look for published articles in one of the common general reference databases available at most college, corporate, and community libraries. Libraries usually pay a per-student fee to gain permission to link to the online database through their computers or Web site. These reference databases, specifically designed for undergraduate researchers, cull citations and articles from a variety of academic journals, periodicals, and newspapers. They can provide a good overview of a topic from popular academic journals and the popular press but generally do not index the more research-oriented technical publications and journals. Three common general reference databases found in libraries are *Expanded Academic ASAP, LexisNexis,* and *Web of Science.*

- *Expanded Academic ASAP* is a highly generalized database that indexes citations, abstracts, and full-text articles from journals, periodicals, and national newspapers in a wide variety of disciplines, including the humanities, social sciences, general sciences, and technology.
- *LexisNexis* is a general topical database that pulls articles, usually in full text, from global newspapers, business and news periodicals, and government publications. It is an excellent resource to investigate current topics or issues that are frequently in the news, such as cloning or global warming.
- *Web of Science* is a metaindex of over 8,500 journals that searches three smaller indexes: *Science Citation Index Expanded, Social Sciences Citation Index,* and *Arts & Humanities Citation Index.* It provides citations and abstracts, as well as the list of works cited for each article. This allows you to search for articles that cite a particular article. Accessing these databases can provide you with an overview of a topic so you can narrow and define your research as you work through a project.

Discipline-Specific Databases. Most academic disciplines have electronic databases that index research and publications in that discipline. Some are published by professional organizations in the field, others by for-profit companies. These databases allow you to search publications and reports specific to the field and often include publications that are not listed in more general databases. Depending on your library's funding, you may be able to find many of these resources in print, CD, and online format. These are only a few of the hundreds of discipline-specific indexes available:

- *AGRICOLA* (AGRI) is a database of the National Agricultural Library, indexing more than 3,300,000 citations in the agricultural literature since 1970.
- *Avery* is an index of architectural periodicals, covering architecture, interior design, and city planning since 1977.
- *Compendex* is a general database for engineering, known in its print version as the *Engineering Index.*
- *MEDLINE* is a general index of the National Library of Medicine, covering over 4000 international biomedical journals.
- *NTIS* (National Technical Information Service) is a database of unrestricted technical reports for U.S.- and non-U.S.–government-sponsored projects.
- *ScienceDirect* is a general, massive, electronic database covering all sciences and some social sciences.

In many cases, several abstracting indexes are available for a single discipline. Because these indexs may not draw from the same sources, you should check the publications to ensure that particular journals are covered. For example, biology abstracts and citations can be found in many indexes, including *Biological Abstracts, Biological Abstracts/RRM, Biological and Agricultural Index, Biology Digest,* and *BioOne.*

Government Documents and Offices. Another source for valuable information on a variety of topics is the U.S. government, which is required by law to publicly post many of its documents, research findings, and proceedings. Virtually all agencies of the U.S. government have public Web sites that are updated regularly, making their information easily and cheaply accessible. For example, the U.S. Census Bureau has the most recent census online (http://www.census.gov), the FBI posts its Most Wanted list (http://www.fbi. gov/mostwant.html), and the Federal Aviation Administration publishes many of its regulations (http://www.faa.gov). Several other government sites are interesting and useful:

- Federal Deposit Insurance Corporation (FDIC) — http://www.fdic.gov
- Fish and Wildlife Service — http://www.fws.gov
- Library of Congress — http://www.loc.gov
- Patent and Trademark Office — http://www.uspto.gov

- Securities and Exchange Commission (SEC) — http://www.sec.gov
- Smithsonian Institution — http://www.si.edu

An extremely useful resource for federally funded government research is supported by the Office of Scientific and Technical Information (http://www.osti.gov), which enables users to search across multiple collections of preprint, noncommercial, and peer-reviewed journal literature with a focus on physical sciences and technology. For example, the *EnergyFiles* collection (http://www.osti.gov/EnergyFiles/) provides information, tools, and technologies to facilitate the use of scientific resources in planning and conducting energy-related research from over 500 databases and Web sites.

While some government documents that may be of value to you may not be available electronically, many national repositories for government documents have electronically indexed their holdings. For example, Iowa State University (ISU) is one of many national repositories; its electronically indexed holdings are searchable through ISU's library catalog.

Which repository of government documents is closest to you?

Web Research. While the terms *Internet* and *World Wide Web* are often used synonymously, they actually refer to two different things. The Internet is the technical infrastructure over which content in the form of Web pages, chat relays, and newsgroups flows.[7] The Web enables anyone with access to a Web server (a computer that connects a group of users to one another and to the Web) to publish documents on it. Because of the Web, very little time is needed to make new information available around the world, with minimal printing or delivery costs.

© Courtesy of NASA

Photographs from the Hubble telescope were made available by NASA on the Web minutes after they were taken. In contrast, most of a day is needed for photographs to be published in newspapers, months to be published in professional journals, and sometimes years to be published in scholarly books.

What are the trade-offs to online publication, which makes information available to users more quickly than traditional print media?

Search engines, both general and discipline-specific, are a good place to start doing Web research. Many engines are available online today, but some of the most reliable and well-respected include www.google.com, www.yahoo.com, www.excite.com, and www.metacrawler.com. Disciplines may also have their own free or subscription-based search engines on the Web, which can usually be located through a discipline's professional organizations. For example, SPIN Web (http://ojps.aip.org/spinweb/) is a subscription search engine for physics journals and research records, and EEVL (www.eevl.ac.uk) is a free search engine for engineering, mathematics, and computer science sites.

Newsgroups allow users to correspond by "posting" and "replying to" messages from other users on the Internet through a special newsreader interface. Lists (e.g., LISTSERV and Majordomo) are automatic mailing list servers that transmit e-mail only to individuals who have subscribed to the list. Newsgroups are separated into more than 25,000 different subjects. Their titles usually begin with a prefix followed by a period and then a suffix such as *alt, soc, comp, misc, rec,* or *sci.* Lists are often associated with professional organizations or particular professions. (See Chapter 13 for netiquette for newsgroups and lists.)

Searching Electronic Resources

The amount of information available in print and online is astounding — and growing exponentially every year. Researchers at Berkeley have estimated that between 1 and 2 exabytes (an exabyte = one billion gigabytes) of unique information is produced by the world every year, and printed documents of all kinds make up only 0.03% of the total.[8] Corporations, universities, research institutions, and even governments are moving more and more print information to online formats, which makes searching for information both easier and more complicated. With these vast electronic resources available, you must develop advanced electronic search techniques to effectively sort and select relevant information. This section explores techniques that can help you more effectively search, sort, and find information in electronic resources.

We often don't think about "the best way" to execute a search as being dependent on the particular preferences of a search engine, but understanding how to effectively word and punctuate a request can save you a great deal of time. Before you begin a search in a new database or electronic resource, take a minute to read the Help or About page that is usually located in the main navigation bar of the site. Nearly all respectable electronic resources have these pages, which are designed to give users suggestions about the best ways to complete effective and targeted searches. The Help page tells you which search strategies and punctuation are most effective in that resource and explain how to limit or broaden your searches using search menu options.

Once you are familiar with a particular electronic resource, you should also consider ways to approach your search. Developing a search strategy helps you focus your ideas so that you get information that is useful, but it also saves you time (and money). Consider the following critical questions:

- What are your key terms? What are synonyms for these terms? What alternative terms are used to refer to the topic?
- What indexing source does the database use?
- How do you want to combine the terms?

Many electronic resources allow to you search in a variety of ways, including by keyword, subject, author, or title. Author and title searches allow you to quickly find pieces by one author or a particular piece. But more often than not, you will not know a specific author or title; therefore, you will use a keyword or subject search. Though they may seem like synonymous terms, keyword and subject searches return very different results.

Before you read the next two subsections, consider why keyword and subject searches return different results.

Keyword Searches. *Keywords* are flexible search terms that can be words, parts of words, phrases, or some combination of these that relate to your topic. You can generate a list of possible keywords when you develop your topic search strategy; consider not only exact topic phrases but also synonyms or related words that might generate hits in the database. Use keyword searches at the beginning of your search when you are trying to get a feel for the type of information that is available about your topic. Because keyword searches search the database for particular words, not specific topics, they often return results that might not be directly related to your topic. At that point you can use the search limiters offered by the resource or add additional keywords to narrow the field.[9]

To complete a keyword search, choose "keyword" from the resource's search menu and type your keywords in the search field. Figure 6.3 illustrates some advanced and Boolean search techniques you can use to combine or eliminate keywords in your search, depending on the preferences of the database as noted on its Help page.

Subject Searches. Subjects are different from keywords because *subjects* are predetermined categories that you can search within the database. Figure 6.4 shows how subject and keyword searches produced different results for Nancy Irish. Many databases use the subject headings defined by the U.S. Library of Congress. These subject headings are listed in alphabetical order in the *Library of Congress Subject Headings (LCSH),* which is available in both print and online editions. The *LCSH* lists both the subject headings and other common phrasings with pointers to the appropriate subject heading. Some disciplines have their own published subject listing guides, so be sure to ask your reference librarian about these resources. To complete a subject search, browse through the *LCSH* or the subject guide provided by the electronic resource, choose Subject Search from the search menu, and type the exacting wording of the subject heading in the search field.[10]

While keyword, subject, author, and title are the most common criteria for general topic searching, many electronic resources offer several ways to further

FIGURE 6.3 **Advanced and Boolean Search Limiters[11]**

Be sure to read the Help or About page for each database to determine which advanced search techniques a particular search engine recognizes.

Search Goal	Strategy	Example	Variation
Search for exact keyword phrase	Group words by enclosing in quotation marks or linking words with underscores	"BT plant proteins"; BT_plant_proteins	
Search for multiple keywords in same document	Type all keywords without punctuation or group terms by Boolean AND	Soymilk Soyfoods Tofu; Soymilk AND Soyfoods AND Tofu	Some engines use + rather than AND
Exclude keywords from search	Use the Boolean AND NOT	Soyfoods AND NOT Tofu	Some engines use − rather than AND NOT
Include spelling or synonym variations	Use the Boolean OR and/or parentheses	Tofu AND (soybeans OR soy-beans)	The combination or elimination of OR or parenthesis depends on the search engine
Search for variation of word endings	Use * after the stem	Soy* will return all results including Soybean, Soymilk, Soyfood, etc.	

limit your hits. For example, many resources allow you to specify important parameters:

Which parameters are important for searches you might conduct?

- *Time frame.* Do you want information from the past ten years? Five years? Only the current year?
- *Types of materials.* Do you want only books? Articles? Articles in refereed publications? Audiovisual materials? Presentations?
- *Language(s).* Do you want materials only in English? Also in German? French? Russian? Spanish? Chinese?
- *Print format.* Do you want only the titles? The full citation? The complete record with the abstract if it's available? The full text?

Developing a search strategy and using these advanced search techniques will help you make effective use of your time.

FIGURE 6.4 Comparison of Search with Subject and Keyword[12]

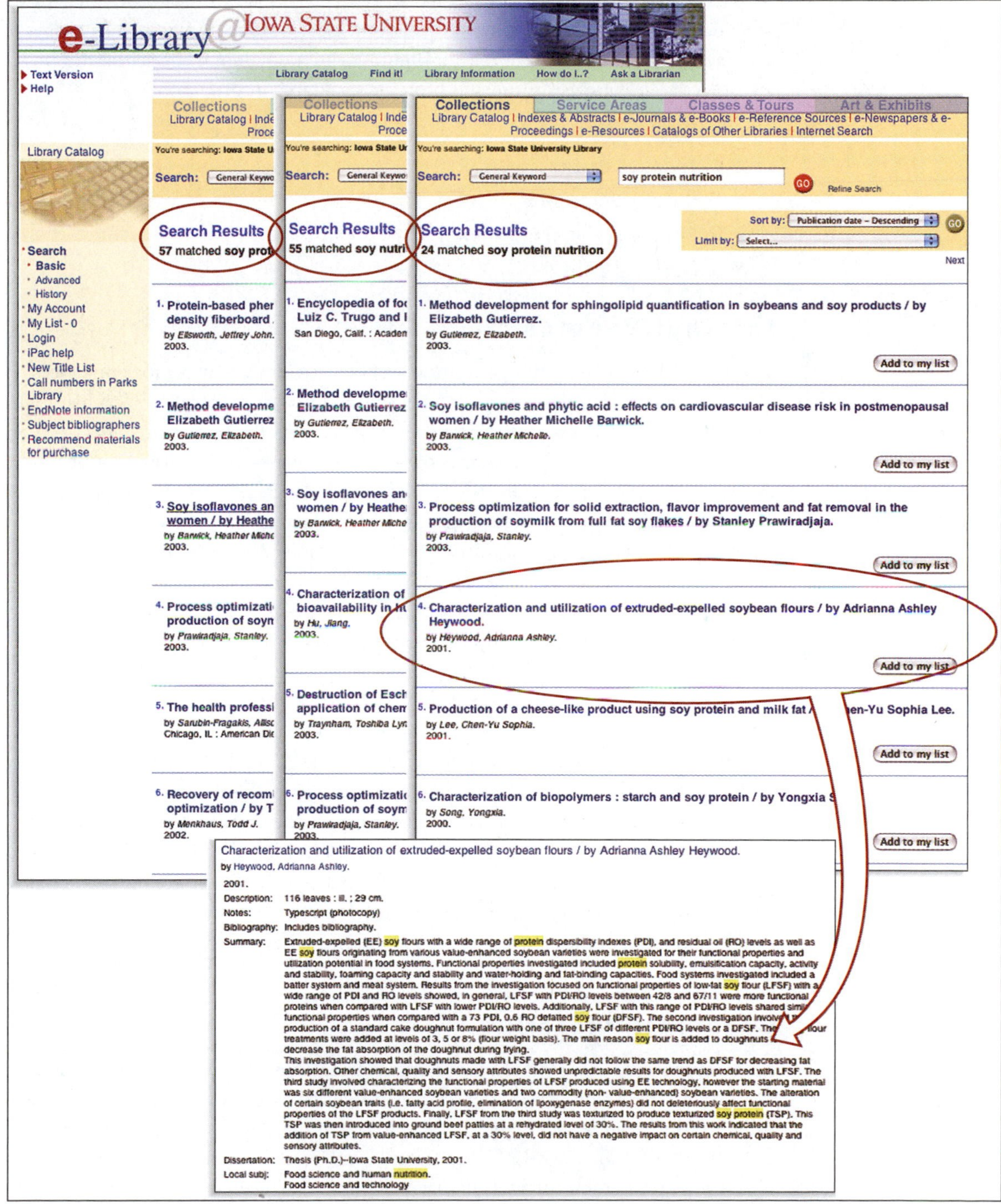

Characterization and utilization of extruded-expelled soybean flours / by Adrianna Ashley Heywood.

by Heywood, Adrianna Ashley.

2001.

Description: 116 leaves : ill. ; 29 cm.

Notes: Typescript (photocopy)

Bibliography: Includes bibliography.

Summary: Extruded-expelled (EE) soy flours with a wide range of protein dispersibility indexes (PDI), and residual oil (RO) levels as well as EE soy flours originating from various value-enhanced soybean varieties were investigated for their functional properties and utilization potential in food systems. Functional properties investigated included protein solubility, emulsification capacity, activity and stability, foaming capacity and stability and water-holding and fat-binding capacities. Food systems investigated included a batter system and meat system. Results from the investigation focused on functional properties of low-fat soy flour (LFSF) with a wide range of PDI and RO levels showed, in general, LFSF with PDI/RO levels between 42/8 and 67/11 were more functional proteins when compared with LFSF with lower PDI/RO levels. Additionally, LFSF with this range of PDI/RO levels shared sim… functional properties when compared with a 73 PDI, 0.6 RO defatted soy flour (DFSF). The second investigation involv… production of a standard cake doughnut formulation with one of three LFSF of different PDI/RO levels or a DFSF. The … flour treatments were added at levels of 3, 5 or 8% (flour weight basis). The main reason soy flour is added to doughnuts … decrease the fat absorption of the doughnut during frying.
This investigation showed that doughnuts made with LFSF generally did not follow the same trend as DFSF for decreasing fat absorption. Other chemical, quality and sensory attributes showed unpredictable results for doughnuts produced with LFSF. The third study involved characterizing the functional properties of LFSF produced using EE technology, however the starting material was six different value-enhanced soybean varieties and two commodity (non- value-enhanced) soybean varieties. The alteration of certain soybean traits (i.e. fatty acid profile, elimination of lipoxygenase enzymes) did not deleteriously affect functional properties of the LFSF products. Finally, LFSF from the third study was texturized to produce texturized soy protein (TSP). This TSP was then introduced into ground beef patties at a rehydrated level of 30%. The results from this work indicated that the addition of TSP from value-enhanced LFSF, at a 30% level, did not have a negative impact on certain chemical, quality and sensory attributes.

Dissertation: Thesis (Ph.D.)--Iowa State University, 2001.

Local subj: Food science and human nutrition.
Food science and technology

Finding Information Using Other Resources

While electronic databases and the World Wide Web provide enormous amounts of valuable (and not so valuable) information, electronic resources cannot always provide specific data to support your argument in a document, presentation, or visual. A variety of other sources also provide valuable information:

- Internal records
- Corporate libraries
- Personal observations
- Interviews and letters of inquiry
- Surveys and polls

Internal Records

The internal records of an organization can be excellent resources for employees and researchers. Many organizations are required by law to keep detailed historical records of their business, either on paper or electronically. Internal records — the data an organization keeps about its own transactions — often serve as supporting material in documents, presentations, and visuals. Such records might provide facts and details about finances, personnel, manufacturing, marketing, or shipping. For example, in submitting a memo proposing the addition of three employees to his sheet metal department, a departmental manager uses a variety of data, including personnel information about salaries and benefits, monthly productivity sheets, backlogged orders, and projected sales figures. His proposal wouldn't have any impact if he merely said his department needed three additional workers. All the evidence he needs to support his proposal is available in the files and internal documents of his own company.

Internal records must be analyzed for information usability and appropriateness before you can use them. Often, historical data need to be interpreted with respect to actual outcomes before they can be used effectively. You must also consider the legal and ethical implications of using internal records, particularly in a document intended for external readers. Many companies require key employees to sign nondisclosure or confidentiality agreements (usually considered legally binding) to restrict the dissemination of proprietary information.

Corporate Libraries

Many organizations, such as law and engineering firms, build extensive resource libraries for their employees. The services offered by large corporate libraries, often called information services, give a clear picture of the importance of librarians to your own professional work. Librarians in a corporate information service often locate the information for their internal clients, saving valuable time for employees. The Information Services division of 3M Corporation, headquartered in St. Paul, Minnesota, estimates that, on average, it saves employees some 28 hours per project. Barbara Peterson, Director of Information Services for 3M, explained in an interview how the 75 professionals in Information Services help meet her division's four distinct responsibilities:[13]

1. Responding to three kinds of inquiries:
   - *What are current facts*? Information Services answers fact-based questions that range from locating potential consultants for a project to identifying regulatory requirements that may affect project development.
   - *What are relevant sources?* Information Services identifies publications related to a particular topic, from state-of-the-art sources to newspaper articles.
   - *What are recent developments?* Information Services monitors ongoing developments with selective dissemination of information so that searches can be tailored to an individual's information needs.
2. Providing 3M professionals with easy access to library materials, ranging from print journals to the Internet.
3. Organizing and maintaining systems that track two kinds of R&D information for 3M:
   - Internal research and development documents, such as research proposals, laboratory notebooks, research reports;
   - An online database of skills of 3M's scientists and engineers so that project teams have a good chance of locating a needed expert from among the company's 8,000 technical employees.
4. Providing ongoing education to help 3M employees use, manage, and share information effectively.

Although corporate libraries in other companies may provide a different range of services, you can depend on a good corporate library to be a tremendous help to you in identifying and collecting information useful for your projects.

Personal Observations

Personal observations are legitimate primary sources if the observer is trained in the area of investigation. For example, botanists can identify a bog by observing and classifying the vegetation in a geographic area, and opticians can make remarkably accurate judgments about the flatness of a mirror surface by observing light refraction.

Personal observations of experts and experimenters were the basis for scientific and technical knowledge long before the days of sophisticated testing and measuring equipment. For instance, in 1628, William Harvey wrote *De Motu Cordis (On the Motion of the Heart),* part of which is quoted in Figure 6.5, based entirely on his personal observations. In fact, modern medicine still uses some of the same observational strategies that Harvey used almost 400 years ago. Notice how Harvey describes the motion of the heart using common but specific terms and draws specific conclusions based on specific examples.

FIGURE 6.5 **Translated Excerpt from Harvey's *De Motu Cordis*[14]**

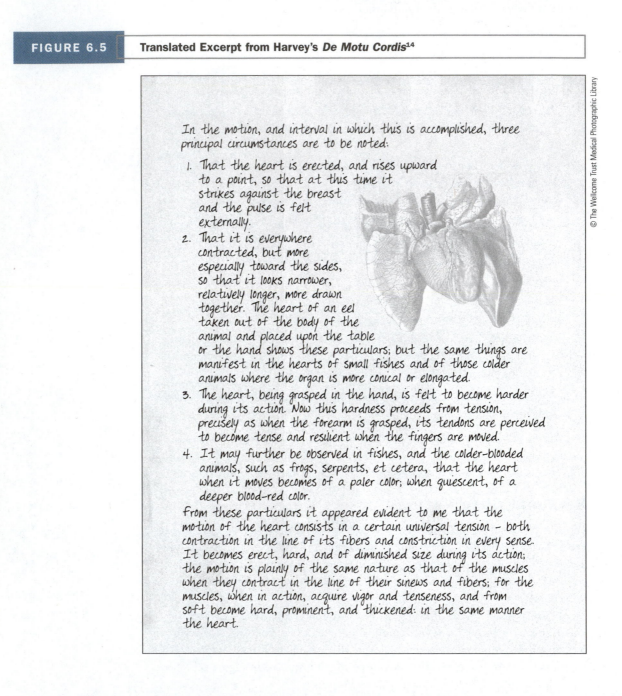

In the motion, and interval in which this is accomplished, three principal circumstances are to be noted:

1. That the heart is erected, and rises upward to a point, so that at this time it strikes against the breast and the pulse is felt externally.
2. That it is everywhere contracted, but more especially toward the sides, so that it looks narrower, relatively longer, more drawn together. The heart of an eel taken out of the body of the animal and placed upon the table or the hand shows these particulars; but the same things are manifest in the hearts of small fishes and of those colder animals where the organ is more conical or elongated.
3. The heart, being grasped in the hand, is felt to become harder during its action. Now this hardness proceeds from tension, precisely as when the forearm is grasped, its tendons are perceived to become tense and resilient when the fingers are moved.
4. It may further be observed in fishes, and the colder-blooded animals, such as frogs, serpents, et cetera, that the heart when it moves becomes of a paler color; when quiescent, of a deeper blood-red color.

From these particulars it appeared evident to me that the motion of the heart consists in a certain universal tension - both contraction in the line of its fibers and constriction in every sense. It becomes erect, hard, and of diminished size during its action; the motion is plainly of the same nature as that of the muscles when they contract in the line of their sinews and fibers; for the muscles, when in action, acquire vigor and tenseness, and from soft become hard, prominent, and thickened: in the same manner the heart.

To learn more about William Harvey's treatise, *On the Motion of the Heart* go to **www.english.wadsworth.com/burnett6e** for a link to Harvey's treatise.
CLICK WEBLINK
CLICK Chapter 6/William Harvey

WEBLINK

Even if you are not a specialist, your own involvement and observations are important and valid sources of primary information. We make mental personal observations, usually subconsciously, all the time; for example, you might speak up without raising your hand in class once you've observed that the instructor prefers a conversational classroom or observe and model a mentor's behavior as you make a transition into a professional environment. Consciously observing relationships, cultural and social factors, behaviors, and repeated processes can provide a great deal of information in your research efforts.

Also, do not underestimate the value of your hands-on experience. For example, technical communicators who can use the computer software they are documenting are better equipped to explain how to accomplish tasks with the software than is a colleague who merely watched someone else use it. Similarly, if you are a supervisor who can actually run the equipment in a machine shop, you are better qualified to train new employees than is a colleague who knows the theory but not the actual operation. A nutritionist experienced in dietary analysis is better equipped to teach such analysis than is an inexperienced nutritionist.

When personal observations are appropriate, consider the following guidelines to increase the likelihood that data will be accepted as credible:

- Identify the situation, the context.
- Indicate the expertise and the disciplinary or social biases of the observer.
- Determine if you are an observer in a situation or if you are what ethnographers call a "participant-observer" — someone who is studying a culture or situation by interacting with it.
- Take extensive descriptive notes so you can identify trends or key points for analysis.
- Look for trends in the data rather than trying to make the data fit your assumptions.
- Select observations for a final report that are representative of the situation as a whole.
- Quantify observations whenever possible and use standard notation and terminology.
- Support generalizations you make based on your observations with specific details and examples from your notes.

Interviews and Letters of Inquiry

© Jose Luis Pelaez, Inc./CORBIS

Interviewing is one of the most important professional skills. Often the information you need is not written in any book or article but is in the mind or indecipherable notes of another person.

What are the advantages and disadvantages of submitting questions you'll ask in an interview in advance to the person you're interviewing?

When you're conducting an information interview, you need to be prepared and persistent about asking the right questions in order to elicit the information you need. Preparing effective questions for an interview requires some homework. You need to prepare in three ways:

- Gather necessary information about the subject: What do you already know? What do you need to know? What do you expect this particular person to be able to tell you?
- Approach the person you want to interview: How should you contact the person? What's the best place for the interview? What's the most convenient time? Be aware when you select the site that time and place can affect the quality of the interview. How should you refer to the person?
- Identify the categories of questions you want to ask, and plan the specific topics that you want to explore. Listing topics rather than specific questions gives you more flexibility.

The answers to these questions provide a focus for the interview. An essential courtesy is learning to pronounce the name of the person you're interviewing. You should also be familiar with the terminology you'll use during the interview.

Questions can be divided into two broad categories: *convergent questions,* which have only one correct answer, and *divergent questions,* which are open-ended and therefore more useful in problem solving. The more prepared you are before you enter an interview, the more likely you are to conduct an effective, focused interview that will uncover the information you are looking for (if, that is, the person being interviewed can answer your questions). Several useful guidelines for asking effective interview questions are illustrated in Figure 6.6 by some of the questions that Nancy Irish revised for her interviews.

What you ask is important. How you ask it is also important. Figure 6.7 identifies actions and attitudes that will make your interviews more efficient and effective. Far more than the content of the questions is important: You also need to consider your behavior and body language, respect conventions of interviewing, and decide ways to manage mechanical considerations.

Interviews can be conducted in a number of ways depending on time and geography. Face-to-face interviews are usually the best because you can easily gauge reactions or clarify meaning. If time is limited or distance is a problem, telephone or e-mail interviews are alternatives. Telephone and e-mail interviews

FIGURE 6.6 Guidelines for Asking Effective Interview Questions

Guidelines	Negative and Positive Examples
Design questions whose answers require a definition or an explanation rather than a simple yes or no.	***Weak:*** Are people's diets improved by using soy protein? ***Effective:*** In what ways are people's diets improved by protein?
If you must ask a yes or no question to establish a position, follow it with a question requiring some explanation. Generally, you should know the answers to basic questions. Don't waste interview time asking questions when you already know the answers except when a confirmation attributed to the person would be useful. Instead, use the basic information you already know to develop more fruitful questions.	***Weak:*** Has there been an increase in the past five years in soy protein used in processed food for human consumption? ***Effective:*** Has there been an increase in the past five years in soy protein used in processed food for human consumption? What has contributed to this increase (or decrease)? ***Better:*** What has contributed to the increase (or decrease) in the past five years in soy protein used in processing food for human consumption?
Ask questions that require a focused response, not a broad, rambling discussion.	***Weak:*** What do you think of the public attitudes about soy use? ***Effective:*** What specific strategies work to persuade the public to incorporate soy into regular meals?
Ask a single question at a time, not combined (complex) questions that require multiple answers.	***Weak:*** Describe the progress that has been made in identifying those components responsible for undesirable flavors and colors and in developing processing technology to eliminate these from food products. ***Effective:*** Describe the progress in identifying components responsible for undesirable flavors and colors in soy products. ***Follow-up:*** What processing technology has been developed to eliminate these from food products?
Use terminology that narrows the area of response.	***Weak:*** In what ways have soybeans been improved? ***Effective:*** In what ways has the nutritive value of soybeans been improved, specifically the amino acid profile and the digestibility?
Prepare questions to tactfully redirect a respondent who has begun to ramble.	***Weak:*** Getting back to the subject, I'd like to ask you again to describe the ways in which soybeans have been improved. ***Effective:*** I'd be interested to hear more of the ways in which the nutritive value of soybeans has been improved.
Research other interviews in the field to avoid asking what have become cliché questions; instead, find a new angle so that the respondent will be interested in what you ask and will know you have done your homework.	***Weak:*** What are the ecological benefits of soy foods? ***Effective:*** How do you think the ecological benefits of soy foods could be used in marketing to increase the American public's interest in them?
Refer specifically to the respondent's published work.	***Weak:*** I liked your recent article on soybeans for human consumption. ***Effective:*** Your article in last month's *Soy Research Journal* effectively supports the importance of increasing human consumption of soybeans.
Prepare questions the respondent might perceive as hostile so that the respondent is not offended but, instead, answers.	***Weak:*** How can you carry out objective research on soy foods when your research is funded by the National Soy Growers of America? ***Effective:*** I understand that your research is funded by the National Soy Growers of America. Does that make it difficult for you to conduct your research objectively?

may allow the possibility of a series of shorter interviews rather than one long personal interview; this is sometimes valuable because it gives both the interviewer and interviewee time to research stronger questions and responses during the interim. These alternatives also present their own challenges, though, such as the lack of visual cues to determine comprehension and the less personal nature of these media.

When a person is simply not available in person or on the telephone, an e-mail request or a letter of inquiry is an appropriate vehicle for collecting information. In some situations, a person may even agree to record answers on an audiotape that you provide so that he or she is spared writing detailed responses or working within a short timeframe. Figure 6.8 displays a letter of inquiry that Nancy Irish wrote to Chris Morgan.

FIGURE 6.7 **Actions and Attitudes for Effective Interviewing**

Category	Actions and Attitudes
Interpersonal behavior	■ Be attentive and courteous. ■ Know how to pronounce the person's name. ■ Generally address the person formally ("Dr. Greenough" rather than "Pat") unless you know him or her personally.
Body language	■ Convey interest through your tone of voice, facial expressions, and body language. ■ Make direct eye contact with the person you're interviewing. ■ Note the person's body language as a signal to how well the interview is going.
Interviewing conventions	■ Confirm arrangements beforehand regarding the time, place, and duration of the interview. ■ Arrive or call on time, and do not exceed the agreed-on length. If you want to record the interview, ask for permission; it's usually granted. ■ Prepare questions in advance. You can modify or extend them if you need to, but don't go into the interview unprepared. ■ Give the person time to respond to your questions. ■ Don't interrupt unless the person strays off track or becomes too long-winded.
Mechanical considerations	■ Be prepared to take notes, paraphrasing the person's comments, in case there are mechanical problems with the recorder or the person prefers that you don't use it. ■ Consider asking the person to prepare responses to statistical questions prior to or after the interview itself.

FIGURE 6.8 **Letter of Inquiry**

112 North Riverside Drive
Ames, IA 50010-5971
October 30, 20—

Make sure to include your return address so the person can contact you.

Chris Morgan, Research Director
National Soybean Institute
731 Jefferson Street
Dubuque, IA 52086

If you don't know the gender of the person (because of a name such as Robin, Pat, or Chris), use the full name.

Dear Chris Morgan:

Identify your role and goal.

As part of my work designing educational programs for members of food co-ops, I am investigating the role of soybeans in an ecologically and nutritionally sound food system. I am particularly interested in the increased use of soybeans in the American diet.

State specifically what you want.

Could you please provide information on the following topics?

1. Percentage of annual U.S. soybean crop consumed by humans vs. livestock; domestic vs. foreign markets
2. Percentage of U.S. arable land (acreage) planted in soybeans
3. Nutrient composition of traditional Asian soy foods
4. Resource efficiency of soybean products vs. grain-fed livestock production
5. Current research to develop new soy products for human consumption

I appreciate your assistance. If you wish, I will be glad to send you a copy of my final report outlining program options.

Thank the person for assisting you.

Sincerely,

Nancy Irish

Nancy Irish
515-555-1234 (office)
515-555-6779 (fax)
nirish@iastate.edu (e-mail)

Provide a telephone number and, if available, a fax number and e-mail address to give the person alternative and perhaps more convenient ways of contacting you.

Should you agree to submit material written from the interview to the person you interviewed for clarification? For approval?

Whether you obtain your information through a personal interview, telephone interview, e-mail interview, written questionnaire with taped response, or written response to a letter of inquiry, take the time to write a thank-you letter, acknowledging your appreciation for the time and effort the person took to assist you. You can give the letter a personal tone by mentioning one or two things that were particularly helpful. If the person is someone you regularly see in your company, you can write an informal note or send an e-mail message; even a person you see every day appreciates a brief thank you. If the person is someone you do not see every day, make the thank you more formal.

Surveys and Polls

Surveys and polls are extremely useful tools for gathering information about opinions and preferences from large groups of people. They are usually based on a predetermined set of questions called a questionnaire, which can be administered on paper, through e-mail or the Internet, or in personal interviews. If you use a random sample of the target audience, you can make statistically acceptable generalizations about the entire population. If you are conducting your own survey, take care in constructing the questions, selecting the test group, and compiling the results in order to generate ethical and statistically significant results. If you are using data from surveys by others, investigate how carefully the surveys were constructed and conducted to ensure the reliability and validity of results.

Survey design is important. The questions should be both valid (the questions really ask what the survey takers intend to ask) and reliable (the questions are likely to be interpreted in a similar way by people completing the survey). Surveys can be designed using any of six different types of questions; each type has advantages and disadvantages you should consider when designing a questionnaire for a survey or poll:

- *Dual alternatives,* the simplest questions to tabulate, offer only two choices: yes/no, positive/negative, true/false, and so on. Such questions adequately address simple issues, but they often unrealistically limit the range of responses needed to portray complex situations. If you want to use this type of question, assure yourself that there really are only two possible answers so that the respondent does not feel as if the choices are inadequate.
- *Multiple choice questions* give respondents alternatives, sometimes limiting them to a single answer, at other times permitting them to check all applicable answers. Because the survey designer provides all the choices, multiple choice questions may not offer an answer with which respondents agree, thus distorting the results. A well-designed multiple choice question does not have two or three choices that are obviously wrong, nor does it have choices in which the distinctions are so slight that the respondent must guess the correct answer.

- *Rank ordering* provides respondents with a series of items and asks them to order the items according to preference, frequency of use, or some other criterion. Tabulating rank ordering is easy, but distortion may occur if the mean (arithmetic average) differs significantly from the mode (most frequently occurring number). If the items in a particular grouping are similar, respondents will have a difficult time ranking them and may assign an arbitrary value rather than an actual preference.
- *Likert scales* provide a method for respondents to express their opinion by rating items either numerically or verbally on a continuum. As with rank ordering, tabulation is easy but may give distorted results if the mean and mode are significantly different. Scales are most effective if they have an even number of choices (usually four or six). If a scale has a middle choice (as occurs when the scale has an odd number of choices), respondents will choose it a disproportionately large percentage of the time.
- *Completions* expect respondents to fill in information to complete an item. The answers are simple to tabulate if the responses are quantitative, such as age, frequency, or amount. But even short, open-ended opinion questions present problems in tabulation if respondents give a variety of answers; the evaluator must decide which terms respondents use are synonymous.
- *Essays or open-ended questions* give respondents the opportunity to fully express themselves, presenting both facts and opinions, but the responses are difficult to tabulate. Essays are much more effective if they ask respondents to focus on specifics rather than request general opinions or reactions.

How much difference do the types of questions really make? Imagine that Nancy Irish read a claim that 23 percent of people questioned agreed they would include tofu in their diets. The statistic doesn't mean much unless she knows the question. That same answer may be a response to several variations of the same question. Dual alternatives, for example, could elicit far more positive responses than completion questions. Figure 6.9 illustrates some of the possible questions that could have elicited such a response. In order to determine the validity and reliability of your questions, test them with a small representative sample. If changes are necessary, you can make them before officially administering the survey.

Designing an effective series of questions is only the first step in preparing a survey. After you have designed and tested your questions, you need to administer the survey to a representative random sample of the target population. Compiling and reporting the results should not be difficult if your questions have been carefully designed. Even when the questions are valid and reliable, the population sample random and representative, and the data compiled accurately, the results may be disappointing, particularly with mail-in surveys, which have a very low return rate.

FIGURE 6.9 **Variations of Questions That Could Produce the Same Response**

Dual Alternative	Would you include tofu in your meals if you knew good recipes? ❑ Yes (23 percent) ❑ No
Single Multiple Choice	Tofu is a traditional Asian soybean food high in protein and low in calories and fat. Considering the high nutrient value of tofu, check the one item you most agree with. ❑ Willing to try tofu if I had a good recipe (14 percent) ❑ Use tofu regularly (19 percent) ❑ Tried tofu and didn't like it ❑ Not interested in trying tofu ❑ Never heard of tofu
Rank Order	Rank order the following soy items in order of preference as additions to your current diet. __ Soy sauce __ Tofu (23 percent ranked tofu second) __ Miso __ Natto __ Tempeh
Continuum (Likert)	Mark your preference for each food item on the scale. Tofu: Would not ever use — Might consider — Would if knew how — Use now (23 percent)
Completion	Which soy product are you most willing to include in your meal planning? ____________ ("tofu"–23 percent)

Using Sources Ethically

As a writer and reader, you are responsible for assessing and using information ethically. The ease with which any person with a computer and server access can post information online means that readers must be much more careful when assessing the reliability and validity of electronic information. And the proliferation of electronic and Web resources has made plagiarizing information increasingly frequent. Ethically using information means understanding ways to assess information on the Web for credibility and understanding ways to incorporate information into your documents, presentations, and visuals without plagiarizing, which is the literary equivalent of stealing.

Assessing Credibility

Many readers assume that information published in professional journals or books is reliable and valid because the peer-review process authors must go through usually eliminates misinformation. Readers of online journals usually assume that these articles have been through similar peer review and editing. But not all online publications are peer reviewed or carefully edited. The Web makes possible the publication of documents with little or no editing or review; thus, you may have difficulty judging the credibility of online information.

The ease of posting information on the Web has changed how you need to approach online information. You need to consider possible biases and inaccuracies in information obtained from the Web and to think critically about a document's origin. For example, documents published by companies may be biased in favor of their own products; documents published by individuals may be partially or entirely based on unsubstantiated personal beliefs or opinions. While a particular bias may not be a sufficient cause to reject a source, you do need to recognize ways in which a bias can influence a source. Consider the following elements when assessing the credibility of Web documents:

1. ***Authorship***
   - Can you identify the author or webmaster on the site?
   - Is the author identified as a person, a corporation, a university?
   - What do you know about the credentials of the author or organization?
   - Does the site have a sponsor that might have a vested interest?
2. ***Timeliness***
   - When was the site originally posted?
   - When was the site last updated?
   - How regularly is the site updated?
   - Is the information current or outdated for your needs?
3. ***Purpose***
   - Can you determine from the design and content if the site is intended to inform, persuade, or sell?
   - Can you determine the intended audience?
   - Does the site have advertising banners? Do they affect the content?
4. ***Content***
   - How detailed and well-researched is the content?
   - Does the new information confirm or disconfirm your prior knowledge?
   - Does the author fully cite sources and link to respected sources to support assertions?
   - Can you verify the author's claims using other resources?
   - Does the content contain spelling and grammatical errors?
   - Can you identify a particular slant or bias to the information?

Avoiding Plagiarism

Avoiding plagiarism is just as important in the workplace as it is in academia. Cutting and pasting information from an original source into a second document has become extremely easy to do with today's word processing programs. But the law in the United States is on the side of the original author. Stealing text from other sources — whether published or unpublished — is unethical and illegal.

WEBLINK

Go to **www.english.wadsworth.com/burnett6e** for a link to learn more about ways to avoid plagiarism.
CLICK ON WEBLINK
CLICK on Chapter 6/plagiarism

What is plagiarism? Using work that is not your own without attribution.

- Using exact quotations from documents, presentations, Web sites, or visual designs and images without attribution
- Using someone else's unique or distinctive ideas without attribution
- Using someone else's unique or distinctive processes without attribution
- Using words, images, or processes, that have only been changed slightly without attribution. What counts as an insufficient change? Changing verbs from past to present tense. Changing nouns from singular to plural. Reordering items. Omitting some information but keeping some of it as it was in the original.

Figure 6.10 provides practical guidelines for avoiding plagiarism. The guidelines can be summarized quite simply: Give credit for ideas, words, and visuals that are not your own, whether they are directly quoted or paraphrased.

Just as Web sites peddling academic term papers for sale are flourishing, so are sites used to combat plagiarism. You can type in a long phrase in the search engine **www.google.com**, and it will return any online document that uses that exact phrasing. Several universities are building searchable databases of student papers for faculty who suspect plagiarism. And many companies are designing software for faculty and research libraries that searches student papers against online materials. The Internet may have made plagiarism easier, but it has also made plagiarism easier to combat.

What's exempt from plagiarism? Using common knowledge, widely known historical facts, or company boilerplate (for company purposes) is usually not considered plagiarism, but a courtesy acknowledgment doesn't hurt. The best guideline you can use is, "When in doubt, include a complete citation."

WEBLINK

The notion of "common knowledge" often causes remarkable problems in deciding what to cite. Go to **www.english.wadsworth.com/burnett6e** for a link to learn more about what constitutes common knowledge.
CLICK ON WEBLINK
CLICK on Chapter 6/common knowledge

FIGURE 6.10 **Ways to Avoid Plagiarism**[15]

▪ To take accurate, usable notes, clearly signal all quoted material with BIG quotation marks, a bold Q (for quotation), and the complete citation information.	. . . and then later	▪ Proofread the final document with clearly marked notes at hand to make sure that all ideas, words, and visuals that are not your own, whether directly quoted or paraphrased, are clearly cited in in-text references, footnotes, or endnotes and the bibliography or works cited.
▪ To avoid plagiarism when you're writing a summary, close the journal or book. When you've finished drafting the summary, check the original source for accuracy. ▪ Write the complete citation information next to the quotation in your notes.	. . . and then later	▪ In the text of your document, give credit to the source. ▪ Use quotation marks for unique words or phrases. ▪ Use parenthetical citations (author, date, page) immediately after quoted and paraphrased information.
▪ To accurately use a direct quotation, use the exact words and quotation marks.	. . . and then later	▪ Refer directly to the person or source in introducing or commenting on the quotation. ▪ Check that the quotation marks are in the correct places.

When integrating information from another source into your document, you have several options, all of which require documentation. You can directly quote a phrase or sentence(s) from the source in quotation marks and support that quote with an introductory phrase such as "According to . . ." or "Jones says . . . ." If the information you want to quote is longer than a few sentences, paraphrase the information into one or two sentences that summarize the information without borrowing exact wording. Paraphrases do not require quotation marks but do require an introductory phrase to identify their source. Both quoting and paraphrasing require that you cite the source in a parenthetical citation at the end of the sentence and that you include a bibliographic reference to the original source at the end of the entire document. Documentation formats vary across disciplines but include MLA, APA, and Chicago style. *Avoiding Unintentional Plagiarism* on the next two pages shows examples of quoted, paraphrased, and plagiarized materials as well as proper documentation forms.

AVOIDING UNINTENTIONAL PLAGIARISM

Part of your professional ethos comes from using information correctly and ethically, which means knowing how to avoid plagiarism by properly citing, attributing, paraphrasing, and integrating information from other sources. Plagiarism is not only an academic offense; it's a legal offense. Companies view plagiarism as unacceptable, unethical, and illegal. The following examples show both plagiarized and acceptable uses of information. The brochure was published by the United Soybean Board and reports a portion of an independent research firm's findings about consumer attitudes toward health and nutrition.

In each box, A–D, decide which example is plagiarized. Check your answers in the key on the next page.

A 1 Organic foods are growing in popularity in the United States as 44% of consumers buy organic food products either frequently or sometimes. More consumers aged 35–54 (20%) frequently buy organic products than do consumers aged 18–34 (13 percent) or 55 and older (14 percent). This suggests that the food industry may not be doing enough to target and increase organic food awareness among younger people.

A 2 According to a report sponsored by the United Soybean Board (USB) (2002), "44% of consumers buy organic food products either frequently or sometimes" (Organic Foods section, para 1). The report also notes that more middle-aged consumers regularly purchase organic food than younger and older consumers (Organic Foods section, para 1). This research suggests that the food industry may not be doing enough to target and increase organic food awareness among younger people.

CONSUMER PERCEPTIONS AND KNOWLEDGE OF ORGANIC FOODS[16]

Forty-four percent of consumers buy organic food products either frequently or sometimes, while 30 percent do so rarely and 25 percent never purchase organics. More consumers aged 35-54 (20 percent) frequently buy organic products than do consumers aged 18-34 (13 percent) or 55 and older (14 percent).

The majority of consumers concerned about biotechnology are not aware of the alternatives they already have. Only three percent of consumers know, without prompting, that organic foods can't contain genetically modified ingredients by law, compared to the much larger 41 percent who are aware that organics must be grown without pesticides or herbicides. Additionally, 33 percent cannot name any requirements for a food to be called "organic."

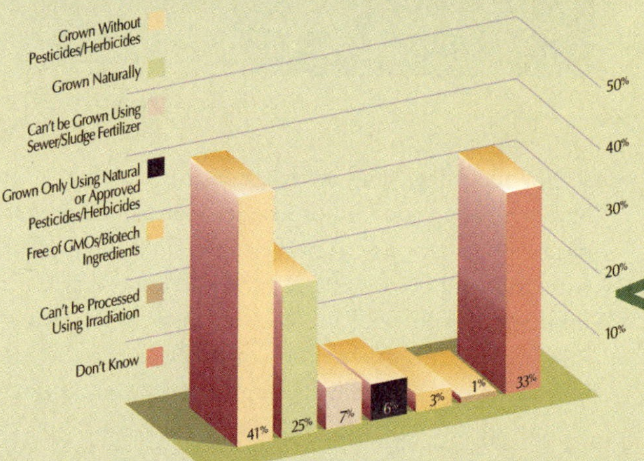

On an aided basis, 48 percent know that organic foods can't contain genetically modified ingredients and 64 percent know that organics must be grown without pesticides and herbicides.

CONSUMER ATTITUDES ABOUT FATS

Consumers continue to show confusion about the health profile of fats and oils. While 90 percent of consumers recognize that saturated fats are unhealthy, most remain uncertain about unsaturated fats with nearly half perceiving mono- and poly- unsaturated fats as unhealthy. Additionally, four in 10 consumers believe that *trans* fatty acids are healthier than saturated fats, while two in 10 believe that saturated fats are healthier and three in 10 do not know which is healthier. In contrast, consumers are much more sure about Omega-3 fatty acids, with 67 percent stating that they are healthy.

On an unaided basis, only three percent of consumers mentioned *trans* fats or hydrogenation as a health concern in relation to soy and fewer than one in 10 perceive any negative health effects of soy or soyfoods. However, when consumers were asked specifically to rate the healthiness of *trans* fats and hydrogenation, seventy-seven percent ranked *trans* fatty acids as very or somewhat unhealthy, compared to the 66 percent who rated hydrogenated vegetable oils as unhealthy. In contrast, 22 percent now feel that *trans* fats are very or somewhat healthy, while 34 percent believe that hydrogenated vegetable oils are healthy.

Sixty-two percent of consumers say they are somewhat or very unlikely to buy products labeled as containing *trans* fatty acids. While this percentage is higher than in 1999, it is significantly smaller than just last year. One-half of consumers (51 percent) feel they are unlikely to buy products featuring hydrogenated vegetable oils on the label.

C1 Even in a very health-conscious society, consumers are still confused about the health benefits and drawbacks of most fats. "While 90 percent of consumers recognize that saturated fats are unhealthy . . . four in ten consumers believe that trans fatty acids are healthier than saturated fats" (USB). This statement shows that consumers have difficulty defining the types of fats and determining which are healthy.

C2 A study sponsored by the United Soybean Board (2002) reported that 90 percent of consumers recognized that saturated fats were bad for their health but were torn about the health value of mono- and poly- unsaturated fats (Attitudes about Fats section, para 1). The study also stated that consumers are unsure of the health value of trans fatty acids but "are much more sure about Omega-3 fatty acids, with 67 percent stating they are healthy" (Attitudes about Fats section, para 1). Credit for the view about saturated and Omega-3 fats can be credited to the extensive coverage these fats have had in the popular media in the last 20 years.

D1 According to a USB report, consumers remain confused about the advantages of fats and oils in one's diet. When speaking about soy, a small number of consumers (3%) thought trans fats or hydrogenation was a health concern. Less than 1 in every 10 people thought soy or soyfoods had negative health effects.

D2 According to a USB-sponsored report (2002), fats and oils still confuse many consumers, most of whom are unsure which fats and oils are healthy for their diets. The report noted that a miniscule number of respondents thought trans fats were unhealthy and that less than 10 percent presumed soy and soyfoods would be harmful to their health (Attitudes about Fats section, para 2).

B1 Consumers cannot agree on the definition of organic, but the majority know it has something to do with how food is grown. Over 50% of consumers think "organic" has something to do with fertilizer, while many have no idea what makes food organic.

B2 According to the USB-sponsored study, 33 percent of consumers cannot define characteristics of organic food, while over 50 percent relate the term to the use of herbicides, pesticides, and unnatural fertilizers during growth (Organic Foods section, graphic). Only three percent were able to determine that organic foods must be free of genetically modified organisms, one of the key requirements for food to be termed organic (Organic Foods section, graphic).

Key

A1 Plagiarism — uses the original source without quotation marks, attribution, or citation

A2 Acceptable — uses quotation marks to signal directly quoted material and appropriately paraphrases other material; uses attributions and citation

B1 Plagiarism — distorts information in the figure and does not attribute or cite the source

B2 Acceptable — accurately interprets information in a way readers can understand, provides attribution, and cites the source

C1 Plagiarism — does not introduce or attribute the partial, and therefore potentially misleading, quotation even though an in-text citation is provided

C2 Acceptable — integrates direct quotation into the text and appropriately attributes and cites quoted and paraphrased information

D1 Plagiarism — uses the paraphrased statistics and conclusions from the original with attribution but without citation

D2 Acceptable — maintains the attribution but now cites the source

Individual and Collaborative Assignments

1. **Learn about your own academic library.** Some academic libraries have cooperative arrangements with local businesses so that the library facilities are available for practicing professionals. Work in small groups with others who share your academic major (or a closely related one).
   (a) Design questions for a librarian about services and resources available to local businesses and industries.
   (b) Select a representative to arrange and conduct an interview with a college librarian.
   (c) Design a hands-on library orientation for local business professionals to develop skills in using library references and resources. Go beyond the usual orientation by including information about the services of reference librarians, database searches, periodical and abstract indexes, technical reference materials, government documents, microfilm and microfiche files, and audiovisual materials. Present the information in a guidebook that also serves as a convenient reference for professionals after the orientation.

2. **Learn about a corporate library.** Learning about corporate library resources is important for technical professionals and for technical writers. Work in small groups with others who share your academic major (or a closely related one).
   (a) Design questions for a corporate librarian about the differences between a university and a corporate library, frequency of use, and the types of professionals who use them.
   (b) Select a representative to arrange and conduct an interview with corporate librarians at different local businesses and industries.
   (c) Design a hands-on library orientation to introduce new professional employees to the references and resources in a corporate library. Include information about services of professional librarians, available books and periodicals, loan privileges, database searches, periodical and abstract indexes, technical reference materials, government documents, microfilm and microfiche files, and audiovisual materials. Present the information in a guidebook that also serves as a convenient reference for employees.

3. **Recommend resources.** Corita Muir of GeoTech, Inc., is investigating the cost effectiveness of her company's instituting a longer lunch hour and flex-time. Its goal is to have a variety of company-sponsored fitness programs available to all employees from 6:00 to 8:00 a.m., 11:00 a.m. to 1:00 p.m., and 4:00 to 7:00 p.m. She plans to use the following resources to prepare her proposal:
   - Company records providing the number of sick days employees used in the past 24 months

- Recent articles in the local newspaper describing the benefits of regular exercise
- Articles in business or science journals about the benefits of company-sponsored exercise

What additional resources would you recommend?

4. **Identify keywords and subject headings.** You are trying to locate information about one of the following topics. Before looking in the Library of Congress Subject Headings (LCSH), brainstorm possible alternative keyword terms for the topic. Determine subject search headings as well.
   - Midwives
   - Medical applications of lasers
   - New coal-mining techniques
   - Proper use of farming equipment
   - Wellness programs
5. **Compare sources in your field.**
   (a) Do a search for information on a particular topic in your field using the library catalog, electronic databases, and the World Wide Web. Use both keyword and subject searches where applicable and compare the different results generated. Determine what you believe are advantages and disadvantages of each type of information source and each search strategy.
   (b) Compile and then annotate a list of reference books, electronic journals and databases, and Internet sources that are particularly helpful to a person in your career field.
6. **Compare Internet sources.** Visit the following Web site, http://www.nist.gov/weblinks.htm, and choose three different government Web sites to visit and analyze. Determine what you believe are the purposes, audience, and advantages or disadvantages of each. In a memo to the class, report your analysis.
7. **Interview a professional, part I.**
   (a) Arrange to interview a professional technical writer or a technical professional in your field who writes frequently. During the first part of the interview, inquire about the person's approach to locating information from a variety of sources. Ask some of the following questions and add additional questions relevant to the specific person and organization:
   - Where/how do you get the information you use?
   - Do you ever have problems getting the information? How do you handle these problems?

- How available and cooperative are your informants?
- What kind of bias tends to be part of the data you collect?
- What's your background? Technical communication? Science? Engineering?
- How does this background affect the way you locate and collect information?

(b) Organize the information you find as part I of a brief oral or written report intended for other members of the class. This report can be posted on the class Web site or made available in print form in a class notebook in the library.

8. **Identify and correct plagiarism.** Several original passages and the related paraphrased and/or quoted passages follow. Read each pair. Then decide how you can revise the paraphrased and/or quoted passages so that the quoted segments and citations are correct.

(a) ***Original passage***[17]

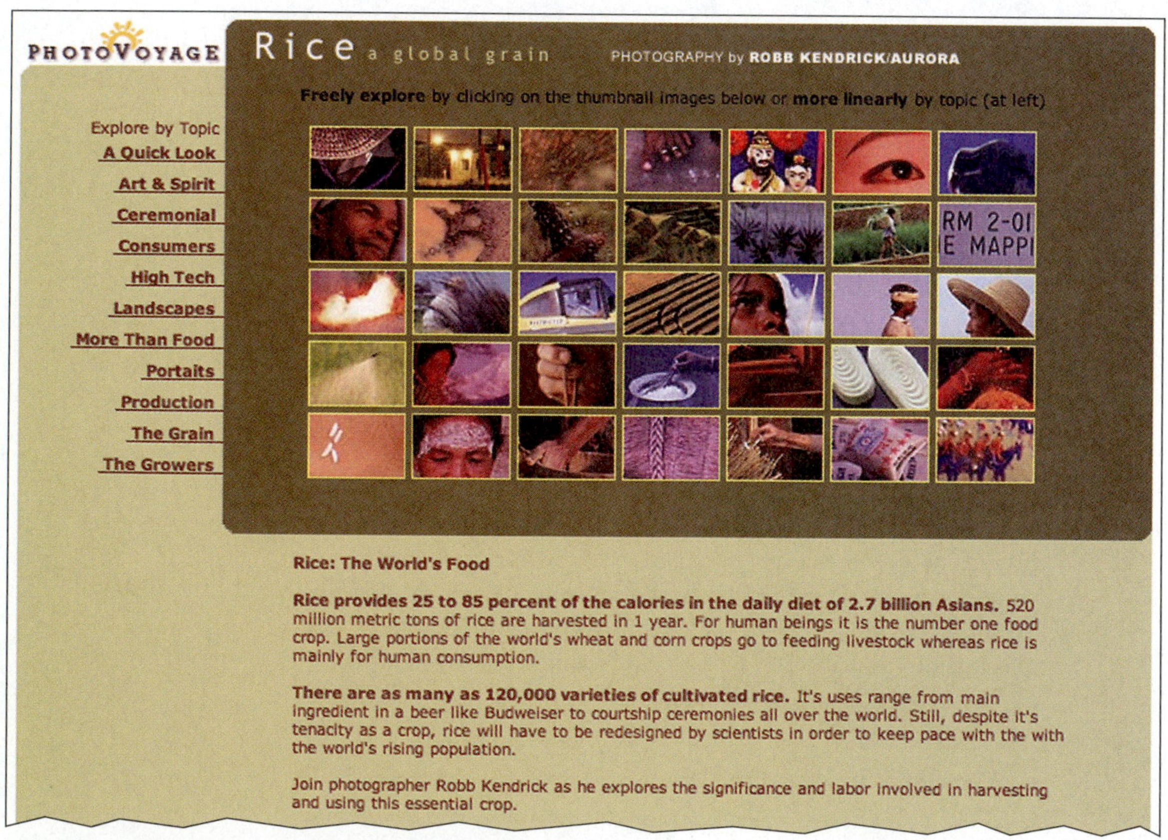
PHOTOVOYAGE

Rice a global grain PHOTOGRAPHY by ROBB KENDRICK/AURORA

Freely explore by clicking on the thumbnail images below or **more linearly** by topic (at left)

Explore by Topic
A Quick Look
Art & Spirit
Ceremonial
Consumers
High Tech
Landscapes
More Than Food
Portaits
Production
The Grain
The Growers

Rice: The World's Food

Rice provides 25 to 85 percent of the calories in the daily diet of 2.7 billion Asians. 520 million metric tons of rice are harvested in 1 year. For human beings it is the number one food crop. Large portions of the world's wheat and corn crops go to feeding livestock whereas rice is mainly for human consumption.

There are as many as 120,000 varieties of cultivated rice. It's uses range from main ingredient in a beer like Budweiser to courtship ceremonies all over the world. Still, despite it's tenacity as a crop, rice will have to be redesigned by scientists in order to keep pace with the with the world's rising population.

Join photographer Robb Kendrick as he explores the significance and labor involved in harvesting and using this essential crop.

Quoted or paraphrased passage

During one year, approximately 520 million metric tons of rice are harvested, making it the number one food crop for people. The more than 120,000 varieties of rice currently being cultivated have many uses, including being the main ingredient in a beer. Food scientists are working to create new ways to use rice, which provides between 25 to 85 percent of the calories in the daily diet of 2.7 billion Asians.

(b) ***Original passage***[18]

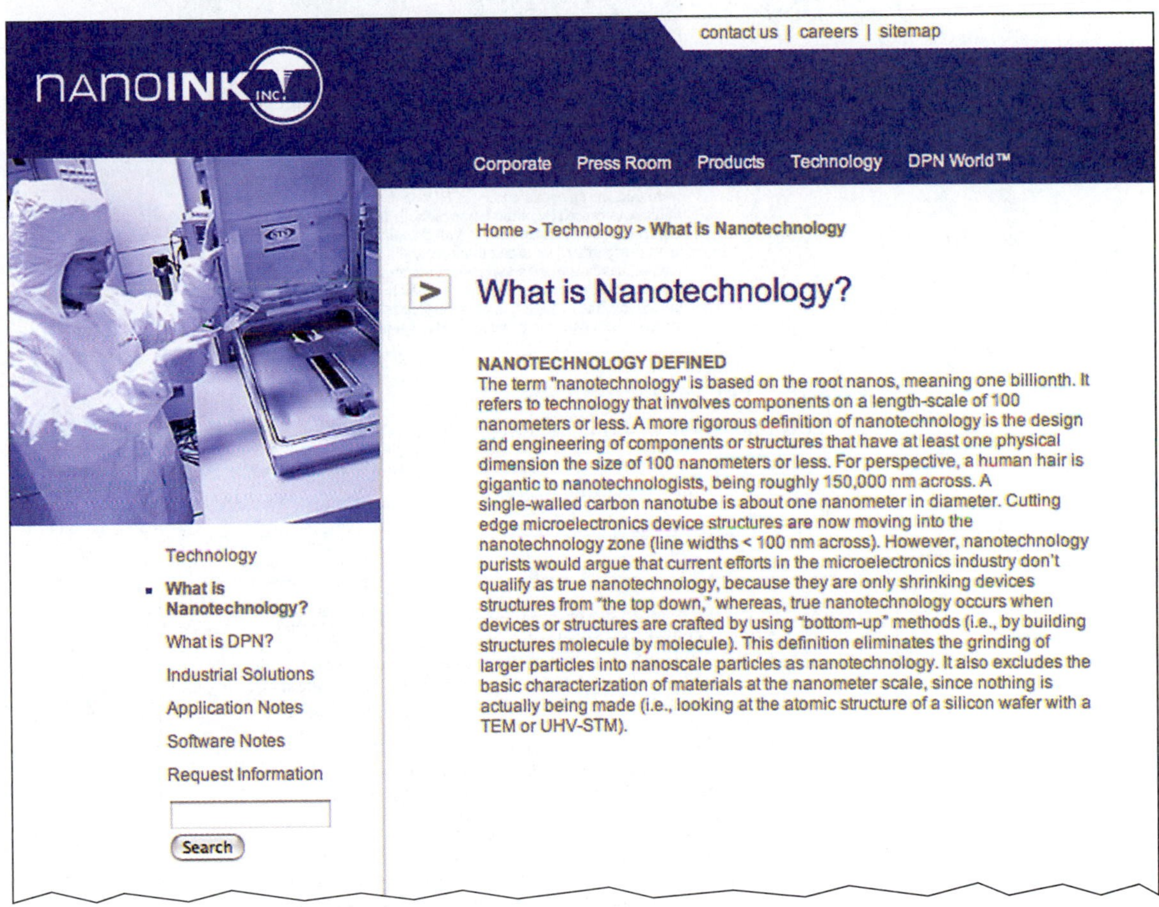

contact us | careers | sitemap

nanoINK INC.

Corporate Press Room Products Technology DPN World™

Home > Technology > **What is Nanotechnology**

> What is Nanotechnology?

NANOTECHNOLOGY DEFINED

The term "nanotechnology" is based on the root nanos, meaning one billionth. It refers to technology that involves components on a length-scale of 100 nanometers or less. A more rigorous definition of nanotechnology is the design and engineering of components or structures that have at least one physical dimension the size of 100 nanometers or less. For perspective, a human hair is gigantic to nanotechnologists, being roughly 150,000 nm across. A single-walled carbon nanotube is about one nanometer in diameter. Cutting edge microelectronics device structures are now moving into the nanotechnology zone (line widths < 100 nm across). However, nanotechnology purists would argue that current efforts in the microelectronics industry don't qualify as true nanotechnology, because they are only shrinking devices structures from "the top down," whereas, true nanotechnology occurs when devices or structures are crafted by using "bottom-up" methods (i.e., by building structures molecule by molecule). This definition eliminates the grinding of larger particles into nanoscale particles as nanotechnology. It also excludes the basic characterization of materials at the nanometer scale, since nothing is actually being made (i.e., looking at the atomic structure of a silicon wafer with a TEM or UHV-STM).

Technology
- **What is Nanotechnology?**

What is DPN?
Industrial Solutions
Application Notes
Software Notes
Request Information

Search

Quoted or paraphrased passage

The word "nanotechnology" comes from the root *nanos,* meaning one billionth, which involves components on a length-scale of 100 nanometers or less. Nanotechnology involves the design and engineering of components with at least one physical dimension of 100 nanometers or less. How small is this? A human hair, which is roughly 150,000 nm across, is gigantic to nanotechnologists. In comparison, a single-walled carbon nanotube has a diameter of about one nanometer. Innovative microelectronics devices are now moving into the nanotechnology zone; however, current efforts don't qualify as true nanotechnology. True nanotechnology occurs when devices are crafted molecule by molecule.

(c) ***Original passage***[19]

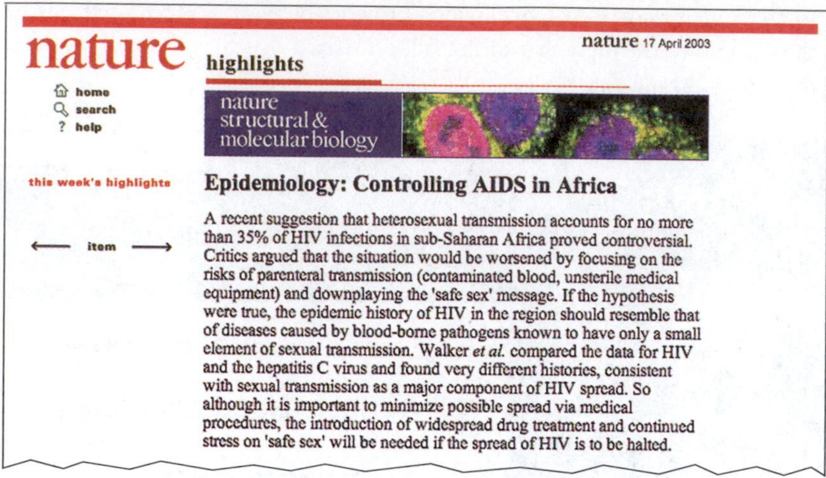

nature

nature 17 April 2003

highlights

home
search
help

nature structural & molecular biology

this week's highlights

← item →

Epidemiology: Controlling AIDS in Africa

A recent suggestion that heterosexual transmission accounts for no more than 35% of HIV infections in sub-Saharan Africa proved controversial. Critics argued that the situation would be worsened by focusing on the risks of parenteral transmission (contaminated blood, unsterile medical equipment) and downplaying the 'safe sex' message. If the hypothesis were true, the epidemic history of HIV in the region should resemble that of diseases caused by blood-borne pathogens known to have only a small element of sexual transmission. Walker *et al.* compared the data for HIV and the hepatitis C virus and found very different histories, consistent with sexual transmission as a major component of HIV spread. So although it is important to minimize possible spread via medical procedures, the introduction of widespread drug treatment and continued stress on 'safe sex' will be needed if the spread of HIV is to be halted.

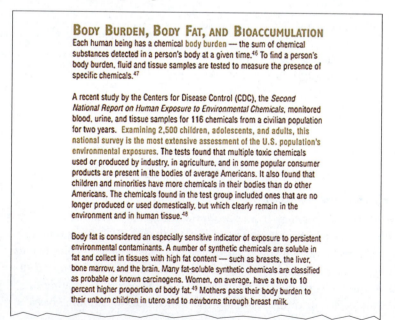

Quoted or paraphrased passage

Less than 35% of HIV infections in sub-Saharan Africa result from heterosexual transmission, so efforts should focus on the risks of parenteral transmission and possible spread from medical procedures.

(d) ***Original passage***[20]

BODY BURDEN, BODY FAT, AND BIOACCUMULATION

Each human being has a chemical body burden — the sum of chemical substances detected in a person's body at a given time.[46] To find a person's body burden, fluid and tissue samples are tested to measure the presence of specific chemicals.[47]

A recent study by the Centers for Disease Control (CDC), the *Second National Report on Human Exposure to Environmental Chemicals*, monitored blood, urine, and tissue samples for 116 chemicals from a civilian population for two years. Examining 2,500 children, adolescents, and adults, this national survey is the most extensive assessment of the U.S. population's environmental exposures. The tests found that multiple toxic chemicals used or produced by industry, in agriculture, and in some popular consumer products are present in the bodies of average Americans. It also found that children and minorities have more chemicals in their bodies than do other Americans. The chemicals found in the test group included ones that are no longer produced or used domestically, but which clearly remain in the environment and in human tissue.[48]

Body fat is considered an especially sensitive indicator of exposure to persistent environmental contaminants. A number of synthetic chemicals are soluble in fat and collect in tissues with high fat content — such as breasts, the liver, bone marrow, and the brain. Many fat-soluble synthetic chemicals are classified as probable or known carcinogens. Women, on average, have a two to 10 percent higher proportion of body fat.[49] Mothers pass their body burden to their unborn children in utero and to newborns through breast milk.

The writer using this report did not verify or read any of the sources cited in its endnotes, which included these in the passage below.

46 *Living Downstream*, 226.
47 Ibid.
48 Department of Health and Human Services, Centers for Disease Control and Prevention, "Second National Report on Human Exposure to Environmental Chemicals," (Washington, D.C.: CDC, 2003) [online]; available from: <http://www.cdc.gov/exposurereport/pdf/SecondNER.pdf>.
49 Margaret Reeves and Terra Murphy, Pesticide Action Network North America and Teresa Calvo Morales, Organización en California de Lídres Campesinas, *Farmworker Women & Pesticides in California's Central Valley* [online]; available from: <http://www.panna.org/resources/gpc/gpc_200304.13.1.08.dv.html>.

Quoted or paraphrased passage

According to a report published by the Centers for Disease Control and Prevention, the body fat of average Americans shows evidence of toxic agricultural and industrial chemicals, including ones that are no longer produced or used in the United States. This recent study examined 2,500 children, adolescents, and adults, monitoring blood, urine, and tissue samples. Many of the tissue samples contained carcinogens.

Chapter 6 Endnotes

1 *Knowledge management glossary.* (n.d.). Retrieved October 30, 2003, from http://www.mccombs.utexas.edu/kman/glossary.htm

2 *Knowledge management.* (n.d.). Retrieved October 30, 2003, from http://searchcrm.techtarget.com/sDefinition/0,,sid1 1_gci212449,00.html

3 Wilson, T. D. (2002). The nonsense of "knowledge management." *Information Research, 8*(1), Paper 144. Retrieved October 30, 2003, from http://informationr.net/ir/8-1/paper144.html

4 EOA national study finds nearly half of American workers engaged in unethical hi-tech behavior last year. (1998). *EOA News,* (Spring), 1, 4–5.

5 Office for Subject Cataloguing Policy, Collection Services. Library of Congress Subject Headings, 15th ed., vol 1 A-C. Washington, DC: Cataloguing Distribution Service. Library of Congress.

6 e-Library@Iowa State University. Collection Search Page. Retrieved on April 7, 2004, from www.lib.iastate.edu

7 Tech Web: The Business Technology Network. (n.d.). *Tech encyclopedia.* Retrieved July 6, 2003, from http://www.techweb.com/encyclopedia

8 Lyman, P., & Varian, H. R. (2000). *How much information?* Retrieved July 15, 2003, from http://www.sims.berkeley.edu/how-much-info

9 Adatped from ISU's *Instruction Commons Guide* by Vega Garcia, S. A. (2000, October). Searching by subject or by keyword. Retrieved July 15, 2003, from http://www.lib.iastate.edu/commons/resources/lcsh.index.html

10 Adapted from ISU's *Instruction Commons Guide* by Vega Garcia, S. A. (2000, October). Searching by subject or by keyword. Retrieved July 15, 2003, from http://www.lib.iastate.edu/commons/resources/lcsh.index.html

11 Adapted from UC Berkley-Teaching Library Internet Workshops. (n.d.). *Finding information on the internet: A tutorial.* Retrieved July 17, 2003, from http://www.lib/berkeley.edu/TeachingLib/Guides/Internet/Strategies.html

12 e-Library@Iowa State University. Collection Search Page. Retrieved on April 7, 2004, from www.lib.iastate.edu

13 Barbara Peterson, Director of 3M Innovation Services (personal communication, 1995, 2001).

14 Harvey, W. (1960). On the motion of the heart and circulation of the blood. In F. R. Moulton and J. J. Schifferes, (Eds.), *The autobiography of science* (2nd ed., pp. 106–107). Garden City, NY: Doubleday.

15 The suggestions include information from these sources:

- Purdue University Online Writing Lab. (n.d.). *Avoiding plagiarism.* Retrieved October 20, 2003, from http://owl.english.purdue.edu/handouts/research/4_plagiar.html
- Baylor College of Medicine. (n.d.). *Plagiary and the art of skillful citation: How to cite!* Retrieved October 20, 2003, from http://www.bcm.tmc.edu/immuno/citewell/artcite.html

- Reference and Instructional Services, Young Research Library, UCLA. (2002, October). *Preventing plagiarism: YRL seminar program.* Retrieved October 20, 2003, from http://www.library.ucla.edu/yrl/referenc/plagiarism.htm

16 *Consumer attitudes about nutrition: Insights into nutrition, health and soyfoods.* (n.d.). Retrieved July 21, 2003, from http://www.talksoy.com/pdfs/Consumer-Attitudes-2002.pdf

17 Robb Kendrick/Aurora. (n.d.). PhotoVoyage: Rice: A global grain. Washingtonpost.com. Retrieved October 30, 2003, from http://wpni01.auroraquanta.com/pv/rice

18 NanoInk, Inc. (n.d.). What is nanotechnology? Retrieved October 30, 2003, from http://www.nanoink.net/4100_whatis.html

19 Epidemiology: Controlling AIDS in Africa. (2003, April 17). *Nature.* Retrieved October 30, 2003, from http://www.nature.com/nature/links/030417/030417-3.html

20 Eshaghpour, T. (2003). *Confronting toxic contamination in our communities.* The Women's Foundation of California. Retrieved October 30, 2003, from www.womensfoundca.org/fullreport10_7.pdf

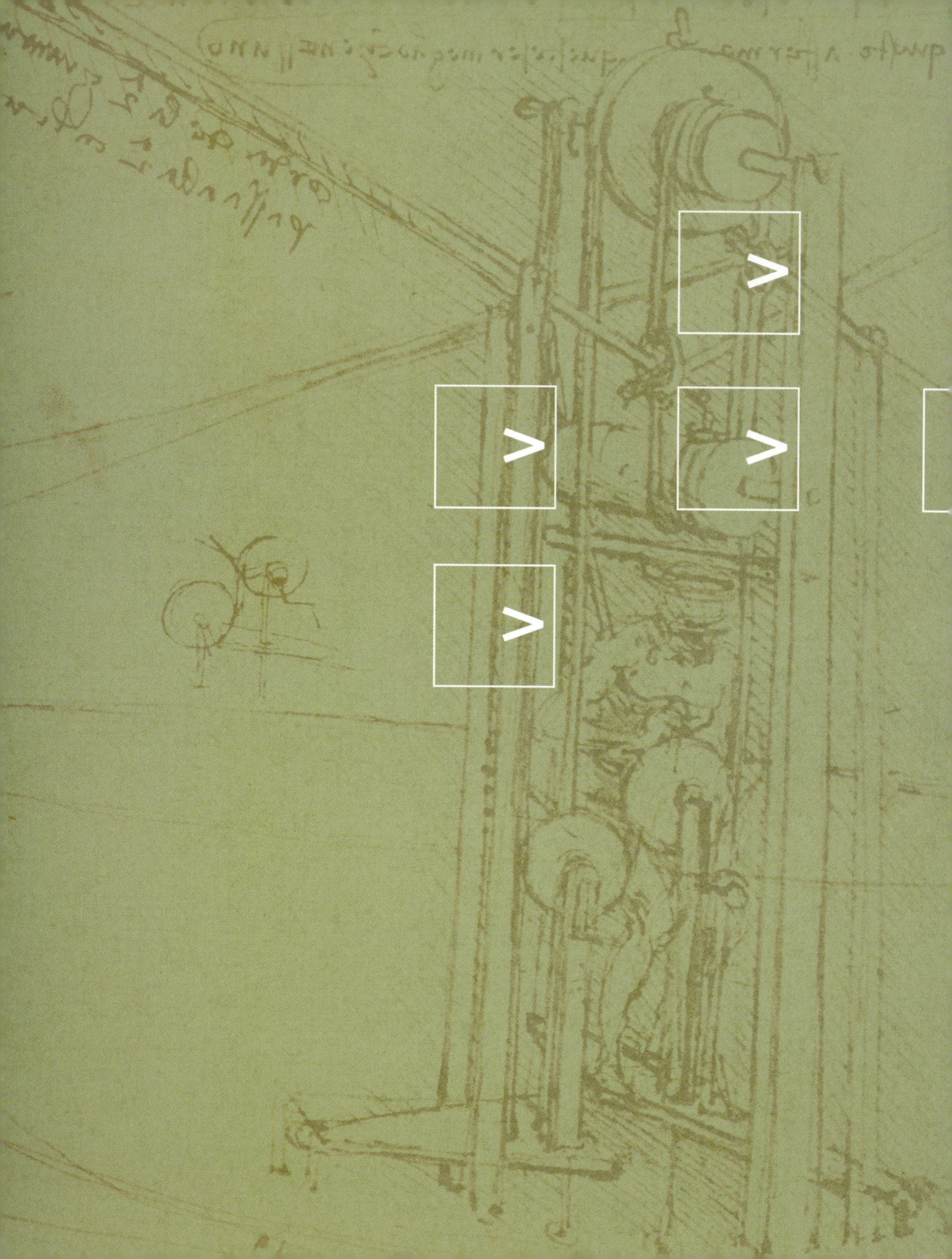

CHAPTER 7

Planning and Drafting

Objectives and Outcomes

This chapter will help you accomplish these outcomes:

- Learn the strategies used by experienced writers as they explore, plan, and draft documents, oral presentations, and visuals
- As part of *inventing and exploring,* use a problem-solving process with proven strategies: brainstorming, 5 *W*s plus *H,* cause-and-effect analysis, and synectics
- As part of *planning,* consider types of planning, use project-management tools, make decisions about rhetorical elements (content, purpose, task, audience, constraints, organization, and design), assess the logic
- As part of *drafting,* select the appropriate person and verb mood, use plain language, avoid density, and use given-new constructions

Research about writing shows that experienced writers recognize the importance of taking time to investigate their subject, plan their approach, and organize the draft of their information. They are concerned with accuracy as well as with the appropriateness and appeal of the information for the intended audience. They typically ask whether their documents, oral presentations, and visuals are accessible, comprehensible, and usable. In contrast, inexperienced writers tend to be concerned with whether they have enough content, minimize their planning, and not think much about the audience.

Let's look at the details of a process that you might use to create a document, oral presentation, or visual. Although the specific steps will change to suit your individual approach and the requirements of a particular project, experienced professionals typically move through these steps during their process:

- Inventing and exploring
- Planning and organizing
- Drafting and designing
- Revising
- Editing

This chapter focuses on what happens in inventing/exploring, planning/organizing, and drafting and suggests a number of strategies to help you become more expert-like in these parts of the process. Revising and editing are the focus of Chapter 8.

Differences Between Writing Processes

What do skillful writers do when they write? The writing process used to be described as three linear stages: prewriting, drafting, and revising. However, as researchers have charted the actions and ideas of actual writers through the writing process, the notion that the process is linear has been dispelled. Writers say things like, "Lots of times while I'm in the middle of writing, I'll get a terrific idea — something I've never thought of before, something that's not part of my outline." Or, "I'll write one word and immediately know it's not quite right, so I'll scratch it out and write a new one." These people are saying that the steps in this creative process do not necessarily happen consecutively; rather, they are typically recursive — that is, steps happen more than once, and more than one step can happen at the same time.

Producing a technical document, oral presentation, or visual isn't a solitary process, either. You collaborate with other people for many different reasons (as you read in Chapter 5): to get the necessary information, to verify your approach, to confirm technical accuracy, and to check readers' reactions. Figure 7.1 identifies some of the major categories you need to consider and some of the questions you need to ask throughout every project. You can learn good questions to ask by starting with any category in Figure 7.1.

FIGURE 7.1 Questions to Ask about Communication Relationships

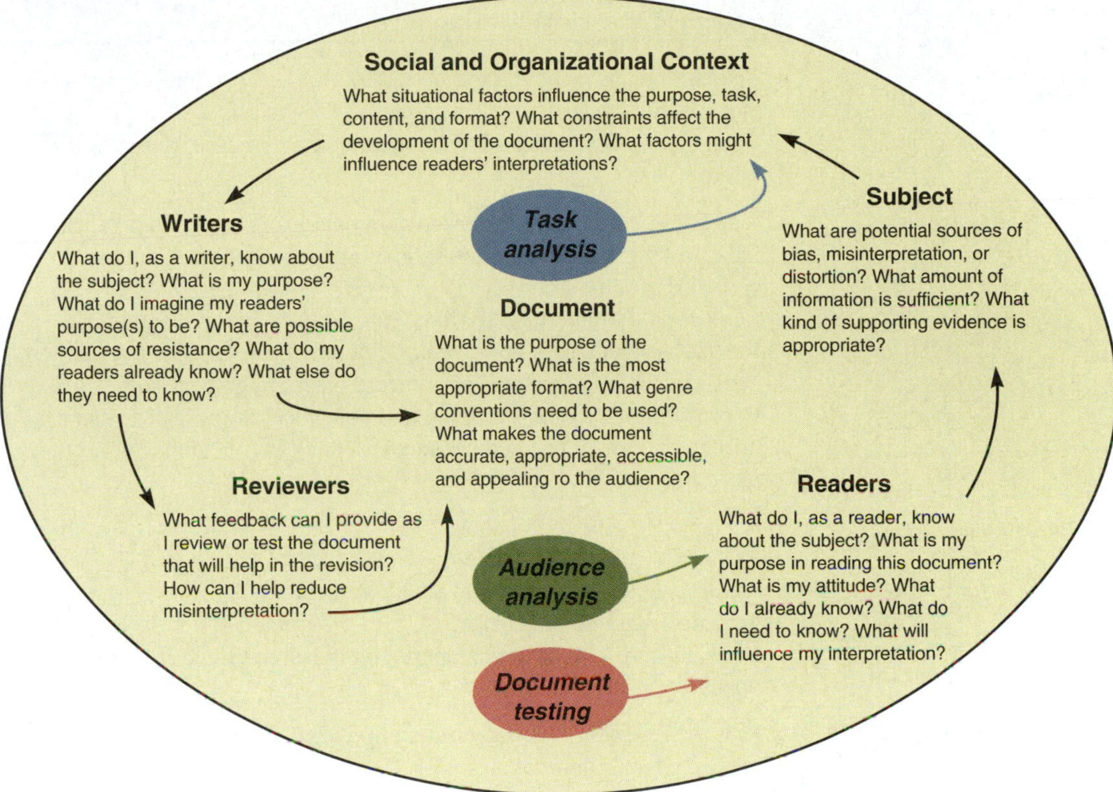
Social and Organizational Context
What situational factors influence the purpose, task, content, and format? What constraints affect the development of the document? What factors might influence readers' interpretations?
Writers
What do I, as a writer, know about the subject? What is my purpose? What do I imagine my readers' purpose(s) to be? What are possible sources of resistance? What do my readers already know? What else do they need to know?
Task analysis
Document
What is the purpose of the document? What is the most appropriate format? What genre conventions need to be used? What makes the document accurate, appropriate, accessible, and appealing ro the audience?
Subject
What are potential sources of bias, misinterpretation, or distortion? What amount of information is sufficient? What kind of supporting evidence is appropriate?
Reviewers
What feedback can I provide as I review or test the document that will help in the revision? How can I help reduce misinterpretation?
Audience analysis
Document testing
Readers
What do I, as a reader, know about the subject? What is my purpose in reading this document? What is my attitude? What do I already know? What do I need to know? What will influence my interpretation?

Many skillful communicators begin by asking about the context, the subject, and the document, oral presentation, or visual itself. But you can also start with yourself, depending on whether your role is as a writer (or speaker or designer), reader (or listener or viewer), or reviewer. Take time to carefully read the questions in Figure 7.1 and get accustomed to asking them every time you plan a document, presentation, or visual. When you're working for a specific organization, you'll add questions to those suggested here based on the people and politics of that organization.

When you find processes for inventing/exploring, planning, and drafting that work for you, examine them, use them, develop them. Once you're confident of an approach, branch out and try another one. If it works, integrate it with your writing practices. Figure 7.2 reviews some of the habits, concerns, and working procedures that are used by experienced workplace writers. You goal should be to develop these expert processes.

FIGURE 7.2 Composing Processes of Experienced Writers

Stage	Experienced Workplace Writers
Inventing and exploring	■ Spend time inventing and exploring. Experienced communicators see value in such activities as cause-and-effect analysis. ■ Spend time considering and contemplating the implications of various alternatives.
Planning and organizing	■ Plan before drafting. Experienced communicators plan and organize ideas in their heads, in conversations with other people, on their computers, and on paper. ■ Consider a lot more than content. Experienced communicators ask themselves questions about their purpose, audience, and organization. ■ Create notes and sketches so a record exists to go back to. ■ Reconsider plans during drafting and designing; revise plans whenever necessary. Experienced communicators don't stop planning when they start drafting.
Drafting/Designing	■ Respond to audience's needs. ■ Frequently stop to rescan, reread, reflect. ■ Respond to all aspects of the situation: content, context, audience, purpose, organization, design. ■ Show concern with accuracy, appropriateness, accessibility, and appeal.
Revising and editing	■ First revise globally for organization and logic, then locally for sentence- and word-level conventions. ■ See revising as ongoing throughout the process of preparing a document, oral presentation, or visual. ■ Proofread with care.

Inventing and Exploring

Communicators really do talk to themselves; they invent an idea and listen to how it sounds. During your inventing and exploring, you may assess your knowledge, read and review available background references, ask questions and discuss ideas, make observations, conduct experiments, and take notes. You may make tentative decisions about your task. Thinking about your options and your approach is part of this stage, one that you will return to many times while preparing a document, presentation, or visual.

Problem-Solving Processes

Inventing and exploring offer an opportunity to investigate problems related to a particular document, oral presentation, or visual; as you consider possible solutions, effective problem solving can make your work easier. For example, you may have problems identifying good alternatives or determining the focus. One way you can be a more skillful problem solver is to go through a problem-solving process (which, like writing, is recursive, with overlapping stages).

Productive individuals and groups in the workplace consciously follow a problem-solving process. They initially (1) identify the context (politics, policies, parameters for solutions) and the problem (which may be defined differently by different people). Then they (2) gather and evaluate information in order to set priorities. Next they (3) typically formulate alternative possible solutions that may differ from conventional approaches, being careful to defer judgment about the preferred solution. Individuals (4) draw on their prior knowledge and experiences with similar problems in order to determine the preferred outcome and the criteria for success. Then they (5) assess the alternatives and select one plan of action. At that point, the group can (6) get started on the most appropriate, efficient solution, considering technical as well as organizational and interpersonal factors. As the solution is implemented, people need (7) to monitor individual and group performance against the outcome. Finally, they can (8) evaluate their performance to determine success of the problem-solving process and final product.

Which of these four problem-solving strategies have you used? How useful have they been? Which ones do you think you'll try in the future?

Problem-Solving Strategies

This section of the chapter applies problem solving to four commonly used and successful strategies: brainstorming, 5*W*s + *H*, cause-and-effect analysis, and synectics.

WEBLINK

To learn more about problem-solving strategies, ones that reinforce a constructive process as well as engage your creativity, go to www.english.wadsworth.com/burnett6e for useful links.

CLICK ON WEBLINK

CLICK on Chapter 7/problem-solving steps

You're probably already familiar with the first problem-solving strategy, *brainstorming,* which encourages you to suggest as many ideas as possible about a given problem or situation without making any judgments until after a number of ideas have been suggested. Despite being remarkably simple, brainstorming is a problem-solving strategy that often works.

© Comstock Images/Getty Images

In a productive problem-solving situation, team members usually follow an established process, which means that along the way they consider lots of alternatives and engage in a considerable amount of substantive conflict (see Chapter 5). People working independently can also follow the same productive process, asking themselves tough questions.

Two other strategies are commonly used by quality control circles.* One of these strategies is also a formula frequently associated with journalism: *5 Ws plus* H. These questions are effective for approaching any problem you are examining:

- WHO — Who is involved? Who should be involved?
- WHAT — What is involved? What should be changed? What should remain the same?
- WHEN — When should it be done? When is the most appropriate or convenient time?
- WHERE — Where should it be done?
- WHY — Why should it be done?
- HOW — How should it be done?

Another strategy recommended by quality control circles is *cause-and-effect analysis,* which focuses on the causes of a particular problem. The initial causes are often separated into four categories: *machine, employee, material,* and *method.* Any of these possible causes can be expanded with additional headings. *Cause-and-effect charts* (also called *fishbone charts*) are a useful way to summarize causal relationships.

*A *quality control circle* is a group problem-solving process used in many companies. Small groups of people doing similar work go through a training period and then meet regularly to identify, analyze, and solve problems connected to the work they do. Many companies have established quality control circles, using research pioneered in Japan, to improve productivity as well as employee involvement.

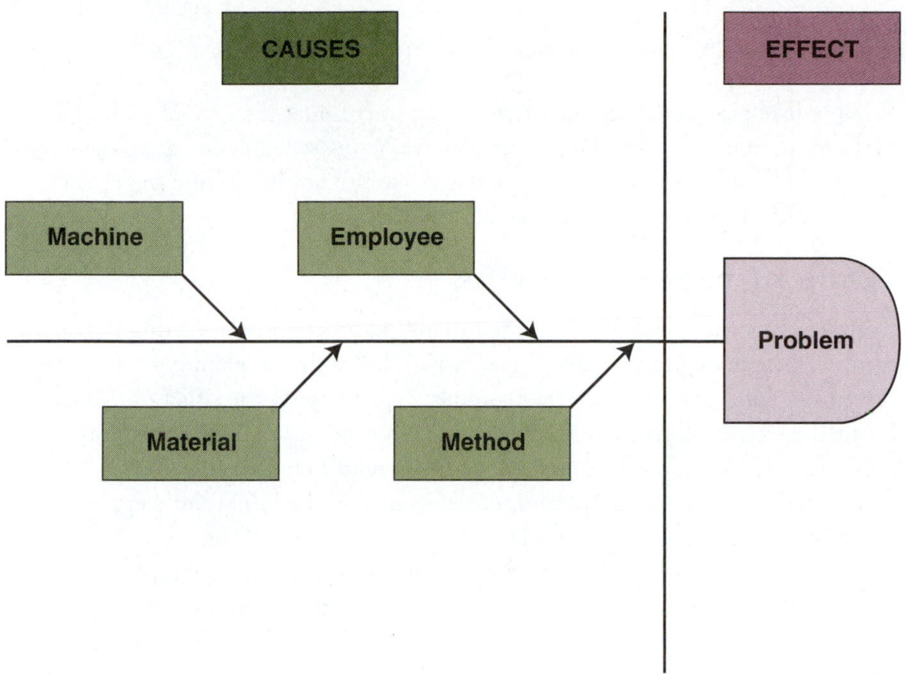

A final problem-solving strategy is *synectics,* often used collaboratively to stimulate creative thinking. *Synectics* is a coined word, from the Greek *syn* (meaning "to bring together") and *ectos* (for "diversity"), which, combined, suggest bringing together diverse people and ideas.[1] This "bringing together" encourages workplace professionals to, for example, combine two unrelated ideas as a means of analyzing a particular problem. The goal is to develop new perspectives and solutions. Using a variety of techniques, including metaphors, analogies, role-playing, and simulations, participants first define a particularly difficult problem. Then they put themselves *into* the problem by creating a metaphor or an analogy or by acting out a component of the problem. In synectics, participants are encouraged to engage in a variety of strategies: search for ideal solutions, imagine that they are the subject, compare the subject to something concrete, and compare the subject to something abstract. You can, of course, use similar problem-solving strategies at any time during the writing process.

To learn more about synectics, a widely used corporate problem-solving process (which developed from William Gordon's book, *Synectics,* published in the 1960s), go to **www.english.wadsworth.com/burnett6e** for useful links.
CLICK ON WEBLINK
CLICK on Chapter 7/synectics

WEBLINK

When you're planning a document, you need to consider the type of planning that's most appropriate for the project, determine the schedule for the project by arranging the sequence of activities, and concentrate on the specific rhetorical elements for the document or presentation.

Types of Expert Planning

Experienced writers usually match their planning process to the writing task. They seem to shift among three types of planning, schema-driven planning, knowledge-driven planning, and constructive planning, depending on the situation. Knowing about these types of planning can help you select the most appropriate kind for a particular task. Inexperienced writers, however, tend to be concerned with simply completing the task and often minimize or even skip the important steps of inventing/exploring and planning and immediately start drafting.

If you create a document according to an existing format or template for a text, you'll follow a *schema.* A schema is your knowledge, your mental image of what is expected or appropriate in a given situation. You already have schema for many texts like memos and business letters. If you use schema-driven planning, you can often complete a document without much thought about format or the kind of information to include. For example, if you have a memo schema, you can quickly produce a memo announcing a meeting. If you have a business letter schema, you can quickly produce a letter requesting product specifications. Writers who use schema-driven planning can usually produce successful documents if their schema is appropriate for the situation.

In another situation, you might not have a schema, but you might have a great deal of knowledge about the subject. In this case, planning a document becomes a matter of your telling what you know about the subject. For *knowledge-driven planning* to work, your information must be well organized and appropriate for the task. For example, a chemist who has a great deal of experience with gas chromatography has the procedural knowledge to write instructions for technicians to operate the equipment fairly easily.

In contrast, if you have a difficult writing problem, you probably can't depend on existing schema or knowledge. Instead, you'll have to do *constructive planning,* which requires careful analysis of the purpose, audience, task, and a variety of constraints. Although constructive planning usually requires more time and effort than other kinds of planning, it helps you set goals and criteria as well as establish procedures for composing your document.

Often you may use a combination of schema-driven planning, knowledge-driven planning, and constructive planning. For example, writers preparing a proposal for municipal recycling of refuse need to be familiar with the format and language of proposals (their schema) and also need to have extensive knowledge about the subject. But the specific problem of preparing a proposal for recycling in a particular community requires constructive planning, so that the writers can solve the problems of the task and establish procedures for completing it.

Project Planning

Inadequate project planning is one of the most common reasons that projects fail to produce excellent results. Early in the process of planning, workplace professionals often manage their time by sketching out what needs to be done and when it needs to be done. Your informal plan may be nothing more than a self-stick note stuck on a bulletin board or on the edge of your computer screen. Or you may prefer notes jotted on a calendar or electronic organizer. Maybe your style is to make a list of tasks that you paper clip to the front of a manila folder containing notes for the project. Whatever the form you choose, though, project planning serves an important time-management function by encouraging you to make a preliminary assessment of the tasks that need to be done, their order, and their deadlines.

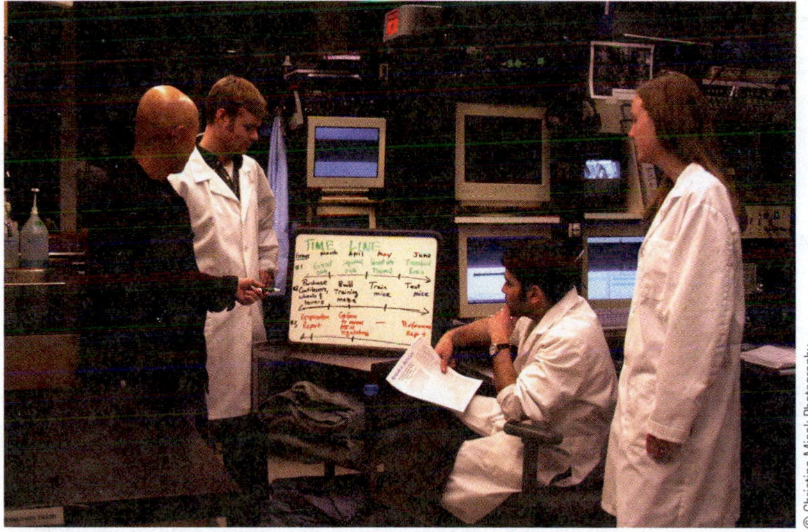

Planning and updating the schedule for a complex project requires input from all the major players on a team. Checking the milestone dates on the schedule with the actual progress should be a regular weekly task.

For larger projects, professionals often turn to more formal time- and project-management tools. Two common tools are Program Evaluation Review Technique (PERT) charts and Gantt charts, both of which keep track of the various project tasks. PERT charts not only track project activities but also show how these activities depend on each other. Gantt charts track project activities on a timeline. Both of these project management tools enable you to plan projects and determine what should be accomplished at each stage. Figure 7.3 (PERT chart) and

Figure 7.4 (Gantt chart) illustrate the same project so that you can easily compare the two tools.

FIGURE 7.3 **PERT Chart Showing Steps in Revision of Mitsubishi TV Manual**[2]

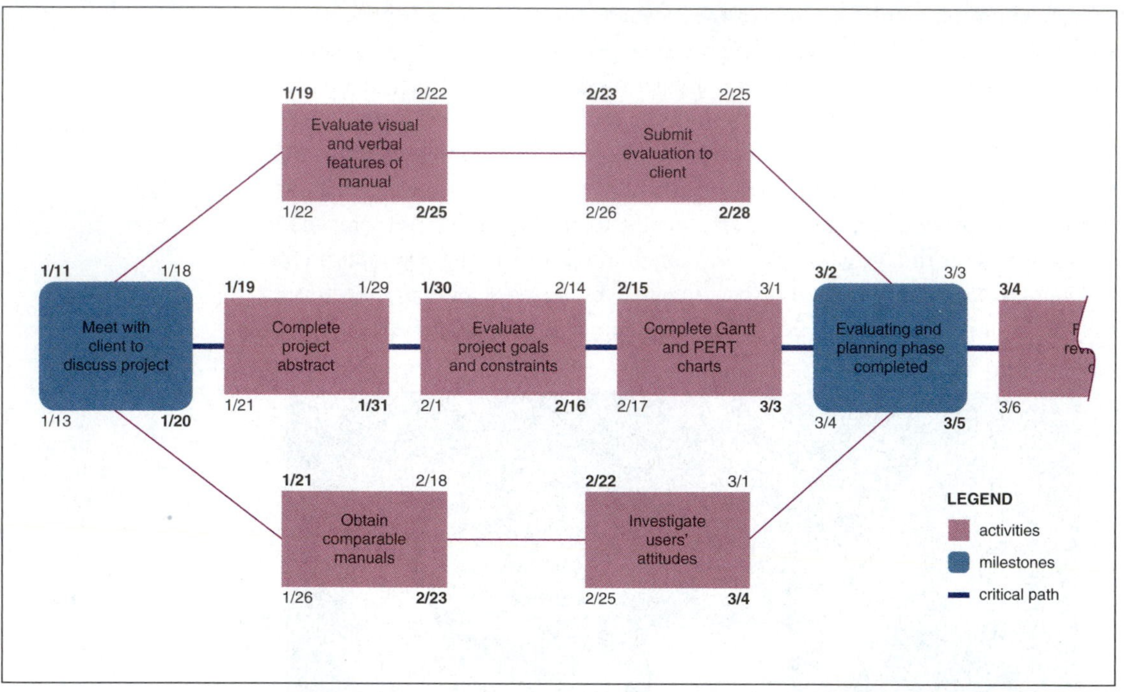

Computer software is available that helps you construct both kinds of time-management charts. Project planning software can make creating PERT and Gantt charts very easy; managers can print up-to-date charts, status reports, and graphs of the project's progress. More sophisticated packages allow users to link the charts for various projects so that managers can see conflicts in deadlines and balance workloads. Project planning software used on a computer network helps managers automatically update the schedule and inform project members. For example, when a manager changes the project schedule, all project members are first notified and then receive e-mail reminding them of changes in upcoming deadlines.

WEBLINK

To see examples of project-planning charts as well as other useful visual displays, go to **www.english.wadsworth.com/burnett6e** for links to these project management tools.
CLICK ON WEBLINK
CLICK on Chapter 7/project management

FIGURE 7.4 **Gantt Chart Showing Phases of Revision of Mitsubishi TV Manual**[3]

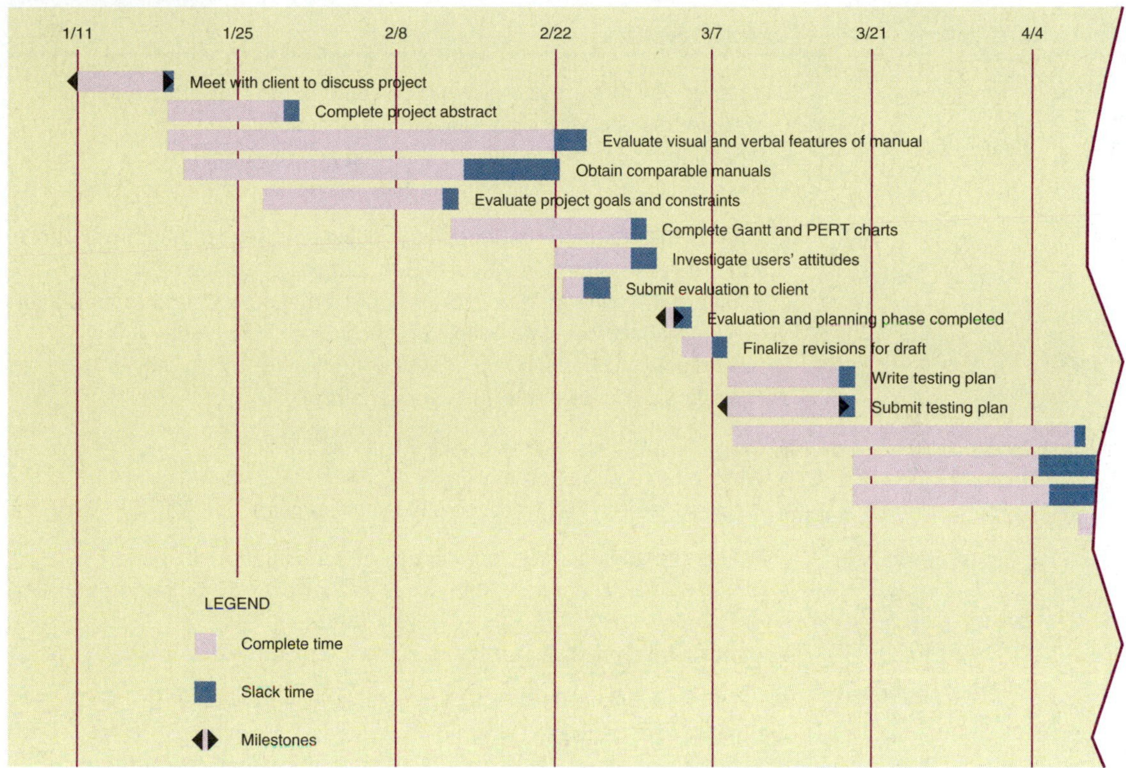

Considering Rhetorical Elements

When you plan, you need to make decisions about a variety of rhetorical elements (which you have read about in Chapter 1). Figure 7.5 lists these rhetorical elements again, this time as areas that skillful communicators consider as they plan their documents, oral presentations, and visuals. The critical questions in Figure 7.5 provide suggestions you can use as you make decisions. As you use these questions, you should jot down additional ones that also help.

What planning questions do you believe will be useful for the kinds of communicating you expect to do in your career?

What you read at the beginning of this chapter about the overall process of writing being recursive also applies to the process of planning: Each element affects the others. The sequence in which you consider these elements has no best order. As your work on a document, presentation, or visual progresses, you can change your plans, but you need to begin with a sense of what you're doing and where you're going, in part because changes in one element influence the other elements. Although skillful communicators plan often and replan as their work progresses, careful planning can minimize the number of major changes that are

FIGURE 7.5 Critical Questions to Ask during Planning

Rhetorical Elements	Critical Questions
Content	■ What do you already know? What do you need to learn to plan and prepare this document, presentation, or visual? ■ What information do you want to include and exclude? ■ What examples and explanations will help achieve the purpose?
Context	■ What is the situation that needs the document, presentation, or visual? How does this situation influence the rest of your planning? ■ What are the constraints — for example, time, subject and format, audience, collaboration, data collection, technology, and noise — that you face in planning? ■ How should you deal with these constraints?
Purpose	■ Do you want to inform? Persuade? Instruct? Train? ■ What argument are you making? ■ What is the best way to accomplish your purpose?
Audience	■ Who is the audience for your document, presentation, or visual? ■ What are their characteristics? What do they expect? What do they need? ■ How will the audience use the document, presentation, or visual?
Organization	■ What is the most effective way to organize the information? ■ How will you ensure that key points are easy to locate? ■ What have you done to ensure that the audience can follow the argument? That the points are coherent?
Visuals	■ Which information should be presented verbally, and which should be presented visually? ■ What visual conventions need to be followed?
Document design	■ How can the design reinforce the purpose of the document, presentation, or visual? ■ What design features will appeal to the audience? ■ What design conventions need to be followed?
Usability	■ How will you determine that the information is accurate? ■ How will you determine that the information in the document, presentation, or visual is usable for all of the intended users?
Language conventions	■ What professional standards are you using? ■ What will the audience expect?

needed later. Simply put, you'll find changing a plan easier than changing a completed document, presentation, or visual.

Planning is often more productive if it is collaborative, even if you are going to draft independently. Working with someone else often generates ideas and approaches that you might not come up with when working alone. Unfortunately, you will almost always be under a deadline, so make time for planning, whether a few minutes alone as you develop ideas for a short letter, or two-minute oral update during a meeting, or a few weeks with others as you collaborate on a complex proposal or an important conference presentation. Despite temptations to jump right into the actual work, you should not skimp on exploring and planning.

One of the most important things you should pay attention to during planning is the ethics of the situation. The ethics sidebar below specifically addresses ethical questions you need to answer early in your planning. Having accurate content or an appealing design doesn't much matter if you're perceived as unethical. The sidebar focuses on achieving a balance between your personal (or private) positions and your public positions.

What issues in your discipline might cause inconsistencies between your private and public positions?

ETHICS SIDEBAR

"Is the communication honest?": Ethics and the Writing Process

Technical professionals face ethical decisions throughout the writing process. From initial brainstorming through the creation of the document, to post-production distribution, technical professionals must continually be aware of the ethical aspects of their communication. Researcher Brenda Sims provides a useful list of questions to ask yourself as you plan, draft, and revise documents and presentations:[4]

- Is the communication honest and truthful?
- Am I acting in the organization's best interest? In the public's? In my own?
- What if everybody acted or communicated in this way?
- Does the action or communication violate the rights of any of the people involved?
- Am I treating similar situations similarly?
- Am I willing to take responsibility for the action or communication publicly and privately?

Asking yourself these questions helps you look at your communication from personal and public perspectives. Ethical problems can occur when we don't look at how communication is perceived privately and publicly.

If you are willing to take responsibility for a communication privately but not publicly, then your behavior may be viewed as unethical by your colleagues or the general public. For example, if you send a private memo to a colleague objecting to company hiring practices but you are unwilling to voice the same concerns publicly, you may be perceived by some colleagues as acting unethically because you will not take a public position.

If, on the other hand, you take a public position but it conflicts with your private view, then your personal ethics may also be in question. For example, if you tell sales representatives that an owner's manual covers all possible problems that owners will encounter, but you neglect to mention that you personally believe that some explanations are very difficult to understand, some people may question your ethics.

How can you revise documents so that they are acceptable privately and publicly?

Ethical problems like these affect your professional standing and credibility. In most companies, organizations, and agencies, you will be held responsible for what you communicate. Using checklists like the one provided by Sims helps you avoid potential ethical problems that could negatively affect your writing and career.

Assessing the Logic

One of the most difficult parts of planning is making sure your document, oral presentation, or visual is logical. You must first recognize potential problems in logic and then know how to correct them. Problems in logic generally fall into four categories:

- Using data from authorities
- Presenting facts without drawing inferences
- Drawing inferences
- Establishing causal relationships

Once you recognize these pitfalls, you are likely to judge your data more critically. The following discussion defines these four categories of logic problems, analyzes examples, and suggests possible ways to eliminate the problems before they find their way into a draft.

What problem(s) have you observed that involved using data from authorities?

Using Data from Authorities Your audience will often respond positively to data from a recognized authority in the field. However, the data lose their effectiveness if the expert's qualifications are suspect or if the expert has a vested interest in the subject. If people or organizations have a vested interest (that is, a proprietary concern in supporting a particular position), their views might be more self-serving than objective. For example, imagine that the New England Soy Dairy prints a flier that states, "People everywhere are excited about the prospect of adding tofu to their diets." The business of the New England Soy Dairy is to market soy products, so it is in their interest to promote positive attitudes about soy products. Before accepting their word about the widespread excitement regarding soy products, you should verify the information with less vested sources.

Presenting Facts without Drawing Inferences You present verifiable information (what we usually call *facts*) in the form of statistics, quotations, or visuals intended to support points in a document or oral presentation. If the facts you present are incomplete, out of context, oversimplified, or distorted, the audience will be misled.

What problem(s) have you observed that involved at least one of the following kinds of presenting facts without drawing inferences?

- ***Omitted or Incomplete Data.*** Omitted or incomplete data, whether deliberate or accidental, affect the accuracy and credibility of a document or oral presentation. Imagine a novice computer user trying to follow instructions that omit a critical piece of information, such as "Press Enter after completing each step" — information that is second nature to anyone who has some computer experience. Having omitted or incomplete data sometimes results from failure to check and recheck facts, but the problem also occurs when a writer hopes to influence the audience by omitting discrepant or conflicting data, as the following example shows:

Statement	Soy-based infant formula is generally the best food for infants with milk allergies.
Comment	The statement fails to mention that breast milk is the best food for infants and greatly reduces the likelihood that an infant will develop milk allergies.
Revision	When breast-feeding is impossible or is not the feeding method of choice, soy-based infant formula is generally the best alternative for infants with milk allergies.

- ***Out-of-Context Data.*** Out-of-context information may be accurate in itself, but if relevant background or related information is omitted, the audience will have a distorted view. Out of context, even indisputable information can be misleading, as seen in this example:

Statement	American farmers often prefer planting corn to soybeans.
Comment	Although the statement is true, taking it out of context ignores both the important reasons that influence farmers' decisions and the relationship between soybeans and other major crops.
Revision	Soybean crops compete with other major crops in the principal production areas, so the amount of land planted with soybeans depends on a complex interaction of factors that affect both soybeans and the competing crops. Most significant appears to be the price relationship between corn and soybeans in the major Midwest area, and between cotton and soybeans in the Mississippi Delta. This relationship, in turn, depends on world market factors, which most recently have tended to favor corn over soybeans and soybeans over cotton.[5]

- ***Oversimplified or Distorted Data.*** Oversimplification reduces a complex situation or concept to a few simple ideas, ignores relevant details, and gives the audience a distorted view, as the following example illustrates:

Statement	Adding soy protein to a person's diet improves the quality of the diet.
Comment	Many factors determine the quality of a diet; the interrelationships of all dietary components must be considered. Adding soy protein to some diets may result in an excess protein intake, limit the intake of other necessary nutrients, and so on.
Revision	The inclusion of soy protein in a person's diet may have economic, nutritional, and sensory benefits.

What problem(s) have you observed that involved at least one of the following kinds of drawing inferences?

Drawing Inferences During your planning of a document, oral presentation, or visual, you obtain data and then draw inferences based on that data. These inferences will be faulty if you make hasty generalizations, assume irrelevant functions, or fall prey to fallacies of composition and division. Data should represent the general condition or situation.

- ***Hasty Generalizations.*** A hasty generalization occurs when a particular case or situation is mistakenly assumed to be typical or representative of a group. This error usually takes place when you have not examined enough cases to warrant a generalization, as in the following example:

Statement	In a survey of members of food cooperatives, 90 percent stated they would like to increase the amount of soy in their diets. Therefore, Americans are willing to eat more soy foods.
Comment	Members of food cooperatives may not be representative of the greater population.
Revision	The willingness of co-op members surveyed to increase soy food intake suggests a potential for increased soy food consumption in at least some segments of the American population.

- ***Irrelevant Functions.*** When you criticize something for not having a characteristic it was never intended to have, you are guilty of assuming an irrelevant function:

Statement	Tofu and tempeh (a traditional Indonesian cultured soybean food) do not taste like meat.
Comment	Tofu and tempeh, traditional foods in several cultures, should be appreciated for their own characteristics. They are not intended to taste like meat.
Revision	Tofu and tempeh are alternative sources of protein for people who choose to eat less meat in their diet.

- ***Composition and Division.*** The fallacy of composition assumes that the characteristics of the individual components can be attributed to the whole. The reverse occurs in the fallacy of division, in which the characteristics of the whole are attributed to each of the individual components composing the whole. The following example illustrates the fallacy of division:

Statement	Peanut butter on wheat bread is a protein dish of high biological value (it is a "complete protein"); therefore, peanut butter alone or wheat bread alone has a high biological value (is a "complete protein").
Comment	In this example of the fallacy of division, the basic assumption is faulty. In fact, peanuts (peanut butter) and grain (bread) proteins complement each other to together provide protein of high biological value.
Revision	Peanut butter combined with wheat bread is a protein dish of high biological value.

What problem(s) have you observed that involved at least one of the following kinds of causal relationships?

Establishing Causal Relationships One of the most frequent relationships employed in technical documents, oral presentations, or visuals establishes causes and resulting effects. Your audience can be misled by several kinds of poorly drawn causal relationships, including a condition not being sufficient cause, variables not being correlated, and the fallacy of *post hoc, ergo propter hoc.*

- ***Condition Not a Sufficient Cause.*** An effect that results from multiple causes must have sufficient causes, and it may also have contributing causes. A *sufficient cause* is one that by itself can produce the effect. A *contributing cause* may help bring about the effect, but if all other causes are eliminated, any one contributing cause by itself cannot produce the effect. The following example demonstrates how poorly established cause–effect relationships can mislead an audience:

Statement	The patient's heart attack was caused because his diet contained too much cholesterol.
Comment	Heart disease is caused by a number of contributing factors, such as diet, obesity, smoking, heredity, hypertension, and amount of exercise. Evidence has not yet shown that cholesterol alone is a sufficient cause, but it is usually a contributing cause.
Revision	High cholesterol may have contributed to the patient's heart disease.

- ***Variables Not Correlated.*** Even though a number of factors frequently occur in sequence, a cause–effect relationship does not necessarily exist:

Statement	In the 1970s, the U.S. production of soybeans rose, and soon after, Americans began eating more soybeans.
Comment	The increase in soy production was primarily to help the balance of trade. Most of the soy was exported and used as animal feed. Americans' increase in soy consumption has resulted from a number of factors such as increased awareness, marketing, and availability.

Revision	In the 1970s, the U.S. production of soybeans rose to help the balance of trade. At the same time, Americans began eating more soybeans in response to a combination of factors, such as increased awareness, marketing, and availability.

- ***Fallacy of* Post Hoc, Ergo Propter Hoc.** This line of reasoning (usually simply called *post hoc*) claims that because one event follows another, the first event must cause the second. The Latin translates as "after this, therefore because of this." The following example demonstrates how such fallacious logic can undermine a document's credibility:

Statement	Japanese women eat a high proportion of soy protein in their diets. Japanese women also have a low incidence of breast cancer; therefore, eating soy lowers the likelihood of breast cancer.
Comment	Even though eating soy precedes the low incidence of breast cancer, the chronology does not necessarily indicate a causal relationship. Japanese women also have a high intake of fish. Does eating fish also lower the risk of breast cancer?
Revision	Japanese women eat a high proportion of soy protein in their diets. Japanese women also have a low incidence of breast cancer. Emphasizing low-fat foods in the diet (such as soy and some types of fish) may be a contributing factor in lowering the risk of breast cancer.

When you're finished revising a document or presentation, checking the statements in Figure 7.6 will help you evaluate whether the information will withstand scrutiny.

FIGURE 7.6 **Questions about the Logic of a Document to Ask Yourself During Revision**

If you answer "I agree" to each of these statements, you can be fairly sure the logic in your document, oral presentation, or visuals will withstand scrutiny.	*I agree*
All the evidence and data are accurate.	❑
No information has been deliberately or accidentally omitted, oversimplified, or distorted.	❑
No sources have special vested interests.	❑
Examples and cases are representative of the general condition(s) or situation(s).	❑
Data are presented with appropriate background information or discussion so that they are meaningful to readers.	❑
Sources have expertise in the area being discussed.	❑
Visuals are presented without distortion of any kind.	❑
Nothing is criticized for something that is not part of its nature or function.	❑
Causes and effects are clearly differentiated and not mistakenly assumed because of a sequence of events.	❑

Drafting involves writing the text and preparing the visuals. Writers, speakers, and designers have many different ways of approaching this stage of the process:

- You may create bits and pieces of a draft during the planning stage, recording key sentences or making preliminary sketches of things you don't want to forget. When your planning seems to be done, you take these ideas, develop them, and fill in the spaces.
- You may sit down to compose or design from beginning to end without interruption.
- You may prepare an outline and notes or a rough thumbnail sketch, then set them aside and start to write or draw, checking your planning materials only when you get stuck.
- You may generate an online outline or a sketch as the framework of your document or visual and then stick to this outline or sketch.
- You may work slowly, pondering and polishing every phrase and sentence or every line and figure as you go along.
- You may work rapidly, encouraging the ideas to flow quickly, tumbling onto each other.

One key to drafting is trying several approaches in order to find out which ways work best for you. A second key is seeing drafting as recursive, something you do throughout the process of preparing a document, oral presentation, or visual. You draft as you think of new ideas, as the nature of the project changes, as sections need to be elaborated, as new information become available. Drafting is seldom a "do it once" part of the process.

Selecting Person

Choosing among first, second, or third person depends on the purpose and audience of the document, oral presentation, or visual. First person (*I, we*) is appropriate if, for example, you are narrating the events in a sequence in which you were involved. If your role in the action was significant, the use of first person emphasizes this. Second person (*you*) is usually reserved for instructions, since readers are being directed to complete a particular action. Third person (*he, she, one, it, they*) is most common, allowing you to emphasize the sequence of action rather than yourself or the readers.

While the guidelines above are appropriate for default decisions, you should follow the conventions of the organization or journal that is the primary audience for your document, oral presentation, or visual. For example, some organizations and journals require third person, while others accept and sometimes even expect first person.

To read about ways to manage gender problems with pronouns, go to www.english.wadsworth.com/burnett6e for a link to a discussion about ways to avoid sexist language.
CLICK ON WEBLINK
CLICK on Chapter 7/sexist language

Verb Mood

Mood refers to the characteristics of verbs that convey a speaker's or writer's attitude toward a statement. During drafting, you should select the verb mood that's appropriate for the situation.

- Indicative mood states facts or opinions or asks questions.

fact	Nurses *use* a rectal thermometer to take a baby's temperature.
opinion	Child-care providers *prefer* a thermometer strip rather than a rectal thermometer for taking a baby's temperature.
question	Why should child-care providers *use* a thermometer strip to take a baby's temperature?

- Imperative mood expresses commands or gives a direction.

command	(You) *Use* a rectal thermometer to take a baby's temperature.
direction	(You) *Get* a thermometer from the drawer labeled "Thermometers" in Utility Room II.

- Subjunctive mood expresses recommendations, wishes, conjectures, indirect requests, and statements of conditions contrary to fact.

recommendation	When children are hospitalized, we recommend that parents *be* with them. [not *are*]
wish	I wish the clinic *were* able to provide more well-baby classes. [not *was*]
conjecture	If the budget *were* not cut, we would have electronic thermometers. [not *was*]
indirect request	If the rectal thermometer *were* to break, a baby could be seriously injured. [not *was*]
condition contrary to fact	The parent asked if the examination *were* almost over. [not *was*]

Because process explanations deal with observable, verifiable information, they are written in indicative mood. But you also need to be able to recognize and use imperative mood (especially for directions) and subjunctive mood (useful in correspondence and reports).

Selecting Active or Passive Voice

Your selection of active or passive voice depends on the subject, the purpose and focus, and the audience. Active voice emphasizes the doer of the action (sometimes called the agent) and de-emphasizes the receiver, which is appropriate in most situations. Such emphasis is appropriate when readers are more interested in the doer or in the action itself than in the receiver or result of the action. Material presented in active voice is also more interesting to read, in part because the subject of the sentence is responsible for the action. Active voice is more direct; the audience does not have to work as hard to figure out who does what. Active voice is less wordy because the "to be" verb form necessary to create a passive construction is omitted. In general, you can use active voice unless you have a specific reason not to. You need to think about active and passive voice while you're drafting a document. The agent (the doer) in active voice does not always have to be a person, or even a living organism, as the following example illustrates:

> Solid-state computer scanners maintain a file of fingerprints. When a fingerprint is put into the file, the computer scanner identifies the characteristic "points" of a fingerprint (up to 150 points, though only 12 are needed for legal identification) and then converts them to a series of numbers that are stored on a computer disk file. When a fingerprint from a crime scene is digitized in a similar manner, the scanner attempts to match the "numbers." Each disk drive stores up to 90,000 fingerprint cards, each card with 10 prints. The computer can riffle through 1,200 such points every second. The computer can increase arrest rates based on fingerprints 10 to 15 percent above the current levels.[6]

In this example, the agent is the computer scanner, which can maintain a file, identify characteristic "points," convert them for storage, and, later, match the numbers.

Passive voice also has appropriate uses. When the receiver is more important than the agent, passive voice is usually correct. In this following example, the agent ("by Hollywood lore") is less important than the concept ("the fingerprint").

Passive voice	The fingerprint is overrated by Hollywood lore as a way to catch criminals.

Passive voice is also appropriate when the agent is unknown or insignificant.

Passive voice	Fingerprints are collected only between 25 and 30 percent of the time, even though they are usually the most prevalent form of physical evidence at the scene of a crime.

In this sentence the agent is not identified ("Fingerprints are collected [by unknown detectives]"), so passive voice is not only appropriate but necessary.

When you are deciding whether to use active or passive voice, consider your purpose and audience. If you want the focus to be on the receiver of the action rather than on the agent, then use passive voice. If, on the other hand, you want to emphasize the agent, use active voice. The following two examples illustrate the difference.

Passive voice masks or minimizes the agent of action and emphasizes the receiver of the action.	**Active voice** emphasizes the agent of action.
Argon lasers and household "super glue" are now used by some dactylography experts to collect fingerprints. A small, highly concentrated, single-wavelength light beam is produced by the laser. Fluorescence is induced by the laser in such fingerprint chemicals as riboflavin. Prints — even old ones — can be lifted from paper, something that has been impervious to traditional dusting, by the laser. The benefits of super glue were discovered by accident. The fumes from the glue interact with the amino acids and outline the print. A fingerprint is raised and preserved by super glue on just about any surface. The glue is used on surfaces such as skin, plastic, coarse metal, and leather.	Some dactylography experts now use argon lasers and household "super glue" to collect fingerprints. The laser produces a small, highly concentrated, single-wavelength light beam that can induce fluorescence in such fingerprint chemicals as riboflavin. The laser can lift prints — even old ones — from paper, something that has been impervious to traditional dusting. Experts discovered the benefits of super glue by accident. The fumes from the glue interact with the print in a way that raises and preserves it on just about any surface, including skin, plastic, coarse metal, and leather.

Using Plain Language

Plain language communicates information — even very complex information — so that it makes sense to most people. As you draft, you should make plain language a goal. Many experts in both law and technical communication agree that plain language has a number of characteristics:

- It eliminates unnecessary or inappropriate formalisms ("wherefore," "whereas"), archaic expressions ("herein," "thereof"), redundancies ("consent and agree," "each and every"), and Latin words.
- It uses strong verbs that eliminate nominalizations ("classify" rather than "classification").
- It is direct and eliminates wordiness.

Plain language is more than a good idea; it's a presidential order that has been in effect since 1998 for all U.S. federal documents (though most people acknowledge that getting all government documents in plain language is not an easy task):

> The Federal Government's writing must be in plain language. By using plain language, we send a clear message about what the Government is doing, what it requires, and what services it offers. Plain language saves the government and the private sector time, effort, and money. . . . Plain language documents have logical organization, easy-to-read design, [and these language features]:
>
> - common, everyday words, except for necessary technical terms
> - "you" and other pronouns
> - the active voice
> - short sentences[7]

To read the full text of the "Presidential Memorandum on Plain Language," go to www.english.wadsworth.com/burnett6e for a link to the original order.
CLICK ON WEBLINK
CLICK on Chapter 7/Presidential Memorandum

The federal agency charged with helping departments and agencies use plain language, the Plain Language Action & Information Network, says that "[p]eople should be able to understand what we write the first time they read it, especially materials that tell people how to obtain benefits or comply with requirements."[8]

To learn more about ways to use plain language, go to www.english.wadsworth.com/burnett6e for a link to sites with guidelines and strategies to make technical documents, including legal documents, more usable.
CLICK ON WEBLINK
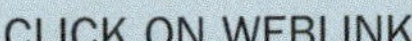
CLICK on Chapter 7/creating plain language

WEBLINK

Efforts in government, business, and industry to increase the use of plain language have been under way for nearly a decade. Does plain language really make a difference? One dramatic situation shows how complex technical language can affect people's lives. The U.S. Court of Appeals for the Ninth Circuit heard the case *Maria Walters and Others v. United States Immigration and Naturalization Service* (INS).

> In this case, the court found that certain government forms were so difficult to read that they violated due process requirements that people be given "notice"

of possible legal actions against them, and of the legal consequences of their own actions. In brief, the 9th Circuit Court of Appeals found that aliens subject to deportation based on INS charges that they committed document fraud did not get due process. The forms used by INS to tell the plaintiffs that they might be deported did not "simply and plainly communicate" legal consequences to the plaintiffs. The court ordered INS to redo the forms to communicate better. The court also ordered INS to refrain from deporting any alien whose case had been processed using the deficient forms.[9]

WEBLINK

To read more about the case *Maria Walters and Others v. United States Immigration and Naturalization Service*, go to **www.english.wadsworth.com/burnett6e** for a link to the complete text of the decision.

CLICK ON WEBLINK

CLICK on Chapter 7/Walters vs. Reno

How else can plain language make a difference? Which of these pairs of sentences would you prefer to read? Which would you prefer to write?

When the process of freeing a vehicle that has been stuck results in ruts or holes, the operator will fill the rut or hole created by such activity before removing the vehicle from the immediate area.	***or this . . .***	If you make a hole while freeing a stuck vehicle, you must fill the hole before you drive away.
The Secretary of the Interior may, in specific cases or in specific geographic areas, adopt or make applicable to off-reservation Indian lands all or any part of such laws, ordinances, codes, resolutions, rules or other regulations of the State and political sub-divisions in which the land is located as the Secretary shall determine to be in the best interest of the Indian owner or owners in achieving the highest and best use of such property.	***or this . . .***	We may apply state or local laws to off-reservation lands. We will do this only if it will help the Indian owners make the best use of their lands.[10]

Using plain language will increase the comprehensibility and usability of your documents, oral presentations, and visuals.

Is plain writing just "dumbing it down"? To learn more about whether using plain language in government, business, and industry really makes any difference, go to www.english.wadsworth.com/burnett6e for links to the results of a study that discusses the cost saving and microeconomic reform that can result from plain writing, as well as a straightforward discussion about the myths and benefits of plain writing.

CLICK ON WEBLINK

CLICK on Chapter 7/It's not "dumbing it down."

WEBLINK

Avoiding Density

One of the ways to draft comprehensible, usable documents is to avoid dense text. For example, sometimes a paragraph you draft focuses on a single topic and has appropriate sentence structure, yet it is difficult to read because you have packed the ideas so tightly that the connections are obscured. As a result, the audience will have trouble following your reasoning. Several techniques can help you draft to reduce the density in your writing:

1. ***Suggestions about verbal elements***
   - Separate information into several sentences rather than a few very long sentences.
   - Develop important points in separate paragraphs.
   - Add examples and explanations to illustrate points.
   - Use direct diction.
   - Add transitions within paragraphs and between paragraphs and sections of a document.
2. ***Suggestions about visual elements***
   - Use headings and subheadings to identify key sections.
   - Illustrate objects and concepts to aid understanding.
   - Use selected visual devices to highlight key ideas: lists, bullets, tables, underlining, italics, boldface.

In Figure 7.7, the notes for the paragraph are so dense that readers have trouble absorbing the information. The sentences are long and wordy, with few transitions to establish relationships among the ideas; in some places, additional information is needed. Careful drafting of this paragraph in Figure 7.8 considers sentence length, separates information into several paragraphs, adds a clarifying figure, and includes transitions to signal relationships.

One of the interesting things about this example is that the comprehensibility has more to do with organization of information than with adherence to mechanical or grammatical conventions.

FIGURE 7.7 Dense Notes

Separate each chunk into sentences to give equal emphasis to each point.

Too much new information in each sentence.

A robot hand can perform rapid, small motions by human-like finger actions—independently of its manipulator arm, be attachable to many different manipulators without any modifications to the arms, and has the potential as a prosthesis for humans. The hand employs three opposed fingers, each one with three joints and a cushioned tip on the outer segment, controlled by four sheathed cables running to motors on the forearm. The motor drives on the forearm impose no loads on the hand actuators, and the drives are more easily accommodated on the forearm and have a much smaller effect on its response than they would if located on the hand. The robot-hand fingers can provide more than three contact areas since more than one segment per finger can contact an object; thus, the robot hand can move objects about, twist them, and otherwise manipulate them by finger motion alone.

Separate into paragraphs to reduce density.

Add a figure so readers can visualize the robot hand.

FIGURE 7.8 Draft of a Paragraph That Avoids Density[11]

Applying given-new principle places "hand" in the subject position.

A robot hand currently under development performs rapid, small motions by human-like finger actions—independently of its manipulator arm. Without modification, the robot hand attaches to many different arms. The hand even has the potential as a prosthesis for humans.

As shown in the figure, the hand employs three opposed fingers. Each has three joints and a cushioned tip on the outer segment. Each finger is controlled by four sheathed cables running to motors on the forearm.

The topic of the paragraph (eliminating problems) is placed first.

The new hand eliminates some problems with previous designs. Its motor drives are on the forearm (rather than on the hand itself), where their weight and inertia impose no loads on the hand actuators. The drives are more easily accommodated on the forearm and have a much smaller effect on its response than they would if located on the hand.

The use of an analogy — comparing the robot hand to a human hand — helps readers understand key distinctions.

Like the fingers on the human hand, the robot-hand fingers provide more than three contact areas since more than one segment per finger contacts an object. Thus, like the human hand, the robot hand moves and manipulates objects by finger motion alone.

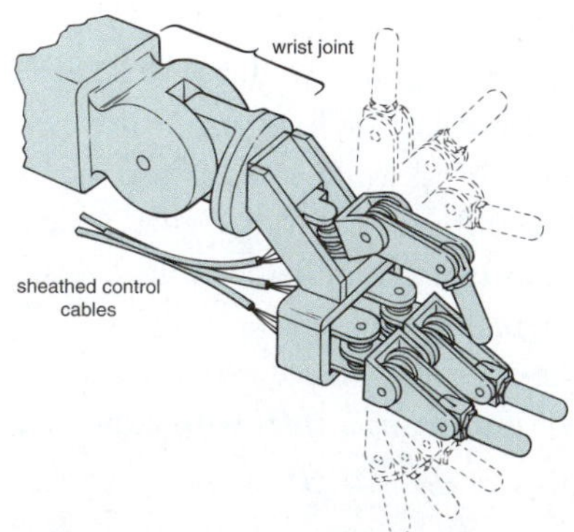

A figure helps readers visualize the robot hand.

Using Given-New Analysis

Another problem that occurs during drafting is working toward document coherence — that is, making sure that one idea logically or clearly relates to the following idea. Connections that aren't obvious to audience must be articulated. This problem often occurs when a writer presumes too much about the readers' prior knowledge. In these situations, a writer forgets (or neglects) to include background and connecting information that readers need.

One way to ensure that you provide the audience with the cues is to use *given-new constructions,* which strengthen the coherence of a document. In such constructions, new information is connected to what the audience already knows, either from background knowledge or from immediately preceding text. Given-new analysis aids drafting by providing a quick, effective way to spotlight the new information you have introduced. These are three of the many variations of given-new that are possible, with *A, B, C,* and so on representing pieces of information:

EXAMPLE 1[12]

sentence	given-new	
(1)	given A: new B	(1) The *ink* of a squid [A] is *brown or black viscous fluid* [B], which is contained in a reservoir.
(2)	given A: new C	(2) The *ink* [A] is *ejected through the siphon when a squid is alarmed* [C].
(3)	given A: new D	(3) This *ink* [A] not only forms an *effective screen* behind which the animal can escape [D], but
	given A: new E	*it* [A] contains *alkaloids that paralyze the olfactory senses of the enemy* [E].

EXAMPLE 2

sentence	given-new	
(1)	given A: new B	(1) All *squids* [A] propel themselves by *taking in and forcibly expelling water* from the mantle cavity through the siphon [B].
(2)	given B: new C	(2) The *force and direction of the water expelled* [B], plus the undulation of the fins and body, determine the *direction and rapid movement* [C] of the animals.
(3)	given C: new D	(3) The *rapid movement* [C] is always in the opposite direction of the *water ejected from the siphon* [D].
(4)	given D: new E	(4) The *ejection of the water* [D] also *oxygenates the gills,* located in the mantle cavity [E].

EXAMPLE 3

sentence	given-new	
(1)	given A: new B	(1) The unusual *coloration* [A] of squid is caused by the presence of *integumental pigment cells* [B].
(2)	given B: new C	(2) These *cells, called chromatophores* [B], contain red, blue, yellow, and black *pigment* [C].
(3)	given A: new D	(3) *Coloration*

(4)	given B: new E	[A] is *specific to each species* [D]. (4) The *chromatophores, color cells,* [B] are controlled by m expand or contract in relation to visual stimuli, thus *changing the color of the*
(5)	given B: new F	(5) The *pigment cells'* [B] release of color into th flesh of a dead animal indicates the *onset of spoilage* [F].

Information in drafts that follow given-new structures is typically easier to read and remember. In addition, given-new structures make the relationships between ideas clear to the audience.

Individual and Collaborative Assignments

1. **Track your writing process.** One way to learn about your writing process is to track your own process as you prepare a document. You will need to maintain careful records. Use one of the following methods.

   (a) Tape-record your planning or drafting. (Note: Trying to track the entire process of preparing the document takes too much time; pick just one part of the process.) Afterward, listen to the tape to identify interesting patterns you can note in your log.

   (b) As you were encouraged to do in Chapter 5, keep a log in which you record every time you work on the document. Record the date and time as well as what happens during the time. Try to be as accurate and complete as possible.

2. **Analyze your writing process.**

   (a) Use the data you have collected in Assignment 1 to write a one- to two-page discussion about one aspect of your writing process. Focus on some specific aspect — for example, the way you defined the task, the way you considered your audience, the way you organized or designed the document. Reread your log, looking for patterns in your oral interaction and your writing. Determine what problem-solving strategies you found useful in your writing. Use quotations from the tape or log to support your decision.

   (b) Write a memo to your instructor in which you describe the productive patterns that you hope to continue and the unproductive patterns that you hope to change. Then speculate about the relationship between your process and the quality of your writing.

(c) Assume the time from the beginning to the end of the writing project in Assignment 1 is 100 percent. Design a chart that will be part of the memo in which you show the typical proportion of time you spend on each stage of your writing process. Explain in which stage(s) you're most effective, in which you're the weakest, and what might you do to improve your process.

3. **Create a rubric.** The only way to make sure that your writing is effective is to analyze it (as in Assignment 2) to determine what areas you need to improve. Use the rubric from Chapter 1 to analyze a document that you created (modify it to fit your document) so that you can assess your own skills as a technical communicator. Each time you create a new document, use the rubric to establish where you have made improvements.

4. **Interview a professional, part II.**

   (a) Arrange to interview a professional technical writer or a technical professional in your field who writes frequently. During the second part of the interview (Part I of the report is an assignment in Chapter 6), inquire about the person's approach to planning and drafting. Ask some of the following questions and add additional questions relevant to the specific person and organization:

   - What kind of planning do you do? How much time does it take? How important is it? Could you reduce or omit planning of your documents with little negative effect?
   - How important is project planning and project management in the successful completion of a project?
   - What factors influence decisions about what content you include? How important are factors such as context, purpose, audience, organization of information, overall argument, design of the document?
   - How important is the creation of a document or presentation that doesn't have errors in logic?
   - How much does a draft change from the initial plan to the penultimate version?
   - How does your background affect your planning and drafting?

   (b) Organize the information you find as part II of a brief oral or written report intended for other members of the class. This report can be posted on the class Web site or made available in print form in a class notebook in the library.

5. **Look at the process of translation.** Because your audience is not always made up of native English speakers, you need to consider how to translate your documents into the language that your audience uses. Translation of technical documents is not as easy as it may sound. In order to test your translation, you should have a native speaker of that language read your original document and its translation to check for errors or miscommunications caused by direct translation. This is called back-translation. In this assignment, you will be asked to look at the process of translation and back-translation. Visit the following Web site: www.babelfish.altavista.com/.
   (a) Type in a short paragraph from a document that you have written in English.
   (b) Ask the "translator" to translate your paragraph to another language.
   (c) Find a native speaker of that language on your campus and ask her or him to read the translation and the English original.
   (d) Write a short memo to your classmates about the process of translation. Consider the effectiveness of the Web site translation program and the value of the native speaker.

6. **Determine the appropriate voice.** Work in a small group with your classmates to determine whether you would use active or passive voice for each situation and audience listed below.

   Example: A document about the grading of pearls could be written in active voice if the document is intended for trainees learning how to grade pearls, because the focus should be on what the person does. However, the document could be written in passive voice if it's for customers who are more interested in the pearls than in the grader.

   (a) blood donation, for a potential donor
   (b) purification of water from sewage treatment plant, for an ecologist
   (c) grading of diamonds for jewelry, for a consumer
   (d) professional dry cleaning of antique clothing, for a dry cleaner
   (e) an operation, for a child in hospital for surgery
   (f) filing a claim, for a new insurance adjuster
   (g) a change in manufacturing procedures, for a customer
   (h) formation of sebaceous cysts, for patient with cysts
   (i) operation of a heat exchanger, for a homeowner

7. **Explain active and passive voice.** Carefully read the following two paragraphs. Explain the context, purpose, and audience that makes active voice appropriate for one paragraph and passive voice appropriate for the other.

active voice The neonatal nurse can decrease the chances that a newborn will suffer from cold stress by following several standard procedures. The nurse should monitor the baby's temperature regularly and should not bathe the baby until the temperature is stabilized. A careful nurse coordinates all medical procedures that expose the baby to cool air and will be certain to remove all wet bedclothes and diapers that cool the skin by evaporation.

passive voice A newborn will not suffer from cold stress if the neonatal nurse follows several standard procedures. The baby's temperature should be monitored regularly. The baby should not be bathed until the temperature is stabilized. A baby should not be exposed to cool air while undergoing medical procedures and should not be left with wet bedclothes or diapers that cool the skin by evaporation.

8. **Revise an explanation to eliminate density.** Read the following dense paragraph and revise it to make it more coherent. You may separate the material into shorter paragraphs, use lists, and add or delete material. This information was part of a memo informing employees about changes in procedures dealing with scrapped parts.

Several steps must be taken to separate and stage inventory to be scrapped. The first step is to locate and stage all excess parts. The white tag on each part will identify the part as either EXCESS, GOOD or EXCESS, DAMAGED. Parts tagged EXCESS, GOOD should be staged in bin #020185. Parts tagged EXCESS, DAMAGED should be staged in bin #020186 for further sorting and staging by code. The code is etched into the frame of each part: 1A — defective consumable, 1B — unidentified damage, 1C — vendor return/unrepairable, 1D — identifiable damage. Parts labeled code 1A, defective consumable, are staged in bin #020188, to be automatically scrapped without further investigation. Parts labeled code 1C, vendor return/unrepairable, are staged in bin #020189; these parts can also be automatically scrapped. If a part is labeled code 1B, additional information is required. Code 1B parts should be visually examined for signs of damage such as broken chips or split jumper cables. Parts with observable damage should be placed on a skid in bin #020187. Parts with no observable damage should be hand carried to the Repair Center where each part will be tested to determine the extent of damage. Parts the Repair Center determines cannot be repaired are staged in bin #020187. Parts the Repair Center believes are repairable are placed in the Repair Stockroom and scheduled for repair. Parts labeled with code 1D are staged in bin #020190. When ten or more skids have accumulated, the manager should be notified to check the accumulated parts and fill out a justification form to scrap each part.

9. Revise using given-new structures. Read the paragraph below, following these steps:

(a) Analyze to identify the current given-new structure, which will help you see why the paragraph is not coherent.

(b) Revise the paragraph to more closely reflect given-new structures.

(c) Identify the given-new structure of your revised paragraphs.

(1) Hot Isostatic Processing makes metal more dense. (2) Dense metal is more durable. (3) Metal parts cannot be cast without unavoidable small cracks and air pockets. (4) Imperfections lead to wear and breakage. (5) These flaws can be eliminated with Hot Isostatic Processing. (6) The parts are heated in special units, then pressurized with gas to minimize flaws. (7) A controlled cooling process ensures that the parts retain their original shape. (8) Hot Isostatic Processing produces stronger parts, able to withstand greater pressures for a longer span of time.

Chapter 7 Endnotes

1 Synectics. (n.d.). http://www.synectics.com/chicago.html. Retrieved October 12, 2003.

2 Downes, N. (n.d.). Mitsubishi Project PERT Chart (Carnegie Mellon University).

3 Downes, N. (n.d.). Mitsubishi Project Gantt Chart (Carnegie Mellon University).

4 Sims, B. (1993). Linking ethics and language in the technical communication classroom. *Technical Communication Quarterly, 2*(3), 285–299.

5 Adapted from Johnson, V. A., et al. (1978). Grain crops. In M. Milner, N. S. Scrimshaw, & D. I. C. Wang (Eds.), *Protein resources and technology: Status and research needs* (p. 244). Westport, CT: AVI Publishing. Used with permission of AVI Publishing Company.

6 Examples modified from McMillan, G. (1984, January 30). Technology finger-printing moves into the computer age. *Boston Globe.*

7 Plain Language Action & Information Network. (n.d.). *Presidential memorandum on plain language.* Retrieved October 18, 2003, from http://www.plainlanguage.gov/

8 Plain Language Action & Information Network. (n.d.). Retrieved October 18, 2003, from http://www.plainlanguage.gov/

9 Plain Language Action & Information Network. (n.d.). *Maria Walters and others v. United States Immigration and Naturalization Service.* Retrieved October 18, 2003, from http://www.plainlanguage.gov/cites/ins.htm

10 Plain Language Action & Information Network. (n.d.). *How to comply with the president's memo on plain language.* Retrieved October 18, 2003, from http://www.plainlanguage.gov/cites/vpguid.htm

11 Three-fingered robot hand. (1983). *NASA Tech Briefs, 8*(Fall), 99.

12 Modified from Bardach, J. E., & Pariser, E. R. (1978). Aquatic proteins. In M. Milner, N. S. Scrimshaw, & D. I. C. Wang (Eds.), *Protein resources and technology: Status and research needs* (p. 461). Westport, CT: AVI Publishing.

CHAPTER 8

Revising and Editing

Objectives and Outcomes

This chapter will help you accomplish these outcomes:

- Differentiate various types of revising and editing and develop the skills required to complete these tasks
- Revise and edit to make documents, presentations, and visuals more accessible, comprehensible, and usable
- Understand that revising is a global function involving complex, interrelated strategies and decisions
- Identify and correct inconsistencies and errors in a draft
- Review a draft of a document that has been carefully edited and then analyze the resulting final version to study the impact of the changes

Revising and editing are critical to the success of your documents. Although these processes often overlap and definitions vary slightly from expert to expert and company to company, their fundamental purposes remain the same: to increase the accessibility, comprehensibility, and usability of documents, presentations, and visuals.

- *Revising* generally refers to the process of changing overall (or global) elements of documents.
- *Editing* generally refers to (1) changing specific (or local) elements of documents and (2) managing administrative details necessary for document publication.

Because of the complexity of editing tasks, a range of skills is necessary. As you can see in Figure 8.1 on the next page, all aspects of revising and editing require attention to detail and an eye for consistency, which can mean a number of things. For example, in revising and substantive editing, these skills can mean recognizing repetitious or discrepant content that may be dozens of pages apart. In proofreading, however, attention to detail and an eye for consistency can mean recognizing vertical alignment or knowing the difference between 0 (the numeral) and O (the letter), between 1 (the numeral) and l (the letter), and between a hyphen (-), an en dash (–), and an em dash (—). This range of tasks also requires remarkably different skills. For example, revising and substantive editing require subject matter expertise related to the topic of the document, oral presentation, or

© Christina Micek Photography

Not even the very best professionals know everything, so they use invaluable print resources such as these for checking facts and language conventions. They also have technical handbooks for the areas they work in, technical communication handbooks, dictionaries in English and other languages as well as references for checking general facts.

visual, which is not a necessity for proofreading or administrative editing. Administrative editing typically requires knowledge of drafting and arranging contracts with a range of vendors, including designers and printers, which is not usually necessary knowledge for copy editors.

FIGURE 8.1 **Types of Revising and Editing**

Types	Necessary Knowledge
Revising focuses on global aspects of the document such as content, organization, argument, and design. ***Substantive editing*** also focuses on global aspects of the document.	▪ Subject matter expertise ▪ Organization/logic of argument ▪ Disciplinary/professional expectations and conventions ▪ Ability to revise a document ▪ Document design conventions ▪ Attention to detail and an eye for consistency
Design review focuses on the overall design of the document. ***Copyediting*** focuses on global aspects such as logic and format and local aspects such as language conventions and consistency.	▪ Organization/logic of argument ▪ Document design conventions ▪ Ability to craft sentences and paragraphs ▪ "Macro" features of a document ▪ Language conventions: grammar, punctuation, capitalization, spelling, symbols, numbers, usage, style ▪ Attention to detail and an eye for consistency
Traditionally, proofreaders compare a manuscript of a document with the typeset version to identify typographical errors. In electronic publishing, ***proofreading*** usually involves eliminating typographical errors in the electronic version (and any paper version) without comparison to the original.	▪ Language conventions: grammar, punctuation, capitalization, spelling, symbols, numbers, usage, style ▪ Citation practices ▪ Proofreading marks to save time and avoid ambiguity ▪ Skill with technology ▪ Attention to detail and an eye for consistency
Administrative editing focuses on the aspects of compliance with the organization's policies and management of a range of tasks related to electronic or print publication.	▪ Organizational policies ▪ Publication and printing processes ▪ Prepress/publication technology and procedures ▪ Intellectual property ▪ Contracts with vendors ▪ Attention to detail and an eye for consistency

WEBLINK

Skillful editors as well as technical professionals and managers who have to edit a great deal make their job easier and faster by having a list of readily available resources to answer questions. Go to **www.english.wadsworth.com/burnett6e** for a list of URLs to a variety of useful professional resources for revision and editing.
CLICK ON WEBLINK
CLICK on Chapter 8/editing resources

Revising

> *Important documents, such as instructions, politically sensitive memos and letters, major reports, and proposals need to be reviewed by experts who can assess factors such as accuracy, audience appropriateness, and design.*

> *Important documents should be tested by actual users who can assess factors such as accessibility, comprehensibility and effectiveness.*

This chapter focuses on two levels of revision — *global revision* (redesigning and reorganizing the overall document) and *local revision* (revising individual paragraphs and sentences). Both kinds of revision depend on chunks of text rather than on individual words and phrases. The information for making revisions can come from various kinds of document testing (see Chapter 9), reviewer feedback, and careful self-assessment.

During revision, you'll read and reread your draft critically to examine choices in content, structure, organization, coherence, logic, and design. You can add, delete, and rearrange material in order to produce a document that is more accurate and effective for the audience. You may be the kind of writer who begins revising almost as soon as you start planning, or you may defer most of your revising until you have drafted a substantial chunk of text.

Ideally, you'll have time to compose the draft and then set it aside for a while before thoroughly revising it. All documents are improved when you check them to verify the accuracy and logic of the content, note needed changes in the examples or organization, and keep an eye open for violations of mechanical and grammatical conventions. In other words, you need to maintain the minimum specifications for good writing identified in Chapter 1.

The Revision Process

Revision is more difficult than most inexperienced writers realize. Part of the difficulty comes from not necessarily understanding what to revise and how to change an acceptable draft into an excellent one. A researcher who studies revision strategies, Barbara Sitko, developed a decision tree that identifies six steps you go through during revision. At each step in the process, you can decide to move to the next step — or bail out. Figure 8.2 shows the steps and provides a

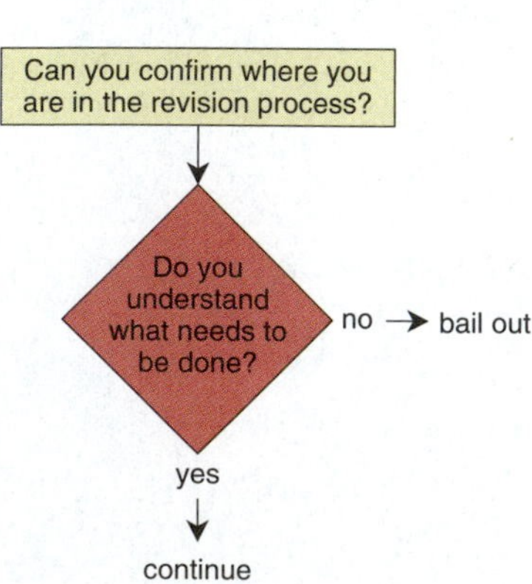

clear explanation of the options you have at each decision point.

Understanding these six steps gives you considerable flexibility in assessing your revision process. Once you realize the kind of knowledge you need in order to revise a document, you are more likely to actually complete the revision.

FIGURE 8.2 **Decision Points Leading to Revision[1]**

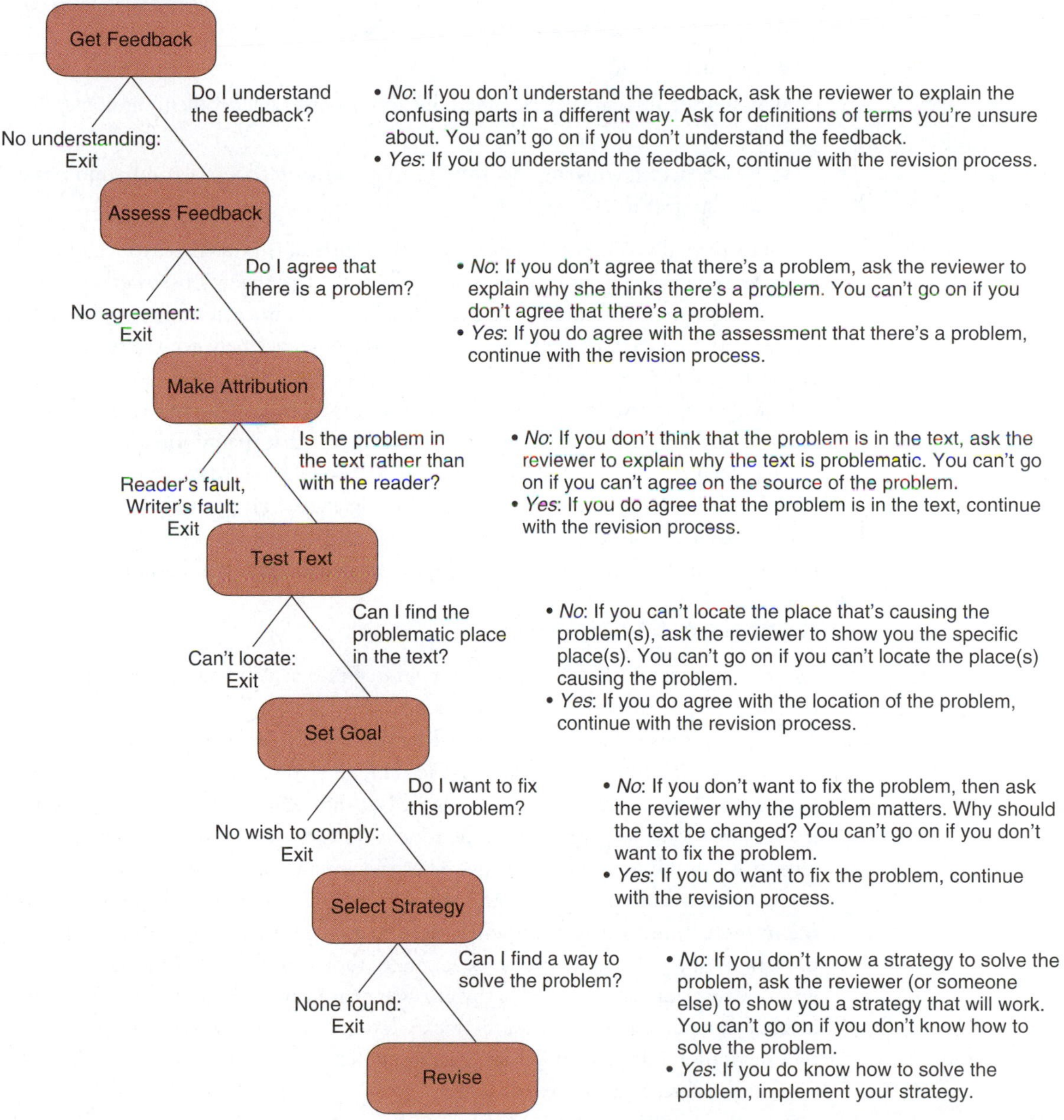

- *No*: If you don't understand the feedback, ask the reviewer to explain the confusing parts in a different way. Ask for definitions of terms you're unsure about. You can't go on if you don't understand the feedback.
- *Yes*: If you do understand the feedback, continue with the revision process.

- *No*: If you don't agree that there's a problem, ask the reviewer to explain why she thinks there's a problem. You can't go on if you don't agree that there's a problem.
- *Yes*: If you do agree with the assessment that there's a problem, continue with the revision process.

- *No*: If you don't think that the problem is in the text, ask the reviewer to explain why the text is problematic. You can't go on if you can't agree on the source of the problem.
- *Yes*: If you do agree that the problem is in the text, continue with the revision process.

- *No*: If you can't locate the place that's causing the problem(s), ask the reviewer to show you the specific place(s). You can't go on if you can't locate the place(s) causing the problem.
- *Yes*: If you do agree with the location of the problem, continue with the revision process.

- *No*: If you don't want to fix the problem, then ask the reviewer why the problem matters. Why should the text be changed? You can't go on if you don't want to fix the problem.
- *Yes*: If you do want to fix the problem, continue with the revision process.

- *No*: If you don't know a strategy to solve the problem, ask the reviewer (or someone else) to show you a strategy that will work. You can't go on if you don't know how to solve the problem.
- *Yes*: If you do know how to solve the problem, implement your strategy.

Redesigning Documents

The organization and format of a report have a strong impact on readers' reactions. Figures 8.3 and 8.4 are different versions of the same memo report, which requested that automatic garage doors be installed in a shop area constantly exposed to harsh weather. Both reports were accompanied by several attachments that are not included here. In the first version of the report (Figure 8.3), readers have to work too hard to get the information they need. This version has been marked for the following revisions:

- Add headings and subheadings.
- Reorder sections.
- Elaborate in some places, especially about personnel problems and energy losses.
- Add expected conventions (for example, the writer's initials and a more specific subject line).

The second version of the memo report (Figure 8.4) is far easier to read because of its headings, organization, details, and adherence to conventions; in other words, it shows a general awareness of readers by making the information more accessible and understandable. Probably the most obvious difference between the two versions is that Figure 8.4 provides design cues in the form of headings and subheadings that save readers time and energy in understanding the relationships among the various sections of the memo. Equally important, though, is that the information in Figure 8.4 has been reorganized and chunked so that following the writer's argument is easy. A third major difference is that financial details are given special attention by being separated from the rest of the text. Another difference is that readers are reminded of the benefits that accrue if the recommendations are followed.

Revision Strategies

You can engage in a number of revision strategies that will help you see your document, oral presentation, or visual in a new way.

The consensus among experienced writers and designers is that revision is much more effective if you focus on one thing at a time. Perhaps begin by focusing on how well the information is adapted to the audience, next focus on how the information is organized, and then focus on whether the information is responsive to the overall context as well as the specific situation. When you're testing whether a strategy works, you don't need to implement it in the entire document; instead, test the revision with a section to see how it works. Rewrite a paragraph or a page using the Track Changes tool in your word-processing program so that you can later reject the changes that don't work and return to your original version. Many writers use Track Changes to create two or three versions of a section of a document before deciding which one is the most accessible, comprehensible, and usable.

FIGURE 8.3 **Report without Carefully Designed Information**

*Add headlines and subheadings to help readers identify issues and recommendation.

MEMO

Add job titles

To: Charles Malatesta
From: Tom Mansur
Subject: Purchase and installation of automatic roll-up doors
Date: September 27, 20—

Make subject line more specific

*Reorder sections to create a stronger argument. (see numbers)

Overview of unacceptable conditions

① The R&D Model Shop moved to its current location—next to the rear shipping and receiving dock of the Colebrook facility—15 months ago. Since then, we have been subjected to excessive heat in summer and extreme cold in winter. Most of the day the loading doors are open, and the heat or air conditioning is lost through the two connecting doorways. This is both expensive and a health risk.

⑤ Previous attempts to solve the problem.

Swinging Metal doors:

One attempt to control the temperature was swinging metal doors, which kept out the weather but were dangerously noisy. These doors banged loudly against the chain link fence outside. Machinists would jump at the sound, presenting the danger of severed fingers. Eventually these doors were knocked of their hinges, leaving us no protection from the outside elements.

Plastic Flap Doors:

We next installed clear plastic strip doors, which flap in the breeze and offer no protection from extreme temperatures. They also present a safety risk to fork truck drivers, in that the flexible strips swing down from the fork truck extensions and slap the drivers on the face and arms.

② Background on current problems

Personnel Concerns: (expand!)

Energy Loss: (add info)

Cost of Lost Energy:

The approximate cost of the lost heat and air conditioning is $5,970.00 per year. (See attached cost breakdown.) In human terms, in the summer when the air conditioners are on, we are sweating; in the winter when the heaters are blowing, we are shivering. These conditions are unacceptable.

③ Recommendation for Automatic Garage Doors

We, therefore, recommend the purchase and installation of two electric garage doors with both wall-mounted and pull-chain openers. These doors would be equipped with pneumatic safety stops at the bottom to prevent injury. In compliance with the fire safety egress requirements, we recommend the installation of two regular doors, with pneumatic closers, as common entry/exit ways. The total cost to purchase and install the equipment to correct the situation is $4,262.00. This would account for $1,708.00 in savings at the end of one year. Please refer to the attached estimates for prices of the garage doors and regular doors. Refer also to the attached building layout that identifes the recommended locations of the proposed doors.

④ Financial Evidence

Purchase and Installation Costs:

Potential Savings:

⑥ Discussion

We need these doors now, since winter is rapidly approaching. (Expand!)

TM/db
Attachments

FIGURE 8.4 **Revised Report**[2]

MEMO

To: Charles Malatesta
Plant/Facilities Administration

From: Tom Mansur *TM*
R&D Model Shop Group Leader

Subject: Justification for the purchase and installation of automatic roll-up doors for the R&D Model Shop

Date: September 27, 20—

A detailed subject line helps orient the reader to the purpose of the memo.

Overview of Unacceptable Conditions

The R&D Model Shop has been in the Colebrook facility, zone 5 (see Figures 1 and 2), for 15 months, adjacent to the shipping and receiving dock. The dock's loading doors are open approximately 8–10 hours a day, so the heat or air conditioning escapes from the R&D Model Shop through the two connecting doorways. The Model Shop machinists are exposed to excessive heat during the summer and extreme cold during the winter.

Establishing the problem—the seriousness and the background—helps the reader see the need for immediate action.

Background on Current Problems

This situation poses personnel concerns as well as energy losses, both of which are influential in analyzing and addressing the current problem.

Personnel Concerns: Machinists working in the R&D Model Shop are exposed to temperature extremes for up to 10 hours a day, causing discomfort and decreasing resistance to colds and flu. Employee satisfaction with the working conditions is low, influencing shop productivity. Additionally, use of sick days by R&D Model Shop employees has increased 7% since we moved to the Colebrook facility, costing the company money.

Energy Loss: Because the loading dock is open to the outside for up to 10 hours a day, the energy expended to heat or air condition the R&D Model Shop is wasted.

Reordering of sections clearly establishes and labels the problems before making a recommendation that addresses the identified problems.

Specific headings and subheadings ease reading.

Recommendation for Automatic Garage Doors

The purchase and installation of two electric garage doors with both wall-mounted and pull-chain openers would eliminate the problem with temperature extremes. These doors would be equipped with pneumatic safety stops at the bottom to prevent injury. In compliance with the fire safety egress requirements, two regular doors (with pneumatic closers) would be installed as common entry/exit ways.

Financial Evidence

This recommendation for automatic garage door openers takes three financial factors into consideration

Cost of Lost Energy: The cost of BTU loss in zone 5—based on heating and air conditioning expenses for the last fiscal year and cost formulas (see Figures 3, 4, and 5)—is approximately $5,970.00.

Reviewing the previous attempts reminds the reader that the problem is long term and that various solutions have been tried.

Purchase and Installation Costs: The total cost to purchase and install the equipment to correct the situation is $4,262.00, based on quotations from Dempsey Door Company and Jackson Lumber Company. (See Figures 6 and 7.)

Potential Savings: The payback period would be less than one year.

Estimated annual cost of problem	$5,970.00
Estimated cost of solution	–4,262.00
Savings the first year	$1,708.00

All the financial information is chunked and labeled so that the implications are easy to see.

The recommended solution would account for more than $1,708.00 in savings at the end of one year. Please refer to the attached estimates for prices of the garage doors and regular doors. Refer also to the attached building layout that identifies the recommended locations of the proposed doors.

Previous Attempts to Solve the Problem

The two previous attempts to address the problem have not been successful.

Swinging Metal Doors: Swinging metal doors were in place for the frst six months that the R&D Model Shop was in the present location. They kept out the weather but were dangerously noisy. When fully loaded fork trucks passed through, these doors banged loudly against the chain link fence that surrounds the shop. Machinists in the area would jump at the sound, presenting the danger of severed fngers or worse, not to mention the rattled nerves. The constant traffic of fork trucks eventually knocked these doors off their hinges.

Plastic Flap Doors: The swinging metal doors were replaced with clear plastic strip doors. These doors flap in the breeze and actually provide no protection at all from extreme temperatures. They also present a safety risk to fork truck drivers. As the fork trucks move through the doorway, the flexible strips swing down from the fork truck extensions and slap the drivers on the face and arms.

Figures, which provide supplementary information, are attached if the reader needs to check any details.

The figures that accompanied the original report are not included here.

Discussion

Implementing the recommendation in this report will benefit both the R&D Model Shop employees as well as the Colebrook operation in general. The comfort of the work environment and job satisfaction will increase for employees in the R&D Model Shop. Sick time should decrease. The impact on the heating and air conditioning expenses should be felt almost immediately.

This report has been approved by Raphael Calvo, Manager of the R&D Model Shop, and Max Freeport, Director of Colebrook Operations.

So that work can begin immediately, a working print (Figure 8) and the completed work request form (Figure 9) are attached.

The brief discussion acts as a conclusion that reminds the reader about benefits that are gained if the recommendations are followed.

Another good way to begin the revision process is to get feedback from the intended audience — an actual or simulated one. For example, you can get feedback from an actual reader, invite feedback from a peer reviewer, or even role-play the intended audience. Various people will have a perspective different from yours, which may give you some insight about what works and what needs to be changed. Writing audience-based prose is difficult if you never check audience reactions.

Sometimes you know something isn't quite right with a document, oral presentation, or visual, but you can't put your finger on the problem. You know it's not the basic grammar, mechanics, or sentence structure that's wrong, but something simply isn't working. If the piece is particularly important, taking extra time to investigate whether some other approach will work is worthwhile. Reading the document aloud is one excellent way to hear problems that you can miss if you're reading silently.

Another way to start the revision process is to play with the draft in unexpected ways. Seeing the information presented in another way gives you opportunities to move forward in a potentially productive manner. Try any combination of these outside-the-box approaches to re-visioning text and design to help you think of your work in new ways:

1. ***Text changes to encourage re-vision:***
   - Put the end section at the beginning . . . and start to revise from that point.
   - Keep the same major argument but make it half the length.
   - Select different supporting evidence — different quotations, different statistics.
   - Change the intended audience (which necessarily changes things such as the emphasis and the kind of details).
   - Change the genre. Try it as a newsletter article, journal article, lab report, letter to a former colleague, after-dinner speech.
2. ***Design changes to encourage re-vision:***
   - Change points in a paragraph into a bulleted list.
   - Change a dramatic example into a sidebar.
   - Change the headings from descriptive labels to questions.
   - Convert critical information in paragraphs to visual displays: charts, diagrams, tables, maps, graphs.
   - Add captions to the visuals (and delete information from the text).
   - Redesign the document. Try it as a two-column piece with callouts, as a Web page with hyperlinks, as a script for a National Public Radio feature report.

A skillful editor pays as much attention to the design of a document, PowerPoint presentation, or visual as to the words. For some very useful guidelines about visual proofreading, go to www.english.wadsworth.com/burnett6e.
CLICK ON WEBLINK
CLICK on Chapter 8/visual proofreading

WEBLINK

Editing

Most people think of editing as having to do with correctness — avoiding errors in mechanics and grammar. You might find a more useful approach is to think of editing as adhering to conventions and assuring consistency. *Conventions* are what a particular group of people agrees is acceptable, expected usage. *Consistency* involves ensuring that a document conforms to those agreed-on conventions. Conventions differ from country to country and from company to company. Skillful editors seldom make absolute decisions; instead, they make situated, contextualized decisions. So, for example, in the United States, *color* is conventional; in the United Kingdom and much of the rest of the English-speaking world outside the United States, *colour* is conventional. Similarly, U.S. currency uses commas ($1,000 = one thousand U.S. dollars) in places where UK currency uses periods (£1.000 = one thousand British pounds).

Excel lets you change the Windows Regional Settings in order to apply foreign currency formats with the appropriate monetary symbols and use of commas and decimals for various national denominations.[3]

Levels of Edit

Effective editing improves a document's technical or textual accuracy as well as its appropriateness and appeal to the audience. The term *editing* as it is used in this text refers to four functions that describe the nine levels of edit identified in Figure 8.5. These functions, which typically take place in the final stages of preparing a document, help writers adhere to conventions and ensure consistency.

If the information in a document is accurate and complete, why does a writer need to revise the document so that it's accessible and adapted to the audience?

In some organizations *content accuracy* is the responsibility of the writer, in others it is the responsibility of an editor, and in still others writers and editors act as safety checks for each other. In situations in which the editor does make changes in content, final approval for the changes usually remains with the writer (unless the changes involve institutional policies and practices more familiar to the editor). Regardless of who controls content, little point exists in completing any other level of edit until the content accuracy is established by a subject-matter expert.

Design review typically includes assuring the consistency of all elements of design, from "macro" elements, such as assuring the same font type, style, size,

READABILITY FORMULAS DON'T DO WHAT THEIR NAME CLAIMS

Readability formulas reputedly assess a text's level of difficulty: How easy the text is to comprehend. Readability formulas are used in government, industry, and education. The question is whether they do what they claim.

How Readability Formulas Work

Readability formulas are based on the relationships between average word length (number of syllables per word) and average sentence length (number of words per sentence). The principle presumes that both the higher the average number of syllables per word and the greater the average number of words per sentence, the more difficult the document is to read. The formulas yield a ratio between word length and sentence length that estimates a document's readability by level of difficulty or grade level.

Problems with Readability Formulas

Because readability formulas are based on a ratio, they do not adequately address several rhetorical elements of writing.

Content complexity plays a critical role in audience comprehension. But readability formulas do not consider difficulty of content. A paragraph may be labeled "seventh-grade readability" because of relatively short words and sentences; but the ideas may be extremely difficult to comprehend. For example, this sentence has only 15 words: "Boards with tight lines and spaces required reworking due to shorts caused by solder bridging." According to some reading scales, the sentence falls between fairly easy and standard because most of the words are short and the sentence has a total of only 21 syllables. However, the terminology is unfamiliar to general readers, and the sentence is not easy to understand — despite formulas that indicate otherwise. Or consider what grade level you would assign for "I think; therefore, I am."

Sentence length can affect material's comprehensibility. Generally, shorter sentences are easier to read than longer ones. Some reading experts assign levels of difficulty to sentence length, as the following examples show. However, the content in these sentences does not vary much in difficulty since all of the sentences are from the same article in *Aviation Week and Space Technology,*[4] a magazine for professionals and experts in aerospace-related industries. The practical difference in difficulty between the sentences is small, a factor ignored by readability formulas.

> The cockpit instrumentation is well designed. (6 words) *very easy, 8 words or less*
>
> This aircraft is equipped with combined dual navigation and communication systems. (11 words) *easy, 11 words*
>
> All controls are within easy reach of the pilot and have readily discernible functions. (14 words) *fairly easy, 14 words*
>
> The DO 228 has easy access to cockpit seats by an optional door located on either side of the cockpit. (19 words) *standard, 17 words*
>
> The aircraft's fuel, electrical, deicing and starting systems controls are color coded and shown in clearly styled flow diagrams on the overhead panel. (23 words) *fairly difficult, 21 words*
>
> The front instrument panel is not cluttered, partly due to the large center console located between the pilots that contains most of the avionics controls. (25 words) *difficult, 25 words*
>
> Dornier's DO 228–200 commuter and utility twin turboprop offers operators a versatile-mission aircraft with short-field performance combined with excellent flight performance characteristics and easily maintained aircraft systems. (28 words) *very difficult, 29+ words*

Variety of sentence length affects comprehensibility. A document with sentences all of the same length quickly becomes repetitious and boring. Variation maintains readers' interest and potentially increases comprehension. Although a single document might not have the tremendous variety shown in the previous examples, effective technical communicators try to vary sentence length, but readability formulas do not account for this.

Subordination of ideas can increase comprehension. Short, separate sentences that treat all ideas equally may be more difficult to comprehend than sentences that have subordinate and main ideas clearly established. Two versions of the same information illustrate this point.

> **Original.** A decayed woven linen robe was found in 1981. The robe dated from c. 1000 B.C.E. It is the oldest known piece of cloth found in Greece. The robe was found by a Greek-British excavation team. The team was digging on the island of Euboeoa. Euboeoa is about 40 miles from Athens.
>
> **Revision.** A decayed woven linen robe from c. 1000 B.C.E., the oldest known piece of cloth found in Greece, was unearthed in 1981 during a Greek-British excavation on the island of Euboeoa, 40 miles from Athens.[5]

Although readability formulas indicate that the original version is easier to read, most adults prefer the revision because relationships are established clearly and repetition is eliminated.

Word length is a central factor in readability formulas, which equate short words with simplicity, without regard to word meaning or audience knowledge. However, shortness is not always a measure of reading ease, for short words do not necessarily have simple meanings. For example, *quark* and *erg,* although short, are more difficult words than *satellite* or *occupation.* And some very long (and sometimes technical) words have a high recognition factor. For example, some multisyllable medical and chemical terms are easily and immediately recognized by all professionals in their respective fields. Likewise, a fourth-grader recognizes the words *Mississippi* or *Massachusetts.*

© Roger Ressmeyer/CORBIS

Editing is done by professional editors as well as by peers who are often willing to provide feedback on drafts. The more specific you are in telling these peer reviewers what needs attention, the more help they're likely to be. Should they look at organization? At adaptation to the audience? At the evidence you use? At the tables and figures?

Such multisyllable, high-recognition words can give a false high grade-level equivalency compared with the difficulty of the content when evaluating readability for adult professionals.

Word length is less important in comprehension if the words are familiar to the audience. Some technical words are long and appear complex to the general reader, but they are easily understood by experts. In many cases a technical term is exactly the correct word; an explanation comprised of shorter, more recognizable terms would be less precise. The following sentence probably won't make sense to general readers, but to plastics engineers it is both accurate and easy to understand.

> Polysulfones comprise a class of engineering thermoplastics with high thermal, oxidative, and hydrolytic stability, and good resistance to aqueous mineral acids, alkali, salt solutions, oils and grease.[6]

Document design is not part of readability formulas. Closely spaced, small printing with little visual relief is more difficult and takes longer to read and comprehend — regardless of the content — than carefully designed, visually pleasing pages.

FIGURE 8.5 **Functions of Editing and Levels of Edit**[7]

Functions of Editing	Levels of Edit
Substantive editing and revising	▪ ***Substantive edit*** — reviews the document globally for accuracy, logic, completeness, coherence, consistency, organization, and tone.
Design review and copyediting *Proofreading*	▪ ***Format edit*** — establishes consistency in "macro" physical elements such as headings, fonts, page design, use of visuals. ▪ ***Integrity edit*** — matches text references to corresponding figures, tables, references, footnotes, and appendixes.
Administrative editing	▪ ***Mechanical style*** — edit establishes consistency in "micro" physical elements ▪ ***Screening edit*** — corrects language and numerical errors. ▪ ***Language edit*** — changes grammar, punctuation, usage, and sentence structure to meet conventions, such as symbols, citations, and numerical copy.
	▪ ***Policy edit*** — ensures that institutional policies are enforced. ▪ ***Clarification edit*** — provides clear instructions to the compositor and graphic artist. ▪ ***Coordination edit*** — deals with the administrative aspects in publishing technical documents.

Which of the nine levels of edit are you skillful at? Which of these levels of edits would help you be successful in your career?

and placement of headings, to details, such as assuring that numbers of figures and their textual references match. Individuals who assume this responsibility need to understand the general principles of document design as well as have a keen eye for inconsistencies in mechanical details.

If "correctness" is often a matter of conforming to conventions, why do people make such a big deal about not following these conventions?

Proofreading encompasses a broad range of responsibilities, including those most typically considered editing. All of the problems and errors in numeracy and language use fall into this category. Individuals who proofread documents need to have a thorough command of language appropriateness as well as the skill and sensitivity to create reader-based documents.

Some writers try to decrease their editing/proofreading burden by using software tools to flag potential problems in style, mechanics, and usage. Most word-processing software comes with built-in spelling and grammar checkers that catch glaring problems. However, such tools cannot guarantee that a document will be error-free or that it will communicate effectively with its audience. Writers should never depend on software tools to identify all the errors in a document, but should consider the following benefits and limitations of software tools for editing:[8]

Benefits

Software can identify

- misspelled words
- passive voice
- complex sentences
- wrong part of speech
- redundancy
- potentially difficult wording (based on word and sentence lengths only)
- slang/colloquialisms
- potentially sexist language
- negative wording

Limitations

Software cannot identify

- correctly spelled words used incorrectly
- inconsistent writing styles
- confusing sentence structures
- what the audience needs to know
- every grammar/mechanical problem
- poorly organized documents
- missing or faulty information
- potentially offensive ideas
- how a document will be used

When preparing a document that will be read online, most professional editors recommend that you print everything for final editing and proofreading. Why is this a good practice to follow?

Finally, a number of *administrative responsibilities* are often considered editorial tasks. An individual who assumes these responsibilities needs to be familiar with corporate policies, understand the production and publication processes, and be a skillful manager of people and tasks.

These nine levels of edit are functions that in a small organization may be performed by a single individual. In very large organizations, several people may be responsible for these editing functions. However, whether one person or several are responsible, the functions are completed in several passes through a document. No editor tries to look for everything at the same time. So, for example, an editor who is doing a substantive edit for content accuracy typically ignores the other areas. Similarly, an editor who is checking for consistency in headings or documentation tends to set other problems aside temporarily.

People who work on a networked computer system often use the network's capabilities to help with *document cycling,* the part of the process that takes a document through several rounds of reviewing, revising, and editing. They can electronically send their document to several reviewers simultaneously and receive the reviewers' comments the same way. Using the network can potentially speed document cycling, which in many organizations is a required part of preparing important documents, especially ones for external readers.

Your responsibilities as an editor are not always clear-cut. Although you often have rules and guidelines to follow (e.g., Does the document have problems in

© Helen King/CORBIS

Editing is done by professional editors and peers. The more specific you are in telling peer reviewers what areas of your draft need attention, the more help they're likely to be. Should they look particularly at organization? At adaptation to the audience? At the kind of supporting evidence you use? At your discussion of the tables and figures?

punctuation? Is all the documentation correct?), an even bigger part of your editorial role requires judgment. You may make recommendations to a writer based on the response you imagine readers will have. Rather than having a rule to enforce, you'll ask, "Is the design accessible?" "Is the document easy to read?" "Is the document easy to understand?"

Common Copyediting Problems

The discussion in the remainder of the chapter identifies four common copyediting problems for writers. They are areas that need a final check before a document is complete: concrete details, directness, positive phrasing, and the elimination of several kinds of wordiness.

Use Concrete Details Concrete words that refer to tangible objects are usually easier for your readers to understand than abstract words. Thus, you can generally make your writing more precise if you use specific details and examples. An effective technique for reducing abstract writing is to insert specific details in place of vague or general statements. The list in Figure 8.6 identifies some of the questions you can ask yourself as you edit a document to make abstract statements precise and concrete, to make your statements quantifiable and verifiable.

Use Direct Language Indirect language often results from inappropriate attempts to be formal or to use inflated or abstract language. In general, direct language is better. Direct language is typically plain and simple. However, no word is inherently right or wrong until you take the context, audience, and purpose into account. For example, the word *feel* would be appropriate in a pamphlet for women on breast self-examination that stated, "If you feel your breast tissue and detect a lump, contact your doctor." However, in a document for medical personnel, the more technical *palpate* would be a better choice than *feel.*

One way to be direct is to use straightforward terms. The revision of the following sentence substitutes simple language for needlessly ornate language.

> *Why might a writer purposely obscure meaning in writing?*

Inflated Language	**Revised: Direct Language**
Your conceptualization of our aggregate capability may enhance our marketing position.	Your ideas about our capability may improve our marketing.

FIGURE 8.6 **Selecting Concrete Details**

Abstract Information		Concrete Details
important client	Who?	Jean Thompson, PPI president
a new development	What?	developed a blight-resistant chestnut
schedule early	When?	schedule before 9 a.m.
ideal location	Where?	Southfield, MA
a substantial profit	How much?	a 37 percent profit
a broken part	Which one?	a broken camshaft
limited leg mobility	What percentage?	lift his leg 40 percent of normal extension
high-temperature environment	What degree?	up to 2,000°F
a small wingspan	What size?	a 7.0-mm wingspan
few changes in the procedure	How many?	three changes in the procedure
corrosion-resistant metal	What kind?	stainless steel

Another guaranteed shortcut to improved writing involves changing abstract nouns into verbs. One example is listed below, but many others appear in technical documents. If you change abstract nouns back to verbs and delete words that add little to the meaning, you'll greatly improve your writing. (Refer to the Usage Handbook on the Companion Web Site for additional examples.)

Abstract Language	**Revised: Direct Language**
The judge provided the required authorization for the search.	The judge authorized the search. (*Authorize* identifies the action.)

Use Positive Phrasing Effective writers often employ positive phrasing for several reasons. Psychologists and linguists know that readers and listeners understand positively phrased sentences more quickly and accurately than negatively phrased sentences, which the human brain takes slightly longer to process. Also, positive phrasing generally is more direct and appealing because eliminating negatives also eliminates words.

The most obvious negative words, which include *no, not, none, never,* and *nothing,* are added to a positive sentence to negate the idea. When you edit, consider replacing a negative word with a less overtly negative term, often an antonym for the word in the "not" phrase:

not many	few
do not accept	reject
do not succeed	fail

Other negative words are created by adding a common negative prefix to a word. The words with the prefixes are more indirect; they're less overtly negative. However, when you want to emphasize a negative point, use *not:*

inefficient	not efficient
impossible	not possible
unreliable	not reliable

In what situations might you choose to word directions positively? Negatively?

While one negative word or phrase only slightly delays a reader, multiple negatives significantly decrease reading speed and inhibit comprehension. Also, possible confusion results from multiple negatives. The following sentences illustrate how replacing negative wording with positive wording lessens the chances of confusion arising from multiple negatives.

Negative Phrasing	**More Positive Phrasing**
The production line changeover will not be ready on time.	The production line changeover will be delayed.
The manager could not approve the decentralization of the tool crib because there would be no inventory control or centralized maintenance checks.	The manager denied decentralization of the tool crib because it provides a regular inventory control and maintenance check.

In some situations, however, you'll find that positive phrasing is inappropriate because it is not emphatic enough. The following set of directions presumes that many users will do the wrong thing unless specific *do nots* are presented, as the following example illustrates.

Failure to follow standard procedures in using a disk will have detrimental results. Follow these guidelines:

- Do not turn drive on or off with disk in the drive.
- Do not bend disk.
- Do not touch disk surfaces with fingers or objects.
- Do not store disk out of protective envelope.
- Do not expose disk to heat, magnets, or photoelectric beams.

Such phrasing cautions computer operators to treat disks carefully. Although the wording is negative, experience has led manual writers to realize that some readers ignore cautions that use positive wording.

Eliminate Wordiness Wordiness can make your writing difficult to understand because readers are forced to plow through unnecessary text to read essential information. Your readers may lose patience. The most effective way to reduce wordiness is to avoid redundancy — the unnecessary repetition of concepts, words, or phrases. Sometimes readers cannot discern the point because it is hidden underneath layers of language. At other times redundancy is so annoying or time consuming that readers give up without completing a document.

The memo in Figure 8.7 has numerous instances of redundancy. Section manager Carl Saunders typed the memo directly into his word processor and read a copy after it was printed. When he proofread his draft, he was appalled by its redundancy, which he eliminated by crossing out unnecessary terms and making minor changes in wording. Figure 8.7 shows the editing he did on his memo. Producing a new copy constituted very little work on his word processor; within a few minutes, he printed the revision (Figure 8.8). You may encounter several redundant elements that you can edit out of the final drafts of technical documents — modifiers similar to those in Mr. Saunders's initial memo.

FIGURE 8.7 **Redundant Language in Memo**

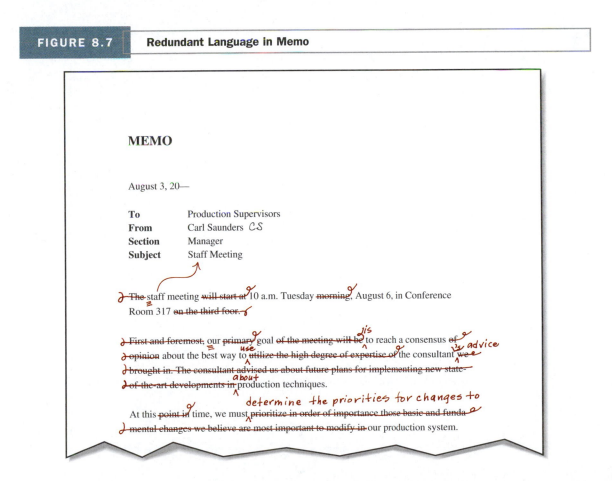

MEMO

August 3, 20—

To Production Supervisors
From Carl Saunders CS
Section Manager
Subject Staff Meeting

~~The~~ staff meeting ~~will start at~~ 10 a.m. Tuesday ~~morning~~, August 6, in Conference Room 317 ~~on the third foor.~~

~~First and foremost,~~ our ~~primary~~ goal ~~of the meeting will be~~ is to reach a consensus ~~of opinion~~ about the best way to use ~~utilize the high degree of expertise of~~ the consultant's advice ~~we brought in. The consultant advised us about future plans for implementing new state-of-the-art developments in~~ about production techniques.

At this ~~point in~~ time, we must determine the priorities for changes to ~~prioritize in order of importance those basic and fundamental changes we believe are most important to modify in~~ our production system.

- ***Redundant Pairs.*** Some common terms contain two words, both of which mean the same thing. Some writers use such redundant pairs to give their writing what they think is an air of elegance. Because both terms mean the same thing, only one is necessary. You could appropriately write *first* or *foremost,* but using both terms is repetitive. Sometimes, as in Mr. Saunders's memo, neither term is necessary because there is only one stated goal. Some other common redundant pairs include *null and void* and *final and conclusive.*

WEBLINK

For additional examples of redundant pairs, go to **www.english.wadsworth.com/burnett6e**.
CLICK ON WEBLINK
CLICK on Chapter 8/redundant pairs

FIGURE 8.8 **Revised Memo**

MEMO

August 3, 20—

To	Production Supervisors
From	Carl Saunders *CS*
Section	Manager
Subject	Staff Meeting 10 a.m., Tuesday, August 6 Conference Room 317

Our goal is to reach a consensus about the best way to use the consultant's advice about production techniques.

At this time, we must determine the priorities for changes to our production system.

- ***Redundant Modifiers.*** Phrases in which one term implies the other are known as redundant modifiers. For example, *collaborate* implies "together," so *collaborate together* is redundant. Some other common redundant modifiers include *initial preparation* and *most unique.*

  Mr. Saunders's memo contains the phrase "high degree of expertise of the consultant brought in," which has three redundant features. A consultant is assumed to have expertise and is also brought in from outside the company; saying a consultant has expertise and is brought in is doubly redundant. Also, expertise assumes a high degree of competence; repeating "high degree" is unnecessary.

For additional examples of redundant modifiers, go to www.english.wadsworth.com/burnett6e.
CLICK ON WEBLINK
CLICK on Chapter 8/redundant modifiers

WEBLINK

- ***Redundant Categories.*** When one term in a phrase is in the general category to which the other term belongs, it's known as a redundant category. For example, the expression *bitter taste* is redundant because bitter is one of four tastes; the word *bitter* is sufficient. In Mr. Saunders's memo, the initial "a.m." implies the category stated in the second term, so "morning" is unnecessary. You could appropriately write "at 10 a.m. Tuesday" or "at 10:00 Tuesday morning." Some other common redundant categories include *reliability factor* and *red in color.* The following examples show how eliminating redundant categories makes sentences less wordy.

Redundant	**Improved**
The reliability factor of the equipment was guaranteed.	The equipment's reliability was guaranteed.

For additional examples of redundant categories, go to www.english.wadsworth.com/burnett6e.
CLICK ON WEBLINK
CLICK on Chapter 8/redundant categories

WEBLINK

Revise Noun Strings Your writing can be improved by eliminating noun strings, which are a series of two or more nouns in which the first nouns modify the later ones. For example, in the string "circulation pump filter," both *circulation* and *pump* modify *filter.* Noun strings are distinguished by the absence of both apostrophes and connecting words that show relationships (such as *of, for, in*).

You'll run into problems if you use noun strings, particularly when the reader is unable to determine exactly how the nouns relate to each other. Strings that are only two words long seldom present difficulties. For example, *data analysis* is easily understood; saying "the analysis of data" is unnecessary. When the strings are three words long, the reader can usually figure out what is meant, but reading may be slowed slightly. When the strings reach four words, extra time is required to figure out the relationship among words. Although your reader probably will accurately interpret the string, this takes unnecessary time and effort, which may distract and annoy the reader. Strings that are five or more words long are open to multiple interpretations and can be indecipherable. The reader may never guess exactly what you intended.

Noun Strings	**Revision**
standardization blending	controller that standardizes blending
test module specification review	the specifications review for the test module
scanner head motion control	control of the scanner head's motion

Revisions of noun strings usually require additional words to clarify the relationships; however, these extra words eliminate ambiguity.

The issues of conventions and correctness addressed in this chapter are often more difficult to apply in workplace situations, which are nearly always complicated by politics and personalities. The ethics sidebar below addresses some of these complications, which often force writers and editors to make difficult decisions. The sidebar suggests that identifying the values that are in conflict will aid decision making.

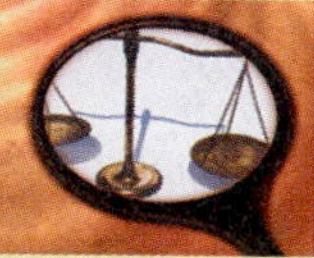

ETHICS SIDEBAR

Author, Audience, and Company: Ethics and Technical Editing

Imagine a technical editor in the following scenario:

Sam is a technical editor, working for a software development company. As she edits the draft of a manual for the company's newest product, she realizes that large sections are very technical, making them unclear and difficult to understand. When Sam talks to the writer, a programmer, she finds him very reluctant to revise the manual; he argues that the technical information, as it is written, is necessary. The deadline for the manual is looming; Sam has little time to revise the manual without the writer's help.

What are the technical editor's responsibilities in this situation? As this scenario indicates, technical editors, because of their unique position in the document production process, must balance obligations to several different groups. First, they have an obligation to the audience to ensure that documents are clear and accurate. Second, they are expected to safeguard the employer's reputation and protect the company from legal liabilities. Finally, they have to respect the author's intellectual independence, especially if the editor is not an expert in the subject. In most situations, each of these different obligations works toward a collective goal of clear, accurate, professional documents.

When these obligations do conflict, however, the editor is faced with an ethical dilemma. To which group does he or she have the greater responsibility: author, audience, or employer? For example, in the situation above, the editor faces several different ethical obligations:

- *To the audience* — they expect the manual to be clear and readable but also expect it to be technically accurate
- *To the author* — the author is an expert on the product but may not appreciate the audience's lack of technical expertise
- *To the employers* — they could lose money if the product is delayed but could also suffer if the documentation is inaccurate or misleading

This situation does not have one simple solution. Researchers Lori Allen and Dan Voss, however, provide a process that allows editors to at least clearly identify the values that are in conflict.[9] The outline of their process includes parenthetical notes about ways in which each step fits the scenario in this sidebar.

- Define what is at issue (producing a clear text) and the groups involved (audience, author, and employer)
- Determine the different groups' interests (clarity, experience, profit)
- Identify the relevant values (usability, integrity, capitalism)
- Determine the values that conflict (usability vs. integrity, capitalism vs. usability)
- Determine which value has greater importance (usability)
- Resolve the conflict for that value (require the writer to revise the document)

Requiring the writer to revise the document is one possible resolution for this dilemma. If the greater value is capitalism, then the ethical decision may be to improve as much of the quality of the manual as time allows. No matter which they choose, editors should heed Allen and Voss's final point: Editors carry a greater ethical responsibility because they are experts with language. Editors have the advantage of looking at a situation from different perspectives, but that advantage comes with weighty ethical responsibilities.

Reread the scenario at the beginning of this sidebar. Using Allen and Voss's six-step process, how would you resolve this conflict? What value(s) would you emphasize?

How would you edit a poorly written technical document that the author does not want altered?

Proofreading

Before any document, accompaniments for an oral presentation, or visuals get approved for final distribution, display, or publication, you need to do careful proofreading — a final check. This isn't a time to add new information or to do rewriting. It's a straightforward check for accuracy and consistency in five broad areas:

- mechanical conventions (e.g., , punctuation, capitalization, spelling)
- grammatical conventions (e.g., grammar, usage)
- design conventions (e.g., typography, visual displays, headings)
- disciplinary conventions, (e.g., abbreviations, citations)
- typographical conventions (e.g., symbols, numbers)

The proofreading marks in Figure 8.9 are widely used by editors as a common system for identifying specific things to change in a document. Learning and regularly using these symbols will enable you to understand editorial marks on documents returned to you as well as to similarly mark documents you are asked to proofread.

Reading aloud helps catch problems — omissions, awkward phrasings, grammatical errors — that you might miss when proofreading silently, especially proofreading on-screen.

Examining Revision and Editing Decisions

Figure 8.10 shows the first few pages of a draft of a feasibility report (pages 288–292). This particular report evaluates two anchor bolts and recommends one for the company. Suggestions for revision and copyediting are marked on this draft. Marginal annotations comment on some of the other problems with the draft. Figure 8.11 shows the final revised and edited version of the same report (pages 293–302).

To assess the effectiveness of this document, you need to know about the purpose of feasibility reports, which is to advocate the best solution to a problem, and their structure. The introduction of the report usually gives an overview of the feasibility study (the work that results in the report) by briefly stating the purpose of the study, describing the problem being investigated, and presenting the scope of the report. An optional background section, which is included when the audience is not likely to be familiar with the situation that necessitated the feasibility study, can contain information such as the history of the problem, the possible long-range impact of not correcting the problem, and relevant work that has already been done in the field.

The criteria and solutions sections are the most important. The criteria allow readers to evaluate the various solutions fairly, so their selection is important. Some of the following criteria are often used:

- cost of purchase, installation, and maintenance
- ease and frequency of servicing
- ease of operation

FIGURE 8.9	Proofreading Marks[10]

Symbol	Meaning	Example
⌇ or ⌇ or ⌇	delete	take it out
⁀	close up	print as o ne word
⌇	delete and close up	cloose up
^ or > or ⌇	caret	insert here (something)
#	insert a space	put onehere
eq#	space evenly	space evenly where indicated
stet	let stand	let marked text stand as set
tr	transpose	change order the
/	used to separate two or more marks and often as a concluding stroke at the end of an insertion	
[	set farther to the left	[too far to the right
]	set farther to the right	too] far to the left
⌒	set as ligature (such as æ)	encyclopaedia
=	align horizontally	alignment
||	align vertically	|| align with surrounding text
X	broken character	imperfect
□	indent or insert em quad space	
¶	begin a new paragraph	
(sp)	spell out	set 5 lbs as five pounds
cap	set in CAPITALS	set nato as NATO
smcap or s.c.	set in SMALL CAPITALS	set signal as SIGNAL

Symbol	Meaning	Example
lc	set in lowercase	set South as south
ital	set in *italic*	set oeuver as *oeuver*
rom	set in roman	set *mensch* as mensch
bf	set in **boldface**	set important as **important**
= or -/ or ⌇ or /H/	hyphen	multi-colored
1/N or en or /N/	en dash	1965–72
1/M or em or /M/	em (or long) dash	Now—at last!—we know.
∨	superscript or superior	2 as in πr^2
∧	subscript or inferior	2 as in H_2O
◊ or X	centered	· for a centered dot in $p \cdot q$
,	comma	
'	apostrophe	
⊙	period	
; or ;/	semicolon	
: or (:)	colon	
" " or ' '	quotation marks	
(/)	parentheses	
[/]	brackets	
OK/?	query to author: has this been set as intended?	
⊥ or ⊥	push down a work-up	an unintended mark
9	turn over an inverted letter	inverted
wf	wrong font	wrong size or style

- kind and duration of operator training
- performance specifications
- dimensions
- compatibility with existing systems
- flexibility for expansion
- environmental impact

Writers should select only criteria that identify significant differences in the solutions, and the solutions offered should be feasible. Including impractical or weak solutions as a way to make a preferred solution look better borders on the unethical and can destroy the credibility of the entire document.

The final section of a feasibility report incorporates the conclusions and recommendations. These sections may be presented separately or together; in either case, the points are often bulleted or enumerated. Sometimes the conclusions in a feasibility report present both advantages and disadvantages of each solution, synthesizing rather than just summarizing the information. The recommendations identify the best solution based on the criteria used in the feasibility study. The recommendations should be clear, precise, unambiguous, and logical.

As you read the draft in Figure 8.10 and the final report in Figure 8.11, consider both global revisions (redesigning and reorganizing the report) and local revisions (revising paragraphs and sentences). Then consider the editing decisions that focus on word- and phrase-level usage that affect the report's accuracy as well as its appropriateness and appeal.

FIGURE 8.10 **Draft of Feasibility Report Marked for Revision and Editing**

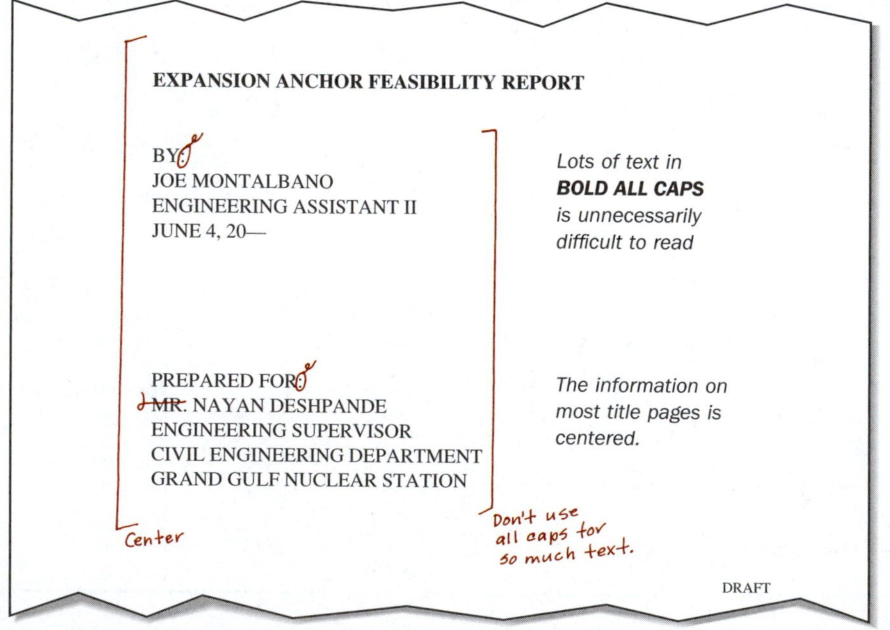

EXPANSION ANCHOR FEASIBILITY REPORT

BY:
JOE MONTALBANO
ENGINEERING ASSISTANT II
JUNE 4, 20—

PREPARED FOR:
MR. NAYAN DESHPANDE
ENGINEERING SUPERVISOR
CIVIL ENGINEERING DEPARTMENT
GRAND GULF NUCLEAR STATION

DRAFT

FIGURE 8.10 **Draft of Feasibility Report Marked for Revision and Editing (continued)**

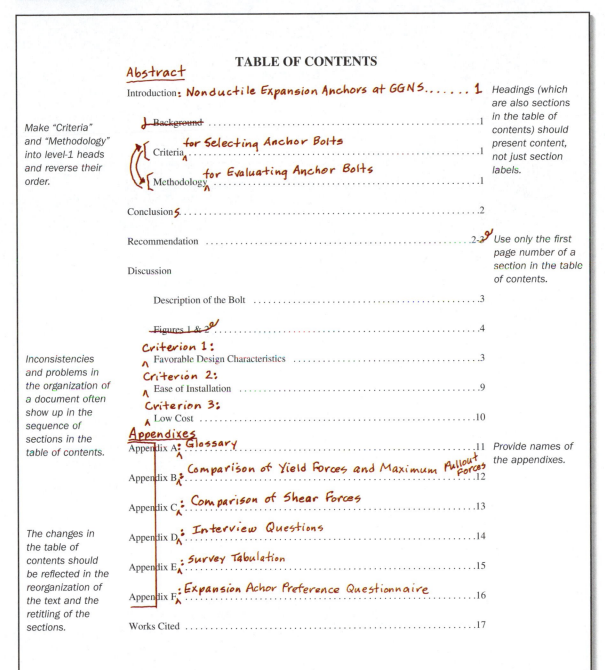

Make "Criteria" and "Methodology" into level-1 heads and reverse their order.

Inconsistencies and problems in the organization of a document often show up in the sequence of sections in the table of contents.

The changes in the table of contents should be reflected in the reorganization of the text and the retitling of the sections.

TABLE OF CONTENTS

Headings (which are also sections in the table of contents) should present content, not just section labels.

Use only the first page number of a section in the table of contents.

Provide names of the appendixes.

DRAFT

FIGURE 8.10 **Draft of Feasibility Report Marked for Revision and Editing (continued)**

Add an abstract

INTRODUCTION : Nonductile Expansion Anchors at GGNS

By convention, the first page of the main body usually doesn't have the page number on it.

Background

The nonductile expansion anchor has a long history of use at Grand Gulf Nuclear Station (GGNS). The nonductile expansion anchor, which has been used almost exclusively, is has been the Hilti Kwik Bolt anchor bolt. This wedge-type anchor is essential for installation of a wide vriety variety of components including conduit supports, pipe supports, instrument supports, and ventilating and air conditioning supports.

However, replacement of the Hilti Kwik Bolt is required because the manufacturer recently discontinued production of it. Therefore, As a result, I conducted a search for a replacement anchor bolt. This search consisted of by evaluating several types of two anchor bolts: they are the Hilti Kwik Bolt II anchor bolt* and the Drillco Maxi-Bolt anchor bolt.*

Double spacing takes up an unnecessary amount of space. Space-and-a-half will maintain the ease of reading but save room as well.

Criteria for Selecting Anchor Bolts

I established three criteria that best describe the desired characteristics of the new anchor bolt. These criteria are listed in order of importance:

- Favorable design characteristics
- Ease of installation
- Lowest cost

Readers may more easily follow the logic of the report if the general approach (the methodology) is presented before the specifics (the criteria for evaluation).

What additional problems have you identified in this report?

Methodology for Evaluating Anchor Bolts

In evaluating the Hilti Kwik Bolt II anchor bolt and the Drillco Maxi-Bolt anchor

*See Glossary for definitions of terms.

All terms marked with an asterisk are defined in the Glossary in Appendix A on page 11.

DRAFT

FIGURE 8.10 **Draft of Feasibility Report Marked for Revision and Editing (continued)**

bolt, I used several methods of gathering information. The main method I used was reading vendor brochures, technical guides, and a technical standard. This provided a substantial amount of information about the two bolt types. Also I gathered information by distributing an expansion anchor preference questionnaire and by interviewing an engineer familiar with the two bolt types.

CONCLUSIONS

The evaluation of the Hilti Kwik Bolt II and the Drillco Maxi-Bolt produced the following results.

Generally single space the items in a list and double space between these items.

- In terms of design characteristics, the Hilti Kwik Bolt II has four of the most important characteristics. The Drillco Maxi-Bolt has only one and it is the least important characteristic.
- In terms of ease of installation, the Hilti Kwik Bolt II satisfied both installation requirements. The Drillco Maxi-Bolt satisfied neither of them.
- In terms of lowest cost, the Hilti Kwik Bolt II satisfied this criterion because it was at least 20% cheaper in cost than the Drillco Maxi-Bolt.

RECOMMENDATION

I recommend that the Civil Engineering Department approve the Hilti Kwik Bolt II for future use at GGNC. This bolt is suitable for use at GGNS because it adheres to all the established criteria, with the exception of load capacity. Although this bolt type does not have the highest load capacity, it is acceptable. The load capacity is very similar to that

The word- and sentence-level editing makes the report easier to read.

DRAFT

FIGURE 8.10 **Draft of Feasibility Report Marked for Revision and Editing (continued)**

of the Hilti Kwik Bolt which was used at GGNS for a long time. Repaginate to eliminate orphan.

Make sure that the visual cues allow readers to easily and quickly distinguish between levels of headings.

DISCUSSION

Description of Bolts

The Hilti Kwik Bolt II and the Drillco Maxi-Bolt represent two different anchor designs with each having an ultimate tensile stress of 125,000 psi. The primary difference between the two bolts is the anchoring mechanism. The Hilti Kwik Bolt II has a wedge anchoring mechanism and is classified as a wedge anchor.* (See Figure 1.) Also it is a nonductile expansion anchor because of its nonductile failure characteristics. The Drillco Maxi-Bolt has a sleeve anchoring mechanism and is classified as a sleeve anchor.* (See Figure 2.) This anchor has ductile failure characteristics and is also called a ductile expansion anchor. Insert Figures 1, 2 here

Incorporate critical figures and tables as close to the text reference as possible after the end of the paragraph.

Criterion 1:
Favorable Design Characteristics close up spacing

The new anchor bolt must have five characteristics. ~~They are~~ listed here in order of importance:

- Largest assortment of bolt diameters and bolt lengths
- Minimum bolt embedment
- Minimum bolt spacing*
- Minimum free edge distance*
- Highest load capacity

Eliminate unnecessary repetition by changing the first mention of the glossary to include references to all succeeding asterisks.

ital

Hilti Kwik Bolt II. The bolt diameters for this bolt type range in sizes from $^1/_4$" to 1" in $^1/_8$" increments excluding $^7/_8$". Also each bolt diameter is available in ~~a~~ at least two lengths. For example, the $^1/_2$" diameter bolt is available in $2^3/_4$", $3^3/_4$", $4^1/_2$", and $5^1/_2$" lengths. The large variety of sizes for this bolt type broadens the number of its applicable uses.

~~*See Glossary for definitions of terms.~~

FIGURE 8.11 **Final Version of a Revised and Edited Feasibility Report**[11]

*The choice to use **BOLD ALL CAPS** draws attention to the title. The brevity of the title makes **BOLD ALL CAPS** acceptable in this situation.*

The title clearly informs readers about the topic (expansion anchors) and the approach (feasibility).

EXPANSION ANCHOR FEASIBILITY REPORT

Joe Montalbano
Engineering Assistant II
June 4, 20—

Prepared for
Nayan Deshpande
Engineering Supervisor
Civil Engineering Department
Grand Gulf Nuclear Station

FIGURE 8.11 Final Version of a Revised and Edited Feasibility Report[11] (continued)

The headings identify the content of each section rather than simply specify the section of the report—that is, Methodology for Evaluating Anchor Bolts rather than just Methodology.

Page numbers are right justified.

TABLE OF CONTENTS

Dot leaders help readers visually track from the topic to the page number.

The organization and typography of the table of contents match what appears in the report itself.

Only the first page number of a section is listed in the table of contents.

FIGURE 8.11 **Final Version of a Revised and Edited Feasibility Report[11] (continued)**

The abstract should preview both the content and the organizational structure of the feasibility study.

The specific headings help readers anticipate the content to be discussed in each section.

ABSTRACT

This report presents the result of a feasibility study of two anchor bolts—the Hilti Kwik Bolt II and the Drillco Maxi-Bolt—and recommends the Hilti Kwik Bolt II for use at Grand Gulf Nuclear Station. The study evaluated both bolt types in terms of three criteria: design characteristics, ease of installation, and low cost.

INTRODUCTION: NONDUCTILE EXPANSION ANCHORS AT GGNS

The nonductile expansion anchor has a long history of use at Grand Gulf Nuclear Station (GGNS). The nonductile expansion anchor used almost exclusively has been the Hilti Kwik Bolt anchor bolt. This wedge-type anchor is essential for installation of a wide variety of components including conduit supports, pipe supports, instrument supports, and ventilating and air conditioning supports.

Replacement of the Hilti Kwik Bolt is required because the manufacturer recently discontinued its production. As a result, I conducted a search for a replacement by evaluating two types of anchor bolts: the Hilti Kwik Bolt II anchor bolt* and the Drillco Maxi-Bolt anchor bolt.*

The report uses first person ("I") because the individual engineer is making a recommendation. By using first person, he accepts responsibility for his recommendation.

Placing the Methodology before the Criteria helps readers understand the overall approach before they learn about the specifics.

METHODOLOGY FOR EVALUATING ANCHOR BOLTS

In evaluating the Hilti Kwik Bolt II anchor bolt and the Drillco Maxi-Bolt anchor bolt, I used several methods of gathering information. The main method was reading vendor brochures, technical guides, and a technical standard. This information provided a substantial amount of background about the two bolt types. I also gathered information by distributing an expansion anchor preference questionnaire to engineers and by interviewing an engineer familiar with the two bolt types.

Glossary terms are easy to identify because they're marked with an asterisk and clearly defined.

What additional strengths have you identified in this report?

---

*All terms with an asterisk are defined in the Glossary in Appendix A on page 9.

CRITERIA FOR SELECTING ANCHOR BOLTS

Bulleting the criteria helps readers see their importance.

I established three criteria, in order of importance, that describe the desired characteristics of the new anchor bolt:

- Favorable design characteristics
- Ease of installation
- Low cost

CONCLUSIONS

The conclusions and recommendation are put up front so that readers know the critical information first. The supporting discussion follows for readers who want more information or who want to understand the rationale for the decision.

My evaluation of the Hilti Kwik Bolt II and the Drillco Maxi-Bolt led me to the following conclusions.

- In terms of design characteristics, the Hilti Kwik Bolt II has four of the most important. The Drillco Maxi-Bolt has only one, and it is the least important.
- In terms of ease of installation, the Hilti Kwik Bolt II satisfies both installation requirements. The Drillco Maxi-Bolt satisfies neither of them.
- In terms of lower cost, the Hilti Kwik Bolt II satisfies this criterion because it is at least 20 percent lower in cost than the Drillco Maxi-Bolt.

RECOMMENDATION

I recommend that the Civil Engineering Department approve the Hilti Kwik Bolt II for future use at GGNS. This bolt meets all the established criteria, with the exception of load capacity. Although this bolt type does not have the highest load capacity, it is acceptable. The load capacity is very similar to the Hilti Kwik Bolt, which was used at GGNS for a long time.

DISCUSSION

In this section, I provide an overall description of the two different bolts and then discuss the three criteria that influenced my recommendation.

FIGURE 8.11 **Final Version of a Revised and Edited Feasibility Report[11] (continued)**

Description of Bolts

The Hilti Kwik Bolt II and the Drillco Maxi-Bolt represent two different anchor designs, each having an ultimate tensile stress* of 125,000 psi. The primary difference between the two bolts is the anchoring mechanism. The Hilti Kwik Bolt II has a wedge anchoring mechanism and is classifiedas a wedge anchor.* (See Figure 1.) Also it is a nonductile expansion anchor because of its nonductile failure characteristics. The Drillco Maxi-Bolt has a sleeve anchoring mechanism and is classified as a sleeve anchor.* (See Figure 2.) This anchor has ductile failure characteristics and is also called a ductile expansion anchor.

Figures 1 and 2 are proportional in size and placed side by side for fair and easy comparison. Note that both are referred to in the text.

FIGURE 1:
Hilti Kwik Bolt II
Source: *Hilti Technical Guide*

FIGURE 2:
Drillco Maxi-Bolt
Source: *Drillco Maxi-Bolt Brochure*

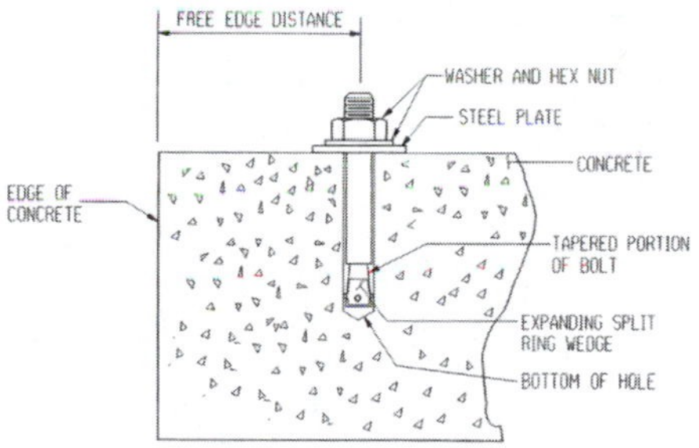

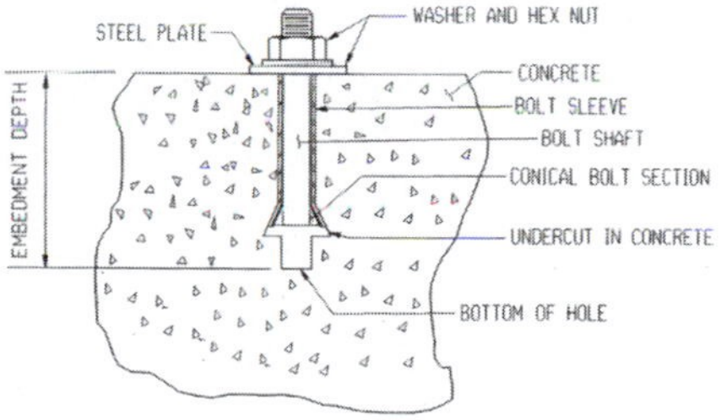

Criterion 1: Favorable Design Characteristics

The new anchor bolt must have five characteristics, listed here in order of importance:

Using a bulleted list to specify the favorable design characteristics helps readers know what to look for in the assessment of each bolt.

- Largest assortment of bolt diameters and bolt lengths
- Minimum bolt embedment
- Minimum bolt spacing*
- Minimum free edge distance*
- Highest load capacity*

Differentiating heading levels — on this page a level-3 heading — helps readers understand the hierarchical relationships in your discussion.

Hilti Kwik Bolt II. The bolt diameters for this bolt type range in size from $^1/_4$" to 1" in $^1/_8$" increments excluding $^7/_8$". Also each bolt diameter is available in at least two lengths. For example, the $^1/_2$" diameter bolt is available in $2^3/_4$", $3^3/_4$", $4^1/_2$", and $5^1/_2$" lengths. The large variety of sizes for this bolt type broadens the number of its applicable uses.

The bolt embedment is the portion of the bolt's overall length that is embedded in concrete (Standard 6). Each bolt diameter has a minimum embedment as indicated in Table 1. In comparing the embedment values given in Table 1, you will notice that the Hilti Kwik Bolt II has at least 33 percent shallower embedment than the Drillco Maxi-Bolt. Shallower embedments allow faster bolt installation.

Table 1 provides readers with an easy-to-read comparison that summarizes information. Note that the table is referred to in the text.

TABLE 1:
A Comparison of Embedments

	Hilti Kwik Bolt II	Drillco Maxi-Bolt
Bolt Diameter (in.)	**Minimum Embedment (in.)**	**Minimum Embedment (in.)**
$^1/_4$	2	3
$^3/_8$	2	4.5
$^1/_2$	2.5	6
$^5/_8$	3	7.5
$^3/_4$	3.5	9.25
1	4.5	12.5

Source: Standard No. SERI-CS-03, "Standard for Design of Concrete Expansion Anchors"

Providing the citation for the information in Table 1 helps readers assess its credibility.

The bolt spacing and free edge distance represent minimum spacing requirements. These design characteristics are smaller for the Hilti Kwik Bolt II than the Drillco Maxi-Bolt. The smaller bolt spacing means the bolts can be installed close to each other. The small free edge distance means that the bolt can be located closer to the edge of a concrete structure. Instances where these design characteristics can be advantageous are on concrete walls or floors of the plant where there is limited room for additional items.

FIGURE 8.11 **Final Version of a Revised and Edited Feasibility Report[11] (continued)**

Readers are referred to the attached appendixes, which increases the likelihood that they'll check that information.

The load capacity of the Hilti Kwik Bolt II is measured in terms of pullout force* and shear force.* For each bolt diameter, Appendix B shows the maximum pullout forces produced at the maximum embedment. Appendix C shows the maximum shear force for each bolt diameter. These two appendixes indicate that pullout forces and shear forces for the Hilti Kwik Bolt II are consistently lower than the yield forces and shear forces for the Drillco Maxi-Bolt. (The pullout forces and shear forces are also lower than for the original Hilti Kwik Bolt.) The lower forces mean that the Hilti Kwik Bolt II is not as strong as the other two bolts.

Drillco Maxi-Bolt. This bolt is available in the same bolt diameters as the Hilti Kwik Bolt II. But for each bolt diameter there are fewer alternate bolt lengths. The lengths of the bolts range from $4^1/_2$" for $^1/_4$" bolts to $16^1/_2$" for 1" bolts.

The embedments for the Drillco Maxi-Bolt are shown in Table 1 on page 4. The embedment for each diameter Drillco Maxi-Bolt is consistently larger than for the Hilti Kwik Bolt II. These deeper embedments are unfavorable because when bolts are installed in reinforced concrete, there is an increased probability of damaging or severing the reinforcing steel.

Like many paragraphs in this report, this paragraph uses given-new principles to provide coherence for readers.

The bolt spacing and the free-edge distance for this particular bolt type are dependent on bolt embedment. Because the embedment of the Drillco Maxi-Bolt is deeper than the Hilti Kwik Bolt II, it follows that the bolt spacing and free edge distance are larger. These larger spacing characteristics are a disadvantage because a larger area is required to produce the rated load capacity of the bolts.

Like many paragraphs in this report, this paragraph uses a clear topic sentence to help readers anticipate what is coming in the next few sentences.

The load capacity of the Drillco Maxi-Bolt is measured in terms of yield force* and maximum shear force. For each bolt diameter, Appendix B shows the yield force corresponding to the maximum embedment. Appendix C shows the maximum shear force for each bolt diameter. Both appendixes indicate that the yield forces and the shear

FIGURE 8.11 **Final Version of a Revised and Edited Feasibility Report[11] (continued)**

forces of the Drillco Maxi-Bolt are consistently higher than the forces shown for the other two bolt types. The higher forces mean that this bolt type is stronger.

The conclusions are clearly labeled and formatted in a bulleted list so they are easy to distinguish from the rest of the text.

Conclusions about Design Characteristics. Based on the information provided for each design characteristic, I have made the following conclusions:

- The Hilti Kwik Bolt II has the largest assortment of bolt diameters and bolt lengths because it provides more alternate bolt lengths for each bolt diameter.
- The Hilti Kwik Bolt II has the desired minimum embedment characteristic because it requires a shallower embedment than the Drillco Maxi-Bolt.
- The Hilti Kwik Bolt II has the desired minimum bolt spacing and minimum free edge distance characteristic because this bolt type provides the smallest distance dimension.
- The Drillco Maxi-Bolt has a higher load capacity than the Hilti Kwik Bolt II.

The Hilti Kwik Bolt II is preferred in terms of design characteristics over the Drillco Maxi-Bolt because it satisfies four of the most important design characteristics. Its load capacity is only slightly lower than the Hilti Kwik Bolt.

A summary conclusion for criterion 1 removes any ambiguity that might have remained if the writer had simply presented the bulleted list of conclusions.

The same structure used to organize the information for criterion 1 is used here to organize the information for criterion 2: required features, compliance of Kwik Bolt II, compliance of Drillco Maxi-Bolt, and conclusion.

Criterion 2: Ease of Installation

Because of the anticipated high frequency of use of the new anchor bolt, it must meet the following installation requirements:

- Easy installation process
- Average installation time of less than 2 hours

Hilti Kwik Bolt II. The process of installing the Hilti Kwik Bolt II consists of three easy steps:

1. Drill a hole in hardened concrete.
2. Insert the bolt in the hole to the desired embedment.
3. Set the anchor by torquing the nut down.

The installation process is so simple that it takes an average of only $1^1/_4$ hours with a few standard tools.

Drillco Maxi-Bolt. The installation of the Drillco Maxi-Bolt is a more involved, four-step process:

1. Drill a hole in hardened concrete.
2. Cut the undercut shape at the appropriate location in the hole.
3. Insert the bolt in the hole to the desired embedment.
4. Set the anchor by torquing the nut down.

This more involved process takes $2^1/_2$ hours and requires special tools such as the Maxi-Bolt undercutting tool and the Maxi-Bolt setting tool.

Conclusion about Installation. The Hilti Kwik Bolt II satisfes both ease of installation requirements, while the Drillco Maxi-Bolt does not satisfy either of them.

Criterion 3: Low Cost

The new anchor bolt must be the less expensive of the two bolt types being evaluated. Since the different bolt sizes are packaged in various quantities, I needed to determine a per unit cost for the Hilti Kwik Bolt II and the Drillco Maxi-Bolt. Table 2, which shows the per unit cost, indicates that for each bolt diameter the cost for the Hilti Kwik Bolt II is a minimum of 20 percent less than the Drillco Maxi-Bolt.

FIGURE 8.11 **Final Version of a Revised and Edited Feasibility Report[11] (continued)**

Table 2, referred to in the text, summarizes the critical information about cost, in a display that is far more efficient and effective than using sentences. Whereas Table 1 came from a single source about industry standards, this table has been constructed by the author from two different sources, both of which are noted.

TABLE 2:

Cost Comparison

	Hilti Kwik Bolt II	Drillco Maxi-Bolt
Bolt Diameter (in.)	**Cost[1]**	**Cost[1]**
1/4	$2.20	$2.65
3/8	3.05	4.00
1/2	4.30	5.55
5/8	5.10	6.35
3/4	6.15	7.40
1	7.50	9.45

[1]cost per bolt

Source: *Hilti Technical Guide* and *Drillco Maxi-Bolt Brochure*

Conclusion about Cost. Based on the available cost information presented in Table 2, I recommend the Hilti Kwik Bolt II.

Because the overall conclusions were presented near the beginning of the report, when the discussion of the criteria is completed, the report simply stops (with appendixes added).

The appendixes that accompanied the actual feasibility report are not included here.

1. **Edit sentences for wordiness.** Work in a small group with a few classmates. For each of the following sentences, prepare a less wordy revision. Explain to your classmates why you think your revision is an improvement given the context, audience, and purpose that you envision.

   - It has come to my attention that the lights in my office have not been working.
   - Subsequent modifications will be disseminated to all users.
   - The deterioration of her condition led to the determination that surgery was necessary.
   - When the inspector did not accept this data, claiming it was unreliable, we were not so inefficient as to reject suggestions for alternative testing methods.
   - A 35mm SLR (single lens reflex) camera is a camera (a light-tight box that uses a physical means of reproducing an image and a chemical means of preserving it) that uses film of 35mm width and allows the photographer to see the image to be photographed through a single lens, the one used to expose the film, rather than through a second one used only for viewing the image.
   - A single-lens method of viewing the image to be photographed is preferable to the viewfinder or rangefinder method, in which the user "finds the view" through a separate lens because it eliminates the problem of parallax error, or not seeing precisely what will be photographed.

2. **Revise law to make more comprehensible.** Read the following state law explaining fishing regulations and then revise to clarify, while maintaining necessary information.[12]

   Lake or Pond Partly in Another State. If, in the case of a lake or pond situated partly in this state and partly in another state, the laws of such other state permit fishing in that part thereof lying within such other state by persons licensed or otherwise entitled under the laws of this state to fish in that part of such lake or pond lying within this state, persons licensed or otherwise entitled under the laws of such other state to fish in the part of such lake or pond lying within such other state shall be permitted to fish in that part thereof lying within this state, and, as to such lake or pond, the operation of the laws of this state relative to open and closed seasons, limits of catch, minimum sizes of fish caught and methods of fishing shall be suspended upon the adoption and during the continuance in force of rules and regulations relative to those subjects and affecting that part of such lake or pond lying within this state, which rules and regulations the director is hereby authorized to make, and from time to time add to, alter and repeal.

3. **Use a software editing program.** Select a document that you have written that still needs careful editing — that is, you still need to clean up stylistic problems as well as eliminate errors in mechanics and usage. Then go to your school (or company) computer lab and select one of the software packages that does this kind of editing. Run the software program to identify problems in style, mechanics, and usage. Make the corrections that you believe will improve your document.
4. **Evaluate a software editing program.** Write a review of the software you used in Assignment 3 and evaluate the effectiveness of the software. Prepare this review for publication in a local Society for Technical Communication newsletter so that other technical communicators and professionals who might use this software may have access to it.
5. **Interview a professional, part III.**

    (a) Arrange to interview a professional technical writer or a technical professional in your field who revises and edits frequently. During the third part of the interview (Part I of the report is an assignment in Chapter 6, and Part II is an assignment in Chapter 7), inquire about the person's approach to revising and editing. Ask some of the following questions and add additional questions relevant to the specific person and organization:

    - What kind of revising and editing do you do? How much time does it take? How important is it? Could you reduce or omit revising and editing of your documents with little negative effect?
    - What factors influence what you decide to revise?
    - How do you use revision and editing to create more accessible, comprehensible, and usable documents, presentations, and visuals?
    - How much does a draft change from the initial plan to the penultimate version?
    - What technology/software do you use for revising and editing?
    - How does your background affect your revising and editing?

    (b) Organize the information you find as part III of a brief oral or written report intended for other members of the class. This report can be posted on the class Web site or made available in print form in a class notebook in the library.

Chapter 8 Endnotes

1 Sitko, B. (1992). Exploring feedback: Writers meet readers. In A. M. Penrose & B. M. Sitko (Eds.), *Hearing ourselves think: Cognitive research in the college writing classroom* (pp. 170–187). New York: Oxford University Press.

2 Mansur, T. (n.d.). Justification for the purchase and installation of automatic roll-up doors for the R&D model shop. *Technical Writing, 42,* 225.

3 Microsoft Press. (n.d.). *Troubleshooting Excel® spreadsheets.* Downloaded October 21, 2003, from http://www.microsoft.com/mspress/books/sampchap/4757b.asp

4 Klass, P. J. (1982, January 11). Software augments manual readability. *Aviation Week & Space Technology, 116,* 106.

5 Adapted from Robe from 1000 BC found on Greek Island. (1983, July 5). *The Boston Globe.*

6 Fried, J. R. (1983). Polymer technology — part 7: Engineering thermoplastics and specialty plastics. *Plastics Engineering, 39*(May), 39.

7 Modified from the work originally done by Buehler, M. F. (1981). Defining terms in technical editing: The levels of edit as a model. *Technical Communication, 28*(4), 13–15. Based on a system developed and used at the Jet Propulsion Laboratory, Pasadena, CA.

8 Modified from a study conducted by Kohut, G. F., & Gorman, K. J. (1995). The effectiveness of leading grammar/style software packages in analyzing business students' writing. *Journal of Business and Technical Communication, 9*(July), 341–361.

9 Allen, L., & Voss, D. (1998). Ethics for editors: An analytical decision-making process. *IEEE Transactions on Professional Communication, 41*(1), 58–65.

10 *Proofreaders' marks.* (n.d.). Downloaded October 20, 2003, from http://www.m-w.com/mw/table/proofrea.htm

11 Montalbano, J. (n.d.). *Expansion anchor feasibility report.* Grand Gulf Nuclear Station, Port Gibson, MS. Used with permission.

12 New Hampshire State Fishing Laws, 1983, 6.

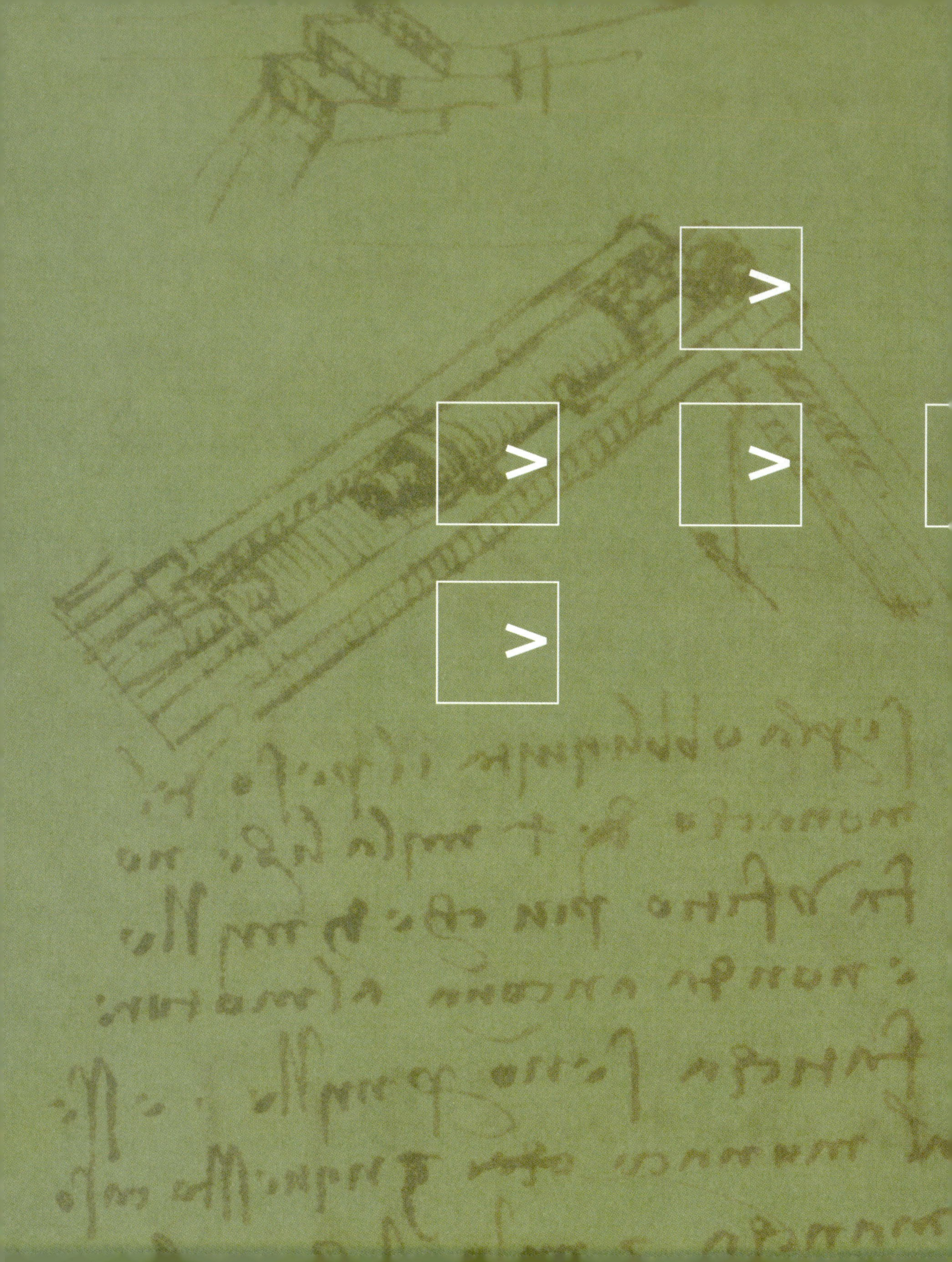

CHAPTER 9

Ensuring Usability

Objectives and Outcomes

This chapter will help you accomplish these outcomes:

- Characterize usability and understand the crucial role that its principles play in allowing people to accomplish tasks
- Understand factors that characterize usability and usability testing
- Differentiate and use text-based testing, expert-based testing, and user-based testing
- Identify, plan, and conduct various types of usability tests based on user and task analyses
- Define accessibility and understand the relationship between accessibility and usability

>

Think about the last time you used directions to assemble or use a piece of equipment. Were the directions clear, well organized, and helpful? Now consider the last time you encountered new computer software or used a Web site to accomplish a task. Did the design of the software or Web site help or hinder your ability to complete your task? Documents and interfaces that are difficult to understand or use or that make your experience unpleasant indicate poor usability. To be *usable,* communication — including documents, software interfaces, Web sites, and other media — must successfully meet the needs of the people using them in complex situations.

Although ensuring usability is an important step in creating documents, designing Web sites, and developing software interfaces, it too often receives inadequate attention or may even be overlooked during product development. Time constraints, tight budgets, and lack of expertise in usability often limit usability testing of print and online texts. In the long run, however, attending to usability throughout the design and development processes saves time and money. Good usability reduces costs incurred through fixing problems later, reduces the need for user support, and increases customer satisfaction. Within a company, high usability also reduces costs by increasing employee productivity.

WEBLINK

Usability is a complex and critical topic in technical communication. For a link to many valuable, even essential, usability sites, go to **www.english .wadsworth.com/burnett6e**.

CLICK ON WEBLINK

CLICK on Chapter 9/usability sites

This chapter begins with a discussion of principles that guide usability and then defines usability testing, identifies various types of usability testing, and considers their importance. The majority of the chapter focuses on conducting usability tests and reporting the results. The final section discusses accessibility, which is an important corollary of usability.

Characterizing Usability

The goal of usability is to place user concerns, rather than text features, at the center of the design and development processes. Usability responds to the needs, goals, skills, and contexts of people who are completing a process or using a product.

This chapter focuses on the way that texts interact with users to affect usability. Texts affecting usability may be displayed in various *materialities* and, thus, necessarily expand the meaning of the term "text." For example, the texts can be printed (usually on paper or plastic) in a manual, on a product package insert, on the packaging itself, on a tag wired to the product, or on a small label glued to a product. Or the texts can be printed or embossed on the products themselves, from face protectors for welders to hypodermic needles. Or the texts can be displayed on an LCD screen on devices ranging from laptop computers to SIDs monitors.

Equally important, texts affecting usability appear in various modes. They may be written, visual, graphic, oral, signed (as in American Sign Language), or a combination, further expanding the meaning of the term "text." For example, texts may be entirely in written language, which is typical of many technical reports and proposals. Texts may be a combination of written language and visual displays, typical in specification sheets. Texts may be largely visual and graphic, such as many Web sites. Texts may be entirely graphic, such as airline evacuation instructions for international audiences. Texts may be entirely oral, which is typical of a considerable amount of workplace training. Texts may be oral and signed, such as a presentation to an audience that includes hearing and deaf people. Texts affecting usability may take any or all of these forms, regardless of the process that users are completing or the product they are using.

© David Binder
Comstock Images/Alamy
© Stephen Simpson/Getty Taxi
© Tony Freeman/Photoedit
Courtesy of US Army Corp of Engineers
© Michael Rosenfeld/Getty Images
Courtesy of Carl Reese/Cabinsafety.com

Definition of Usability

How is usability defined in your discipline or profession? Why is usability important in your discipline or profession?

In technical communication, *usability* is the degree to which texts, regardless of their materiality or mode, effectively and easily enable people to accomplish their goals. Usability is one aspect of the larger concern of usefulness. Usefulness also encompasses *utility*, or the degree to which a product and related texts are adequate for the intended purpose. Thoughout this chapter, you will see that usability, usefulness, and utility are interdependent. As researcher Barbara Mirel suggests about software interfaces, "Ease of use involves being able to work the program efficiently and easily; usefulness involves being able to use the program to do one's work in context effectively and meaningfully. Stripping usefulness from ease of use and focusing primarily on the latter is an incomplete recipe for usability or user-centeredness."[1] The same can be said of all types of communication.

WEBLINK

What's the truth behind usability testing? Is it expensive? Time consuming? Stifling? Unnecessary? What are the definable benefits? To differentiate fact from fiction, go to www.english.wadsworth.com/burnett6e for links.
CLICK ON WEBLINK
CLICK on Chapter 9/usability facts

Critical Principles

Usability is guided by five critical principles that can be applied to many types of print and electronic texts, oral presentations, and visuals. These principles apply equally to the usability of texts, objects, and processes with which people interact.[2]

- ***Learnability.*** How easily can people learn to use the text and the product? How quickly can they become productive using them?
- ***Efficiency.*** How productive are people using the text and the product?
- ***Memorability.*** How well do people remember how to use the text and the product from one use to the next?
- ***Error Recovery.*** How many errors do people make using the text and the product? How serious are the errors and how quickly can users correct them?
- ***Satisfaction.*** How satisfied are people with the performance of the text and the product? Do they enjoy using them?

These five principles can be used to assess the usability of different types of texts that technical communicators and technical professionals typically develop. Figure 9.1 provides representative questions you can ask when you use these principles to assess usability of software interfaces, Web sites, and print information.

FIGURE 9.1 Representative Questions for Applying Usability Principles to Different Genres

Usability Principle	Interfaces and Web Sites	Print Information
Learnability	■ How quickly can users begin using the software to accomplish tasks? ■ Must users sift through a manual to find information about using the software, or is the interface intuitive?	■ Do textual cues help readers determine the key terms they need to learn? ■ Does the sequence of information simplify the task of learning it?
Efficiency	■ How are menu items on a software interface organized? ■ Are menus and links on a Web site labeled in a way that enables users to readily determine what content is available? ■ How many "clicks" are required on a Web site for users to access the information and services they need?	■ How quickly can readers find information using an index or table of contents? ■ Do headings help readers use the text? ■ What elements of the design (e.g., tabs, screened sidebars, index) help readers use the text?
Memorability	■ Can users remember the calculation for completing a complex task in Excel? PowerPoint? PhotoShop? Flash? ■ Can users easily remember keystroke shortcuts rather than using menus?	■ Are safety warnings clustered at the beginning of the manual or placed with the appropriate steps? ■ Are instructions written so that readers can remember steps?
Error Recovery	■ If users make a mistake, how easily can they fix it? ■ If users click to the wrong place on a Web site, how easily can they find their way back?	■ If readers lose their place in the text, how easily can they find it again? ■ Does the format assist readers in avoiding errors while performing a task?
Satisfaction	■ What will prompt users to return to a Web site? ■ Do users report that their needs are effectively met by a Web site or a software interface?	■ What will prompt readers to return to a text? ■ Do readers report that their needs are effectively met by a text?

Interpreting the effectiveness of each principle is highly situational; it is influenced by the genre and media, by the type and complexity of the text, and by the users' prior knowledge, experience, needs, and expectations. Consider the complexity of assessing learnability. For example, most people learn to use an automatic teller machine (ATM) fairly quickly and then learn to use other ATMs even more quickly because the task is simple, the instructions are short, the sequence of steps is usually intuitive, and the process is very similar from one situation to the next. What determines learnability in this situation? A simple measure could be that a person could make a withdrawal from virtually any ATM, whether in the United States or in another country.

Select any one of the other usability principles and explain how it might be interpreted in various situations, both simple and complex and with both print and electronic documents.

However, people typically take much longer to learn more complex tasks such as manipulating digital images. When faced with a new software for manipulating digital images, they may be faster than the first time they ever tried this type of software, but they probably won't be speedy. The task is complex, the instructions are usually long, the sequence of steps is not necessarily intuitive, and the process is usually idiosyncratic from one situation to the next. What determines learnability in this situation? Unlike using an ATM, using software for manipulating digital images is a complex process, so measuring its learnability is more complex.

Characterizing Usability Testing

Iterative *is an adjective that means "repetitious."*

- *In computer programming,* iterative *describes a sequence of instructions that can be performed multiple times. One pass is called an* iteration. *If the sequence is executed repeatedly, it is called a* loop.
- *In software development,* iterative *describes heuristic planning and development processes; an application is developed in small sections called* iterations. *Each iteration is reviewed and critiqued by the software team and potential end-users.*
- *In usability testing,* iterative *describes the cycle of testing, rewriting, and retesting.*

Now that you have a basic understanding of usability principles and the complexity of measuring them, this section focuses on defining usability testing, identifying the types of such testing, and considering the importance of *iterative* testing.

Definition of Usability Testing

As you've learned, usability is the degree to which texts (regardless of their materiality or mode) effectively and easily enable people to accomplish their goals. *Usability testing* is a structured process that gathers information about specific use from people similar to the intended users. This user-centered approach to testing allows people involved in all aspects of design and development to assess usability by observing people interacting with products and texts. Usability testing provides a realistic context (whether in natural workplace settings or in testing labs) for measuring learnability, efficiency, memorability, error recovery, and user satisfaction.

Purposes. Usability testing has both immediate and long-term purposes. The immediate purpose is to identify problems prior to the release of a text so the problems can be fixed. A long-term purpose is to maintain a historical record of test results and benchmarks that can be referenced during development. This is particularly helpful for future editions of documents, for the development of additional materials, or for new versions of Web sites or interfaces.

In usability testing, test administrators elicit feedback from subject matter experts, design and development professionals, and either actual users or testers who represent various user groups. With effective planning, implementation, and evaluation, usability testing provides specific and useful information about how well texts meet usability principles. Choices of testing procedures and selection of test participants provide information about the ways in which users think, believe, feel, and perform:

- ***Cognition.*** How are users thinking about the text?
- ***Perception.*** How do users understand the text? Envision results of use?
- ***Affect.*** What are users' attitudes toward the text and the process they're engaged in?
- ***Performance and behavior.*** What do users actually do with the text (and, of course, with the product)? How do they try to solve problems? What do they do when faced with a problem?[3]

Quality Standards. Usability testing is one method of ensuring high-quality texts. When a text is not usable, various kinds of minor, moderate, or severe problems result. Minor problems probably irritate users, but they do not actually delay the completion of the task. Moderate usability problems actually hinder completion of the task, but users usually can develop workarounds. However, severe usability problems prevent the completion of the task or "result in catastrophic loss of data or time."[4]

Many types of communication benefit from careful testing: Web sites, software interfaces, online help and user manuals, form letters, advertising copy, product descriptions, spec sheets, and standard sections of proposals (such as the boilerplate sections about an organization). Usability testing should not be limited to the language sections of communication; other components should be tested as well, including visuals and the overall design. Your goal is to determine the effectiveness of communication quality for the intended audience.

WEBLINK

The International Organization for Standardization (ISO) defines usability as "the extent to which a product can be used by specified users to achieve specified goals with effectiveness, efficiency and satisfaction in a specified context of use."[5] ISO 9241, *Guidance on Usability,* is one of several ISO standards that are central to quality and usability. For a link to sites dealing with ISO and usability, go to www.english.wadsworth.com/burnett6e.

CLICK ON WEBLINK

CLICK on Chapter 9/ISO and usability

Limitations

Testing results are only as good as the tests; they can be accurate but not necessarily representative. Usability tests, whether conducted in usability labs or in work sites, can't be conducted on every potential user and in every likely situation, so when you're designing a test, you need to consider these limitations:

> *Test validity is the extent to which tests measure what they claim to measure.*[6] *Test reliability is the extent to which results on tests are repeatable and yield consistent scores.*[7] *Valid tests are always reliable, but reliable tests are not necessarily valid. In what ways might weak reliability or weak validity of usability tests distort the results and lessen their usefulness?*

- ***Test participants.*** The usefulness of test results depends in part on choosing test participants who reflect the needs and attitudes of the actual end users.
- ***Test situation.*** Testing doesn't guarantee a communication's usefulness. In order to be generalizable beyond the specific situation, tests need to be carefully designed to reflect a variety of situations in which the communication will be used.
- ***Test techniques.*** Test techniques must be appropriate at various stages of text development. Different stages require different procedures. For example, user testing does not replace evaluations such as expert or technical reviews; it supplements them. Your goal should be to design appropriate testing for each stage of development.
- ***Testing procedures.*** Testing procedures can also affect results. However, with effective planning, implementation, and evaluation, testing can provide more specific and useful information than other forms of feedback (such as postpublication satisfaction surveys).

Benefits

Everyone involved in the design and development of technical communication — writers, designers, technical experts, managers, designers, editors, prepress technicians, print/production managers — agrees that testing takes time. The

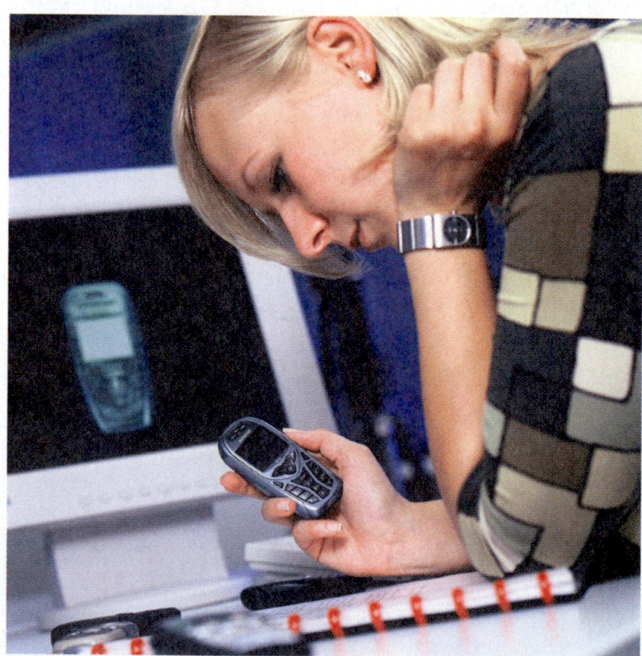

Testing in Siemens' ICM Usability Lab.[8] One of the benefits of testing is eliminating expensive design changes at late stages of development. This is one of the reasons that many companies have a usability lab. The test participant in this photo is in the Siemens' ICM Usability Lab, testing a cell phone to determine how easy it is to operate. Other typical users test the menus and keys on cell phones and cordless phone in all stages of development to determine their ergonomic comfort. In various scenarios, users in the lab might use the phones themselves, print directions, or access online help.

Courtesy of Siemens

benefits, though, are undeniable, because a well-planned usability testing program can provide invaluable information for revising and editing communication. Successful testing looks for unrecognized problems so that they can be eliminated; testing communication for flaws you've already identified wastes time and energy.

WEBLINK

Does usability really matter? To read some success stories about the benefits of usability, go to www.english.wadsworth.com/burnett6e for links.
CLICK ON WEBLINK
CLICK on Chapter 9/usability benefits

Types of Usability Testing

Organizations must make testing part of their process development and production schedules, or it simply will not get done. As the value of testing becomes more widely recognized, more organizations are employing some kind of testing. In the long run, testing during development and revision is much more cost effective than dealing with the results of poor communication. Testing throughout the development process increases the chances that serious usability problems can be avoided. Waiting until the project is completed to do some kind of testing usually means that no time will exist to incorporate necessary changes.

What important information about usability might you miss if you just asked users to recall the path they took in moving through a series of steps to activate online banking? To sort materials into recycling bins? To store hazardous products according to OSHA guidelines?

Three broad categories of testing — text-based, expert-based, and user-based testing — are distinguished by the ways the information is collected and by the nature of the feedback. Each type of test provides a different kind of feedback that can be used to improve communication quality. When you have access to information from several kinds of testing, you can suggest changes to improve the functionality as well as the technical and textual accuracy of communication. Perhaps more important, you can make communication more accessible, comprehensible, and usable for your intended audience.

WEBLINK

When you decide to conduct a usability test, you can choose from quite a number of tests. The links at www.english.wadsworth.com/burnett6e provide specific information about tests you can administer, ranging from card sorting and focus groups to surveys and user profiles, from think-aloud protocols to observations of task performance.
CLICK ON WEBLINK
CLICK on Chapter 9/test choices

Most kinds of testing can be *structured* (that is, planned to include specific questions and tasks) or *unstructured* (that is, asking for feedback without specific questions or tasks). And most kinds of testing can take place in an actual workplace or in a usability lab.

Text-Based Testing

People doing text-base testing examine a range of local-level language, visual, and design features and then draw conclusions about changes that are necessary to improve the text's accessibility and comprehensibility. Text-based testing can incorporate a variety of guidelines and checklists, readability tests, and computer programs to assess structural and stylistic features (for example, types of sentences or active versus passive voice). It can also incorporate analyses of the visuals and the design, including comparisons between two (or more) design alternatives. Figure 9.2 identifies the general functions, benefits, and limitations of four common types of text-based testing.

Several simple text-based tests for consistency can be performed at any time in the development process on selected pages or individual documents using word-processing software such as Microsoft Word®. After checking the spelling and grammar in a text, Microsoft Word displays two types of readability test scores of your text: Flesch reading ease and grade-level ratings.

- ***The Flesch Reading Ease*** score claims to indicate how easy the text is to read. It rates chunks of text on a 100-point scale; the higher the score, the easier the text is presumed to be. Standard documents aim for a score of approximately 60 to 70. Lower scores indicate that the text might be more difficult.
- ***The Flesch-Kincaid Grade Level*** score rates text on a U.S. grade-school level. For example, a score of 8.0 means that an eighth-grader should be able to understand the document. Standard documents aim for a score of approximately 7.0 to 8.0. However, the purpose(s) and audience(s) must be considered in determining the grade level.

> *Imagine a technically accurate text that is assessed as having a Flesch-Kincaid score of 8.0, but the lab technicians (all with two-year or four-year college degrees) consistently make errors in implementing the new process. What factors in the text might contribute to the problem even though the reading level says "eighth-grade"?*

Unfortunately, readability tests are not a useful measure of the difficulty that actual users will have reading the information. Why? These readability scores are based on various ratios between word length and sentence length. While these scores do little to reflect the actual difficulty of the text because they omit consideration of critical factors such as content complexity, use of explanatory examples, use of visuals, and document design, these scores can be useful to determine the relative difficulty and the internal consistency of text. For example, if you're testing pages for a Web site and some pages score 80 in the Flesch test and others score 30, the range of scores indicates inconsistency in the writing across the site. Consistency is an issue that should be resolved before the site is tested with actual users.

Text-based tests are useful for identifying specific errors and for establishing the presence or absence of common textual, visual, and design features. Quick and inexpensive, text-based testing reminds writers and designers about things they need to check. However, such tests do have drawbacks:

FIGURE 9.2 **Text-Based Testing**

Test Types	Phases in Development				Functions	Benefits	Limitations
	Early	Middle	Late	Post Release			
Document Design Review	✔	✔			Assesses consistency, correctness, and conventions of design	Provides data for revision; can be formal or informal	Reviewers need to know generic and disciplinary conventions of design
Comparison Test	✔	✔	✔	✔	Compares design alternatives for the overall text	Provides data for decision making; can be formal or informal, quantitative or qualitative	Best results are achieved when substantively different alternatives are used
Visual Review		✔	✔	✔	Assesses consistency, correctness, and conventions of visuals	Provides data for revision; can be formal or informal	Reviewers need to know generic and disciplinary conventions of visuals
Editorial Review		✔	✔	✔	Assesses consistency, correctness, and conventions of language	Provides data for revision; can be formal or informal	Reviewers need to know generic and disciplinary conventions of language

- They provide generalized information without criteria for assessing how effectively a feature has been used.
- They ignore contextual factors.
- They do not assess the technical accuracy of a text.
- They do not indicate how actual users will respond to a text.

Can you think of any circumstances in which text-based testing alone would be sufficient?

Despite the obvious limitations, text-based testing can help assess a whole range of factors related to language, visuals, and design, such as adherence to language conventions, consistent terminology, sentence structure, documentation, verbal and visual coherence, adherence to visual conventions, and consistent design features. Since writers and designers have a difficult time seeing their own errors, this kind of testing can be very helpful.

Why isn't technical accuracy sufficient for a text?

Expert-Based Testing

Expert-based testing includes several kinds of reviews: technical reviews by subject-matter experts, substantive editorial reviews, and design reviews. Expert-based testing is particularly useful for assessing technical accuracy and also for selecting supporting evidence and for identifying the level of detail for the intended audience. Recent workplace research supports the practice of using several kinds of experts in testing. Figure 9.3 identifies the general functions, benefits, and limitations of two common types of expert-based testing.

What kinds of texts do you think should be reviewed and tested by experts in several areas? What areas might benefit from expert testing in addition to those in identified in Figure 9.3?

"Expert" does not just mean technical subject-matter expert (SME). For example, consider the expert testing conducted for a user manual accompanying impendance spectroscopy software for a multinational market. The manual benefited from review by three kinds of experts: experts in engineering, in information design, and in nonnative readers of English.[9]

FIGURE 9.3 **Expert-Based Testing**

Test Types	Phases in Development				Functions	Benefits	Limitations
	Early	Middle	Late	Post Release			
Expert Review	✔	✔	✔	✔	Reviews various components such as design accessibility, international audiences, and legality	Provides expert feedback to developers about factors such as design, readability, legal compliance, and safety	Provides expert perspective without necessarily understanding what users need
Technical Review	✔	✔	✔	✔	Reviews text or presentation for accuracy as well as adherence to professional standards	Provides detailed feedback about accuracy of content and compliance with professional standards	Provides peer perspective without necessarily understanding what users need

Many large companies and agencies have formal review boards that read every publication that goes out of the company, both to ensure technical accuracy as well as to ensure that writers are not unintentionally revealing proprietary information (that is, guarded or secret information such as production processes). Such committees are also sometimes charged with ensuring that texts adhere to the organization's policies and practices regarding publication, which might include anything from checking the correct bibliographic format to assuring that the texts are coherent.

Formal review is a process that is mandated by some organizations. It is a sequence or cycle that texts go through to be assessed for consistency and appeal in format and design. Documents may move back and forth between writer, editor, designer, and SME and then back two or three times, or perhaps a dozen or more times. Similarly, a text may cycle back and forth between experts, for example, the SME and a graphic designer. In many organizations, texts cycle through a review-revise-review sequence several times, with different reviewers at different levels in the organization who identify problems in textual and technical accuracy.

Informal expert-based testing doesn't follow such a rigid process. Rather, informal expert-based testing generally means stopping by the office of a colleague and saying, "Sandy, I'd appreciate it if you had the time to read through this to check *X*." You fill in the *X*, asking for a review of technical content, logic of your argument, organization of information, and so on. As you develop a significant text or presentation, you may see that it receives dozens of these informal reviews by various kinds of experts.

Although experts can examine both specific and overall features of a text, they cannot predict how actual users will respond, largely because they know so much more than the intended users. However, feedback from experts is important because they can identify areas that need improvement in accuracy, completeness, coherence, consistency, and appeal for the verbal and visual aspects of a text. Revising and editing are much easier when you have expert feedback.

Experts know an immense amount about their area of specialization, certainly much more than average users. Why can't they assume the role of users and identify problems in a text?

User-Based Testing

Because text-based testing and expert-based testing aren't always sufficient predictors of usability, user-based testing elicits information directly from users. Revisions based on user feedback is so important that the U.S. government urges federal departments and agencies to user-test their documents. Figure 9.4 identifies the general functions, benefits, and limitations of five common types of user-based testing.

Information collected from users as they read and use a text is called *concurrent testing*. Information collected from users after they have finished using a text is called *retrospective testing*. Generally, writers have their choice of two kinds of concurrent testing.

Concurrent Testing. One of the most effective kinds of concurrent testing involves creating brief, realistic scenarios to which users can respond. Scenarios are

FIGURE 9.4 User-Based Testing

Test Types [All can be either *concurrent* or *retrospective.*]	Phases in Development: Early	Middle	Late	Post Release	Functions	Benefits	Limitations
Exploratory Test	✔				Assesses user assumptions, needs, skills, goals, concepts	Provides user feedback about conceptual design	Deals with abstractions; may require high level of user/tester contact
Assessment Test	✔	✔			Assesses implementation of conceptual model	Provides data about lower-level skills in time to modify design and content; focuses on user performance; easy to implement	May require high level of user/tester contact
Validation Test		✔	✔		Assesses usability; evaluates the ways components in the product or process work together	Provides data about the text for revision; provides data to benchmark future documents	Requires retesting of the changes at the last minute
Beta Test			✔		Provides limited review of initial published version of document	Provides feedback from actual users about a product in limited release	Customer already has the product; may influence their future attitude toward the company and product
User Feedback				✔	Provides postpublication review	Provides feedback that may help solve problems in later versions; provides feedback to strengthen future documents; adds information about customer base	Customer already has the product; difficult, costly, and inconvenient to fix problems; involves resources in customer service that might be spent in product development or marketing

what usability researcher Ginny Redish calls an "intersection of user analysis, task analysis, [and] context (environment) analysis" that capture a story of who, what, why, and how.[10] For example, imagine that an online airline reservation system is testing whether a new site design is accurate, fast, and hassle-free for users. Based on profiles of typical users, notes from observing other users, and conversations with individual users about situations they encounter, the test designer creates this scenario:

> Amelia James, a 26-year-old engineer who lives in Jackson, Mississippi, is considering a job offer in Houston. The company has asked her to come to Houston for an on-site interview next week. While the company will pay for her travel expenses, she has been asked to provide an airline itinerary, arriving Wednesday evening or very early Thursday morning and leaving Friday evening. Amelia prefers nonstop flights. She has less than 15 minutes before her next meeting to find the information and e-mail it to the company in Houston so the Human Resources Department can purchase the ticket she prefers.

Or imagine that a professional association offers its members the choice of receiving its quarterly publication in print or electronic format. The association wants to assess the accuracy and ease of completing an online form for switching to the online journal. The test designer creates a scenario based on file information about members, observation notes from previous testing, and conversations with individual members:

> You're busy and have little storage space in your office, so you've decided to shift from a print to an electronic journal for one of your primary professional organizations. You want to be able to receive upcoming publications electronically as well as access back issues electronically. You also want to be able to forward citations about interesting articles with your comments to colleagues and to bookmark articles you want to refer to again. You want the site to keep track of what you've already accessed. Complete the form provided by the association.

Concurrent testing often involves watching users' behaviors as they perform a task (in a scenario or not). Whether in a natural workplace setting or in a usability testing lab, as you observe the users, you can identify places where they have difficulty interpreting or completing the information. For example, you can observe the amount of time or apparent ease or difficulty they have in locating information or performing individual tasks or sequences of tasks. You can keep notes on any number of factors, from the number and type of errors they make to their attitudes while using the document.

What do you see as the benefits and disadvantages of conducting tests in a natural workplace setting or in a usability testing lab?

Another kind of concurrent testing asks users to read and think aloud — that is, not only to read aloud but also to say aloud all their comments, reactions, and opinions. Users' comments are tape-recorded as they read a document, browse a Web site, or use a software interface. By reviewing the tape, you can identify specific places where information is confusing or tasks are difficult for users.

Why is user testing often considered the most important kind of usability testing?

If you conduct user-based testing with several individuals, you will begin to see patterns of problems signaling areas in the text that need revising and editing or features of design and navigation that need improvement. As few as five to eight test participants gives you usable data for revision. Concurrent testing takes time, requires trained testers, and involves costs that are not incurred in text-based and expert-based testing. But the advantage is that writers, designers, and developers receive critical feedback directly from actual users.

During retrospective testing, why would some test participants make up what they think they probably did while completing a task rather than saying, "I don't remember"?

Retrospective Testing. Retrospective testing includes such methods as questionnaires, interviews, focus groups, and audience feedback cards. Although retrospective testing does provide information from actual users, it should be used cautiously because readers' memories are not necessarily accurate, and the information they provide is often vague. Despite the challenges of retrospective testing, it provides invaluable information. Using this feedback, you can revise and edit communication so that it is more accessible, comprehensible, and usable for the intended audience.

Conducting Usability Testing

In order for testing to ensure usability, you must carefully plan tests and incorporate them at various stages of the project. This section suggests preliminary steps for conducting testing. These generic steps will need to be modified to fit specific situations.

Creating a Usability Testing Plan

Before starting the testing process, define the broad goals and scope of your testing plan and determine what types of usability tests will best meet your goals, when they should be conducted, and how the results will be reported and used. Consider the following issues as you plan your testing.

1. ***Goals.*** Establish the goals for your testing. What do you want to find out? Ways to reduce time on task? To increase accuracy? To reduce calls to the helpline?
2. ***Criticality.*** Assess the importance of the text you plan to test and decide what kind of testing feedback you need. For example, Web sites and user manuals require text-based, expert-based, and user-based testing. However, this degree of testing would be excessive for a document such as a monthly progress report.
3. ***Constraints.*** Identify the constraints that you have to work with. They typically include available expertise (someone to conduct and interpret the tests), available test participants, time, and budget.

4. ***Schedule.*** Build the time for testing into the project schedule, and provide for that testing in the project's budget. Establish a schedule that, if at all possible, includes testing of different kinds as an ongoing part of the project.
5. ***Involvement.*** Explain the purpose and procedures of usability testing to all key personnel on the project so that they understand the goals and cooperate with the testing.
6. ***Timing.*** Test products and texts at various points in the development cycle, including the initial stages. Feedback on early design is essential for test results to become part of the design process.
7. ***Goodness of fit.*** Test procedures should elicit data that is appropriate and relevant to your specific goals. You may, for instance, need to test separately for information retrieval and for reading.
8. ***Ease of use.*** Test procedures must be easy for test participants to understand and for testers to facilitate.
9. ***Usable form.*** Manage the production of test data so that usable results are achieved.
10. ***Updating.*** Determine how the test results will be used as part of the development and revision process.[11]

Which of these issues will be most challenging for you? Which will be the easiest?

Many organizations expect a testing plan to be presented and approved as part of a project. Which of these issues would you choose to address in a testing plan in your discipline or field?

Are you ready to develop a test plan? To learn more about usability tools and strategies that you'll find useful, go to www.english.wadsworth.com/burnett6e for several links.

CLICK ON WEBLINK

CLICK on Chapter 9/usability tools

WEBLINK

A well-planned program of testing for a complex document, Web site, or software interface generally incorporates text-based, expert-based, and user-based testing at various points in the project cycle. When user-based testing is part of the design and development process, decisions about usability are based on observed use and problems rather than drawn from untested speculation. In other words, instead of working from guesses (even educated guesses) about how a product works with actual users (e.g., "When the users perform a particular task under certain circumstances, the result is probably going to be this."), user-based testing helps product developers get answers to practical questions about the ways in which the product will function ("What happens when the users do this?"). User-based testing can answer questions at various stages during iterative development while changes can be made relatively easily.

Testing processes vary considerably, depending on the kind of test you're conducting; the following steps are typical, although their specific form differs from situation to situation:

Give examples of how the complexity and time requirements of these steps will change in different situations.

- Analyze the users and tasks for which you are preparing communication.
- Locate representative test participants.
- Develop strategies for inquiry.
- Identify and prepare a test location and test materials.
- Explain the test procedures to participants.
- Conduct the tests.
- Report test results so the information can be used to improve design.

Analyzing Users and Tasks

When you design documents, Web sites, and software interfaces, you often focus attention on content, design preferences, and perceptions of the task or activity that the communication is meant to facilitate. This focus may mean you don't adequately account for the needs of actual users, the demands of their particular situations, and the tasks and limitations of the users in those situations.

Three major components affect the effectiveness of communication: the people, the activity, and the context. Testing helps you focus on users and their tasks so you can better deal with important usability issues. For example, when you develop a Web site, you probably have in mind a target audience of people who are most likely to use your site. Since you can't know about everyone who will visit your site, focus your analysis on your perception of your intended audience or representatives of actual users.

User Analysis. Many users have significantly different skills, attitudes, expectations, and experiences than the writers and designers. What do you need to know about your users, and how do you find out? Usability experts Joann Hackos and Janice Redish, in *User and Task Analysis for Interface Design,* recommend that writers and designers investigate the ways that users think about what they do ("jobs, tasks, tools, mental models") and "how they differ individually (personal, physical, and cultural characteristics, as well as motivation)."[12] When analyzing your users, consider the user characteristics listed in Figure 9.5.

The questions in Figure 9.5 are representative, not inclusive. What additional questions can you add?

Try to answer as many of the questions in Figure 9.5 as possible about your intended audiences. Use the information to write brief user profiles that you can use to guide your design and development. To address the many different needs and situations for which you develop communication, you need to recognize that usability is determined by how well the text works for users who want to accomplish specific tasks.

FIGURE 9.5 Characteristics for User Analysis[13]

Characteristic	Questions
Learning styles	■ Do Web site users read help files and manuals, or do they prefer to experiment when they use software? ■ Do Web site users use menus, site maps, or search functions when they encounter a new site?
Physical differences	■ Do users have visual, hearing, or motor disabilities, color blindness, or other characteristics that will affect their perception, interpretation, and use of the information?
Language and culture	■ How do users from different cultures respond to different metaphors, concepts, or organizational structures? ■ Does your information need to be provided in several languages?
Reading preferences and abilities	■ Do users read material on computer screens? Scan it? Print it out? ■ Do they prefer large print? Illustrations?
Motivation	■ How motivated are users to use your information? Motivated users might be more tolerant of design flaws than users who are less motivated.
Attitudes toward the type of communication you're preparing	■ Is the communication positive or negative? ■ How will target audiences react to the communication you're preparing?
Environment in which the communication is used	■ Will your audiences use your communication at work, at home, at school, or in a public location?
Users' job titles and job responsibilities	■ Do your audiences include managers? Professionals? Craftspeople? Students? ■ Do they often use the type of communication you're preparing?
Extent, purpose, and comfort using the type of communication	■ How do your audiences perceive their level of expertise? Do they consider themselves to be novices? Experts?
Types of tasks performed using the communication	■ What tasks will your audiences use your communication to accomplish?
Levels of knowledge provided in the communication	■ What do users already know about the information? About the task?
Types of training users have received in using the type of communication	■ Have users been trained to use similar communication? Will training be provided? How much additional help or documentation will you need to provide?
Users' mental models of communication	■ How do the audiences conceptualize the tasks that your communication is designed to facilitate? ■ What metaphors do they comprehend? Use?
Users' vocabulary	■ What terms do users recognize for various tasks and products?
Tools available for using the communication	■ What technology — for example, modem speeds, screen sizes, browsers, etc. — is available to the users?

Task Analysis. Once you have completed a preliminary user analysis, select representative users who fit the profile of your target audiences and begin assessing the tasks in which they are involved. Whether you are writing a technical manual or designing a Web site, understanding the tasks people will accomplish with your communication is a critical part of assessing and improving usability. Several strategies are available for gathering information about workplace tasks:

1. Interview users about their tasks and, if possible, observe them working. Ask them to show you what they typically do and what they do most frequently.
2. Prepare a list of important and frequent tasks people do with the type of communication you're developing. Note the important tasks that your communication should support.
3. Focus on the most important tasks in detail. Several aspects of tasks are particularly important:

   - ***Task Steps.*** What individual actions are required to complete the task? Do some users use shortcuts? Conflate steps? What variations in steps and methods are possible?
   - ***Resources.*** What tools do users need to complete the tasks (documents, computers, databases, search tools)? What information do they need? What processes do they use?
   - ***Constraints.*** What limits, such as time or knowledge, affect the task?
   - ***Task Environment.*** Where is the task performed — an office, retail store, factory? Does the communication need to be portable? What aspects of the environment, for example, noise or interruptions, affect task performance?
   - ***Problems.*** What problems are common? How can they be avoided? If problems occur, what information can best be given to help the users resolve the problem?
   - ***Frustrations.*** What aspects of the task or task environment irritate users? What aspects of the type of communication you're preparing confuses users?

Once you understand the audience for your communication and the tasks in which they're involved, you'll be better able to prepare tests and identify test subjects.

Implementing the Test Plan

Once your plan has been completed and you have carefully analyzed your users and their tasks, you're ready to implement the plan. This involves selecting test participants, developing and then piloting your test questions or scenarios, preparing the setting and materials, introducing the procedures to the test participants, and conducting the tests themselves.

Locate Representative Test Participants. Locate and schedule participants for your test. If you have completed user and task analyses, your user profile should identify test participants who are representative of the intended users. Successful user-based testing doesn't require the participation of large numbers of people. In fact, according to usability expert Jakob Nielsen, you can identify 80 percent of usability problems by testing your site with five to eight people.[14] Make sure, though, that your test participants have various levels of experience with the type of communication you are developing.

Why should you not have all novice users or all expert users?

Develop Strategies and Pilot Test for Inquiry. Develop a list of questions, scenarios, and key points that you want to explore during testing. For example, you may ask users to participate in a think-aloud protocol in which they respond orally to questions such as these, about a Web site:

- Describe the first items you notice on the page.
- Identify which elements on a page are actionable/clickable.
- Explain what you expect to find behind this link.
- Given the situation of *X,* show what steps you'd take to resolve it.
- Please describe your experience when trying to complete *X* task.

Before testing with participants, "test the test" with a trial or pilot run of the test procedures with other people in your class, department, or company. Pilot tests can help you identify any adjustments that you need to make to the test before testing with participants.

Conducting pilot tests adds time, effort, and expense to a project. What do you lose when you skip them?

Identify and Prepare a Test Location and Materials. Find an appropriate place to conduct tests that is free from serious distractions or interruptions (no one works in complete silence, so a little background noise will not interfere with your test) and keep the setting of the evaluation informal and comfortable. Make sure the space you choose is large enough to accommodate both test participants and test administrators without making anyone uncomfortable.

Regardless of where the test is held, prepare and inspect the room the day prior to the test. Organize all the required supplies, including copies of documents you're testing; pencils, pens, and forms for recording test activities; computers and software if applicable; tables, chairs, and sufficient lighting. Prior to your test procedure, make sure that everything is working properly.

Explain the Test Procedures to Participants. You should consider the value of writing an introductory script summarizing the test purposes and procedures that can be explained by a test administrator or read individually by the participants. A script ensures that each participant hears or reads the same introductory information about what a usability test is, how it works, and what can be done to help you get the best information.

When you are administering a usability test, are you more comfortable using a script, a full outline, notes, or your own good memory? What are advantages and disadvantages of each?

Observation Room in IBM Lotus Usability Testing Lab.[15] This corporate usability testing lab provides a room for the test participants and one for the text administrators to observe the participants.The 42-inch monitor in the observation room shows two inputs: each user's computer screen and each user's face. The large monitor enables test administrators to see exactly what the user is clicking on. The audio speakers are strategically placed so test administrators can hear what the user is saying.

Test administrators should put participants at ease, so before testing starts, tell participants the following information:

- Please be honest in your responses. I didn't design the [document, Web site, interface, manual . . .] you are testing, so your comments won't hurt my feelings.
- The test is meant to evaluate the text's performance, not yours. We expect you to encounter some problems, but if the text were perfect, we wouldn't need to test it.
- Do things just as you would if you were at home or at work. (Participants may try too hard to complete tasks to please the test administrator. For example, make sure they don't spend more time reading instructions than they normally would.)

Remind participants that they are helping you by testing the text, helping you to identify places that need to be revised in some way — reworded, defined, more fully explained, or illustrated. And remember to thank them.

Conduct the Tests. While the duration of user-based tests depends on the complexity of the communication you're testing, individual tests generally range from about 20 to 90 minutes. Test administrators should avoid asking questions too quickly and should give test participants sufficient time to frame their responses. User feedback is usually more useful when the test is run like a conversation rather than an interview. However, test administrators should focus on the user's perspective and refrain from giving their opinions about the features of the communication being tested.

Why are writers and designers typically bad choices to act as test administrators for their own work?

During the tests, record test results by taking notes, by audio or video taping, or both. Note-taking and taping both have benefits and limitations. If you choose to take notes, you can determine in advance how you will record information and you can focus on specific aspects of the testing. On the other hand, note-taking requires you to divide your attention between observing the participants' activities and writing down their reactions and comments as well as your own observtions. Recording frees you to focus on the user, but yields a considerable amount of "unfiltered" data from which you'll need to take notes later.

What are you likely to choose as ways to record information during testing? Why?

Test administrators who keep these additional tips in mind will generally obtain more useful data:

- Ask participants for clarifications when necessary, but avoid leading questions. If you're not sure about a participant's response, paraphrase what you think the user is saying and ask if your interpretation is correct. For instance, you might ask, "Just to clarify, I think I hear you saying that following the directions on page 10 was difficult because you're not told what results to expect. Is that accurate?"
- If participants become confused or frustrated, request that they move on to the next task or question. Determine whether the participant can complete the task in a reasonable amount of time under typical conditions. Make sure to remind participants that their performance is not "wrong."
- When participants identify problems, document them clearly. Ask participants what they would recommend to eliminate the problem. Also, record information about aspects of the communication that users find particularly effective or satisfying.
- When most participants notice the same deficiency or encounter the same problem repeatedly, note the pattern and eliminate the question or task from the test. Once the presence of a problem is clear, you don't need to continue testing that aspect of the communication.
- Pay attention to participants' actions, facial expressions, and body language. What users do during a test is as important as what they say. Nonverbal communication also conveys significant information. You may be able to discern when participants are confused, frustrated, satisfied, or confident.

Arrange practice testing sessions with classmates or colleagues to try out these suggestions as you test drafts of your own documents.

During the test itself, be flexible and depart from your planned questions when necessary. If users testing a text or Web site bring up issues that you didn't anticipate, encourage them to talk about what they notice.

Reporting Test Results

When you follow a testing plan that includes text-based, expert-based, and user-based testing, you will collect quite a bit of raw data to review, organize, and report. Your first task will be to review your data from each test to identify key quotes, recurring comments, and interesting findings. Then determine which findings are most critical.

Organize critical findings, using categories that will be helpful in determining revision priorities and goals. For example, you might organize your findings by the assessment criteria presented in Chapter 1 of this text: accessibility, comprehensibility, and usability. Or you might focus on types of usability areas such as learnability, efficiency, memorability, error recovery, and user satisfaction.

You may also need to determine ways to quantify some results. Decisions about quantification should be decided prior to administering the test, because the way you decide to evaluate the tests may affect the type of test tasks and questions you include. Consider these possibilities:

- ***Time*** — how long test participants take to complete tasks
- ***Error rate*** — how many errors occur during various tasks
- ***Number of similar comments*** — how often different test participants note the same usability issues

User satisfaction is more difficult to quantify. However, you might ask questions about user satisfaction with various aspects of the communication using a Likert scale.

Once key findings are identified and prioritized, prepare a test report. Clearly identify your test procedures, participants, and findings. Also provide analyses of your findings that help you and others make judgments about what changes should be made to the communication you tested. Your goal for the written report is to prepare a useful and readable document for the designers and often for clients, not just a compilation of data without conclusions.

Well-written testing reports are easy and interesting to read. Figure 9.6 presents an excerpt from a testing report that describes usability tests conducted by the National Cancer Institute (NCI) on two Web-based booklets that they created for patients. Notice that the report provides readers with background information about goals, materials, test participants, and methodology before presenting results and recommendations. By conducting tests with a variety of users, including patients and clinicians, the developers were able to identify and report on specific usability problems with the booklets.

FIGURE 9.6 Excerpt from National Cancer Institute Web Testing Report[16]

Goals of the Usability Tests

We wanted to find out more about issues that may make an online booklet more usable:

- Will users have more success using a "detailed" table of contents (a set of links showing primary and secondary headings) or a "minimal" table of contents (a set of links showing only primary headings)?
- What makes a heading informative enough to help users find the information they are looking for?
- Do anchor links (links at the top of a page that take users to information further down that page) help users?
- Will users scroll through long pages of text? Can they easily find what they are looking for by scrolling?
- How much of a booklet do users want to print at one time?
- How much explanatory text is needed to go with icons that indicate other ways of getting a booklet? Icons include "print," "pdf," and "order" (for a paper copy).

Clear identification of testing goals phrased as questions

What Was Tested

We tested two online booklets: *Chemotherapy and You* and *Facing Forward: Life After Cancer Treatment*. Primary audiences for the booklets include patients, their family, and friends. The secondary audience is the health care professionals who provide the booklets to the patients.

Both online booklets are written at the eighth-grade reading level.

Chemotherapy and You has information on what to expect during chemotherapy and what patients can do to take care of themselves before and after treatment. *Facing Forward: Life After Cancer Treatment* addresses issues faced by patients who have completed treatment.

Specification of online documents being tested: titles, primary and secondary audiences, content

Who Participated in the Usability Tests

Fourteen participants tested the booklets:

- 9 clinical nurses
- 1 clinical assistant
- 2 cancer patients
- 2 family members of cancer patients

Identification of the test participants: number, role, general sense of prior knowledge

Most clinicians were very Web savvy. The patients and their family members were not as Web savvy.

Methodology Used

Each participant used the two online booklets for about an hour spending one half-hour on each booklet. During this time they:

- Answered questions on who they think the booklets are for, what they think the booklets are about, and what they would do with them
- Used each online booklet to search for specific information.
- Answered questions about their impressions of the online booklets
- Answered questions specifically about the format of the icons in the booklets, the format they prefer, and reasons for their preference.

Process used to test each online document

Figures 9.7 and 9.8 show two of the recommendations included in the report — what the NCI report calls "Lessons Learned." Each separate "lesson" includes the same categories of information: a recommendation, an explanation of that recommendation, a suggestion about how to implement it, and a visual example and text summary of the problem. This deductive organization of each subsection makes understanding the argument easy: the point is made in the recommendation; the rest of each subsection provides support. The evidence is credible, in large part because readers are presented with an annotated screen capture that illustrates the problem. As a result of the organization and the evidence, the recommendations are easy to accept.

FIGURE 9.7 **Excerpt from National Cancer Institute Web Site Redesign: Classification Problem[17]**

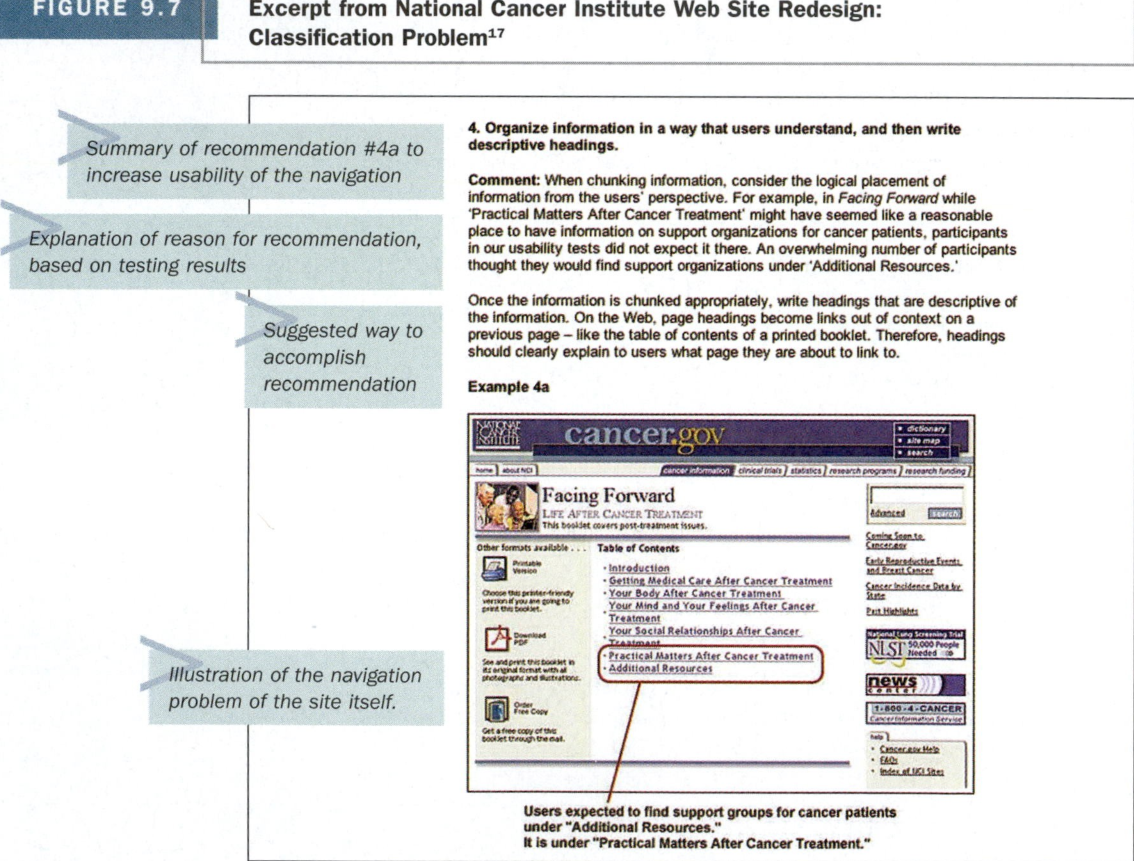

4. Organize information in a way that users understand, and then write descriptive headings.

Comment: When chunking information, consider the logical placement of information from the users' perspective. For example, in *Facing Forward* while 'Practical Matters After Cancer Treatment' might have seemed like a reasonable place to have information on support organizations for cancer patients, participants in our usability tests did not expect it there. An overwhelming number of participants thought they would find support organizations under 'Additional Resources.'

Once the information is chunked appropriately, write headings that are descriptive of the information. On the Web, page headings become links out of context on a previous page – like the table of contents of a printed booklet. Therefore, headings should clearly explain to users what page they are about to link to.

Example 4a

Users expected to find support groups for cancer patients under "Additional Resources."
It is under "Practical Matters After Cancer Treatment."

FIGURE 9.8 **Excerpt from National Cancer Institute Web Site Redesign: Vocabulary Problem**[18]

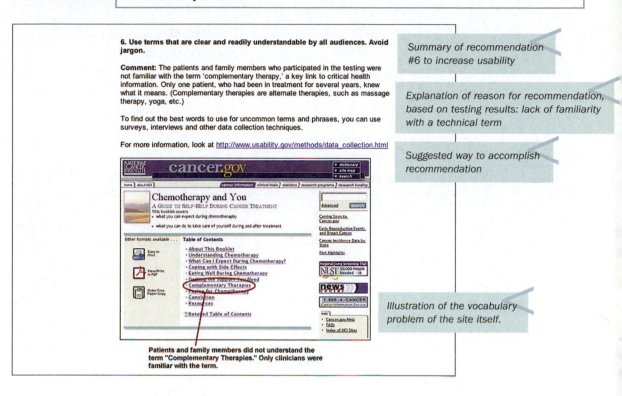

6. Use terms that are clear and readily understandable by all audiences. Avoid jargon.

Comment: The patients and family members who participated in the testing were not familiar with the term 'complementary therapy,' a key link to critical health information. Only one patient, who had been in treatment for several years, knew what it means. (Complementary therapies are alternate therapies, such as massage therapy, yoga, etc.)

To find out the best words to use for uncommon terms and phrases, you can use surveys, interviews and other data collection techniques.

For more information, look at http://www.usability.gov/methods/data_collection.html

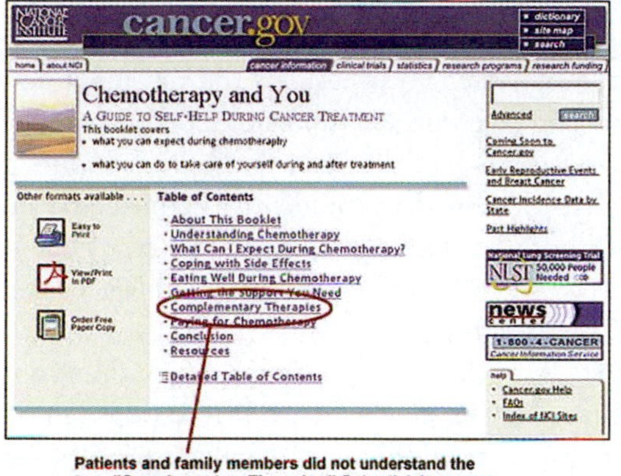

Patients and family members did not understand the term "Complementary Therapies." Only clinicians were familiar with the term.

Summary of recommendation #6 to increase usability

Explanation of reason for recommendation, based on testing results: lack of familiarity with a technical term

Suggested way to accomplish recommendation

Illustration of the vocabulary problem of the site itself.

Ensuring Accessibility

As part of improving usability, designers of communication, whether print or electronic documents, oral presentations, or visuals of various kinds, should also consider *accessibility,* an important aspect of usability that focuses on providing access to information and services, especially to people who have disabilities. Despite the growth of the Internet and advancements in computer hardware and software, barriers to information and to technologies used to communicate exist for many people. Technical communicators and technical professionals should promote access to information and services for all people. In thinking about audiences for all types of documents, presentations, and visuals, consider differences in users' vision, hearing, dexterity, cognition, and other learning and perception areas as well as limitations in their economic, educational, and physical environments.

Do you know someone in the United States who has a disability that affects communication?

- Nearly 3 million people in the United States have speech impairments.
- Approximately 3 million people in the United States are color blind.
- Over 8 million people in the United States have visual impairments.
- Approximately 22 million people in the United States are deaf or hard of hearing.
- More than 40 million people in the United States are affected by dyslexia.[19]

Mobility

- difficulty reaching equipment
- difficulty manipulating physical objects or parts

Vision

- color blindness
- low vision
- total blindness

Hearing

- selective hearing loss
- total deafness

Cognition

- memory disabilities
- perception disabilities (e.g., dyslexia)
- problem-solving disabilities
- conceptualization disabilities

Barriers to communication are created when designers don't consider the full range of users' abilities.

Sometimes you design and write for very specific audiences with whom you are familiar, such as safety engineers using a tutorial about hazardous waste disposal or pharmacists reading about possible complications with a new medication. However, much of the time, you write for a broad audience, one that is so diversified that it doesn't have an identity defined by a singular purpose, education, ability, or technological expertise. The greater the variability in the audience, the greater the likelihood that some members of the audience will not be able to access the information.

How can users vary? Purpose. Education. Ability. Access to technology. How much of the audience in the United States have various kinds of impairments that affect their ability to access information? 1 million people? 5 million? 20 million? Check the marginal annotation for demographic data.

WEBLINK

Many agencies, businesses, and organizations provide accessible information for various audiences. For a link to examples, go to www.english.wadsworth.com/burnett6e.

CLICK ON WEBLINK

CLICK on Chapter 9/accessible information

Principles of Accessibility

In an effort to address the differences and reduce barriers to communication, a group of architects, product designers, and environmental design researchers has identified seven principles for evaluating existing designs, guiding the design process, and educating both designers and consumers about the characteristics of accessibility, in products themselves and in the environments in which they're used. These researchers and advocates have collaboratively established universal design principles to guide a wide range of design disciplines, including communication. Some of these principles overlap the general principles of usability, although not all the principles are applicable in all instances:

- ***Equitable use.*** The design is useful and marketable to people with diverse abilities.
- ***Flexibility in use.*** The design accommodates a wide range of individual preferences and abilities.
- ***Simple and intuitive use.*** The design is easy to understand, regardless of the user's experience, knowledge, language skills, or current concentration level.
- ***Perceptible information.*** The design communicates necessary information effectively to users, regardless of ambient conditions or the user's sensory abilities.
- ***Tolerance for error.*** The design minimizes hazards and the adverse consequences of accidental or unintended actions.
- ***Low physical effort.*** The design can be used efficiently and comfortably and with a minimum of fatigue.
- ***Size and space for approach and use.*** Appropriate size and space are provided for approach, reach, manipulation, and use, regardless of user's body size, posture, or mobility.[20]

In what ways do universal design principles overlap with usability principles?

Identify a text or a piece of software that you have used recently that violates one or more of these universal design principles. How might the application of one or more of the principles make the particular text or software more accessible?

WEBLINK

To learn more about universal design principles, go to www.english.wadsworth.com/burnett6e for interesting links.
CLICK ON WEBLINK
CLICK on Chapter 9/universal design

Accessibility and Electronic Communication

With people's increasing reliance on computers and the Internet for information and services, hardware, software, Web sites, and Web-based applications pose challenges for designers because of the diversity of the audiences. For example, to design Web sites that are usable by and accessible to the largest number of potential users, designers need to understand accessibility concepts in a number of areas including hardware, software, and design principles.

WEBLINK

Eliminating biased or stigmatizing language is always important. Some organizations offer guidelines for unbiased language. To learn ways to avoid using biased language, go to www.english.wadsworth.com/burnett6e for a link.
CLICK ON WEBLINK
CLICK on Chapter 9/unbiased language

FIGURE 9.9 **Technology Affects Accessibility**[21]

Disability	Technology Aids Access
Mobility	Special-purpose hardware and software — such as on-screen keyboards, eyegaze keyboards, and sip-and-puff switches systems — provide alternate ways of creating keystrokes for people with severely limited dexterity.
	Voice recognition is used both by people with some physical disabilities or temporary injuries to hands and forearms as well as by users interested in greater convenience. It's also an input method in some voice browsers.
	Extended keyboard and mouse capabilities provide extended keyboard, mouse, and sound access. For example, the "sticky key" function solves the problem of having to press two or more keys at once (like performing CRTL-ALT-DEL). One key at a time does the job.
	Alternative keyboards and alternative mice allow people with limited or no ability to use their hands to perform keyboard activities.
	Word prediction software predicts the word a person is typing and the next word based on word frequency and context. This software, which may include spell checking, speech synthesis, and hot keys for frequently used words, is useful for slow typists, probe or pen users, and for people with visual disabilities or dyslexia.
Vision	Scanning reading systems allow printed information to be scanned into a system and read to a person using a speech synthesizer.
	Screen readers, used by people who are blind or have learning disabilities, interpret what is displayed on a screen and direct it either to speech synthesis for audio output or to a refreshable braille reader for tactile output.
	Screen magnifiers, used primarily by individuals with low vision, magnify a portion of the screen for easier viewing. Some screen magnifiers offer two views of the screen: one magnified and one default size for navigation.
	Dynamic or refreshable braille involves a mechanical display that raises and lowers dots to allow any braille words to be displayed. Some braille printers allow simple graphics to be drawn.
Hearing	Visual notification — rather than sound notification — allows people who are deaf or hard of hearing to receive a warning about a computer error.
	Real-time chat capabilities are possible with assistive technology that helps people who are deaf or hard of hearing to participate in real-time online chats.
Cognition	WYNN software reads, spells, and defines electronic text aloud so that people with dyslexia or other reading challenges can perform day-to-day work more easily. People can simultaneously view the text and hear it read aloud, thus taking advantage of both aural and visual input.

While accessibility can be restricted because of physical or cognitive barriers, technology provides a remarkable variety of tools to help overcome such barriers. Figure 9.9 summarizes technology that helps address problems with mobility, vision, hearing, and cognition. While technology is powerful and effective, it may be expensive, incompatible with existing hardware and software, too complex to use easily, infrequently accompanied by easy-to-use and thoroughly tested instructions, and insufficiently supported by knowledgeable technicians and usable online help.

Why are accessibility principles so important to Web-based products? Technical communicators are increasingly involved in the design of Web sites and Web-based applications, as you'll read in the ethics sidebar below; thus, they must be aware of all aspects of usability, including accessibility. In addition, recent regulations pertaining to disability and information technologies require that certain Web sites, software, and other communication products meet accessibility standards. Most important, meeting the needs of people who use information technologies is ethically responsible as well as practical and profitable.

WEBLINK

Designing accessible, usable online help for blind and low-vision users of text-based readers is important and challenging. This article provides very helpful suggestions that are clearly explained and illustrated. For a link to the article, go to www.english.wadsworth.com/burnett6e.

CLICK ON WEBLINK

CLICK on Chapter 9/low vision

SIDEBAR

"What You Say!"

"All Your Base Are Belong to Us!" is an enigmatic phrase that has become a phenomenon on the Web,[22] which is littered with sites that tell its story. It comes from the strange Japanese-to-English translation of the introduction to Sega Genesis' 1989 arcade game *Zero Wing,* part of the story of a hapless flight crew facing off against a villainous enemy called "Cats." In 2101, war was beginning:

Captain: What happen?
Mechanic: Somebody set up us the bomb.
Operator: We get signal.
Captain: What!

What arguments could you make to colleagues who say these things:

- *"I had three years of Spanish in high school, so I can probably translate this product description; it's not very long."*
- *"Let's just use a machine translation."*
- *"It's an American product. Let's keep it in English."*
- *"I think Gemma in Public Relations has a degree in French literature from the Sorbonne. Let's ask her to translate these instructions about biohazardous waste disposal into French for the Haitian facilities workers."*
- *"Translation is simply too expensive; let them read it in English."*

Operator:	Main screen turn on.
Captain:	It's You!
Cats:	How are you gentlemen! All your base are belong to us. You are on the way to destruction.
Captain:	What you say!
Cats:	You have no chance to survive make your time. HA HA HA HA . . .
Captain:	Take off every 'zig'! You know what you doing. Move 'zig'. For great justice.[23]

Would you know how to take off your 'zig'?

Of course, poor translations from English to other languages occur as well. For instance, Coca-Cola first introduced its brand-name drink in China with a translated name that means "bite the wax tadpole." Coca-Cola corrected the unfortunate mistake, renaming the drink with a word that roughly translates to English as "happiness in the mouth." Liz Elting of Transperfect Translations notes that, "One of the biggest failures of a product was for the Chevy Nova in Latin American countries. 'No va' in Spanish means 'no go,' not a good name brand for a car."[24]

The results of bad translations may be funny, and they can be costly from a marketing perspective, but imagine poorly translated information on package insert with prescription medicine; the unusable results could be disastrous.

And what happens if no translation is provided at all? According to Andrea Perera of the *Philadelphia Inquirer,* "Nearly 30 percent of people surveyed in three major metropolitan areas in [2003] guessed at the proper dosage of their drugs because they weren't sure what their prescription said. More than half of the Spanish-speaking participants said they found it 'impossible' to fully understand their prescriptions because of language difficulties."[25] Pharmacies throughout the United States are seeking ways to rectify the problem by translating prescription directions into various languages.

When technical translations are poor or nonexistent, products become unusable and even dangerous. What should you do if you need to have a text translated?

- Don't rely on computer-generated translations, which can be particularly poor.
- Do hire a reputable translator with a track record who is fluent in both the original language and the target language of the translation.
- Make sure the translator is conversant with the subject matter. The translator's understanding of technical terms in areas such as medicine, law, and engineering is critical as well.
- Don't depend on literal, word-for-word translation to convey information. Good translations reflect cultural contexts and norms as well as correct lexical and grammatical aspects of text.
- Don't skimp on translation costs. The real costs in terms of usability will be higher in the long run.

Understandable text — in any language — is a usability issue.

Accessibility and Government Regulations

"Accessibility" often refers to the Americans with Disabilities Act. In the case of technical communication, this means Section 508 of the statute, which requires that the electronic and information technology of federal agencies, vendors, and contractors be accessible to people with disabilities, including employees and members of the public.

As a result of this far-reaching federal act, government agencies have been reducing these barriers so that their Web sites are accessible to broad audiences. Many of them have statements that announce their efforts, such as the one in Figure 9.10, which provides recourse to users who believe that the site is not accessible to them.

What example of compliance with accessibility do you see on your campus or workplace, both in the physical environment and in the electronic environment?

WEBLINK

Creating accessible Web sites is boring? Expensive? Too difficult? Unnecessary? Too many people have the wrong idea about what Web accessibility means. To differentiate fact from fiction, go to www.english.wadsworth.com/burnett6e for a link.

CLICK ON WEBLINK

CLICK on Chapter 9/accessibility myths

FIGURE 9.10 **Public Statement of Accessibility**[26]

Home ▪ Search ▪ Subject ▪ Organization ▪ Project ▪ Facilities ▪ Resources ▪ People

NASA Web Accessibility Statement

The Web page you just visited has been reviewed to be accessible to individuals with disabilities in accordance with provisions of Section 508 of the Workforce Investment Act and the Rehabilitation Act. If you experience problems accessing this information please contact the Web site curator directly. If your initial request is not acknowledged within five business days, contact the Discrimination Complaints Program Manager at 216-433-2323, or write to the Office of Equal Opportunity Programs, NASA Glenn Research Center, Mail stop 500-311, Cleveland OH 44135.

ACCESSIBILITY AND ATTITUDE

Accessibility in the workplace has to do with building access as well as with screen resolution and font size. But it also has to do with attitude and the way you communicate that attitude.

When you prepare communication for others, one of the most important responsibilities you have is to represent the needs and concerns of the people in your audience. Another is to recognize how language can either exclude and stigmatize people or include and empower them.

Read the following scenarios and consider how language can be as much a barrier to full participation as other factors in the workplace environment. As you read them and formulate a position, decide how you would behave in each situation.

SCENARIO 1

You use a wheelchair to get around. Your friend is visiting you in your home and invites you to go out to her favorite restaurant for lunch. You accept the invitation but say, "I assume the restaurant is accessible."

The friend then decides to call to make a reservation, using your speakerphone so you can listen in. The friend begins by saying, "I want to make a reservation for lunch but I also need to find out if your establishment has handicapped facilities since my friend, who is a victim of polio, is wheelchair-bound."

The manager of the restaurant responds, "Oh yes, we comply with all of the handicapped requirements. We have handicapped parking, a new wooden handicapped ramp at the back door and a newly remodeled restroom with grab bars. It has a handicap sink with a sloped mirror. We are proud to serve our friends with special needs."

Your reaction

You are disappointed that your friend decided to make the call in your behalf, especially now that you know how the conversation went. You resent the demeaning tone of the conversation on both ends of the phone.

Both your friend and the manager of the restaurant used the term "handicapped." You know the connotations of the word "handicap," which has its origin from beggars who put their caps in their hands on street corners. You take offense at the term, especially when it is used in a condescending manner to describe you. You do not like your friend's reference to you as a "victim" of polio. You are also not "bound" to your wheelchair; it is simply a device you use to get from place to place.

As for the restaurant manager, you question why everything, including parking, restrooms, sinks and mirrors

© Jean Dobbs

need to be described as "handicapped." What is so special about being given access to a building and a restroom? You also do not understand why the ramp is at the back door. Why should you have to have a separate, stigmatizing experience to enter the building, and why did they choose to install a wooden ramp? You also question why they selected a special sink with a sloped mirror. Special sinks are unnecessary and expensive and sloped mirrors do not work well for standing people. An open counter with a mirror that extends down to the top of the counter works well for everyone, regardless of a person's height or standing or seating position. You believe that the choice of a "special fixture with a sloped mirror" reflects an attitude of the owner/manager, who apparently thinks of accessibility accommodations as providing extras, rather than serving everyone on an equal basis.

You do not feel "handicapped" by old or poorly remodeled buildings nearly as much as by people's attitudes.

SCENARIO 2

You use a wheelchair to get around. Your friend is visiting you in your home and invites you to go out to her favorite restaurant for lunch. You accept the invitation but say, "I assume the restaurant is accessible."

The friend asks you if you think it is a good idea to call the restaurant to find out if reservations are needed and to make sure the restaurant is accessible. You agree that it is a good idea. Your friend asks you if you would like to make the call. You say that you would like to make the call.

When the owner/manager answers, you begin by asking if there is a wait. You also ask if the restaurant is accessible. The manager responds, "Our restaurant is fully accessible. Do you have any specific questions for me regarding access?" You answer, "I use a wheelchair. Will I have any difficulty getting in, getting seated, or using the restroom?" The manager says "No, I don't believe so, but if you do experience any problems with our building, our service, or our menu items, I welcome your suggestions on how we can make improvements."

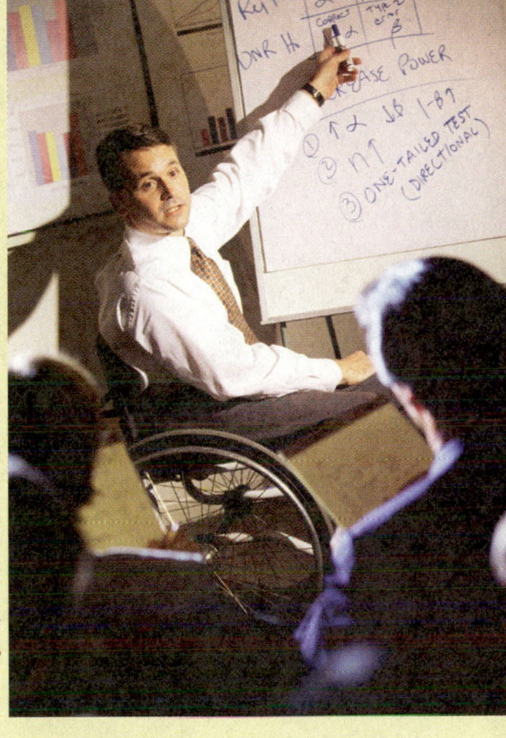

Anton Vengo/Superstock

Your reaction

You appreciate your friend's invitation to lunch and the concern about the restaurant being accessible. You appreciate your friend asking you if you would like to make the call, instead of the friend assuming that the call needs to be done for you. You appreciate the restaurant owner/manager's candid responses and the fact that you were given the opportunity to make suggestions for improvements to the building, services, and menu.

Most of all, you appreciate being treated with respect by both your friend and the restaurant owner/manager without any patronizing terms being used in the conversation.

The barriers people face can be as specific as the lack of access to the front door of a building or the content of a Web site or as general as attitudes expressed through communication. When you imagine yourself in the position of others in the workplace, you can begin to assess the ways that barriers interfere with people's autonomy and participation in workplace tasks and activities. In your communications with and for others, identify and eliminate as many barriers as possible.

Dr. Arvid E. Osterberg is a professor of architecture at Iowa State University where he teaches and conducts research in the areas of safety, accessibility, historic preservation, and design.

He also conducts building assessments and does consulting on safety and accessibility issues. He is coauthor (with Donna J. Kain) of *Access for Everyone: A Guide to Accessibility with References* to ADAAG.

Individual and Collaborative Assignments

1. **Evaluate a software application.** Write a review of a software application you've recently used. Evaluate the usability of the software. Prepare this review for publication in a local Society for Technical Communication (STC) newsletter so that other technical communicators and professionals who might use this software will have access to your assessment.
2. **Plan usability testing.** Select a print or Web document that you are working on that would benefit from usability testing. Prepare a usability test plan, drawing on the questions in Figure 9.5 to help you focus on the critical issues. Write the test plan as a memo to your instructor or to the client for the project.
3. **Conduct usability testing.** Select a print or Web document that you are working on that would benefit from usability testing. Prepare a testing report that includes these three parts:

   (a) Your best shot before usability testing — that is, what you consider the very best you can do in preparing the document.

   (b) The usability testing results. You must include at least one kind each of text-based, expert-based, and user-based testing. Identify the tests you selected and justification for the selection, the test results, and your analysis/interpretation of the results.

   (c) The revision of your best-shot document based on the testing results and your analysis/explanation of the changes.
4. **Assess and redesign a Web site.** Go to http://uaweb.arizona.edu/resources/makeovers/demo.shtml to study two versions of the same Web site, one inaccessible and the second one revised to be compliant with section 508.

   (a) Carefully read the features that made the old site inaccessible. Then review the revision.

   (b) Select a site to analyze that you believe is inaccessible for a portion of the intended audience. List the problems (as in the example above) and write specific recommendations for revision.
5. Do all Web sites need to be accessible to people with vision impairments? Yes, if it's a federal Web site or a site provided under contract to a federal agency. Should other sites be accessible? Yes, if they want to reach the broadest audience. To learn more about access guidelines developed by the Web Accessibility Initiative of the World Wide Web Consortium, go to these Web sites:

   http://usability.gov/accessibility/index.html#understanding

   http://usability.gov/accessibility/508.html

(Alternatively, go to www.english.wadsworth.com/burnett6e and then Chapter 9/accessibility for direct links.) After you have reviewed the information, prepare a presentation that could be used in a workplace workshop introducing accessibility.

Chapter 9 Endnotes

1 Mirel, B. (2002.) Advancing a vision of usability. In B. Mirel & R. Spilka (Eds.), *Reshaping technical communication: New directions and challenges for the 21st century* (pp. 167–189). Mahwah, NJ: Lawrence Erlbaum.

2 Adapted from Nielsen, J. (1993) *Usability 101* [electronic version]. Boston: Academic Press. Retrieved December 10, 2003, from http://www.useit.com/alertbox/20030825.html

3 Adapted from Nielsen, J. (1993) *Usability 101* [electronic version]. Boston: Academic Press. Retrieved December 10, 2003, from http://www.useit.com/alertbox/20030825.html

4 Artim, J. M. (n.d.). *Usability problem severity ratings.* Retrieved December 10, 2003, from http://www.primaryview.org/CommonDefinitions/Severity.html

5 *Usability Partners. (n.d.). ISO standards.* Retrieved December 18, 2003, from http://www.usabilitypartners.se/usability/standards.shtml

6 Brown, J. D. (2000). What is construct validity? *Shiken: JALT Testing & Evaluation SIG Newsletter, 4*(2): 7–10. Retrieved January 2, 2004, from http://www.jalt.org/test/bro_8.htm

7 Neill, J., Campbell, H., & Dalby, S. (2003, August 4). *Essentials of a good psychological test.* Retrieved January 2, 2004, from http://www.wilderdom.com/personality/L3-2EssentialsGoodPsychologicalTest.html

8 Siemens. (n.d.). *Testing in Siemens' ICM usability lab.* Retrieved January 2, 2004, from http://www.siemens-mobile.com/cds/frontdoor/0,2241,hq_en_0_3911_rArNrNrNrN_img%253A1124297,00.html

9 For further information, see Molitor, K. (1995). *How participants' expertise influenced expert testing of a technical user manual.* Unpublished master's thesis, Iowa State University, Ames, Iowa.

10 Redish, G. (2001). *Storytelling: The power of scenarios.* Retrieved January 3, 2004, from http://www.redish.net/content/handouts.html

11 Schumacher, G., & Waller, R. (1985) Testing design alternatives: A comparison of procedures. In T. Duffy & R. Waller (Eds.), *Designing usable texts* (pp. 377–403). Orlando, FL: Academic Press.

12 Hackos, J. T., & Redish, J. C. (1998). *User and task analysis for interface design* (p. 35). New York: Wiley.

13 Hackos, J. T., & Redish, J. C. (1998). *User and task analysis for interface design* (p. 35). New York: Wiley.

14 Nielsen, J. (1998). *Cost of user testing a website.* Retrieved October 10, 2003, from http://www.useit.com/alertbox/980503.html

15 IBM. (n.d.). *Usability.* Retrieved December 10, 2003, from http://www-10.lotus.com/ldd/use.nsf/308c971706adfdef8525640500696fa8/b89473efab094bbe852565e50070168f?OpenDocument

16 National Cancer Institute, Office of Communications. (n.d.). *Designing educational booklets for the web.* Retrieved December 16, 2003, from http://usability.gov/lessons/chemo.html

17 National Cancer Institute, Office of Communications. (n.d.). *Designing educational booklets for the web.* Retrieved December 16, 2003, from http://usability.gov/lessons/chemo.html

18 National Cancer Institute, Office of Communications. (n.d.). *Designing educational booklets for the web.* Retrieved December 16, 2003, from http://usability.gov/lessons/chemo.html

19 Meyertons, J. (n.d.). *Accessibility of online materials: Assistive technologies.* Retrieved December 10, 2003, from http://www.willamette.edu/~jmeyerto/Accessibility/workshops/technologies.htm

20 Connell, B. R., Jones, M., Mace, R., Mueller, J., Mullick, A., Ostroff, E., Sanford, J., Steinfeld, E., Story, M., & Vanderheiden, G. (1997). *Principles of universal design.* Version 2.0. Raleigh NC: North Carolina State University. Copyright © 1997 NC State University. The Center for Universal Design. Retrieved December 10, 2003, from http://www.design.ncsu.edu:8120/cud/univ_design/princ_overview.htm
For additional information, also see *General concepts, universal design principles and guidelines* from the Trace Research & Development Center, College of Engineering, University of Wisconsin-Madison, http://www.trace.wisc.edu/world/gen_ud.html

21 Meyertons, J., & Bowers, N. (2000, October 27). *Accessibility of Online Materials: Assistive Technologies.* Portland State University. Retrieved on December 10, 2003, from http://www.access.pdx.edu/workshops/technologies.html

22 Benner, J. (2001, February 23). When gamer humor attacks. *Wired News.* Retrieved January 6, 2004, from http://www.wired.com/news/culture/0,1284,42009,00.html

23 Story of All Your Base. (n.d.). *All your base are belong to us.* Retrieved January 6, 2004, from http://www.planettribes.com/allyourbase/story.shtml#game; also reprinted in Pearrow, M. (2002). *The wireless web usability handbook* (p. 100). Hingham, MA: Charles River Media.

24 Transperfect Translations, Press Room. (2003, January 6). *Bad translations make for a good laugh but are bad for business.* Retrieved January 6, 2004, from http://www.transperfect.com/tp/eng/badxlate.html

25 Perera, A. (2003, October 7). Directions for prescriptions in English a problem for some. *In the News.* Retrieved January 6, 2004, from http://www.transperfect.com/tp/eng/phillyinq100703.html

26 NASA Glenn Research Center. (n.d.). *NASA web accessibility statement.* Retrieved December 10, 2003, from http://www.grc.nasa.gov/Doc/access.html

PART III

Shaping Information

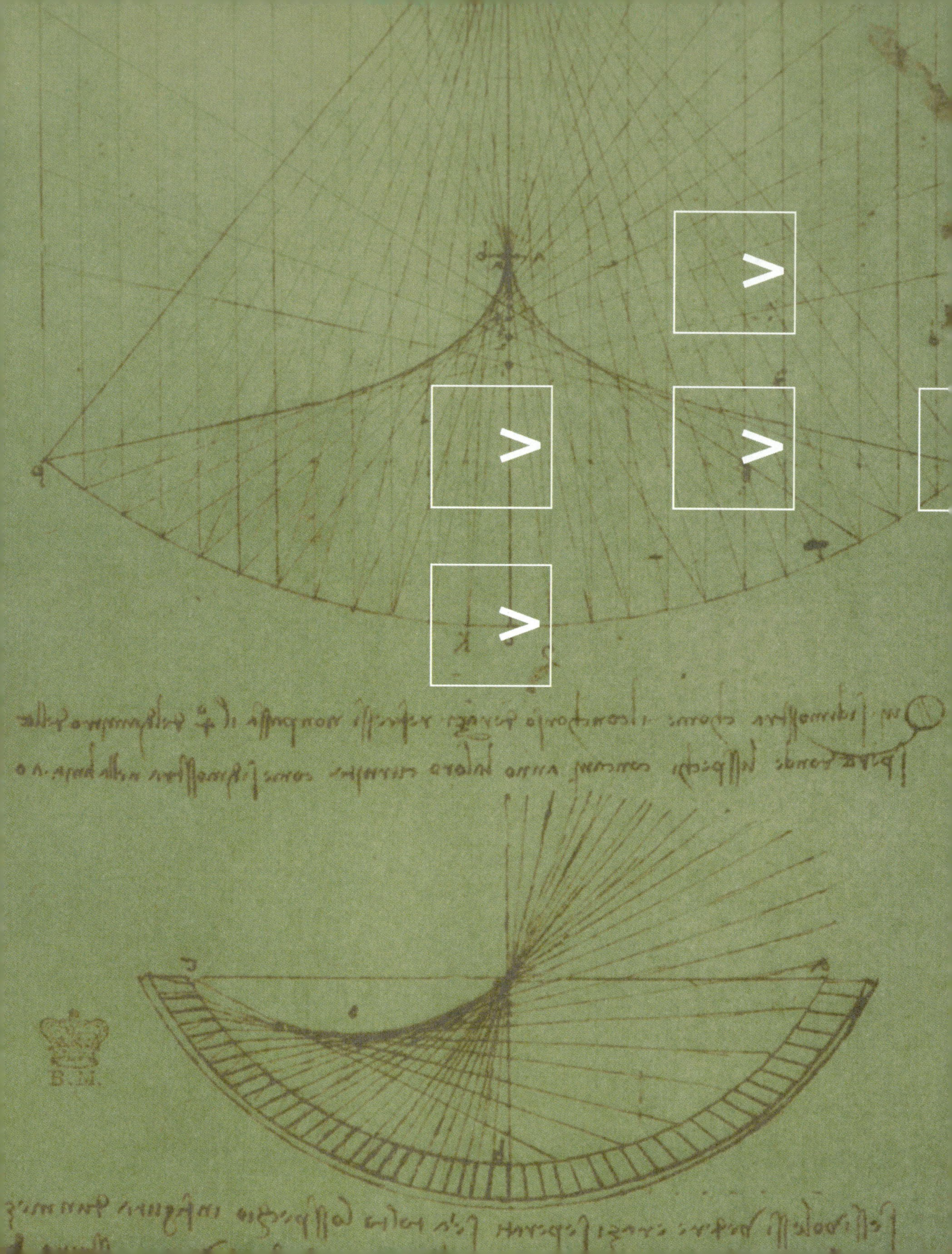
B.M.

CHAPTER 10

Organizing Information

Objectives and Outcomes

This chapter will help you accomplish these outcomes:

- Organize information as part of knowledge management
- Use outlines, storyboards, and tables as tools to test various ways to organize information
- Use topic sentences and transitions to signal organization
- Use conventional organizational patterns — whole/parts, chronology, spatial order, ascending/descending order, comparison/contrast, cause and effect — to present information verbally and visually

>

The way you choose to organize information affects the way you interpret it and, necessarily, the meaning you make from it. While Chapter 6 discussed the beginning steps of knowledge management (finding and selecting information), this chapter is about organizing and transforming that information. If you skimp on this part of the communication process — organizing information — you will have difficulty producing high-quality documents, oral presentations, and visuals. Outlines, storyboards, and tables are valuable tools for organizing information in various ways. Additionally, six conventional patterns enable you to organize verbal information (for both documents and oral presentations) and visual information.

Transforming Information into Knowledge

Decontextualized information doesn't carry meaning, but once information is transformed into knowledge, that information does carry meaning. This transformation requires you to make sense of the information, organizing it so that you can compare it to your prior knowledge, consider the implications and consequences, and talk about it with informed colleagues.

The process of transforming information into knowledge — and, thus, making meaning — is messy, sometimes unpredictable, often difficult, and always contextualized. When you change the context, you change the meaning. The organization of your information affects the rest of the project, so if you're involved in an important or complex project, you should allow time during your planning to organize your information in more than one way because you'll see different things every time you rearrange it. Why? The meaning you construct from information is contingent on a range of things, including national and organizational

Informal Settlements in Durban, South Africa.[1] This is housing in an "informal settlement." These overcrowded residential areas present complex social, economic, and engineering challenges since these areas typically do not have sanitation systems, fresh water, or electricity.

William L. Jeffries © 2002

culture, the local context and specific situation, and the perspectives of you and the audience(s). If you do nothing to organize and interpret the information you present, other people will organize it to suit themselves, perhaps seeing meanings different from what you might have presented.[2]

The strategies for organizing information are widely recognized, but you should apply them purposefully rather than mechanically — that is, have reasons for what you do rather than just following convenional practices. The purpose of a document, oral presentation, or visual is to communicate information to an audience; thus, making your information understandable is critical. As you initially explore a topic, you may initially create a *writer-based* document (or presentation or visual), one that helps you examine and organize the information. However, it doesn't consider whether the information will be effective for your audience. Your goal should be to create an *audience-based* document, presentation or visual, one that considers the needs and reactions of your audience and organizes information so that these readers, listeners, and viewers can understand the issues.

WEBLINK

Can you have too much relevant information? Probably not. But you can have too much information that's irrelevant, badly organized, and inaccessible. To read a short version of how one company handled this problem, go to http://english.wadsworth.com/burnett6e.

CLICK ON WEBLINK

CLICK on Chapter 10/information silos

William L. Jeffries © 2002

Informal Settlements in Context. Audience reactions to the same housing may be different when the informal settlement is put in context — adjacent to more elaborate housing. Even though the problems with sanitation systems, fresh water, and electricity don't change, the context influences interpretation.

The overall persuasiveness of your documents, oral presentations, and visuals depends on their organization. You can use strategies such as outlines, storyboards, and tables to experiment with and test various ways to arrange and rearrange your information to determine the most effective organization.

Outlines

You may decide to outline your information simply because (as you read in Chapter 6) changing an outline is usually easier than changing the draft of a text. Any outline you develop should be flexible and easy to change as you arrange and rearrange ideas, add new information, and delete unnecessary material. Outlines are not intended to restrict you; rather, they are tools to help you manage the material for a document. Think of them as blueprints that show the overall structure and primary features. As with buildings that exist only as blueprints, changes in documents or oral presentations are easier to make in their preliminary stages.

Outlines can help you arrange and examine and then rearrange collected information. They do not have to be formal, complete-sentence outlines. Initially, you can just jot down information and then rethink, rearrange, and reorganize it

FIGURE 10.1 **Online Outline Showing Problems in Completeness, Organization, and Parallelism**

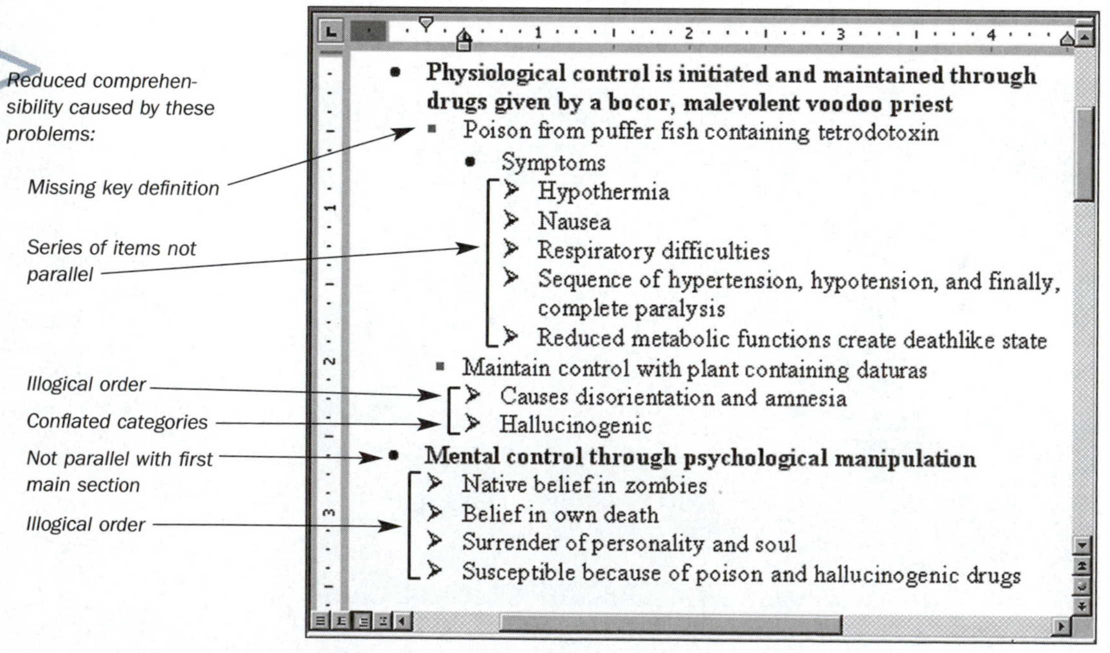

in an outline. For example, a simple list of unorganized points about native beliefs and practices reports the work of Wade Davis, a Harvard ethnobotanist, who used pharmacology to discuss the zombies of traditional Haitian voodoo.[3]

Figure 10.1 shows a computer screen of an outline that was developed from this preliminary list. Outlining is an option with most word-processing software. An electronic outline offers you a different view of your document rather than creating a separate document. The electronic changes you make in the outline automatically become part of the document, and vice versa. You can easily switch back and forth between an outline view of your document and a full text view.

The writer could examine the electronic outline in Figure 10.1 to determine whether the information is complete and parallel — equivalent in importance, sequence, and wording. In this case, the outline has incomplete information, needs reordering, and isn't parallel. But without the outline (whether paper or electronic), the writer might not see these inadequacies.

Figure 10.2 shows a revised electronic outline in which the writer has corrected the problems. The changes are more than cosmetic. First, expressing ideas in parallel structure demonstrates that the writer intends to treat them equally, as shown in the revision. Next, the sequence must be logical, as shown in

FIGURE 10.2 **Online Outline Showing Revisions**

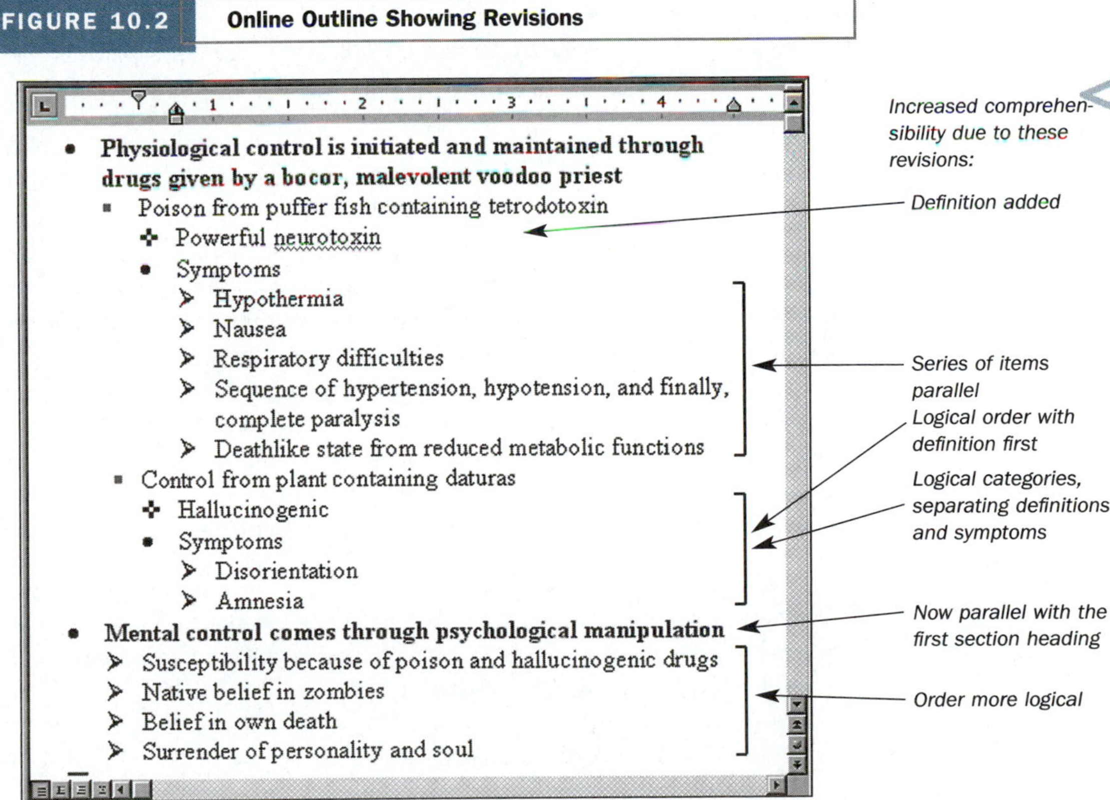

the revision in the order of entries. Here, indicating susceptibility logically comes first; instances of beliefs resulting from this susceptibility follow. Finally, essential information must not be inadvertently omitted, as shown by the omission of the type of poison in the first section. In each case, changing the outline is easier than changing the draft of the paper.

The original and revised outlines about zombies (Figures 10.1 and 10.2) show how helpful outlines can be to organize and reorganize information before starting to draft a document. These outlines identify gaps in data, inconsistencies in the relative importance of various segments, and problems in sequencing information. If you use outlines as tools for planning and revision, you may save yourself a great deal of frustration and time.

WEBLINK

Ethnobotonists study the connections between plants and people. Harvard ethnobotonist Wade Davis has "traveled to Haiti to investigate legends of a 'zombi poison.' The so-called poison was supposedly made from human bones and parts of lizards, poisonous toads, sea worms, puffer fish, and other items; it was said to lower the metabolism of anyone who swallowed it and paralyze his or her vital functions, leaving the individual in a condition that could easily be mistaken for death. Davis's supporters believed that the drug might have important applications for anesthesiology and artificial hibernation,"[4] which might control neurological diseases. To read more about Wade Davis and the work of ethnobotonists, go to http://english.wadsworth.com/burnett6e for a series of links.

CLICK ON WEBLINK

CLICK on Chapter 10/ethnobotony

As you plan your material, you can ask yourself the questions in Figure 10.3 to check whether your outline is likely to result in a successful document. If you can answer yes to all the questions in Figure 10.3, your outline will probably be useful in planning, organizing, and drafting your document.

Outlines have additional uses besides helping to organize information. In a long document, an outline's main headings can provide ready-made headings for the document itself and also serve as a table of contents. For example, Figure 10.4 — a U.S. Food and Drug Administration online report, "Improving Innovation in Medical Technology" — shows the entire three-level heading structure as a numeric outline.

WEBLINK

Sometimes the most useful way to organize information is visually. For a link that provides examples of ways to organize using tables, diagrams, lists, and maps, go to http://english.wadsworth.com/burnett6e.

CLICK ON WEBLINK

CLICK on Chapter 10/visual strategies

FIGURE 10.3 **Questions to Ask about Your Outline**

Content and context	▪ Is all important information included and all unnecessary information omitted? ▪ Are contextual factors that will influence the document's interpretation acknowledged?
Audience	▪ Do the headings give readers an accurate overview of the document? ▪ Is the level of detail appropriate for the purpose and audience?
Purpose and key points	▪ Are the main headings and subheadings logically arranged? ▪ Is the organization in the outline appropriate to the content, purpose, audience, and genre?
Organization	▪ Is all information in the outline in the appropriate places — both in the main sections and the subsections?
Professional standards	▪ Do the details in each subsection reflect the emphasis for that heading? ▪ Are main headings in the outline written in parallel grammatical form (so that you can use them as headings in the document)?

Storyboards

Storyboards are a second way of experimenting with and testing your organization of information. They have two broad purposes and audiences: (1) a powerful organizing tool for the writers and designers who create them; and (2) a short, dramatic visual summary showing the gist of the final project for clients or customers. Storyboards are a related sequence of hand-drawn or electronic sketches of pages or screens that organize their critical points. Basically, storyboards create shorthand stories that have a setting, characters, and actions for a particular audience. They also convey the purpose, hierarchy, and sequence of the information as well as the navigation. Using them enables you to plan highly visual projects such as tutorials, training videos, ads, and Web sites.

Storyboards have been around for a long time; even Leonardo da Vinci put ideas up on a wall and examined the layout. Modern storyboards can trace their origins to the beginnings of cinema with Sergei Eisenstein. Then, in 1928, Walt Disney and his staff developed a storyboard system allowing writers and designers to trace the overall story and see connections between the various parts.[5]

Some storyboards are passive — simply a sequence of thumbnail sketches or drafts of screenshots of a Web site. With a *passive* storyboard, the writer or designer usually walks a client through the sequence, explaining what happens.

Other storyboards are *active* (animated on a timed program) or *interactive* (requiring the user to be engaged in the process, for example, using PowerPoint so a client can independently walk through the sequence of a presentation).[6] And,

FIGURE 10.4 **Outline Headings in FDA Report**[7]

Improving Innovation in Medical Technology

1.0 Background
 1.1 FDA Product Approvals and Challenges
 1.2 Factors contributing to decline in new product applications.

2.0 Reducing Delays and Costs in Product Approvals by Avoiding Multiple Review Cycles
 2.1 Factors that cause unnecessary delays in new product approvals
 2.2 Reducing avoidable delays in time to approval
  2.2.1 Addressing the root causes of longer review times for human drugs and biologics
  2.2.2 Reducing Delays in Medical Device Reviews
  2.2.3 Reducing Delays in Animal Drug Reviews

3.0 Improving the review process through a quality systems approach to medical product review
 3.1 Instituting quality systems in review of new drugs and biologics
 3.2 Implementing medical device quality initiatives

4.0 Improving Product Development and FDA Review Process through Developing Clear Guidance
 4.1 Better Guidance for Targeted Clinical Areas
 4.2 Better Guidance for Emerging Technology Areas
  4.2.1 Cell & Gene Therapy
  4.2.2 Pharmacogenomics
  4.2.3 Novel Drug Delivery Systems

4.2.2 Pharmacogenomics

Certain new therapies will b… identify the responding sub… find people who are prone… combinations must be faci… benefits while minimizing… uncertainty. Over the nex…

- Issue guidance on … FDA during drug d… pharmacogenomic… information would… next 6 months)
- Hold a workshop … test and a drug.(this year)
- Issue joint guidance (CDRH-CDER) on the regulatory pat… combinations. (In 18 months)

4.2.3 Novel Drug Delivery Systems

Novel drug delivery systems present a wide diversity of technologies and applications, e.g., infusion pumps, drug-eluting stents, lasers for photodynamic therapy, hyperthermia devices, etc. Most of these products require application submissions to, and reviews by multiple centers, and multiple offices and divisions within those FDA centers. The complexity of the review issues vary dramatically from product to product, ranging from simple device/complex drug combinations to simple drug/complex device products. The novel technologies and regulatory uncertainties can present challenges for product innovators.

finally, some storyboards have evolved to *animatics,* which are drafts of animated presentations that are "produced by photographing storyboard sketches on a film strip or video with the audio portion synchronized on tape."[8]

WEBLINK

Animatics are similar in function to storyboards. They are animated presentations — for example, short Quicktime videos — that give clients a sense of the focus, tone, and content of the final product. For some examples of animatics, go to http://english.wadsworth.com/burnett6e.

CLICK ON WEBLINK

CLICK on Chapter 10/animatics

Typically, storyboards identify the context or setting and display the relationships between the players and their activities. The individual frames in storyboards often have annotations or captions, especially if they are active or interactive. If you include a separate text, you should cross-reference each chunk of text with the number of the appropriate storyboard frame. In general, storyboards have one of two layouts:[9]

- A storyboard might be in a linear layout when you want users to visit each page sequentially without skipping around. What kinds of projects might use linear storyboards? Introductory computer-based tutorials or training modules. Procedural task instructions. Slide shows. PowerPoint presentations.
- A storyboard might be in a hierarchical layout to structure Web documents. This layout can begin with a homepage and show links that lead to other pages, each with additional links to other pages. For the usability of your site, create no more than three or four levels of linked pages. What kinds of projects might use hierarchical storyboards? Complex Web sites. Interactive tutorials. Training modules.

What are the benefits of storyboarding? A large portion of the project planning and revising takes place up front, which saves time and money. Why? The revisions are done to a plan, not to a product. This reduces changes later in the process. People involved in the project can review and revise storyboards as a way to explore alternative ideas.

Some companies have templates for storyboards, which often include a box for information such as the name and purpose of the project, people who are involved (including client, subject matter experts, writers, and designers), the projected start and completion dates, and the media that will be used. Other kinds of information on (or accompanying) most storyboards include the text that goes with each frame; typographic specifications; and the nature of the visuals, animation, audio, and video. Storyboards let you try out various designs without the effort of a fully developed project. Many professionals see storyboards as a way to increase efficiency while encouraging creativity in content, structure, and design.

WEBLINK

Storyboarding is an easy way to accomplish a lot of planning. For links to some useful sites about storyboarding, go to http://english.wadsworth.com/burnett6e.

CLICK ON WEBLINK

CLICK on Chapter 10/storyboarding

Storyboards are tools, so don't make too big a production out of them. You can use them frequently to plan/organize and to get user feedback. These guidelines will save your time and money:

- Don't invest too much time and effort in storyboards. Keep storyboards sketchy.
- Make storyboards easy to modify. If you don't change anything, you don't learn anything.
- Whenever possible, make storyboards interactive. The user's experience will generate more feedback and will elicit more new requirements than passive storyboards.
- Create storyboards early and often. Always use storyboards on projects that have new or innovative content.[10]

Tables and Spreadsheets for Organizing Information

Tables and spreadsheets are a third way to experiment with and test ways to organize information. They enable you to classify information into comparable groups and then identify categories of details (features, functions, applications, and so on) about each group. As you collect information, you can organize it in the table or spreadsheet.

Color Systems	CMYK~ Subtractive	RGB~ Additive
Identifying Primary Colors		
Mixing Primary Colors		
Using Color Systems		
Creating Other Colors		

If you create the tables and spreadsheets on a computer, you can rearrange the rows and columns to consider additional comparisons. Electronic spreadsheets have a further benefit for numeric data: You can easily organize and interpret the data using various formulas, both standard equations and ones that you create for special purposes.

An example illustrates how useful tables can be for organizing information. The use of color is critical for a number of professionals, including scientists, designers, and printers. Differentiating these color systems is sometimes confusing, especially if you learned in elementary school that red, yellow, and blue are the only primary colors, and now you realize that scientists and artists talk about the CMYK Subtractive Color System and the RGB Additive Color System. So you sketch the information you minimally need to organize in a table, as shown on the previous page: What are the primary colors in both the CMYK and RGB systems? What happens when you mix the colors in each system? What are the applications for each color system? What are the specific ways of colors can be combined in each system? This rough sketch can then be refined into a table for organizing the information you collect about the CMYK and RGB systems. Figure 10.5 shows this simple table for organizing information.

FIGURE 10.5 **Table for Organizing Information**

Color Systems	CMYK ~ Subtractive Color	RGB ~ Additive Color
Identifying primary colors	CMYK = cyan, magenta, yellow, black 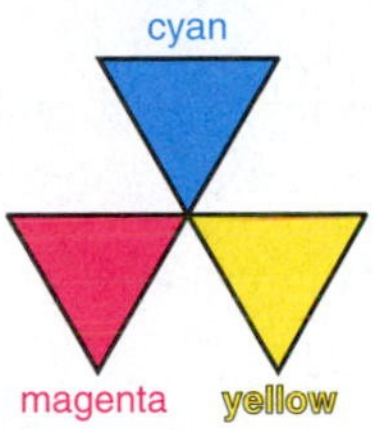 cyan magenta yellow	RGB = red, green, blue red green blue 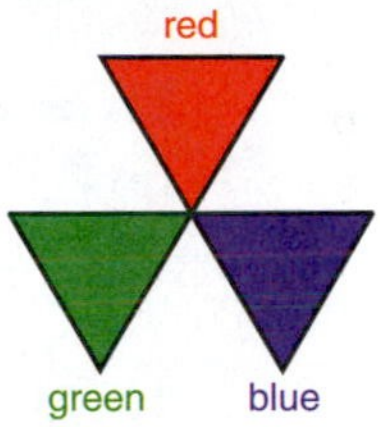
Mixing primary colors	In their purest form, cyan, magenta, and yellow cannot be made by mixing other pigments. When any two subtractive primary colors come together, they form one of the additive primary colors.  cyan yellow magenta	When additive primary colors — beams of light of dots of colored light — are combined, they form different colors than mixing subtractive colors.  red green blue
Using color cystems	Subtractive color applies to print documents, photography, lithographs, and paintings by combining physical pigments such as paint, dyes, or inks.	Additive color applies to stage lighting, computer monitors, and television by combining colored light such as spotlights or the dots of color on computer monitors and TV screens.
Creating other colors	• Mix paint together. • Combine clear color liquids. • Place one color glass in front of another. • Print several color inks on top of each other on the same paper during several press runs. • Use several transparent color layers to make a photo, a slide, or movie film.	• Shine colored lights or spotlights on the same area. • Visually combine the separate dots of colored light on a television screen or computer monitor.

WEBLINK

Color can be a critical part of documents, oral presentations, and visuals. Understanding subtractive and additive color is important. You might also be wondering why school children are usually introduced to red–yellow–blue as primary colors. To learn more about these issues and to see how the colors work together, go to http://english.wadsworth.com/burnett6e for useful links.

CLICK ON WEBLINK

CLICK on Chapter 10/color

Implementing the Organization of Information

The way information in a print or electronic document, oral presentation, or visual is organized affects the meaning that the audience constructs.

- *Print documents* chunk information into paragraphs that readers can see; the paragraphs can be signaled by indentation or extra line spacing as well as by topic sentences and transitions.
- *Electronic documents* chunk information into paragraphs as well; they have the additional benefit of hyperlinks (usually signaled visually by color and underlining), so users can organize personal sequences of information.
- *Oral presentations* also chunk information, often in similar ways to the chunks in documents; the audience has to listen for cues about shifts to another topic, so changes in vocal pacing, pitch, and inflection take the place of indentation and line spacing, and the topic sentences and transitions need to be much more explicit than in documents.
- *Technical visuals* also chunk information; labels and cues direct movement through the visual.

Sometimes information can be organized in lists and continuums, sometimes in paragraphs, and sometimes in visuals. These various options are explained and illustrated in the rest of the chapter.

Alphabetical Order, Numeric Order, and Continuums

Information can be organized in alphabetical order, numeric order, or continuums.[11] One of the most basic organizational patterns is alphabetical order, which is useful for quite a range of documents, such as dictionaries, encyclopedias, glossaries, indexes, and phone books. Another common organizational pattern is *numeric order,* which arbitrarily associates a particular number (or range of numbers) with some category, as in the Dewey Decimal System categories:

000	general reference	500	science
100	philosophy and psychology	600	technology
200	religion	700	arts and recreation
300	social science	800	literature
400	language	900	history and geography

A numeric system is effective because each item — whether a book or an automotive part — is identified by a single number. However, the assigned number doesn't indicate a rank ordering.

WEBLINK

If you want to learn more about the complexities of an enormous numeric classification system, go to http://english.wadsworth.com/burnett6e for a link. On the home page for the Dewey Decimal System, select the link *Introduction to the DDC,* which takes you to a 37-page introductory list related to managing such complex systems.

CLICK ON WEBLINK

CLICK on Chapter 10/dewey decimal

A third way of organizing information is to use a *continuum,* which ranks or rates the objects or practices being organized, such as runs batted in (RBIs), an airline's on-time record, a company's safety record, a division's manufacturing performance. Figure 10.6 displays an edited excerpt from an online list of changes in several updates of a computer program ("Common Gateway Interface [CGI], a set of rules that describe how a Web server communicates with other software on the same machine and how the other piece of software, the CGI program, talks to the Web server").[12] Referring to this list of earlier changes helps CGI users trace the product's history and also identify current capabilities. Because the releases are on a continuum, the numbers — 1.01 through 2.01 — show that version 2.0 was a major revision.

Topic Sentences and Transitions to Signal Organizations

Sometimes alphabetical or numeric lists and continuums are not sufficient because you need to explain the relationship between the ideas or objects you're organizing. Once you've decided how to organize the information, you can help the audience by using signals that identify the organization. Two tools help signal the organization of information, whether written or oral:

- A *topic sentence* identifies both the content and organization of a paragraph so that the audience anticipates what forthcoming information is about and how the information will be sequenced.

FIGURE 10.6 **Continuum: Changes in CGI Program Releases**[13]

The continuum makes the chronology clear.

The newest release is version 2.01.

A big change in features and capabilities came in release 2.0. The release prior to that had been 1.07.

The first revision after the initial release was 1.01.

What's new in version 2.01?

- Makefile supports "make install"
- Compiles without warnings under both C and C++ with strict warnings and strict ANSI compliance enabled . . .

What's new in version 2.0?

1. CGIC 2.0 provides support for file upload fields. User-uploaded files are kept in temporary files, to avoid the use of excessive swap space. . . .

What's new in version 1.07?

A problem with the cgiFormString and related functions has been corrected. These functions were previously incorrectly returning cgiFormTruncated in cases where the returned string fit the buffer exactly.

What's new in version 1.06?

1. A potentially significant buffer overflow problem has been corrected. . . .

What's new in version 1.05?

Non-exclusive commercial license fee reduced to $200.

What's new in version 1.04?

For consistency with other packages, the standard Makefile now produces a true library for cgic (libcgic.a).

What's new in version 1.03?

Version 1.03 sends line feeds only (ascii 10) to end Content-type:, Status:, and other HTTP protocol output lines, instead of CR/LF sequences. . . .

What's new in version 1.02?

Version 1.02 corrects bugs in previous versions. . . .

What's new in version 1.01?

Version 1.01 adds no major functionality but corrects significant bugs and incompatibilities . . .

For a reader who is an expert in the subject, why are transitions necessary? Won't an expert understand the relationships among the ideas?

- *Transitions* are words, phrases, and sentences that act as the glue connecting ideas and sentences within a single paragraph, linking one paragraph to another, and relating one section of a document or presentation to the next section.

Technical communicators organize information to make it clear and accessible to an audience. Some common ways to organize information are whole/parts organization, chronological order, spatial order, ascending/descending order, comparison/contrast, and cause and effect. With few adaptations, these ways of organization apply to written documents, oral presentations, and visuals.

No communicator settles down to write and says, "I'm about to plan a document or presentation using chronological order." Instead, the communicator asks questions like these: "What's the situation or problem I'm responding to?" "What's my purpose?" "Who's the audience, and what are their expectations?" "How can I help my audience understand this information?" "What's the appropriate genre?" And then the communicator may ask, "What's the most appropriate way to organize the information given the situation, purpose, audience, and genre?" In organizing written and oral information, technical communicators use topic sentences and transitions to help make the information more accessible and appealing to the audience.

In what ways do topic sentences make technical information easier to read?

Does every paragraph need a topic sentence? Not necessarily. Some paragraphs are transitional, connecting one main paragraph or section to the next. Occasionally, an excessively long paragraph in a document is separated into two or more to make the information easier to read and give readers a chance to breathe. Usually, however, most of the paragraphs in a well-constructed document have clear topic sentences that can be listed together as a summary. If the message from this topic sentence summary is clear and logical, then the information is probably well organized.

These points about organizing information for documents naturally extend to Web sites. However, Web sites not only have available the strategies that work for written documents, they also have additional organizational strategies because hyperlinks can be incorporated. Jakob Nielsen, an expert of Web usability, reports on some of the problems that result from poorly organized Web sites:[14]

- In one recent study of 15 large commercial sites, users went to the correct home page before they started the study's test tasks; even with that help, they found information they needed only 42 percent of the time.
- In a study of electronic shoppers, 62 percent gave up looking for the item they wanted to buy because they couldn't find it.
- A third study analyzed 20 major sites to determine whether these sites followed simple Web usability principles: Is the site organized by user goals? Does a search list retrievals in *order* of relevance? Only slightly more than half of the sites — 51 percent — complied with basic principles of organization.

What's the impact of such poorly organized Web sites?

- Sites lose approximately 50 percent of potential sales because people can't find what they want.
- Sites lose repeat visits from up to 40 percent of users who do not return to a site after an initial negative experience.

What's the most effective solution? Jakob Nielsen recommends usability testing of all Web sites. If a site is well organized, users should be able to find what they want. The strategies in this chapter can be part of your repertoire for organizing print and Web documents. These strategies are useful tools for making both print and electronic documents comprehensible and usable, but in some remarkably different ways. For example, print documents can be organized chronologically; however, you can be quite sure that users of a Web site will follow links in different ways and thus necessarily create different sequences of information. Therefore, a Web author needs to ensure that information makes sense in the various ways that users might access it. If a specific sequence is critical for comprehension, a Web author needs to tell users to follow the prescribed sequence.

WEBLINK

Organizing information for your Web site gives you more options than organizing an effective print document. You should create logical chunks, determine a hierarchy, establish relationships, implement your plan, and analyze its effectiveness. For a link to a more detailed discussion, go to http://english.wadsworth.com/burnett6e.

CLICK ON WEBLINK

CLICK on Chapter 10/organizing information for the Web

Whole/Parts Organization

A document that uses *whole/parts organization* presents readers with a relationship between the whole (whether an idea, object, or entire system) and parts of that whole (whether on a micro level or a macro level). Sometimes whole/parts organization involves separating a single item into individual components. At other times, it involves identifying related types of an item. It may also involve identifying the broad category to which something belongs.

A whole/parts organization to present the categories of integrated circuits is shown in Figure 10.7. The paragraph shows the relationship between the whole

FIGURE 10.7 **Whole/Parts Organization**

A clear topic sentence identifying the whole increases comprehensibility.

Bullets signaling the three types of circuits make the information more accessible.

Transitions signaling the parts of the whole make the information more comprehensible.

A simple visual (referred to in the text) reinforces the whole/parts relationship.

Integrated circuits are divided into three categories, depending on their function and capability in the final product. (See the diagram below.)

- The first type is a programmable integrated circuit, a multifunctional component designed to be programmed by the supplier or by in-house technicians.
- The second type of integrated circuit is the memory device, used to store memory in an end product.
- The third type is a linear device designed to do many specific predetermined functions and used in conjunction with other components to operate computers.

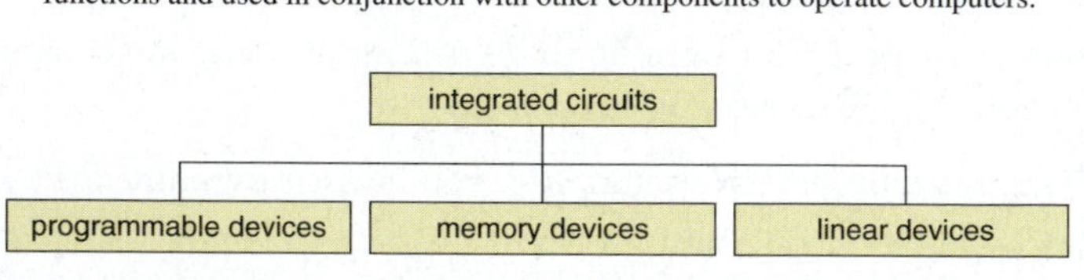

(the category of devices called integrated circuits) and the parts of that whole (three types of circuits: programmable devices, memory devices, and linear devices). This example (and the following related ones) was written by the testing line supervisor in an inspection group of a plant that manufactures and assembles integrated circuits. This supervisor was required to train the unskilled entry-level workers assigned to her area. She decided to provide these workers with short, easy-to-read explanations about the inspection process, believing that helping workers understand how their specific job fit into the larger process would increase the quality of their work. Although she hoped they would learn this information during their initial job orientation, she knew she would need to review it periodically in ongoing training sessions.

What additional examples of using a whole/parts organization for information are likely in the workplace?

A new researcher is organizing a report about a six-month project he just completed. He wants his readers to follow his experiments step by step, not being able to know the results until they read his conclusion and recommendations at the end. He says he worked hard; he wants readers (especially his manager) to appreciate his efforts. What do you think of his decision? What would you say to persuade him to reconsider his plan?

Chronological Order

Chronological order presents readers with material arranged by sequence or order of occurrence. When the purpose is to give instructions, describe processes, or trace the development of objects or ideas, chronological order is appropriate. For example, chronologically organized information about CDs might include a description identifying each step of the manufacturing process or an explanation of the historical development of electronic storage media. Information can be presented chronologically in visual forms, as shown in Figure 10.8, which also lists an example of each form.

FIGURE 10.8 **Visual Forms for Chronological Order**

Form	Example
Schedules	Plane, train, or bus schedule
Flow chart	Sequence of manufacturing process
Time line	Development of synthetics for medical use
Genealogy chart	History of family with Huntington's disease
Sequential photos	Embryo development
Sequential drawings	Steps in resuscitation of drowning victim
Storyboard	Public service ad urging water conservation
Time-lapse photos	Emergence of butterfly from cocoon
Line graph	Increase in eye movement during REM sleep
Calendar	Production schedule for new product
Chart	Stratification of rock layers according to geologic periods

A topic sentence conveying chronology includes words or phrases that indicate a process or a sequence of actions, as illustrated in the following examples:

- Seven operations are necessary to fabricate sheet metal.
- Most poultry farms are vertically integrated, from breeding to egg to packaged product.

Readers can justifiably expect the paragraph that follows the first topic sentence to identify the seven steps of sheet-metal fabrication. The paragraph that follows the second topic sentence should identify each stage in the process of poultry production.

Transitions in chronological paragraphs can indicate the sequence of events or the passage of time. Figure 10.9 presents an example that explains the chronology of incoming inspection that integrated circuits go through. Readers can easily follow the process because the chronological transitions highlight each step of the five-step process.

FIGURE 10.9 **Information Organized Using Chronological Order**

Chronological transitions:
- *the first stage*
- *then*
- *Next*
- *After this*
- *The final step*

Integrated circuits received at incoming inspection go through a five-step process. As the flowchart shows, the first stage of inspection ensures that the parts have been purchased from a predetermined qualified vendor list. The parts are then prioritized according to daily back-order quantities and/or line shortages assigned by the production floor. Next the parts are moved into the test area, where a determination is made as to which lots will be tested at 100 percent and which will be sample tested. After this, the parts are electrically tested for specific continuity and direct current parameters using MCT handlers. The final step of incoming inspection is distributing the parts according to need on the production floor.

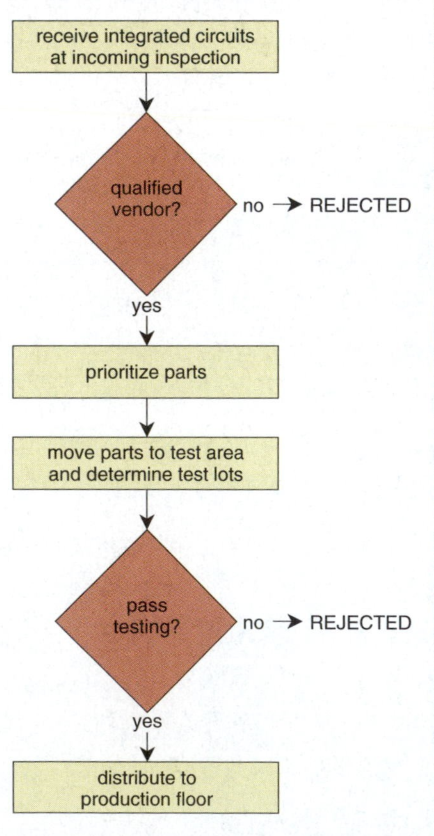

What additional examples of using chronological order for organizing information are likely in the workplace?

Spatial Order

Spatial order — arrangement by relative physical location — describes the physical parts of nearly anything, from cellular structures to the orbital path of a satellite. Spatial organization could explain parts of a CD or the location of the hard drives in relation to the other parts of the computer. Figure 10.10 presents examples of several visual forms. You may find that visuals are particularly effective for spatially arranged material because they help your readers see the actual physical relationships.

Because spatial arrangement deals with the relative physical location of objects, the topic sentence suggests their placement, as seen in these examples.

- Unnecessary or damaged inventory that is scheduled to be scrapped is placed on skids in one of six bin locations in the Defective Stockroom.
- Sound is a ripple of molecules and atoms in the air that travels from its source to our ears.

Readers of the first sentence anticipate the identification of each bin's location according to type of scrap material. The second topic sentence indicates to readers that the paragraph will track the sound as it moves through the air from source to listener.

Transitions in spatial paragraphs suggest the relative physical location of components or objects. Figure 10.11 presents an example that uses spatial order to describe the incoming inspection of an integrated circuit.

FIGURE 10.10 **Visual Forms for Spatial Order**

Form	Example
Map	Identification of migration stopovers
Blueprint	Specification of dimensions for machined part
Navigational chart	Location of sand bars and buoys
Celestial chart	Sequence of moons around Jupiter
Exploded view	Assembly of disk brake
Cutaway view	Interior components of pool filter
Wiring diagram	Wiring of alarm system
Floor plan	Workflow in busy area
Set design	Arrangement of furniture/props for *Hamlet*
Architectural drawing	Appearance of building with solar modifications

FIGURE 10.11 **Information Organized Using Spatial Order**

Spatial transitions:

- *through the test area from the supplier*
- *into a removable channel within a clear tube*
- *in the same direction*
- *into the MCT handler*
- *in the upper left through a slot*
- *to the manufacturing area*
- *to another engineering station*

The movement of the IC (integrated circuit) chip through the test area is very efficient. The chips arrive from the supplier, already set—24 at a time—into a removable channel within a clear tube. The chips are aligned in the same direction within the tube. This tube is inserted by the operator into the MCT handler so that pin 1 of the first chip, marked by a small dot, is in the upper left. The tube slides through a slot, into the testing compartment, where each chip is tested individually. Automatically, the good chips are placed in one channel, the rejects in another. The channels are moved so the operator can slip on the protective tubes. The good chips are sent to the manufacturing area; the rejects are sent to another engineering station for further testing.

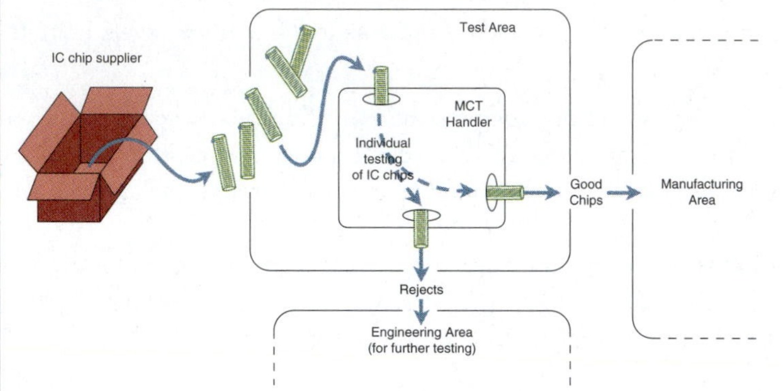

> *What additional examples of using spatial order for organizing information are likely in the workplace?*

Ascending/Descending Order

Ascending and *descending orders* present readers with information according to quantifiable criteria.

appeal	durability
authority	ease of manufacture, operation, repair
benefit	frequency
cost	importance
delivery	size

Descending order uses a most-to-least-important order; ascending order, a least-to-most. Descending order is found in workplace writing more frequently than ascending order because most readers want to know the most important points first. Readers generally form opinions and make decisions based on what they read initially; they expect descending order in nearly all technical documents. Either descending or ascending order would be appropriate for organizing the relative convenience of various forms of electronic data storage or for identifying the CD specifications in various price ranges. If you wanted to arrange information visually, one of the forms illustrated in Figure 10.12 would work.

FIGURE 10.12 **Visual Forms for Ascending/Descending Order**

Form	Example
Numbered list	Priority of options for treating breast cancer
Bull's-eye chart	Population affected by nuclear explosion
Percent graph	Percent of different economic groups receiving balanced nutrition
Pareto diagram (bar graph of ranges of data arranged in descending order)	Productivity using different methods
Line graph	Increasing success for breeding endangered species in captivity over 20-year period

Unlike topic sentences for paragraphs organized in other ways, those beginning descending or ascending paragraphs do not give an immediate clue about the subsequent organization. The reader understands that the paragraph presents a series of related ideas, but the specific relationship is not clear until the second sentence. The topic sentence in the next example begins a paragraph about the master satellite station in Beijing by presenting the characteristics of the Beijing antenna in descending order of importance.

> The largest earth station in China is a 15-to-18-meter-diameter dish antenna in Beijing for domestic satellite communications.

A paragraph using a descending organization to identify the various sized antennas in China begins with a general statement before going on to identify each type of station.

> Three types of earth stations are planned for domestic satellite communication in China. The largest is a 15-to-18-meter-diameter dish antenna of the master station in Beijing. . . . Regional stations are equipped with a 10-to-13-meter-diameter antenna. . . .[15]

Readers would expect the remainder of the paragraph to identify a series of satellite dish antennas, arranged by size, from largest to smallest.

Transitions in ascending/descending paragraphs indicate the relative priority of points in the paragraph or document. Figure 10.13 uses *descending order* to identify the priorities for testing circuits.

FIGURE 10.13 **Information Organized Using Descending Order**

Cues to the organizational pattern:
- . . . *the most important integrated circuits . . .*
- . . . *first priority . . .*
- . . . *priority parts are further separated . . .*
- . . . *All other integrated circuits are then prioritized . . .*

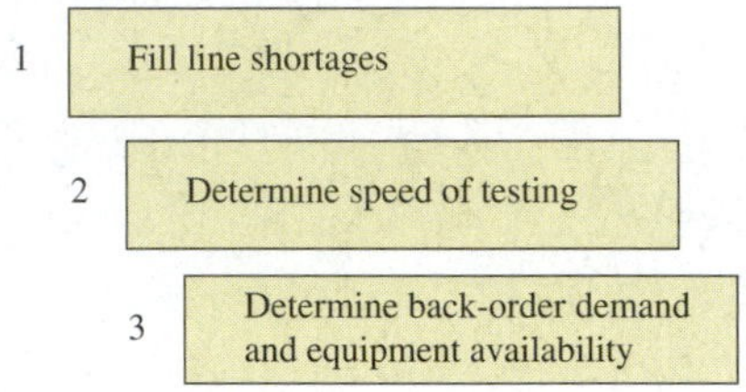

What additional examples of using ascending and descending order for organizing information are likely in the workplace?

Integrated circuits received at incoming inspection are processed according to priorities. As the following figure shows, the most important integrated circuits are those that fill line shortages on the production floor. These parts take first priority at incoming inspection and are handled according to frequency of use and critical demand. These priority parts are further separated according to how fast they can be accurately tested and sent to the production floor. All other integrated circuits are then prioritized by back-order demand and the availability of open test equipment.

INSPECTION PRIORITIES

1 Fill line shortages

2 Determine speed of testing

3 Determine back-order demand and equipment availability

Comparison/Contrast

Comparison and contrast tell readers about similarities and differences. *Comparison* identifies the similarities of various ideas, objects, or situations; *contrast,* the differences. A comparison or contrast organization could present the advantages and disadvantages of certified versus uncertified computer disks or the ease or difficulty of various methods of storing electronic data. Any of the techniques in Figure 10.14 could visually present information that you want to compare and contrast.

Readers expect comparison and contrast topic sentences to present ideas dealing with similarities and differences, or both, as illustrated by the following two sentences:

- The Honey Bee Lens, with three telescopic lenses for each eye, is patterned after the compound eye of a bee.
- Computers can now analyze measurable differences between the cries of healthy newborns and high-risk infants.

The first sentence introduces a paragraph that compares new lenses to the compound eyes of bees. The second sentence leads readers to expect the paragraph to deal with characteristics that differentiate cries of healthy and high-risk infants.

Transitions in comparison and contrast paragraphs identify various similarities and differences. Figures 10.15 and 10.16 use comparison and contrast to differentiate the responsibilities of incoming inspection and in-process inspection of integrated circuits. Figure 10.15 shifts between incoming inspection and in-

FIGURE 10.14 Visual Forms for Comparison and Contrast

Form	Example
Paired photos or drawings	Before/after of patient treated for scoliosis
Multiple or paired bar graphs	Expenditures for utilities for each quarter of the fiscal year
Multiple or paired percent graphs	Utilization of nutrients with and without coconut oil to increase absorption
Line graph	Changes in toxicity of emissions since installation of scrubbers
Multiple or paired gauges	Illustration of danger/no-danger in training manual for pilots
Table	Data collected on size of bats according to age, sex, and location
Dichotomous key	Distinction of edible wild plants
Pareto diagram	Distinction between major and minor causes of shipping delays
Histogram	Women, grouped by age, affected by lung cancer
Columned chart	Physical symptoms of substance abuse

process inspection, explaining how each deals with specific responsibilities. Figure 10.16 takes the same paragraph and rearranges it to present all the information about incoming inspection first and discuss the in-process inspection second.

Cause and Effect

The *cause-and-effect* organization of information focuses on precipitating factors and results. You can move from cause to effect or from effect to cause. For example, you could carry a CD through an electronic surveillance scanner and then trace the effects — the various disk errors that appear. Or beginning with the effect, a damaged disk, you could investigate the causes of the damage. Figure 10.17 identifies and illustrates various visuals that are effective for presenting cause-and-effect relationships.

Understanding inductive and deductive reasoning is critical in constructing effective cause-and-effect arguments. Making strong arguments involves considering the ways to organize the information and avoiding errors in logic.

One type of cause and effect — *inductive reasoning* — moves from specific instances to broad generalizations, forming the basis for the *scientific method* used in much research and experimentation. You begin by collecting data in support of

FIGURE 10.15 Information Organized Using Comparison and Contrast

Organization of information:

- *incoming inspection*
- *in-process inspection*
- *incoming inspection*
- *in-process inspection*
- *mutual goal*

Integrated circuits are inspected and/or tested by two separate quality control departments: incoming quality control and in-process quality control. Incoming quality control is responsible for ensuring that all integrated circuits sent to the production floor meet all electrical standards set by the Component Engineering Department. In-process quality control is only responsible for ensuring that the parts are properly mounted on the printed circuit board. Incoming quality control also has to verify the markings on the integrated circuits in order to do proper testing and make certain that the company has purchased a qualified product. In contrast, in-process quality control only has to do random inspections of the circuit markings to ensure that qualified parts are being used in the manufacturing process. Although incoming and in-process inspection are two different areas, they do share the same goal of building a quality product. The following lists summarize each department's responsibilities.

Component Engineering Department responsible for *incoming inspection*

- verify parts meet standards
- verify IC markings
- confirm qualified vendors

Quality Control Department responsible for *in-process inspection*

- ensure proper mounting
- do random inspections

FIGURE 10.16 Information Organized Using Comparison and Contrast

Organization of information:

- *incoming inspection*
- *in-process inspection*
- *in-coming inspection*
- *in-process inspection*
- *mutual goal*

Integrated circuits are inspected and/or tested by two separate quality control departments: incoming quality control and in-process quality control, as shown in the lists below. Incoming quality control is responsible for ensuring that all integrated circuits sent to the production floor meet all electrical standards set by the Component Engineering Department. Incoming quality control also has to verify the markings on the integrated circuits in order to do proper testing and make certain that the company has purchased a qualified product. In-process quality control is only responsible for ensuring that the parts are properly mounted on the printed circuit board. In-process quality control only has to do random inspections of the circuit markings to ensure that qualified parts are being used in the manufacturing process. Although incoming and in-process inspection are two different areas, they do share the same goal of building a quality product.

Component Engineering Department responsible for *incoming inspection*

- verify parts meet standards
- verify IC markings
- confirm qualified vendors

Quality Control Department responsible for *in-process inspection*

- ensure proper mounting
- do random inspections

What additional examples of using comparison and contrast for organizing information are likely in the workplace?

FIGURE 10.17 **Visual Forms for Cause and Effect**

Form	Example
Paired photos or drawings	Effects of two different treatments for removing facial birthmarks
Weather map	Impact of cold front on majority of Midwest
Bar graph	Efficiency of various methods for harvesting cranberries
Line graph	Destruction of American chestnut by blight during the 1900s
Cause-and-effect diagram	Identification of multiple contributing factors to contamination of drinking water
Pareto diagram	Identification of major causes of low birth weight

an unproved hypothesis. After you have organized and examined a sufficient body of data, you draw a conclusion. When your conclusion proves consistently to be valid, it is considered a generalization. Most scientific principles and theories are based on this method of inquiry.

Because no way exists to test every instance, inductive reasoning has a certain risk, and professionals must be careful to avoid basing their conclusions on invalid assumptions. Two kinds of errors will put your cause-and-effect thinking at risk: You cannot assume chronology is the same as causality, and you need to examine a large sample before drawing a conclusion.

The first problem equates chronology with causality. Just because B follows A does not mean that A causes B. Because inductive reasoning moves from specifics to a generalization, an investigator should not assume that sequence of events alone causes the effect. (Such an error in reasoning is called *post hoc, ergo propter hoc,* Latin for "after this, therefore because of this.") For example, a donor may become ill the day after donating blood to the Red Cross, but she cannot logically conclude that donating blood caused her to become ill. Guard against fallacious reasoning by examining all possible causes. A closely related problem is using *non sequiturs,* or arguments that begin with an unwarranted assumption, so its inference isn't correct.

The second problem deals with sample size. You need to examine a large enough number of instances before drawing a conclusion. For example, before a new drug is allowed on the market, the Food and Drug Administration requires extensive tests with a broad segment of the target population. As a result of testing, a powerful painkiller such as propoxyphene, when taken according to directions and under a physician's care, is certified as safe even though all propoxyphene tablets and capsules have not been individually tested. Unfortunately, errors occur, although rarely, causing some people to suspect all inductive reasoning. So make sure your methodology is sound, your sample large, and your analysis free from bias or distortion.

When a generalization is widely accepted, you can use it as a base from which to predict the likelihood of specific instances occurring. This process is called *deductive reasoning* — moving from general premises to specific causes. A patient taking propoxyphene trusts that the pills are safe even though those particular ones have not been tested.

In paragraphs that use cause and effect as a way to organize information, both the cause and the effect should be identified in the topic sentence, as seen in the next examples:

- One hypothesis about the formation of mineral-rich marine nodules suggests that marine bacteria break down organic material into free-floating minerals that eventually collect to form nodules.
- The Zephinie Escape Chute (ZEC) can rapidly evacuate people from 10-story burning buildings because of its unique construction.

Readers of the first sentence expect the information in the paragraph to explain how minerals form marine nodules. The second sentence develops a paragraph that explains how the ZEC's unique construction aids rapid evacuation.

Transitions in cause-and-effect paragraphs signal the relationship between an action and its result. The example in Figure 10.18 uses cause and effect to organize information about integrated circuit (IC) testing. In this example, the cause-and-effect transitions indicate the descending order of the reasons an IC chip can be rejected.

FIGURE 10.18 **Information Organized Using Cause and Effect**

Causal transitions:
- *because*
- *therefore*
- *so*

Reasons for rejection of IC chips during incoming inspection fall into three categories (see figure). Most often, rejections occur because of some flaw in the chip itself. For example, a chip may have a short in the circuitry or fail to perform at the specified voltage or current. A second reason for rejection occurs when the supplier sends the wrong parts or a mixed batch of parts; therefore, assembly is delayed and production schedules slip. The final reason, which happens infrequently, occurs when the automatic test equipment has the wrong program or a program with a bug, so the contacts for the electrical testing are misplaced.

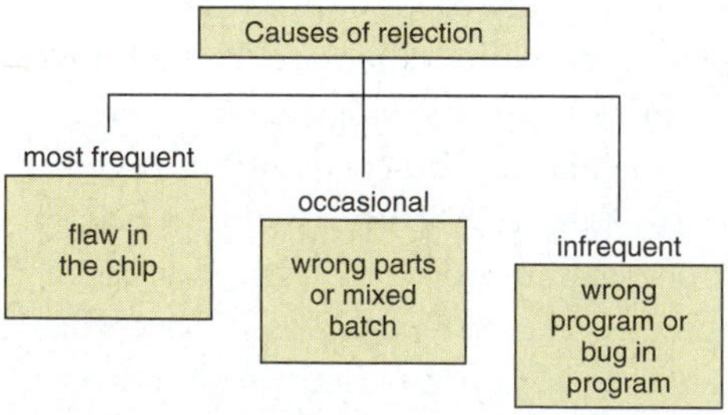

What additional examples of using cause and effect for organizing information are likely in the workplace?

These ways of organizing information are frequently used in technical documents, both as short, self-contained segments (like the examples about integrated circuits) and also combined in longer pieces of writing.

By organizing information so that it meets the needs of the content, purpose, and audience, you can make a paragraph or document more understandable, as well as overcome some of the noise that interferes with readers' acceptance or comprehension of information. For example, processes, procedures, and directions are best organized chronologically for all audiences. Descriptions of physical objects, mechanisms, organisms, and locations frequently make the most sense if the information is organized spatially. Reasons and explanations are usually presented in descending order so that the audience reads the most important information first. Explanations of problems and their solutions often make the most sense if the information is organized using comparison/contrast and cause and effect.

You can also use organization of information to adapt your material to readers' attitudes by taking advantage of what you know about induction (specific to general) and deduction (general to specific). If you think readers might be reluctant to accept your conclusions or recommendations, you can organize the material inductively, moving from various specifics to your conclusion. Thus they can follow your line of reasoning and, perhaps, be persuaded by your analysis. If you think readers will agree with your conclusions, you can organize the information deductively, presenting the conclusion initially and then following it with the specifics that led to it. Deductive organization is more common in technical documents.

Sometimes decisions about organizing information aren't obvious or easy. The ethics sidebar on this page focuses attention on unethical design choices, ones that distort information and thus mislead readers because of the way information is organized. Decision making might be easier if workplace professionals could agree about what is unethical, but, as you'll read in the sidebar, opinions are divided.

ETHICS SIDEBAR

"Is This Ethical?": Ethics and Document Design

Consider the following scenarios:

Scenario 1: "You have been asked to evaluate a subordinate for possible promotion. In order to emphasize the employee's qualifications, you display these in a bulleted list. In order to de-emphasize the employee's deficiencies, you display these in a paragraph. Is this ethical?"

Scenario 2: "You are designing materials for your company's newest product. Included is a detailed explanation of the product's limited warranty. In order to emphasize that the product carries a warranty, you display the word 'Warranty' in a large size of type, in upper and lower case letters, making the word as visible and readable as possible. In order to de-emphasize the details of the warranty, you display this information in smaller type and in all capital letters, making it more difficult to read and thus more likely to be skipped. Is this ethical?"[16]

Are these ethical or unethical document design choices? If you are uncertain, then you are not alone. Technical professionals and teachers, responding to these same scenarios (and others) in a survey, revealed differing opinions about the ethics involved. A majority of respondents (54.6%) considered the design choices of the evaluation ethical, although a significant number were uncertain (26.7%) or found the evaluation unethical (18.7%). The warranty divided opinions even more: ethical (33.3%), unethical (44.1%), uncertain (22.6%).

> *Follow Dragga's advice for self-examination by reviewing the scenarios. Is the reader deceived by the different formats in the evaluation? Does the difference in font size in the text of the warranty increase the likelihood of deception for the reader?*
>
> *What can help you decide whether a document's design is misleading?*

Researcher Sam Dragga, who coordinated the survey, believes it indicates the thin line separating rhetorically savvy design from deceptive design. He did find, however, that the survey revealed a general guideline most technical professionals follow for producing ethical document design: "The greater the likelihood of deception and the greater the injury to the reader as a consequence of that deception, the more unethical is the design of the document."[17] Determining the degree of deception and injury requires a technical communicator to weigh different items, "including typical communication practices, professional responsibilities, explicit specifications and regulations, as well as rhetorical intentions and ideals."[18] Dragga recognizes that workplace pressures can sometimes lead us to choose convenience over ethics. To avoid making unethical document design choices, Dragga advises "periodic self-examination" for technical professionals to help them create ethical communications.

Individual and Collaborative Assignments

1. **Identify expectations based on topic sentences.** Read the following topic sentences and identify expectations that readers might have if they read them. What content would readers expect in each paragraph? One example is provided.

   Example The use of a sulfur-asphalt mixture for repaving the highway will result in several specific benefits.

   Analysis The sentence presents a cause-and-effect relationship between using a sulfur-asphalt mixture and specific benefits. The paragraph will identify these benefits.

   a. The CAT scan creates an image resembling a "slice" that clearly visualizes anatomical structures within the body.

b. The human fetus is in the birth position by the ninth month of a normal pregnancy.
c. The routine use of drugs in labor and delivery sometimes has adverse effects on otherwise healthy, normal infants.
d. The stages of normal labor and delivery begin at term when the fetus reaches maturity and end with the expulsion of the placenta.
e. Two major forms of leukemia — chronic myelocytic and acute myelocytic — have distinct differences.
f. Improper downstroke and follow-through can cause the golf ball to either hook and fade to the left or slice and fade to the right.
g. Three main types of parachutes are used for skydiving. The most widely used parachute has a round, domelike canopy.
h. A square parachute provides more maneuverability and a better overall ride than does a conventional round parachute.
i. Two methods of disinfecting treated wastewater are chlorination and ozonation.
j. The chlorinator room contains the evaporators, chlorinators, and injectors, three of each. The evaporators are used only when liquid chlorine is being drawn from the containers.

2. **Revise a memo by improving organization.** Read the memo below. Revise it so that the subject line in the heading, the topic sentences, and the paragraphs are well organized, unified, and coherent. The intended readers are interested in information that will help decrease rejects during manufacturing. They are not particularly concerned with personnel or cost.

Stanford Engineering, Inc.

February 14, 20—

To Quality Control Supervision
From T. R. Hood, Engineer TRH
Subject Increased scrap and customer rejects

The Engineering Department is recommending the purchase of an International glue line inspection system to strengthen standard visual inspection. Machine operators will not be slowed down by the addition of this new glue line inspection system. The system will detect breaks in the glue line and eject a carton from the run before it reaches shipping. If the system detects more than five consecutive rejects, it will automatically stop the machine.

Purchasing this International system is a better solution than purchasing a new glue pot for $6,500. The savings in purchasing the $4,000 International system will allow us to rebuild the existing glue pot.

3. **Revise paragraphs by adding topic sentences.** Read the two paragraphs below and write topic sentences that help readers anticipate the content and organization.

   Paragraph 1 Buss bars, the smallest of the parts the Sheet Metal Fabrication Shop produces, are made of grade-A copper and are tin-plated before being used for internal grounding. Paper deflectors, used in printers, are made of stainless steel and do not require any plating or painting. Deflectors guide the paper through the printer and usually measure 4″ in width and 15″ in length, depending on the size of the printer. A larger box-like structure, made of aluminum and requiring plating in the enclosure chassis, is designed to hold a variety of electronic devices within an even larger computer main frame. A steel door panel requiring cosmetic plating and painting is the largest of the parts produced by the Sheet Metal Fabrication Shop.

   Paragraph 2 The fetal causes for spontaneous abortion are infectious agents: protozoa bacteria, viruses, particularly rubella virus. Drugs such as thalidomide cause fetal abnormalities. When radiation is given in therapeutic doses to the mother in the first few months of pregnancy, malformation or death of the fetus may result.

4. **Characterize and evaluate the organization of a Web site.** Work with a small group of people in your major or a closely related field of study.

   (a) Visit several Web sites that are related to your field of study. How would you characterize the organizational strategies used: whole/parts, chronological, spatial, ascending/descending, comparison/contrast, or cause and effect?

   (b) Establish criteria for evaluating the effectiveness of Web site organization based on your responses to Assignment 4(a).

   (c) Create a rubric for these criteria, and evaluate one Web site from your list.

   (d) Share your rubric and evaluation with your classmates.

5. Tour and evaluate the organization of a Web site. Visit one of the following Web sites:

   www.ashastd.org/
   www.hhs.gov/
   www.ostrich.ca/index-2.htm
   www.hc-sc.gc.ca/pphb-dgspsp/sars-sras/index.html
   www.centerforfoodsafety.org/
   www.avert.org/condoms.htm
   www.ercim.org/
   www.epa.gov/

   Follow the links to learn as much as you can about the Web site's audience and purpose. Pay special attention to the overall organization of the site — that is, the site architecture. Write a one-page review for your instructor about the comprehensibility and usability of the site's organization, given the audience and purpose.

Chapter 10 Endnotes

1 Jeffries, W. L. (2003). *Informal settlements* and *Informal settlements in context.* © William L. Jeffries. Durban, South Africa.

2 Allee, V. (n.d.). A delightful dozen principles of knowledge management. In *The knowledge evolution: Expanding organizational intelligence.* Retrieved November 8, 2003, from www.vernaalee.com/library%20articles/A%20Delightful%20Dozen%20Principles%20of%20Knowledge%20Management.pdf

3 Modified from Jordon, N. (1984). What's in a zombie. [Review of the book *The serpent and the rainbow*]. *Psychology Today,* (May), 6.

4 H. W. Wilson Co. (2003, January). Wade David. In *Current biography.* Retrieved November 8, 2003, from http://www.hwwilson.com/currentbio/cover_bios/cover_bio_1_03.htm

5 *Storyboarding.* (n.d.). Retrieved November 8, 2003, from http://members.ozemail.com.au/~caveman/Creative/Techniques/storyboard.htm

6 HM Customs and Excise. (2002, January 25). *Business change lifecycle — storyboarding.* Retrieved October 18, 2003, from http://www.ogc.gov.uk/sdtkdev/examples/HMCE/Guidance/RequirementsManagement/Storyboarding.htm

7 U.S. Food and Drug Administration. (n.d.). *Improving innovation in medical technology: Beyond 2002.* Retrieved November 7, 2003, from http://www.fda.gov/bbs/topics/news/2003/beyond2002/report.html

8 Arens, W. F. (n.d.). Glossary. In *Contemporary advertising* (8th ed.). Retrieved November 7, 2003, from http://highered.mcgraw-hill.com/sites/0072415444/student_view0/glossary.html

9 *Storyboarding.* Retrieved November 8, 2003, from CalPolyWeb Authoring Resource Center Web site: http://www.calpoly.edu/warc/plannning/layout/storyboarding.html

10 HM Customs and Excise. (2002, January 25). *Business change lifecycle — storyboarding.* Retrieved October 18, 2003, from http://www.ogc.gov.uk/sdtkdev/examples/HMCE/Guidance/RequirementsManagement/Storyboarding.htm

11 The discussion expands the examples provided by Shedroff, N. (1994). Information interaction design: A unified field theory of design. Retrieved November 8, 2003, from http://www.nathan.com/thoughts/unified/index.html

12 1001homepages.com (n.d.). CGI. In *Glossary of terms.* Retrieved November 9, 2003, from http://1001resources.com/hosting/glossary.htm1#C

13 Boutell, T. (2003). *CGIC 2.02: An ANSI C library for CGI programming.* Retrieved November 9, 2003, from http://www.boutell.com/cgic/

14 Nielsen, J. (1998). Failure of corporate websites. *Jakob Nielsen's Alertbox for October 18, 1998.* Retrieved November 9, 2003, from http://wwwuseit.com/alertbox/981018.html

15 Modified from Lenorovitz, J. M. (1983, November 21). China plans upgraded satellite network. *Aviation Week & Space Technology, 65,* 71–75.

16 Dragga, S. (1996). 'Is this ethical?': A survey of opinion on principles and practices of document design. *Technical Communication, 43*(1), 29–38, p. 30.

17 Dragga, S. (1996), p. 35.

18 Dragga, S. (1996), p. 35.

CHAPTER 11

Designing Information

Outcomes and Objectives

This chapter will help you accomplish these outcomes:

- Understand the relationship between the design of information and the critical goals of accessibility/legibility, comprehensibility/readability, and usability
- Use chunking and labeling to group topically related information
- Arrange visual and verbal information using a page grid
- Emphasize information by appropriate use of typographic and design elements such as lists, boxes, and color

Regardless of your technical specialty, from aviation to biology to computers, from mechanical engineering to nursing to oceanography, from writing to xerography to zymurgy, part of your job will involve the design of information in both print and electronic documents, for both presentations and Web sites. Your knowledge of ways to effectively combine visual and verbal elements to communicate to audiences may distinguish you from other professionals. And if you are lucky enough to have designers and graphic artists to work with, you need to be able to communicate your vision of a document to them.

Whether information is in print form or electronic form, a number of general principles about design apply. This chapter presents principles that affect accessibility/legibility, comprehensibility/readability, and usability for paper documents and also notes variations and adjustments that are often necessary when technical information is presented in other ways, such as PowerPoint presentations, technical and scientific posters, and Web sites.

Information design is concerned with the ways in which you organize and present information to increase audience comprehension. *Document design,* a term you're probably more familiar with, is part of information design. As you design the information in your documents, you manage five categories of elements:[1]

> *Take a quick look at this page and the facing page. Identify the textual, spatial, graphic, color/textural, and dynamic design elements. Then open some other book, journal, or Web site that's close at hand and identify the design elements.*

- ***Textual elements*** — letters, numbers, and symbols (for example, the characters that form the words in a document, as well as headings, labels, and page numbers)
- ***Spatial elements*** — the spaces between elements (for example, the spacing between letters, sentences, paragraphs, and margins, sometimes called *white space* or *negative space*), as well as the size and placement of textual and graphic elements
- ***Graphic elements*** — punctuation marks, typographic devices (for example, bullets and icons), geometric forms (for example, lines and arrows; boxes on flowcharts), and visual images (for example, tables, graphs, diagrams, drawings, photographs, maps)
- ***Color and textural elements*** — the hue (what the color is), saturation (how pure the color is; the amount of color), and value/brightness/luminescence (how bright the color is; the color's lightness or darkness), which are called HSV, HSB, or HSL color model; texture is the tactile or virtual features of a surface (for example, smoothness, glossiness, hardness, and so on)
- ***Dynamic elements*** — the motion that is implied in a print document (by, for example, layout, typography, placement of images); often actual motion in an electronic document that uses various kinds of animation

For a better understanding of hue, saturation, and value in color, go to www.english.wadsworth.com/burnett6e for a link to useful examples.
CLICK ON WEBLINK
CLICK on Chapter 11/HSV

WEBLINK

These five categories of design elements are important whether you're designing print documents or Web pages. Regardless of how carefully you design information, though, reading electronic documents is about 25 percent slower than reading from paper. Furthermore, many people simply do not like to read extended chunks of text on a computer screen. And, finally, many people don't like to scroll, so brief electronic pages are likely to get more attention. To compensate for these reading preferences, when you're designing an electronic document, you should consider writing "50 percent less text and not just 25 percent less since it's not only a matter of reading speed but also a matter of feeling good."[2]

This chapter begins by illustrating basic principles of information design in an excerpt from an online report and then discussing strategies of information design that use the following principles for managing text, space, graphic elements, color, and movement in both print and electronic documents:

- Chunking and labeling information by effective use of white space and headings
- Arranging information by appropriate integration of visual and verbal chunks
- Emphasizing information by effective use of typographic devices and typefaces

Chunking and Labeling Information

You can apply the design principles in your print document or Web site by grouping, or chunking, topically related information and then labeling that information for the audiences. In simple terms, chunking makes the information in undifferentiated text accessible to audiences. Decisions about chunking information involve two factors:

- Logical topical relationships
- Audience needs for the information

PRINCIPLES OF DESIGN IN ACTION

You can use a number of widely accepted principles that enable designers to effectively integrate text, space, graphics, color, and implied or actual movement in print and electronic documents. The example here applies the most critical of these principles in a typical document — in this case, the beginning of an online technical report about tropical deforestation[3] — to show that design principles apply even in everyday print and online documents.

Direction. *Where should people begin reading? In what direction should they move? People begin a document where the design directs them; the upper left part of the page or screen contains the most important information — the title and the primary text. The entry is prompted by column width and placement, typography, and visual emphasis.*

Proportion/Scale. *How big are elements of the design in relation to the audience? In relation to each other? Scale helps the audience understand how big various elements are in relation to human size. Proportion helps the audience understand the relationship between the various elements of the design.[4] The photo lets readers know the scale of the forest.*

Contrast. *What creates contrast — size, color, shape, or placement? Is the contrast obvious — even dramatic? The primary contrast here is provided by differences in size: the wider column of the primary text contrasts with the narrower column of the supplemental text, not only letting the audience know which is more important but also creating visual variety.*

Rhythm. *How dynamic is the design? Do the elements have a pace or rhythm? The elements of a design — text, space, graphics, color, and implied or actual movement — are repeated to help the audience move through the design and better understand and interpret information. Selected elements are usually repeated. When people access the information online, the right-hand information is repeated on every screen, so the links are always accessible.*

Balance. *How are the various elements in the design balanced?*

- *Symmetrical — each half of the design (horizontal or vertical) is a mirror image*
- *Asymmetrical — design has uneven number of elements and/or elements that are not placed evenly*
- *Radial — design moves out from the center (or near the center) of the page or screen*

Most designers envision horizontal and vertical grids for a design. The example uses a three-column grid, but the site looks like it has two asymmetrical columns. Why? The primary text takes up two columns in the grid (making it appear as a single, wider column). The third column serves as a site menu, so it has a large amount of white space that balances the text.

Alignment. *How do various chunks of the text line up with each other? How does the text line up with the visuals? Text can align in columns on a page or screen:*

- *Left aligned text (also called flush left, left justified, or ragged right) lines up evenly on the left.*
- *Right aligned text (also called flush right, right justified, or ragged left) lines up evenly on the right.*
- *Centered text has equal white space on both sides.*
- *Justified text lines up evenly on both sides.*

The primary text in this example has left justified/ragged right margins, which is easy for readers.

You can read the full report about tropical deforestation by going to www.english.wadsworth.com/burnett6e for a link.
CLICK ON WEBLINK
CLICK on Chapter 11/tropical deforestation

WEBLINK

earth observatory

home • data & images • features • news • reference • missions • experiments • search

glossary on ○ off ◉

REFERENCE

TROPICAL DEFORESTATION

By Gerald Urquhart, Walter Chomentowski, David Skole, and Chris Barber

The clearing of tropical forests across the Earth has been occurring on a large scale basis for many centuries. This process, known as deforestation, involves the cutting down, burning, and damaging of forests. The loss of tropical rain forest is more profound than merely destruction of beautiful areas. If the current rate of deforestation continues, the world's rain forests will vanish within 100 years-causing unknown effects on global climate and eliminating the majority of plant and animal species on the planet.

Why Deforestation Happens

Deforestation occurs in many ways. Most of the clearing is done for agricultural purposes-grazing cattle, planting crops. Poor farmers chop down a small area (typically a few acres) and burn the tree trunks-a process called Slash and Burn agriculture. Intensive, or modern, agriculture occurs on a much larger scale, sometimes deforesting several square miles at a time. Large cattle pastures often replace rain forest to grow beef for the world market.

Commercial logging is another common form of deforestation, cutting trees for sale as timber or pulp. Logging can occur selectively-where only the economically valuable species are cut-or by clearcutting, where all the trees are cut. Commercial logging uses heavy machinery, such as bulldozers, road graders, and log skidders, to remove cut trees and build roads, which is just as damaging to a forest overall as the chainsaws are to the individual trees.

The causes of deforestation are very complex. A competitive global economy drives the need for money in economically challenged tropical countries. At the national level, governments sell logging concessions to raise money for projects, to pay international debt, or to develop industry. For example, Brazil had an international debt of $159 billion in 1995, on which it must make payments each year. The logging companies seek to harvest the forest and make profit from the sales of pulp and valuable hardwoods such as mahogany.

"THE CAUSES OF DEFORESTATION ARE VERY COMPLEX."

Emphasis. *What's the focus? What draws audience attention? What's important? The hierarchy visually differentiates the title and headings from the primary text by placement (set off from text) and by typestyle (small caps and boldfacing).*

Gestalt. *How do all the elements work together? This report has a coherent gestalt (also called unity, consistency, variety, harmony, wholeness, oneness). All the elements look as if they belong together in their current location and sequence.*

Proximity. *Where are the elements — text, space, graphics, color, and implied or actual movement — placed? How are elements grouped? The primary text is placed close to the heading and photograph. The supplementary information is grouped in a narrower column.*

© Tui De Roy/Minden Pictures

Hornbills are an exotic bird found in Africa and Asia. The female nests in a tree cavity or rock crevice and then seals the nest entrance, leaving only a narrow slit for receiving food from the male.

Figure 11.1a–c shows the sequence of chunking and labeling information. The first chunk, Figure 11.1a, includes information about several topics related to hornbills. Unfortunately, the information is in no particular order and is without any design elements to aid audiences.

The second chunk, Figure 11.1b, groups all the information about the various subtopics and then visually separates these topical chunks into paragraphs. The third chunk, Figure 11.1c, labels the subtopics and uses a bulleted list to make reading easier.

Decisions about the most appropriate way to chunk and label information are not necessarily as simple as grouping topically related information, as in Figures 11.1a–c. For example, if you are preparing a fact sheet about the rivets your company sells, you could chunk the information by such characteristics as types of materials, strength of materials, applications, dimensions, resistance to corrosion, distinctions from competitors, price, and so on. Information from all these categories could appear on the fact sheet, but you could make a decision about how to chunk the information based on the needs of your audience. For example, the top list on the right chunks information about rivet material according to type; the second list chunks information about material according to strength.

What other information could you chunk in two (or more) different ways, similar to these lists categorizing rivets?

Types of Rivet Materials
Steel
Aluminum
Brass
Plastic

This topical chunking would be useful for audiences interested in marine applications, where anticorrosion is a critical factor.

Strength of Rivet Materials
Tensile strength
Compressive
Flexure
Torsion
Shear

This topical chunking would be useful for audiences interested in bridge construction, where strength is a critical factor.

FIGURE 11.1 **Sequence from (a) Undifferentiated Text to (b) Chunked Text to (c) Labeled Text[5]**

(a) Twenty-three species of hornbills inhabit the savannas and forests of Africa, and twenty-seven more species are found in Asia, mostly in tropical rain forests. The Tangkoko red knobs are a species bothered by few dangerous predators; their unusual nesting behavior seems a holdover from earlier times. The nesting behavior of the Tangkoko red knobs is thought to have evolved from one or more of these reasons: protection from predators; defense against intruding hornbills; ensurance of mate fidelity (female can't escape to the nest of another; or, variously, male is too exhausted feeding one female and chick to seek an additional mate). Tangkoko red knobs weigh more than five pounds and produce a variety of honks, croaks, squawks, and barks, some of which can be heard more than 300 yards away. The female Tangkoko red knob nests in a tree cavity or rock crevice and then seals the nest entrance — with mud delivered by the male and her own fecal matter — leaving only a narrow slit for receiving food from the male. One of the Borneo species of hornbills, the Tangkoko red knobs, is large and loud. Only four species are found to the east of the biogeographical boundary between Borneo and Sulawesi.

The information about hornbills is raw, run-together text. It is undifferentiated, except for capital letters and periods to mark the beginning and end of sentences. The information is unordered and is in a sans serif font, without any spatial or graphic elements to aid audiences.

(b) Twenty-three species of hornbills inhabit the savannas and forests of Africa, and twenty-seven more species are found in Asia; mostly in tropical rain forests. Only four species are found to the east of the biogeographical boundary between Borneo and Sulawesi.

One of the Borneo species of hornbills, the Tangkoko red knobs, is large and loud. They weigh more than five pounds and produce a variety of honks, croaks, squawks, and barks, some of which can be heard more than 300 yards away.

The female nests in a tree cavity or rock crevice and then seals the nest entrance — with mud delivered by the male and her own fecal matter — leaving only a narrow slit for receiving food from the male. This behavior is thought to have evolved from one or more of these reasons: protection from predators; defense against intruding hornbills; ensurance of mate fidelity (female can't escape to the nest of another; or, variously, male is too exhausted feeding one female and chick to seek an additional mate). However, the Tangkoko red knobs are bothered by few dangerous predators, so their unusual nesting behavior seems a holdover from earlier times.

All the information about the three subtopics, now in a serif font, is topically chunked and then visually separated into paragraphs with a ragged-right margin.

(c) **Hornbill Territory.** Twenty-three species of hornbills inhabit the savannas and forests of Africa, and twenty-seven more species are found in Asia, mostly in tropical rain forests. Only four species are found to the east of the biogeographical boundary between Borneo and Sulawesi.

Tangkoko Red Knobs. One of the Borneo species of hornbills, the Tangkoko red knobs, is large and loud. These birds weigh more than five pounds and produce a variety of honks, croaks, squawks, and barks, some of which can be heard more than 300 yards away.

Unusual Nesting Behavior. The female nests in a tree cavity or rock crevice and then seals the nest entrance — with mud delivered by the male and her own fecal matter — leaving only a narrow slit for receiving food from the male. This behavior is thought to have evolved from one or more of these reasons:

- protection from predators
- defense against intruding hornbills
- ensurance of mate fidelity (female can't escape to the nest of another; or, variously, male is too exhausted feeding one female and chick to seek an additional mate)

However, the Tangkoko red knobs are bothered by few dangerous predators, so their unusual nesting behavior seems a holdover from earlier times.

The subtopics in each paragraph are labeled with run-in, boldfaced headings. A bulleted list makes reading easier.

FIGURE 11.2 **Clearly Chunked Table of Contents**[6]

What principles of design do you see at work in this home page?

Not only do you need to chunk small amounts of information (like categories of rivets), but you also need to show the overall organization of information. One of the most efficient ways to inform your audience about the way you've chunked information is in a table of contents. Figure 11.2 shows a clearly chunked home page for Nikon, in which every bulleted item is a link.

White Space to Chunk Information

Once you have determined the chunks of information, you can separate these chunks by using white space, and then you can label them with headings. *White space* (or *negative space*) is the part of any page or screen that is blank, without print or visuals. Not only does white space signal chunks of information, it also makes documents more appealing.

No hard and fast rules exist concerning the use of white space. Several conventions, which apply both to print pages and electronic pages, suggest that white space be used for margins, between lines within a paragraph (called *leading* or *line spacing*), between paragraphs and sections of a document, and around visuals. The amount of white space is also affected by the justification and length of the lines of text. The impact of white space on a page is illustrated by the two segments in Figure 11.3a and 11.3b — the first with crowded printing and skimpy margins, the second with wider margins and more space between lines.

FIGURE 11.3A **Skimpy White Space**

This paragraph is more difficult to read because little attention is given to physical features of the presentation. Narrow leading and full justification create greater eye strain. The small margins give an impression of crowding. The proportion of text is too great for the amount of space.

FIGURE 11.3B **Appropriate White Space**

This paragraph is easier to read because the physical features consider the reader. The material has sufficient space between the lines and is surrounded by margins. The white space and text are balanced.

Margins. The four margins on a print page, which usually have different widths, are also used to chunk information. The top (head) margin is the narrowest. The inner margins (gutters) are wider to ensure that no words are lost in the binding; the outer margin is wider still. Type that runs nearly to the edge of the paper not only is unattractive but also leads readers' eyes off the page. Sometimes, outside margins are even wider to provide space for note taking or a running gloss, which are marginal notes that emphasize particular points. The widest margin is usually at the bottom of the page. The thumbnail sketches in Figure 11.4 show the flexibility you have in adjusting margins: the typical 1-inch margins in version a, the wide equal margins in version b, the extra-wide left margin in version c.

FIGURE 11.4 **Thumbnail Sketches Showing Changes in Margins**

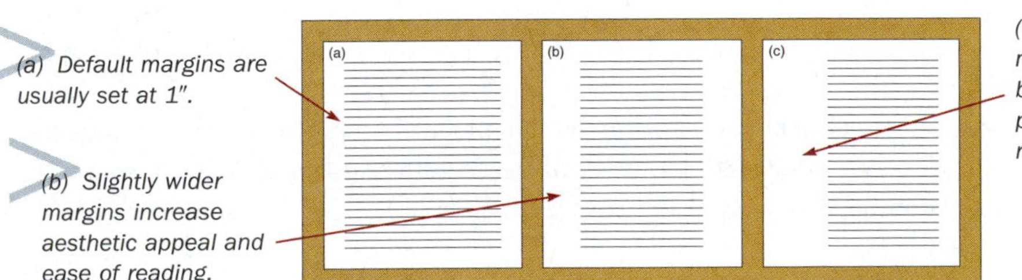
(a) Default margins are usually set at 1".
(b) Slightly wider margins increase aesthetic appeal and ease of reading.
(a)
(b)
(c)
(c) An extra-wide margin can be left blank, or it can be a place for annotations or references.

Explain which you think is more important: a neat, clean appearance or an easy-to-read document.

Alignment. Another element that affects chunking and, thus, accessibility/legibility is alignment. If all the lines of type on a page are exactly the same length, the lines have been justified, or adjusted to equal length by proportional spacing between words on each individual line. Fully justified lines give a document a neat, clean appearance, as Figure 11.5a illustrates. If the lines all begin at the same left-hand margin and the right margin is ragged, the right margin is unjustified, as Figure 11.5b shows.

FIGURE 11.5 **Thumbnail Sketches Showing Alignment**

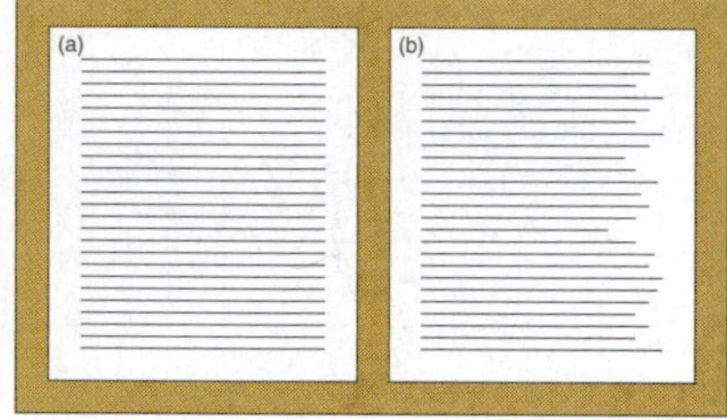

(a) Fully justified text has a formal, neat look, but reading many pages of fully justified text is difficult, especially if the print is small and the lines are long.

(b) Most ragged-right text (justified only on the left) is easier to read than fully justified text.

Although some editors and designers prefer that both left and right margins be justified, both because of tradition and because more words will fit on a page or screen, fully justified text is not as easy to read as text with a ragged right margin. Because all the lines in a fully justified text are the same length, readers can easily lose their place — and readers lose their place even more easily in on-screen reading than in paper reading. In addition, a long-cited survey in the journal *Technical Communication* indicates that a majority of both managers and nonmanagers prefer documents with ragged-right margins.[7]

Generally, readers find texts with shorter line lengths and ragged-right margins easier to read; however, these generalizations need to be applied with an awareness of the type of document you're producing. For example, a formal corporate annual report might have one column that is fully justified, whereas a monthly newsletter for the same company might use a two-column format with ragged-right margins.

Alignment can also apply to elements beyond lines of text in a paragraph. The example in Figure 11.6a, the poster for the Human Genome Project, is a masterpiece of data compression and alignment in its list of selected genes, traits, and disorders associated with each of the 24 different human chromosomes. Even in a thumbnail version of the poster, which is 24″ × 36″ in its original size,

FIGURE 11.6 **(a) Thumbnail of Human Genome Landmarks Poster; (b) Human Chromosome 5 in Detail[8]**

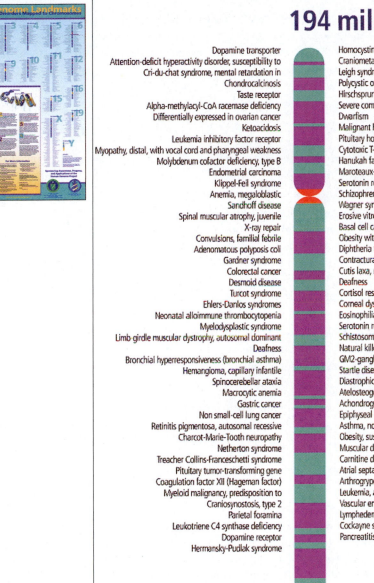

A

194 million bases

5

Dopamine transporter
Attention-deficit hyperactivity disorder, susceptibility to
Cri-du-chat syndrome, mental retardation in
Chondrocalcinosis
Taste receptor
Alpha-methylacyl-CoA racemase deficiency
Differentially expressed in ovarian cancer
Ketoacidosis
Leukemia inhibitory factor receptor
Myopathy, distal, with vocal cord and pharyngeal weakness
Molybdenum cofactor deficiency, type B
Endometrial carcinoma
Klippel-Feil syndrome
Anemia, megaloblastic
Sandhoff disease
Spinal muscular atrophy, juvenile
X-ray repair
Convulsions, familial febrile
Adenomatous polyposis coli
Gardner syndrome
Colorectal cancer
Desmoid disease
Turcot syndrome
Ehlers-Danlos syndromes
Neonatal alloimmune thrombocytopenia
Myelodysplastic syndrome
Limb-girdle muscular dystrophy, autosomal dominant
Deafness
Bronchial hyperresponsiveness (bronchial asthma)
Hemangioma, capillary infantile
Spinocerebellar ataxia
Macrocytic anemia
Gastric cancer
Non small-cell lung cancer
Retinitis pigmentosa, autosomal recessive
Charcot-Marie-Tooth neuropathy
Netherton syndrome
Treacher Collins-Franceschetti syndrome
Pituitary tumor-transforming gene
Coagulation factor XII (Hageman factor)
Myeloid malignancy, predisposition to
Craniosynostosis, type 2
Parietal foramina
Leukotriene C4 synthase deficiency
Dopamine receptor
Hermansky-Pudlak syndrome

Homocystinuria-megaloblastic anemia, cbl E type
Craniometaphyseal dysplasia
Leigh syndrome
Polycystic ovary syndrome
Hirschsprung disease
Severe combined immunodeficiency
Dwarfism
Malignant hyperthermia susceptibility
Pituitary hormone deficiency
Cytotoxic T-lymphocyte-associated serine esterase
Hanukah factor serine protease
Maroteaux-Lamy syndrome
Serotonin receptor
Schizophrenia susceptibility locus
Wagner syndrome
Erosive vitreoretinopathy
Basal cell carcinoma
Obesity with impaired prohormone processing
Diphtheria toxin receptor
Contractural arachnodactyly, congenital
Cutis laxa, recessive, type I
Deafness
Cortisol resistance
Corneal dystrophy
Eosinophilia, familial
Serotonin receptor
Schistosoma mansoni infection, susceptibility/resistance to
Natural killer cell stimulatory factor-2
GM2-gangliosidosis, AB variant
Startle disease, autosomal dominant and recessive
Diastrophic dysplasia
Atelosteogenesis
Achondrogenesis
Epiphyseal dysplasia, multiple
Asthma, nocturnal, susceptibility to
Obesity, susceptibility to
Muscular dystrophy, limb-girdle, type 2F
Carnitine deficiency, systemic primary
Atrial septal defect with atrioventricular conduction defects
Arthrogryposis multiplex congenital, neurogenic
Leukemia, acute promyelocytic, NPM/RARA type
Vascular endothelial growth factor receptor
Lymphedema, hereditary
Cockayne syndrome
Pancreatitis, hereditary

B

the alignment of the chromosomes is remarkably clear. Figure 11.6b shows chromosome 5 and a number of the genes, traits, and disorders associated with it.

WEBLINK

Go to www.english.wadsworth.com/burnett6e to find a link to the online chromosome viewer, which gives you an opportunity to click on any chromosome and see selected genes, traits, and genetic disorders associated with each chromosome.

CLICK ON WEBLINK

CLICK on Chapter 11/chromosome

> *How do leading and line length affect accessibility of information for users of online documents?*

Leading and Line Length. The spacing between lines of type (called *leading* in paper documents; more often called *line spacing* in Web documents) is another way of chunking information that improves accessibility/legibility and, thus, increases ease and speed of reading. Generally, text that is easiest to read has line spacing that is one-and-a-half times the letter height. (See Chapter 3 and Chapter 13 for additional discussion about reading electronic documents.)

Lines that are too short annoy readers; lines that are too long are difficult to read. But "short" and "long" are relative, related to font type and size rather than to absolute line length. In general, font size larger than 12 points has a "primer" look and reminds people of their elementary school reading. (Exceptions to this, of course, are materials for readers who are visually impaired.) And a font size smaller than 7 points is usually too tiny to read easily. However, changes in leading and adjustments in justification and line length can have a great impact on ease of reading, as the following two examples show:

8-point Geneva, 8-point leading, full justification, and long lines

The whole world ocean extends over about three-fourths of the surface of the globe. If we subtract the shallow areas of the continental shelves and the scattered banks and shoals, where at least the pale ghost of sunlight moves over the underlying bottom, there still remains about half the earth that is covered by miles-deep, lightless water, that has been dark since the world began. (Rachel Carson, from "The Sunless Sea" in *The Sea Around Us*)

8-point Geneva, 12-point leading, ragged-right margin, and shorter lines

The whole world ocean extends over about three-fourths of the surface of the globe. If we subtract the shallow areas of the continental shelves and the scattered banks and shoals, where at least the pale ghost of sunlight moves over the underlying bottom, there still remains about half the earth that is covered by miles-deep, lightless water, that has been dark since the world began. (Rachel Carson, from "The Sunless Sea" in *The Sea Around Us*)

Does line length really make a difference? Read a short study that shows that line length and margin width interact to affect the ease of reading a document. This is an experiment that you could replicate. To access it, go to www.english.wadsworth.com/burnett6e for a link.

CLICK ON WEBLINK

CLICK on Chapter 11/line length

WEBLINK

Headings to Label Chunked Information

Headings and subheadings can label chunked information and identify the relative importance of these chunks in a document, whether print or online. Headings not only establish the subject of a section; they also give readers a chance to take both a literal and a mental breath while previewing the upcoming content. Some writers try to use a heading or subheading every three to five paragraphs to avoid visual monotony and keep the reader focused. Although you may find such breaks too frequent, the concept is important. As noted in Chapter 10, a well-designed outline can serve as the structure for the table of contents and also provide headings and subheadings that make a document easier to read.

The thumbnail sketches in Figure 11.7 show various ways you can incorporate headings into your documents. Relative importance is signaled by capitalization, type size, and typeface. When you test the draft of a document with your audience, check that they are helped, not confused, by your titles, headings, and subheadings.

FIGURE 11.7 **Thumbnail Sketches of Heading Placement**

(a) Headings can be pulled out of the text and placed in a separate narrower column used solely for headings and annotations.

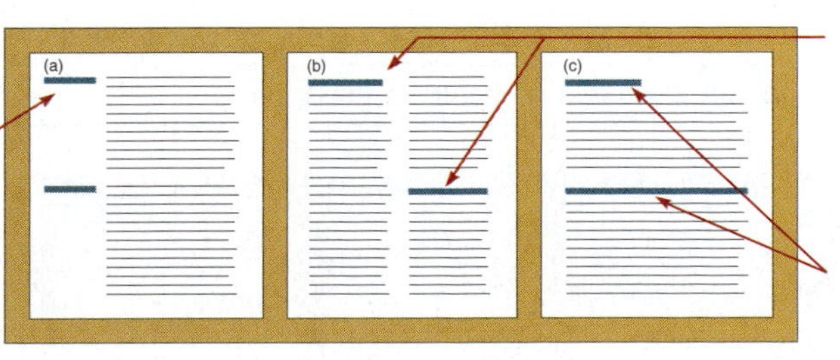

(b) Headings can be used to signal a change of topic as well as to reduce visual monotony and create visual interest for the reader.

(c) Headings can partially or fully extend the width of the grid column.

Arranging Related Chunks of Verbal and Visual Information

In arranging verbal and visual information, you can draw on design conventions as well as avoid design problems.

Using Design Conventions

Two practices used by professional designers will help you produce more effective documents:

- Selection of appropriate grids
- Placement of visuals near related text

Why should subject matter experts spend time learning about the design of documents?

The easiest and most efficient way to design a page (or an entire document) is to see the page or a screen as a *grid* — columns and rows that help you organize the textual and visual chunks. In most situations, you will use one-column, two-column, or three-column grids, illustrated in thumbnail sketches in Figure 11.8. For more detailed grids, you'll probably be working with a designer or graphic artist. When you are working independently, you'll be able to create thumbnail sketches of various grids and consider which one is most appropriate.

Different grids are appropriate for different purposes and audiences. Imagine, for example, that the one-column grid in Figure 11.8a is for an in-house technical report about reconfiguring computer workstations; the two-column grid in Figure 11.8b is for an operator's manual for office workers using desktop publishing; the three-column grid in Figure 11.8c is for a corporate newsletter.

Readers of print documents appreciate not having to constantly turn back and forth between the page they're reading and visuals placed in an appendix. A long-accepted survey concerning the format of NASA technical reports indicated that 80 percent of management and nonmanagement engineers and scientists preferred visuals integrated into the text rather than placed in an appendix. The only exceptions noted were if several consecutive pages of visuals interrupted the flow of the text and thus distracted the reader.[9] Figure 11.9 shows several acceptable possibilities for incorporating visuals into your text.

FIGURE 11.8 **Thumbnail Sketches of Columns for Grids**

(a) (b) (c)

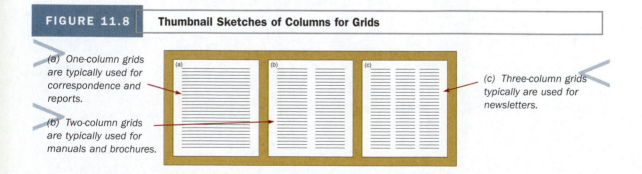

(a) One-column grids are typically used for correspondence and reports.

(b) Two-column grids are typically used for manuals and brochures.

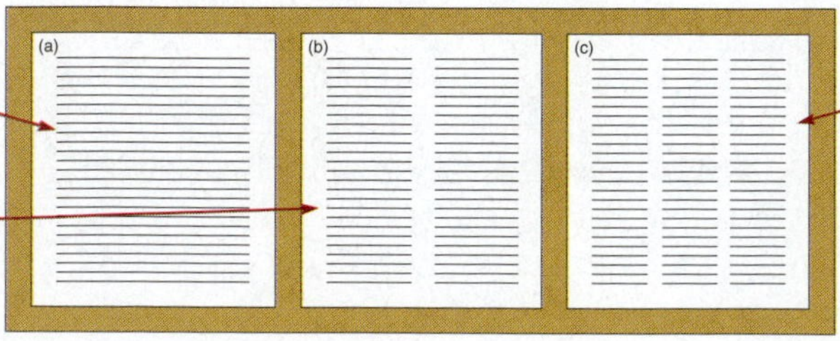

(c) Three-column grids typically are used for newsletters.

FIGURE 11.9 Thumbnail Sketches Showing Placement of Visuals

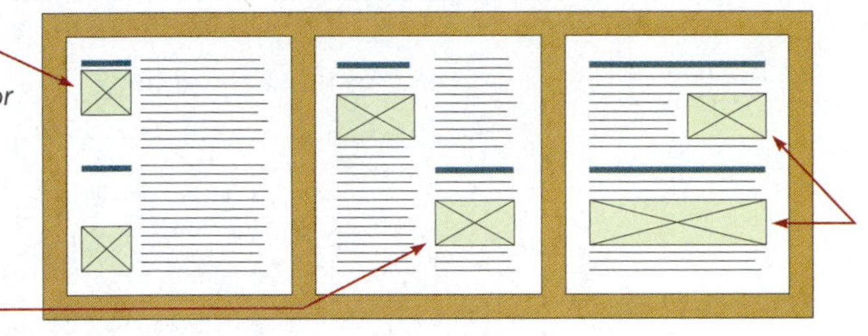

Visuals can be pulled out of the text and placed in the margin (or on a separate page).
Visuals can be incorporated in the text, fitting into one column on the grid.
Visuals can also take up more or less space than a grid's column.

Readers of online documents appreciate hyperlinks that allow them to move back and forth between the page they're reading and a linked document or visual. Readers working on a relatively small screen will toggle back and forth; however, as monitor size increases and the price of high-resolution flat-panel screens decreases, users often have access to monitors that can display two full pages at the same time.

Avoiding Problems in Arranging Information

Before you design a print or online document, you need to be aware of four potential problems that distract readers:

- Chartjunk
- Tombstoning
- Heading placement
- Widows and orphans

One problem involves the temptation to clutter visuals with unnecessary *chartjunk,* miscellaneous graphic junk that does nothing to help people understand the information. This temptation is multiplied when you are using computer software that entices you to add fancy features to visuals. Figure 11.10 contrasts a graph with unnecessary and distracting chartjunk and one with less chartjunk and more white space. Whatever is included on visuals should contribute to the meaning and make the information more accessible and appealing.

Another problem, called *tombstoning,* involves aligning headings so that readers mistakenly chunk the text when they look at the page. In Figure 11.11a, readers could easily believe that the top half of the page was one section and the bottom half was another. Figure 11.11b shows that you can rearrange the headings to avoid this potential confusion.

A third problem, *heading placement,* comes from leaving too few lines after a heading or subheading at the top or bottom of a column or page. Figure 11.12a shows the problem of not leaving at least three lines in a column or on a page before beginning a new heading or having at least three lines following a new heading at the bottom of a column or page. Figure 11.12b shows the adjustment, which included more lines.

FIGURE 11.10 **Chartjunk versus White Space**

This version has too much chartjunk and too little white space, so the data are obscured.

This version, without the chartjunk, has more white space, so the changes in monthly production are easier to see.

What arguments can you think of for and against using chartjunk?

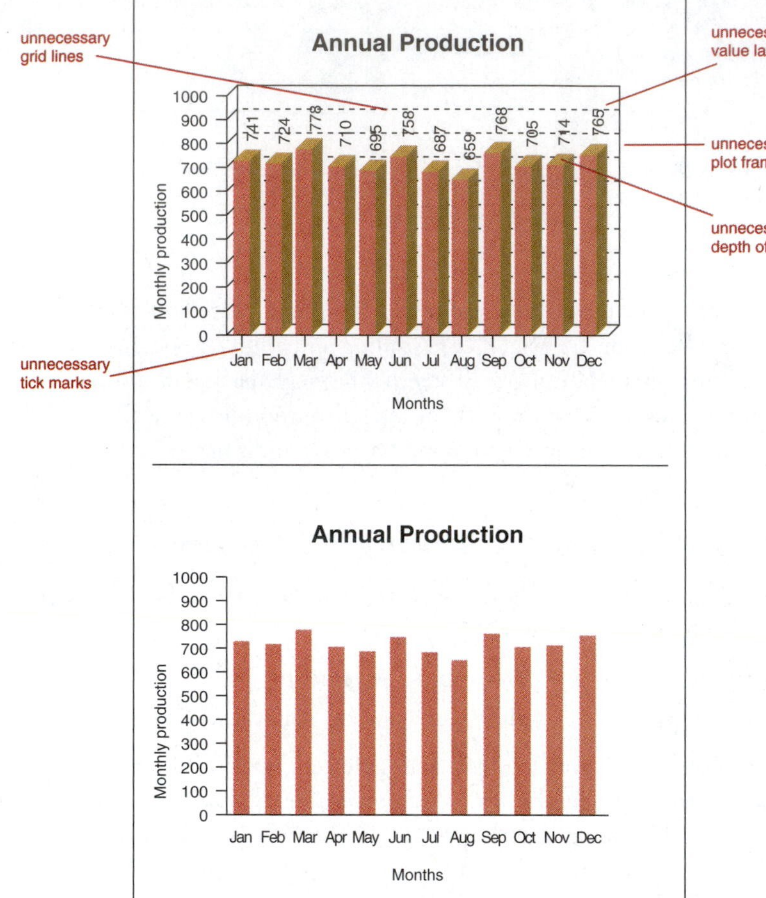

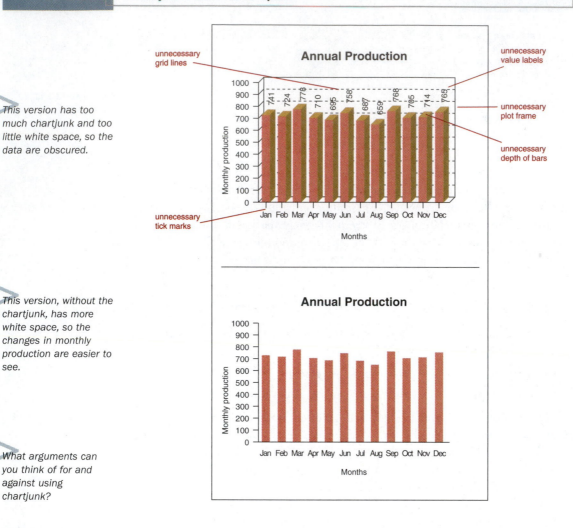

FIGURE 11.11 **Tombstoning**

(a) Because of the aligned headings (arranged like tombstones), the top two columns could be mistakenly seen as one chunk of the text.

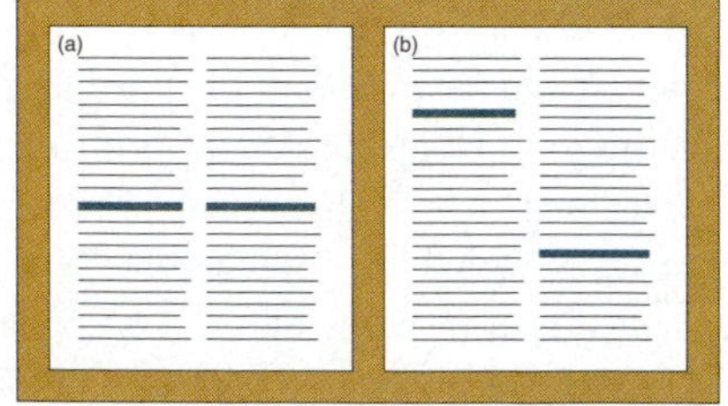

(b) The placement of the headings makes much clearer that readers should move through the first column and onto the second. Any temptation to mischunk the text is lessened.

FIGURE 11.12 **Placement of Headings**

(a) Avoid placing a single line or two at the top of a column or page when followed by a new heading.

Avoid a heading at the bottom of a column or page that has room for only one or two lines of text.

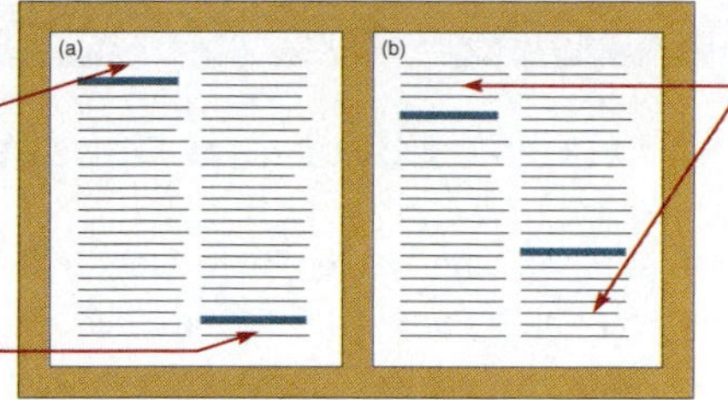

(b) Have several lines of text preceding and following a heading.

A final problem deals with widows and orphans. *Widows* are leftover words — one or two words hanging on awkwardly as the last line of a paragraph. You can avoid them by revising the sentences in the paragraph to add or delete a few words so that one or two are not left on a line alone. *Orphans* occur when a column or page break occurs in a paragraph after the first line of the paragraph (see Figure 11.13). Try to arrange column or page breaks after a few lines rather than at the very beginning or end of a paragraph.

FIGURE 11.13 **Widows and Orphans**

WIDOWS
Widows occur when one or two words hang on as the last line of a paragraph.

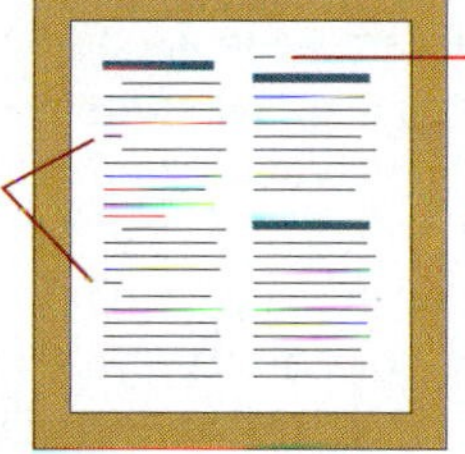

ORPHANS
Orphans occur when a column or page breaks after the first line of the paragraph.

Sometimes problems in arranging information are more serious than simply inconveniencing readers whose time may be wasted or understanding restricted because of careless or inadequate attention to information design. The ethics sidebar on the next page discusses the document design of a government document and the ways that type size, white space, and boldface type potentially can bias readers.

What are some techniques you could use to avoid using font sizes deceptively?

A short, easy-to-read article summarizing the "seven deadly sins of information design" will help you avoid some basic errors. To access the article, go to www.english.wadsworth.com/burnett6e for a link.
CLICK ON WEBLINK
CLICK on Chapter 11/deadly sins

WEBLINK

ETHICS SIDEBAR

"A Wound to the Hand": Ethics and Document Design

Emphasizing information is an important part of document design in technical communication, but technical professionals must be aware of the ethical line between emphasis and deception. Where is this line for you? How much does document design, in elements like type size or spacing, affect how a document is interpreted?

Can technical professionals provide verifiable facts in a technical document but still be at fault for presenting an unethical document? Yes, they can — if those facts are found in a document that is designed to deceive and mislead the audience. So found researcher TyAnna Herrington in her review of research on the relationship between ethics and document design.[10] Herrington's research indicates the way facts are presented in a document is as important as the facts themselves. Design elements like headings, titles, spacing, and font size guide readers in determining the inherent value of the information and the relative importance of that information when compared to other information inside (and outside) the document. These design elements can intentionally or unintentionally lead readers to interpretations and conclusions that may not be supported by the facts or that hide more credible or relevant explanations.

Herrington believes an example of unethical document design can be found in a government report about the siege at the Branch Davidian compound in Waco, Texas, in 1998. The Branch Davidians, a small religious group, and the U.S. Department of Alcohol, Tobacco, and Firearms (ATF) clashed in February of that year when the ATF attempted to confiscate weapons they believed the Branch Davidians had obtained illegally. The violent confrontation between the ATF agents and the Branch Davidians at the Davidians' housing compound led to multiple deaths on both sides.

What are some techniques you could use to avoid using font sizes deceptively?

Herrington reviewed the tables in the government report and found that type size, white space, and boldfacing emphasized the ATF injuries and de-emphasized the Branch Davidian injuries, even though more Davidians than agents died during the attack. For example, the titles for the ATF tables are in boldface and are a larger type size (14-point vs. 12-point) than the Branch Davidian titles, which are not boldfaced. She also found additional white space around the titles in the ATF tables, which helped these tables stand out more in the overall document. These visual differences, Herrington believes, help accentuate the ATF casualties.

The descriptions in tables listing ATF and Davidian injuries also accentuate the ATF casualties. In the ATF table, the wording of the descriptions is more understandable to the untrained reader than the descriptions in the Davidian tables. For example, the ATF table lists a "gunshot wound to the hand" while the Davidian table lists a "craniocerebral trauma-gunshot wound." Herrington believes these differences are unethical because they may lead to unsupported and misleading interpretations of the events in Waco.

When you get really involved in the design of information, you often want to read about various practices and issues. For a list of remarkably varied and valuable links, go to www.english.wadsworth.com/burnett6e. You'll be able to access one of the special interest groups (SIGs) of the Society for Technical Communication that focuses on information design, as well as access the tables of contents of *Document Design: Journal of Research and Problem Solving in Organizational Communication.*

CLICK ON WEBLINK

CLICK on Chapter 11/design references

WEBLINK

Emphasizing Information

Once information is chunked, labeled, and arranged, you may still need to emphasize selected portions of the text to make the information more accessible, comprehensible, and usable.

On printed pages, chunking, labeling, and arranging are fixed. However, readers can alter the way they view Web pages by changing browser preferences. For example, readers can choose to "view pictures and text," "pictures only," or "text only" by turning graphics on or off. They can choose specific fonts for their browser to use as a default and turn off the fonts specified in the document, disable dynamic fonts, and so on. Readers can also specify link and background colors.

Despite a designer's careful and considered choices, some readers may choose to override the designer's decisions. Most readers, though, do use the design on the Web page; thus, typeface and typographic devices remain important for creating emphasis for both print and Web pages.

Explain whether knowing a great deal about making documents appealing to readers increases the likelihood that carelessly done work — perhaps inaccurate or unsupported information — could be presented in an effective "package" that might disguise its inadequacies.

Typography in the Western world dates from the early- to mid-15th century, depending on which historical records you read. To learn more about the history of typography, including examples of how typography evolved, discussions about the politics of typography, and critical features of particular typefaces, go to www.english.wadsworth.com/burnett6e.

CLICK ON WEBLINK

CLICK on Chapter 11/typographic history

WEBLINK

Typefaces

Typeface (also called *font*) affects readers' attitudes and reactions to a print or electronic document, as well as their ability to access, comprehend, and use information easily and quickly. The desktop publishing revolution makes imperative your knowledge of basic characteristics about typefaces:

- Serif or sans serif
- Typeface variations
- Type size
- Style choices

WEBLINK

Skillful designers can distinguish various typefaces and determine which ones are appropriate for various purposes. Part of being able to make informed decisions about typefaces involves learning to recognize the fonts that are available. To access sites illustrating various typefaces, go to www.english.wadsworth.com/burnett6e for links.

CLICK ON WEBLINK

CLICK on Chapter 11/typography

Whether you're selecting typefaces for a print document or for a Web page, you have the same choices, but the constraints are considerably different. For a Web page, follow these general guidelines:

- Select conventional fonts that you can be fairly sure users will have installed on their computer.
- Remember that in print the relationship between typographical elements is fixed on a page. However, on a Web site, users scroll and link in different (and sometimes unanticipated) ways, so reaction to typographical elements is never to a fixed page.
- Web sites can use technology to build in interaction — for example, overlays or pop-ups — to engage users.

Serif or Sans Serif. Most typefaces can be classified as *serif* or *sans serif.* Serifs are tiny fine lines usually at the top or bottom of letters.

These are examples of lowercase and uppercase versions of the letters *b, p,* and *x* in common serif and sans serif typefaces:

serif: b B (10-point Times Roman)	sans serif: b B (9-point Helvetica)
serif: p P (10-point Palatino)	sans serif: p P (9-point Arial)
serif: x X (10-point New York)	sans serif: x X (9-point Geneva)

Sans serif typefaces have a neat, appealing appearance and are often used for short documents. Because sans serif typefaces are simpler, the letters don't have as many distinguishing features, making them slightly less accessible in print documents for some people; thus, documents for children, the elderly, people with visual impairments or disabilities, and people with learning disabilities are often printed in serif typefaces. Similarly, long documents (for example, long technical reports and journal articles) for all readers often use a serif typeface so that readers won't tire so quickly. These generalizations don't apply, however, to electronic documents, which many people believe are easier to read in sans serif typefaces, especially those designed for on-screen reading.

The information architecture and typography for Web sites requires special attention to a range of factors ranging from platforms to screen sizes. For useful information about Web design, go to www.english.wadsworth.com/burnett6e for links.

CLICK ON WEBLINK

CLICK on Chapter 11/web design

WEBLINK

Typeface Variations. Your selection of typeface should be influenced both by the kind of document and your sense of how the readers will react. Readers are discouraged if typefaces are inappropriate for the document or situation or are difficult to read.

- *This typeface is called Edwardian Script and is usually used only for social announcements.*
- This typeface is called Helvetica and is often used for standard business letters.
- This typeface is called Adobe Garamond and is often used for typesetting lengthy reports and books, including this one, because it is easy to read.

Figure 11.14 illustrates ways in which typographical conventions, including typeface, are used in a Dell Computer user's manual as visual cues to help readers. Each typeface has a specific meaning. For example, all items on the manual's menu screens are presented in Helvetica; all commands lines are presented in Courier. Associating meaning with various fonts and other typographic cues helps readers because it gives them textual content as well as visual cues to help construct meaning.

Type Size. You can also affect readers by the size of the type. You can discourage, insult, or alienate readers by using an inappropriate type size. For example, although tiny type reduces the number of pages in a user manual, it also makes the manual difficult for readers to use. Unnecessarily large type can make a document seem elementary because many people associate large type with children's books. You generally use 10- or 12-point type for the text of business documents. Headings may use larger type. The type used for PowerPoint

What federal mandates exist for making information accessible to visually impaired readers? What guidelines exist to help design documents for elderly readers?

FIGURE 11.14 **Typographical Conventions Used in a Dell Computer User's Manualiv[11]**

__Bold__ signals interface components such as menu names.

<Angle brackets> signal keycaps and key combinations.

__Lower case bold__ signals filenames, directory names, information for reference, and commands.

[Brackets] indicate optional information.

`Courier New` signals command lines and text to be typed.

Italics signals variables.

Typographical Conventions

The following list defines (where appropriate) and illustrates typographical conventions used as visual cues for specific elements of text throughout this document:

- *Interface components* are window titles, button and icon names, menu names and selections, and other options that appear on the monitor screen or display. They are presented in bold.

  Example: Click **OK**.

- *Keycaps* are labels that appear on the keys on a keyboard. They are enclosed in angle brackets.

  Example: <Enter>

- *Key combinations* are series of keys to be pressed simultaneously (unless otherwise indicated) to perform a single function.

  Example: <Ctrl><Alt><Enter>

- *Commands* presented in lowercase bold are for reference purposes only and are not intended to be typed when referenced.

  Example: "Use the **format** command to . . . ."

  In contrast, commands presented in the Courier New font are part of an instruction and intended to be typed.

  Example: "Type `format a:` to format the diskette in drive A."

- *Filenames* and *directory names* are presented in lowercase bold.

  Examples: **autoexec.bat** and **c:windows**

- *Syntax lines* consist of a command and all its possible parameters. Commands are presented in lowercase bold; variable parameters (those for which you substitute a value) are presented in lowercase italics; constant parameters are presented in lowercase bold. The brackets indicate items that are optional.

  Example: **del** *[drive:] [path] filename* **[/p]**

- *Command lines* consist of a command and may include one or more of the command's possible parameters. Command lines are presented in the Courier New font.

  Example: `del c:\myfile.doc`

- *Screen text* is a message or text that you are instructed to type as part of a command (referred to as a command line). Screen text is presented in the Courier New font.

  Example: The following message appears on your screen:

  `No boot device available`

  Example: "Type `md c:\programs` and press <Enter>."

- *Variables* are placeholders for which you substitute a value. They are presented in italics.

  Example: DIMM_x (where x represents the DIMM socket designation).

presentations or transparencies (vu-graphs) should be larger (18-point, 24-point, or 32-point type works well for making visuals to accompany oral presentations) so that the information can easily be read from the back of the room.

Variation in type size can also be used to capture your readers' attention. This warning, printed in 20-point Berkeley Old Style, is intended to grab readers.

WARNING! Ingestion of
this chemical could be fatal!

However, the warning would not be nearly as effective if it were printed in 8-point type. Reducing the type size reduces the impact.

WARNING! Ingestion of this chemical could be fatal!

Typeface itself influences the size; all 10-point type does not look the same size. Combinations of type size and typeface provide visual appeal and variety. For example, the body of this book is set in 10.5-point Adobe Garamond with 12-point Franklin Gothic Demi main headings; 9-point Franklin Gothic Book is used for the examples. The part headings are in 18-point Franklin Gothic Book, as are the chapter titles.

If particular material must fit into a prescribed and limited amount of space, you can use type size to make minor adjustments. For example, information for a résumé could fit attractively on a single 8½-by-11-inch sheet using a 10-point font, whereas the information might take up one and a quarter sheets using a 12-point font.

Flexibility in type size depends on the equipment you use. Some printers may restrict your choices of typeface and type size. However, most inkjet and laser printers offer tremendous flexibility. Many inkjet printers offer the option of color cartridges so that you can print in color for a relatively low cost. Word-processing packages let you select from 8-point (or smaller) to 72-point (or larger) type in a great variety of fonts.

And what about fonts for Web sites? The same fonts are displayed differently on Macintosh and Windows operating systems. In general, fonts on Windows-based browsers look two to three points larger than the equivalent fonts on Macintosh browsers. The difference in the way fonts are displayed can have a major impact on your page design.[12]

Style Choices. The style of type you select can also influence the audience. Depending on the computer software you're using, you have the capability of using CAPITALIZATION, SMALL CAPS, **boldface,** and *italics,* as well as fancier variations such as shadow or outline.

Using ALL CAPS for occasional emphasis can be effective; however, your use of ALL CAPS should be limited to headings and single words or short phrases in the text. WHEN YOU USE ALL CAPS FOR ENTIRE SECTIONS OF TEXT, THE

READER IS NOT ABLE TO RAPIDLY DIFFERENTIATE THE WORDS BECAUSE ALL THE LETTERS ARE THE SAME HEIGHT. THUS, READING IS SLOWED. YOUR INTENT TO EMPHASIZE A POINT IS LOST IN THE VISUAL MONOTONY OF CONSISTENT CAPITALIZATION.

Visual emphasis can be created by using **boldface** or **BOLD SMALL CAPS** or **BOLD ALL CAPS.** These techniques are usually reserved for signaling warnings, cautions, and dangers and for calling attention to important points and terms. Use all these type variations with restraint. Try to work within a general guideline of using no more than two typefaces and a total of four variations of typeface, type size, or style on a single page or screen.

Typographic Devices

Sometimes you need to emphasize information by separating it or visually distinguishing it from the text. Effective devices include numbered lists, bulleted lists, underlining, boxes, shading or tints, and colors. As with any device, their impact diminishes with overuse. Too many visual devices make a page look so cluttered that the reader cannot concentrate on content.

Numbered lists are particularly common elements in sets of instructions, and they also appear frequently in reports and proposals. Numbered lists suggest one of three things to readers:

- Sequence or chronology of items is important.
- Priority of items is important.
- Total count of items is important.

When all items in the list are equivalent, a bulleted list is preferable, as illustrated in the preceding lines. The bullets draw attention to each item in the list but infer no priority to the sequence. You can create bullets with most word-processing software by using one of the option keys or menu items.

Underlining words or phrases is a holdover from typewriters that didn't have italics or boldfacing to create emphasis; instead, underlined text was converted to italics when the material was typeset. A summary of underlining conventions appears in the *Usage Handbook* on the Companion Web site at **www.english.wadsworth.com/burnett6e**. And, as the preceding sentence illustrates, underlining now has another function: to signal hyperlinks in Web documents. In addition to being convenient for all readers, underlined hyperlinks increase accessibility for colorblind users as well as those with monochrome monitors. Unless you are signaling a hyperlink, avoid underlined text.[13]

Boxed and shaded information is emphasized. Information that is boxed should be surrounded by white space so that the text does not run into the box. Sometimes the boxed material relates directly to the text; other times it is supplemental. Too much shading diminishes its impact. Note particularly the use of shading for boxed figures in this text.

Boxes are often effective in the following situations:

- Identify major headings
- Highlight key terms
- Emphasize formulas or equations
- Separate anecdotal material

Shading highlights and emphasizes material. Sometimes shaded areas are used in conjunction with boxes. Readers are drawn to shaded material because it is differentiated from the rest of the text.

Color is an especially appealing visual device, which often contributes significantly to the effectiveness and clarity of a document. Some technical materials require color. For example, as you'll see in Chapter 12, anatomical diagrams need shading and color. Similarly, color-coded electronic components should be accompanied by color-coded troubleshooting diagrams. In documents with difficult material, color can create visual interest, highlight section headings, identify examples, and emphasize important points. In this text, color is used to help distinguish parts of visuals and to highlight important textual elements more distinctively than if a gray-scale were used.

Shading highlights and emphasizes material. Sometimes shaded areas are used in conjunction with boxes. Readers are drawn to shaded material because it is differentiated from the rest of the text.

Using color instead of a gray-scale calls even more attention to the information. The color should not decrease legibility (e.g., blue text on a black background) nor be so flamboyant that the content is ignored.

WEBLINK

Color is a powerful design element. You can learn a great more about color by going to **www.english.wadsworth.com/burnett6e** for a link to a considerable number of references.

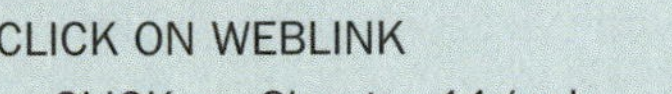
CLICK ON WEBLINK
CLICK on Chapter 11/color portal

Color has a number of specific benefits, as illustrated in the thumbnail sketches in Figure 11.15. (The use of color in specific visuals is discussed in greater detail in Chapter 12.)

FIGURE 11.15 **Use of Color for Creating Emphasis in Text**

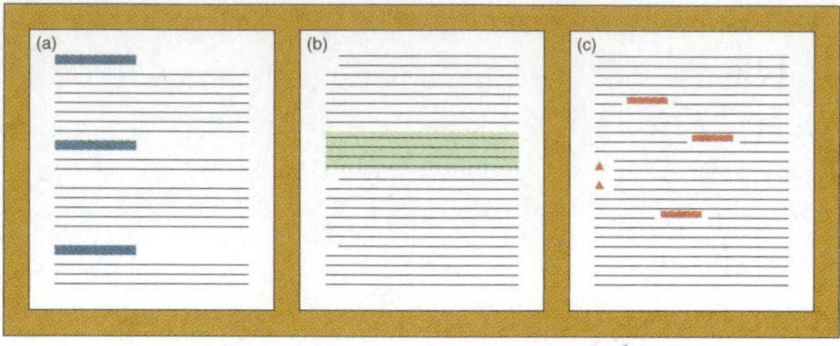

(a) Highlight text hierarchy

(b) Highlight important parts

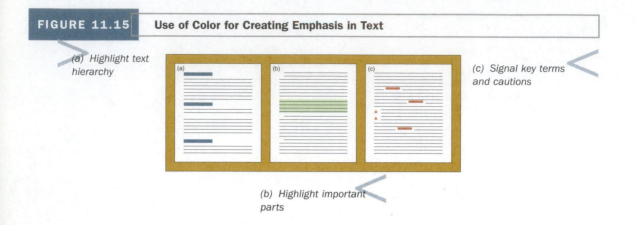

(c) Signal key terms and cautions

- *Identify text hierarchy.* Color can help readers locate the main sections in a text. For example, Figure 11.13a shows how color can be used to highlight headings, but it can also be used for section dividers.
- *Chunk information.* Color can effectively chunk related information for readers. For example, Figure 11.13b shows how color can be used to highlight an important part of a paragraph.
- *Emphasize key points.* Figure 11.13c shows how color can be used to highlight terms, which helps readers recall and remember that information. Similarly, color can be used to emphasize critical parts of the text — for example, to signal cautions and warnings.

Despite its value, the use of color is often restricted by cost if you are using a commercial printer because a separate print run is needed for each color added to a page. Often, a well-designed document employing a variety of visual devices other than color can be equally effective.

WEBLINK

If you are particularly interested in the theory of design, go to **www.english.wadsworth.com/burnett6e** for a link to an interesting and provocative article.

CLICK ON WEBLINK

CLICK on Chapter 11/design theory

Individual and Collaborative Assignments

1. **Choose the appropriate format for a document.** Four common presentation formats have been explained in this chapter:

   - textual presentation alone in paragraph format
   - textual presentation with visuals
   - visual presentation with text explanations
   - visual presentation alone

   Choose one of these formats for each of the following situations and explain your choices:

   - Instructions for first-aid emergency
   - Preparation of a car for repainting
   - Explanation of a solution for a manufacturing problem
   - Presentation of computer printer components
   - Application of fertilizer to golf course

2. **Evaluate effective page design in a technical document.** Work with a small group of classmates in your academic discipline or a related one.

   (a) Based on what you have learned in this chapter and your own observations of page design, create a rubric that you can use to evaluate several documents.

   (b) Look for examples of page design in technical documentation and evaluate several.

   (c) Choose a page that is a particularly good example of effective page design and discuss the elements that make the page design effective and appealing.

   (d) Choose a page that is difficult to read because of poor design or poor visuals and redesign its layout.

3. **Use design principles for analysis.** Work with a small group of classmates to carefully study the following page from the quarterly publication *Conservation Frontlines,* published by Conservation International. Identify the design elements and then analyze the design principles that are at work in this page. Specifically, note the way information is chunked and labeled, the way it is arranged, the integration of textual and visual information, and the way information is emphasized.

dispatches

Biological survey uncovers marine marvel

A CI marine Rapid Assessment Program survey of reefs off the northwestern coast of Madagascar uncovered a remarkably diverse ecosystem, recording at least nine coral and three fish species new to science. Researchers documented 304 coral species, one-third of the world's known total and nearly double the number known to exist in Madagascar. CI and local groups are using survey results to help establish protected areas in this extraordinary marine environment. ■

[LEFT] Resplendent goldie *(Pseudanthias pulcherrimus)*, a species found in Madagascar's diverse coral reefs.

Rangers trained to save imperiled "man of the forest"

CI is providing antipoaching and law enforcement training to park rangers in Sumatra's Gunung Leuser National Park, part of a new program aimed at saving the endangered Sumatran orangutan. The program also is building a cadre of local monitors within the park to enhance protection efforts. The shaggy red orangutan, whose name translates as "man of the forest," is one of the world's most imperiled primates. ■

[ABOVE] Orangutan, one of the world's most endangered primates.

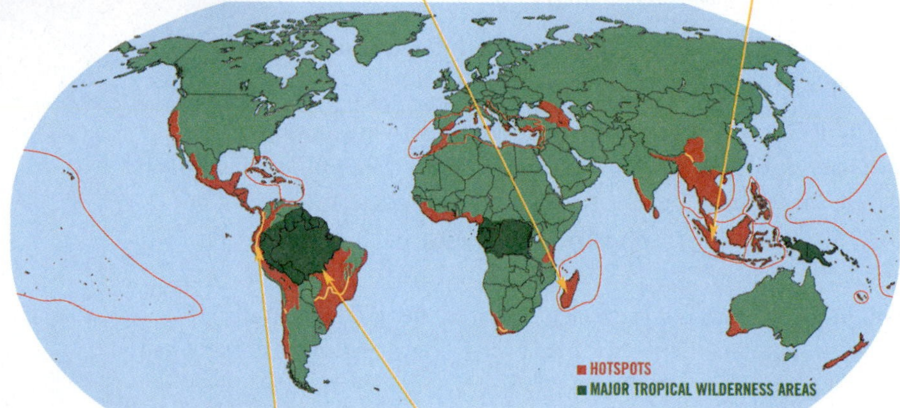

[LEFT] Nangaritza River, located in Ecuador's newly formed Nangaritza Protected Area.

Brazilian government backs Kayapo accord

In its first-ever agreement with an environmental organization, Brazil's Funai, the government agency responsible for indian affairs, has agreed to partner with CI to help the Kayapo indigenous community safeguard its 25-million-acre ancestral home in the Amazon wilderness area. Under the agreement, CI is providing equipment and training to support Kayapo efforts to protect the reserve. ■

[RIGHT] Men from the Kayapo reserve in Brazil wearing ceremonial headdress.

Protected area in Ecuador doubles in size

Collaboration among CI, local partners and the government of Ecuador has led to the creation of the 295,000-acre Nangaritza Protected Area adjacent to Podocarpus National Park, doubling the size of the protected region. Nangaritza is located in southeast Ecuador, near the border of Peru and within the biologically important Condor-Cutucú conservation corridor. ■

frontlines | summer 2002 | 1

4. **Assess the differences between print and electronic journals.** Work with a group of classmates who share your major or a related field of study.
   (a) Locate an academic journal in your field in the library.
   (b) Search the Web for its corresponding electronic version.
   (c) Evaluate the differences between the two versions of the journal. Write your findings in a memo to your instructor.
5. **Transfer a print document to the Web.** Electronic publication of existing print documents involves different design and organizational considerations. Use a print document that you have already completed for this or another class. Change its format, design, and organization and publish it on the Web. Write a memo to your instructor in which you detail the changes that you made in your print document in order to make it Web accessible.
6. **Evaluate the effectiveness of the design of a technical Web site.** Locate several Web sites that deal with technical information in your field of study. Compare the effectiveness of the design of these Web sites.
   (a) Use (and modify) the rubric you created in Assignment 2 or use (and modify) the rubric. Consider the following questions:
      - Which presentation formats are most popular in Web-based documents?
      - Why do you think this is so?
      - What suggestions do you have for technical communicators designing Web-based documents?
   (b) Present your findings and your final rubric in a memo to your classmates.

Chapter 11 Endnotes

1 Modified from (1) Kostelnick, C. (1989). Visuals rhetoric-oriented approach to graphics and design. *The Technical Writing Teacher, XVI*(1), 77-88; (2) Allen, J. T., & Chance, B. *Formal aspects of design.* Retrieved November 1, 2003, from http://desktoppub.about.com/gi/dynamic/offsite.htm?site=http%3A%2F%2Fs9000.furman.edu%2Fcs16g%2Fresources$2Felements.html

2 The information about elements of graphic design has been drawn from a variety of print and electronic sources, all of which are worth checking for further details: (1) Bear, J. H. (2004). *What you need to know about desktop publishing* (and the many links available). Retrieved January 20, 2004, from http://desktoppub.about.com/cs/designprinciples/; (2) Goin, L. (2002). *Graphic design basics.* Retrieved November 1, 2003, from http://www.graphicdesignbasics.com/article1043.html/; (3) Jirousek, C. (n.d.). *Art, design, and visual thinking.* Retrieved November 1, 2003, from http://char.txa.cornell.edu/language/principl/rhythm/rhythm.htm; (4) Schriver, K. A. (1997). *Dynamics in document design: Creating text for readers.* New York: John Wiley & Sons. (5) Williams, R. (1994). *The non-designer's design book.* Berkeley, CA: Peachpit Press.

3 Urquhart, G., Chomentowski, W., Skole, D., & Barber, C. (n.d.). *Tropical deforestation.* Retrieved November 1, 2003, from http://earthobservatory.nasa.gov/Library/Deforestation/

4 Allen, J. T., & Chance, B. (n.d.). *Formal aspects of design.* Retrieved November 1, 2003, from http://desktoppub.about.com/gi/dynamic/offsite.htm?site=http%3A%2F%2Fs9000.furman.edu%2Fcs16g%2Fresources%2Felements.html

5 Kinnaird, M. F. (1996, January). Indonesia's hornbill haven. *Natural History, 105*(1), 40–44.

6 Nikon. (n.d.). Home page. Retrieved November 1, 2003, from http://www.nikonusa.com/home.jsp

7 Pinelli, T. E., Cordle, V. M., McCullough, R. (1984). Report format preferences of technical managers and nonmanagers. *Technical Communication, 31,* 6–7.

8 U.S. Department of Energy Human Genome Project Information Web site. (n.d.). *Human genome landmarks poster.* Retrieved November 2, 2003, from http://www.ornl.gov/sci/techresources/Human_Genome/posters/chromosome/index.shtml

and http://www.ornl.gov/sci/techresources/Human_Genome/posters/chromosome/chromo05.shtml

9 Pinelli, T. E., Cordle, V. M., McCullough, R. (1984). Report format preferences of technical managers and nonmanagers. *Technical Communication, 31,* 6–7.

10 Herrinton, T. (1995). Ethics and graphic design: A rhetorical analysis of the document design in the "Report of the Department of the Treasury on the Bureau of Alcohol, Tobacco, and Firearms investigation of Vernon Wayne Howell also known as David Koresh." *IEEE Professional Communication, 38*(3), 151–157.

11 Notational Conventions. (n. d.). *Preface: HP OpenView Network Node Manager Special Edition 1.3 With Dell OpenManage™ HIP 3.3 User's Guide.* Retreived May 12, 2004, from http://support.ap.dell.com/docs/SOFTWARE/smnnmse/nnmse13.preface.htm

12 Lynch, P. J., & Horton, S. (2002). Type size. In *Web style guide* (2nd ed.). Retrieved November 3, 2003, from http://www.webstyleguide.com/type/size.html

13 Lynch, P. J., & Horton, S. (2002). Emphasis. In *Web style guide* (2nd ed.). Retrieved November 3, 2003, from http://www.webstyleguide.com/type/size.html

CHAPTER 12

Using Visual Forms

Objectives and Outcomes

This chapter will help you accomplish these outcomes:

- Understand that visuals not only attract attention and create appeal but also benefit cognitive processing and learning
- Adapt visuals by varying the complexity of content, presentation, color, and size to different audiences and different situations (limited technical knowledge, limited time)
- Make effective decisions about textual references and labeling and placement of visuals
- Carefully design or select visuals to fulfill specific functions:
    - Provide immediate visual recognition
    - Organize numeric or verbal data
    - Show relationships among numeric or verbal data
    - Define or explain concepts, objects, and processes
    - Present chronology, sequence, or process
    - Illustrate appearance or structure
    - Identify facilities or locations
- Use color appropriately and productively

Technical visuals are not a recent addition to technical communication. Leonardo da Vinci is only one of many scientific investigators who have produced a wide variety of technical art. His work, some of which is reproduced in chapter dividers of this text, exemplifies the accuracy and attention to detail found in effective technical visuals.

Like Leonardo da Vinci's visuals from the late fifteenth and early sixteenth centuries, visuals in contemporary technical documents should have a specific purpose and convey specific content. Why use visuals instead of or in addition to text? Not only do visuals attract attention and add appeal, they also strengthen documents in other ways.

> *In what situations might visuals be distracting or minimally helpful to a reader?*

- *Visuals can be more specific than text.* The word *tugboat,* for example, could represent anything from "Tommy Tug" in a children's story to barge tugboats, but a visual of a specific tugboat is easily identifiable.

© Chris Hallbeck Illustration

© Alan Schein Photography/CORBIS

- *Well-designed visuals can usually be understood more easily than text.* Visuals can be particularly effective when dealing with numeric data. Consider the following text and the corresponding graph. While the text presents precise numbers, the graph makes the overall trend in the figures far easier to understand. Most readers are better able to process and remember trends that are presented in graphs and charts.

Quickly read the following two versions that present the same information. Which one is faster to comprehend?

Verbal Version: Text

Staff growth. In the past four years, our staff size has changed. For full-time staff, numbers have changed from 24 in 2002 to 26 in 2003 to 29 in 2004 to 30 in 2005. For temporary staff, numbers have changed from 5 in 2002 to 6 in 2003 to 8 in 2004 to 7 in 2005.

Visual Version: Graph

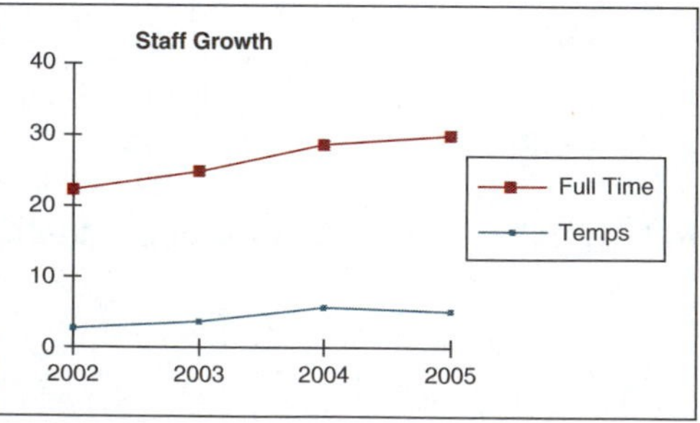

Virtually all readers can comprehend the visual version faster and can recall the visual information more easily. Why?

- *Visuals can be processed more quickly than text.* Visuals can be processed at a glance in as little as one-third of a second (although obviously more time is required to get the full meaning of a complex chart), while text must be consciously scanned.
- *Visuals help readers learn.* Readers who are given documents containing visuals consistently comprehend and retain information better than readers given only text. On average, readers who are given documents containing text and visuals learn 36 percent more than those given text-only documents.

Given these benefits, visuals can be extraordinarily important in increasing the accessibility and usability of your documents. However, you also need to understand the parameters in which you're operating. For example, how much can you alter and adjust a visual in order to make your point, such as changing the color on a photo or shifting the scale on a graph? The ethics sidebar below addresses some of the issues involved in deciding whether manipulation of a visual is ethical or unethical.

Is any alteration an unethical manipulation? Where do you draw the line?

ETHICS SIDEBAR

"With no sacrifice to truth": Ethics and Visuals

Consider this scenario:

> You are working on a report that includes illustrations and photographs. When reviewing the photographs, you find that certain important details are hard to see. To draw out these details, you use visual design software to alter the intensity and hues, making the colors richer. The revised photos seem more appealing.
>
> After you turn in the report, however, your supervisor calls you, claiming you unethically manipulated the photos. You defend yourself, arguing that the changes were artistic and didn't fundamentally alter the truth of the photographs. But your supervisor's response makes you wonder: Did you create unethical photos?

Are changes in intensity or color hues unethical manipulation of visuals? The editors of *Time* magazine didn't think so, at least not in the summer of 1994. That summer, the nation was captivated by the O. J. Simpson murder trial. After Simpson's arrest in June, both *Time* and *Newsweek* used versions of Simpson's police mug shot on their covers. *Time's* version, however, drew public criticism because the photo had been altered. The image, referred to by *Time* as a photo-illustration, artificially

How could you alter a visual without deceiving the audience?

To avoid creating unethical visuals, Kienzler offers some questions we can ask ourselves as we work with documents:

- *What are possible consequences of our visuals?*
- *How would we feel if we received the visual?*
- *What would the world be like if everyone used these visual techniques?*

Do you agree with Kienzler that the altered photo was unethical?

darkened Simpson's face, although it still looked like the official police mug shot.

Critics argued that the altered photo was racist, making Simpson look more menacing, especially when compared to *Newsweek's* nearly identical but unaltered cover. *Time's* managing editor, James Gaines, replying to the criticism, argued that the changes "lifted a common police mug shot to the level of art, with no sacrifice to truth." While Gaines recognized that "altering news pictures is a risky practice," he stated that "every major news outfit routinely crops and retouches photos to eliminate minor, extraneous elements, so long as the essential meaning of the picture is left intact."[1]

But was the "essential meaning" intact? Not according to researcher Donna Kienzler, who identifies the altered photo as unethical. Kienzler believes the image is unethical because readers did not immediately know that the image had been altered. Kienzler argues that the altered image should be seen as art, which "filters reality," and not as truth. The acknowledgment that the photo had been altered was not placed on the cover; it was instead placed on the contents page. According to Kienzler, placing the acknowledgment within the magazine violated the "readers' informed consent to be persuaded by art, rather than themselves evaluating a photograph" (180).[2]

Your professional success will be greatly influenced by your ability to understand and use visuals. Among the things you need to learn are factors that affect the way you incorporate visuals into a document, specific functions of visuals in technical documents and oral presentations, and conventions in using color.

Incorporating Visuals

In what situations might visuals be more appropriate and useful than words for conveying technical information?

Your initial decisions about incorporating visuals involve balancing and integrating verbal and visual information, adjusting visuals for different audiences, and knowing when to choose visuals instead of text.

Visual/Verbal Combinations

In technical communication, visuals work by themselves and in combination with text to create stories for the audience. While visuals should make sense by themselves, they should also illustrate, explain, demonstrate, verify, or support the text. In deciding about appropriate visual/verbal combinations, you can choose from several choices, displayed in the thumbnail sketches in Figure 12.1.

You need to select visual/verbal combinations that communicate effectively to your audience. In many situations, visuals are more efficient than words. For example, a troubleshooting manual could begin with a verbal table identifying the problems along with analyses and solutions. Such a table would be far more useful

FIGURE 12.1 **Various Combinations of Visual/Verbal Integration**

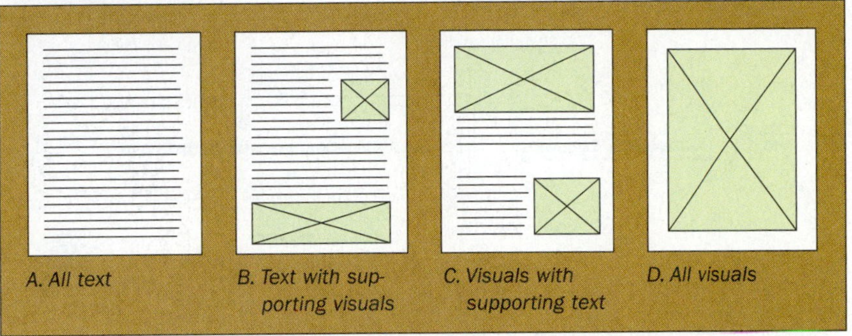

than a series of paragraphs detailing potential problems. You should consider the value of effective, well-designed visuals in the following situations:

- When the audience's understanding of the technical content is limited
- When speed is critical, and reading text would slow the process
- When the process is more clearly illustrated visually

Some concepts or processes are so complex that one visual is insufficient. The three visuals in Figure 12.2, originally published in *NASA Tech Briefs,* all help to illustrate the uses of the Fabry-Perot Fiber-Optic Temperature Sensor. This photonic temperature sensor monitors and controls temperature in highly sensitive areas such as aircraft engines, conventional power plants, and industrial plants, where electronic temperature sensors pose a sparking hazard.

- The top visual in Figure 12.2 shows a cutaway view of the sensor head that is clearly labeled with noun phrases, while arrows point to specific components. Crosshatching distinguishes the nickel-alloy sheath from the Fabry-Perot interferometer that it holds.
- The middle visual in Figure 12.2 diagrams the entire sensor system; the blue arrow indicates the placement of the sensor head. A source of white light flows into one of the pair of optical fibers (represented by black lines) that flow into (and out of) the fiber-optic coupler to the fiber-optic connector and finally to the sensor itself. Shading, spot color, and labels distinguish one component from the other.
- The bottom visual in Figure 12.2 depicts a reflected light spectrum that is characteristic of the temperature in the sensor head. The colored line of this line graph is easily discernible against the white background.

Taken together, this series of visuals tells a story about a particular piece of hardware, the system it connects to, and the work that it does.

FIGURE 12.2 Series of Related Visuals to Tell a Story[3]

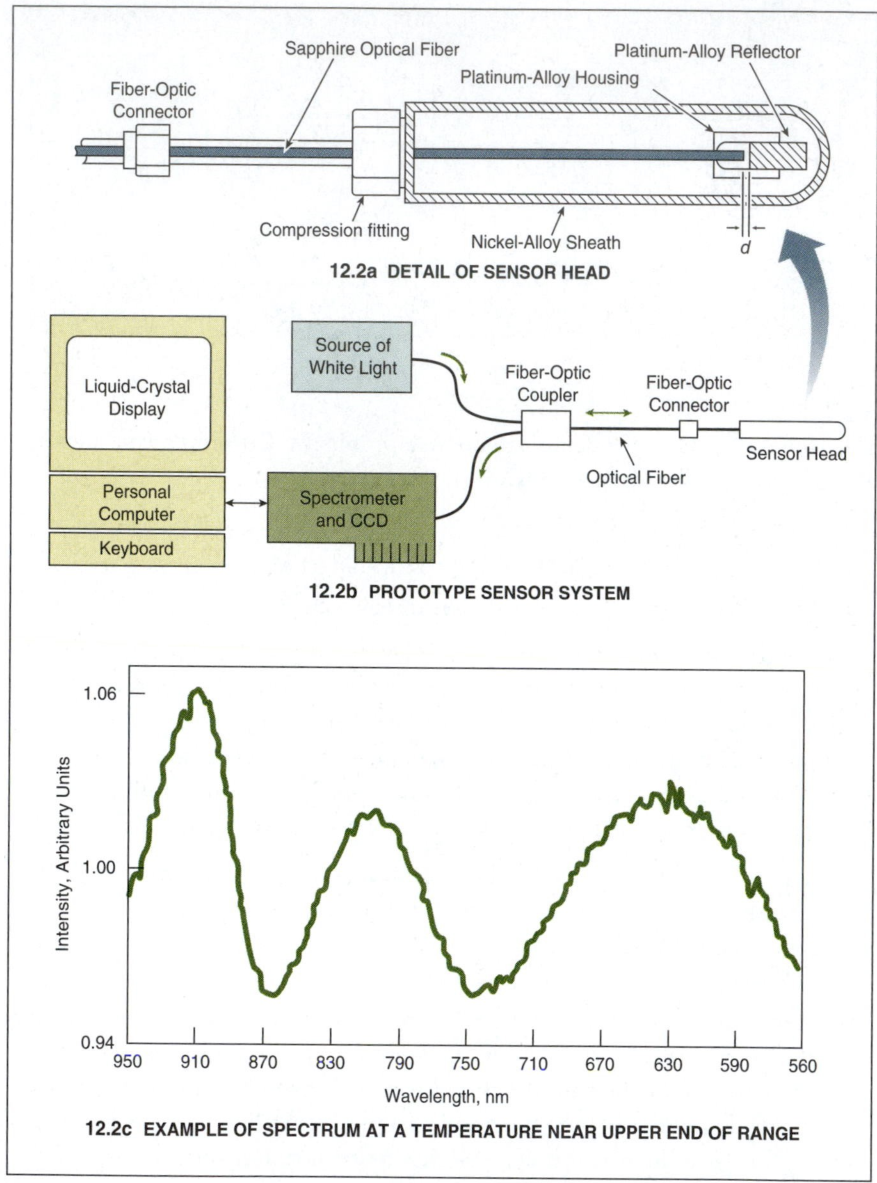

12.2a DETAIL OF SENSOR HEAD

12.2b PROTOTYPE SENSOR SYSTEM

12.2c EXAMPLE OF SPECTRUM AT A TEMPERATURE NEAR UPPER END OF RANGE

Various kinds of visuals — for example, graphs, maps, photographs, and drawings — can often be successfully used in combination to tell a story far more effectively than any single kind of visual. For an example of a technical fact sheet that is highly dependent on multiple visuals, go to www.english.wadsworth.com/burnett6e for a link.

CLICK ON WEBLINK

CLICK on Chapter 12/multiple visuals

WEBLINK

Adapting Visuals to Audiences

What's the difference between "dumbing down" and adapting to the audience?

Visuals can be adapted to different audiences by the complexity of content, presentation, and sometimes color and size. Audience members who are not experts need more frequent and simpler visuals than experts. Since nonexperts also may not understand visual conventions that experts readily recognize, they may need additional explanations beyond the standard level of titles, legends, and captions. For example, whereas an expert would know that the bars in a histogram represent ranges of data, nonexperts may need to have that explained.

Figures 12.3 and 12.4 illustrate one way that the same information can be adapted to two very different audiences. Both figures are about the same discoveries and hypotheses in the study of human evolution. Figure 12.3 is more easily understood by educated nonexperts interested in science; Figure 12.4 updates science professionals about new research.

Conventions in Referencing and Placing Visuals

Virtually all visuals have widely accepted conventions that accompany their use. Although you may sometimes choose to ignore a specific convention for good cause in a particular situation, generally you should follow these guidelines, which will help you to reference, label, and place visuals in ways that will be most useful to readers.

Textual Reference As a general practice, you should refer to visuals in the text, rather than simply including visuals and expecting the audience to see them and connect them to the appropriate section of the text. Do not assume that the audience will check a visual unless you refer to it. Include adequate information in your text reference such as the figure number and title. Textual references may be embedded references in sentences or parenthetical references:

- "The incision should be made just above a joint with a bud, as illustrated in Figure 2: Grafting."
- "The effectiveness of the antitoxins tested is presented in Figure 4: Antitoxin Response."
- "Table 5 shows the rapid increase of gas prices during a five-year period."
- ". . . (see Table 3)."

FIGURE 12.3 Visual for General Readers[4]

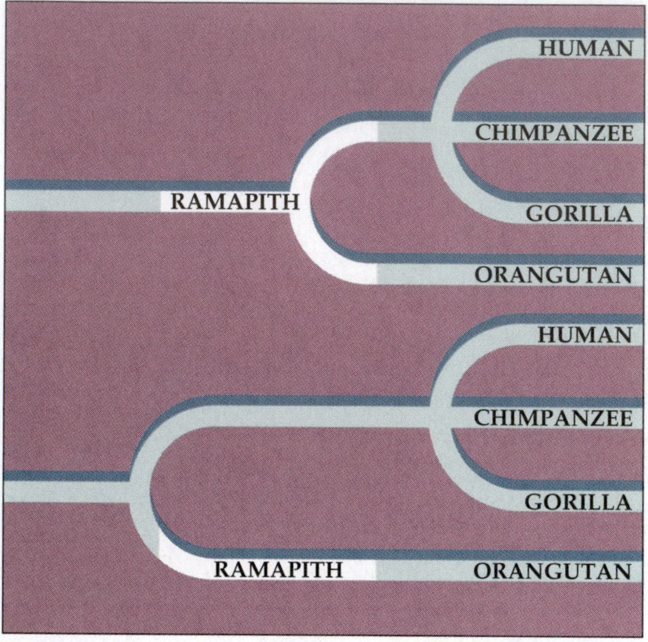

Rival interpretations of the ramapith's role in human evolution: If the top family tree is correct, ramapiths were the common ancestors of orangutans, African apes, and humans. An alternative interpretation, below, has ramapiths ancestral only to orangs. In this view, the common ancestor of humans and African apes remains unknown.

FIGURE 12.4 Visual for Professionals[5]

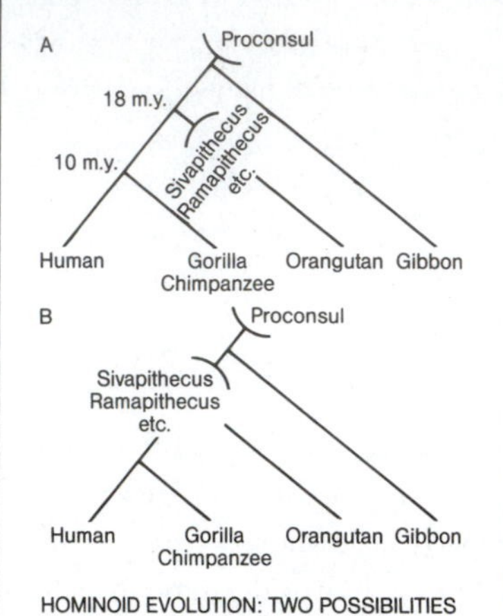

(A) *If the facial features of* Sivapithecus *and* Ramapithecus *are specialized characters shared with the modern orangutan, the group is not ancestral to the later apes and hominids. The dates of the divergences would then be as shown.* (B) *If the* Sivapithecus *orangutan face is primitive, then the group could be the common ancestor to the later hominids.*

In your reference, you can also suggest the particular focus or interpretation you expect the audience to apply when examining a visual. Without an explanation describing its significance, people may not understand a visual's purpose.

Describe how you would counter this statement: "The visual is right there on the page. I'll insult my readers if I not only tell them to look at the visual but also tell them what to pay attention to."

Labeling Complete and accurate labeling of visuals makes them much easier to use. Complete labeling includes identification, title, and caption.

Identification — Title

Table 1: Worker Fatigue Using Wire Cutters
Worker fatigue was compared using three different models of ergonomic cutters during a two-week period.

Caption

The following conventions are generally followed to help readers locate, interpret, and verify visuals the way you intend:

- If a formal report has more than five visuals or includes visuals that readers would need to access independently from the text, include a list of figures or list of tables at the beginning of the document.

- Include the complete dimensions of objects in each visual, making sure to specify the units of measure or scale.
- Whenever possible, spell out words rather than using abbreviations. If abbreviations are included, use standard ones and include a key.
- Identify the source of the data as well as the graphic designer.

Placement Generally, place visuals as close as possible following the text reference. Surround visuals with white space to separate them from the text of the document.

If a visual requires an entire page in a document that's printed on both sides of the paper, place the full-page visual on the page facing the text that refers to it. If a visual requires an entire page in a document printed only on one side of the paper, place the visual on the page following the text reference. Visuals that readers need to refer to repeatedly can be placed near the end of the document. For example, they can be located after the final text reference or in the first appendix on a fold-out page, as shown in this thumbnail sketch.

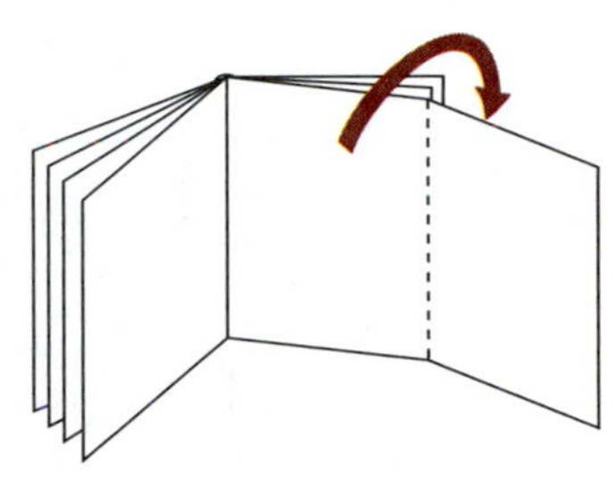

WEBLINK

The National Cancer Institute has published guidelines in both English and Spanish for developing publications that are heavily dependent on visuals for limited-literacy audiences. You can review the entire set of guidelines, or you can go directly to the subsection about integrating visual and verbal information. To access these links, go to www.english.wadsworth.com/burnett6e.

CLICK ON WEBLINK

CLICK on Chapter 12/NCI complete guidelines

CLICK on Chapter 12/NCI visual and verbal integration

Visual Functions

Visuals of different types — tables, graphs, diagrams, charts, drawings, maps, and photographs — all fulfill one or more functions in technical documents. Seven major functions of visuals are the focus of this section:

Locate a figure in this textbook that is particularly useful in helping you understand the information it conveys. Share your preference in class. Note distinct preferences that different readers have.

- Function 1: Provide immediate visual recognition
- Function 2: Organize numeric or textual data (for example, tables and diagrams)
- Function 3: Show relationships among numeric or verbal data (for example, tables, graphs, and diagrams)
- Function 4: Define or explain concepts, objects, and processes (for example, drawings, photographs, and diagrams)
- Function 5: Present chronology, sequence, or process (for example, line graphs, flow charts, organizational charts, and milestone charts)

- Function 6: Illustrate appearance or structure, which may include describing objects or mechanisms (for example, drawings, photographs, maps, and diagrams)
- Function 7: Identify facilities or locations (for example, maps, charts, schematics, and blueprints)

WEBLINK

Visuals are more than decorative. For a useful discussion about the many ways that illustrations can aid comprehension, go to www.english.wadsworth.com/burnett6e.

CLICK ON WEBLINK

CLICK on Chapter 12/cognitive benefits

Function 1: Provide Immediate Visual Recognition

Some things need rapid visual recognition; they range from the convenient (restrooms) to the critical (radiation). Typically, visual recognition can be provided by symbols that are used by most countries:

- A solid blue circle with a white symbol signals a safety precaution.
- A yellow triangle with a black band and black graphic warns about whatever is displayed in the triangle.
- A red circle with a slash and black graphic prohibits whatever is under the black slash.[6]

Quickly skim the common workplace safety symbols in Figure 12.5 to determine if you know what they mean. They are representative of the dozens of safety symbols used by business and industry around the world. Confirm your opinion by reading the marginal annotations.

All workplace organizations are responsible for educating and training their employees about the meaning of safety symbols. All individual employees have a responsibility to learn the safety symbols associated with their profession or industry.

WEBLINK

Symbols are important on safety labels, though they are more commonly used in Europe than in North America. One manufacturer of safety labels says that "European standards recognize that symbols have the ability to communicate across language barriers."[7] To access a link with more information about using symbols to communicate, go to www.english.wadsworth.com/burnett6e for a link.

CLICK ON WEBLINK

CLICK on Chapter 12/ISO safety symbols

FIGURE 12.5 **Safety Symbols**

Safety Symbols in Figure 12.5
1. *ear protection necessary*[8]
2. *biological hazard*[9]
3. *low temperatures*[10]
4. *lasers*[11]
5. *electrical hazard*[12]
6. *flammable, combustible*[13]
7. *crushing or pinch point*[14]
8. *explosive material*[15]
9. *hot surface*[16]

Symbols designed for immediate recognition can be representational (a symbol that looks similar to the actual object or situation) or abstract (an arbitrary symbol that becomes associated with a particular meaning). Why do these categories exist? What contributes to the effectiveness of each?

Function 2: Organize Numeric or Textual Data

Numeric and textual information identifying the characteristics of ideas, objects, or processes can be displayed in tables. The rows and columns of a table provide a system for classifying data and showing relationships that might be confusing if presented only in sentences and paragraphs.

Tables are often incorrectly seen as simply straightforward presentations of data. How can the design of a table influence the way readers interpret the data?

While the information in a table is usually organized in a way that gives readers the sense of reaching their own interpretations, the display itself shapes these interpretations. For example, readers are influenced by the sequence of the rows and columns, by the column and line heads, by the inclusion of footnotes, and so on. The text accompanying a table should discuss the information, providing readers with a further direction for their interpretation.

If the data in a table are self-contained and self-explanatory, they are usually boxed as well as surrounded with white space to set them apart from the text. Such tables are labeled with a number and a title. Established conventions for designing an effective table are listed here and illustrated in Figure 12.6.

- Place columns to be compared next to each other.
- Round numbers if possible.
- Limit numbers to two decimal places.
- Align decimals in a column.
- Label each column and row.
- Use standard symbols and units of measure.
- Use footnotes for headings that are not self-explanatory.
- Present the table on a single page whenever possible.

FIGURE 12.6 Model Format for Tables

Model of an Effective Tabular Format
Subtitle is an Optional Addition

rule

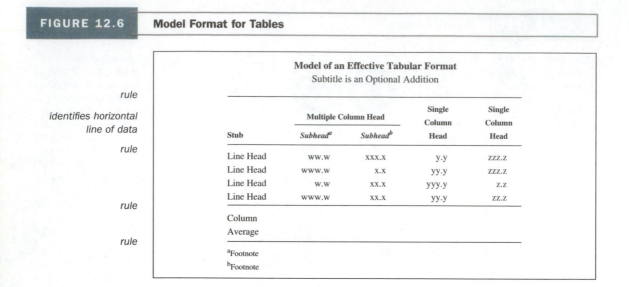

identifies horizontal line of data

rule

rule

rule

Stub	**Multiple Column Head**		**Single Column Head**	**Single Column Head**
	Subhead[a]	***Subhead***[b]		
Line Head	ww.w	xxx.x	y.y	zzz.z
Line Head	www.w	x.x	yy.y	zzz.z
Line Head	w.w	xx.x	yyy.y	z.z
Line Head	www.w	xx.x	yy.y	zz.z
Column Average				

[a]Footnote
[b]Footnote

Less formal tables are integrated into the text. These shorter tables depend directly on the surrounding text to provide a context for the data. They are not numbered or titled, but they usually do have clear row and column headings.

Numeric Tables That Organize Data In addition to their obvious organizational benefits, tables can present large amounts of numeric data in an accessible format that takes far less space than the verbal presentation of the same information would. For example, see Figure 12.7, which shows body weight and blood alcohol percentages. The same information presented in a paragraph would be difficult to follow and would force readers to work unnecessarily hard to get the information.

FIGURE 12.7 Table with Visual Cue for Interpreting Data[17]

Multiple column heads help readers chunk appropriate information.

Two visual cues — the shading and the right-hand column indicating the level of influence — make the table easier to read.

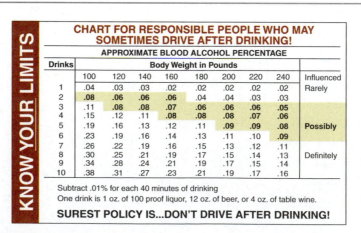

KNOW YOUR LIMITS

CHART FOR RESPONSIBLE PEOPLE WHO MAY SOMETIMES DRIVE AFTER DRINKING!

APPROXIMATE BLOOD ALCOHOL PERCENTAGE

Drinks	**Body Weight in Pounds**								
	100	120	140	160	180	200	220	240	Influenced
1	.04	.03	.03	.02	.02	.02	.02	.02	Rarely
2	**.08**	**.06**	**.06**	**.06**	.04	.04	.03	.03	
3	.11	**.08**	**.08**	**.07**	**.06**	**.06**	**.06**	**.05**	
4	.15	.12	.11	**.08**	**.08**	**.08**	**.07**	**.06**	
5	.19	.16	.13	.12	.11	**.09**	**.09**	**.08**	**Possibly**
6	.23	.19	.16	.14	.13	.11	.10	**.09**	
7	.26	.22	.19	.16	.15	.13	.12	.11	
8	.30	.25	.21	.19	.17	.15	.14	.13	Definitely
9	.34	.28	.24	.21	.19	.17	.15	.14	
10	.38	.31	.27	.23	.21	.19	.17	.16	

Subtract .01% for each 40 minutes of drinking
One drink is 1 oz. of 100 proof liquor, 12 oz. of beer, or 4 oz. of table wine.

SUREST POLICY IS...DON'T DRIVE AFTER DRINKING!

Tables can use visual cues to help readers understand how to interpret the data. For example, Figure 12.7 effectively uses shading to help readers interpret the connection between alcohol consumption and intoxication and, thus, make an informed decision about safe driving. Originally printed on a business card, this table is designed for people to carry in their wallets or pockets for quick reference.

Textual Tables That Organize Data. Although tables generally present numeric data, the tabular format is also appropriate for some primarily and even completely textual material. Textual tables can be economical and effective. Explaining the information in Figure 12.8 in sentences and paragraphs would be time-consuming, and the resulting text would be difficult to read. The columns in Figure 12.8 are clearly labeled. The rows are presented in alternating shades so that readers can easily see related information.

Function 3: Show Relationships

Visuals can be used to depict relationships in very different ways. First, they can show spatial relationships, such as proportion, proximity, size. Second, they can show quantitative relationships between sets of data.

Spatial Relationships. Spatial relationships are often depicted in various kinds of maps, although drawing and photographs are also frequently used. Typically, a large-scale map, photo, or drawing is presented, with a small area

FIGURE 12.8 Textual Information Organized in a Table[18]

MEET THE BUGS

Name	Possible Symptoms *(from most to least common)*	Foods that Have Caused Outbreaks	How Soon it Typically Strikes	How Soon it Typically Ends
Campylobacter (bacteria)	diarrhea (can be bloody), fever, abdominal pain, nausea, headache, muscle pain	chicken, raw milk	2 to 5 days	7 to 10 days
Ciguatera (toxin)	numbness, tingling, nausea, vomiting, diarrhea, muscle pain, headache, temperature reversal (hot things feel cold and cold things feel hot), dizziness, muscular weakness, irregular heartbeat	grouper, barracuda, snapper, jack, mackerel, triggerfish	within 6 hours	several days (neurological symptoms can last for weeks or months)
Clostridium botulinum (bacteria)	marked fatigue; weakness; dizziness; double vision; difficulty speaking, swallowing, and breathing; abdominal distention	home-canned foods, sausages, meat products, commercially canned vegetables, seafood products	18 to 36 hours	get treatment immediately
Cyclospora (parasite)	watery diarrhea, loss of appetite, weight loss, cramps, nausea, vomiting, muscle aches, low-grade fever, extreme fatigue	raspberries, lettuce, basil	1 week	a few days to 30 days or more
***E. coli* O157:H7** (bacteria)	severe abdominal pain, watery (then bloody) diarrhea, occasionally vomiting	ground beef, raw milk, lettuce, sprouts, unpasteurized juices	1 to 8 days	get treatment immediately
Hepatitis A (virus)	fever, malaise, nausea, loss of appetite, abdominal pain, jaundice	shellfish, salads, cold cuts, sandwiches, fruits, vegetables, fruit juices, milk, milk products, infected food handlers	10 to 50 days	1 to 2 weeks

circumscribed; that enclosed area is then enlarged so that it can be shown in greater detail. Each individual map, photo, or drawing must be interpreted in the context of the other(s). The maps in Figure 12.9 depict the May 13, 2002, Gilroy, California, earthquake, which was recorded by Advanced National Seismic System (ANSS) instruments "designed to detect strong ground shaking from urban earthquakes." The small map of California has an area marked in red, which is then enlarged to provide the primary visual focus (map A). Identifying the area where the ANSS instruments recorded the highest ground motion on map A is easy: It's the red bull's eye. Curiously, though, this area is not the earthquake's epicenter, but is about 10 kilometers east-northeast of the epicenter. "The absence of higher ground motion near the epicenter and the portrayal of localized areas of anomalously high or low ground motion marked by bull's eye contour patterns are indicative of too few ANSS stations in the area." This statement is reinforced by map B, an enlargement of the boxed section of map A. Map B shows that a dense array of instruments makes obvious the variability in ground motion. The maps are making a clear argument: "The number of ANSS-quality seismic stations in the San Francisco Bay Area is insufficient to map out the variability in ground motion from earthquakes."[19]

FIGURE 12.9 **Spatial Relationship**[20]

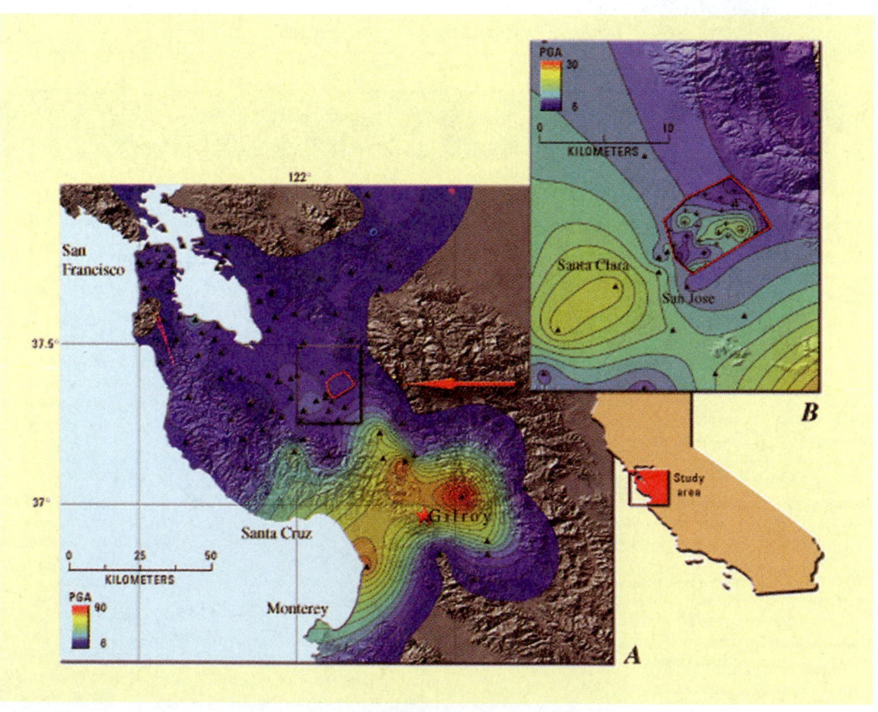

Quantitative Relationships. Relationships between two or more sets of data can be displayed using several types of visuals, but the most frequently used are graphs, including line graphs, scatter graphs, pie graphs, bar graphs, and pictorial graphs.

What graphs are most commonly used in your discipline or profession? What are ways to distort data with these graphs?

Line graphs show the relationship between two values, represented by intersecting values projected from the *abscissa* (horizontal) and *ordinate* (vertical) axes on a coordinate grid. They usually plot changes in quantity, showing the exact increases and decreases over a period of time, whether minutes, days, decades, or centuries, or other quantifiable variables. Since line graphs are one of the most commonly used forms of displaying relationships, most readers are familiar with them. In constructing line graphs, these conventions are usually followed:

- Use the horizontal axis to depict time, some event occurring over time, or some other quantifiable variable.
- Limit the number of lines on a graph to those easily interpreted by readers.
- When using more than one line, differentiate the lines by design or color and use a key or label to identify each line.
- Add notes or labels to make information clear to readers.
- Keep the vertical and horizontal axes proportionate.

The line graph in Figure 12.10 was produced by an energy company for a major medical center to show its long-term trends in energy use in a new building and factors that affect these trends. The audience can see the beginning of each year (January) and then every other month (March, May, July, September, November) for a 12-year period. Five lines are plotted: current and expected gas use, current and expected electrical use, and square footage of the building. Without any energy conservation, both electric and gas use grew at approximately 3 percent and 5 percent respectively each year. Referred to as "creep," this growth results from increased amounts of medical equipment and the normal deterioration of heating and air conditioning systems.

One of the features that makes this graph particularly useful is that milestones are marked. For example, the very end of 1989 shows a dramatic increase in energy consumption; the callout notes this is when initial occupancy began, with full occupancy a little more than two years later. The text accompanying the graph provides this explanation:

> occupancy is not an instantaneous process. Not every department can move in on the same day and heating and air conditioning systems must be turned on for the whole building even though it is not yet fully occupied, and full "occupancy" is also not immediate in terms of patients, staff and medical and other equipment.[21]

The callouts also note the beginning and termination of an energy conservation program, which may account for otherwise unexplained increases and declines. And finally, the callouts indicate projected trends.

FIGURE 12.10 **Multiple Line Graph**[22]

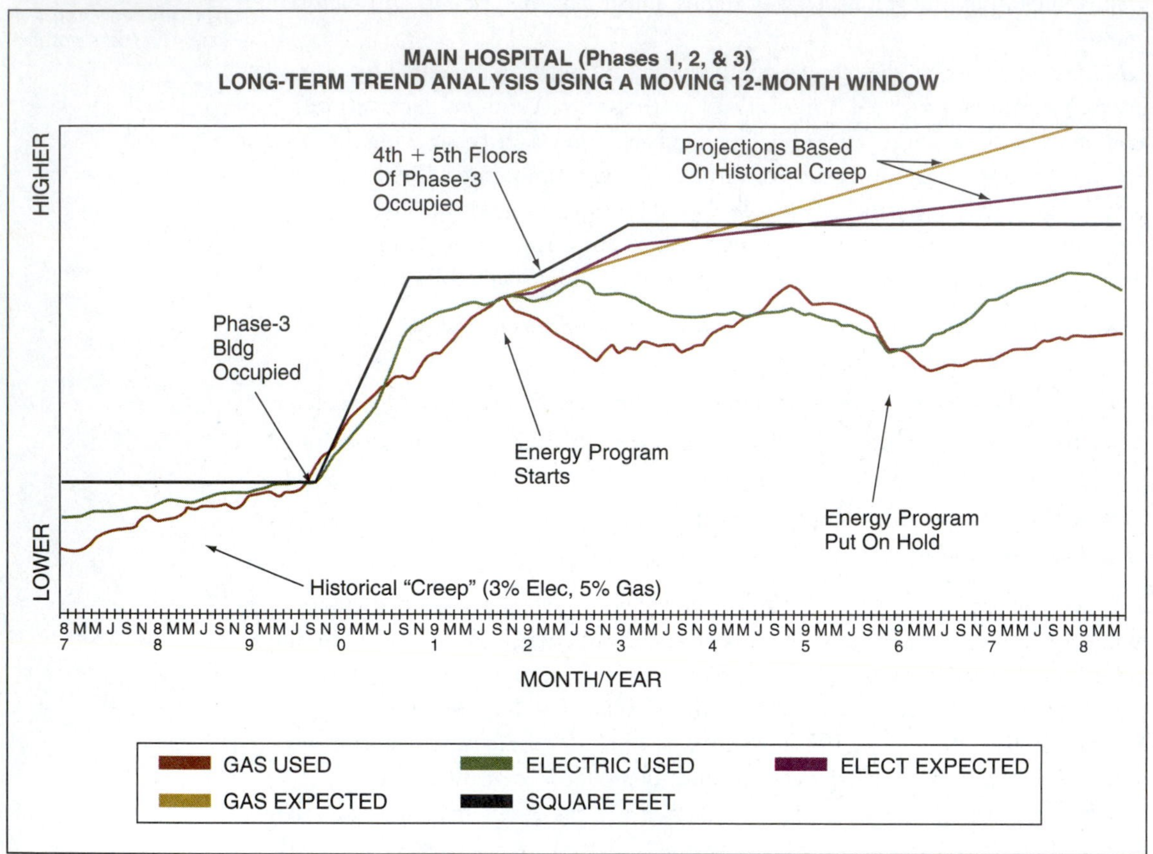

Scatter graphs use single, unconnected dots to plot instances where two variables (one on each axis) meet. Usually, scatter graphs are plotted on graphs where the x-axis (horizontal) and y-axis (vertical) are proportionate. However, if the range of data is very large, the data can be charted logarithmically to show more clearly the direction of the correlation.

The pattern of the dots expresses the relationship between variables, as illustrated in Figure 12.11. If the dots are randomly scattered, the two variables have no correlation. If the dots are primarily on a diagonal running from the lower left to the upper right, the correlation is positive. If the diagonal runs from the upper left toward the lower right, the correlation is negative.

The correlation between the variables is sometimes highlighted by shading an area on the graph. However, the statistical significance of the correlation must also be discussed in the text. Because interpreting scatter graphs is often difficult, their use is generally limited to professional and expert audiences.

FIGURE 12.11 Relationships Displayed in a Scatter Graph[23]

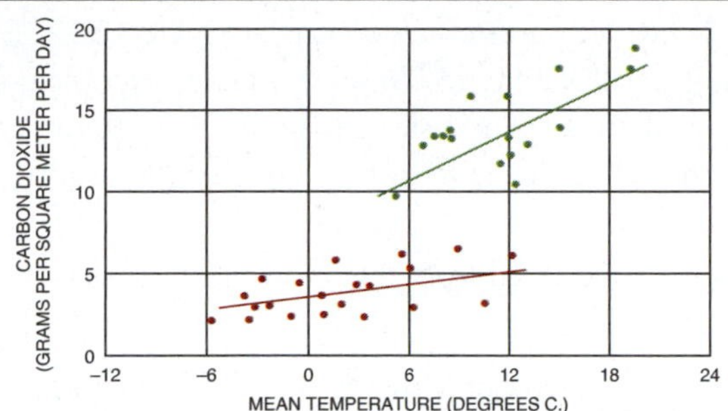

RESPIRATION OF THE FOREST, plotted against temperature, is seen to proceed at a higher rate in summer (colored curve) than in winter (black curve). Annual respiration was calcualted in grams of carbon dioxide, then converted to yield the total respiration, 2,100 grams.

Adding a plot curve and using two colors, one for summer rates and the other for winter rates, helps readers see the relationship between the variables.

The clear caption helps readers interpret the information in the graph

One kind of distortion and manipulation with graphs can be avoided if you understand the impact of linear and logarithmic scales. For a link to a useful discussion, go to www.english.wadsworth.com/burnett6e.

CLICK ON WEBLINK

CLICK on Chapter 12/tricky graphs

WEBLINK

FIGURE 12.12 Relationships Displayed in a Pie Graph[24]

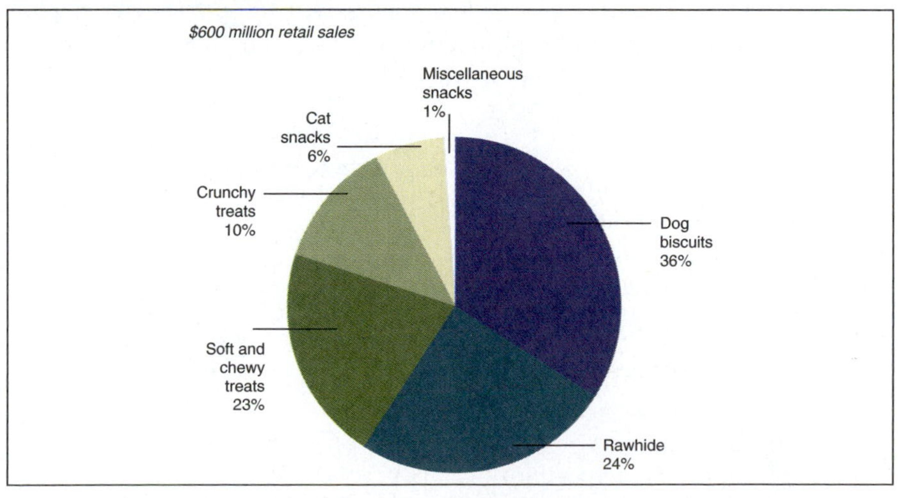

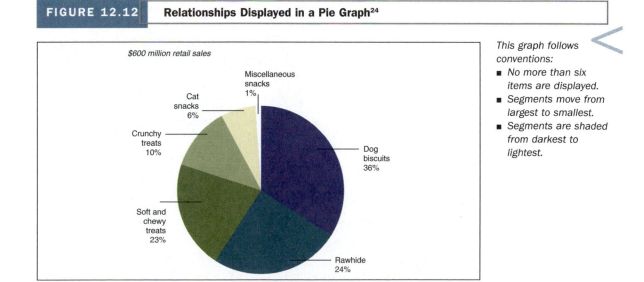

This graph follows conventions:

- *No more than six items are displayed.*
- *Segments move from largest to smallest.*
- *Segments are shaded from darkest to lightest.*

Pie graphs — also called pie diagrams, pie charts, percent graphs, or divided circle graphs — emphasize the proportionate distribution of something, frequently money or time. Pie graphs total 100 percent, with each percent representing 3.6 degrees of the circle, as shown in Figure 12.12. Even though software gives you great flexibility in the way you construct pie graphs, following these conventions will make them easier for the audience to understand:

- Slices of the pie are arranged from largest to smallest, starting at "noon" and moving clockwise.
- Slices of the pie are colored from darkest to lightest, starting at "noon" and moving clockwise.

Pie graphs can make a striking visual display, but the significance of the presented information must be discussed in the accompanying text. Although they are popular attention-getting devices that focus the reader's attention for examining more detailed data, they are generally unsuitable for the comparison of more than five or six items. The primary problem is the impossibility of comparing areas. Additionally, the visual difference between areas representing similar percentages is minimal.

Bar graphs can show several kinds of relationships, including comparisons, trends, and distributions. Like line graphs, bar graphs are drawn from a series of values plotted on two axes, but the values are represented by vertical or horizontal bars instead of points joined by a line. Because each bar represents a separate quantity, bar graphs are especially appropriate when the data consist of distinct units, such as tons of grain or megawatts of hydroelectric power produced over a specified period.

Commonly used bar graphs include a simple bar graph, subdivided bar graph, and subdivided 100-percent bar graph. Figures 12.13, 12.14, and 12.15 illustrate how the same data plotted on these three kinds of bar graphs can create quite a different appearance for readers.

Other variations of bar graphs, such as a subdivided 100-percent area graph, multiple bar graph, sliding bar graph, and floating bar graph, shown in Figures 12.16, 12.17, 12.18, and 12.19, are appropriate for somewhat more complex displays of information.

In the *simple bar graph* in Figure 12.13, all bars represent the same type of information, so the differences are stressed. These guidelines apply to any bar graph:

- Make bars the same width.
- Make the space between bars one-half the bar width.
- Label each bar.

FIGURE 12.13 **Simple Bar Graph**

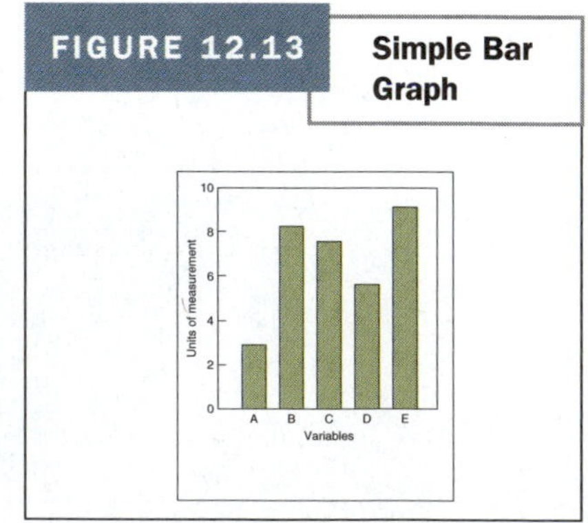

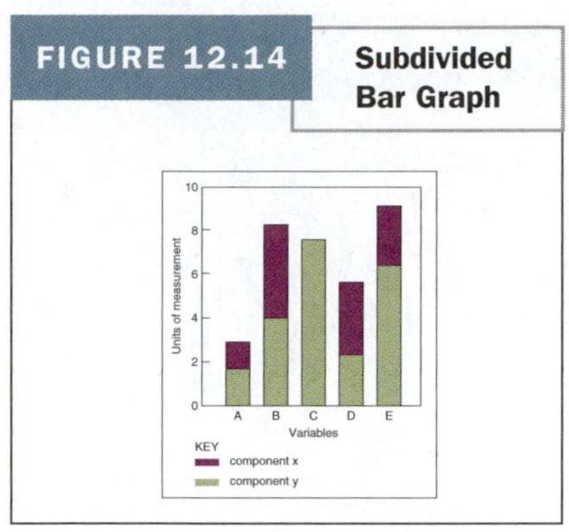

FIGURE 12.14 Subdivided Bar Graph

In the *subdivided bar graph* in Figure 12.14, each bar is subdivided to represent the magnitude of different components. Parts are differentiated by shading or cross-hatching. Although the total magnitude of each bar can be compared, as in a simple bar graph, the individual components are not easily compared.

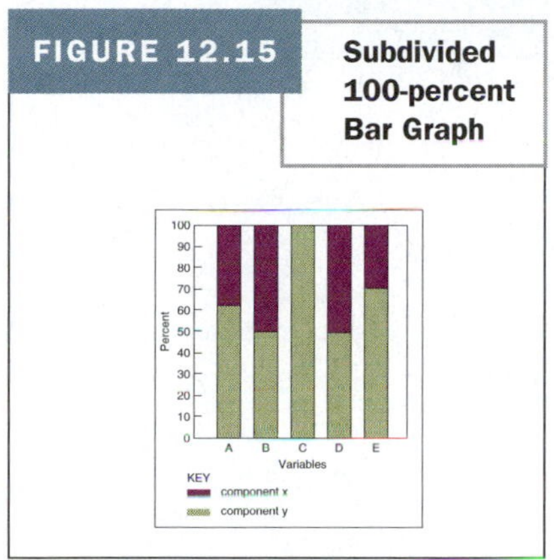

FIGURE 12.15 Subdivided 100-percent Bar Graph

In the *subdivided 100-percent bar graph* in Figure 12.15, each bar extends to 100 percent, and the components of the bar are separated by percentage. Unlike simple bar graphs and subdivided bar graphs that enable you to compare total magnitude, subdivided 100-percent bar graphs enable easy comparison of the individual components.

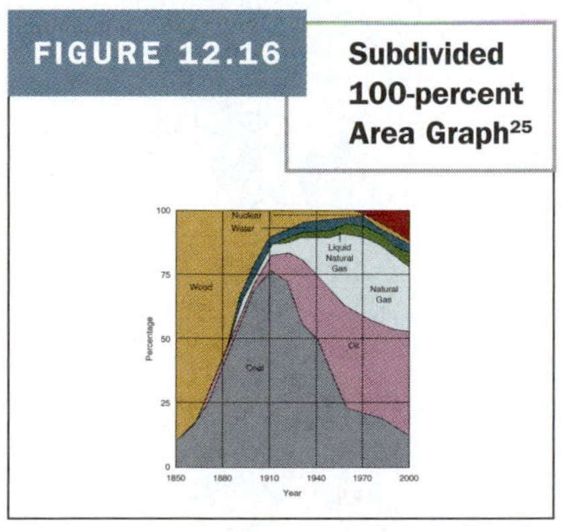

FIGURE 12.16 Subdivided 100-percent Area Graph[25]

Figure 12.16 illustrates a variation of a subdivided 100-percent bar graph — a *subdivided 100-percent area graph.* Imagine a series of 150 bars, each extended to 100 percent and each bar subdivided to show the percentage of, say, types of fuel used during one year. The bars are then pushed together so the overall effect is a continuous line for each category. The area under each bar is shaded to make the distinctions clear.

A *multiple bar graph* groups two or more bars to present the magnitude of related variables. The example in Figure 12.17 compares "Feature X" and "Feature Y" over an eight-year period.

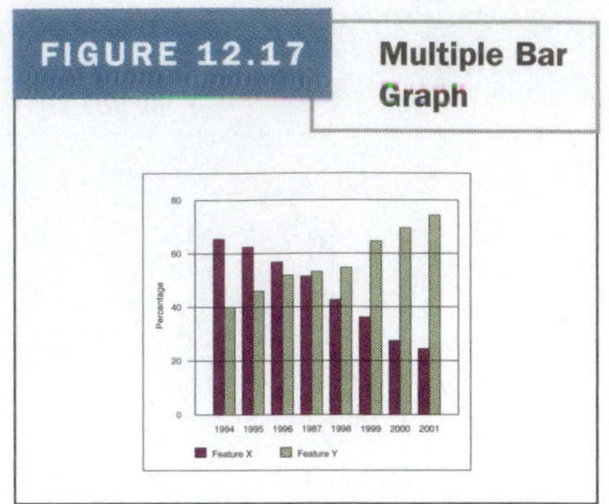

FIGURE 12.17 Multiple Bar Graph

The bars in the *sliding bar graph* in Figure 12.18 move along an axis usually marked in opposing values (active/passive, hot/cold) that extend on either side of a central point, such as values on a temperature scale.

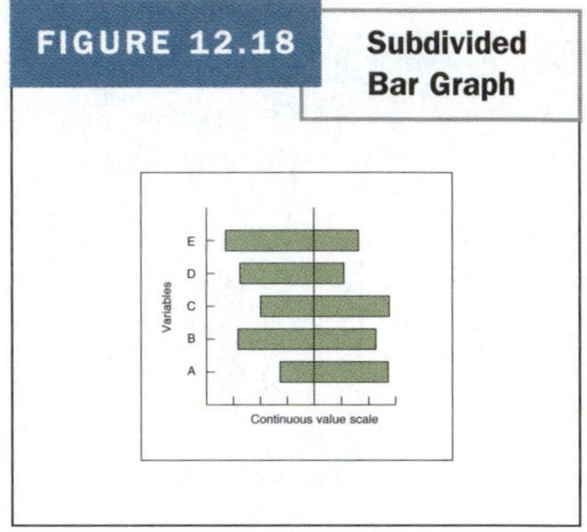

FIGURE 12.18 Subdivided Bar Graph

The *floating bar graph* in Figure 12.19 has bars that "float" in the area above the x-axis, which may extend below zero.

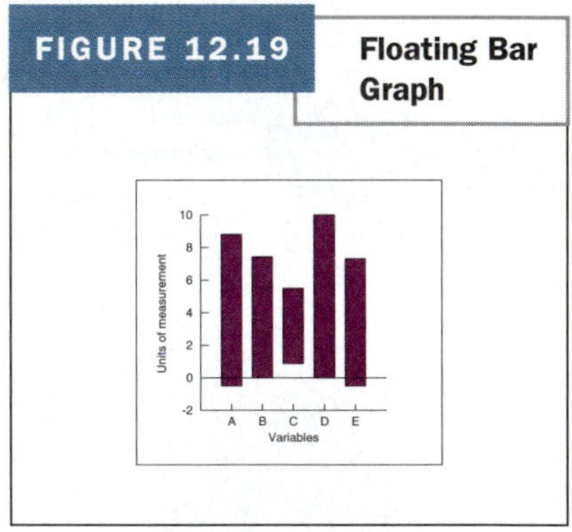

FIGURE 12.19 Floating Bar Graph

Pictorial graphs use actual symbols to make up each bar. Each symbol (*isotype*) represents a specific number of people or objects. Pictorial graphs are very appealing and are widely used with many audiences. Problems arise, however,

FIGURE 12.20 **Problems of Pictorial Graphs**

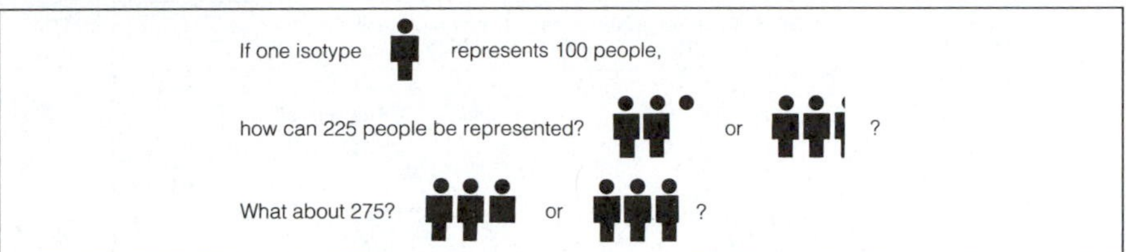

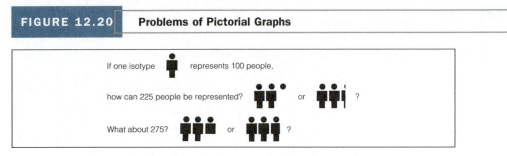

when depicting fractions (see Figure 12.20). Following these guidelines when creating pictorial graphs generally avoids problems:

- Round off numbers to eliminate fractions.
- Make all symbols the same size and space them equally.
- Select symbols that are clearly representative of the object.

Another version of a pictorial graph uses single isotypes of different sizes to represent the quantity or magnitude of each variable, as shown in Figure 12.21. In this graph, the increasing number of children allowed in day care groups is represented by successively larger isotypes representing children in each age group. Such a graph is appropriate for attracting reader attention, but it should not be used to present technical data.

What's the difference between simply ignoring graph conventions and purposely flouting those conventions in relation to the information being accessible, comprehensible, and usable? How might each approach affect your credibility?

Flouting Graph Conventions. Adhering to conventions for designing graphs is generally a good idea. Flouting those conventions, however, may be done for good cause. For example, Figure 12.22 shows that renewable energy consumption made up six percent of total U.S. energy consumption, a total of 96.6 quadrillion BTUs. If the graph were created conventionally, the pieces of the pie would be in descending order (starting at "noon") from the largest to the smallest. However,

FIGURE 12.21 **Relationships Displayed in Pictorial Graphs[26]**

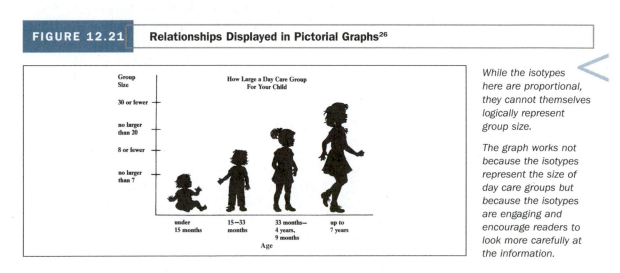

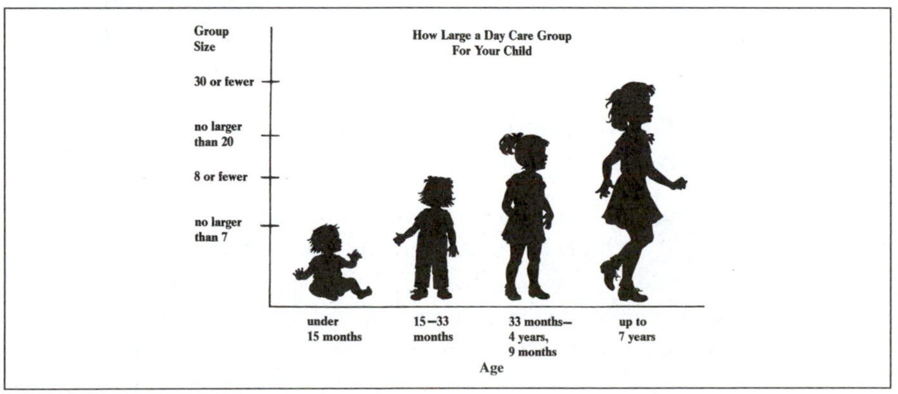

While the isotypes here are proportional, they cannot themselves logically represent group size.

The graph works not because the isotypes represent the size of day care groups but because the isotypes are engaging and encourage readers to look more carefully at the information.

FIGURE 12.22 **Purposely Flouting Design Conventions**[27]

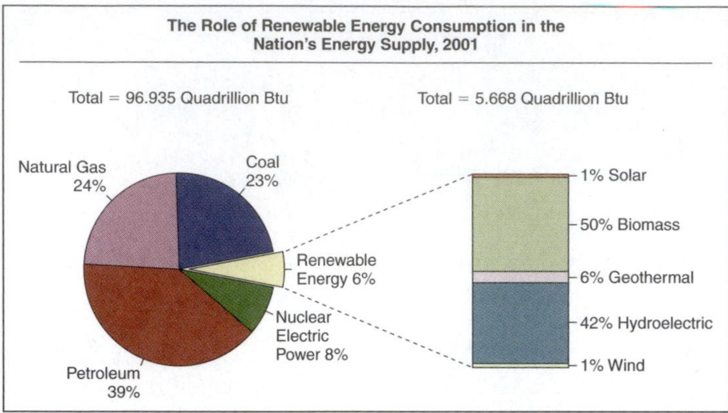

for the overall energy consumption to be connected to the sources of renewable energy, the conventions needed to be flouted, so the pieces of the pie aren't in descending order. Instead, the smallest slice is placed at about 3:00 rather than at about 11:00, where it would be conventionally. As a result, the audience can easily see that of the overall six percent devoted to renewable energy sources, biomass provided 50 percent, hydroelectric energy provided 42 percent, and other sources collectively provided the remaining eight percent.[28]

WEBLINK

One of the most frequent visual forms for professional publication is technical or scientific posters. The primary appeal of posters is their limited text and their dependence on visuals to convey critical information. For a link to several useful sets of guidelines for creating posters, go to www.english.wadsworth.com/burnett6e. (Also see an example of a scientific poster in Chapter 18.)

CLICK ON WEBLINK

CLICK on Chapter 12/posters

Function 4: Define Concepts, Objects, and Processes

How does your discipline or profession use visuals for definition?

Visuals can be exceedingly valuable as definitions. The drawings of types of screw heads in Figure 12.23 are more efficient and useful than textual descriptions. Visuals can illustrate details that are difficult to describe, such as the types of heads and slots on the top of screws. Additionally, the angle of a flathead screw is easily depicted in a drawing. Explaining the same information in words would not be nearly as effective.

FIGURE 12.23 **Visual Definition of Screws**[29]

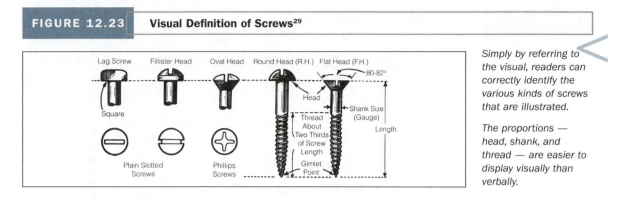

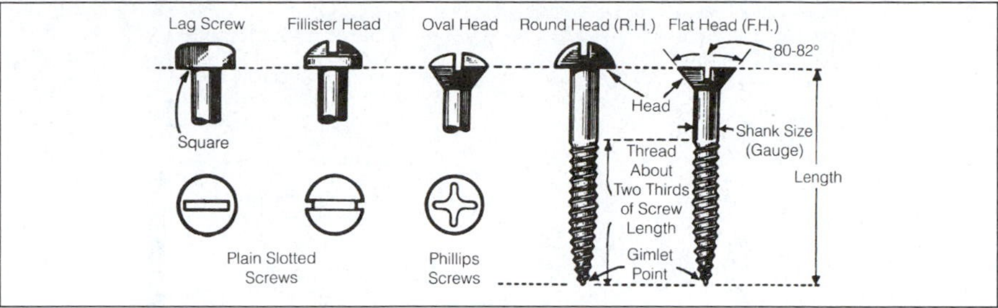

Simply by referring to the visual, readers can correctly identify the various kinds of screws that are illustrated.

The proportions — head, shank, and thread — are easier to display visually than verbally.

Figure 12.24 shows a chromosome with attention in the drawing to the structure of the DNA in one gene. Each chromosome contains many genes, the basic physical and functional units of heredity. Genes include specific sequences of bases that encode instructions about how to make proteins. All the instructions needed to direct their activities are contained within the chemical DNA. In this figure, the structure of the DNA is elegantly and simply communicated through the use of color in the spiraled double helix. This structure determines the exact instructions required to create a particular organism with its own unique traits. The figure clearly defines the DNA sequence through the side-by-side arrangement of bases along the DNA strand (e.g., ATTCCGGA).[30]

Visuals may be more appropriate than text when readers need a definition of an unfamiliar or complex object or process. Figure 12.25 provides a good example. In this visual, the toilet on the space shuttle is explained clearly, with a discrete, non–gender-specific human figure that helps readers understand how the system works. Sufficiently large font and high figure-ground contrast make the

FIGURE 12.24 **Chromosome with DNA Structure Defined**[31]

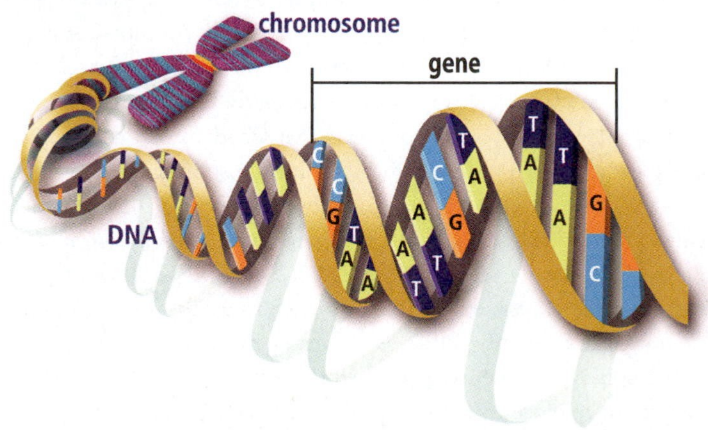

FIGURE 12.25 Visual Definition of NASA Commode[32]

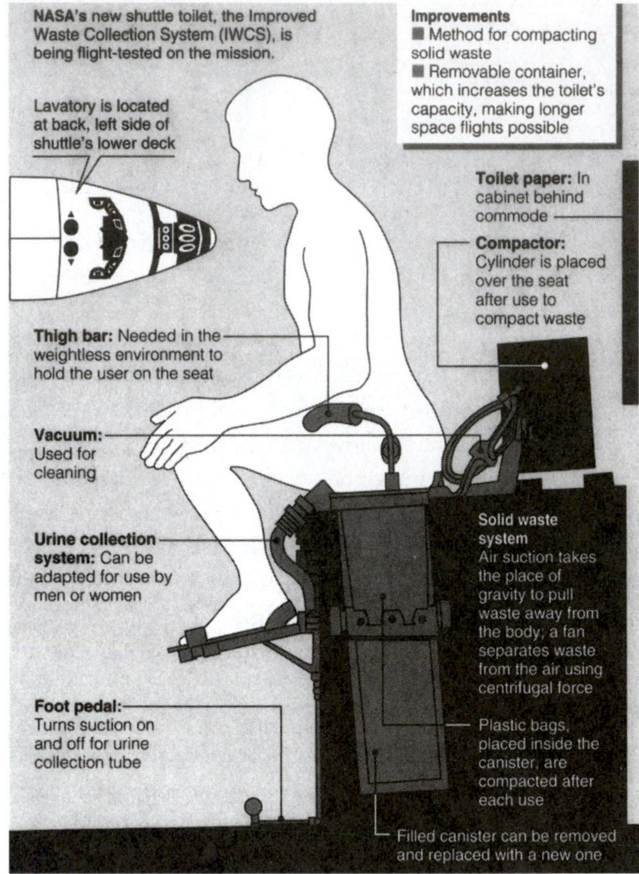
NASA's new shuttle toilet, the Improved Waste Collection System (IWCS), is being flight-tested on the mission.
Improvements
■ Method for compacting solid waste
■ Removable container, which increases the toilet's capacity, making longer space flights possible
Lavatory is located at back, left side of shuttle's lower deck
Toilet paper: In cabinet behind commode
Compactor: Cylinder is placed over the seat after use to compact waste
Thigh bar: Needed in the weightless environment to hold the user on the seat
Vacuum: Used for cleaning
Urine collection system: Can be adapted for use by men or women
Solid waste system
Air suction takes the place of gravity to pull waste away from the body; a fan separates waste from the air using centrifugal force
Foot pedal: Turns suction on and off for urine collection tube
Plastic bags, placed inside the canister, are compacted after each use
Filled canister can be removed and replaced with a new one

information legible. Readers' comprehension can sometimes be faster and more complete when visuals illustrate overall concepts as well as details.

Color is important in visually defining certain objects and organisms. For example, color patterns help to identify insects and to distinguish between insects that look identical except for color patterns, particularly butterflies. Figures 12.26 and 12.27 show two brilliantly colored insects that are identified in large part by their color. Figure 12.26 is a photograph of a common short-horn grasshopper (*Phymateus saxosus*), which entomologist and photographer Tom Myers took when he stopped along a roadside in Madagascar. The particular combination of colors is unique to grasshoppers in Madagascar. Figure 12.27 is a photograph of a wasp moth (*Orcynia calcarata*) that Myers took while in French Guiana. The black and yellow of this moth mimics the coloration of a wasp, warning away birds and other predators. Myers's photographs are frequently published in magazines, but he also uses them in presentations about rainforest insects to professional, school, and community groups.

FIGURE 12.26 Photograph of Short-Horn Grasshopper

© Tom Myers

FIGURE 12.27 Photograph of Wasp Moth

© Tom Myers

Function 5: Present Action or Process

Visuals are particularly appropriate for action views and processes. While visuals vary widely according to the process being presented, actions are particularly easy to depict in a sequence, and processes are easy to depict in various kinds of charts.

Action Sequences. Figure 12.28 shows the breach of a whale. The sequence of five drawings with arrows to indicate the direction of the movement enables readers to understand the breach without referring to the text. The visual tells a story by itself.

Charts. Charts can represent the components, steps, or chronology of an object, mechanism, organism, or organization. The most common charts are block charts, organizational charts, and flowcharts.

How does your discipline or profession visually present actions or processes?

FIGURE 12.28 Drawing Showing Action: Whale Breaching[33]

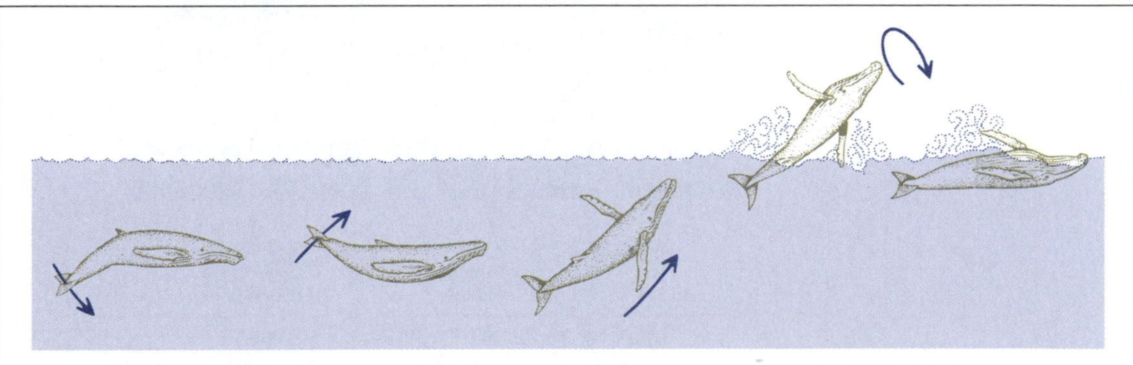

A *block chart* (also called block diagram or classification chart), illustrated in Figure 12.29, uses blocks to represent the components or subdivisions of the whole object, system, or process. This example uses shapes, colors, and arrow placement to increase the likelihood that the audience will access, comprehend, and use the information about the components of Multidisciplinary Design Optimization (MDO), a computer software analysis procedure used to streamline the complicated aircraft design process.

At top, the purple ovals represent the various disciplines necessary to aircraft design, such as aerodynamics and acoustics. At center, the ovals and arrows combine to illustrate the cyclical and complex design process. The MDO networks all of these disciplines throughout each phase of the process. At bottom, the green boxes depict the variety of aircraft products, such as commercial aircraft, launch vehicles, that can be designed, as in Figure 12.31.

Figure 12.29 uses both arrows and color to draw your eyes from the top of the block diagram to the bottom, while the use of a gradient purple color in the

FIGURE 12.29 **Block Chart**[34]

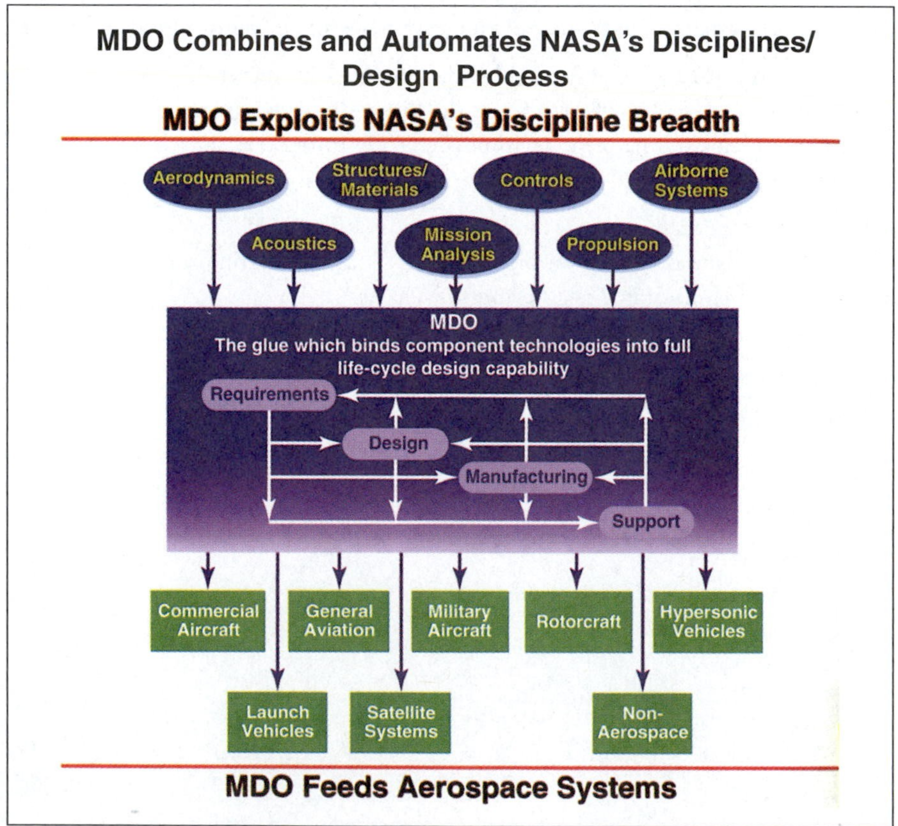

center box also draws your eyes from the top to the bottom. The purple arrows against the white background lead your eye from the purple ovals through the center box to the green "aircraft products" boxes.

An *organizational chart* portrays the hierarchy of an organization by putting each position in a separate block, as in Figure 12.30. This chart shows the vertical and horizontal relationships in the U.S. Department of Energy's Office of Environmental Restoration and Waste Management. In addition to the basic information, the chart also provides icons and brief descriptions of the responsibilities associated with each office.

A *flowchart* (also called a route chart) depicts the sequence of steps in a process. Conventional symbols make flowcharts easy to comprehend. Such charts sometimes also indicate the amount of time each step takes. Conventional flowcharts use standard symbols — for example, ▭ = a step in the process and ◇ = a decision in the process — that make flowcharts easy for professionals to understand regardless of their specialization or language, as in Figure 12.31.

However, some charts designed for general audiences use the principles of flowcharts but substitute small diagrams or drawings for the conventional

FIGURE 12.30 **Organizational Chart**[35]

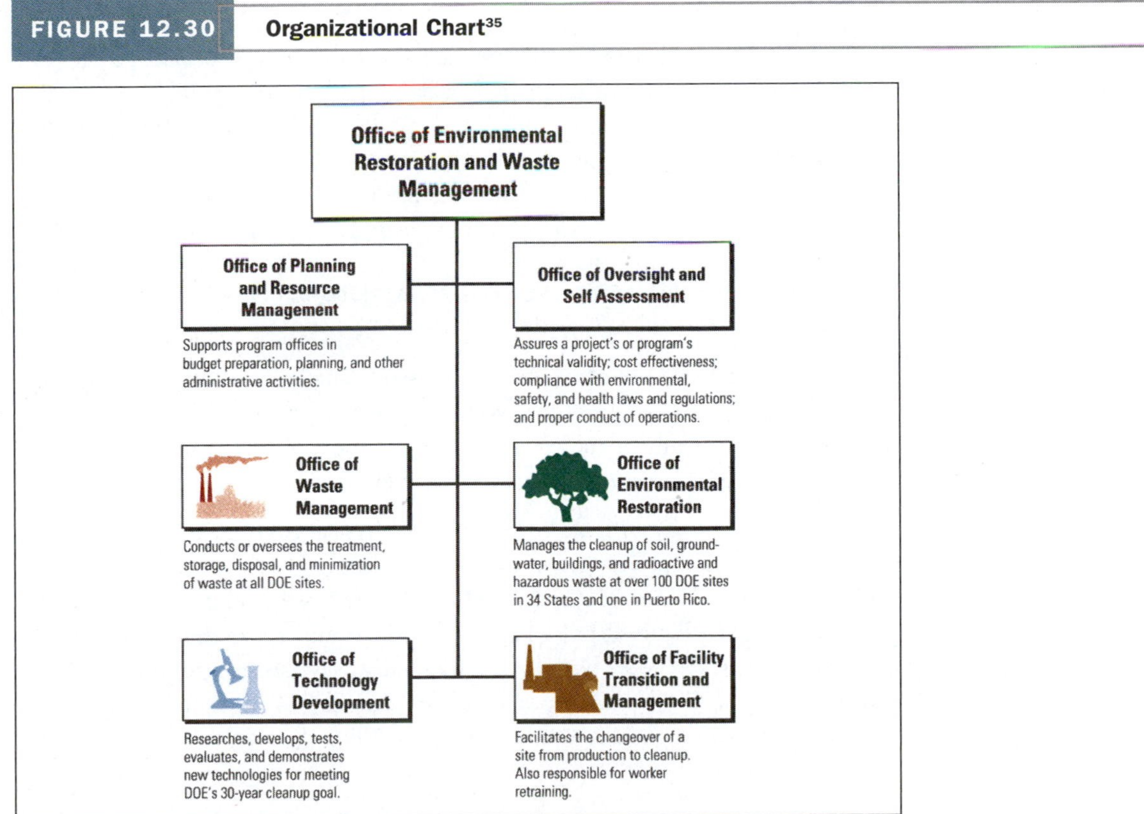

FIGURE 12.31 **Flowchart**[36]

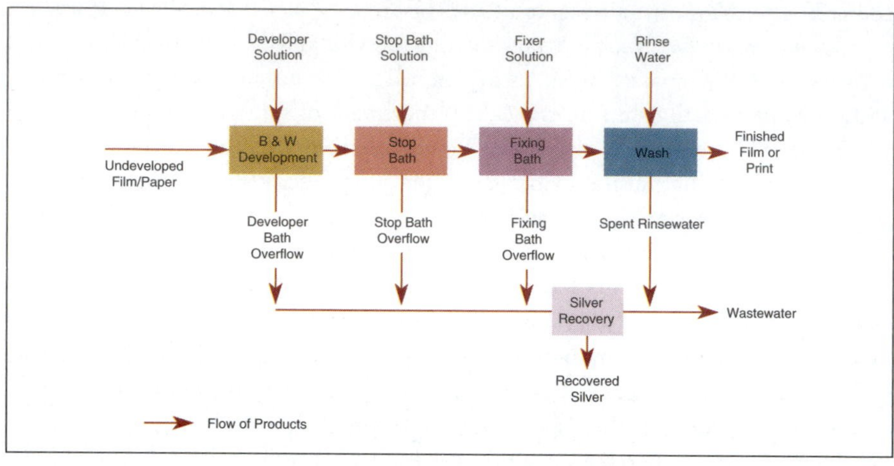

symbols. Figure 12.32 is an example of such a modification. It provides a simple visual overview of the manufacture of Therban®, Bayer's original high-performance specialty elastomer, which has been specially designed for demanding applications requiring exceptional durability in aggressive environments. This visual overview increases audience comprehension by having clear beginning and ending points as well as by labeling each major stage in the process.

Function 6: Illustrate Appearance, Structure, or Function

> *How does your discipline or profession visually depict parts of objects, mechanisms, or organisms?*

Physical characteristics are often easier to present visually than verbally. Diagrams and drawings are especially effective ways to show the parts of objects, mechanisms, or organisms and the relationships among those parts. Only the parts readers need to know about are represented.

Diagrams. *Diagrams* illustrate the complex physical components and structures of objects, mechanisms, or organisms. Indeed, they are often easier to understand than photographs or representative drawings because readers are not distracted by unnecessary details. The diagram in Figure 12.33 shows a system comprised of a motor, heat exchanger, tank, and valves used to cool the space shuttle's cryogenic liquid rocket propellants. The system works to continuously circulate warmer cryogenic liquid with subcooled liquid.

FIGURE 12.32 **Visual Overview of Process**[37]

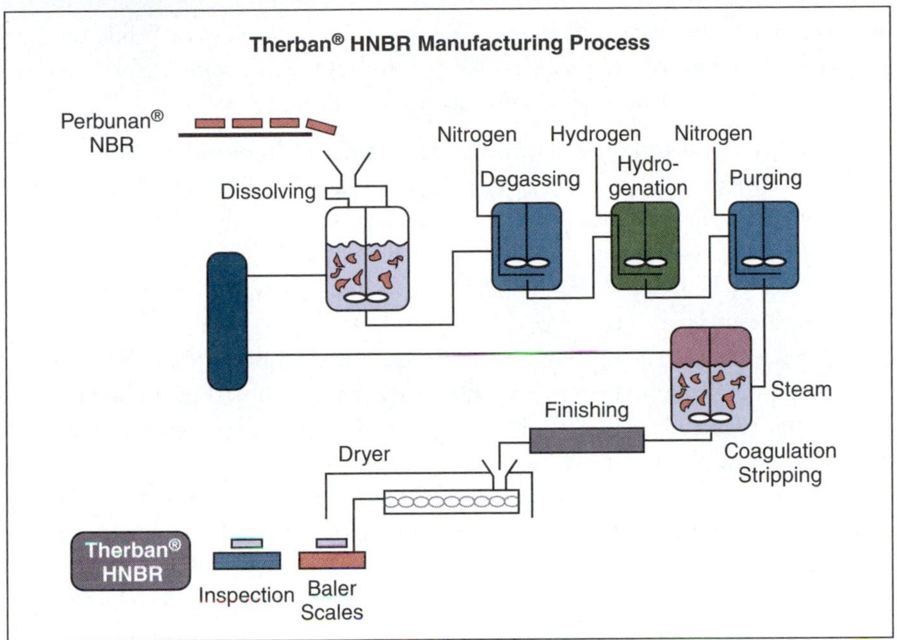

FIGURE 12.33 **Diagram**[38]

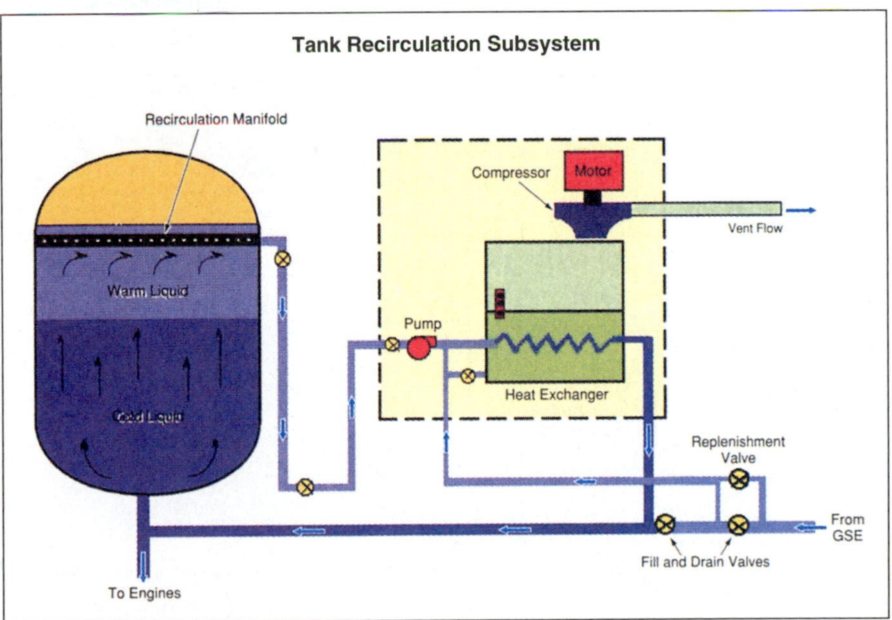

Color distinguishes the warmer liquid (dark purple) from the cooler liquid (light purple), while the blue arrows signal the direction of the flow. The white outlines of these blue arrows make them easier to distinguish. The green of the heat exchanger also helps to illustrate the flow of liquid through it. While the diagram's components are clearly labeled, a consistent use of color enables you to easily identify the valves (yellow) and the motor and pump (red).

Drawings. *Drawings* depict the actual appearance of an object or organism. Unlike a photograph, a drawing can delete details and emphasize more important portions. Drawings are appropriate when you want to focus on specific characteristics or components of a subject. A drawing does not have to be complicated to be effective. Figure 12.34 shows just how well a simple drawing can illustrate an abstract concept.

Various components and aspects of objects, mechanisms, and organisms can be shown by different drawing views. Drawings that a technical professional might refer to include perspective drawings, phantom views, cutaway views, exploded views, and action views.

Figure 12.35 shows a three-dimensional view of the Nebula GNX car. This drawing was produced using a special computer-aided design (CAD) software that allows the user to create realistic images of a variety of different shapes and surfaces. While Figure 12.35 shows a three-dimensional view, the view itself is not realistic — that is, the play of light and shadow on the car allows you to see both the outlines of the car's exterior and the shape of the car's engine and frame.

The car's exterior is deep red, while its interior is a shadowy, gunmetal gray. The car's frame and suspension is yellow, a color picked up in parts of the engine (including the Nebula GNX logo on the engine itself). The yellow and red emphasize the engine and immediately draw your eyes to this area.

FIGURE 12.34 **Simple, Effective Drawing**[39]

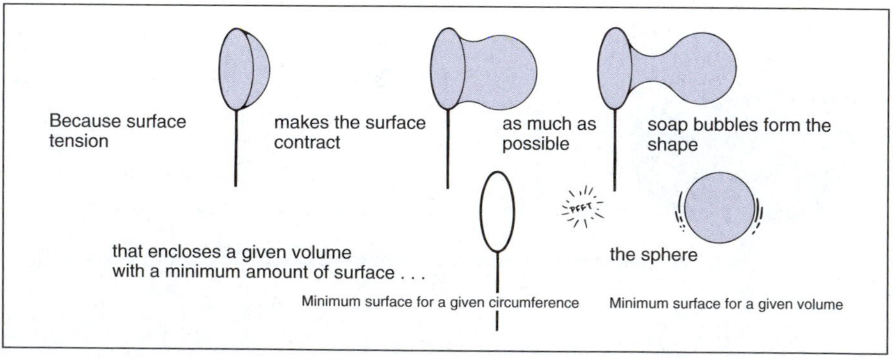

Because surface tension
makes the surface contract
as much as possible
soap bubbles form the shape
that encloses a given volume with a minimum amount of surface . . .
the sphere
Minimum surface for a given circumference
Minimum surface for a given volume

FIGURE 12.35 **Three-Dimensional Drawing**[40]

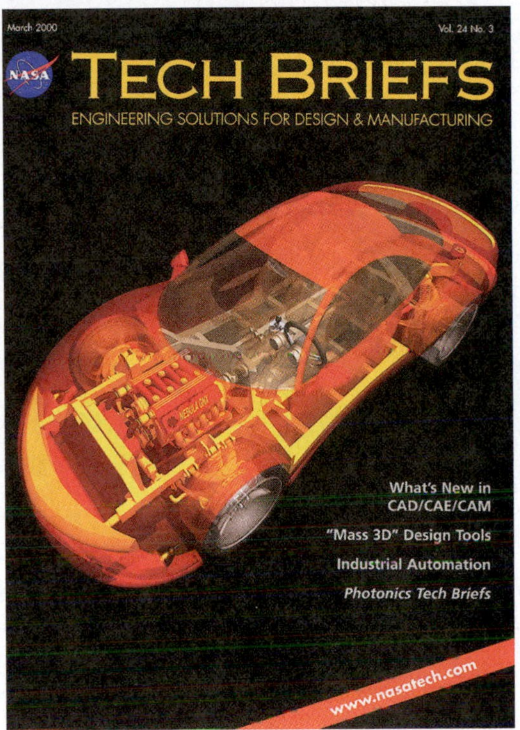

What do you see as the benefits and limitations of photographs, drawings, and diagrams of the same object? How would photographs or diagrams of the Nebula GNX respond to the limitations but also have their own limitations?

WEBLINK

No one would argue with the idea that visuals often communicate to international audiences when words fail. However, that doesn't mean that visuals (and their related text) are not bound by culture. For several links to Web sites that suggest ways to adapt visuals and text for an international audience, go to www.english.wadsworth.com/burnett6e.
CLICK ON WEBLINK
CLICK on Chapter 12/international

The versatility of drawings is demonstrated in a book for farriers. The series of three drawings in Figure 12.36 shows an external view — a representational drawing — of a hoof, followed by a phantom view and a cutaway view to reveal the internal structure. Because these drawings have the same scale and are placed close together, you can easily make comparisons.

What kinds of problems might occur from errors in proportion and scale in technical drawings?

Another common type of technical drawing is an exploded view, illustrated in Figure 12.37. An exploded view shows an entire mechanism or organism by separating (exploding) the whole to provide a clear view of each component.

FIGURE 12.36 External View, Phantom View, and Cutaway View[41]

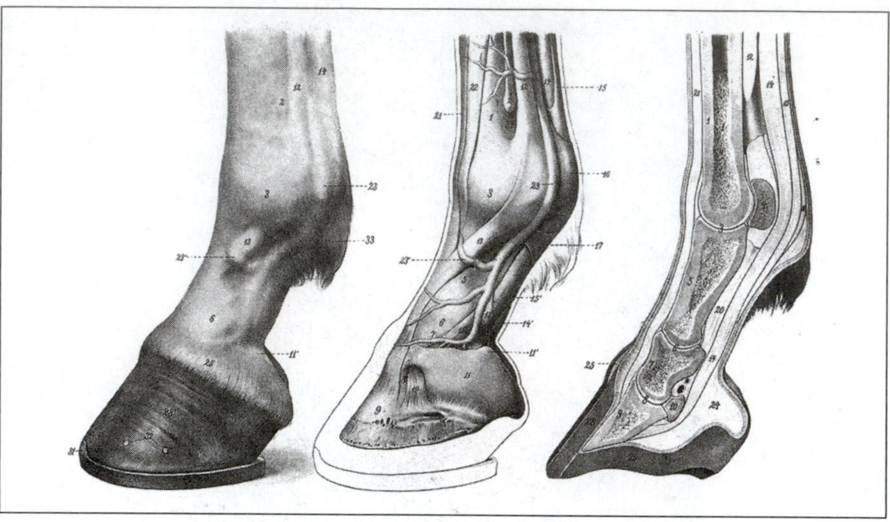

Exploded views are useful as part of an overall description of a mechanism or organism, but they are most frequently used in assembly and repair manuals. Your understanding of the compression faucet in Figure 12.37 is increased by the clear labels next to the appropriate parts and the concise process explanation of the way the faucet works.

FIGURE 12.37 Exploded View[42]

What would be the differences between a drawing like this one designed to show the components and explain the overall operation and a drawing of the same faucet designed to accompany instructions to repair a faucet leak?

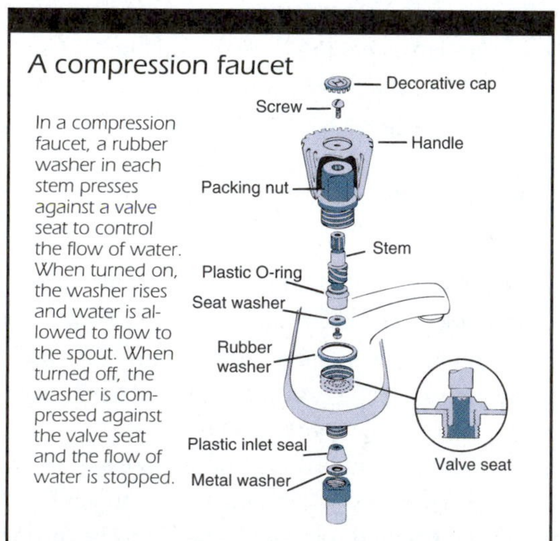

A compression faucet
In a compression faucet, a rubber washer in each stem presses against a valve seat to control the flow of water. When turned on, the washer rises and water is allowed to flow to the spout. When turned off, the washer is compressed against the valve seat and the flow of water is stopped.
Decorative cap
Screw
Handle
Packing nut
Stem
Plastic O-ring
Seat washer
Rubber washer
Plastic inlet seal
Metal washer
Valve seat

Function 7: Identify Facilities or Locations

Identifying facilities and locations traditionally has meant maps and photographs. Now, however, *map* also refers to a navigational tool used on the Web, and workplace photographs are made as often with digital cameras as with traditional film cameras.

Maps. Geographic information is displayed on maps (called charts, not maps, for air or water). To many people, maps mean road guides. But maps also display topographic, demographic, agricultural, meteorological, and geological data. And now, *maps* also refers to Web sites, providing users with an overview of the electronic terrain.

How does your discipline or profession use maps?

Maps show features of a particular area, such as land elevation, rock formations, vegetation, animal habitats, crop production, population density, or traffic patterns. Statistical maps can depict quantities at specific points or within specified boundaries. Data can be presented on maps in a number of ways: dots, shading, lines, repetitive symbols, or superimposed graphs. One of the most common maps shows political boundaries, for example, state boundaries in the United States and provincial boundaries in Canada, as shown in Figure 12.38.

FIGURE 12.38 **Map Showing Political Boundaries**[43]

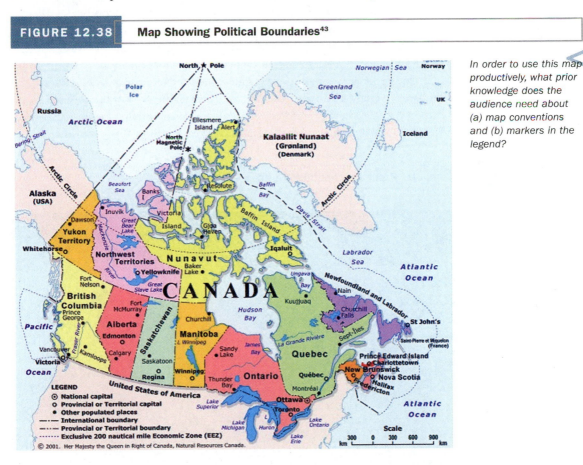

In order to use this map productively, what prior knowledge does the audience need about (a) map conventions and (b) markers in the legend?

Another kind of map is presented in Figure 12.39, which shows a computer model — a map — depicting possible underground paths of water flowing toward the Columbia River in the state of Washington. This groundwater may contain contaminants leaking from tanks at the Hanford Reservation, used to store waste radioactive materials from the manufacture of nuclear weapons. Tracking the groundwater helps the U.S. Department of Energy establish the severity of the problem, determine concentrations and radioactivity of contaminants, and show the feasibility of slowing the movement of the groundwater toward the Columbia River.

The computer model, designed by a geophysicist at Bechtel-Hanford, Inc., in Hanford, Washington, uses color to help viewers interpret the information. The map model shows the bottom and top boundaries of the region being studied and modeled: (1) The purple mountain depicts the impermeable layer of bedrock far underground. (2) The yellow grid (only the top plane of which shows on the model) depicts the actual ground level. Groundwater can flow between purple mountain and yellow grid.

Another way to depict environmental changes is illustrated in Figure 12.40, a map that documents deforestation in Brazil. According to NASA's Earth Observatory, "If the current rate of deforestation continues, the world's rain forests will vanish within 100 years — causing unknown effects on global climate and eliminating the majority of plant and animal species on the planet."[44] Most of the 21,000 square miles deforested annually in South America are in the Amazon Basin. So far, due to the isolation of forest fragments and the increase in forest/clearing boundaries, a total of 16.5% of the Amazon River Basin (an area nearly the size of Texas) has been affected by deforestation. Audiences interpreting Figure 12.40 are helped by a key that shows that the darker the area, the more forest remains.

FIGURE 12.39 **Computer Rendition of an FEM Surface Grid**[45]

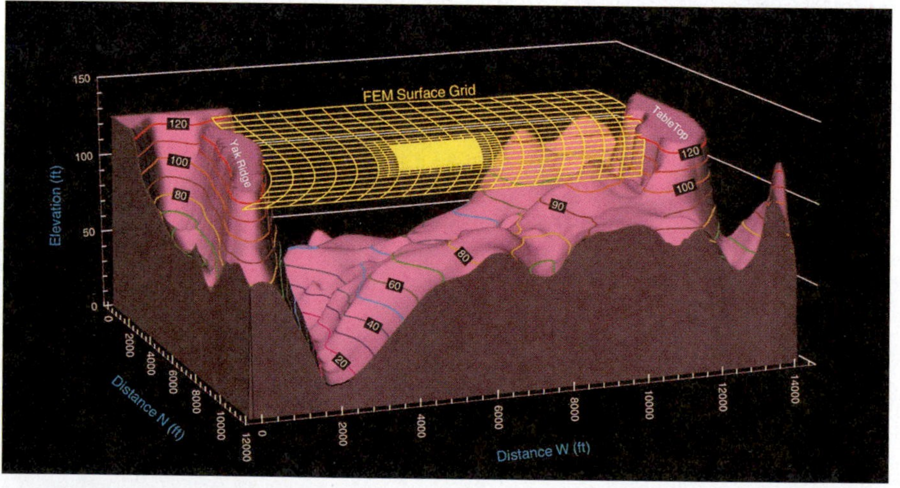

FIGURE 12.40 **Map Documenting Environmental Changes**[46]

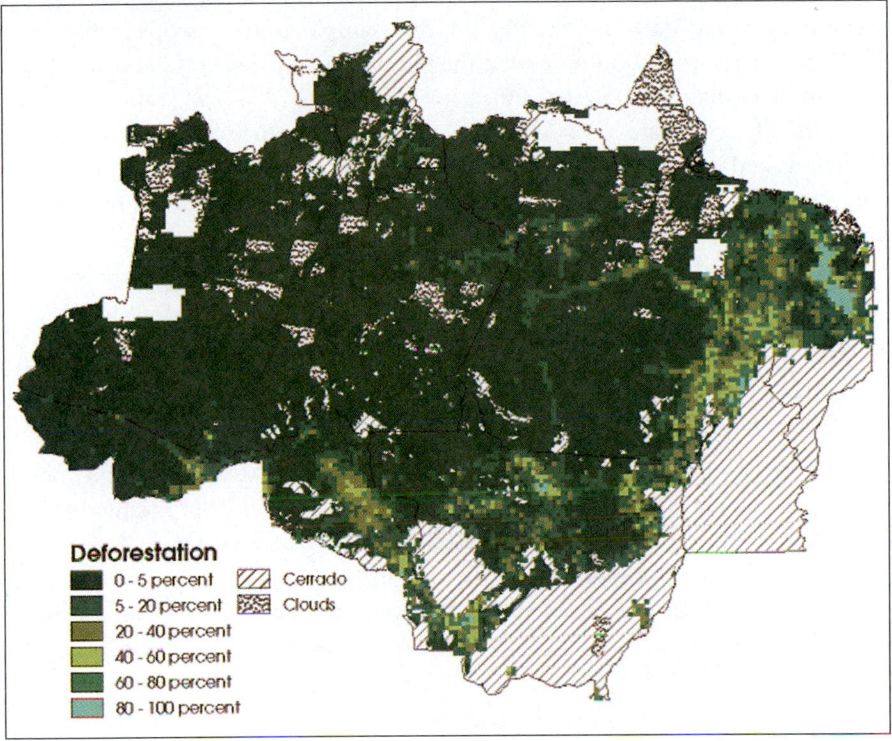

This map shows deforestation at one point in time. How could you effectively display the increasing deforestation over a 20-year period? Over a 50-year period?

WEBLINK

The *Atlas of Cyberspaces* is a collection of cybermaps and graphic representations that help us access, comprehend, and use digital landscapes available in global communication networks and online information resources. Some of the maps are appear familiar, using the cartographic conventions of real-world maps; others are abstract representations of electronic spaces. For a link to these cybermaps, go to www.english.wadsworth.com/burnett6e.

CLICK ON WEBLINK

CLICK on Chapter 12/cybermaps

Photographs. Because a photograph displays an actual view of a subject, it's appropriate when you want to emphasize realism, particularly the natural features of a setting. However, even though photographs accurately depict locations, they often show too much detail. For this reason, *callout arrows* (small arrows superimposed on a photo) can be used to draw attention to main features.

How does your discipline or profession use photographs?

When a photo is printed, its appearance can be altered so that the primary subject becomes more prominent than the background, thus giving emphasis that would not be possible if the photo were printed normally. Photos can also be reduced, enlarged, or cropped to emphasize a particular portion of the subject.

Photographs can be remarkably effective in displaying a range of subjects, including very tiny objects. Quickly skim the examples of microscopy in Figure 12.41 to determine if you know what the photos actually depict. Confirm your opinion by reading the marginal annotations.

Virtually every state has both public and private photo archives containing thousands and in some cases millions of aerial photographs that provide a complete record of features in the state. Routinely taken, aerial photos record various kinds of information:

- Agricultural information (crop data, erosion management, forestry management)
- Municipal information (property lines, utilities, streets)
- Transportation information (major and minor roads and highways, bridges, waterways, traffic patterns, railroads, airports)

Figure 12.42 is an aerial photograph. It shows a water treatment facility and county maintenance shed. The hydrogeologist who used this photo wanted confirmation of land features and building locations as he tracked an underground hazardous waste spill.

Photographs can provide an alternative perspective. Before you read further, take a look back at the map in Figure 12.40, which shows the deforestation in the

FIGURE 12.41 **Microscopy (Can you identify these objects?)**

Microscopy subjects
1. *Geranium leaf*
2. *Coral*
3. *Fruit fly*
4. *Tomato*
5. *Mouse embryo*
6. *Algae diatom*

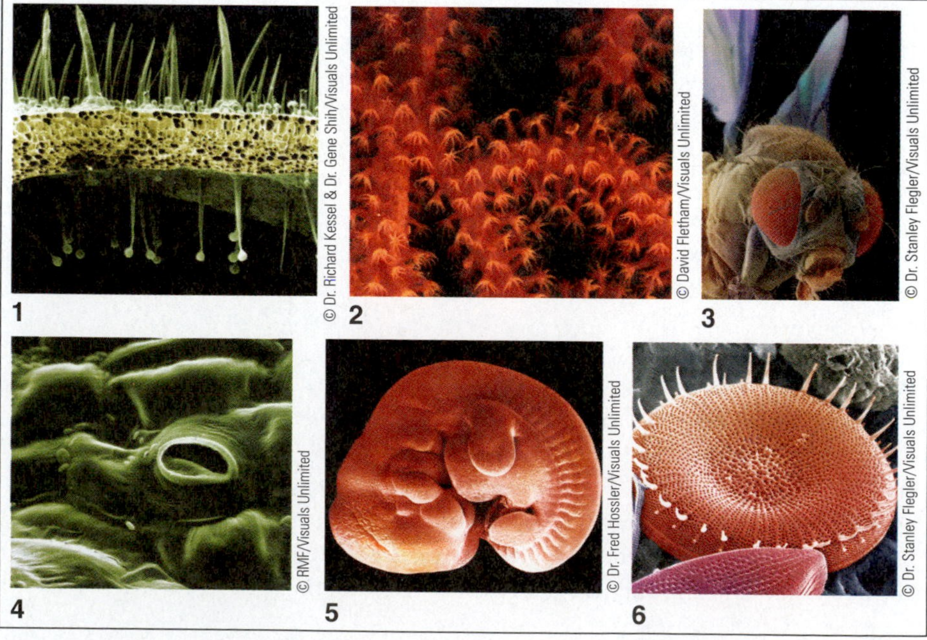

FIGURE 12.42 **Aerial Photograph from Low-Flying Airplane**

Iowa Department of Transportation

FIGURE 12.43 **Satellite Photo Documenting Environmental Changes**[47]

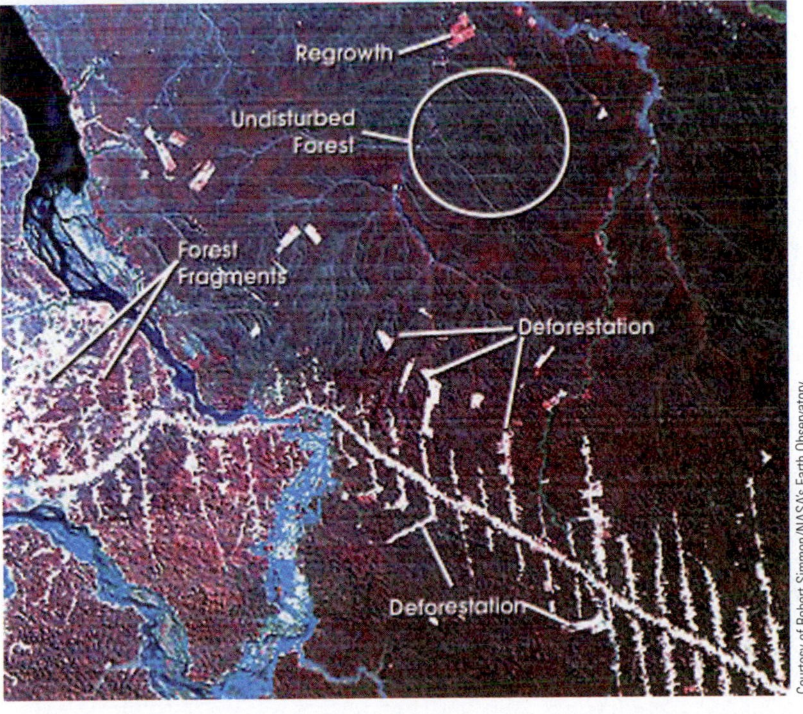

Courtesy of Robert Simmon/NASA's Earth Observatory

What would be possible different audiences and uses for Figures 12.40 and 12.43?

Amazon Basin. Now look at Figure 12.43, which is a photograph of the same phenomena. In this satellite image of deforestation in the Brazilian state of Para, the dark areas are forest, the white is deforested areas, and the gray is regrowth. The pattern of deforestation spreading along roads is obvious in the lower half of the image. Scattered larger clearings can be seen near the center of the image.[48]

Another kind of photograph is presented in Figure 12.44, which shows four views of the Long Valley region of California using synthetic aperture radar (SAR) technology. SAR allows all-weather mapping of topographic and geographic features of land surfaces. The four views illustrate different SAR processing steps (clockwise from upper left): the initial SAR image, the interferogram, the perspective view, and the contour map.

Color is important in each of these views. The initial SAR image uses white, black, and gray shadow to illustrate the features of the land; the interferogram uses a range of colors — violet, yellow, blue, and green — to highlight these surface features. The contour map relies on shades of green, yellow, and gray to highlight land surfaces, and black contour lines further distinguish these features. The perspective view relies more heavily on light and shadow rather than contrasting colors — to distinguish the features of the land surface.

FIGURE 12.44 **Satellite Photographs**[49]

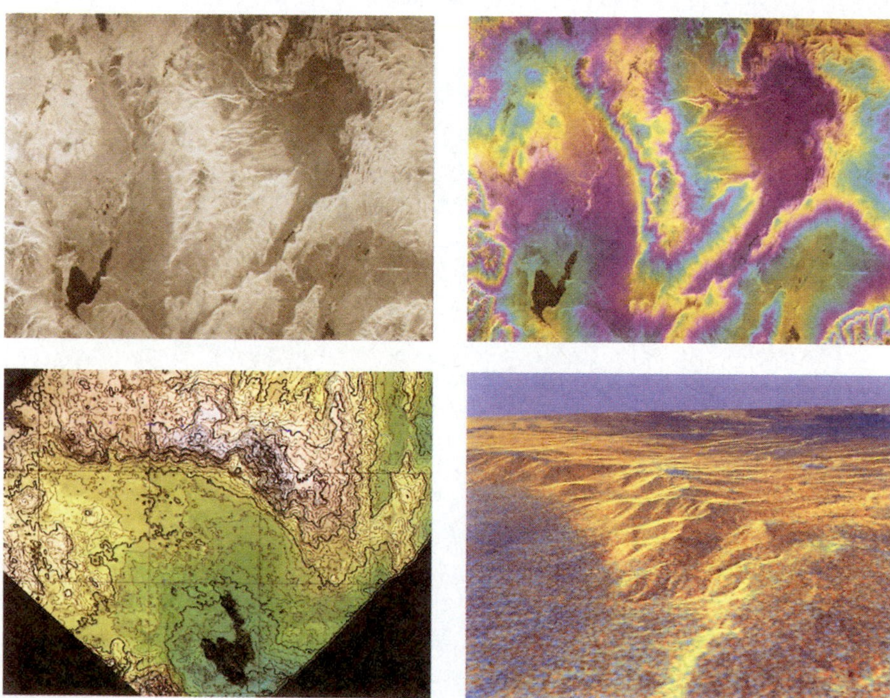

Figure 10-1. Long Valley region of east-central California acquired by SIR-C/X-SAR interferometer, illustrating processing steps from SAR image (upper left) to interferogram (upper right) to contour map (lower left) to perspective view (lower right).

With the wide availability of relatively inexpensive color printers and color photocopiers, color in technical visuals is commonplace. It can add to the meaning or help interpretation, but it is sometimes unnecessary and too often badly done or inappropriately used. Color can be an extraordinarily powerful tool to help create more effective visuals when the appropriate conventions are followed.

Cautions against Misuse of Color

Because color is so easy to use, the temptation to overindulge is great. These are some of the problems to avoid:

In what situations might black-and-white visuals be more appropriate and more useful than ones in color?

- *Overuse of decorative color.* Using color simply as decoration contradicts the basic premise that color in technical documents should be functional. It should be associated with meaning or function so that when readers see a color used in a particular document, it triggers appropriate associations:

  The printout of the stress test shows the weak spots in red, the questionable spots in yellow, and the unstressed parts in shades of blue and green.

  or

  The troubleshooting guide for each section is signaled with a red strip on the edge of the page so these pages are immediately identifiable.

- *Too much color.* Using too many colors, too much color, or inappropriate color distracts or annoys readers. Select a few colors rather than trying to pack in as many as possible. Similarly, the intensity of the color should match the context, purpose, or audience. For example, an industrial catalog for laboratory safety supplies can appropriately use seven colors in a table that matches them to OSHA and ANSI color-coded safety guidelines, but in most tables using that many colors would be unnecessary and, therefore, inappropriate.
- *Cultural insensitivity.* Violating cultural expectations can mean that the use of color contradicts the expectations in either the workplace culture or the broader social culture. For example, in industry neon orange is usually used in small triangles to signal warnings, so readers of an instruction manual for a precision lathe might feel disconcerted if all the manual's headings were in bright orange.

Suggestions for Appropriate Use of Color

Used effectively in visuals, color can help readers with issues such as consistency, emphasis, and organization.[50] Most documents are printed in one color — black — on white paper. Some documents have a second color added, which can be printed in a variety of intensities from pale to dark, giving the appearance of a range of colors. In most of the print documents you produce, cost and time will dictate that you use a single color (black) or use black and a second color selected to increase the appeal and accessibility of the document.

Sometimes the content, context, purpose, or audience require that you produce four-color documents. A few documents, this book included, are printed in four colors that combine to create all the rest of the colors in the book. So the rest of the discussion in this section focuses on some of the important uses of four-color visuals.

Whenever color is used in technical documents, it should be an integral part of the information that readers need. Even though the discussion of the following examples emphasizes a particular use of color for each visual, in practice, most well-designed color visuals use color to accomplish multiple purposes. These are among the most important purposes:

- Signal safety
- Attract attention
- Enable accurate identification
- Show structure or organization
- Highlight components and their process or movement
- Aid comprehension
- Influence interpretation

Signal Safety. One of the most important uses of color is to signal safety. The most widespread international agreement about the use of color is probably is with traffic lights: Green = go, yellow = caution, and red = stop, regardless of the country. Additionally, a number of government agencies and international organizations specify the use of particular colors to increase attention to safety: Occupational Safety and Health Administration (OSHA), American National Standards Institute (ANSI), American Public Works Association (APWA), and International Organization for Standardization (ISO) all use color to signal various conditions as well as levels and kinds of dangers. For example, OSHA guidelines indicate that U.S. workplaces should code safety equipment and hazardous areas with specific colors. Similarly, APWA guidelines indicate that U.S. public agencies, utilities, contractors, and others involved in excavation should use color to prevent accidental damage, service interruption, or injury. ANSI provides national standards for U.S. safety signs and labels, while ISO governs international safety labeling requirements.

Some colors such as yellow, orange, and red are mandated by OSHA; other colors such as blue and green are used by common practice but not mandated. Figure 12.45 highlights the colors specified by OSHA to signal various levels of safety or risk in the U.S. workplace and by APWA to label various kinds of cables, wires, and conduits that could be damaged or broken during a construction project.

FIGURE 12.45 OSHA and APWA Safety Colors[51]

OSHA	APWA
Black / White = TRAFFIC MARKINGS, signaling things such as passing and no passing, exit ramps, and shoulder areas.	
Green = SAFETY, signaling instructions about safe work practices, proper safety procedures, safety equipment location (including first aid equipment other than for fire-fighting).	**Green** marks surveying and general construction markings.
Blue = NOTICE, signaling need for attention to safety of personnel or protection of property. Must not used in place of CAUTION, WARNING, or DANGER.	**Blue** marks water, irrigation and slurry lines.
Yellow = CAUTION ▪ Signals a potentially hazardous situation that, if not avoided, may cause minor or moderate injury. ▪ Alerts users to unsafe practices or the potential for property damage.	**Yellow** marks gas, oil, steam, petroleum or gaseous material.
Orange = WARNING ▪ Signals a potentially hazardous situation that, if not avoided, has some probability of death or serious injury.	**Orange** marks telephone, communication, alarm or signal cables or conduit.
Red = DANGER ▪ Red indicates danger or a hazardous situation that, if not avoided, has high probability of death or severe injury. Used in extreme situations. ▪ Indicates STOP (e.g., for emergency bars, buttons, and switches). ▪ Identifies the location of fire protection equipment and apparatus.	**Red** marks electric power lines, cables, conduit and lighting cables.
Fluorescent orange or **orange-red = BIOLOGICAL HAZARD**	
Purple = RADIATION HAZARD	

Courtesy of Ron Doll/Falk Corporation

This orange coupling guard complies with ANSI safety standards.[53]

Documents and signs are not the only places where color plays a critical role in workplace communication. ANSI Standard Z535 specifies that hazardous machinery parts themselves must be marked. For example, protective machine guards covering sections of equipment that have an intermediate level of potential risks that is not self-evident or visible must have a visible, legible, warning-level orange safety label to signal a potentially hazardous situation that, if not avoided, could result in serious injury or death. Beyond the warning label, when a safety color is used to identify hazardous machinery parts having an intermediate level of risk, the machine guards themselves should be safety orange, as shown in the photo here.[52]

Unfortunately, some people are color-blind, which means they need to have compensatory skills. For example, color blindness does not impede safe driving because traffic lights are cued by location as well as color. As long as a person knows that the top light is red, the ability to perceive red the same way that other people do is not necessary. This leads to an important principle when designing visuals: Do not have color be the only distinguishing cue in anything you design.[54] Furthermore, because all organizations are responsible for educating and training their employees about the relationship between colors and various levels of risk, employees need to learn additional ways to identify these risks.

WEBLINK

If you want to know more about color blindness, go to www.english.wadsworth.com/burnett6e for links.
CLICK ON WEBLINK
CLICK on Chapter 12/color blind

Attract Attention. A second use of color is to attract readers so that they are drawn toward the topic. Figures 12.46a and 12.46b show a German brown trout in two versions. The technical illustrator, Dean Biechler, created the color version (Figure 12.46a) for the folder cover for a series of fact sheets produced by the Coldwater Stream Program of the Iowa Natural Heritage Foundation. The strong color drawing on the cover attracts attention. The fact sheets focus on conservation and resource management. Biechler created the black-and-white version (Figure 12.46b) for one of the fact sheets. (This black-and-white version was also used on the Coldwater Stream Program t-shirt.)

FIGURE 12.46A **Color Drawing of Fish**[55]

FIGURE 12.46B **Black-and-White Drawing of Fish**[56]

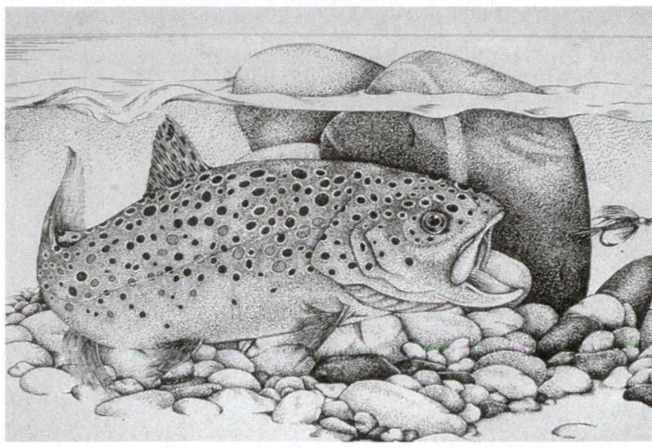

Color attracts attention, ensures accurate identification, and provides contextual details important to those who are interested in and provide support for the program.

Enable Accurate Identification. A third purpose for using color is to help readers focus on critical features of the object. Figure 12.47 represents the human lymphoid system. Color is useful in this drawing in several important ways. First, color helps readers to identify key features of the lymphoid system by highlighting two kinds of lymphoid tissues. By drawing the diffuse lymphoid system in green (obviously not the true color of the lymphoid system), the artist immediately draws readers' attention to this key feature. The artist did not select blue or red to represent the lymphoid system because these are colors conventionally used to represent arteries (red) and veins (blue). Despite the fact that the encapsulated organs are also in color, they are de-emphasized when compared to the green diffuse lymphoid system. Second, the artist uses other colors as well as colored arrows to represent the development and maturation of the T lymphocytes and the B lymphocytes in the detail drawing in the lower right-hand corner. Finally, the artist uses the green to highlight the relationship between the primary illustration and the detail drawing. Because green is used to represent the lymph node in the detail drawing, readers are likely to recognize that this information is directly related to the green lymph nodes in the primary drawing.

Show Structure or Organization. A fourth purpose for using color is to enable readers to better understand the structure displayed in the visual. It can also organize those parts, chunking them so that readers can more easily see critical relationships. Information of a similar type is presented in the same color so that readers immediately see the likenesses and thus more easily understand and recall information.

Figure 12.48 shows a computer-generated stress analysis of a tooth from the front of a tractor bucket. Test engineers at John Deere first drew the tooth and then used a computer modeling system to apply loads at the positions and from the directions the tooth would be stressed in actual situations. The system then

FIGURE 12.47 **Drawing of Lymphoid System**[57]

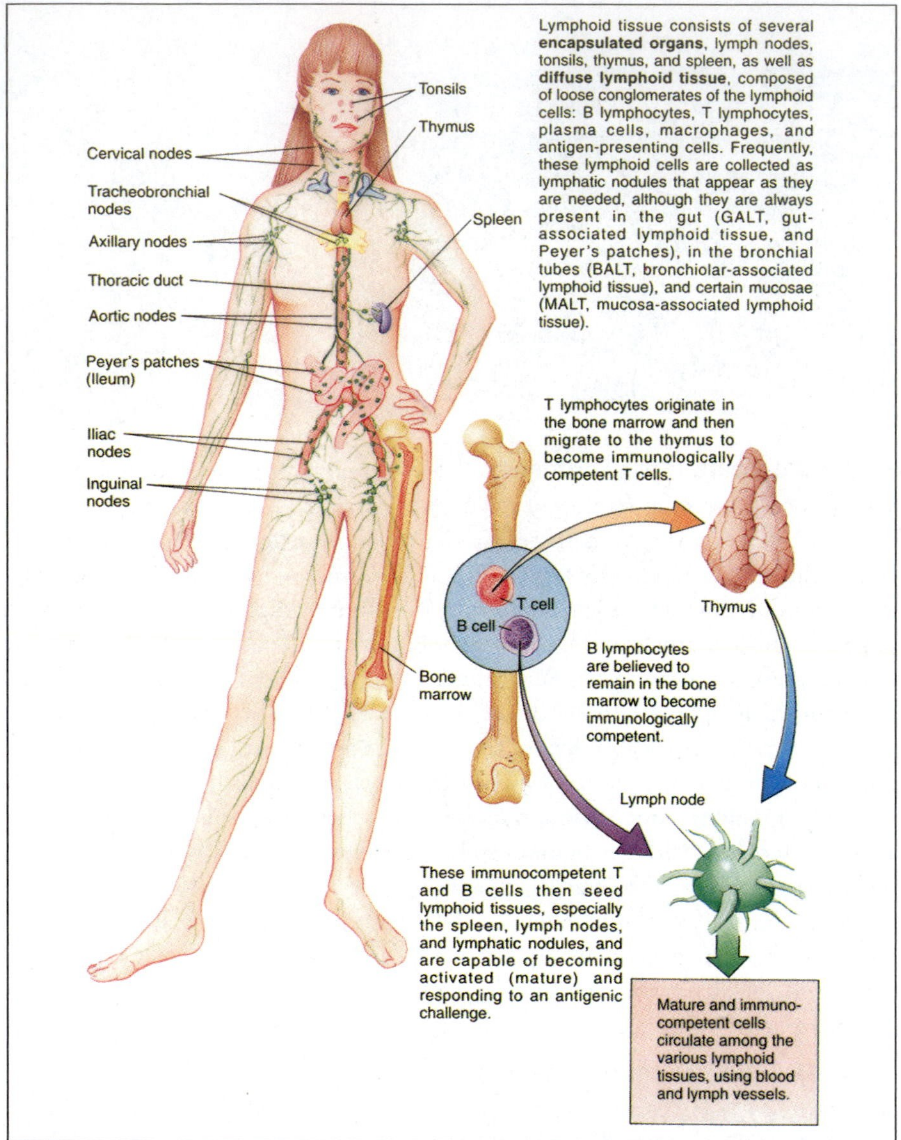

Lymphoid tissue consists of several **encapsulated organs**, lymph nodes, tonsils, thymus, and spleen, as well as **diffuse lymphoid tissue**, composed of loose conglomerates of the lymphoid cells: B lymphocytes, T lymphocytes, plasma cells, macrophages, and antigen-presenting cells. Frequently, these lymphoid cells are collected as lymphatic nodules that appear as they are needed, although they are always present in the gut (GALT, gut-associated lymphoid tissue, and Peyer's patches), in the bronchial tubes (BALT, bronchiolar-associated lymphoid tissue), and certain mucosae (MALT, mucosa-associated lymphoid tissue).
Tonsils
Thymus
Cervical nodes
Tracheobronchial nodes
Spleen
Axillary nodes
Thoracic duct
Aortic nodes
Peyer's patches (Ileum)
Iliac nodes
Inguinal nodes
Bone marrow
T lymphocytes originate in the bone marrow and then migrate to the thymus to become immunologically competent T cells.
T cell
B cell
Thymus
B lymphocytes are believed to remain in the bone marrow to become immunologically competent.
Lymph node
These immunocompetent T and B cells then seed lymphoid tissues, especially the spleen, lymph nodes, and lymphatic nodules, and are capable of becoming activated (mature) and responding to an antigenic challenge.
Mature and immuno-competent cells circulate among the various lymphoid tissues, using blood and lymph vessels.

calculated the stresses that would occur on the tooth. Changes in color show regions of stress: dark blue shows the lowest levels of stress, then light blue, green, and orange show successively higher levels of stress. Red shows the highest level.

Highlight Components and Their Function or Movement. A fifth purpose of color is to show readers a path for moving through a visual. Color can signal a change or draw attention to the nature and direction of the change.

FIGURE 12.48 Stress Test for Tractor Bucket Tooth[58]

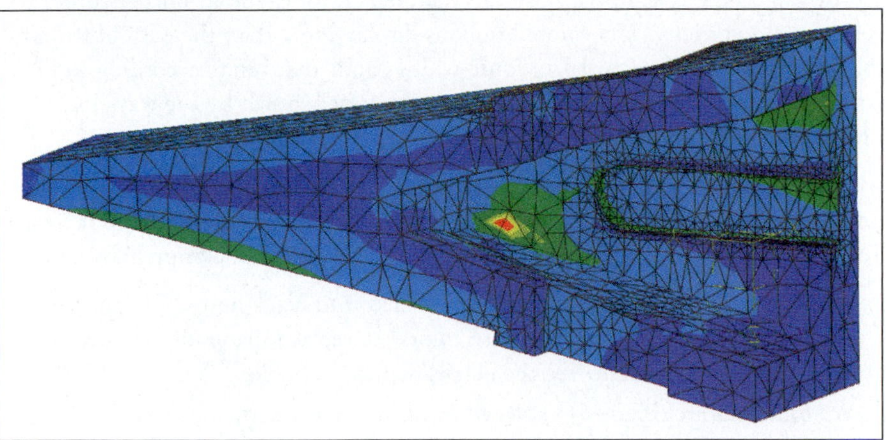

Consistency in the use of color helps readers track the changes an object or organism goes through during a process.

Figure 12.49 shows a computer model created by AmTec Engineering in Bellevue, Washington, using software called TecPlot. The figure is a 3-D depiction of flow over an airfoil (airplane wing), which is useful for aeronautical engineers who want to increase the stability and efficiency of airfoils they are designing or analyzing. The model shows an airfoil, the air flow around the airfoil, and a

FIGURE 12.49 Computer Rendition of 3-D Turbulent Flow over a Wing[59]

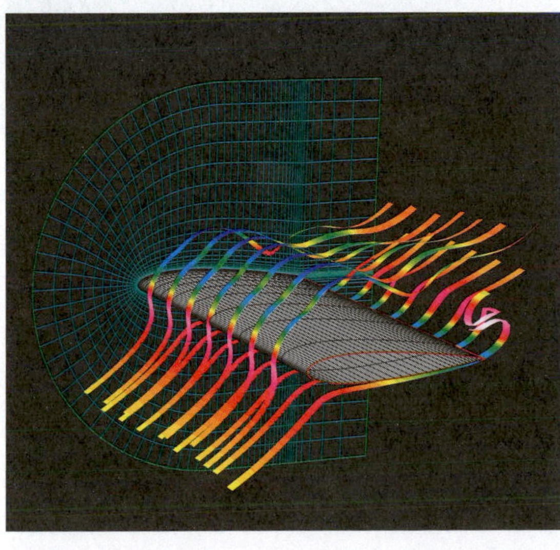

computational grid (only the last plane of this green grid is visible) for analyzing the flow. The model enables aeronautical engineers to collect and analyze huge sets of data as they create visualizations that depict, for example, air pressure, velocity, and vorticity. The colored ribbons display more than the path of the air; they show how the flow twists in finite-width bands that rotate according to the vorticity. In the figure, you can see a flow separation, shown by the ways the ribbons flow up and over the top.

Aid Comprehension. A sixth purpose for color is to make an image easier to understand. For example, the photographs taken by cameras on the Hubble Space Telescope have often been translated into color images to aid comprehension.[60]

- Some Hubble data are originally translated into black-and-white photos. Scientists arbitrarily choose "false colors" to replace shades of gray because people can more easily see the details depicted in color.
- Some data are originally captured in color; scientists then enhance or intensify selected colors to emphasize particular features.
- Some data are captured in true color by taking photographs through separate red, green, and blue filters and then combining the images into a realistic photograph.

The eerie, dark, pillar-like structures in Figure 12.50 are actually columns of cool interstellar hydrogen gas and dust that are incubators for new stars. The pillars protrude from the interior wall of a dark molecular cloud like stalagmites from the floor of a cavern. They are part of the "Eagle Nebula" (also called M16,

FIGURE 12.50 **Pillars of Creation in a Star Forming Region**

Jeff Hester, Paul Scowen/NASA

the sixteenth object in Charles Messier's eighteenth-century catalog of "fuzzy" objects that aren't comets), a relatively close star-forming region 7,000 light years away in the constellation Serpens.

The picture was taken with the Hubble Space Telescope. The color image is constructed from three separate images taken in the light of emissions from different types of atoms. Red shows emission from singly ionized sulfur atoms. Green shows emission from hydrogen. Blue shows light emitted by doubly ionized oxygen atoms.

Influence Interpretation. A final purpose for using color is its ability to influence the way viewers interpret information in visuals such as phase diagrams, which are familiar to chemists, physicists, chemical engineers, and materials scientists.

Phase diagrams are important for the design of chemical separations equipment such as absorbers and distillation columns. These diagrams show the phase behavior of a complex system at a glance and often eliminate the need for in-depth study of detailed, numerical equilibrium data. For example, the phase diagrams in Figure 12.51 were produced at Iowa State University using Animate, an interactive computer graphics program for the study of multicomponent, fluid-phase equilibria.

Phase diagrams for fluid mixtures containing four chemical species (quaternaries) use tetrahedral models to show the results of boiling or condensation. In the computer-generated drawings in Figure 12.51, the pure components A (acetonitrile), B (benzene), C (ethanol), and D (acetone) are designated by the vertexes of the tetrahedron, while all intermediate quaternary mixtures are located in the space within. The compositions of boiling liquids are shown by the red surfaces and those of condensing vapors in green, with the specific, red-green pairs that coexist together (that is, in equilibrium) connected by tie-lines, several of which have been drawn in white.

In the series of images I–IV, the movements of the surfaces, both through the tetrahedron and also relative to one another, show how the equilibrium compositions change in response to increasing fluid pressure while the temperature remains fixed. The rotation of the model itself lets viewers study the diagram from all angles.

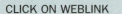

Visuals can be very helpful during preparation for a criminal trial as well as during the trial itself. Visuals can go a long way in aiding comprehension of juries and judges, who must make sense of testimony about complex technical topics. For examples of successful visuals used during trials, go to www.english.wadsworth.com/burnett6e for a link.

CLICK ON WEBLINK

CLICK on Chapter 12/visuals persuasion

WEBLINK

FIGURE 12.51 **Composition Tetrahedron for a Quaternary System in Vapor-liquid Equilibrium (Temperature, Pressure Fixed)**[61]

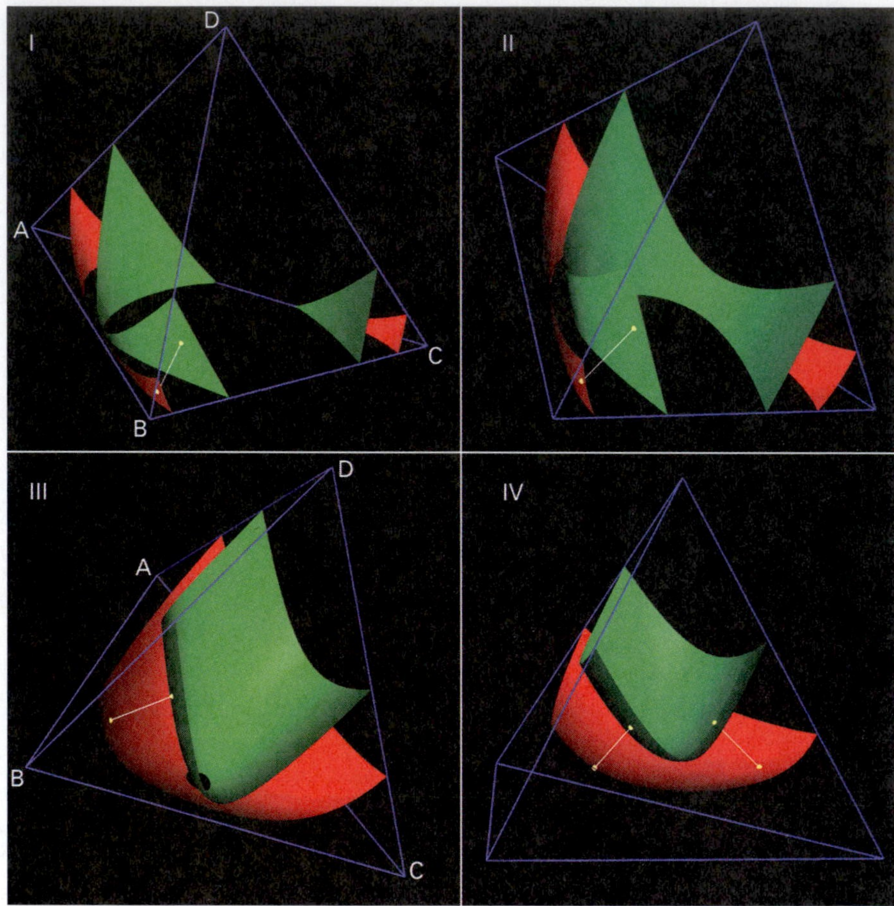

The software was written by Eric Cochran and Professor Kenneth R. Jolls, ISU Chemical Engineering Department, and the sequence was designed by Chad Sanborn, ISU Engineering Publication and Communication Services.

Color in Designing Electronic Documents

In print documents, color can be a valuable, or even essential, element. Color is just as important (and is perhaps even expected) in many electronic documents. Because color is produced differently on a screen than it is on paper, technical communicators must think about it differently. On-screen colors, like paper colors, are affected by the environment in which they are viewed — the reflections off the monitor face, the light in the room, and the combinations of colors on the screen all affect the way you see on-screen colors, and the effects work differently than on paper documents.

Because of the variations in the ways viewers interpret on-screen colors, you need to consider color and viewers' reactions to it when you're designing electronic documents.

- Will the document be viewed on a screen of lower or higher resolution than the one you're using? If so, you must consider whether the graphics will look grainy or poorly defined.
- Will the document be viewed in a room with bright lights (such as a fluorescent-lit office) or dim lights (such as a private home)? The reflections from the screen can have a major impact on how color is perceived.
- Will the document be printed? If so, the differences between screen color and paper color must be considered. If readers might print the document without color, you should look at black-and-white hard copy to see if what you thought was functional color becomes a gray blob when viewed in black and white.

What additional questions could be important in your decision about using color in electronic documents?

The issues raised by these questions are discussed in Chapter 13: Designing and Using Electronic Media.

Individual and Collaborative Assignments

1. **Design a table.** Conduct the appropriate research and then design a speed comparison table that presents the equivalents for kilometers per hour, miles per hour, and knots per hour. For whom would this comparison table be useful?
2. **Transform a pictorial graph.** The following pictorial graph[62] shows the per-capita income disparity for developing, transition, and well-developed countries. Well-developed countries (such as Canada, the United States, Japan, and Israel) show the greatest per-capita income disparity over a 30-year period. That is, over the next 30 years, the rich will become richer while the poor will become poorer. Design another visual depiction of this information. Identify the audience and purpose of your visual.

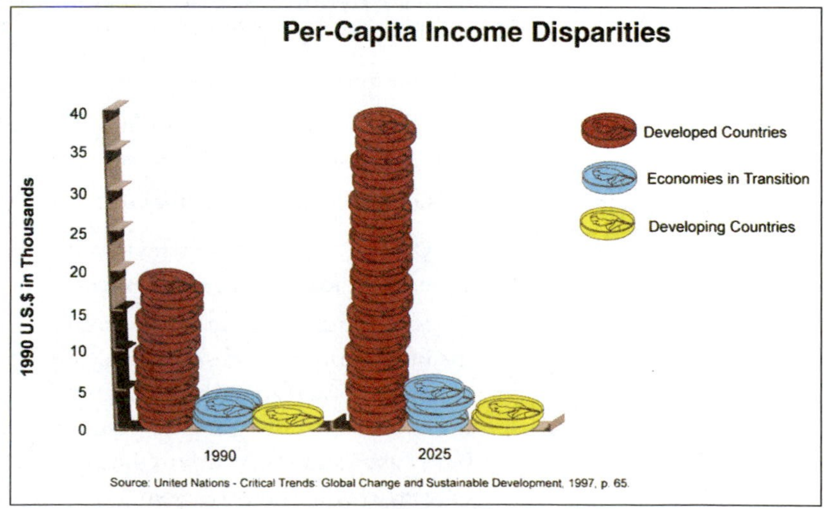

3. **Write a description based on visual information.** Refer to any of the flowcharts or process overviews in this chapter. Write a description of the process based on the information in the visual. Make sure the information is accurate. After you have completed the paragraph, examine it to identify the textual features you used to help readers follow the process. Then decide what's missing from your explanation.
4. **Transform a bar graph.** The following subdivided 100-percent bar graph shows the proportions of saturated, polyunsaturated, and monounsaturated fatty acids in different vegetable oils.[63] All dietary fats comprise mixtures of fatty acids, but lower amounts of saturated fatty acids (like those found in sunflower and safflower oil) are healthier for your body than saturated fatty acids. Develop two additional ways to depict this same information. Identify the audience and purpose of your two new visuals.

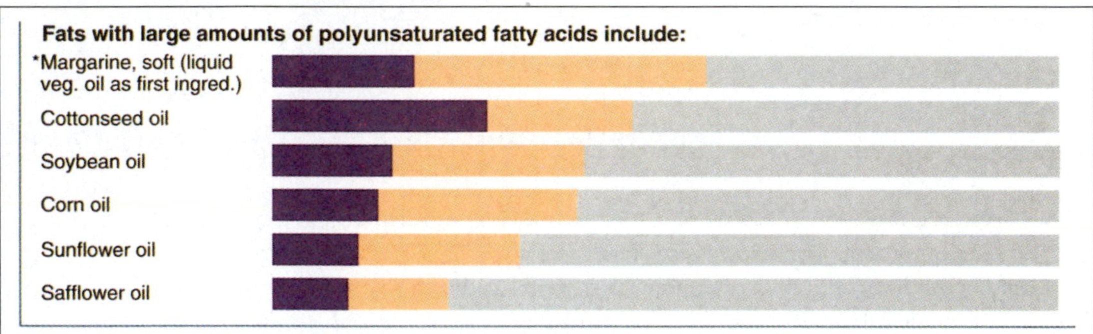

Fats with large amounts of polyunsaturated fatty acids include:
*Margarine, soft (liquid veg. oil as first ingred.)
Cottonseed oil
Soybean oil
Corn oil
Sunflower oil
Safflower oil

* The amounts of unsaturated fatty acids in these foods vary. Chart from Home and Garden Bulletin No. 232-8, *Foods and Planning Menus*, USDA, Human Nutrition Information Service.

5. **Evaluate a published visual.** (a) Select a published visual and examine it according to the following criteria:
   - Is the appearance appealing?
   - Is the accompanying description/discussion complete?
   - Is the type of visual appropriate for the data and purpose?
   - Is the presentation free from distortion?

   (b) Use these and other criteria that you identify to create a rubric for evaluating the effectiveness and accuracy of visuals.
6. **Assess a visual.** (a) Read this explanation of the following visual:[64] It is used in veterinary classes to show changes to blood as it moves through a capillary. In capillaries (the smallest blood vessels of the body), bright red blood containing oxygen loses oxygen to the tissues and turns slightly blue. Blood then goes back to the heart and on to the lungs where it is oxygenated and turns red again.

(b) Explain whether the use of color is effective. Is it necessary?

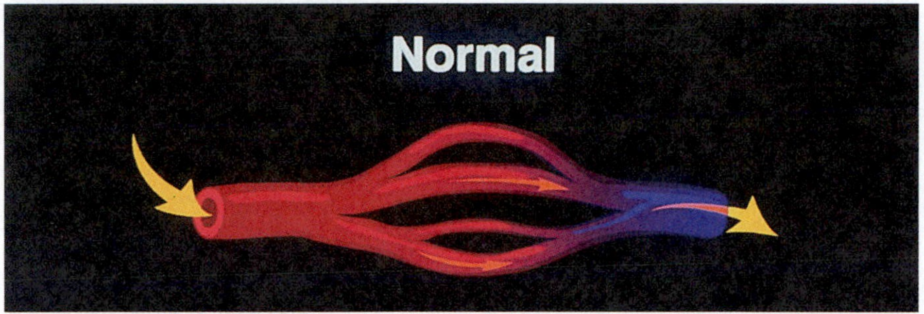

7. **Revise a document by adding visuals.** Select a document that you have previously completed (in this or another course, or in the workplace). Revise it using appropriate visuals to clarify or support the points you made in the document.
8. **Analyze the use of visuals on a Web site.** Visit the following Web sites:

   www.benjerry.com
   www.dell.com
   www.pbs.org/wgbh/nova
   www.jnj.com
   www.discover.com
   www.nationalgeographic.com
   http://cfa-www.harvard.edu
   http://natzoo.si.edu/default.cfm

   (a) Evaluate the visuals on two of the above Web sites according to the rubric you developed as well as the following criteria:

   - Do the visuals effectively and accurately explain or enhance the text?
   - Are visuals used for navigational cues? Are they effective?
   - What are some positive attributes of Web site visuals?
   - What are some negative attributes of Web site visuals?
   - What suggestions do you have for Web site technical communicators regarding the use of visuals?

   (b) Prepare your analysis in a five-minute oral presentation in which you display selected screen captures from the Web sites to illustrate your points.
9. **Assess the quality of visuals in a set of guidelines.** Work in a small group to read and then access the visuals in the document *Scientific and Technical Information: Simply Put,* a PDF file that's available at www.cdc.gov/od/oc/simpput.pdf. Use everything you've learned in Chapter 11 and this chapter to assess the visuals. Consider their comprehensibility and usability for the audience and purpose. Consider the image they create. Work with your group to categorize your assessment, using explanations and examples. Present your assessment in a detailed memo to your instructor.

10. Design a safety sign. Specialized situations require specialized warnings. For example, the Hawaiian Lifeguard Association includes dramatic visuals on warning signs, as this example illustrates.[65] Such a visual indicates a potentially life-threatening situation, with waves reaching heights of 15 to more than to 25 feet.

Carefully observe your local environment (for example, lab, campus, community, surrounding region) to identify a potentially dangerous situation or area. Use either of the two standard warning templates sketched below to create your own effective sign. You need to decide on an effective signal word, message, and pictorial symbol.

SYMBOL PICTORIAL PANEL

SIGNAL WORD

WORD MESSAGE PANEL

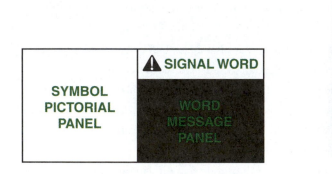

HORIZONTAL LAYOUT

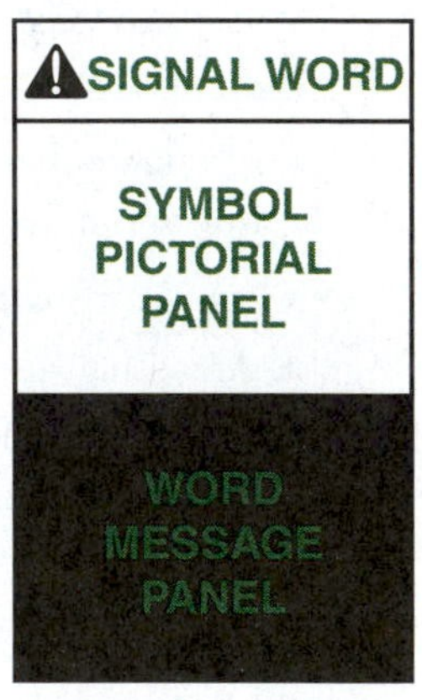

VERTICAL LAYOUT

Chapter 12 Endnotes

1 Gaines, J. R. (1994, July 4). To our readers. *Time,* 144(1).

2 Kienzler, D. (1997). Visual ethics. *Journal of Business Communication,* 34(2), 171–187.

3 (1999, January). *NASA Tech Briefs, 23*(1), 34.

4 Illustration by Ellen Cohen in Hammond, A. L. (1983, November). Tales of an elusive ancestor. *Science 83,* 43.

5 Lewin, R. W. (1983, December). Is the orangutan a living fossil? *Science, 222,* 1223. Copyright © 1983 by the AAAS. Reprinted with permission of the AAAS.

6 Peckham, G. (n.d.). *On your mark: A primer on symbols.* Retrieved November 30, 2003, from http://www.ce-mag.com/archive/02/03/peckham.html

7 HCD safetylabel.com. Are symbols necessary? *Standards FAQ: Symbols.* Retrieved on November 30, 2002, http://www.safetylabel.com/safetylabelstandards/iso-ansi-symbols.php

8 Health and Safety Executive. (n.d.). *Signpost to the health and safety (safety signs and signals) regulations 1996.* Retrieved November 30, 2003, from http://www.hse/gov.uk/pubns/safesign.htm

9 Lehtola, C. T., Brown, C. M., & Becker, W. J. (2000). Biological hazard symbol. Retrieved November 30, 2003, from University of Florida, Institute of Food and Agricultural Sciences Web site: http://edis.ifas.ufl.edu/BODY_OA091

10 EUROport. (n.d.). *International labels, standards, and publications.* Retrieved November 30, 2003, from http://www.europort.com/labels.htm

11 Peckham, G. (n.d.). *On your mark: A primer on symbols.* Retrieved November 30, 2003, from http://www.ce-mag.com/archive/02/03/peckham.html

12 Peckham, G. (2001, January/February). On graphical symbols. *Compliance Engineering.* Retrieved November 30, 2003, from http://www.ce-mag.com/archive/2001/janfeb/Peckham28.html

13 EUROport. (n.d.). *International labels, standards, and publications.* Retrieved November 30, 2003, from http://www.europort.com/labels.htm

14 EUROport. (n.d.). *International labels, standards, and publications.* Retrieved November 30, 2003, from http://www.europort.com/labels.htm

15 Health and Safety Executive. (n.d.). *Signpost to the health and safety (safety signs and signals) regulations 1996.* Retrieved November 30, 2003, from http://www.hse.gov.uk/pubns/safesign.htm

16 EUROport. (n.d.). *International labels, standards, and publications.* Retrieved November 30, 2003, from http://www.europort.com/labels.htm

17 New Hampshire DWI Prevention Council. (n.d.). *Know your limits.* Dover, NH: Author. Reprinted with permission of the New Hampshire DWI Prevention Council.

18 DeWaal, C. S., Alderton, L., & Liebman, B. Food safety guide. *Nutrition Action Healthletter,* 26, 9.

19 McCarthy, J., Hartzell, S., & Peterson, M. (n.d.). *ANSS — Reducing the devastating effects of earthquakes* (U.S. Geological Survey Fact Sheet 046-03). Retrieved November 30, 2003, from http://pubs.usgs.gov/fs/fs-046-03/fs-046-03.html

20 McCarthy, J., Hartzell, S., & Petersen, M. (n.d.). *ANSS — Reducing the devastating effects of earthquakes* (U.S. Geological Survey Fact Sheet 046-03). Retrieved November 30, 2003, from http://pubs.usgs.gov/fs/fs-046-03/fs-046-03.html

21 Energy Resource Associates. JMMC Energy Analysis Graphs. Retrieved on December 1, 2003, from http://www.eraenergy.com/energy/_graphs.html

22 Energy Resource Associates. (n.d.). *JMMC energy analysis graphs.* Retrieved December 1, 2003, from http://www.eraenergy.com/energy_graphs.html

23 Woodwell, G. M. (1970). The energy cycle of the biosphere. In *The biosphere* (p. 72). San Francisco: W. H. Freeman & Company. Copyright © 1970 by Scientific American, Inc. All rights reserved. Reprinted with permission of Scientific American, Inc.

24 H. J. Heinz Company. (1993, Second Quarter). *Heinz quarterly report of earnings and activities,* 13. Used by permission.

25 Singer, S. F. (1970). Human production of energy as a process in the biosphere. In *The biosphere* (p. 184). San Francisco: W. H. Freeman & Company. Copyright © 1970 by Scientific American, Inc. All rights reserved. Reprinted with permission of Scientific American, Inc.

26 MassPIRG. (1985, April). Consumer alert: How to choose day care for your child. *Masscitizen,* 11. Reprinted with permission of the Massachusetts Public Interest Research Group.

27 U.S. Department of Energy, Energy Information Administration. (n.d.). *Renewable energy annual 2001 highlights.* Retrieved December 1, 2003, from http://www.eia.doe.gov/cneaf/solar.renewables/page/rea_data/rea_sum.html

28 U.S. Department of Energy, Energy Information Administration. (n.d.). *Renewable energy annual 2001 highlights.* Retrieved December 1, 2003, from http://www.eia.doe.gov/cneaf/solar.renewables/page/rea_data/rea_sum.html

29 Sears. (1969). Screws. In *Sears Craftsman Master Shop Guide* (sheet 3). Hearst Corporation. Reprinted with permission.

30 U.S. Department of Energy, Genome Image Gallery. (n.d.). *DNA with features.* Retrieved December 1, 2003, from http://www.ornl.gov/sci/techresources/Human_Genome/graphics/slides/scidnafeature.shtml

31 U.S. Department of Energy, Genome Image Gallery. (n.d.). *DNA with features.* Retrieved December 1, 2003, from http://www.ornl.gov/sci/techresources/Human_Genome/graphics/slides/scidnafeature.shtml

32 (1993, January 19). *Des Moines Register,* p. 4A. Reprinted by permission of Tribune Media Services.

33 Whitehead, H. (1985). Why whales leap. *Scientific American,* 252(3), 87. Copyright © 1985 by Scientific American, Inc. All rights reserved. Reprinted with permission of Scientific American, Inc.

34 (1999, August). MDO exploits NASA's discipline breadth. *Insights: High Performance Computing and Communications, 10,* 14. Retrieved 2000 from www.hpcc.nasa.gov

35 U.S. Department of Energy. (1993). *Environmental restoration and waste management: An introduction* (DOE/EM-0104).

36 Environmental Protection Agency. (1991, October). *Guides to pollution prevention: The photoprocessing industry.*

37 Bayer Polymer. (n.d.). *Therban® manufacturing process.* Retrieved November 25, 2003, from http://www.therban.de/intertherban/c1standard_en.nsf/LPSNavigationLUByContentID/CHAR-5C3H4M?OpenDocument&nav=CHAR-5C3H8S

38 (1999, November). *NASA Tech Briefs, 23*(11), 54.

39 Eames, C., & Eames, R. (n.d.). Surface tension. *Mathematics IBM Exhibit Catalog.*

40 (2000, March). *NASA Tech Briefs, 24*(3).

41 Canfield, D. M. (1968). *Elements of farrier science* (2nd ed., p. 10). Albert Lea, MN: Enderes Tool Co. Reprinted with permission of Donald Canfield.

42 (n.a.). (n.d.). *Easy Home Repair, Packet #2* (p. 2). Pittsburgh, PA: International Masters Publishers.

43 Natural Resources Canada. (2001). *The atlas of Canada.* Retrieved November 30, 2003, from http://atlas.gc.ca/rasterimages/english/maps/reference/national/canada_eng.jpg

44 Urquhart, G., Chomentowski, W., Skole, D., & Barber, C. *Tropical deforestation.* NASA Earth Observatory. Retrieved November 30, 2003, from http://earthobservatory.nasa.gov/cgi-bin/texis/webinator/printall?//Library/Deforestation/index.html

45 AmTec. (n.d.). Bellevue, WA.

46 Urquhart, G., Chomentowski, W., Skole, D., & Barber, C. *Tropical deforestation.* NASA Earth Observatory. Retrieved November 30, 2003, from http://earthobservatory.nasa.gov/cgi-bin/texis/webinator/printall?/Library/Deforestation/index.html

47 Urquhart, G., Chomentowski, W., Skole, D., & Barber, C. *Tropical deforestation.* NASA Earth Observatory. Retrieved November 30, 2003, from http://earthobservatory.nasa.gov/cgi-bin/texis/webinator/printall?/Library/Deforestation/index.html

48 Urquhart, G., Chomentowski, W., Skole, D., & Barber, C. *Tropical deforestation.* NASA Earth Observatory. Retrieved November 30, 2003, from http://earthobservatory.nasa.gov/cgi-bin/texis/webinator/printall?/Library/Deforestation/index.html

49 NOAA. (1996, July). Four views of California using computer technology, SAR image, interferogram, contour map, and perspective view. *Operational Use of Civil Space-based Synthetic Aperture Radar (SAR)* 10–8.

50 For additional discussion, see these two standard sources: White, J. V. (1990). *Color for the electric age.* New York: Watson-Guptill Publications; and Keyes, E. (1993, November). Typography, color, and information structure. *Technical Communication, 40,* 638–654.

51 InCom Supply. (n.d.). *OSHA & APWA colors: The right colors for the right reasons.* Retrieved November 30, 2003, from http://www.incomsupply.com/customers/specdata/osha.html

RO-AN Corporation. (n.d.). *OSHA and AWPA safety colors.* Retrieved November 30, 2003, from http://www.roancorp.com/techtips/tisafetycolors.html

Superior Graphix. (n.d.). *OSHA color codes.* Retrieved November 30, 2003, from http://www.superiorgraphix.com/osha.html

52 Doll, R. (2001, September 29). *Avoiding equipment hazards: Ensuring machine-guard safety compliance.* retrieved November 30, 2003, from http://www.chemicalprocessing.com/Web_First/cp.nsf/ArticleID/DPIC-562MWY/

53 Doll, R. (2001, September 29). *Avoiding equipment hazards: Ensuring machine-guard safety compliance.* Retrieved November 30, 2003, from http://www.cehmicalprocessing.com/Web_First/cp.nsf/ArticleID/DPIC-562MWY/

54 Firelily Designs. (n.d.). *Color vision, color deficiency.* Retrieved November 30, 2003, from http://www.firelily.com/opinions/color.html

55 Biechler, D. (n.d.). Chichaqua Bend Studios, Ames, IA.

56 Biechler, D. (n.d.). Chichaqua Bend Studios, Ames, IA.

57 Garter, L.P., & Hiatt, J. L. (Eds). *Color atlas of histology* (graphic 9.1, lymphoid tissues). Baltimore: Williams & Wilkins. Reprinted by permission.

58 John Deere.

59 AmTec. (n.d.). Bellevue, WA.

60 Amazing science of space photography. (1997, October 24–26). *USA Weekend,* p. 14.

61 (1999, October). *Science, 286,* 30.

62 U.S. Commission for National Security in the 21st Century. (1999, September 15). Graph showing per-capita income disparities. *New World Coming: American Security in the 21st Century* (p. 38).

63 U.S. Department of Agriculture Food and Nutrition Service. (1992). Bar graph of a dietary fat chart showing fat and fatty acid proportions. *Building for the Future: Nutrition Guidance for the Child Nutrition Programs* (p. 40). (FMS-279).

64 Tyler, D. E. School of Veterinary Medicine, University of Georgia.

65 *Ocean safety signs.* (n.d.). Retrieved November 30, 2003, from http://www.aloha.com/~lifeguards/bsigns.html#waves

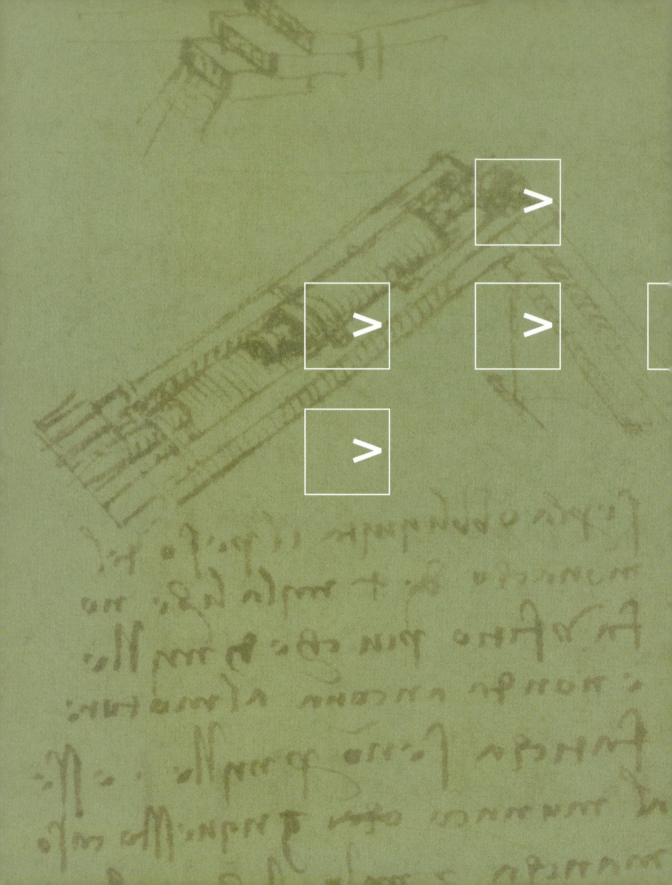

CHAPTER 13

Designing Electronic Communication

Objectives and Outcomes

This chapter will help you accomplish these outcomes:

- Identify the characteristics and features of effective electronic communication
- Understand the principles of effective design for various electronic media
- Analyze key aspects of information architecture: organizing, labeling, and navigating
- Analyze key aspects of effective Web page/screen design: layout, color, graphics
- Understand the standards and tools for designing electronic communication
- Understand the iterative design process
- Assess Web sites for usability and accessibility

>

Electronic tools and processes that allow people to share information will affect many aspects of your professional success. No matter what role you have, when you contribute to the development of electronic communication, you need to consider the virtual environment from a number of perspectives: stakeholders', developers', and especially users' perspectives.

Because most people using electronic information access the World Wide Web, this chapter focuses on Web-based content accessed from personal computers (PCs) as well as from small-screen devices such as personal digital assistants (PDAs) and cellular telephones. The nature, types, and functions of electronic communication affect the ways you read and construct information on the Web. You need to understand the principles and practices for effective site, page, and content design; important aspects of developing usable content for delivery via a range of technologies; and the iterative design process used for that development.

Characterizing Electronic Communication

Users and designers often have different perspectives about electronic communication. Users tend to think about tasks they want to accomplish, such as finding information, purchasing a product, or playing a game. Designers tend to think in terms of design and functionality.

Despite these different perspectives, both users and designers see electronic communication as interactive and nonlinear, virtual and open, complex and dynamic:

> *In what ways do the electronic tools and processes you have used exemplify the characteristics in this list?*

- ***Interactive and nonlinear.*** Electronic communication environments are interactive and nonlinear, established by multiple possibilities for interactions among users, computers, software, interface components, and developers. The goal of electronic communication is for users to accomplish tasks, sometimes something as simple as accessing information on static Web pages.
- ***Virtual and open.*** Electronic communication environments are virtual and open spaces. Virtual spaces do not have a material, face-to-face reality. "Open" means two things, both related to consistency and user expectations: (1) The virtual spaces allow users to move beyond boundaries at will. (2) Standards and conventions are fluid, leading to varied designs and functionality, and often uneven experiences for users.
- ***Complex and dynamic.*** Electronic communication environments are complex and dynamic development efforts that integrate diverse components. Complex development efforts include managing both static and dynamic content, hundreds of individual text and graphic files, multimedia components, and databases. Because development technology changes rapidly, becoming more dynamic and multidimensional, designers must plan for differences in users' available technology.

While these characteristics of electronic communication pose challenges for information designers, people who regularly use electronic communication take its availability for granted, at least until they run into difficulty accomplishing their goals or completing their activities. That's why companies and organizations commit significant resources to create and maintain environments and capabilities for electronic communication, which usually require the collaboration of people with different sets of skills.

Types of Electronic Communication

The chances are high that you are one of the millions of people who access electronic information and services via the Internet and World Wide Web using a variety of PCs, a vast array of software, and different types of Internet service providers. People visit Web sites using browsers. People search libraries online to locate and read articles and books from the comfort of dorm rooms and offices. People use computers to check weather and bank balances, to make airline reservations and order flowers, even to take courses. People send e-mail messages and documents and share music and graphics. People instant message, blog, participate in synchronous and asynchronous chats, and participate in virtual conferences.

> *What different types of electronic devices do you currently use? What kinds of activities do you engage in with each? What differences have you noticed about the design of information for each?*

In recent years, the capabilities of Web resources and PCs have changed dramatically. In addition, people are now using handheld devices, such as PDAs and cell phones, to access Web-based resources and engage in tasks that they once could only accomplish with desktop computers. People can now use cell phones, paging devices, and PDAs to communicate, store addresses, browse headlines, and keep calendars.

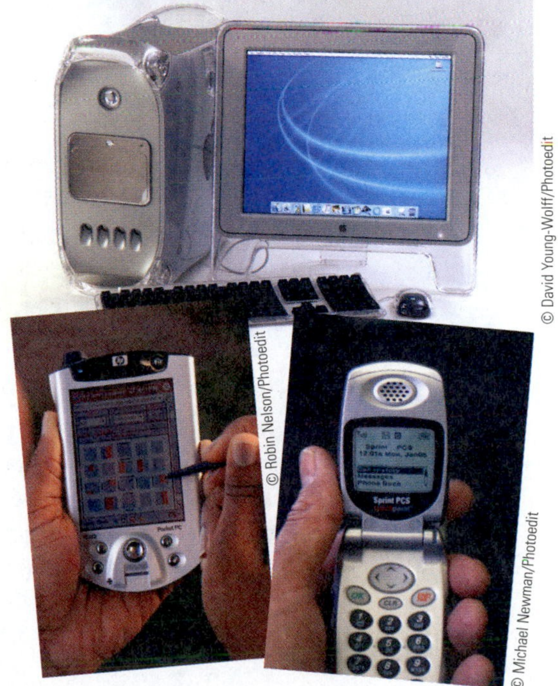
© David Young-Wolff/Photoedit
© Robin Nelson/Photoedit
© Michael Newman/Photoedit

Each of these examples of electronic communication results from the convergence of hardware and software, engineering and programming, connectivity and content. Each type of device places different demands on both technology users and information designers. One obvious example of difference is screen size. The monitor of a desktop or laptop computer obviously displays more information at once than the display screen on a PDA or cell phone. Other less immediately visible differences include memory, bandwidth, connectivity, and the types of standards and protocols required to display information. Designers must account for such differences in designing and developing electronic communication, including handheld and wireless devices.

FIGURE 13.1 Current and Projected Statistics on Web Access by Various Media[1]

Web Access via PCs and Wireless Devices, 2000, 2002, and 2005

Year-end	2000	2002	2005
USA (in millions)			
Web appliances in use	3.2	23.6	115.4
Web appliance share of Internet users	2.3%	14.2%	55.4%
PCs in use	153.2	178.9	221.9
Internet users	135.7	165.7	208.3

Global Internet and Wireless Users, 2001, 2004 and 2007

Subscribers	2001	2004	2007
Internet users (in millions)	533	945	1,460
Wireless Internet users as a percentage of all Internet users	16	41.5	56.8

The use of all electronic devices has increased dramatically in a short time (see Figure 13.1). This rapid expansion of electronic resources and environments means that we need to be concerned with security, especially protecting personal information. How much of a concern this should be is the subject of the ethics sidebar on the next page.

WEBLINK

The statistics show that mobile commerce is booming, especially in Europe, Japan, and the United States. The data in Figure 13.1 are just a hint of its enormous activity. For more details, go to www.english.wadsworth.com/burnett6e for a link. Select what for you is the most surprising statistic in the information presented. Compare your opinion with others in your class.

CLICK ON WEBLINK

CLICK on Chapter 13/mCommerce

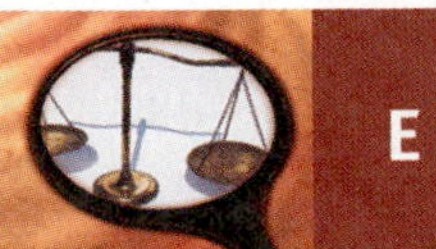

ETHICS SIDEBAR

We Know Who You Are and Where You've Been

Imagine it's Saturday afternoon and you're browsing at your local bookstore. A salesperson approaches you:

> "Can I help you? I noticed that you came in at 1:14 pm through the Main Street entrance, which makes sense since you were just shopping in Baskets and Bows next door. You went right to the bestsellers and considered the latest romance for 56 seconds and then the top-rated mystery for 32 seconds. Now you're browsing self-help books related to depression. Are you depressed? If so, we have several other titles that might interest you. We're also providing information about you to some other businesses with whom we share information — they'll get in touch with you soon. By the way, Jane — your name is Jane Smith, isn't it? — we could have your selections sent right to your home at 123 Elm Street, if you prefer. And, since today is the twentieth time you've been in the store, we have some specials that might interest you. . . ."

Hard to imagine? Not on the Web. Many Web sites collect information about users. Existing technologies for tracking visitors allow Web administrators to collect various kinds of information:

- IP addresses
- Information about operating systems and web browsers
- Time and duration of site visits
- Individual pages accessed, order in which users access them, and amount of time on each
- Other sites users visited, including the link from which a user was referred
- Where users go after their visit

Information about visitors to Web sites is collected a number of ways:

- **Cookies,** small files that servers place on computers, contain information, such as passwords, which frequently makes visiting sites more convenient. Cookies can also track user movements. Session cookies are only used during visits, and then they terminate. Persistent cookies remain on computers until removed or until they expire at a preset time. Cookies are the most common way to collect and use information about users.[2]
- **Forms** users fill out voluntarily ask for personal information that could not otherwise be collected, for example, sex, nationality, age, and e-mail address. The information, often stored in cookies, can assist the site in providing services and offers for users. Some companies sell this information to others.

How can users become better informed about issues of security and privacy?
What responsibilities do schools have to protect the privacy of their students?
What responsibilities do organizations have to protect the privacy of their employees?

- **Hit counters** collect information about visiting IP addresses, users' computers, and the pages users access in a site. Web managers use this information to determine how often their sites are visited, by whom, and for what information.
- **Spyware** (and adware) programs can be placed on users' computers deceptively. While these programs are not illegal, they should cause concern because users often don't know these programs have been installed. Some spyware can track every keystroke that a user makes. Collected information can be sent to a server while the user is online and sold to marketers.[3]

Many companies collect information to make their sites more useful to users. For example, by tracking the most visited pages, companies can provide more of the information users want. Some collected information helps users. For instance, when users check airline flights, they go through several screens of information and make a number of choices that the site server must record to return flight data. Unfortunately, not all data collection is so benign.

If you are developing a Web site that will collect information about users, you should create clear privacy policies and post them on your site. At a minimum your policy should do these things:

- Inform people about the information you collect and the ways you collect it
- Explain the purposes for information collection
- Disclose any information you share and with whom you share it
- Allow users to "opt out" of information collection
- Provide contact information so that users can register any complaints or concerns about your policy

Information collection is a concern to many people including the World Wide Web Consortium,[4] which is drafting policies aimed at making privacy statements on Web sites more thorough, clear, and consistent for users.

Web Sites and Web-Enabled Environments

The World Wide Web is the largest part of the Internet, a huge network comprised of other networks and millions of individual computers. Internet traffic is routed along a number of *backbones,* which are primary networks owned by organizations and companies. These backbones are connected by *hubs* that can move traffic from one backbone to another. The Internet uses a variety of *protocols,* rules and standards that people have agreed to use when developing for the Internet and the Web. For instance, Internet participants use a protocol called TCP/IP (Transmission Control Protocol/Internet Protocol) that allows computers to locate and communicate with each other. This cooperation is essential to keep traffic on the Internet moving and to provide relatively unlimited access to Internet resources. Various organizations and consortiums develop and maintain the shared protocols.

The Web itself is comprised of networks of servers and users' computers that exchange Internet resources using Internet protocols. Hypertext transfer protocol (HTTP), the primary protocol of the Web, facilitates the exchange of hypertext documents. Hypertext documents are plain-text documents that contain hypertext markup language (HTML), a system of tags placed within the documents that are interpreted and implemented by *browsers,* programs that allow users to interact with servers to request, view, and use Web pages. At the most basic level, HTML tags control the appearance of documents in browsers and allow pages to be linked to other pages.

What do you capitalize? When "web" refers to the World Wide Web, it's a proper noun, so it's capitalized. "Internet" is also a proper noun, so it is capitalized as well.

WEBLINK

Lots of people use computer terms carelessly or inaccurately. You have easy access to a number of online resources to help you, from the very simple (TekMom with simple example sentences) to the carefully illustrated (TechEncyclopedia). Go to **www.english.wadsworth.com/burnett6e** for links.

CLICK ON WEBLINK

CLICK on Chapter 13/definition of terms

The individual Web sites that comprise the Web are collections of files that include individual Web pages and graphics and sometimes databases and programs that facilitate sophisticated interactivity, such as searching, placing orders, and playing games. Basically, all the files for a site are stored on a computer with special software that allows the computer to act as a *server.* As the name suggests, what a server does is "serve" users by providing resources that they request by interacting with their browsers. Users' computers are referred to as *clients.*

The relationship between servers and users' computers is called a *client/server relationship.* The concept of a client/server relationship is important to understand because some functions of Web-based materials are performed by the client, and others can only be performed by the server. For example, when you mouse over a menu link on a Web page, the link may change appearance. A small script in the HTML page that is read and acted upon by the client's browser triggers this change in appearance on your computer. On the other hand, when you fill out an online form on a Web site, the information you provide must be handled by a program on the server.

To use Internet resources, people must have access to one of the networks on the Internet. This is generally accomplished by establishing a connection from the client computer and a modem to an Internet service provider (ISP) server that is connected to a network. Examples of ISPs include America Online (AOL), Earthlink, or a point-to-point (PPP) dial-up connection provided through an organization or company. Clients can then request Web pages by accessing the uniform resource locator (URL) of Web pages via browser programs, such as Netscape Navigator, Mozilla, Microsoft Internet Explorer, Mosaic, Opera, and Lynx (which is a text-only browser). Figure 13.2 explains how to interpret a URL.

FIGURE 13.2 **Understanding URLs**

The names and locations, or more simply, the addresses, of electronic resources that you visit or create are referred to as URLs, Uniform Resource Locators. A typical URL looks like this Web address:

http://www.agency.gov/news/archives/project087690.html#participants

Each part of this fictional Web page address, which is highlighted and explained below, contains a considerable amount of information that allows users to find and access resources on the Internet and Web.

http://www.agency.gov/news/archives/project087690.html#participants

The first part of the URL is called the *scheme,* which contains information about the protocol that the computers and servers need to use to interact. These are other familiar schemes:

- https:// (secure http server)
- ftp:// (file transfer protocol)
- mailto: (e-mail address)

http://www.agency.gov/news/archives/project087690.html#participants

The second part of the URL is the actual address or *domain name* of the server on which the Web page is located. URL names correspond to numeric IP addresses assigned to every computer using the Internet (for example, http://128.10.004.1) but are much easier for people to remember and use.

- *WWW* indicates that the resource type is a Web site.
- "agency" is the domain name of the specific Web site. Domain names must be registered so that duplications in names are avoided.
- *.gov* designates the top-level domain in which the resource is included; in this case *.gov* indicates a government agency. Top-level domains are based on the type and geographical location of the resource. The most prevalent top-level domain is *.com.* These are some of the other top-level domains:
- .net (provides Internet services)
- .org (organization)
- .uk (United Kingdom)
- .ca (Canada)

http://www.agency.gov/news/archives/project087690.html#participants

The third part of the URL indicates the folders or *directories* in which the Web page resides on the server. Directory locations are important for providing a complete path to an individual page or resource.

http://www.agency.gov/news/archives/project087690.html#participants

The fourth part of the URL is the *file name,* or the individual Web page or other resource. The *file extension,* in this case *.html* for hypertext mark-up language, indicates the type of document that the user is accessing. These are other common file types:

- asp (active server page)
- .pdf (Adobe portable document file)
- .wml (wireless mark-up language, used by PDAs and cell phones)

http://www.agency.gov/news/archives/project087690.html#participants

The final part of the URL locates a specific place on a Web page. The pound sign in this example signifies a *named anchor* that is identified by the word "participants."

- When an anchor is included in a URL, users' browsers will take them to a specific spot on a page.
- Anchors are particularly helpful on pages that contain a lot of text that users would otherwise need to scroll through.

Audiences and Electronic Communication

Meeting the needs of audiences should be a primary goal of writers and designers of electronic communication. Unlike other types of professional communication, however, electronic environments offer audiences unique opportunities to co-construct the environment and information each time they enter it. For example, because of the Web's hypertext nature, individual users can create unique sequences of information as they move around a Web site or among different Web sites. At a minimum, the sequence of links that users select leads to individual interpretations of information.

Reading and Navigating Electronic Communication. Reading in electronic communication environments is different from reading on paper because it involves an *interface.* In other words, reading electronic media is not simply reading an electronic display of information; instead, it is interactive and brings with it a number of complications. A number of researchers, including Christina Haas at Kent State University,[5] Karen McGrane at Razorfish,[6] and Jakob Nielsen at useit.com,[7] have identified factors that are particularly important in online reading:

- **Screen and page size.** How much text can a reader see at one time? Larger screens (and larger windows on those screens) enable readers to see more. But readers sometimes have difficulty even with large screens, because reading on a computer monitor reduces their awareness of where they are in relation to the whole document.
- **Legibility.** How easy is it to read what's on the screen? Factors such as screen flicker, spacing, and background and text color affect legibility. Readers also sometimes have difficulty because visual cues such as boldface and italics may not show on the screen (as in some e-mail systems), and spacing is sometimes difficult to judge; thus, proofreading is often more difficult to do on a monitor.
- **Responsiveness.** How quickly should a system respond to users' actions? Human factors research shows that users believe a system is reacting instantaneously if the response time is about 0.1 second or less. Users will stay focused and on task if the response time is no more than about 1.0 second, but that's usually not possible on the Web. If the response time is more than ten seconds, people typically lose attention and wander to other tasks.[8]
- **Navigation.** How easily can readers navigate the Web site — that is, how easily can readers move through and locate places in the text? Web readers/users are influenced by images and icons that affect their ability to navigate on the Web, by color, links and backgrounds, and, of course, by typography and layout. But even with navigational aids, it's easier to get lost in an electronic document than in a paper document.
- **Equipment and service.** How much are readers constrained by physical realities? Research at Georgia Tech suggests that even though fast modems and higher bandwidths are available, the modems and lines used by most people are too slow for decent Web response times.[9]

Which of these factors are important to you when reading in an electronic environment? Are some of these factors more important to you than others? Does the purpose or situation change what's more important?

Navigating Electronic Communication. When electronic communication is audience-focused, readers can easily construct paths through resources by choosing their own sequence of links. Researcher Paul Levinson calls this "empowerment of the author through the empowerment of readers."[10] Simply put, hypertext documents allow writers and audiences to link related concepts and restructure knowledge. Different readers can arrive at the same point in a document having discovered different things along the way, either because they activated different links or because they activated the same links but in a different sequence. A document that is read interactively is rarely experienced the same way twice. Electronic communication environments that offer readers this empowerment respect audience needs.

Figures 13.3 through 13.5 show pages from the Food Safety Project Web site. Depending on a user's prior knowledge, needs, and interests, this site could be interpreted in a number of different ways. Visitors to the Food Safety Project Web site can select their own sequence of information.

When readers arrive at the Project's home page, they find that the menu on the left side gives them choices for linking to information about consumer issues, food safety education, foodborne pathogens, and other relevant topics. To the right

FIGURE 13.3 **Food Safety Project Home Page**[11]

Clicking on the link to Food Safety Education brings users to several additional content choices. They can get lessons about food safety and pointers about kitchen safety.

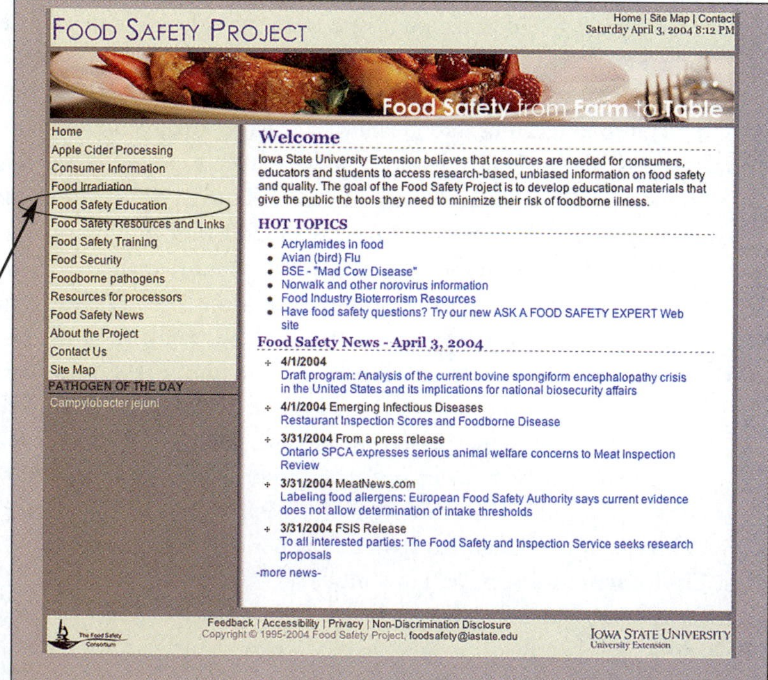

Menus and links allow users choose their own paths through a Web site. For instance, they might be interested in food safety education.

FIGURE 13.4 Food Safety Project Consumer Control Point Kitchen[12]

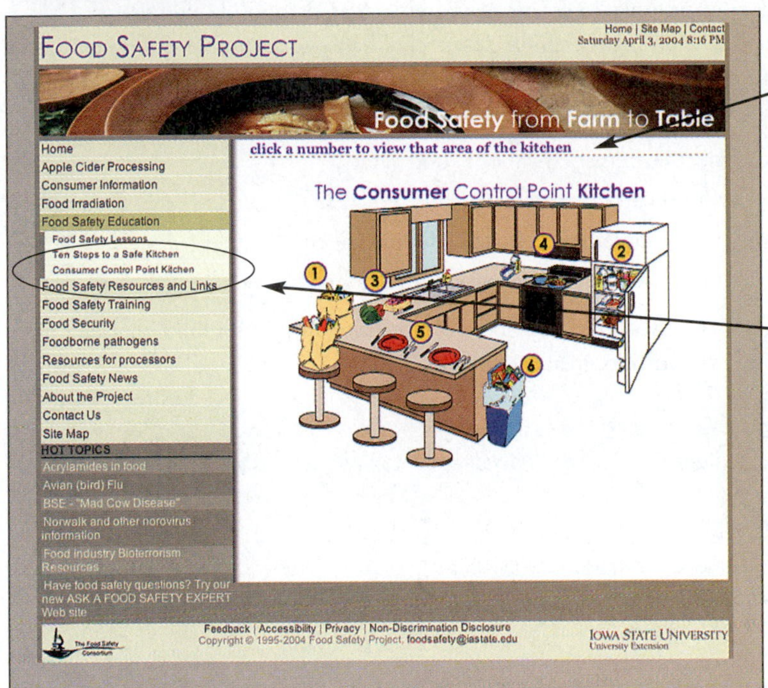

A simple instruction above the kitchen graphic informs users about how to access information.

Users can refine their path by selecting from additional menu choices. What is the Consumer Control Point Kitchen?

The Consumer Control Point Kitchen is a Flash presentation that includes hotspots allowing users to further explore the page's content. Users can click on the numbered items in the kitchen in any order in order to receive safety information specific to that kitchen, or topic, area.

FIGURE 13.5 Food Safety Project Consumer Control Point Kitchen Hotspot[13]

Clicking on Item 1 in the kitchen graphic (shown in Figure 13.4, above) allows users to access information about safe food storage.

Clicking a number triggers a simple animation that provides a close-up of the section of the kitchen that users want to learn about and a text dialog box with specific related information. Users can return to the kitchen for additional information by clicking on "Back to Kitchen."

If users don't find the information they want in the kitchen, they can quickly and easily change their paths through the site because the navigational menu is continuously available. One user may leave the kitchen and visit Food Safety News; another may look for resources under Food Security. In this way, users create their own individual paths through the information.

of the menu, the main section of the home page provides a very brief overview of the site's purpose and links to recent news items related to food safety issues.

After quickly scanning these links, users could choose a number of different and equally productive directions depending on their needs. For example, by following different links, users could do any of these things in whatever order they choose:

- Read the most current news about food safety.
- Learn how food can become hazardous
- Discover the characteristics of a specific pathogen
- View an online video presentation about kitchen safety

And, of course, each link they go to might have additional links that enable them to explore the idea in greater depth.

WEBLINK

To view the complete Food Safety Web site, go to www.english.wadsworth.com/burnett6e for a link.

CLICK ON WEBLINK

CLICK on Chapter 13/food safety site

Principles and Practices of Effective Design

Anyone with time, knowledge, and resources can create a Web site. But designing and developing Web sites that function well for a variety of audiences and a range of technologies requires understanding of users' needs, careful planning, thorough testing, and coordinated design activities. Designers need to consider three important factors:

- ***Information architecture*** is the framework that structures content. The structure should meet the goals and expectations of the user. Structure can be *sequential* (for example, Web pages that link to the next in a linear style), *hierarchical* (outline format), or *interlinked* (less structured, liberally linked).
- ***Page/screen design*** is the look and feel of the information in the space on the screen, another mechanism to help users understand information organization and context.
- ***Content*** is organized and written differently for electronic communication than for traditional print documents. The style of content should match users' ways of finding and reading information using electronic devices.

As you read about the specifics of each factor, remember that the relationships among them determine the degree to which your communication will be usable. For example, you can create beautifully designed Web pages, but if the navigation of the site is poor, your audience will quickly go somewhere else.

Designing Web sites for international audiences brings additional challenges in areas including language, conventions in graphics and design, and expectations about the organization of information. Go to www.english.wadsworth.com/burnett6e for links to interesting discussions about a range of related issues.

CLICK ON WEBLINK

CLICK on Chapter 13/international Web issues

WEBLINK

Information Architecture

Two of the most significant factors of context creation in electronic communication are the design of information that users access directly through sight, touch, and hearing and the design of the components and pathways that allow user access and navigation. Together, these design factors contribute to the architecture of information — the visible areas, the structures beneath, and the design, engineering, programming and electronic media used to create them. This section focuses on three factors of information architecture common in all types of electronic communication: organizing, labeling, and navigating information.

Organizing Information. You can think of information architecture as the relationships between the content pages and the other components comprising an electronic communication environment. A site map is like a blueprint that shows the site architecture. The site map shown in Figure 13.6 depicts the architecture for a corporate intranet, which is a Web site that provides information to individuals in an organization. Generally, Internet users outside the organization cannot access a company's intranet.

In the site map in Figure 13.6, the blue rectangle in the lower left corner represents the intranet home page. Each rectangle in the rest of the figure represents an HTML page that is linked to the home page via one or more hyperlinks. Boxes and cylinders represent Web-enabled software applications (e.g., billing software) and databases (e.g., personnel information), respectively.

As the example illustrates, organizing the components of an electronic communication environment structures information in several ways that affect users:

- Categories of information available to users are determined.
- Relationships of categories of information are established.
- Pathways through information are created based on judgments about the relationships among categories by the designers (and based on user testing as well).
- Points of interaction are established, for example, where certain functions such as forms are included.

FIGURE 13.6 Intranet Architecture[14]

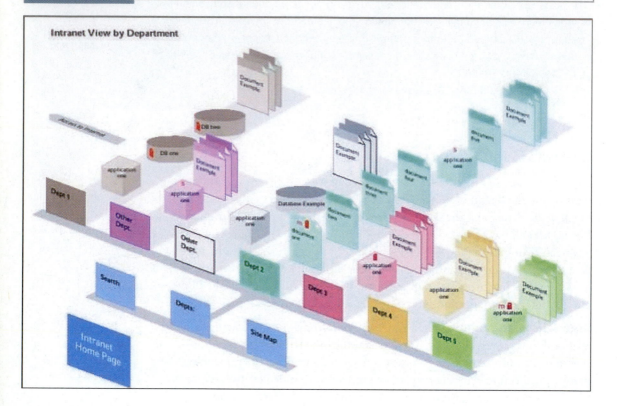

Intranet View by Department

Ultimately, organizing information is a negotiation among users' needs, stakeholders' purposes, the types of media used to provide information, and the capability of the systems supporting the media.

Each of these factors must be carefully considered when organizing information. While no hard and fast rules exist about the best way to organize information, you can begin to think about organizing it by considering several general organizational schemes, illustrated in Figure 13.7, that are frequently used.

WEBLINK

Maps of cyberspaces are often remarkably innovative, much different from the maps you're accustomed to using because the spaces are virtual, not geographic. Go to www.english.wadsworth.com/burnett6e for links to interesting examples.

CLICK ON WEBLINK

CLICK on Chapter 13/maps of cyberspaces

FIGURE 13.7 **Generic Organizational Schemes**

Hierarchical structures are frequently used on the Web. A main entry point, such as a homepage, provides access to sections of related information. Sections serve as discreet collections of content and lower-level pages may only be accessible from other pages within the same section.

Which of these organizational schemes do you prefer when you're trying to locate information? Is your preference related to the subject matter? To the situation?

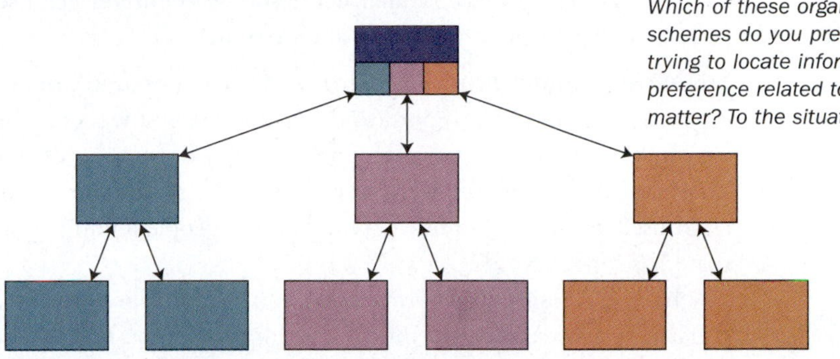

Sequential structures exert more control over paths the users can take through pages by providing minimal navigation through content. For example, a multi-page document may provide only sequential links in a prescribed order.

Interlinked structures create multiple relationships among various pages of information and allow significant flexibility for users to choose paths. This type of structure is often used for parts of sites in conjunction with hierarchical structures.

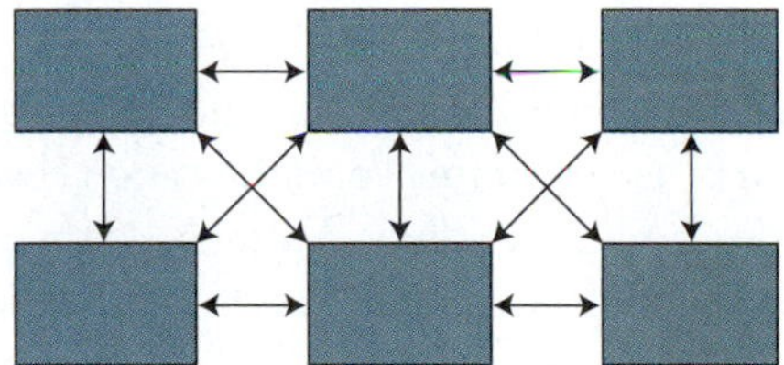

What advantages do Web pages have in comparison to print documents? What disadvantages?

Structure on demand is becoming more prevalent, particularly in e-commerce. These structures are often parts of sites that contain some static elements. Forms for user input are linked to programs and databases that allow people to create information unique to their needs. For example, when you use a search engine to locate particular information, the page of links you receive in response to your query is unique, based on your request and the information available. However, the appearance of the information usually predetermined by the designers, may also be changed by users on some sites.

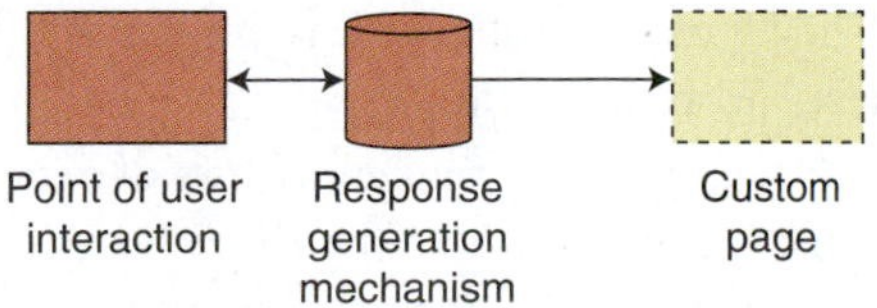

Organizing information and creating a structure for its use are the first steps in developing an environment for electronic communication. As you work on organizing information for electronic environments, you may not know all the individual pieces of information that your text will include, so plan for flexibility so that information can be added. Flexibility is important because electronic information is frequently updated and expanded.

Labeling Information. As you develop categories of information and begin organizing them, you also need to determine the best ways to label the information so that users are able to find and use the resources they need. Labeling information effectively is one of the most important ways that you can assist users of electronic information, because people tend to scan electronic materials for key words.

Before choosing specific words for labels, think about the categories of labels you'll need to describe your information and that users will need to identify it. Claire Harrison, president of Cando Career Solutions, Inc., has developed one effective classification system for the types of links that are often used on Web sites, as shown in Figure 13.8.

> *Think about your department's Web site. Make a list of the category labels that users might expect/want/need. Then check the site to see if your expectations match what's actually on the site.*

Labels are pervasive in electronic communication. Labels appear as page titles, menu items, links, headings within electronic documents, buttons and controls on forms, and other functions features. Because labels are necessary for identifying information and the users' location within the information, labeling is a significant aspect of information architecture.

Navigating Information. Another important factor in information architecture is navigation. As the name suggests, navigation concerns ways that users move through electronic information. Once you have determined the organization for your information and decided how to label it, you need to provide tools to help users move around. The tools should help users develop a mental map of the electronic communication space. Consistent navigational elements (for example, toolbars, buttons, site maps, arrows, menus) let users know where they are in relation to the rest of the information. Users consistently want answers to these questions: Where am I on the site? Where have I been on the site? Where else can I go on the site?

> *Look at your department's Web site. What navigational elements let you answer these three questions?*

How you navigate information depends considerably on your purposes and audiences. For example, why do people visit a particular Web site? What will they look for, and how will they find it? Your navigation strategy also depends on the size and scope of your information. Navigation is usually managed several ways:

- *Menus* are generally horizontal or vertical lists of links to sections or individual pages within a Web site. Users often prefer menus that look like indexes or are in vertical lists.[15]
- *Breadcrumb trails* are sequential lists of pages that let users know where they are on the site and where they have been in relation to either the site's home page or their entry point onto the site.

FIGURE 13.8 A Classification of Links According to Primary Function[16]

Link	Primary Function	Examples on Web Sites
Authorizing	Authenticates the site and its content by describing the organization's legal status, its formal policies, contact information, and so on.	■ About Us ■ Customer Service Policies ■ Archives
Commenting	Provides opinions about the site and/or its content.	■ Press Releases ■ Testimonials ■ Links to reviews
Enhancing	Provides factual information about site content by offering more details or painting a bigger picture.	■ Site Map ■ Guidelines for Membership ■ Site Index; Site Help
Exemplifying	Provides specific examples of content within a broader category.	■ Future Events ■ Job Postings ■ Available Products/Services
Mode-Changing	Moves users from reading to some kind of interaction: an activity or decision making.	■ Online Survey ■ Shopping Cart ■ Taking a Quiz
Referencing/Citing	Provides background or supplemental information about the site's content.	■ Bibliography ■ Related Links ■ Useful References
Self-Selecting	Allows users to narrow searches based on age, sex, geographical location, life situation, personal interests, and so on.	■ E-commerce Sites ■ Your Local Chapter ■ Professional Associations

- *Embedded links* are links within text that take users to another page or site. Be judicious about including links within text. Too many embedded links make reading more difficult.

WEBLINK

The EServerTC Library has a remarkably useful collection of links to articles about information architecture, ranging from usability guidelines to tutorials, from research studies to detailed descriptions of design and development processes. Go to www.english.wadsworth.com/burnett6e for a link.
CLICK ON WEBLINK
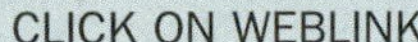
CLICK on Chapter 13/EServer TC Library information architecture

You can help visitors navigate information by taking a few other simple steps that typically make the site more accessible, understandable, and usable:

- Put the organization logo, name, or other identifying information on every page of information. This helps users know where they are relative to the Web as a whole by letting them know when they've left your site and entered another one.
- Include a site map, index, and/or search function so that users can navigate directly to information they want.
- Use the same types of menus, buttons, links, and other navigational elements for all of your information. For example, if you include a back button to return users to the previous page, every page should contain a back button that looks and operates the same way. The colors, sizes, and metaphors of icons, graphics, and text features (such as underlined links) used for navigation should provide a scheme that users can follow.
- Provide dialog boxes and other forms that are similar in appearance and function. Make sure that the location of interactive features makes sense in the overall scheme of the site.

Organizing, labeling, and navigating are critical factors in information architecture. Well-designed electronic information has clues about the architecture built into each screen. For example, the redesigned Fermi Web site in Figure 13.9 contains several embedded clues about site architecture.

Was the redesign a big change? Yes. Figure 13.10 shows what the Fermi site looked like before the 2001 redesign. The 2001 site redesign was a response to the accumulation of electronic documents that the Fermi National Accelerator Laboratory had collected over the time they had been maintaining a Web site. The information needed to be reorganized to provide better access for users. In addition, developing Web technologies made new strategies available for creating content quickly and for providing additional navigational elements, such as the pop-out submenus on the main page links to information hubs.

The Fermi Lab has maintained a site on the Internet since the days of text-only document posting in the early 1990s, when the Internet served scientists who wanted to share information quickly. The Fermi Lab's original site is shown in Figure 13.11. Though the site might look arcane on the modern Web, its simplicity would not be out of place on some PDA or cell phone screens today.

WEBLINK

Based on the information you've read about Fermi National Accelerator Laboratory, you might find viewing the entire site interesting. For a link, go to www.english.wadsworth.com/burnett6e.

CLICK ON WEBLINK

CLICK on Chapter 13/Fermi Lab

FIGURE 13.9 **The Old Home Page of the Fermi Lab Web Site**[17]

In 2001, the Fermi National Accelerator Laboratory revised its Web site. As part of their effort, the designers created "information 'hubs.' Each current public page has a spot in one of these hubs and new content pages will join one of these hubs."

The "hubs" into which information is organized are listed on the main page and include sub-menus.

The Fermi designers note that, "Each hub has a unique page header that may be included on pages within the hub. This header includes navigation buttons to each of the other hubs."

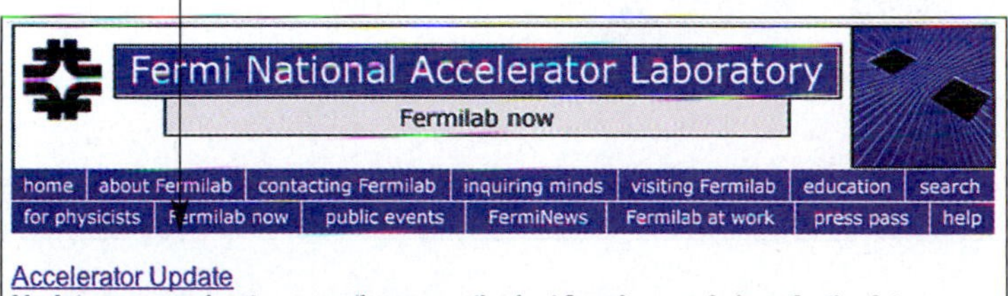

One of the primary purposes of the Fermi site is to distribute news from this national lab, which the designers have made a focus of the site.

The Fermi Web designers point out that, "A news box is prominently featured on the home page. Headlines link directly to the featured news, sections, articles or releases. Members of the Fermi lab community are encouraged to submit news items here. Simply click 'Got news?' from the news box."

FIGURE 13.10 The Old Home Page of the Fermi Lab Web Site[18]

Consider the information categories and labeling of the Fermi Lab's previous site. How would you characterize the organization, labeling, and navigation?

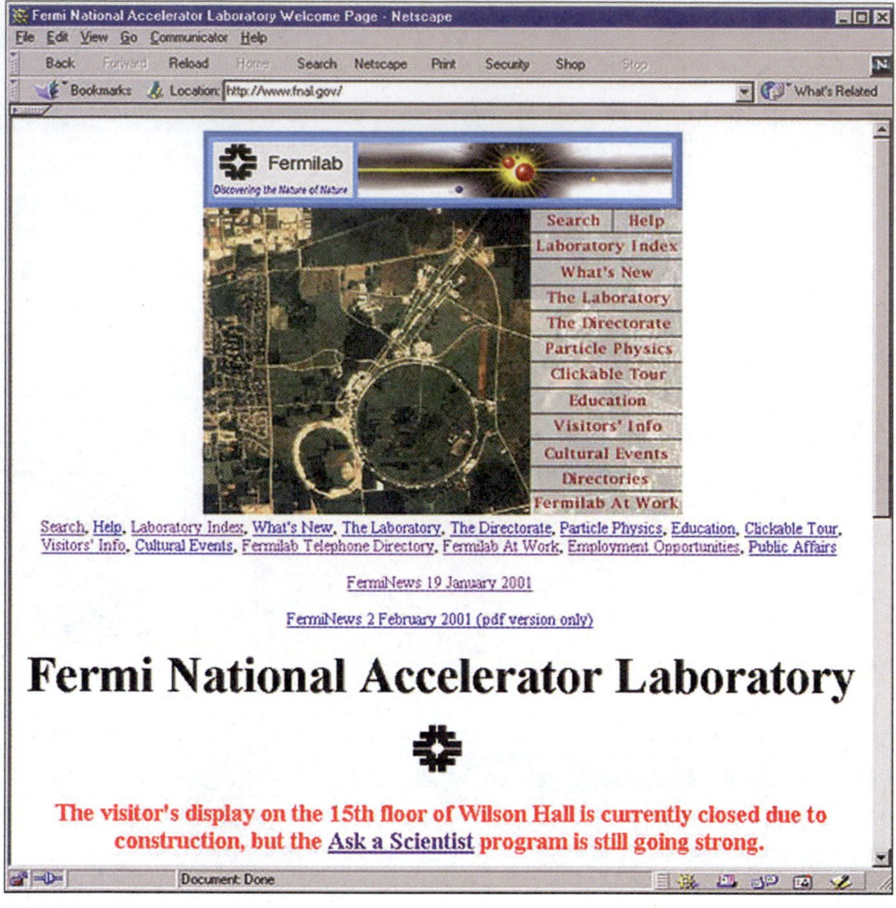

FIGURE 13.11 Fermi's First Web Site[19]

How new is the Web? The World Wide Web was born at CERN in Europe in 1991 as a tool for exchanging particle physics data. When the Fermi Lab Web site was established in June 1992, it was either the second or third Web site in the United States; one at MIT was established at about the same time as Fermi. The first U.S. Web server was created at Stanford Linear Accelerator Center in December 1991.

FERMILAB COMPUTING DIVISION

Computing[1]	Documentation: General[2], Offline Products[3], Online (DOCDB)[4]
INFO[5]	Access to Fermilab and Computing Division announcements.
H E P[6]	World Wide Web services provided by other High Energy Physics Laboratories, CERN[7], SLAC[8]
Spires[9]	Access to Spires preprint database.
Help[10]	Fermilab Help Page for WWW
WWW[11].	Information on the World-Wide Web project.

Page/Screen Design

The aspects of information architecture — organizing, labeling, and navigating — come together in the display of electronic information. Pages and screens are the visible means by which users access content. To bring information architecture to users, you'll need to consider several aspects of screen and page design including layout, color, and graphics.

Layout. The layout of screens and pages is an important aspect of designing electronic information. Electronic materials and the ways people use them are different from print pages and their use. Screen displays combine interface elements that people click, mouse over, scroll, type into, and otherwise manipulate with navigation, written content, graphics, and other elements so that users can easily access resources.

You can get a good start on screen and page design by choosing not to move texts formatted for print directly into an electronic environment. Instead be guided by the architecture of your information and apply a consistent grid when deciding how information should be displayed on the screen. Though one set of rules about page layout does not exist, using an established grid may be helpful to beginning designers. Established grids rely on what is known about the ways people look for information and the types of design conventions with which users are already familiar. On the Web, for example, people have become used to navigation menus at the top and sides of the screen. However, conventions are like habits, which can be good or bad, and decisions about information design should be driven by the types of information and user needs within a specific communication context.

Figures 13.12 through 13.16 illustrate the screen/page designs of the current NASA Web site. The pages reflect the complex architecture and variety of user interests that an organization such as NASA must take into account when creating a design. The grids for the information are tailored for NASA's specific goals, but also incorporate some conventional choices.

The content for each type of page on the NASA site is adapted to enhance the information and the users' experiences with the site. Despite the remarkable amount of information available, the NASA site maintains a high level of consistency in the layout of the main site components across all types of pages. When the grids of various page types are stacked on top of each other, a basic pattern emerges. This grid is illustrated and explained in Figure 13.16. The consistency provided by grids is essential for learnability and memorability — two important usability principles discussed in Chapter 9.

Sizing and positioning content on the page are also important considerations of page and screen design because of variations in display sizes. Small PC monitors can generally display information that is 640 pixels wide by 480 pixels high on one screen, for a total of 307,200 pixels. (A "pixel," short for picture elements, is the unit of measurement for Web page elements and monitor screen

FIGURE 13.12 NASA Site Home Page[20]

Carefully study this home page for the NASA Web site. What can you tell from this page about the organization, labeling, and navigation of the site? What can you tell about the layout, color, and graphics?

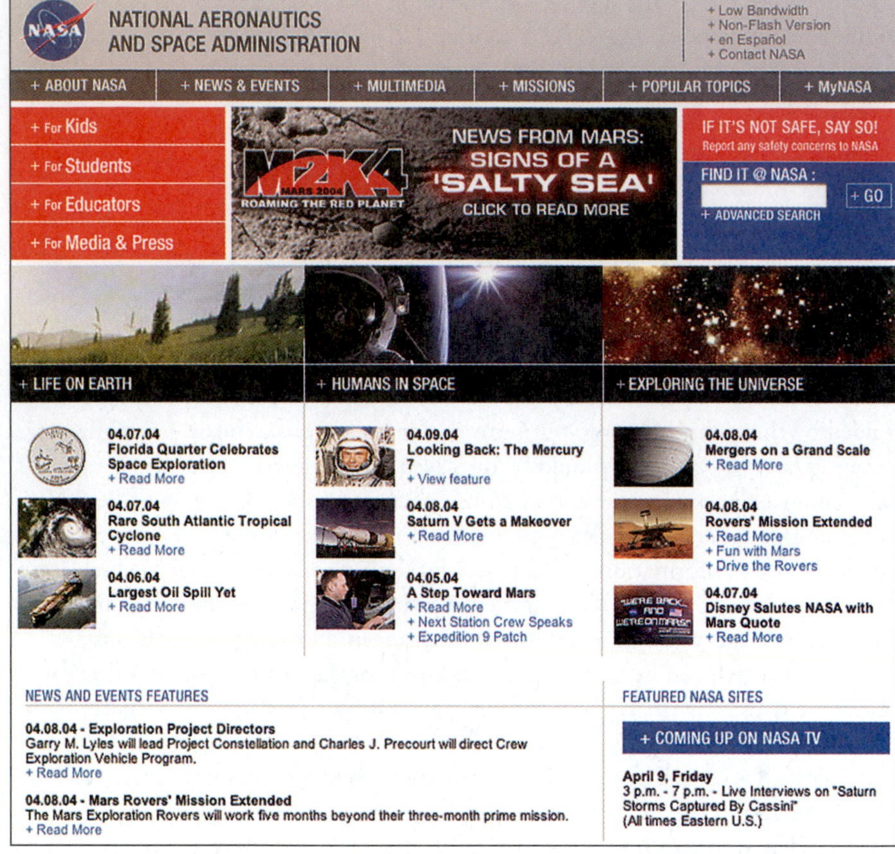

What resolution do you typically use on your computer monitor? What is the range of possible resolutions?

size. The number of pixels affects resolution; the higher the number of pixels, the higher the screen resolution.) Larger monitors can display 800 by 600 pixels or more. Cell phones, PDAs, and other handheld devices typically have much smaller displays. In addition, the way that users set the preferences for their browsers can change the display.

Browsers, by default, display Web page content horizontally to fit within browser windows. The length of a Web page, or the number of screens that you need to scroll down to reach to bottom of the page, is determined by the amount of content that fits on a screen.

You cannot control all the variations in users' hardware and software. If you're designing primarily for the Web, consider designing for small displays or designing pages that display correctly regardless of the way they are resized by browsers. This type of design is called *resolution independent,* which allows the users' browsers to manage the screen space. If you choose this option, make sure

FIGURE 13.13 **NASA Home Page with Grid Overlay**[21]

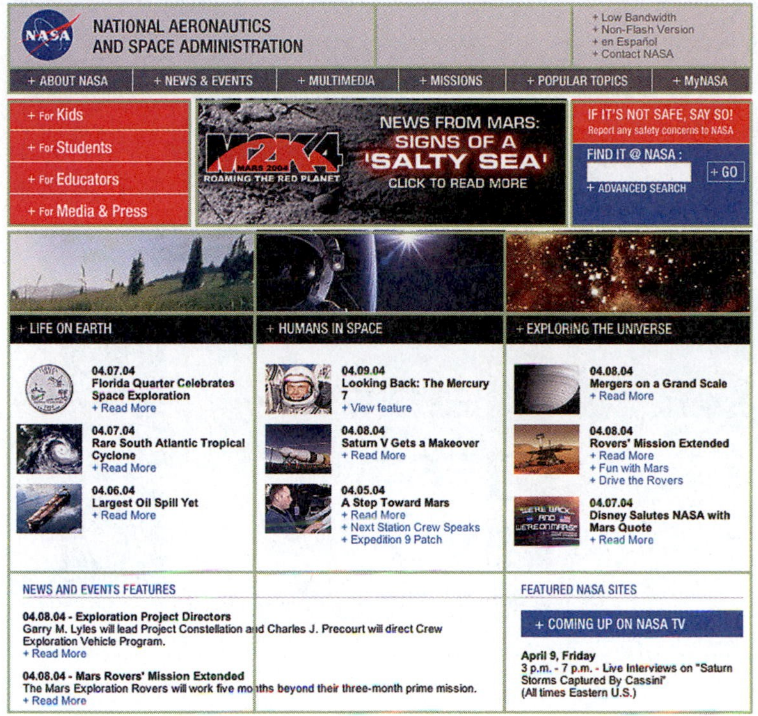

NASA's site uses top navigation to main sections of the site. Top navigation has become a popular choice. The top of the site is also used for identification. Content is displayed below the navigation.

Be careful that your page doesn't become top heavy, with excessive graphics at the top that do little to help the reader situate themselves, navigate, or receive information. Top heavy banners often push important content too far down the page, below the visible screen. If users don't scroll down, they may miss important information.

The green lines on the image of the Web site here and in Figures 13.14, 13.15, and 13.16 show the grid for each page. The thumbnail sketch for each figure highlights the grid for that particular page.

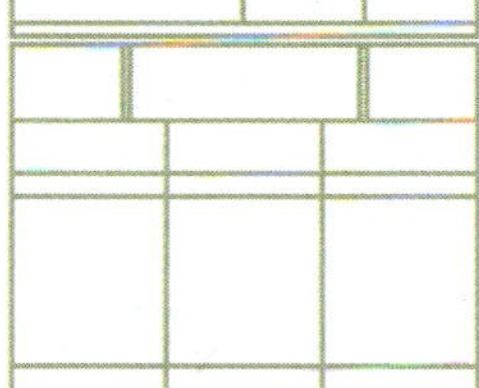

your design organizes the space in such a way that long lines of text don't stretch across the width of large monitors. If you are designing for a variety of information devices, for instance, Web materials that will also be displayed on PDAs, tools are available to help manage and adjust the same information for different modes of delivery. The amount and complexity of information often makes preparing and maintaining separate sets of materials for different systems costly and time consuming.

The following general guidelines will help you design (or revise) usable Web pages:

- ***Content.*** Provide more content than navigation on each page. Users come to your site for the content, so at least 75 percent of your pages beyond the home page should be content.
- ***Identification.*** Include critical information and site identification at or near the top of each page. The top left of a page is an important focal area of Web pages as well as print pages.
- ***White space.*** Don't fill every pixel of your page. White space is important to help guide the user's eye around the page. Leaving space is often preferable to creating divisions with graphics or color bars that detract from your

FIGURE 13.14 NASA News and Events Section Home Page with Grid Overlay[22]

Some pages include navigation and other elements on the left or right sides of the page (and occasionally on the top and both sides, as with a number of the search engine and news sites). The NASA site includes right-side navigation for second-level content within sections of the site.

When you provide navigation and other elements in various locations on the page, it's important to be consistent about the type of information you include in those areas.

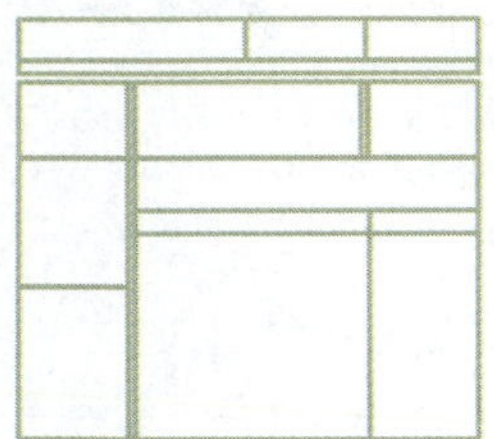

FIGURE 13.15 NASA MyNASA Utility Page with Grid Overlay[23]

The NASA site provides a number of utilities to enhance users' experience of the site. These include links to tools that adjust delivery of the site depending on users' bandwidth, a search engine, and — as illustrated to the right — the opportunity to customize the site.

This screenshot also includes the footer, which appears on every NASA page. Footers can be used to repeat site identification and mirror site navigation at the bottom of pages. Other information commonly found at the bottom of pages includes the site author's name, Web manager's name, last date updated, and/or contact information.

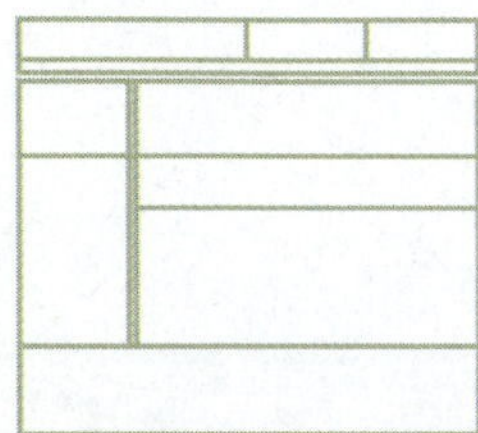

FIGURE 13.16 Basic Grid for NASA Web Pages

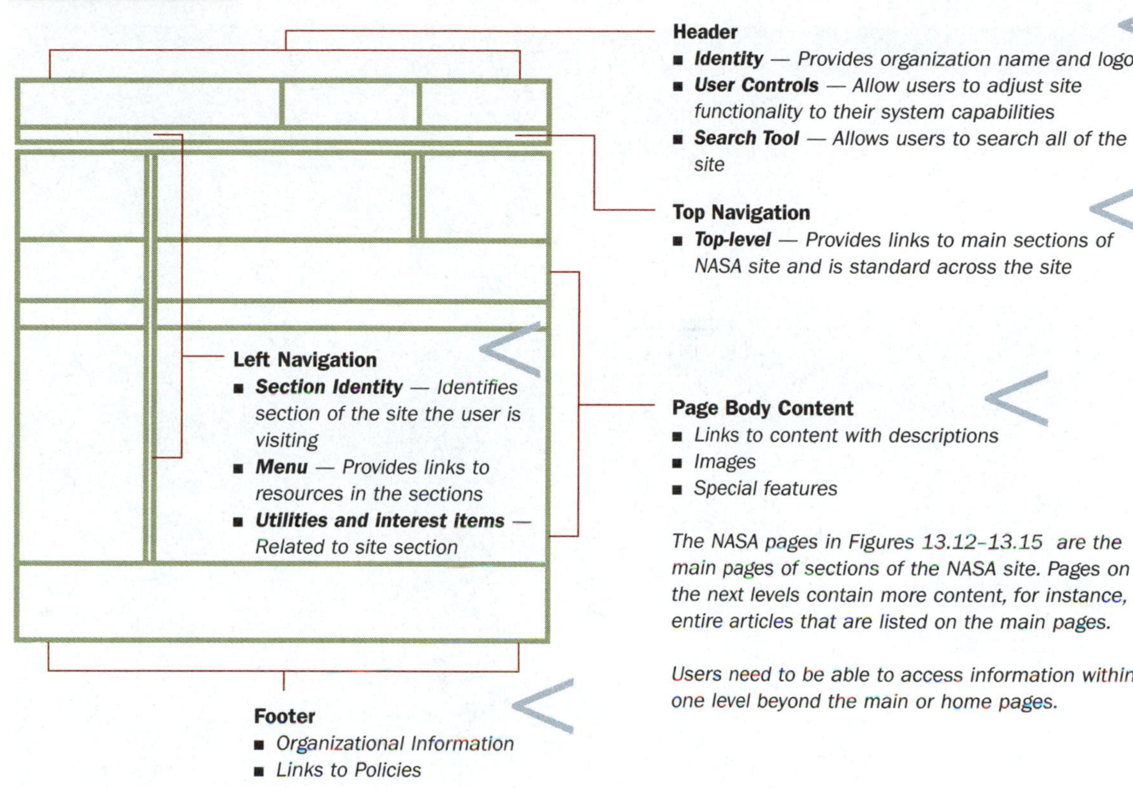
Header
■ Identity — Provides organization name and logo
■ User Controls — Allow users to adjust site functionality to their system capabilities
■ Search Tool — Allows users to search all of the site
Top Navigation
■ Top-level — Provides links to main sections of NASA site and is standard across the site
Left Navigation
■ Section Identity — Identifies section of the site the user is visiting
■ Menu — Provides links to resources in the sections
■ Utilities and interest items — Related to site section
Page Body Content
■ Links to content with descriptions
■ Images
■ Special features
The NASA pages in Figures 13.12–13.15 are the main pages of sections of the NASA site. Pages on the next levels contain more content, for instance, entire articles that are listed on the main pages.
Users need to be able to access information within one level beyond the main or home pages.
Footer
■ Organizational Information
■ Links to Policies

content and navigation. ("White space," of course, may not be white on the Web but will be your background color.)

- ***Scrolling.*** Prevent horizontal scrolling. General consensus among user-testing experts is that people accept a reasonable amount of top-to-bottom page scrolling, which is actually preferable to excessive "click through." Left-to-right scrolling in almost all cases is not tolerated well at all.

Figure 13.17 illustrates the differences in screen real estate for various monitors and devices.

Color. Use color both to highlight different areas of screen displays and to unify the site. Generally, designers choose a palette of a few colors for an individual site. Color can orient users on the site, identify navigation, and signal types of content.

Color should do more than just create interest; it should function as an integral part of the information that users need. Color lets users know where they are on your site and differentiates functions or elements on pages. For instance, in-text links on Web pages are generally all the same color so that people can recognize what text contains links.

FIGURE 13.17 Display Sizes

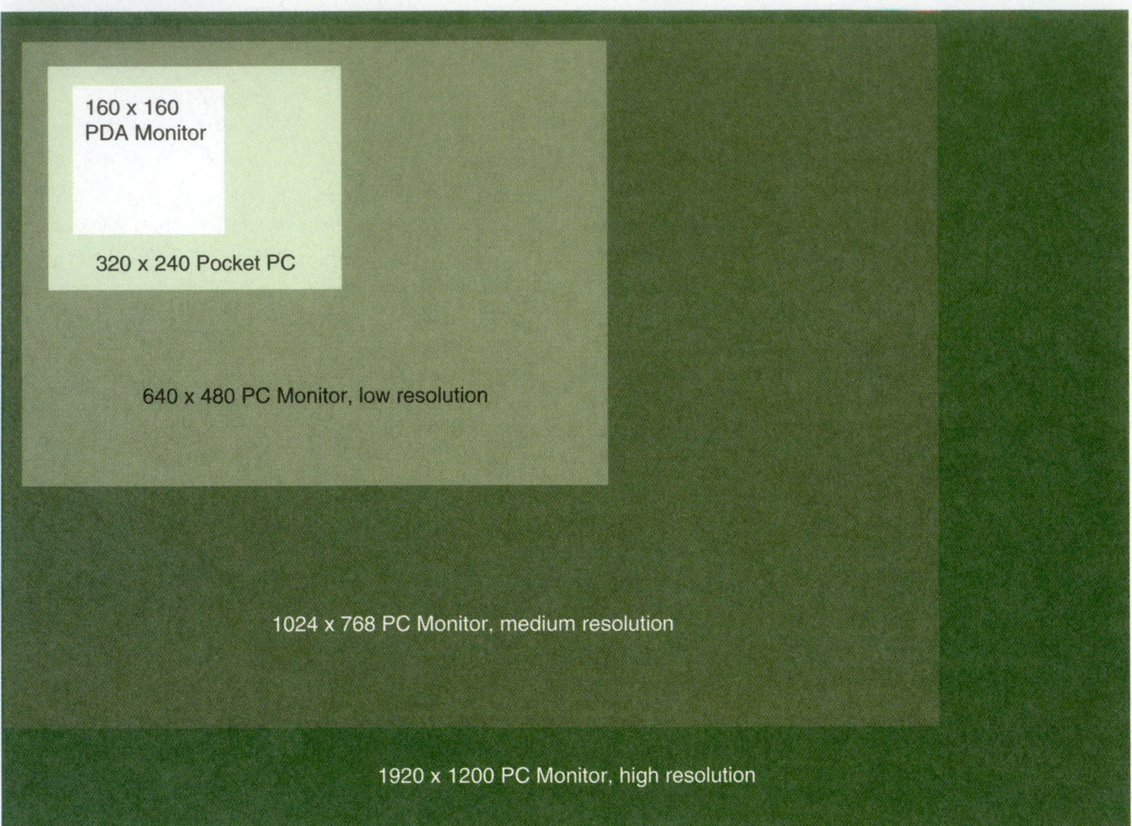

Think about color as you are planning your development project. Establish a coordinated color palette before putting your site together so that the colors are consistent throughout and complement the purpose of your site. The palette should include background colors for various elements of the site (for instance, navigation areas and text areas), text colors (for body text, heading texts, links, and other types of text), and accent colors (for arrows, buttons, dividers, etc.). Coordinate your palette with the site's graphics. If you are unfamiliar with color coordination, get advice from someone with graphic design experience. If you are designing for a company or organization, follow its identity guidelines, which typically include the use of specific colors.

In choosing background and text color, ensure high contrast between your background color and text color so that the text is easy to read. Although black text on a pale background may not seem exciting, it is still the most readable choice for large areas of text. Avoid backgrounds containing patterns, designs, or images that reduce the legibility or obscure your text and compete with other graphics.

Avoid problem color combinations. Red on green, for instance, causes difficulty for people with some types of color blindness and makes reading more difficult in general.

Avoid placing text over patterned backgrounds. Reading information on computer screens is more difficult than reading in print because of the lower screen resolution. Busy backgrounds place additional demands on readers.

Color on a computer screen is produced by combining three colors (red, green, and blue). These three colors make up all colors viewable on a computer monitor. All three colors at 100 percent of their value make white, and the complete absence of all three makes black. Various percentages of each make up 16.7 million colors. Until a few years ago, computers were very restricted in the number of colors they could display. In addition, different computer platforms also used different color palettes; both Macs and PCs had 256 colors in their system palettes, but only 216 of them were the same. To ensure that colors displayed the same way across platforms, Web designers generally restricted themselves to using a Web-safe color palette of 216 colors common to both Windows and Mac computers.

By 1999, however, almost 90 percent of computers that were in use could display colors into the thousands.[24] Even now, though, because the problem of browser compatibility persists to some degree and because Web pages are being served to handheld devices, WebTV, and other platforms, you will get consistent results by continuing to use a Web-safe browser palette. Colors are associated with codes made up of combinations of 6 letters and numbers. For instance, the code for white is #FFFFFF, and the code for black is #000000. Codes are important because they tell the browsers what colors to display. You should use color numbers in your HTML documents to indicate where you want specific colors to appear. Keep a list handy of the color numbers you are using for each project.

Selecting Web-safe colors is easy. Many sites provide the hex/decimal codes. To see variations of these sites, including one that automatically identifies the code when you mouse over a color, go to **www.english.wadsworth.com/burnett6e** for links.

CLICK ON WEBLINK

CLICK on Chapter 13/Web colors

WWW

WEBLINK

Graphics. People include graphics on Web sites for several reasons:

- Enhance the design and usability of the information
- Help establish identities for Web sites and create excitement and interest
- Help users navigate by calling attention to links, sections of sites, special features, and different areas of pages
- Provide information to illustrate products to consumers or as integral parts of Web-based articles, reports, or other documents

However, you should not rely only on graphics to convey information about your site because that may affect usability. Browsers can be set to display text only, a feature that some people use to lower the load time of pages. Some browsers do not display graphics at all, and people who use screen readers because of a visual disability do not benefit from graphics unless additional information is provided.

The size of graphic files is an important consideration for consistency because large graphics can slow the load time of information. The larger the file, the longer it takes to load into the browser window. Graphic files can be large, and most of the time are larger than the file for the page that contains the text information. File size doesn't necessarily indicate big images, but rather large files sizes (indicated by the number of kilobytes (K), or computer space they use). In general, you should reduce the file size of graphics as much as possible, balancing the quality loss against the load time.

You can control the size of graphics files through your choice of image format. Graphics for electronic display come in a wide variety of file types for different purposes. Three types of image formats are especially useful for display on monitors:

- **JPG image (pronounced "j-peg")** — named for the Joint Photographic Experts Group that created the type — works well for photographic images, images with a high number of colors (above 265), and images with graduated color.
- **GIF image (pronounced "gif" or "jif")** — short for Graphics Interchange Format developed by Compuserve — works well for images with "flat" color areas, transparent images (when you want to have the background of your site show through part of the image), and for small images such as icons. Gifs manage up to 256 colors.
- **PNG image (pronounced "ping")** — for Portable Network Graphics — is a new graphics type that promises to offer qualities of both .jpgs and .gifs with smaller file sizes. However, older browsers do not support this image type, and some graphic production software packages do not include the capability of saving images in the .png format.

The examples on the next page illustrate some of the qualities of .jpgs and .gifs, the two most used types of graphics.

FIGURE 13.18 **Web Graphic Examples**[25]

The photo of the rubber duck **(9 KB)** and the word "Ducky" **(2 KB)** are .jpg images.	The photo of the rubber duck **(5 KB)** and the word "Ducky" **(1 KB)** are .gif images.	The big masks **(25 KB)** and the small masks **(7 KB)** are .jpg images.	The big masks **(5 KB)** and small mask **(2 KB)**, are .gif images.
The .jpg files render photographic images more smoothly. However, the file sizes of the .jpg images are larger. You need to decide when you can sacrifice load time for clarity in photographic images. Small pictures of people, for instance, might be worth the extra wait.		Here, large .jpg image file size is quite large compared to .gif image file size — five times larger. However, in the case of flat colors, the .gif is just as clear as the .jpg, so the extra size of the .gif image is not worth the space.	

Graphics files, particularly .jpg files, can be optimized for the Web, which means reducing the file size and, thus, the quality. Generally, programs for creating and adjusting graphics, such as Adobe Photoshop, have tools for optimizing graphics. For graphics that enhance the design and usability of Web sites, follow several useful rules of thumb:

- ***Balance graphics and text on a page.*** Graphics should not overwhelm your text.
- ***Coordinate graphics throughout a site.*** Too many different types of graphics confuse rather than help users.
- ***Keep size of image files as small as possible to reduce load time.*** The size of image files (the K) depends on the type of image used and the procedures to create and save the image.
- ***Notify users when you must include a slow-loading graphic.*** A message such as "Please wait for image to load" tells users that you are sensitive to users' needs and that the image may be worth waiting for.

- ***Use larger images on lower-level pages.*** If people are satisfied with the design and navigation of your home page, they are more likely to tolerate additional load times on lower-level pages if the graphics are informative and useful.
- ***Don't rely on images alone to convey information.*** People sometimes set browsers to display text only because of excessive load times. Some people use text-only browsers, like Lynx, and people with visual disabilities may use screen readers that rely on text information only. These users rely on alternate text information to be included with graphics.

You can obtain graphics by using software to create your own or by converting them from images you draw or photos you take. Software packages often include generic graphics (clip art) that you can use, although clip art is generally overused and often becomes a visual cliche. A number of sites on the Web also offer free images that you can download and use. When you use graphics from clip art files or from Web sites, read any rules of use or permissions that protect the images. For instance, some permissions allow you to use graphics if you credit the creators on your site. Others allow you to use graphics for nonprofit reasons but not to resell or to use on an e-commerce site. You may not just take graphics from sites and use them on your own. That is copyright infringement.

When designing information for small-screen devices such as PDAs and cell phones, consider that the graphic capabilities for these devices are more limited than PCs and laptops. Small-screen device displays have a much lower resolution and fewer colors. When — and if — you use graphics in content targeted for these devices, use simple graphics that contain only a few basic colors and ensure that the graphics are necessary.

Developing Effective Content

People must sift through massive amounts of information daily, particularly on the Internet, which means that writing and managing content are critical to the usefulness and usability of electronic information. The best screen layout in the world and the most interesting bells and whistles are useless if content is missing, inaccurate, inadequate, poorly written, or hard to use. Two aspects of content to consider when developing for electronic communication include writing effectively and ensuring credibility.

Writing for Electronic Communication. Writing for electronic media is different than writing for print. People read less and skim more online. Because of the possibilities offered by hyperlinking, people also move quickly among various pages and sites. Your content should facilitate the ways that users locate and read electronic information. Author Steve Outing, writing in the online publication *Editor & Publisher Interactive,* summarizes some of the advice given by Crawford Kilian in his book, *Writing for the Web.*[26]

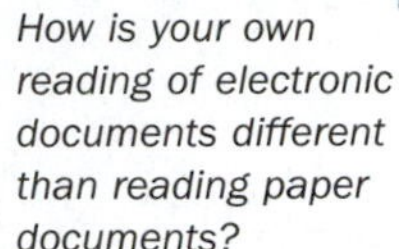

- ***Be concise.*** Low-resolution computer screens make reading from a monitor more difficult than reading on paper. Reading from a monitor can be 25 percent slower than reading on paper.
- ***Keep chunks of text short.*** Readers of electronic documents need minimal text. Write chunks of text between 150 and 200 words, then edit so you get each chunk to about 60 words. Kilian advises that writers "pack the maximum meaning into the minimum text, so your readers will get the message in the shortest possible time."
- ***Use headings and bulleted lists.*** Long paragraphs don't work well on computer screens. Instead, use headings and bulleted lists so users can scan the text easily.
- ***Use active voice.*** Active voice identifies the doer of the action, so the action is clearer and readers are more engaged. Say "You need to review the budget" or "The committee needs to review the budget" rather than "The budget needs to be reviewed."
- ***Consider international readers.*** The whole world may read your message. Your text may be read in Australia, Germany, Israel, Japan, Saudi Arabia, South Africa, and Venezuela; therefore, avoid culture-specific idioms and metaphors.
- ***Use an "inverted pyramid" structure for organizing information.*** Similar to news reporting, your text should include the important information up front, followed by the details. People are going to scan your text, so make the critical information easy to find.
- ***Limit in-text links to other sites and provide information about the links you do include.*** Content on the Web is sometimes littered with links and long "hotlists" that contain links with no explanation of what users can expect on the other end. Many users find themselves linking from a page of links with little content to other pages of links with little content.

How is your own reading of electronic documents different than reading paper documents?

What ways of presenting information do you prefer when you're reading electronic documents?

Sites often contain reports, white papers, and articles that are essential for users of information. Consider archiving longer documents for download in .pdf or other formats and providing brief summaries of these materials, using guidelines for Web writing.

If you're interested in how people read on the Web, go to www.english.wadsworth.com/burnett6e for a link to what usability expert Jakob Nielsen has to say.

CLICK ON WEBLINK

CLICK on Chapter 13/reading electronic documents

WEBLINK

The need for extreme brevity is essential for information accessed via handheld devices. Web expert Mark Pearrow explains that the content for many types of PDAs "is based on decks of small cards, each of which is tiny enough to be just about displayed on a single screen."[27] When the information is more than can be managed on a card, information may be condensed automatically to accommodate the display. Since you can't know how the adjusted content might be read or what information might be excluded, you need to ensure that your content is designed to be efficiently and effectively managed on a PDA. Also important in the small device environment is management of linking. More linking and more site-level depth helps keep the page length short.

Building Credibility. In all situations involving information, people want to be able to trust the accuracy of what they read. Although print media doesn't always guarantee the accuracy of information, the fluidity of electronic information make determining and developing credibility challenging. While publishing information in print form often includes review processes that help separate accurate information from junk, widespread access to the Internet means that just about anyone can author and disseminate information.

Designers and users of information need to critically consider how to establish credibility of electronic information. B. J. Fogg, a researcher at Stanford University's Persuasive Technology Lab, has developed guidelines for building information credibility in electronic environments. The guidelines in Figure 13.19 can also be used to evaluate the credibility of sites you visit.

Standards and Tools

Information designers use standards and other tools to create content and maintain consistency across multiple-page Web sites. Standards include definitions of site elements and processes for implementing design decisions. Though good standards may take time to develop initially, they can improve efficiency and reduce the costs of future development and maintenance because the products of many design tasks can be reused. Some standards, such as the scripting languages that allow electronic information to appear in browsers, are industry standards that have been developed and adopted over time. Documentation, such as style guides and code libraries, are used during development and for maintenance after development.

Markup Languages, Scripts, and Programming

Documents on the World Wide Web are "marked up" with HTML. HTML is not programming. It is a system of tags that, when inserted into plain-text documents, tell Web browsers how to display documents. Writing documents using HTML enables your documents to be displayed on any computer that has a Web browser and an Internet connection, whether a Mac, PC, or UNIX-based workstation. HTML tags also control the appearance of pages on a screen.

FIGURE 13.19 Guidelines for Building Credibility[28]

Guidelines	Comments
1. Simplify the process of verifying the accuracy of information on your site.	Provide third-party support (citations, references, source material) for information you present, especially if you link to this evidence. Even if people don't follow these links, you've shown confidence in your content.
2. Show the actual organization sponsoring your site.	Show that your Web site represents a legitimate organization. The easiest way to do this is by listing a physical address. You can also post a photo of your offices or list membership in the Chamber of Commerce or other credible organizations.
3. Highlight the expertise of people in your organization and of the content and services you provide.	Provide information about in-house expertise. Do you have experts on your team? Are your contributors or service providers authorities? Be sure to give their credentials. Are you affiliated with a respected organization? Make that clear. Conversely, don't link to outside sites that are not credible; your site becomes less credible simply by association.
4. Show that honest and trustworthy people stand behind your site.	Show the actual people behind your site and in your organization. Find ways to convey their trustworthiness through images or text. For example, some sites post employee bios that tell about community involvement, family, or hobbies.
5. Make contacting you easy.	Make your contact information clear: phone number, physical address, and e-mail address.
6. Design your site so it looks professional (or is appropriate for your purpose).	Pay attention to factors such as layout, typography, images, and consistency. Some people quickly evaluate a site by visual design. The visual design should match the site's purpose.
7. Make your site easy to use — and useful.	Make sites easy to use and useful. Some site operators neglect users by self-aggrandizement or using excessive animation.
8. Update your site's content often; review it regularly.	Indicate the date the site was updated. People believe sites are more credible if they have been recently updated or reviewed.
9. Limit promotional content such as ads or special offers.	If possible, avoid having ads (especially pop-up ads) on your site. If you must accept advertising, clearly distinguish the sponsored content from your own.
10. Avoid errors of all types, no matter how small.	Avoid all errors. Typographical errors and broken links severely damage a site's credibility.

HTML has been revised several times and is being supplemented (and in some cases replaced) by newer markup languages. Other types of scripting tools and programming languages now work with HTML to deliver rich graphic and interactive electronic content. In addition, the development of new types of hardware and connectivity has required markup and programming languages to function with the new tools. Figure 13.20 lists several types of markup languages, scripting tools, and programming languages that you will encounter. You may not need to be a computer programmer to design electronic communication, but you should know the tools and their use.

Pages that display electronic content are plain-text documents that can be created in any text editor. However, programs are available that streamline the design process, help manage site structure and files, and allow designers with little HTML knowledge or experience to create Web pages. Some popular development programs for Web pages include Macromedia's Dreamweaver, Microsoft's Frontpage, and Adobe's GoLive.

Style Sheets and Templates

Cascading style sheets manage the appearance of Web pages across a site. Style sheets contain information about elements such as fonts, heading levels, colors, and backgrounds. One cascading style sheet can be used to coordinate the appearance of hundreds of Web pages. If, for instance, you want the font on all your pages to be the same type style, size, and color, using a cascading style sheet ensures consistency. In addition, if you want to change the appearance of an element managed by a style sheet across the whole site, you can change one style sheet, rather than hundreds of individual Web pages.

Templates can also be used to manage the layout of Web pages. Greeked templates, which are examples filled in with "fake" content, can be used early in the development process to test the usability of a design before content is added. Greeked content can also contain styles used for the actual content.

Code libraries contain reusable scripts and codes that can be "plugged into" new pages and sites. Reusing programming for forms and other functional aspects of Web sites saves development time and ensures consistency in function.

Style Guides

Style guides include information about the way that particular information is designed and should be maintained. Style guides can include a variety of information, from color and text choices to guidelines for writing and incorporating graphics. Though creating a style guide takes a little time, the effort is usually worthwhile, particularly in maintenance time and expense.

FIGURE 13.20 Summary of Widely Used Tools

Tags, Scripts, Styles, and Programming	Description	Uses			
		Web/Internet	Wireless	Client Side	Server Side
HTML	**HyperText Markup Language** is a system of tags that make up the basic markup language for the Web.	✔	✔	✔	
HTML 4.0	**HTML 4.0** is the newest version of HTML currently in use. This standard offers more types of tags but is a stricter form of HTML, which means that designers must be more careful about using tags. Browsers are moving towards accommodating this standard as a replacement for HTML.	✔	✔	✔	
DHTML	**Dynamic HTML** combines HML with other programming and scripting languages to provide interactivity and animation.	✔		✔	
XML	**eXtensible Markup Language** is a "metalanguage" for development. It does not include specific tags like HTML; instead, it contains rules for creating tags that can be used to format many types of documents. For example, XML helps information designers create content once and then format it for a variety of media.				
XHTML 1.0	**eXtensible HyperText Markup Language** combines properties of HTML and XML. As with HTML 4.0, this transitional standard offers more design capabilities for designing content and may be used in place of HTML.	✔	✔	✔	
JavaScript	**JavaScript** is a scripting language that shares features of some programming languages. JavaScript enables some functionality on Web pages, such as roll-over images and drop-down menus.	✔		✔	
CGI	**Common Gateway Interface** applies to several types of programming languages — for example, Perl is particularly popular for Web development projects — that process interactive Web content between clients and servers. For instance, CGI is required to receive and display information from Web-based forms.	✔			✔
PHP	**Hypertext Preprocessor** is a scripting language that works with HTML. It resides on the server and allows the creation of dynamic content. PHP works well with databases and can interact with various types of resources across the Internet, not just the Web.	✔			✔
WML	**Wireless Markup Language,** which derives from XML, is used to display information on wireless phones and some other types of handheld devices.		✔		
WAP	**Wireless Application Protocol** allows users to access information on handheld devices.		✔		
CSS	**Cascading Style Sheets** are used with HTML and XHTML to control the appearance of Web pages. CSS can be stored in a separate file from content pages and linked to them.	✔			

The consistency of all aspects of an electronic communication environment is important for usability. Establishing and maintaining consistency requires coordination during the design process. In addition, maintenance procedures must be documented to ensure that consistency is continued as the site is updated. Consistency is a basic usability principle because it reduces the time users need to learn your site and the number of problems they have using your site.

Planning the Iterative Process

Creating a project plan is essential, whether you are developing a site alone or as part of a team. Before you start developing a site, you need to know its purpose and scope, the overall look and feel, the resources you'll need, limitations you must contend with, and the schedule, all of which you will define and explain in the project plan.

During development, the project plan helps the development team stay on track by providing guidelines that everyone can consult. If you are developing a Web site for your company, another company, or other organization, a project plan often becomes part of your proposal. Of course, you will need to make adjustments during the project, for instance, as a result of user responses in testing. Maintain the plan throughout the project and include any changes that you make. Your project plans should address some or all of the issues identified in Figure 13.21, depending on the purpose and scope of the site.

Analyzing Existing Sites

Begin the development process by analyzing existing electronic communication. The analytical process allows you to identify sites that provide information, products, or services similar to yours and provides an opportunity to test the usability of others' sites before developing your own. Analyzing other sites can cut your planning time by allowing you to identify what works and what doesn't. The goal is not to copy other sites but to learn from them. Find sites that interest you and see how well they work. As you analyze sites, identify specific design features that caused problems and those that worked well. The worksheet in Figure 13.22 can also be used to assess your own site during and after development. To complete a fair assessment, you need to actually use a site rather than just look at it.

- What's available for user response or feedback?
- How many clicks from the home page do you need to read actual content? In general, three clicks is reasonable, five is acceptable, and beyond five difficult to manage and very annoying.
- How much time is needed to download? Is the maximum download time within acceptable limits, that is, 10 seconds?

In addition to developing your own analysis guided by the questions in Figure 13.22, you may want to conduct some preliminary usability testing of these sites. This testing may be minimal, involving only three or four users, to give you a preliminary sense of how people interact with various sites. Of course, as your project develops, you will need to regularly conduct usability testing sessions on various portions of your site. This testing lets you make changes as you go along, so each iteration of the site eliminates problems identified by test participants.

Creating Prototypes of Your Web Site

Creating prototypes, which occurs early in the design process, involves developing mock-ups of ideas for your Web site and brainstorming those ideas with team members, clients, and possibly potential users. This allows you to get feedback about your ideas before a significant amount of time, resources, and effort go into development.

Initially, you should develop low-tech storyboards as prototypes of the site. For example, try using simple index cards as "pages" of your site. Draft a scenario explaining the site and its goals. Write page names on index cards (and/or a description of site content). Meet with your team as well as clients and representative users if possible. Using a large board, tape or pin the cards to it, creating a simple display of the proposed site content and architecture. This storyboarding task helps you answer several questions:

- Is the content sufficient and useful?
- Are the names of site areas and pages helpful?
- Does the organization make sense?
- Are the pages of information in the appropriate categories and areas of the site?

Ask team members, clients, and users to comment on and make suggestions about proposed names, content, and architecture. Changes are easy to complete on the spot by writing new index cards and moving others around the board.

Why is using a low-tech storyboard often more productive when done collaboratively rather than when done individually?

From developing low-tech prototypes, you can go on to develop more complete outlines. Keep the initial designs simple. Sketch the page design and possible site structure by creating a small number of linked pages. You don't need actual content and graphics at this point; you'll be testing primarily for overall functionality and usability of the page design. Develop two or three prototypes from your best ideas and then get feedback from team members, clients, and users before proceeding.

Coordinating the Process

Coordination requires both a team effort and a central authority. Early in the design process, the results of testing prototypes help developers decide which features and design elements work best. The team should have a shared vision of the ultimate look and feel of the site. Particularly on a large project, a project manager or steering committee should coordinate all aspects of a project so the site is coherent and consistent.

FIGURE 13.21 Questions and Tasks for Web Site Development

AREA	QUESTIONS AND TASKS
DEVELOPMENT	
Purpose(s)	■ What is the purpose of the site? ***TASK:*** Write a brief statement articulating as clearly as possible the primary and secondary purposes of the Web site.
Intended Audience(s)	■ Who is the primary audience for the Web site? Secondary audiences? ■ What information will they need to use your site? What tasks should they be able to complete? ***TASK:*** Develop user and task analyses and attach them to the project plan.
Project Requirements	■ What are the client's project requirements? ■ Can these project requirements be met? Which will need to be adjusted and in what ways? ***TASK:*** List any project requirements from the client. Categorize the requirements by area (i.e., content, design, graphics, technology, reporting, etc.).
REQUIREMENTS	
Project Resources	■ What hardware is required for development (computers, scanners, printers, servers)? ■ What software is required for development (for example, designing and creating pages, managing site structure and links, designing and creating graphics, project management, mounting site to server)? ■ What server capabilities are required/available? Where is the server on which the site will be mounted? Who can access the server, and mount and manage files? ***TASK:*** Identify what hardware and software are available and where they are located. Determine the necessary server information and supervision.
User Technology	■ What platform/browser combinations are you designing for? ***TASK:*** Create a matrix and indicate the combinations that you anticipate designing for. Your goal should be to accommodate users' technology to the maximum degree possible. However, if you choose not to accommodate older browsers or some platforms, provide a rationale.

	Netscape Nav 4.x	Netscape Nav 6.x	IE 4.x	IE 6.x	Other
Mac					
PC					
Handheld					
28.K modem					
56.K modem					
Broadband					
Wireless					
Other					

FIGURE 13.21 Questions and Tasks for Web Site Development (continued)

AREA	QUESTIONS AND TASKS
REQUIREMENTS	
Project Timeline and Milestones	▪ What is the deadline for delivery of the project? ***TASK:*** Moving backwards from the completion date, schedule all activities and milestones.
Teams/Team Member Responsibilities	▪ What are the human resource requirements for the project? ***TASK:*** List teams and responsibilities. List all specialties required for the project and the number of people needed to complete tasks: Graphic designers? Programmers? Database developers? Writers? Site designers? Multimedia experts? Match available personnel to responsibilities and teams.
PROJECT MANAGEMENT	
Communication	▪ How will the project be managed? How will team activities be coordinated? ▪ What are the proposal requirements (if any)? To whom will it be delivered? ▪ What status updates are required and for whom? ▪ What materials will be provided to client? Archived? ▪ How will change management be handled (process by which changes are documented, reported, approved, and implemented)? ***TASK:*** Provide an organizational chart to identify the team members and their responsibilities. Write the project specifications. Create a Gantt or PERT chart.
Documentation	▪ What documentation is required on the site (FAQ, help, etc.)? ▪ What documentation is required for site maintenance (e.g., style guide, instructions, templates)? ***TASK***: Identify the documentation required.
Quality Assurance	▪ What testing will be completed? When will testing be completed? ▪ What resources are required for testing? ***TASK:*** Create a testing plan.
Site Design/ Navigation	▪ What does the site architecture look like? ▪ What do the pages look like? ▪ What are some key paths users might take? ***TASK:*** Create the site architecture on paper. Create a storyboard of the key pages and paths. Develop names for projected section, page of the site. Prototype the site.
Content	▪ What content is to be provided on the site? How will content be acquired? ▪ What subject matter experts (SMEs) are needed to develop site content? ▪ What writing conventions will be followed? ***TASK:*** List content provided by client and content to be developed. Identify SMEs. Determine appropriate tone and style as well as spelling and naming conventions. Prepare a style guide. Write sample content to review and test.
Visuals	▪ What types of visuals are required? For site design? For content? ▪ How will visuals be created/acquired? ***TASK:*** Select and/or design the visuals.

FIGURE 13.22 Worksheet for Website Assessment

Ratings: (4) Excellent, (3) Acceptable, (2) Weak, (1) Unacceptable, (N/A) Not Applicable

Site address: ____________________

Type/purpose of site: ____________________

Site Features	Rating
Context	
Sponsorship and affiliation evident and visible	
Information easily available	
Perspective/persuasion (news, opinion, etc.)	
Advertising	
Audience	
Reader benefits obvious	
Appropriate for intended audience	
Responses to customer queries/feedback	
Information Architecture	
Clear navigation	
Type(s) and usability of navigation (list)	
Names of pages consistent	
Clicks between home page and information	
Use of lists, headings	
Inclusion of relevant links	
Page Design	
Logical balance: content, graphics, navigation	
Effective use of white space	
Consistent use of color	
Use of headings aids navigation	
Appropriate for rhetorical situation	
Site design consistent	

Site Features	Rating
Content	
Up to date	
Accurate and error-free	
Verifiable	
Reliable	
Research-based	
Sources listed	
Contact information provided	
Visuals	
Relevant	
Consistent	
Aid navigation	
Load time	
Usability/Accessibility	
Maximum download acceptable	
Names of links consistent	
Text and links readable	
Functions on multiple platforms/browsers	
Alt tags included with graphics	
Title tags included with links	
Help and use information provided	
All pages identified; site ID on all pages	
Site map, index, or search functions	

People involved in professional communication must be aware of accessible design concepts as they relate to information development and management. Although tools like special hardware and software can help users who have disabilities, they will not make up for inaccessible information design. Developers need to follow these useful practices:

- Understand the opportunities and limitations of the virtual environment and its potential users.
- Know something about the assistive hardware and software available and be aware of how the design of electronic information could impact the technology your audience may be using.
- Concentrate on good design principles and integration rather than on what "cool" things you can do with programming languages, unless those functions help the majority of your audience receive the information and services you are offering.
- Use various methods for providing information so that you accommodate the greatest number of visitors.

WEBLINK

The more we learn about the ways that users interact with Web sites, the more that older, static sites need to be redesigned. The lessons learned from CancerNet, a Web site that provides current and accurate cancer information from the National Cancer Institute (NCI), the federal government's principal agency for cancer research, is very useful. It summarizes and illustrates data collection and user analysis, iterative prototype development, usability testing, and site launching. Go to www.english.wadsworth.com/burnett6e for a link.

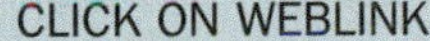
CLICK ON WEBLINK

CLICK on Chapter 13/Web redesign success

Features of Accessible Electronic Communication

Developers can do a number of very simple things to help audiences better access their electronic communication environments:

- ***Provide alternative representations of information.*** Give people choices of page displays, such as with or without frames; graphic and text or text only; with or without audio; or large-font text displays. This doesn't mean just taking the page and stripping out the "cool stuff." This is about considering each rendition on its own merits.

MEETING THE CHALLENGES OF A WEB ACCESSIBILITY POLICY

University Web sites have a challenging array of audiences, including former, current, and potential students. The Emporia State University (ESU) Web presence originated in the mid-1990s as a way to share the wealth of knowledge at ESU with enrolled students. It developed without much oversight regarding design or accessibility into a massive compilation of sites that form a Web presence that reaches worldwide.

When the state of Kansas mandated that all agencies in the state incorporate accessibility guidelines into their Web development policy, ESU realized the complex implications of displaying information on the Internet.

Identifying the Challenges. Implementation of the Kansas Web Content Accessibility Guidelines (KWAG) was met with fierce resistance because current Web development practices would have to change. These changes presented four areas of challenges affecting the production of high-quality Web sites:

1. ***Institution environment:*** Balance was needed between administrative autonomy and technical experience.
   - Politics — From the inception of the university Web site, departments were given absolute control over their respective sites.
   - Technical reality — Many departments assigned office staff who had little or no skill in implementing any type of guidelines.
2. ***Perceptions about the significance of the policy:*** Balance was needed between the mandated changes and information about the importance of the changes.
   - Communication gap — Many stakeholders did not understand the significance of producing accessible Web pages.
3. ***Resistance to change.*** Institutional compliance and consistency needed to be balanced by creativity and academic freedom.
   - Ongoing resistance — Many stakeholders believed the policy detracted from creativity and academic freedom.

4. ***Support for implementation.*** Balance was needed between people's Web design and site management skills and their need to understand the impact of the policy change.
   - Knowledge gap — Many department Web managers lacked the basic skills to maintain a quality web site.
   - Policy problems — Many individuals at the university were unable to determine the ways in which the policy affected them.

To skim through the Kansas Web Content Accessibility Guidelines, go to www.english.wadsworth.com/burnett6e for a link.
CLICK ON WEBLINK
CLICK on Chapter 13/KWAG

WEBLINK

Meeting the Challenges. Implementing new policy guidelines requires a common knowledge base, so ESU developed a four-part process:

- Develop personas to illustrate the legitimacy of following Web development standards such as the KWCAG.
- Develop templates that meet both state and university Web standards.
- Hold workshops that cover basic Web page management activities.
- Provide an open forum where information technology development and strategies can be discussed freely and without consequence.

Organizations regularly must decide to take the time to develop an information system the right way or to create a quick fix, which more than likely ignores multiple issues that will need to be addressed in the future.

One way to do things right from the beginning is to develop *accessibility personas,* like the one for Stephanie in the sidebar. Accessibility personas are composites representing individuals who are members of specific groups of ESU Internet users. These personas are created by grouping critical user characteristics (much like you'd do in writing usability scenarios). The likely users of one site can be represented by creating a series of personas. Such personas help personalize the legitimacy of conforming to Web standards, as the example of Stephanie shows.

Do they work? Yes. Development teams that write accessibility personas tend to create more accessible sites that are responsive to user needs.

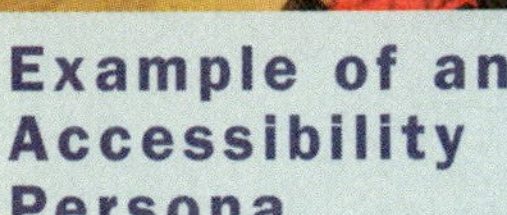
© Christina Micek

Example of an Accessibility Persona

Making the ESU Web Accessible to Stephanie

Stephanie is an 18-year-old who is scheduled to attend ESU in the fall. She is deaf and has low vision. She uses a screen magnifier to enlarge the text on Web sites to a font size that she can read. When screen magnification is not sufficient, she also uses a screen reader to relate to a braille display.

Stephanie has been able to enroll at ESU through the admissions Web site because the online enrollment form uses the appropriate form labels specified by KWCAG.

Stephanie is able to use much of the Web site because of accessible design in a number of important areas:
• style sheets • accessible multimedia • device-independent access • labeled frames • appropriate form labels • table markup

Departments that don't follow the KWAG are inaccessible to Stephanie; thus, she is unable to use those university services.

Zachary Lavicky is the webmaster at Emporia State University. He has a seat on the State of Kansas Web Accessibility Subcommittee, where he is involved in developing and implementing Web Accessibility Standards for the state of Kansas.

- ***Use alternative tagging ("ALT" tags).*** When a user places the mouse on a graphic, the alternative tag enables a pop-up box with a simple text explanation of a function, link, or graphic.
- ***Add transcripts and captioning to audio.*** If people can't hear the audio, they won't get your message.

While these suggestions won't solve all the accessibility problems with Web-based information, they will go a long way in making that information available to a much larger audience.

Checking Web Sites for Accessibility

In addition to paying attention to accessible design principles, you should ensure that your Web sites are accessible by conducting usability tests through the development process. As you develop information for electronic media, ensure that your design serves people with varying needs. Design and implement a testing program that includes diverse users. Also consider using online testing services such as the World Wide Web Consortium's free HTML Validator Service.

WEBLINK

A number of useful guides to developing usable Web sites are available on the Web. Go to www.english.wadsworth.com/burnett6e for a list of links.
CLICK ON WEBLINK
CLICK on Chapter 13/Guides to usability

Individual and Collaborative Assignments

1. **Identify accessibility, comprehensibility, and usability**

   (a) Work in a small group to visit any three of the following Web home pages, which represent a range of national and multinational corporations as well as not-for-profit organizations.

   www.admworld.com
   www.fedex.com
   www.dupont.com
   www.bayerus.com
   www.nestle.com
   www.oracle.com
   www.birdseye.com
   www.lucent.com
   www.ups.com
   www.cocacola.com
   www.merck.com
   www.whymilk.com

   Consider these questions:

   - ***Who's the sponsor?*** Can users easily identify a name, logo, and tag line on the home page describing what the organization does?
   - ***Who are the intended audiences?*** Who are the probable users? What are their purposes for using the site?

- ***How accessible is the site?*** To whom is the site accessible? To users reading other languages? To users with visual disabilities? To users with limited or old computer resources? Does the home page have distracting or competing elements?
- ***How easy is the site to understand?*** Can users easily determine what's most important on the home page? What prior knowledge is assumed? What conventions does the home page use? What is the technical level of definitions and explanations?
- ***How easy is the site to use?*** How easily can users determine the site's organization from the home page? Can users find a clear place to start? How are links to other parts of the site indicated? How easy is the site to navigate?

(b) Summarize your findings in an online guide for Web site designers who are interested in identifying what users see as successful practices. Provide specific examples (correctly documented) from your visits to the Web sites.

(c) Summarize your findings in a collaborative memo written to the sponsor of one of the Web sites, highlighting the strengths and recommending improvements. Provide specific examples from your visits to the Web site.

2. **Compare impact of screen size.** Work individually to identify several Web sites that provide information in formats for both PCs and for PDAs. Compare the ways that information is displayed in each medium.
   - What are the differences in screen display?
   - What are the differences in the ways content is presented?
   - What is gained or lost in the different screen sizes? In content? In convenience?
3. **Credibility of Web sites.** Individually select three companies as your top choices for employment and locate their Web sites. Then use the criteria in Figure 13.19, Guidelines for Building Credibility, to assess each Web site for credibility. Create an effective way to summarize your findings for each organization (for example, matrixes, paragraph descriptions, checklists). Rank order the three organizations according to your findings. Then prepare a 5–8 minute oral presentation for your class in which you summarize and support your findings.
4. **Credibility of nonprofit organizations.** Work with a small group to select the Web sites of four or five nonprofit organizations to assess their credibility. Use the criteria in Figure 13.19, Guidelines for Building Credibility, to assess each Web site for credibility. Do the task collaboratively, discussing the ways in which each criterion can be applied.

Create an effective way to summarize your findings for each organization. Rank order these nonprofit organizations according to your collective decisions. Then prepare a 5–8 minute oral presentation for your class in which you summarize and support your findings.

5. **Assess the usability of your university's Web site.** Work with a small group to assess your own university's Web site.

   (a) Use the criteria in Figure 13.22, Worksheet for Website Assessment, to consider these seven categories: context, audience, information architecture, page design, site features, visuals, and usability/accessibility. Rate each item as (4) excellent, (3) acceptable, (2) weak, (1) unacceptable, or (N/A) not applicable. Do the task collaboratively, discussing what contributes to the rating in each category. Be able to point to at least two aspects of the site that contribute to the rating in each category.

   (b) Based on your assessment, create a list of ways to improve the site.

   (c) Prepare a memo about your assessment and recommendations to submit to the university's webmaster.

6. **Create a prototype for a Web site.** Work with a small group to select a small organization that would like to create or significantly update a Web site. This organization will be your client. Interview the key people in the organization to learn what they believe their users need/expect/want on the site. Two tools from this chapter will be especially useful: Figure 13.21, Questions and Tasks for Web Site Development, and Figure 13.16, Basic Grid for NASA Web Pages. First, complete the tasks in Figure 13.21, which includes creating a prototype of the site with index cards to create a storyboard. As part of this process, sketch the grid. See if that grid can be used with only slight modifications for each page on the site, so that you have consistency and coherence in the architecture. Prepare a memo about your prototype, including the prototype, for your client.

7. **Consider privacy policies.** Visit eight to ten Web sites to locate, read, and assess their privacy policies.

   (a) Figure out a way to record and display your findings. You may need to ask additional questions. In order to have an equitable review of the sites, you need to answer the same questions on each site. Use these questions as a starting point:

      - How easy is locating the privacy policy?
      - Does the site appear to collect information about visitors? If so, what information is collected? How is it collected?
      - Does the site appear to share the collected information? With whom and why?
      - How can users "opt out" of information collection?

(b) Based on your findings, create what you consider to be an excellent privacy policy.

8. **Assess Web writing style.** Visit five Web sites in order to assess their writing style.

(a) Use the principles of writing for the Web that are presented and discussed in this chapter to evaluate the writing on the sites. Figure out a way to record and display your findings, making sure to include examples that illustrate your points. Consider these questions as a starting point for your analysis:

- In what ways does the writing conform to principles of effective writing for the Web?
- In what ways does it fail to conform?

(b) Choose a passage of content from one of the sites that does not conform to the guidelines. Revise the passage to improve the readability for the Web.

Chapter 13 Endnotes

1 Compiled from data available from ePaynews.com. (n.d.). *Statistics for mobile commerce.* Retrieved December 21, 2003, from http://www.epaynews.com/statistics/mcommstats.html#39

2 Winker, M. A., Flanagin, A., Chi-Lum, B., White, J., Andrews, K., Kennett, R. L., et al. (posted 3/17/2000). Principles governing AMA web sites. *Guidelines for Medical and Health Information Sites on the Internet.* Retrieved January 6, 2004, from http://www.ama-assn.org/ama/pub/category/1905.html

3 Post, A. (2003). *The dangers of spyware* (White Paper). n.p.: Symantec Corporation. Retrieved January 6, 2004, from http://www.symantec.com/avcenter/reference/dangers.of.spyware.pdf

4 W3C Technology and Society Domain. (n.d.). *Privacy activity statement.* Retrieved January 6, 2004, from http://www.w3.org/Privacy/Activity

5 Haas, C. (1996). *Writing technology: Studies on the materiality of literacy.* Mahwah, NJ: Lawrence Erlbaum.

6 McGrane Chauss, K. (1996). Reader as user: Applying interface design techniques to the web. Kairos, 1(2). Retrieved December 21, 2003, from http://english.ttu.edu/kairos/1.2/features/chauss/bridge.html; Retrieved January 3, 2004, from http://www.asis.org/Conferences/Summit2000/Information_Architecture/mcgrane.html

7 Nielsen, J. (n.d.). *useit.com: Jakob Nielsen's website.* Retrieved January 3, 2004, from http://www.useit.com/

8 Nielsen, J. (1994). "Response times: The three important limits." *Usability engineering.* San Francisco: Morgan Kaufmann. Retrieved January 3, 2004, from http://www.useit.com/papers/responsetime.html

9 Nielsen, J. (1997, March 1). *The need for speed.* Retrieved January 3, 2004, from http://www.useit.com/alertbox/9703a.html; and GVU's WWW user survey.(n.d.). Georgia Institute of Technology, GVU Center, College of Computing. Retrieved January 3, 2004, from http://www.cc.gatech.edu/gvu/user_surveys/

10 Levinson, P. (1997). *Soft edge: A natural history and future of the information revolution* (p. 146). New York: Routledge.

11 Iowa State University Extension. (1997–2003). *Food safety project.* Retrieved December 16, 2003, from http://www.extension.iastate.edu/foodsafety/

12 Iowa State University Extension. (n.d.). The consumer control point kitchen. *Food safety project.* Retrieved December 16, 2003, from http://www.extension.iastate.edu/foodsafety/educators/ccp.cfm?articleID=62&parent=2

13 Iowa State University Extension. (n.d.). The consumer control point kitchen. *Food safety project.* Retrieved December 16, 2003, from http://www.extension.iastate.edu/foodsafety/educators/ccp.cfm?articleID=62&parent=2

14 Kahn, Paul. (n.d.). *Mapping web sites.* Dynamic Diagrams Seminar on Web Site Planning Diagrams, Visualizing and Analyzing Existing Web Sites. Copyright © 2000 by Dynamic Diagrams, Inc.

15 Bernard, M., & Hamblin, C. (2003). Cascading versus indexed menu design. *Usability News,* 5(1). Retrieved January 7, 2004, http://psychology.wichita.edu/surl/usabilitynews/51/menu.htm

16 Adapted from Harrison, C. (2002, October). Hypertext links: Whither thou goest, and why. *First Monday,* 7(10). Retrieved December 16, 2003, from http://firstmonday.org/issues/issue7_10/harrison/index.html

17 Fermi National Accelerator Laboratory. (n.d.). *Home page.* Retrieved December 10, 2003, from http://www.fnal.gov/; Fermi National Accelerator Laboratory. (n.d.). *Fermilab now.* Retrieved December 10, 2003, from http://www.fnal.gov/pub/now/index.html

18 Fermi National Accelerator Laboratory. Old Home Page. Retrieved December 10, 2003, from http://www.fnal.gov/pub/help/oldhomepage.jpg

19 Fermi National Accelerator Laboratory. (n.d.). *History of the FNAL website.* Retrieved December 10, 2003, from http://www.fnal.gov/pub/help/history.html

20 National Aeronautics and Space Administration. (2004). *NASA news highlights.* Retrieved January 4, 2004, from http://www.nasa.gov/home/index.html

21 National Aeronautics and Space Administration. (2004). *NASA news highlights.* Retrieved January 4, 2004, from http://www.nasa.gov/home/index.html

22 National Aeronautics and Space Administration. (2004). *NASA news and events.* Retrieved January 4, 2004, from http://www.nasa.gov/news/highlights/index.html

23 National Aeronautics and Space Administration. (2004). My NASA. Retrieved January 4, 2004, from http://mynasa.nasa.gov/portal/site/mynasa/index.jsp?bandwidth=high

24 Nielsen, J. (2000). *Designing web usability* (p. 29). Indianapolis: New Riders Publishing.

25 Ducky images: Adobe clip art from Adobe Photoshop.

Mask images: Microsoft clip art from MSOffice 2000.

26 Adapted from Outing, S. (1999, June 18). Some advice on writing, web-style. *E&P Online.*

27 Pearrow, M. (2002). *The wireless web usability handbook* (p. 205). Hingham, MA: Charles River Media.

28 Adapted from Fogg, B.J. (2002, May). *Stanford guidelines for web credibility.* Retrieved November 15, 2003, from Stanford University, Stanford Persuasive Technology Lab Web site: www.webcredibility.org/guidelines

PART IV

Understanding the Communicator's Strategies

V

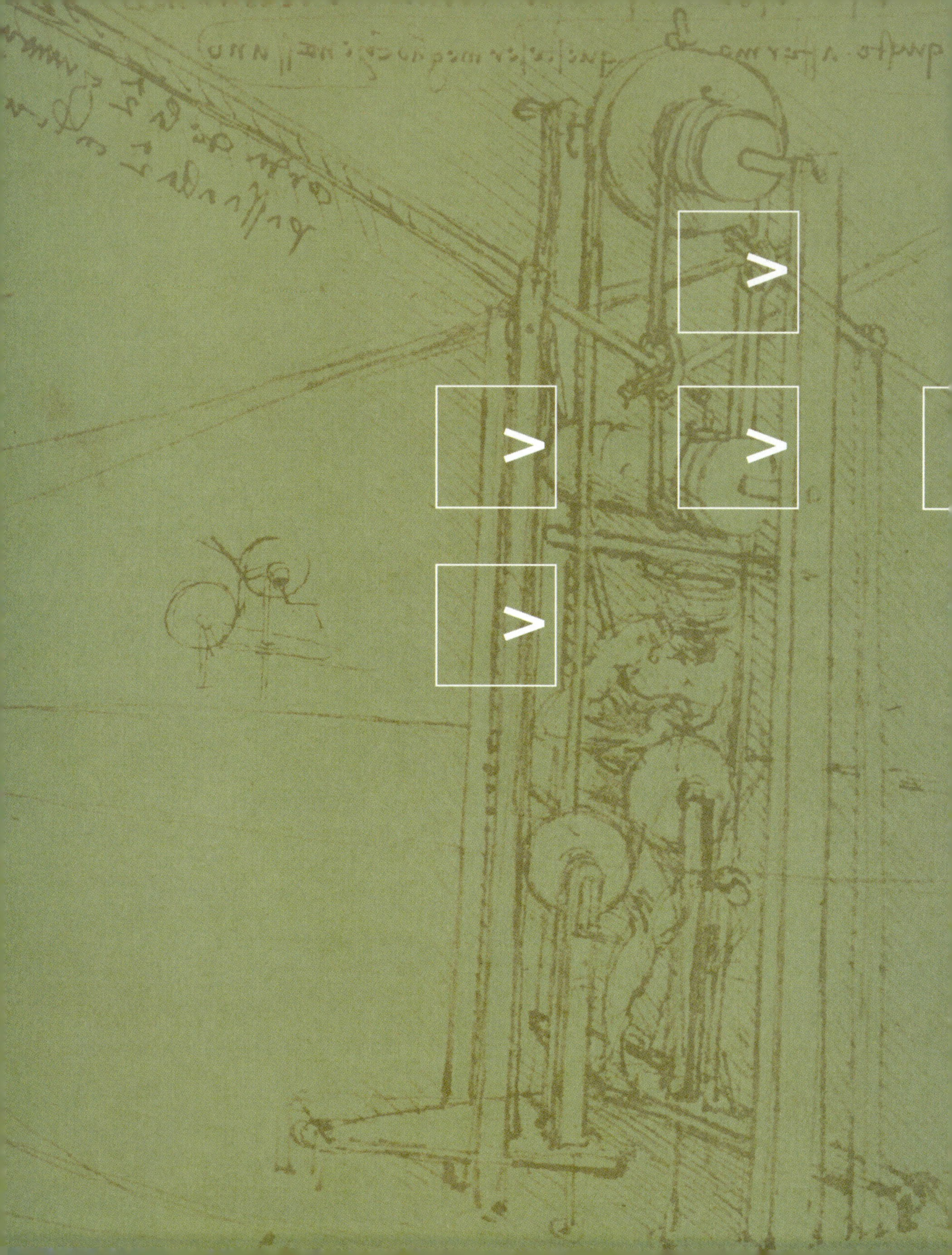

CHAPTER 14

Creating Definitions

Objectives and Outcomes

This chapter will help you accomplish these outcomes:

- Avoid problems caused by multiple meanings, complexity of meanings, technical jargon, and symbols
- Create several categories of definitions:
  - Informal definitions: synonym, antonym, stipulation, negative, analogy, and illustration
  - Formal definitions (*species* = *genus* + *differentia*)
  - Operational definitions summarizing steps involved in a function
  - Expanded definitions using etymology, history, and examples
- Make appropriate decisions about using definitions in glossaries, information notes, and appendixes

Not only do the meanings of words change, but new words come into our language as well. For example, in helping readers to understand the meaning of *hypertext,* a word that has recently entered our common academic and workplace vocabulary, you could use many forms of definition. The following brief explanation of hypertext incorporates many of the types of definition that you'll read about in this chapter.

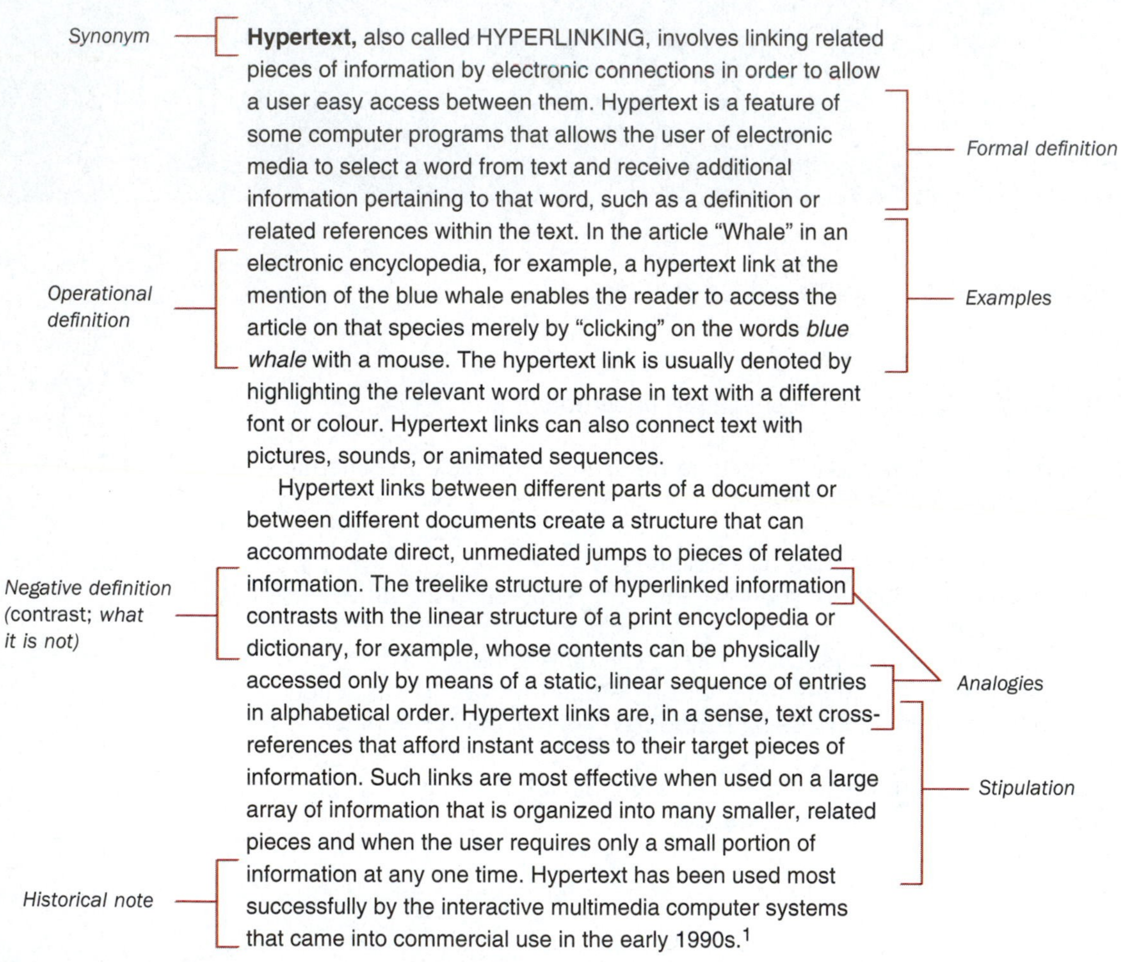

Hypertext, also called HYPERLINKING, involves linking related pieces of information by electronic connections in order to allow a user easy access between them. Hypertext is a feature of some computer programs that allows the user of electronic media to select a word from text and receive additional information pertaining to that word, such as a definition or related references within the text. In the article "Whale" in an electronic encyclopedia, for example, a hypertext link at the mention of the blue whale enables the reader to access the article on that species merely by "clicking" on the words *blue whale* with a mouse. The hypertext link is usually denoted by highlighting the relevant word or phrase in text with a different font or colour. Hypertext links can also connect text with pictures, sounds, or animated sequences.

Hypertext links between different parts of a document or between different documents create a structure that can accommodate direct, unmediated jumps to pieces of related information. The treelike structure of hyperlinked information contrasts with the linear structure of a print encyclopedia or dictionary, for example, whose contents can be physically accessed only by means of a static, linear sequence of entries in alphabetical order. Hypertext links are, in a sense, text cross-references that afford instant access to their target pieces of information. Such links are most effective when used on a large array of information that is organized into many smaller, related pieces and when the user requires only a small portion of information at any one time. Hypertext has been used most successfully by the interactive multimedia computer systems that came into commercial use in the early 1990s.[1]

When you prepare technical documents, oral presentations, and visuals, you need to define critical terms by using vocabulary and concepts within the audience's grasp. You can tailor a definition for different audiences by adjusting details, vocabulary, and types of examples and explanations. This chapter discusses the need for definitions, examines the construction of various types of definitions, and presents specific places to use definitions in various documents.

Writers, speakers, and designers often understand the need to provide definitions of various kinds. As a result, you can find definitions included in many documents, oral presentations, and visuals that provide the audience with information they need to interpret critical terms.

At other times, much like Humpty Dumpty in *Alice in Wonderland,*[2] people use language in whatever way they want . . . and members of the audience are left to figure out the meanings for themselves. Inadequate or missing definitions cause a variety of problems. Readers may become confused by multiple meanings, complexity of meanings, technical jargon, and symbols.

When I use a word, it means just what I choose it to mean, neither more nor less

Multiple Meanings

Some words have *multiple meanings,* different definitions in other contexts, which might mislead readers. The definition of even simple terms sometimes changes entirely when the term is applied in a different field. For example, a biologist, geologist, and naval gunner would probably react differently to the everyday word *focus:*

- In biology — the localized area of disease or the major location of a general disease or infection[3]
- In calculus — one of the points that, with the corresponding directrix, defines a conic section[4]
- In earth science — the location of an earthquake's origin[5]
- In photography — the adjustment of a camera lens to a particular image to ensure a sharp and clear picture[6]
- In physics — the small area of a surface that light or sound waves converge upon[7]
- In naval gunnery — the rotation and elevation of a gun to accurately hit a target[8]

Indeed, definitions are frequently necessary for common words if their meaning is ambiguous or unclear to the audience. Examples include such words as *base, cell, cover, limit, positive, stock,* and *traverse.*

What other words can you think of that are used in multiple disciplines or professionals and have multiple meanings?

Words with multiple meanings are especially treacherous for one of the largest audiences — nonexpert professionals. Imagine a manager with a background in agronomy reading a memo from a graphic artist that includes the phrase "insufficient crop." The context may make the meaning clear, but only after a moment or two of hesitation. To eliminate problems caused by multiple

What problems might occur if you introduce and define key terms near the beginning of a training manual and then throughout the manual substitute a variety of synonyms for these key terms to add some variety to your writing?

meanings, assess your audience and decide whether any of your terms has a meaning that members of that audience might think of before discerning the intended technical definition. If they might be confused, include an unobtrusive parenthetical definition.

Complexity of Meaning

Complexity can be considered in two broad ways. First, definitions can be simple or detailed, depending on the intended audience. The following two definitions of *volt* illustrate two levels of complexity. The first definition comes from a general dictionary like those in public schools and homes; the second comes from a technical dictionary.

This first definition would satisfy most general readers.

- *volt* — standard unit of electromotive force; after Alessandro Volta, an Italian electrician[9]

The second definition requires some technical knowledge because of the vocabulary (ampere, watt, electromagnetic units), *kind of details, and amount of information.*

- *volt* — the derived SI unit of electric potential defined as the difference of potential between two points on a conducting wire carrying a constant current of one ampere when the power dissipated between these points is one watt. Also the unit of potential difference and electromotive force. 1 volt = 108 electromagnetic units. Symbol V (= W/A). Named after Alessandro Volta (1745–1827).[10]

Your decision about the complexity of the definition depends on your assessment of both the intended audience and the situation. You need to decide what information is needed in order for the audience to comprehend and be able to use the term in the context of the document, oral presentation, or visual.

A second way to think about complexity involves the factors that contribute to definitions themselves. For example, *case definitions* are "sets of criteria used by public health agencies in the surveillance, or monitoring, of disease syndromes. In the United States, case definitions are established by the Centers for Disease Control and Prevention (CDC).[11] An especially controversial case definition is the one for AIDS. In 1981, the CDC started to track the cluster of opportunistic infections, which eventually came to be called AIDS. It is now recognized as "a syndrome . . . characterized by more than two dozen different illnesses and symptoms as well as by specific indications on blood test findings."[12] The initial case definition was established by the CDC in 1982 and then underwent major revision in 1985 (to include 20 conditions, three of which were types of cancer) and again in 1987 (to include three more conditions).

The complex definition of AIDS evolved as medical knowledge expanded and as social and political pressures brought attention to two problems: an inadequate case definition and new uses of the definition beyond its original epidemiological purposes. For example, by the early 1990s, the case definition being used to assess the effects of AIDS was inadequate, in part because it didn't sufficiently acknowledge "certain populations, namely women, injecting drug users, and communities of color. . . . Physicians, activists, and, in particular, HIV-positive women, began a movement to force the CDC to expand the surveillance definition of AIDS."[13] So in 1993, the case definition evolved a third time to include three additional conditions and a system of categorization that more accurately reflects the all affected populations.

This example illustrates that definitions — especially complex definitions — need to be contextualized. They work as long as they are accepted by the majority of people in a particular community as accurate, but when they cease to be usable, they can be challenged and revised.

WEBLINK

For a link to the case definition of AIDS, go to **http://english.wadsworth.com/burnett6e**. You'll learn about the evolution of a case definition as knowledge changes and about the legal and financial implications of case definitions.

CLICK ON WEBLINK

CLICK on Chapter 14/AIDS

Technical Jargon

Definitions are frequently needed when technical rather than everyday terms are used. You should assess whether readers are familiar with the terms; if so, no definitions are necessary. As the Peanuts cartoon shows, you can easily misjudge the level of technical information an audience can handle.[14]

The following paragraph contains technical vocabulary that the writer could be fairly certain the intended readers already know; the magazine is a trade publication for the plastics industry, and the intended readers are professionals knowledgeable about plastics. The specialized technical terms, italicized here, were not italicized in the original.

> *Ratios* are indexed via a digital thumbwheel. *Reinforcement* is loaded into the emptied tank (by means of a *positive vertical auger conveyor*) and activated to achieve a uniform density; *polyol* is then added from bulk storage tanks and mixed with *reinforcement*. With initial production now under way, no requirement for *heat-tracing* of mixing tanks has been observed, and this is probably due to the effective but essentially *shearless mix action* of the *orbital auger principle* employed.[15]

Under what circumstances and in what types of documents might a writer take the time and effort to define terms for audiences who have technical expertise?

Sometimes, however, technical terminology unfamiliar to the audience is necessary. For instance, a process may be introduced that requires a technical explanation employing new terms. In such cases, the writer should define the terms. The following sentence, from a short article about safety precautions necessary to protect workers who use industrial robots, defines the unfamiliar term *dwell-time.*

> Perhaps the most dangerous condition exists when people are unaware of a robot's *dwell-time.* (Dwell-time is the temporary period of inactivity between motions.)

Figure 14.1 illustrates longer definitions incorporated into an informational sales brochure for *diamond laps,* instruments used to make metal surfaces flat. The

FIGURE 14.1 **Incorporated Definitions**[16]

Accessibility

- *The figures are placed to closely follow the textual references.*

Comprehensibility

- *The broad characteristics are identified in the opening paragraph.*
- *Informal parenthetical definitions help readers move quickly through the information.*

Usability

- *The features of surface texture — "roughness, waviness, lay, and flow" — are (1) forecast in a sentence, (2) defined in an accessible bulleted list, and (3) illustrated in figures.*

FINISH/FLATNESS CONCEPTS

SURFACE FINISH

The surfaces produced by machining and other methods of manufacturing are generally irregular and complex. Of practical importance are the geometric irregularities generated by the machining method. These are defined by height, width and direction, and other random characteristics not of a geometric nature.

The general term employed to definethese surface irregularities is *surface texture,* the repetitive or random deviation from the nominal surface (Figure 1) that forms the pattern of the surface. It includes roughness, waviness, lay, and flaw.

- Roughness consists of fineirregularities in the surface texture produced by the machining process (Figure 2).
- Waviness is the widely spaced component of surface texture. It has wider spacing than roughness (Figure 2). It results from cutting tool runout and deflection.

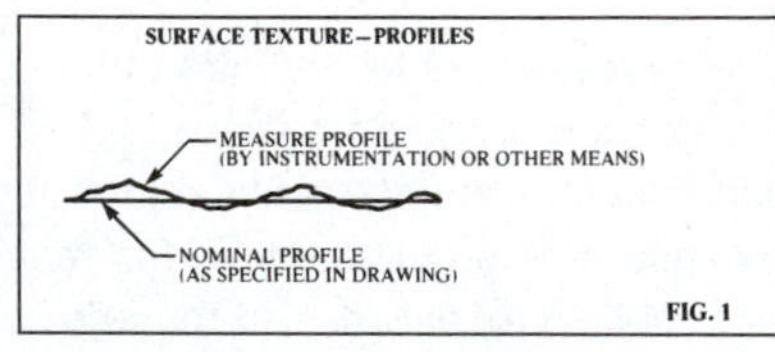

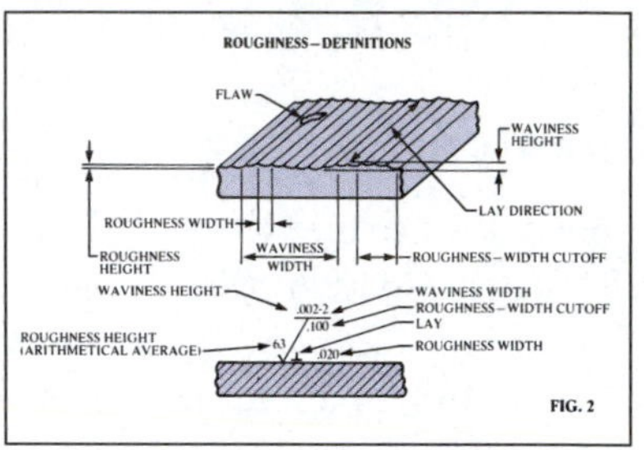

- Lay is the direction of the predominate surface pattern and it is determined by the machining process used in producing the surface (Figure 2).
- Flaws are irregularities that occur at scattered places, without a predetermined pattern. They include cracks, blow holes, checks, ridges, scratches, etc. (Figure 2).

Roughness is definedas the arithmetical average (AA) deviation of the surface roughness expressed in microinches from a mean line or roughness centerline (Figure 3). AA has been adopted internationally and is often referred to as CLA or c.l.a. (centerline, average). Many instruments still in use employ an average deviation from the roughness centerline, which is the root mean square average (RAMS) deviation of surface roughness, also expressed in microinches. RMS, while used frequently, has actually been obsolete since definitions are given credibility because they are taken from *Machining Data Handbook,* a standard machinists' reference book. The company includes the definitions to educate its customers about the comparative benefits of its product.

Symbols

Technical language can be nonverbal, as in the symbolic language of mathematics, chemistry, and physics. For example, the equation explaining conservation of matter and energy ($E = mc^2$) is completely understandable to a physicist and, in fact, to virtually all physics students, who would know that E = energy, m = mass, and c = velocity of light and could translate the symbolic statement into a

FIGURE 14.1 **Incorporated Definitions[16] (continued)**

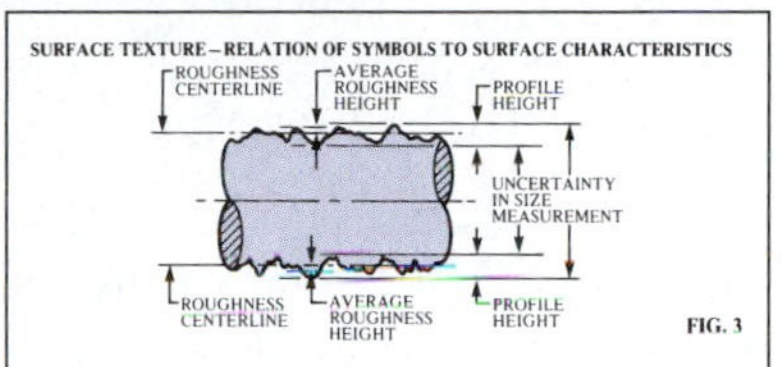

about 1950. Roughness measuring instruments calibrated for RMS will read 11 percent higher, on a given surface, than those instruments calibrated for AA. The difference is usually much less than the point-to-point variations on any given machined surface.

The commercial ranges of surface roughness produced by various machining processes are shown in the table. A range of finishescan be obtained by more than one process; however, the selection of a surface finishinvolves more than merely designating a particular process. The ability of a process operation to pro-duce a specificsurface roughness or surface finish depends on many factors. In turning, for example, the surface roughness is geometrically related to the nose radius of the tool and the feed per revolution. For surface grinding, the finalsurface depends on the type of grind-ing wheel, the method of wheel dressing, the wheel speed, the table speed, cross feed, down feed, and the grinding fluid.A change in any one of these factors may have a significant effect on the fnish of the fnal surface produced.

Type of Surface	Roughness Height (Microinches)
Honed, lapped, or polished	2 4 8
Ground with periphery of wheel	4 8 16 32 63
Ground with flat side of wheel	4 8 16 32 63
Shaped or turned	32 63 125 250 500
Side milled, end milled or profile	63 125 250 500
Milled with periphery of cutter	63 125 250 500

Comprehensibility

- *The figures are carefully labeled with callouts.*
- *The figures are explained in the accompanying text.*

Usability

- *Types of surfaces produced by various machining processes and the corresponding roughness are presented in an accessible table, with the degree of roughness listed in ascending order in microinches.*
- *Even though the information is technical, it is presented so that it can be understood by professional nonexperts — in this situation, probably managers who have to approve decisions that may in part be based on the flatness that can be achieved with certain materials and equipment.*

verbal statement. Yet the symbols might be confusing even to highly educated people not trained in physics or a related field.

Simply defining the symbols will not ensure that nonexperts comprehend the content of either the symbolic or verbal statement. The definitions you construct need to do more than identify the unfamiliar terms; you must consider the audience's knowledge and adjust the definition to the appropriate level. These definitions illustrate two of the possibilities:

- $E = mc^2$ means that *energy* is equivalent to *m*ass times the square of the constant *velocity of light*.
- $E = mc^2$ means mass-energy is conserved. The energy produced directly from the loss of mass during a nuclear fission or fusion reaction is equal to that mass loss times the square of the constant velocity of light.

You need to understand the concepts behind the symbols in order to provide an explanation appropriate for the audience's education and experience.

Construction of Definitions

Effective definitions answer questions considered by members of the audience, *before* they verbalize those questions. Recognizing the nature and variety of possible questions helps you construct your definitions, recognize when a definition is appropriate, and determine the effectiveness of existing definitions when editing. After the initial "What is it?" you can ask yourself some common questions:

Physical Characteristics	■ What does it look like?
	■ What are its physical features?
Comparison	■ How is it classified?
	■ What is it similar to?
	■ How does it differ from similar objects (theories, procedures, situations)?
Whole/Parts	■ What are its distinguishing characteristics?
	■ What are its components (structural parts and functional parts)?
Function	■ What does it do?
	■ How does it work (function, operate)?
Operation	■ Who uses it?
	■ What are examples of its use?
	■ What is its value?

A variety of techniques will aid you in answering these questions when constructing definitions, depending on the category: a formal, informal, operational, or expanded definition.

Formal Definitions

Because dictionaries use *formal definitions* in many entries, people often believe that is the only way to define a term — identifying the broad category to which a term belongs as well as its distinctive characteristics. As a writer, you may be expected to construct clear and accurate formal definitions for new products and processes when no definition exists and for existing products and processes when current definitions are inadequate. The format of formal definitions is always the same.

Species	**equals**	**Genus**	**plus**	**Differentia**
term being defined		Class or category to which the term (species) belongs		distinguishing characteristics that differentiate this species from other species in the same genus

A simple example illustrates the structure and demonstrates the application of guidelines that you should follow when constructing effective formal definitions.

Species	**equals**	**Genus**	**plus**	**Differentia**
A robin	is	a bird	with	a red breast and yellow beak.

You should make the genus as narrow as possible. A robin is a bird, but can the category be more specific? Yes. A robin is a type of thrush, so the formal definition can be revised.

Species	**equals**	**Genus**	**plus**	**Differentia**
A robin	is	a thrush	with	a red breast and yellow beak.

You should make the differentia as inclusive as possible to eliminate the possibility of mistakenly identifying one species with another. Do robins have additional characteristics that differentiate them from other thrushes? Again, yes. A robin has a distinctive black back and wing tips. A more complete formal definition for robin can be constructed:

Species	**equals**	**Genus**	**plus**	**Differentia**
A robin	is	a thrush	with	a red breast, yellow beak, and black back and wing tips.

Formal definitions answer questions such as these: How is it classified? How does it differ from similar objects? What distinguishes this from related objects? What are the identifying characteristics? The need to construct your own formal definitions arises when dictionary definitions are inadequate or nonexistent. The following example of a formal definition was constructed for a specific report. Notice that in this example, the genus is purposely kept narrow and the differentia are inclusive.

Species	**equals**	**Genus**	**plus**	**Differentia**
Hypertext	is	electronically linked pieces of information	with	connections that allow users easy access between them.

Informal Definitions

Informal definitions tend to be the type we insert in communication without realizing that we're defining a term. We integrate informal definitions casually and comfortably and frequently out of necessity into our normal writing and speech. The six types of informal definitions presented in Figure 14.2 are particularly useful for technical communicators: synonym, antonym, negative, stipulation, analogy, and illustration.

Three of these types — stipulation, analogy, and illustration — deserve additional discussion because they are so useful in technical documents and presentations. They are especially good at providing the audience with information to differentiate a specific term from similar ones. For example, the following definition begins by providing a synonym, presenting an additional definition, and then offering an example to help the audience understand the concept of *domain name.* But perhaps the most helpful part of the definition is the set of the stipulations of specific extensions for particular situations. The stipulations are what the audience is likely to remember.

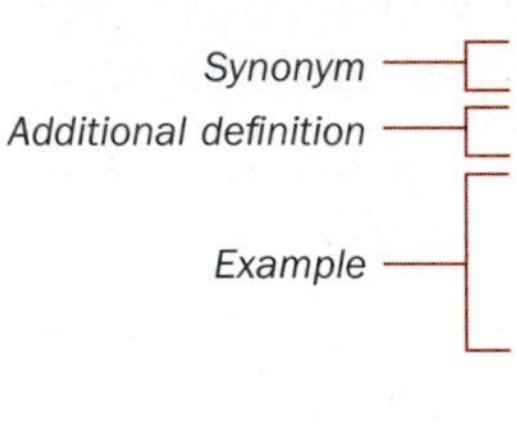

domain name. The address or URL of a particular Web site, it is the text name corresponding to the numeric IP address of a computer on the Internet. For example: www.netlingo.com is the domain name for the numeric IP address "66.201.69.207." There is an organization called InterNIC that registers domain names for a fee, to keep people from registering the same name. As of February, 2002, here is the approximate number of registered names for each top-level domain:

Stipulations

* .com - 22.5 million
* .net - 3.9 million
* .org - 2.5 million
* .info - 700,000
* .biz - 500,000
* .name - 150,000
* .coop - 5,000
* .museum - 2,000

To register a domain name, you can contact a company (such as Network Solutions, Inc.) or you can ask your ISP or hosting company to register names for you. In addition to the suffixes listed above, there is also .edu, .gov, .mil, and the list of country codes, as well as the following:

* .arts for arts and cultural entities
* .firm for businesses
* .pro for professional
* .rec for recreation and entertainment
* .store for merchants
* .web for Web services[17]

FIGURE 14.2 Types of Informal Definitions

Term	Definition	Examples	Comment
Synonym	A word that means essentially the same thing as the original term is asynonym.	microbe = germ helix = spiral Corrugated paperboard is the technical term for what is popularly known as cardboard. The bellows in a thermostatic element is made of a paper-thin, hardened (heat-treated) copper to make it strong, elastic, and corrosion resistant.	Synonyms usually answer such questions as "What is it similar to? What do I know that it resembles?"
Antonym	A word that is opposite in meaning to the original term is an antonym.	deviating - direct indigenous - foreign	Antonyms answer the obvious question, "What is the opposite?"
Negative	Explaining what something is not provides readers with useful information.	Machine rivets, usually made from metals such as aluminum or titanium, are unlike the rivets used in iron work in that they do not need to be heated before insertion.	Negatives respond to such questions as "What similar things should I not equate with this object? What similar things might mislead me?"
Stipulation	Stipulative definitions specify the meaning of a term for a particular application or situation.	When the term *x* [not necessarily mathematical, just any term] is used in this paper, it means . . .	Stipulations respond to such questions as "What are the limitations of use?"
Analogy	An analogy directly compares the unfamiliar to the familiar to identify major characteristics of the unfamiliar term.	A kumquat is a citrus fruit about the size and shape of a pecan. The skin is much thinner than that of a tangerine, and entirely edible. When fully ripe, the kumquat is a motley orange-green, much like an unprocessed orange.	An analogy responds to the same questions as a synonym: "What is this similar to? What related object has characteristics I'm already familiar with?"
Illustration	An actual drawing or diagram can illustrate a term.	[diagrams, drawings]	Visuals respond to the question, "What does it look like?"

Analogies, another kind of informal definition, are particularly powerful when explaining unfamiliar concepts or objects, especially for audiences that lack expert knowledge of the topic. They link the familiar and the unfamiliar. The following example begins with a visual display of how the word *modem* was coined and then provides a formal definition. The analogy that follows, though, is probably what will help an audience unfamiliar with a modem remember its primary function.

modem

short for: MOdulator, DEModulator

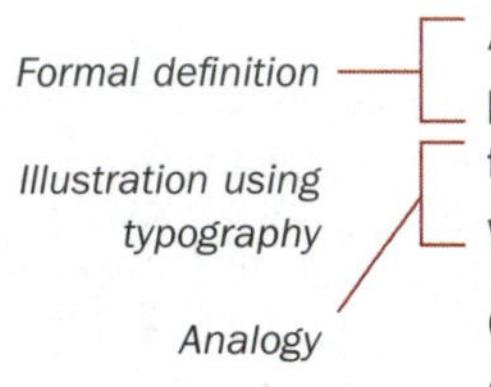

A hardware device you connect to your computer and to a phone line. It enables the computer to talk to other computers through the phone system. Basically, modems do for computers what a telephone does for humans.

Generally, there are three types of modem: external, PC card, and internal.

Most computers now have internal modems so you can plug the telephone cord directly into the back of the computer.[18]

An actual drawing or diagram can illustrate a term. In many cases, a visual definition is far more efficient, accurate, and easy to understand than a verbal definition. In Figure 14.3, annotated drawings accompany a definition so readers can readily understand the explanation of the terms *capillary attraction* and *capillary repulsion.*

Sometimes visual definitions are not merely desirable but essential. Visuals respond to the question, "What does it look like?" For example, both acetone and propionaldehyde molecules contain three carbon atoms, six hydrogen atoms, and one oxygen atom. Although each molecule has the same number and kind of atoms, they are arranged differently, resulting in two compounds with different characteristics. The chemical formulas in Figure 14.4 state the differences verbally, but the addition of the diagrams brings those differences into focus.

FIGURE 14.3 **Illustration to Increase Comprehension[19]**

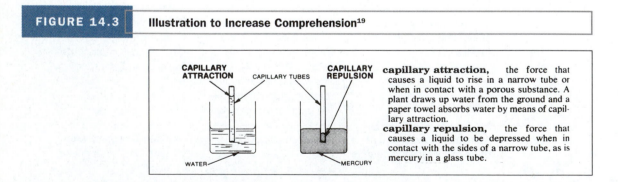

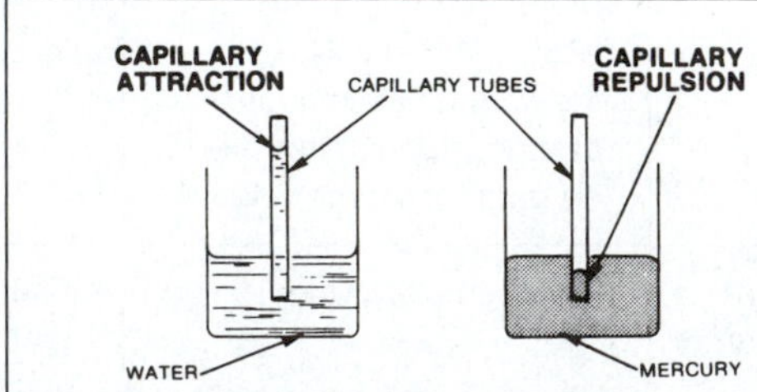

capillary attraction, the force that causes a liquid to rise in a narrow tube or when in contact with a porous substance. A plant draws up water from the ground and a paper towel absorbs water by means of capillary attraction.
capillary repulsion, the force that causes a liquid to be depressed when in contact with the sides of a narrow tube, as is mercury in a glass tube.

FIGURE 14.4 **Visuals Essential to Show Critical Differences**

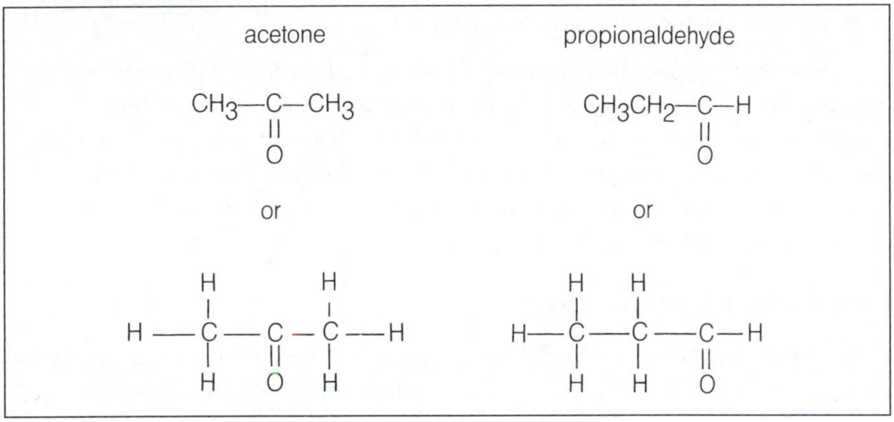

acetone	propionaldehyde
$CH_3—C(=O)—CH_3$	$CH_3CH_2—C(=O)—H$
or	or

```
      H         H                  H    H
      |         |                  |    |
  H — C — C  —  C — H          H — C —  C — C — H
      |   ||    |                  |    |   ||
      H   O     H                  H    H   O
```

Operational Definitions

The term *operational definition* means different things to different technical professions. For example, for experimental researchers, operational definitions specify the activities (the operations) that researchers use to measure a variable. In contrast, for engineers, operational definitions specify the functions or workings of an object or process. In fact, although these two uses of operational definition are different, they both depend on a definition to identify the key steps that make up a process, either to clarify the process or to measure it.

Situations that lend themselves to operational definitions require answers to questions such as these: How does it work? How can I measure or test it? How can I determine if its function is successful? What are the steps in its operation?

An operational definition summarizes or outlines the primary steps involved in the function, usually in chronological order. An effective operational definition can form the basis for a detailed process explanation (see Chapter 16). Whereas an operational definition usually outlines the major steps in a procedure, a process explanation provides specific details of each step, often describing the relationships between steps as well as offering theoretical background. The following example shows how useful an operational definition can be when defining a term:

> A thermostatic element with a remote bulb is a temperature-sensitive instrument that converts a temperature change into a mechanical force. The instrument consists of three copper parts (bulb, capillary tube, and bellows) soldered together, with a liquid sealed inside. The bulb contains liquid that turns to a gas when heated. Since the gas requires more volume per unit weight than the liquid, the pressure in the bulb increases, forcing gas through the capillary tube and into the bellows. This increase in the volume of gas causes the bellows to expand. Just the opposite happens when the bulb is cooled. This loss of heat in

Why is a formal definition a useful way to begin? Why is identifying the physical structure useful? Where does the actual operational definition start?

What are the primary steps in this process? How could you illustrate this process?

the bulb causes some of the gas in the bulb to condense into its liquid state. Since the liquid has a much smaller volume per unit weight, the pressure drops, which pulls gas back from the bellows. A decrease in the volume of gas in the bellows causes the bellows to contract.[20]

Sometimes operational definitions move beyond specifying physical features or processes. The ethics sidebar below discusses definitions of acceptable professional behavior — that is, codes of conduct. Virtually every profession has a code of conduct that you should be familiar with. In the workplace, you'll also find operational codes of conduct in an organization's policy manuals, which define legal and ethical activities and behaviors.

Expanded Definitions

Expanded definitions explain and clarify information. They also maintain audience interest and can adapt a document, oral presentation, or visual for a wider audience. In fact, many of the documents you prepare will contain expanded definitions. What are the most common forms of expanded definitions? Etymology, history, and examples. To determine which form is most appropriate to use in a particular situation, you need to analyze both the audience and the task.

- Presenting the *etymology* (the linguistic origin) of a term is appropriate for general audiences.
- The *history* of a term is also appropriate for general audiences; audiences with technical expertise may need recent historical information as well.
- Relevant *examples* have value for all audiences.

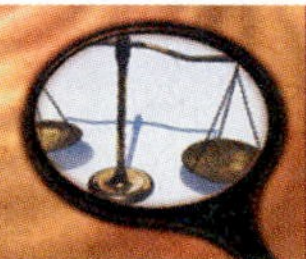

ETHICS SIDEBAR

What would you include in an operational definition of professional conduct for a new organization?

Professional Codes of Conduct

Many professional organizations provide their members with ethical guidelines, sometimes referred to as codes of conduct — that is, definitions of what constitutes acceptable professional behavior. Codes of conduct have many purposes, including providing members of professional organizations with possible responses to ethical dilemmas. They also help build the credibility of the organization (and its profession) to the outside community.

A useful code of conduct for technical professionals is the Society for Technical Communication (STC) code of conduct, available at this Web address:

www.stcconsultants.org/ethics.html

This code considers legality, honesty, confidentiality, quality, fairness, and professionalism.

The STC code, like almost all professional codes of conduct, is designed to influence the way technical communicators act both toward others within their profession and toward those outside the profession. For example, the STC code of conduct requires that its members avoid conflicts of interest in fulfilling professional responsibilities and activities. If a conflict of interest does arise, STC members are expected to disclose it to people who are interested or concerned and obtain their approval before moving ahead with a project.

An online listing of codes of conduct, sponsored by the Illinois Institute of Technology Center for the Study of Ethics in the Professions, is available at this Web address:

www.iit.edu/departments/csep/

In additional to an enormous number of actual codes of ethics from a large array of professional organizations, you'll also find discussions about ways to use codes of ethics, the process of authoring codes of ethics, and links to other sources about codes of ethics.

- *Are you aware of a code of conduct in your field of study? Should you be?*
- *How much of a code of conduct is common sense?*
- *Codes of conduct do not always clearly indicate how or even if an organization will enforce the code. How should codes of conduct be enforced? Can they be? Should codes of conduct be viewed as rigid rules or as advice?*

Etymology. Etymologies anticipate questions such as these: How did this object get its name? How old is this word? Where did this word come from? What are its historical precedents? Presenting the linguistic origin of a term sometimes gives insight into its current meaning(s). Etymological information is found in dictionaries or specialized reference books. In a dictionary entry, this information is most frequently presented in abbreviated form inside square brackets. Every dictionary contains a page near the front of the volume that will help you understand its abbreviations.

Etymologies are a useful part of a definition if knowledge of the original meaning will increase your audience's understanding of the modern meaning and usage. Such explanations are particularly appropriate if you are writing for general audiences, although etymologies also add interest to definitions for more technical audiences. The following example demonstrates how etymologies can be used effectively as a technique for definition.

> The word rivet comes from the Vulgar Latin word *ripare,* which means to make firm, or from the Middle French words *rivet* or *river,* which means to clinch.[21]

History. Presenting historical background about the development and use of the term or subject puts its current meaning into perspective. History can cover several thousand years (if the subject is chemistry, which began at least as early as the magicians in the Egyptian pharaohs' courts) or decades (if the subject is lexan, which was invented in the 1960s). The use of historical background anticipates questions such as these: What are the subject's origins? How long have such objects (concepts) existed? How has the history affected modern development? How were the original objects (concepts) different from modern ones? The example in Figure 14.5 begins with the history of anthrax in 70 BC and takes readers quickly up to the present day.

WEBLINK

Anthrax has been in the news, but not many people know more than what they hear or read in dramatic headlines. Figure 14.5 presents the first of six screens that define anthrax by presenting a brief history of the disease. For a link to the other five screens, go to http://english.wadsworth.com/burnett6e.

CLICK ON WEBLINK

CLICK on Chapter 14/anthrax

FIGURE 14.5 **Example of Definition Using History[22]**

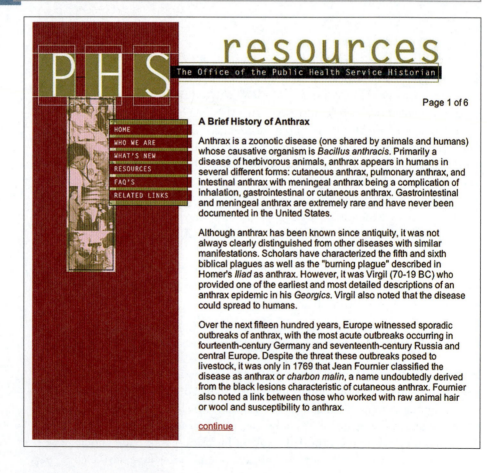

resources

The Office of the Public Health Service Historian

Page 1 of 6

A Brief History of Anthrax

Anthrax is a zoonotic disease (one shared by animals and humans) whose causative organism is *Bacillus anthracis.* Primarily a disease of herbivorous animals, anthrax appears in humans in several different forms: cutaneous anthrax, pulmonary anthrax, and intestinal anthrax with meningeal anthrax being a complication of inhalation, gastrointestinal or cutaneous anthrax. Gastrointestinal and meningeal anthrax are extremely rare and have never been documented in the United States.

Although anthrax has been known since antiquity, it was not always clearly distinguished from other diseases with similar manifestations. Scholars have characterized the fifth and sixth biblical plagues as well as the "burning plague" described in Homer's *Iliad* as anthrax. However, it was Virgil (70-19 BC) who provided one of the earliest and most detailed descriptions of an anthrax epidemic in his *Georgics.* Virgil also noted that the disease could spread to humans.

Over the next fifteen hundred years, Europe witnessed sporadic outbreaks of anthrax, with the most acute outbreaks occurring in fourteenth-century Germany and seventeenth-century Russia and central Europe. Despite the threat these outbreaks posed to livestock, it was only in 1769 that Jean Fournier classified the disease as anthrax or *charbon malin*, a name undoubtedly derived from the black lesions characteristic of cutaneous anthrax. Fournier also noted a link between those who worked with raw animal hair or wool and susceptibility to anthrax.

continue

Examples. Using specific examples to illustrate the application of a term effectively expands a definition. Defining a concept with an example can be particularly effective, as in the following example, "Rayleigh-Taylor Instability and ICF," taken from a sidebar in the article "Diagnostics for Inertial Confinement Fusion Research," from *Energy and Technology Review,* a publication of Lawrence Livermore National Laboratory.

> We are all familiar with the Rayleigh-Taylor instability. Consider, for example, what happens if a layer of water is carefully laid on top of a lower-density liquid, such as alcohol, in a container. The heavier water will find its way through the lighter alcohol to the bottom of the container if the container is disturbed. The mechanism that initiates this fluid interchange is the Rayleigh-Taylor (RT) instability: fingers of the heavier fluid start poking into the lighter fluid, and bubbles of the lighter fluid rise through the heavier fluid until eventually the interchange is complete.[23]

Placement of Definitions

Writers of technical material have five basic choices for placing and incorporating definitions, although the choices are not mutually exclusive:

- glossary
- information notes and sidebars
- incorporated information
- appendixes
- online help

Where are definitions placed so they are most convenient for you as a reader? Where are you most likely to use them? Where are they least obtrusive?

Glossary

A *glossary* is a mini-dictionary usually located at the beginning or end of a technical document. Sometimes glossary definitions are located on the page where a term initially appears, either as a footnote or as a marginal annotation. A glossary at the beginning is particularly useful when readers are unfamiliar with the information and must know the terminology in order to comprehend the document. The disadvantage of an initial glossary is that, without having read the document, readers may lack a frame of reference and may not be able to judge which terms to focus on. A glossary at the end of a document provides definitions and explanations readers can refer to as they need the information. Probably the most useful place for definitions is close to the initial use of the term. When a terminal glossary is used, the terms should be marked in some way (boldface, italics, asterisks) as they occur in the body to let readers know that they can find the definition in the glossary.

Individual entries can employ many of the forms of definition: formal, informal, operational, or expanded. Figure 14.6 shows an excerpt from excerpts from two glossaries. Like most workplace documents that use definitions,

FIGURE 14.6 **Excerpts from Two Glossaries**[24]

National Geographic.com	*ST*atistical *E*ducation through *P*roblem *S*olving (STEP)
Adaptive radiation the evolution of a single ancestor species into several new species within a relatively short period of time and in a certain geographic area. The plants and animals of the Galápagos Islands are a result of adaptive radiation, where one plant or one animal species diversified into many species that fill a variety of ecological roles. For example, more than a dozen species of finches evolved from a single founding species that colonized the islands from the mainland of South America. *Various kinds of examples increase readers' understanding.* **Atoll** a ring-shaped coral reef or string of coral islands, usually enclosing a shallow lagoon **Biodiversity** the variety of life on Earth and the interconnections among living things **Biogeography** the study of living systems and their distribution. Biogeography is important to the study of the Earth's biodiversity because it helps with understanding where animals and plants live, where they don't, and why. **Biotic** refers to the living components of the environment (such as plants, animals, and fungi) that affect ecological functions **Boreal** pertaining to the north *Entries sometimes begin with an informal definition.* **Brackish** slightly salty or briny. Brackish water is saltier than fresh water but less salty than seawater.	**Box and Whisker Plot (or Boxplot)** A box and whisker plot is a way of summarising a set of data measured on an interval scale. It is often used in exploratory data analysis. It is a type of graph that is used to show the shape of the distribution, its central value, and variability. The picture produced consists of the most extreme values in the data set (maximum and minimum values), the lower and upper quartiles, and the median. *Entries often begin with a formal definition that is sometimes explained.* A box plot (as it is often called) is especially helpful for Indicating whether a distribution is skewed and whether there are any unusual observations (outliers) in the data set. Box and whisker plots are also very useful when large numbers of observations are involved and when two or more data sets are being compared. 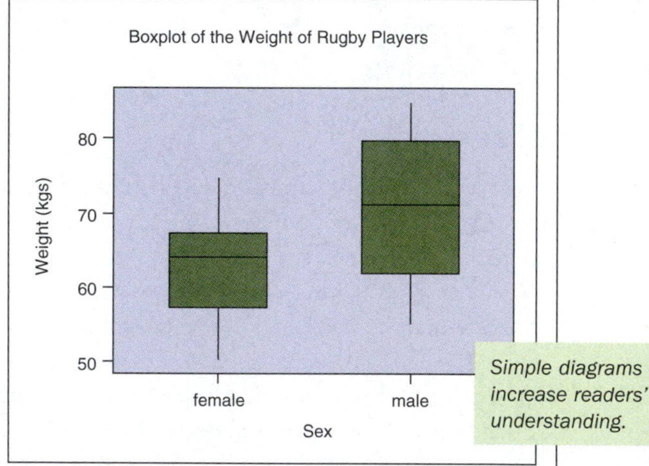 *Simple diagrams increase readers' understanding.* See also 5-Number Summary. *Cross references that help readers locate additional information are signaled, here by underlining and color.*

these glossaries do not restrict themselves to one kind of definition; instead, they incorporate a variety of types of definitions to address the readers' anticipated needs.

Information Notes and Sidebars

When readers need extended information, it may interrupt the flow of the text if included in the main discussion. Presenting this information as information notes or sidebars gives readers the option of reading the additional information if they need it.

Information notes may simply define a term or concept; they also enable writers to provide examples, cite related studies, explain tangential concepts, present possible explanations, and so on. They can be placed on the bottom of a page or collected at the end of a document along with source or reference notes. Technical reports for decision making usually contain information notes that offer brief definitions or explanations, since their audience usually has little time. Technical documents for research or academic purposes may include more detailed information notes for readers who might want to investigate an idea in greater depth.

Sidebars usually provide more elaborated information than footnotes or endnotes. The sidebar and accompanying illustration in Figure 14.7 are taken from an article about balloon angioplasty in the *Harvard Health Letter.* This newsletter, published by the Harvard Medical School, says its overall goal is to "interpret medical information for the general reader in a timely and accurate fashion." The sidebar defines the categories of blockages that occur in arteries. The accompanying illustration helps readers visualize both the blockages and the tools surgeons use to unclog arteries.

Appendixes

Lengthy documents intended for readers with widely varying backgrounds often have difficulty appealing to the entire range of readers. For example, nonexperts can be confused if a document jumps into the subject without sufficient explanation. Technical experts can be bored or even offended if the documents include too much elementary material. One way to resolve this dilemma is to include *appendixes* that provide both operational and expanded definitions of critical concepts. Readers already familiar with the material can glance at the reference to the appendix in the text and continue reading, virtually uninterrupted. Readers who need to review the background material will appreciate the detailed, illustrated definitions and discussions.

Online Help

Virtually all software companies provide users with critical information electronically via online help systems rather than in print manuals. *Online help* systems are designed to provide users with information immediately, in several

FIGURE 14.7 Illustrated Sidebar from a Health Newsletter[25]

WHAT TYPE OF BLOCKAGE?

Physicians use *coronary angiography* to determine the location, size, shape, and to some degree the composition of atherosclerotic lesions that narrow the coronary arteries. This information helps predict who is a good candidate for balloon angioplasty, who is likely to have a better result with a more spe-cialized procedure, and who should go straight to coronary artery bypass graft surgery.

A popular classificationsystem devised by the American College of Cardiology and the American Heart Association divides blockages into three categories. (*See illustration.*)

Type A: These are simple lesions in easily accessible locations that are concentric, less than 10 mm long, and contain little or no calcium and no clots. They are ideal for treatment with balloon angioplasty.

Type B: These somewhat complex, irregular narrowings are 10–20 mm long, with moderate-to-heavy calcifcation and some *thrombus* (clot), and are situated in a bend or vascular junction. Balloon angioplasty is slightly less likely to restore blood flowand slightly more likely to have complications.

Type C: This category includes heavily calcifed lesions, total blockages more than three months old, and narrowings located in a sharp turn or in a deteriorated bypass graft. The success rate for balloon dilation is signifcantly lower so bypass surgery, stenting, or atherectomy may be a better choice.

Balloon angioplasty is most successful for patients who are male, less than 70 years old, with normal left ventricular function (pumping ability), who have no more than two blocked arteries and no history of diabetes, heart attack, or bypass surgery.

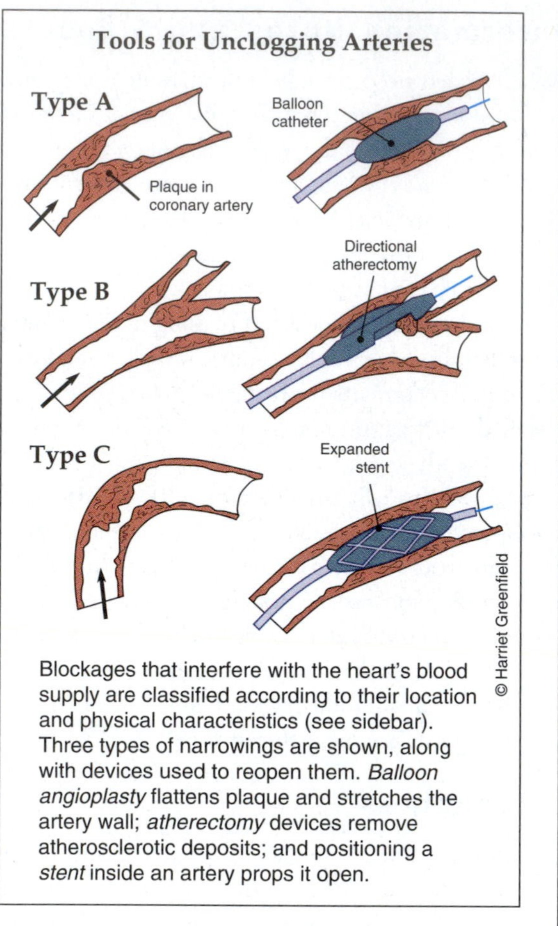

Blockages that interfere with the heart's blood supply are classified according to their location and physical characteristics (see sidebar). Three types of narrowings are shown, along with devices used to reopen them. *Balloon angioplasty* flattens plaque and stretches the artery wall; *atherectomy* devices remove atherosclerotic deposits; and positioning a *stent* inside an artery props it open.

different formats. For example, the word processing program used to prepare the manuscript for this textbook has a "balloon help" option; when activated, it provides the user with a callout that includes brief definitions and explanations of many features and functions on the computer desktop.

In addition to the balloons, this same software has a number of other online help features, which is typical of most software:

- an alphabetic index of all help topics available to users of this software
- a list of frequently used topics that often saves time

- an on-screen box that provides hints and wizards (shortcuts to common practices)
- a searchable database to answer questions

Virtually all of these online help options include various kinds of definitions to assist users who are confused or stuck.

Individual and Collaborative Assignments

1. **Identify types of definitions.** Read the following examples and identify each type of definition. Some of the examples have more than one type of definition embedded. Discuss the appropriateness of each definition for the audiences.

   (a) From a teacher's manual: The word for chemistry comes from the Middle East; long ago the sacred name for Egypt was *Chemia,* which means black and probably referred to the fertile black soil of the Nile valley. In those ancient times almost any kind of change was mysterious, such as the change of wood to ash when it burns, or the transformation of sand into glass. Men who understood how to make some things change were considered to be magicians (at least we might call them that today). From their mysterious abilities to cause changes, the word for the study of change was derived.[26]

   (b) From a dictionary: Chemistry — The study of the composition of substances and of the changes of composition which they undergo. The main branches of the subject are inorganic chemistry, organic chemistry, and physical chemistry.[27]

   (c) From a Web site defining network terms: Hypertext — A system of writing and displaying text that enables the text to be linked in multiple ways, to be available at several levels of detail, and to contain links to related documents. The term was coined by Ted Nelson to refer to a nonlinear system of information browsing and retrieval that contains associative links to other related documents.[28]

   (d) From a technical report: The reduction in speed of the steam turbine engine is necessary to achieve efficiency. The efficient speed for operating the turbine is higher than the speed of the shaft. The transmission of your automobile functions similarly by converting a high engine speed to a lower wheel speed.

   (e) From a scientific journal in an article about lipid barriers in biological systems: The current model views the mammalian stratum corneum as a two-component system ("bricks and mortar") consisting of protein-rich cells (bricks) embedded in a lipid matrix (mortar), with these intercellular lipids being primarily responsible for the integumental water barrier.[29]

(f) From a Web site defining network terms: Search engine — A Web site (actually a program) that acts as a card catalog for the Internet. Search engines attempt to index and locate desired information by searching for the keywords a user specifies. The ability to find this information depends on computer indices of Web resources (maintained in a database) that can be queried for these keywords. These indices are either built from specific resource lists (as is the case with a search directory) or created by Web programs with strange-sounding names that seem to be inspired by insects: bots, spiders, crawlers, and worms.

From a surfer's point-of-view, search engines can be quite tiresome and not very efficient if you don't know how to use them correctly. On top of that, different engines are good for different kinds of searches, so it's a good idea to read the engine's advanced search section before you do a search. In many cases, using Boolean logic will help narrow down the results for you and better optimize your search results.

SEARCH TIP: The best way to get the most out of a search engine is to understand its features. Always check the site's help page or advanced search page when you arrive, to find out what features are available. Then use them — they really help. Be specific and enter all of the words you are looking for. Pick your search site smartly: If you are looking for a bio of a football figure, use a directory, but if you want to see every instance where his name appears, use an engine.[30]

2. **Determine three levels of definitions for the same term.** Work with a classmate who is in the same discipline or field (if possible) and identify a concept or term that you would like to research. Check dictionaries, texts, encyclopedias, handbooks, professional journals in the reference and periodical section of your library, and various Web sites. Make copies of the definitions you select, and document your sources, using the appropriate formats from the Usage Handbook. Identify the major distinctions among the various definitions, and evaluate the effectiveness of each definition for the intended audience. Prepare a brief oral report for your classmates.

3. **Incorporate definitions into a document.** Imagine that your audience is managers of small businesses who have been resistant to converting their businesses to use computers. They need some elementary information about computers. You are revising a newsletter article dealing with computer basics that will contain a complete glossary at the end of the article, but you believe that brief definitions should be incorporated into the article itself so the readers don't have to regularly check the glossary while they're reading. Rewrite this basic paragraph, incorporating the following definitions (or parts thereof) as you believe is appropriate for this resistant audience.

Every computer system has two types of components, hardware and software. Most visible is the hardware, typically a central processing unit (CPU) and peripheral equipment. Less visible, but usually even more critical, is the software. Too often computer buyers and sellers focus on stylish, competent-looking hardware, rather than on the software that determines what a system can or can't do for a particular business. Another often overlooked expense is the backup storage equipment that generally is needed in addition to basic system components.

- *Backup storage.* Copies of data files, used as a safeguard against damage or loss.
- *Computer system.* A computer plus software plus one or more pieces of peripheral equipment.
- *CPU (central processing unit).* The part of a computer that performs calculations and processes data according to the instructions specified by the software. *CPU* is sometimes used interchangeably with *computer.* See also *microcomputer.*
- *Hardware.* The computer itself or any item of peripheral equipment.
- *Peripheral equipment.* Input-output and data storage devices: printers, keyboards, CRTs, remote terminals, and tape and disk drives.
- *Software.* The programs, or instructions, that tell the computer how to respond to specific user commands.[31]

4. Identify audience for a document that uses definitions.

(a) Carefully read the following paper, "The Fugue." [32]

(b) Marginally annotate to identify all the types of definition the writer uses.

(c) Identify probable audiences for this paper.

The Fugue

A fugue is a polyphonic form of music that is most easily recognized by several reappearances of a short melodic theme. Fugues are usually written in from two- to five-part harmonies. Each part blends with the other parts to form harmony, but it is also a melodic line as well. This characteristic of a fugue makes it unique. Writing a short melody with simple harmony is easy; however, if the supporting parts do not form a melody very similar to the melodic theme first stated in the fugue, then the composition cannot be called a fugue.

A fugue is similar to a canon or a round, but a canon usually has only two parts that are always exactly alike. One part begins the canon and the second part enters later, duplicating the first.An example of a canon is shown in Figure 1.

Figure 1 *Canon*

(By the way, can you name that tune?)

The fugue, however, always has the second part entering in a different key. This new key is called the dominant key and is fivetones higher than the original key, the tonic key. (See Figure 2.)

Figure 2 *Tonic-Dominant Relationship*

There are, then, two important characteristics of a fugue:

1. The melodics are imitative.
2. The imitation occurs in related keys.

The fugue is commonly divided into three distinct sections: exposition, development, and stretto.

The exposition consists of the statement of the melodic theme and its answer in the dominant key. These statements and answers may occur several times, depending upon the number of parts written into the composition. Most frequently, fugues are written in four parts and require two statements and two answers.

The development is the most exciting and improvisational part of the fugue. Indeed, the fugue is so-named because of the action within the development. Fugue is derived from the Latin *fugere,* meaning to fee or run away. The development runs away from the straightforward statements and answers of the exposition and adds color and variety to the composition. The development uses the following techniques to achieve interest:

1. Countersubject—playing a new theme with the original theme
2. Augmentation—playing the theme more slowly
3. Diminution—playing the theme more quickly
4. Inversion—playing the theme upside down

The stretto section of the fugue is a restatement of the original theme by all the parts. They often overlap unexpectedly and reach a climax to provide a dramatic end to the fugue.

The fugue was a popular form of composition during the Baroque era. Most of the major composers have written fugues or parts of fugues in larger compositions. However, Bach remains the most famous and prolifc writer of fugues.

5. Identify an audience, context, and purpose.

(a) Carefully study the diagram, titled "It Does a Body Bad,"[33] that defines immediate and long-term effects of stress.

(b) Identify audiences for whom this presentation would be appropriate and effective; what would be possible context(s) and purpose(s)?

IT DOES A BODY BAD

■ **These are some** immediate and long-term effects of stress. If long-term symptoms persist for more than a few weeks and interfere with your ability to work or perform normal daily activities, see a physician:

Brain

Instant: Hypothalamus and pituitary glands, which stimulate other organs and release stress hormones (primarily adrenaline, cortisol and norepinephreine), kick in.
Long-term: Mental and emotional problems, including insomnia, depression, anxiety, personality changes, irritability, sleeping problems, exhaustion.

Face/Head

Instant: Face "flushes" as blood flow increases.
Long-term: Muscles in the head, neck, jaw can tighten, causing chronic headache, neck ache, jaw pain, tics.

Lungs/Breathing

Instant: Lungs speed up to take in more air and deliver more oxygen to muscles, producing a breathless feeling.
Long-term: Stress can worsen chronic lung problems, such as asthma or emphysema.

Muscles

Instant: Muscles become more efficient and stronger, sometimes producing seemingly superhuman strength, speed, reaction.
Long-term: Muscle aches, soreness, pain, tension, upper and lower back, shoulder, neck problems.

Kidneys

Instant: A rush of anti-diuretic hormone (adh), decreased blood supply and tightened muscles decrease urine output and temporarily shut down the kidneys. Defecation and urination are prevented.
Long-term: Diarrhea or uncontrolled urination may occur as muscles relax, blood supply increases and hormones level out.

Blood

Instant: Blood clotting mechanisms improve with stress.

Senses

Instant: Pupils dilate. Smell is heightened.

Salivary glands

Instant: Salivary glands stop secreting saliva, making the mouth feel dry.
Long-term: Mouth ulcers, sores.

Heart

Instant: Heart pumps more blood to muscles, heart rate increases, producing a pounding, flutteringfeeling in the chest. Blood pressure increases.
Long-term: Heart disease, heart attack, and high blood pressure.

Liver

Instant: Releases more sugar into the blood to provide muscles with instant energy. May also release more cholesterol into the blood.

Stomach/Digestive system

Instant: Stomach secretes more acid, producing a feeling of "butterflies" or gurgling. Digestion shuts down temporarily.
Long-term: Ulcers, colitis, irritable bowel syndrome, gastritis, heartburn.

Skin

Instant: Skin turns pale as blood is drawn toward internal organs. Palms and feet sweat to cool the body.
Long-term: Outbreaks of rashes or other skin problems, including eczema, psoriasis, hives.

SOURCES: *The Complete Manual of Fitness and Well Being* (Reader's Digest; 1984); Dr. Howard Schertzinger Jr., of Queen City Sports Medicine Rehabilitation; The American Institute of Stress, Yonkers, N.Y.

GANNETT NEWS SERVICE

Then identify inappropriate or ineffective contexts, purposes, and audiences.

(c) Describe (and, if appropriate, do thumbnail sketches of) other ways to define the psychological effects of stress.

6. **Write a paper incorporating definitions.**

   (a) Work with a classmate who is in the same discipline or field (if possible) and identify a concept or term that you would like to define.

   (b) Write a multiparagraph paper incorporating various forms of definitions for that single term. Your paper should resemble the samples in Assignments #3 and #4. Make sure to include visuals if appropriate for illustrating or clarifying the term. Document any sources used in preparing the paper.

7. **Analyze definitions.** Visit the following Web site: www.search.eb.com.

   (a) Choose a term from your discipline and search for it in the online *Encyclopedia Britannica* and *Merriam-Webster's Collegiate Dictionary* (www.m-w.com).

   (b) Identify additional Web sites that provide useful definitions of your term.

   (c) Create a rubric that lists the criteria for analyzing the effectiveness of each definition, and then use your rubric to assess each definition by identifying the different purposes and audiences.

8. **Define Web structures.** Create a table in which you define and illustrate various ways Web sites can be structured — for example, sequentially, hierarchically, and so on. Imagine your table will be in a book about Web design for managers who need to understand options for their company's Web site and make decisions about the most effective structures for various purposes and audiences.

9. **Revise glossary entries.** As you carefully read the following entries, you'll see that they are not parallel in sentence structure or specificity. Revise the glossary definitions so they're consistent, comprehensible, and usable for interested nonexperts.

Turbine Glossary[34]

Anemometer: Measures the wind speed and transmits wind speed data to the controller.

Blades: Most turbines have either two or three blades. Wind blowing over the blades causes the blades to "lift" and rotate.

Brake: A disc brake that can be applied mechanically, electrically, or hydraulically to stop the rotor in emergencies.

Controller: The controller starts up the machine at wind speeds of about 8 to 16 miles per hour (mph) and shuts off the machine at about 65 mph. Turbines cannot operate at wind speeds above about 65 mph because their generators could overheat.

Gear box: Gears connect the low-speed shaft to the high-speed shaft and increase the rotational speeds from about 30 to 60 rotations per minute (rpm) to about 1200 to 1500 rpm, the rotational speed required by most generators to produce electricity. The gear box is a costly (and heavy) part of the wind turbine and engineers are exploring "direct-drive" generators that operate at lower rotational speeds and don't need gear boxes.

Generator: Usually an off-the-shelf induction generator that produces 60-cycle AC electricity.

High-speed shaft: Drives the generator.

Low-speed shaft: The rotor turns the low-speed shaft at about 30 to 60 rotations per minute.

Nacelle: The rotor attaches to the nacelle, which sits atop the tower and includes the gear box, low- and high-speed shafts, generator, controller, and brake. A cover protects the components inside the nacelle. Some nacelles are large enough for a technician to stand inside while working.

Pitch: Blades are turned, or pitched, out of the wind to keep the rotor from turning in winds that are too high or too low to produce electricity.

Rotor: The blades and the hub together are called the rotor.

Tower: Towers are made from tubular steel (shown here) or steel lattice. Because wind speed increases with height, taller towers enable turbines to capture more energy and generate more electricity.

Wind direction: This is an "upwind" turbine, so called because it operates facing into the wind. Other turbines are designed to run "downwind," facing away from the wind.

Wind vane: Measures wind direction and communicates with the yaw drive to orient the turbine properly with respect to the wind.

Yaw drive: Upwind turbines face into the wind; the yaw drive is used to keep the rotor facing into the wind as the wind direction changes. Downwind turbines don't require a yaw drive, the wind blows the rotor downwind.

Yaw motor: Powers the yaw drive.

Chapter 14 Endnotes

1 Hypertext. (2000). In *encyclopedia Britannica* (15th ed.). Chicago: Encyclopedia Britannica.

2 Carroll, L. (1871). Chapter 6: Humpty Dumpty. Illustrator John Tenniel. *Through the Looking Glass.* Retrieved November 12, 2003, from http://www.sabian.org/Alice/lgchap06.htm

3 Focus. (1940). In *American pocket medical dictionary.* Philadelphia: Saunders.

4 Thomas, G. B., Jr. (1972). *Calculus and analytic geometry* (5th ed., p. 479). Reading, MA: Addison-Wesley.

5 Earthquake. (2000). In *Encyclopedia Britannica* (15th ed.). Chicago: Encyclopedia Britannica.

6 Polaroid Corporation. (1974). *The square shooter* (p. 3). Cambridge, MA: Author.

7 Semat, A. (1966). *Fundamentals of physics* (p. 96). New York: Holt, Rinehart, & Winston.

8 Bureau of Naval Personnel. (1965). *Principles of naval ordnance and gunnery* (p. 4) (NAVPERS 10783-A). Washington, DC: U.S. Navy.

9 Volt. In *Webster's new school and office dictionary.*

10 Volt. In *Penguin dictionary of science.*

11 McGovern, T., & Smith, R. A. (1998). Case definition of AIDS. In *Encyclopedia of AIDS: A Social, Political, Cultural, and Scientific Record of the HIV Epidemic.* Retrieved November 12, 2003, from http://www.the body.com/encyclo/aids.html

12 McGovern, T., & Smith, R. A. (1998). Case definition of AIDS. In *Encyclopedia of AIDS: A Social, Political, Cultural, and Scientific Record of the HIV Epidemic.* Retrieved November 12, 2003, from http://www.thebody.com/encyclo/aids.html

13 McGovern, T., & Smith, R. A. (1998). Case definition of AIDS. In *Encyclopedia of AIDS: A Social, Political, Cultural, and Scientific Record of the HIV Epidemic.* Retrieved November 12, 2003, from http://www.thebody.com/encyclo/aids.html

14 © 1999, *Peanuts* reprinted by permission of United Features Syndicate.

15 Snealler, J. (1982). Mass production of RRIM parts start up on automotive lines. *Modern Plastics, 59* (January): 48.

16 Source: *Machinability Data Center* (p. 3). Cincinnati, OH: Metcut Research Associates. Reprinted with permission.

17 NetLingo.com. (n.d.). *Domain name.* Retrieved November 12, 2003, from http://www.netlingo.com/inframes.cfm

18 NetLingo.com. (n.d.). *Modem.* Retrieved November 12, 2003, from http://www.netlingo.com/inframes.cfm

19 Capillary attraction and capillary repulsion. In *Scott Foresman advanced dictionary.* Reprinted with permission of Scott Foresman.

20 Drake, C. (n.d.). Operational definition of a thermostatic element. *Technical and Scientific Writing, 42,* 225.

21 Rivet. (1980). In *Webster's new world dictionary.*

22 The Office of the Public Health Service Historian. (n.d.). *a brief history of anthrax.* Retrieved November 12, 2003, from http://lhncbc.nlm.nih.gov/apdb/phsHistory/resources/anthrax/anthrax.html

23 Rayleigh-Taylor instability and ICF. (1992). In Diagnostics for inertial confinement fission in research. *Energy and Technology Review,* (July). Livermore, CA: Lawrence Livermore National Laboratory.

24 *Wild*World Glossary. *National Geographic.* Retrieved April 1, 2004, from http://www.nationalgeographic.com/wildworld/glossary.html

Easton, V. J., & McColl, J. H. (1997). *Statistics Glossary,* vl.1. Retrieved April 1, 2004, from http://www.stats.gla.ac.uk/steps/glossary/sampling.html

Copyright Statement and Licence Agreement http://www.stats.gla.ac.uk/steps/licence.html

25 Source: Bittl, J. A., & Thomas, P. (1996, January). Opening the arteries: Beyond the balloon. *Harvard Health Letter, 21,* 4–6. Copyright © 1996, President and Fellows of Harvard College. Reprinted by permission.

26 Mandell, A. (1974). *The language of science* (p. 23). Washington, DC: National Science Teachers Association.

27 Chemistry. In *Chamber's technical dictionary.*

28 Reproduced by permission. Copyright © 1995 by Netlingo, Inc. The Online Computer Dictionary, www.netlingo.com

29 Hadley, N. F. (1989). Lipid water barriers in biological systems. *Progress in Lipid Research, 28,* 23.

30 NetLingo.com. (n.d.). *Search engine.* Retrieved November 12, 2003, from http://www.netlingo.com/inframes.cfm

31 Adapted from Computerspeak glossary. (1980 May). *Inc.,* pp. 102–103. Copyright © 1980 by Inc. Publishing Corporation, 38 Commercial Wharf, Boston, MA 02110. Reprinted with permission of *Inc.* magazine.

32 Source: Stanhope, C. The fugue. *Technical Communication EN 4676,* Northern Essex Community College.

33 MacDonald, S. (1995, November 27). Treating stress. *Des Moines Register,* p. 3T. Reprinted with permission of the Des Moines Register. Copyright © 1995.

34 U.S. Department of Energy/Energy Efficiency and Renewable Energy. (2004). Inside the wind turbine. *Wind Energy Basics.* Retrieved January 2, 2004, from http://www.eere.energy.gov/windandhydro/wind_how.html

CHAPTER 15

Creating Technical Descriptions

Objectives and Outcomes

This chapter will help you accomplish these outcomes:

- Understand that technical descriptions can be used to organize specific details about objects, substances, mechanisms, organisms, systems, and locations for an identified audience
- Summarize physical characteristics, answering questions you expect your readers to have about appearance, acceptability, and impact
- Use technical descriptions in observation notes, manuals and training materials, proposals and reports, marketing and promotional materials, and public information and education
- Prepare technical descriptions:
  - Meet audience needs by answering their questions
  - Partition your subject into structural parts and/or functional parts
  - Adjust diction to audience needs, choosing accurate terms, and using appropriate metaphors
  - Choose from a variety of visuals: photographs or realistic drawings, topographic and contour maps, phantom views, overlays, schematics and wiring diagrams, cross-section maps, exploded views, blueprints
  - Typically use spatial order to give a clear view of appearance and structure

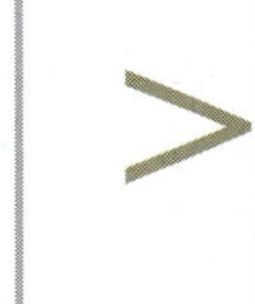

Since the beginning of the seventeenth century, scientists have been describing four of Jupiter's moons — Io, Europa, Ganymede, and Callisto — although over nearly 400 years, the level of detail has changed considerably. The inquiry started in 1610, when Galileo Galilei announced "the occasion of discovering and observing four planets, never seen from the very beginning of the world up to our own times, their positions, and the observations made during the last two months about their movements and their change of magnitude. . . ."[1]

In his journal, Galileo carefully recorded his observations of these four satellites, establishing size as "greater than [another that was] exceedingly small," luminescence as "very conspicuous and bright," and location as "deviated a little from the straight line toward the north."

Courtesy NASA/JPL-Caltech

"This 'family portrait,' a composite of the Jovian system, includes the edge of Jupiter with its Great Red Spot, and Jupiter's four largest moons, known as the Galilean satellites. From top to bottom, the moons shown are Io, Europa, Ganymede and Callisto."[2]

> *January 11.* [My observations] established that there are not only three but four erratic sidereal bodies performing their revolutions round Jupiter. . . .
>
> *January 12.* The satellite farthest to the east was greater than the satellite farther to the west; but both were very conspicuous and bright; the distance of each one from Jupiter was two minutes. A third satellite, certainly not in view before, began to appear at the third hour; it nearly touched Jupiter on the east side, and was exceedingly small. They were all arranged in a straight line, along the ecliptic.
>
> *January 13.* For the first time four satellites were in view. . . . There were three to the west and one to the east; they made a straight line nearly, but the middle satellite of those to the west deviated a little from the straight line toward the north. The satellite farthest to the east was at a distance of 2′ from Jupiter; there were intervals of 1′ only between Jupiter and the nearest satellite, and between the satellites themselves, west of Jupiter. All the satellites appeared of the same size, and though small they were very brilliant and far outshone the fixed stars of the same magnitude.

These same four moons have been the object of intense scrutiny by the National Aeronautics and Space Administration (NASA). The mission of *Voyager 1* and *Voyager 2,* beginning in 1977, was to collect data about Jupiter's "miniature solar system," including the four Galilean satellites. In vivid contrast to Galileo's brief comments, the *Voyager Bulletin* (a mission status report that regularly reported the discoveries) described the moons in considerable detail:

> Of all the satellites, Io generated the most excitement. As *Voyager 1* closed in on Io, the puzzle was why its surface, so cratered and pocked when viewed from a distance, began to look smoother and younger as the spacecraft neared. . . . But

the mystery was solved with the discovery of active volcanoes spewing sulfur 160 km (100 mi) high and showering it down on the crust, obliterating the old surface. Infrared data indicated hot spots at the locations of the plumes identified in the photographs, confirming the find. Io is undoubtedly the most active known surface in the solar system, surpassing even the Earth.[3]

After *Voyager1* came *Galileo,* whose mission, from 1989–2003, was to study Jupiter in even greater detail. NASA's Space Science News Web site regularly reported on the descriptions coming from the *Galileo* project, noting that "since the first volcanic plume was discovered by Voyager in 1979, Io has remained under intense scrutiny." Based on data from *Galileo,* earlier generalizations were replaced with more detailed descriptions.

For a world dominated by fiery volcanoes, it's curious that Io is also very, very cold. The ground just around the volcanic vents is literally sizzling, but most of Io's surface is 150 degrees or more below 0°C. The moon's negligible atmosphere traps little of the meager heat from the distant Sun. As soon as volcanic gases spew into the air, they immediately begin to freeze and condense. The plumes of Io's sizzling volcanoes are very likely made up of sulfur dioxide snow.[4]

Like the observations of Galileo Galilei and the *Voyager* spacecraft, the *Galileo* mission provided valuable data about Jupiter and its moons, including Io, until the spacecraft was destroyed in 2003 (Figure 15.1). The excerpts from Galileo

NASA put Galileo on a collision course with Jupiter for two reasons. First, Galileo's propellant was nearly depleted. "Without propellant, the spacecraft would not be able to point its antenna toward Earth or adjust its trajectory, so controlling the spacecraft would no longer be possible." Second, NASA wanted to eliminate any chance of an impact between Galileo and Jupiter's moon Europa, which Galileo discovered may have a subsurface ocean. "The possibility of life existing on Europa is so compelling and has raised so many unanswered questions that it is prompting plans for future spacecraft to return to the icy moon."[6]

FIGURE 15.1 **End of Galileo Mission[5]**

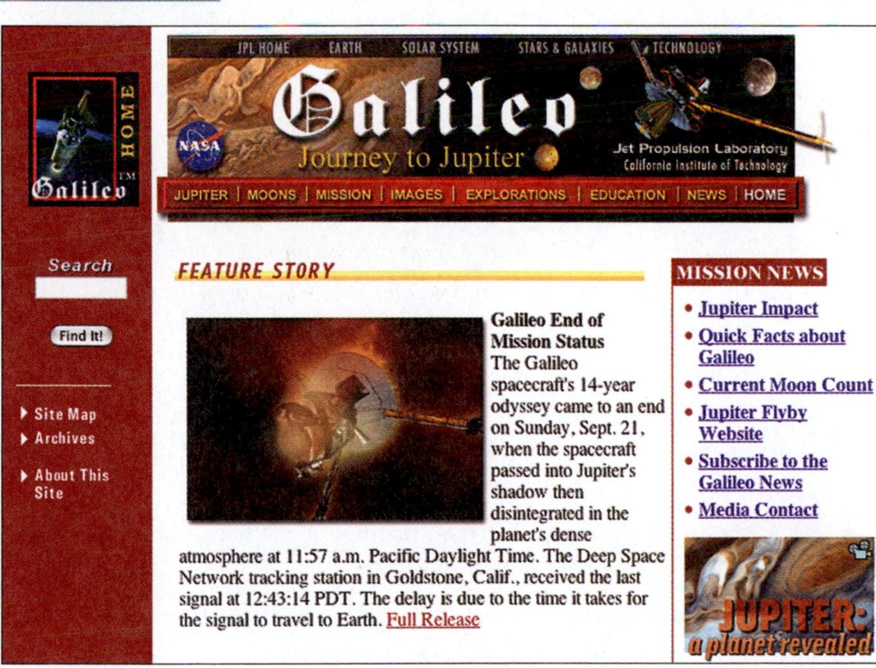

Galilei's journal, from the *Voyager Bulletin,* and from the NASA Space Science News Web site illustrate several characteristics of description that are discussed in this chapter. Most important, the descriptions include specific details, presented in an organized manner to meet the needs of an identified audience.

WEBLINK

For a link to more information about the inventor Galileo Galilei (1564–1642), including his remarkable inventions, go to http://english.wadsworth.com/burnett6e.
CLICK ON WEBLINK
CLICK on Chapter 15/Galileo Galilei

WEBLINK

For a link to more information about NASA's 14-year *Galileo* mission to Jupiter, which ended in September 2003, go to http://english.wadsworth.com/burnett6e.
CLICK ON WEBLINK
CLICK on Chapter 15/Galileo Mission

Defining Technical Description

Descriptions summarize physical characteristics, answering questions you expect your readers to have about the appearance or composition of an object, substance, mechanism, organism, system, or location. Regardless of a description's length or subject, it is characterized by verifiable information that responds to assumed questions.

1. What is it? How is it defined? By whom?
2. What is its purpose? What is its importance or impact?
3. What are the characteristics of the whole? What is "normal" or "typical"? What are within acceptable tolerances or specifications?
   - What does it look like (size, shape, color)?
   - What are its characteristics (material or substance, weight, texture, flammability, density, durability, expected life, method of production or reproduction, and so on)?
4. What are its parts? What is "normal" or "typical"? What are within acceptable tolerances or specifications?
   - What is the appearance of each part (size, shape, color)?
   - What are the distinctive characteristics of each part?

What additional characteristics might be important for descriptions in your professional field?

5. How do the parts fit together? How do they work together? What defines effective function?

Which of these questions are answered depends on the depth of detail required by the description. Complex descriptions clearly answer more questions.

Sometimes technical description constitutes an entire document, oral presentation, or visual. However, in most situations a technical description is limited to a single segment of a longer document or presentation. For example, descriptions could range from a few lines included in a one- or two-page memo to several paragraphs in a longer report. So, a description of equipment could be just one part of a report about monitoring airport noise that is interfering with animals in an adjoining wildlife refuge. Or the description of equipment could take up one section in a proposal to purchase a new X-ray machine.

Do technical descriptions matter? Do they have any impact beyond providing accurate information about physical features? The ethics sidebar below shows that if descriptions are inadequately presented and explained, readers may underestimate their importance or even neglect them entirely. Sometimes the results can be deadly.

ETHICS SIDEBAR

"The decision was flawed": Ethical Responsibility and the Challenger Explosion

If you were in a similar situation, with millions or even billions of dollars and lives on the line, could you have the ethical strength to speak up? How would you do it? See Chapter 16 for a discussion about whistleblowers.

One of the most tragic accidents in NASA history was the space shuttle *Challenger* explosion in 1986. The explosion occurred when fuel leaked out of the fuel tanks and mixed with engine exhaust. The leaks were traced to defective seals, called O-rings, on the fuel tanks. During launch, the O-rings were charred by the hot engines enough that they allowed fuel to leak out.

Researcher Paul Dombroski believes the explosion occurred in part because of an ethical failure of the technicians and administrators involved to act on the information they had, not from a lack of information about the O-rings.[7] Technical descriptions of the fuel tanks, written after each previous shuttle flight, revealed that the O-rings were being charred during launch. Although technicians and administrators knew the O-rings were burned, the burning was described in the reports as "allowable erosion" and "acceptable risk." While some engineers questioned these descriptions (and the integrity of the O-rings) the night before

Courtesy of NASA

In February 2003, the space shuttle Columbia *was destroyed in a disaster that claimed the lives of its seven-person crew. Follow the WEBLINK below to help you form an opinion about the ways in which various kinds of communication problems contributed to the* Columbia *disaster.*

the launch, they were unable to provide strong enough arguments to persuade administrators to cancel the launch. The technical descriptions were not in question; everyone agreed that O-rings were charring. However, the contextual pressures to launch influenced the way that technical information was interpreted. The O-rings were deemed safe enough, and seven lives were lost.

Dombrowski uses this example to illustrate that technical professionals cannot necessarily rely on conventional procedures to deal with ethical dilemmas. Technical professionals must recognize when the information in a document is not enough, even when it is entirely accurate. They should be willing to take personal responsibility to make sure they have done all that is possible to convey the full picture and explain the implications of descriptions that would otherwise go unnoticed.

WEBLINK

Following the space shuttle *Columbia* disaster, a Columbia Accident Investigation Board (CAIB) was formed. In August 2003, the CAIB released a report that said, "unless the technical, organizational, and cultural recommendations made in this report are implemented, little will have been accomplished to lessen the chance that another accident will follow." For a link to more information about the *Columbia,* including the final CAIB report, transcripts and videos of the press conferences, and background information about the missions and crew, go to **http://english.wadsworth.com/burnett6e**.

CLICK ON WEBLINK

CLICK on Chapter 15/Columbia

Using Technical Description

How could you write a description that would be sure to get the attention of managers who needed to know the information for decision making?

How do you know when to use a technical description? You can decide whether to include a description and what kind of details to incorporate by examining the context, purpose, and task of your document.

- Will a description help accomplish *your purpose* of providing information, persuading readers or listeners, or helping them complete a task?
- Will a description help members of the audience accomplish *their purpose* of gathering information, making a decision, or completing an action or activity?
- Will a description help prevent problems?

The following discussion identifies and illustrates some common applications for technical descriptions: observation notes, reference and training materials, reports and proposals, marketing and promotional materials, and public information and education.

Observation Notes

Many situations require accurate first-hand descriptions, particularly in medicine, field study, and scientific research. Technical experts observe, select, and record relevant data, often employing abbreviations and jargon specific to the field. The initial purpose of observation notes is to maintain an accurate record. Later, the notes may be extended or transcribed so others can read them, or they may be used as the basis for a more formal document.

The *Manual of Pediatric Therapeutics,* a reference volume for pediatric practitioners, outlines the longstanding and widely used criteria for immediate evaluation of newborns. This evaluation, based on observation by medical professionals, provides a detailed physical description of the newborn. The excerpt in Figure 15.2 provides information for the delivery room observation and evaluation.

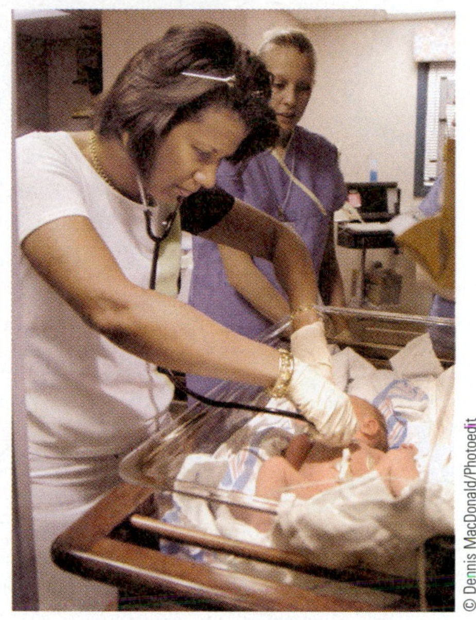

© Dennis MacDonald/Photoedit

In the delivery room, a newborn infant is screened with an Apgar test. Here, the heart-beat is being checked.

Training Materials

Both student interns and new employees often need descriptive overviews of the tools and systems with which they'll work, so technical descriptions can be an appropriate part of initial training. For example, the short catalog description, as

FIGURE 15.2 **Apgar Score: Description Based on Observation**[8]

Accessible

- *The outline makes the hierarchy clear.*
- *The five critical signs are emphasized with italics.*

Comprehensible

- *The purpose of the Apgar score is clearly stated.*
- *Technical vocabulary restricts the information to experts.*
- *The embedded reference to the table helps readers link the text and table.*

Usable

- *The explanation precedes the application of the system.*
- *Time-sensitive instructions are presented.*
- *Presenting critical criteria and rating in a table make information easy to use.*

I. EVALUATION OF THE NEWBORN

A. **Delivery room.** Immediate assessment of the newborn infant by the Apgar scoring system should help to identify infants with severe metabolic imbalances. At 1 and 5 minutes after delivery (the times at which feet and head are both first visible), the infant is to be evaluated for five signs — *heart rate, respiratory effort, muscle tone, reflexes* and *irritability* and *color* — and given a rating of 0, 1, or 2 (as defined in Table 5-1). In the extremely compromised infant, prompt and efficient resuscitation is far more important than his exact Apgar score.

TABLE 5-1

Apgar Score *(Score infant at 1 and 5 minutes of age)*

Sign	0	1	2
Heart rate	absent	slow, less than 100	100 or over
Respiratory effort	absent	weak cry, hypoventilation	crying lustily
Muscle tone	flaccid	some flexion, extremities	well-flexed
Reflex irritability	no response	some motion	cry
Color	blue, pale	blue hands and feet	entirely pink

shown in Figure 15.3, provides only sketchy details about a circular inspection mirror — sufficient for ordering the tool but not for learning about it.

An intern or new employee would not know why this tool is used, how it's used, or what special features it has that make it important. The description of a circular inspection mirror in Figure 15.4, which presents a more detailed description than the catalog does, could be used as background reading during initial orientation to introduce this basic piece of inspection equipment. Notice that both the text and the figure are essential for a complete description; either by itself provides only partial information. Notice also that while the information is precise and carefully organized, the language is easy to understand and the tone is friendly, so the audience is more likely to read the information sheet and understand why this particular small tool is important.

Technical Manuals

Most manuals include a technical description of the mechanism or system that the manual deals with. The description usually appears in one of the manual's early sections, often providing a general overview followed by more detailed information. It introduces the user, operator, technician, or repairperson to the physical characteristics of the mechanism or system. Technical descriptions in manuals are usually accompanied by a variety of visuals: the entire mechanism or system, exploded views, blowups, and phantom and cutaway views of individual parts and subparts.

FIGURE 15.3 **Catalog Description of a Circular Mirror**[9]

Accessible

- *Features and details are differentiated by headings.*
- *Information is presented in easy-to-read lists.*

Comprehensible

- *Vocabulary is basic.*
- *A photo depicts the overall appearance.*

Usable

- *Dimensions are clearly identified.*

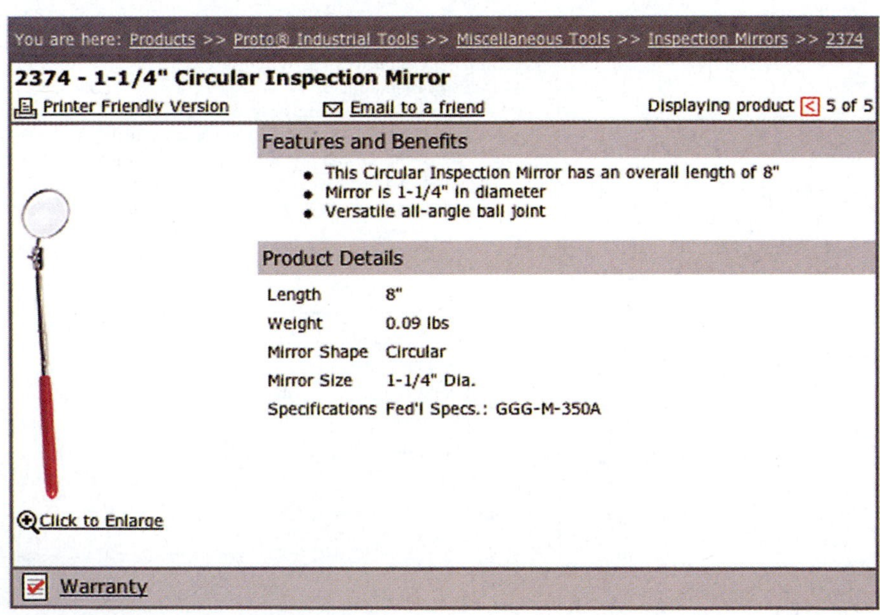

FIGURE 15.4 **Circular Inspection Mirror[10] Information Sheet**

CIRCULAR INSPECTION MIRROR

Everyone working in our group uses a circular inspection mirror. It looks similar to a hand mirror used by a dentist to inspect teeth, with one exception: this mirror swivels separately from the handle.

Function. The circular inspection mirror is one of the most important tools you'll use to visually inspect general electrical and mechanical equipment for production flaws. The mirror helps you observe areas that — because of the angled displacement within the unit — are normally hidden from view.

Components. The inspection mirror shown in Figure 1 consists of three main parts: mirror, universal swivel joint, and handle.

An especially important feature is the lack of distortion in the 1⅛"-mirror, which reflects identical size figures. The mirror's durable stainless steel casing adds ⅛" to the overall diameter, making the total diameter of the mirror 1¼".

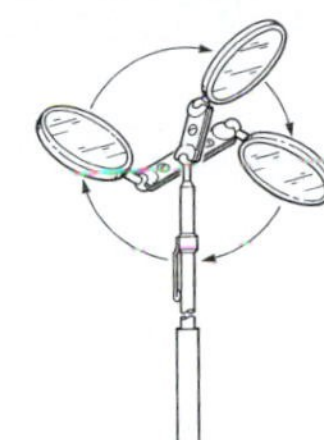
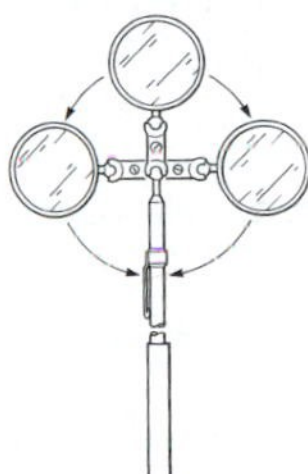

Figure 1. Circular Inspection Mirror

The mechanics incorporated in the swivel design allow a complete 360° spherical positioning of the mirror with no movement of the handle.

Attached by spot welding to the inside of the casing back is a small stem extending ⅜" and concluding in a round bearing. This bearing is positioned inside a two-bearing universal joint.

The simple universal joint uses two encloser plates held together by a nut and screw. Impressed in the plates are four concave pockets that prevent the bearings from leaving the joint but allow maximum rotation to the attached handle or mirror. By tightening and loosening the screw, you can adjust the mirror to the desired tension.

Also located inside the universal casing and opposite the bearing attached to the mirror is the second bearing, which connects to a hard tempered-steel handle that is approximately 6" long.

Safety. The rough surface of the metal handle is covered for 3" with orange plastic insulation that protects you from electrical shock and possible electrocution.

Convenience. An additional feature of the circular inspection mirror is a pocket clip, located toward the middle of the handle, which allows the tool to be carried like a pen in your shirt pocket.

Accessible

- *Run-in headings identify the four topics the audience should understand and remember.*
- *The serif font is sufficiently large to read easily.*
- *The paragraphs are very short, which is appropriate for this kind of quick reference document.*

Comprehensible

- *Second-person "you" lets readers know this information is related to them.*
- *Vocabulary is technically accurate but easy to understand.*

Usable

- *The figure is embedded in the text in the appropriate place and referred to directly.*
- *Critical information — lack of distortion, adjustability, safety — is clear.*

Figure 15.5 provides users with information about CGI. Because this information is available electronically, users can select the path that best matches their own needs and experience. For example, users are told, "If you have no idea what CGI is, you should read this introduction." And if they click that hyperlink, they get Figure 15.6. Users are also told that is they have a basic idea of what CGI is, they can select a primer. And if they click that hyperlink, they get Figure 15.7.

Proposals and Reports

If a description helps the audience understand and approve a proposal, it should be included. This type of description usually gives an overview and then provides details appropriate for the primary reader(s). For example, a proposal from a manager of research and development (R&D) to the company comptroller about an equipment purchase would logically include a description of the equipment. However, the description would not be detailed because the equipment's technical specifications and capabilities are not relevant for that audience. If secondary

FIGURE 15.5 Common Gateway Interface[11]

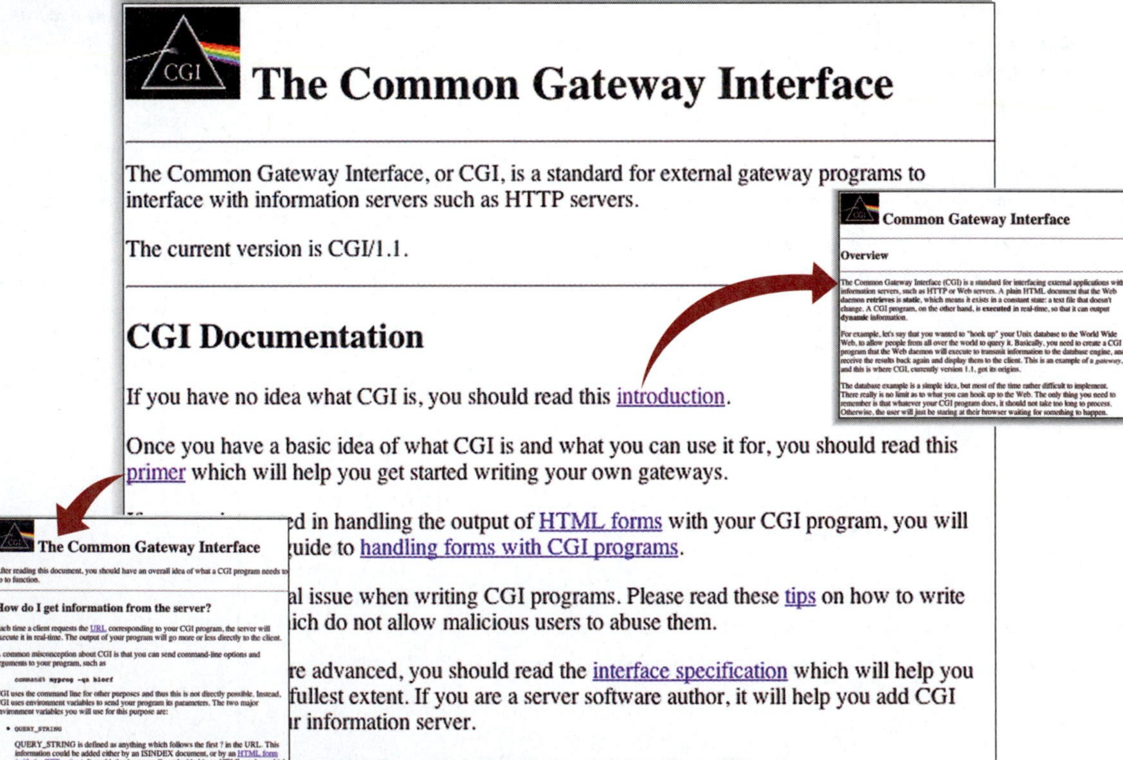

FIGURE 15.6 Common Gateway Interface for Novices Who Need an Introduction[12]

Common Gateway Interface

Overview

The Common Gateway Interface (CGI) is a standard for interfacing external applications with information servers, such as HTTP or Web servers. A plain HTML document that the Web daemon **retrieves** is **static**, which means it exists in a constant state: a text file that doesn't change. A CGI program, on the other hand, is **executed** in real-time, so that it can output **dynamic** information.

For example, let's say that you wanted to "hook up" your Unix database to the World Wide Web, to allow people from all over the world to query it. Basically, you need to create a CGI program that the Web daemon will execute to transmit information to the database engine, and receive the results back again and display them to the client. This is an example of a *gateway*, and this is where CGI, currently version 1.1, got its origins.

The database example is a simple idea, but most of the time rather difficult to implement. There really is no limit as to what you can hook up to the Web. The only thing you need to remember is that whatever your CGI program does, it should not take too long to process. Otherwise, the user will just be staring at their browser waiting for something to happen.

FIGURE 15.7 Common Gateway Interface for Beginners Who Need a Primer[13]

The Common Gateway Interface

After reading this document, you should have an overall idea of what a CGI program needs to do to function.

How do I get information from the server?

Each time a client requests the URL corresponding to your CGI program, the server will execute it in real-time. The output of your program will go more or less directly to the client.

A common misconception about CGI is that you can send command-line options and arguments to your program, such as

```
command% myprog -qa blorf
```

CGI uses the command line for other purposes and thus this is not directly possible. Instead, CGI uses environment variables to send your program its parameters. The two major environment variables you will use for this purpose are:

- QUERY_STRING

  QUERY_STRING is defined as anything which follows the first ? in the URL. This information could be added either by an ISINDEX document, or by an HTML form (with the GET action). It could also be manually embedded in an HTML anchor which readers for the same proposal are familiar with R&D operations, an appendix could discuss technical details. In contrast, a proposal to a state's environmental control commission from a local community to preserve a wetlands area would include a detailed description of the geographic area as a main part of the proposal. The members of the commission would need the details to make an informed decision about the validity of the preservation plan.

Several types of reports incorporate descriptions. For example, a report about changes in the work flow in an assembly area because of new automatic insertion equipment could logically include a description of this equipment. A supervisor writing to the division manager would emphasize features of the equipment that have affected the work flow. Generally, any report justifying or recommending acquisition or modification of equipment or facilities should include a description.

Marketing and Promotional Pieces

Technical descriptions in marketing materials are usually both informative and persuasive. Positive (and, of course, subjective) terms are often incorporated into the initial description. The information presents an overview, identifying major components and characteristics. Additional information is often condensed on specification sheets (specs). Promotional and marketing materials frequently include visuals that first display the entire object or mechanism and then highlight its special features.

Figure 15.8 shows a technical description that balances text and visuals. It's an excerpt from a four-page, four-color, 11" × 17" glossy marketing brochure for a Ruud Achiever 90 Plus Modulating Gas Furnace with Contour Comfort Control. The brochure defines and describes a modulating gas furnace in several ways because it will be unfamiliar to many homeowners who are considering buying a furnace. Figure 15.8 is one of the critical parts of the brochure, describing the furnace with a cutaway view of the furnace itself and succinct explanations of each of the major components.

Public Information and Education

Much of the technical and scientific information presented to the public in newspapers, general-interest magazines, and Web sites includes a substantial amount of description, simply because people need to know *what* something is before they make decisions about its value. Sometimes the presentation of this information follows the same general organization as presentations in more technical documents (that is, as text supported by visuals). At other times, though, the presentation to general readers uses a small amount of text to support dramatic visuals.

An example of a useful Web site providing valuable public information — much of it technical description — is one sponsored by the Danish Wind Industry Association. Because wind energy is an international technology and

FIGURE 15.8 **Technical Description of a Modulating Gas Furnace**[14]

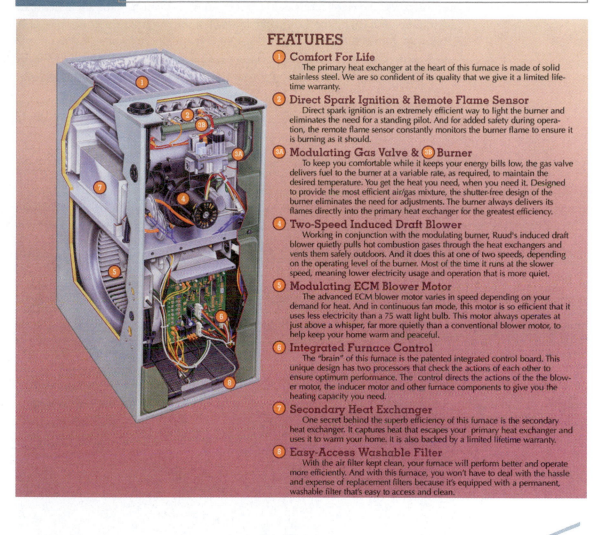

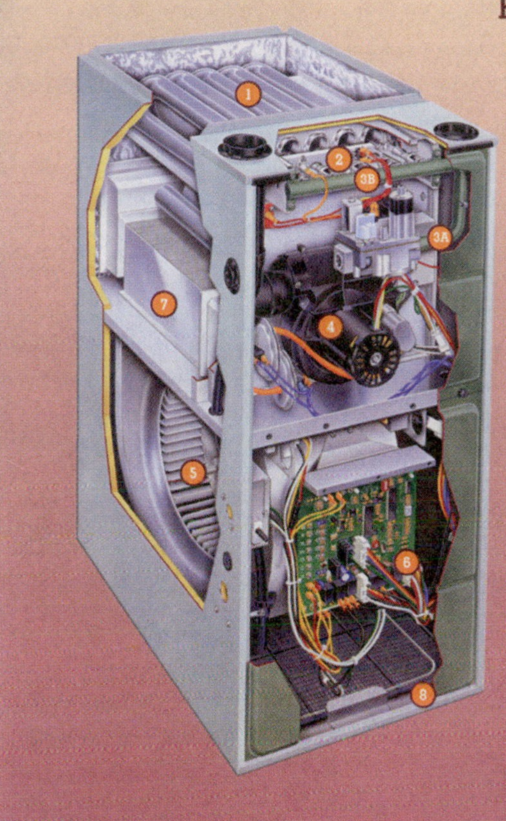

FEATURES

① Comfort For Life

The primary heat exchanger at the heart of this furnace is made of solid stainless steel. We are so confident of its quality that we give it a limited lifetime warranty.

② Direct Spark Ignition & Remote Flame Sensor

Direct spark ignition is an extremely efficient way to light the burner and eliminates the need for a standing pilot. And for added safety during operation, the remote flame sensor constantly monitors the burner flame to ensure it is burning as it should.

3A Modulating Gas Valve & 3B Burner

To keep you comfortable while it keeps your energy bills low, the gas valve delivers fuel to the burner at a variable rate, as required, to maintain the desired temperature. You get the heat you need, when you need it. Designed to provide the most efficient air/gas mixture, the shutter-free design of the burner eliminates the need for adjustments. The burner always delivers its flames directly into the primary heat exchanger for the greatest efficiency.

④ Two-Speed Induced Draft Blower

Working in conjunction with the modulating burner, Ruud's induced draft blower quietly pulls hot combustion gases through the heat exchangers and vents them safely outdoors. And it does this at one of two speeds, depending on the operating level of the burner. Most of the time it runs at the slower speed, meaning lower electricity usage and operation that is more quiet.

⑤ Modulating ECM Blower Motor

The advanced ECM blower motor varies in speed depending on your demand for heat. And in continuous fan mode, this motor is so efficient that it uses less electricity than a 75 watt light bulb. This motor always operates at just above a whisper, far more quietly than a conventional blower motor, to help keep your home warm and peaceful.

⑥ Integrated Furnace Control

The "brain" of this furnace is the patented integrated control board. This unique design has two processors that check the actions of each other to ensure optimum performance. The control directs the actions of the the blower motor, the inducer motor and other furnace components to give you the heating capacity you need.

⑦ Secondary Heat Exchanger

One secret behind the superb efficiency of this furnace is the secondary heat exchanger. It captures heat that escapes your primary heat exchanger and uses it to warm your home. It is also backed by a limited lifetime warranty.

⑧ Easy-Access Washable Filter

With the air filter kept clean, your furnace will perform better and operate more efficiently. And with this furnace, you won't have to deal with the hassle and expense of replacement filters because it's equipped with a permanent, washable filter that's easy to access and clean.

Accessible

- *Strong figure-ground contrast makes the diagram easy to see.*
- *The font is sufficiently large to be legible in home lighting.*
- *Each bulleted and numbered item is in a contrasting color for easy legibility.*

Comprehensible

- *The description presumes readers' purposes: (1) consider the purchase of a new furnace, (2) understand the distinctive features.*
- *Definitions of critical features are embedded in the text.*
- *Analogies help nonexperts.*

Usable

- *Numbers in the bullets refer to critical features on the diagram in ascending order of importance to furnace function.*
- *Each bulleted and numbered item has a key phrase heading to make content easy to locate.*
- *The cutaway view enables readers to see the interior of the furnace.*

because Denmark is part of the European Union, the home page (Figure 15.9) gives users the choice of accessing the information in Danish (naturally), German, English, Spanish, and French. If you choose Danish, you see a page with several choices, each briefly described: a guided tour, a manual for calculations, experiments, class activities for schools, and a Quicktime video. Figure 15.10 shows part of the initial Danish page and the English page, with descriptions of the options the site offers.

WEBLINK

The Danish Wind Industry site (Figures 15.9 and 15.10) can be useful to you in two ways. First, you can learn a great deal about wind energy. Additionally, you can study a tutorial designed for a broad audience and assess whether you like the approach. For a link, go to http://english.wadsworth.com/burnett6e.
CLICK ON WEBLINK
CLICK on Chapter 15/wind energy

FIGURE 15.9 **Danish Wind Industry Association Home Page Offering Access in Five Languages[15]**

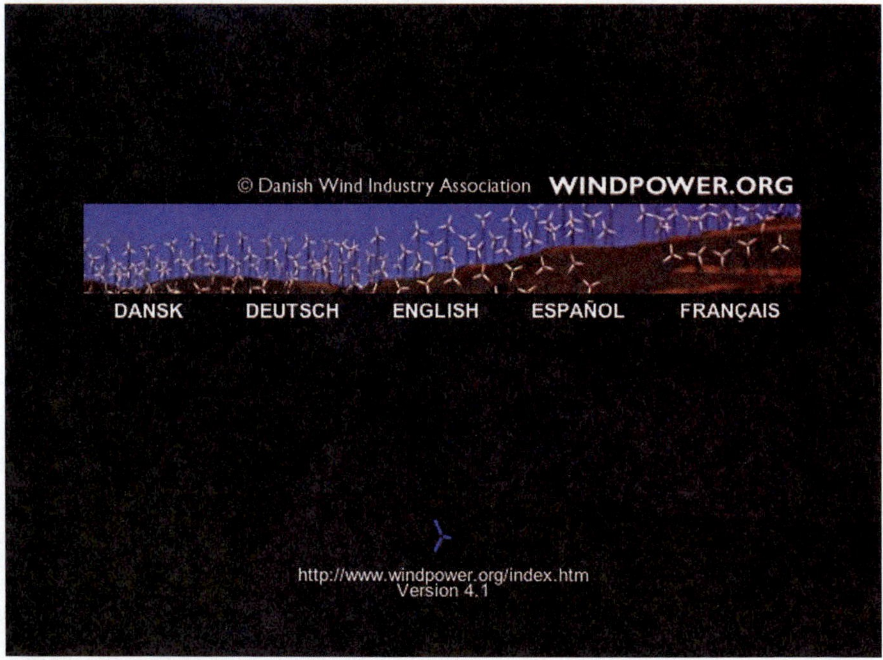

FIGURE 15.10 Danish Wind Industry Association Contents Page in Danish[16] and in English[17]

< > Søg Print zoom

© Vindmølleindustrien WINDPOWER.ORG

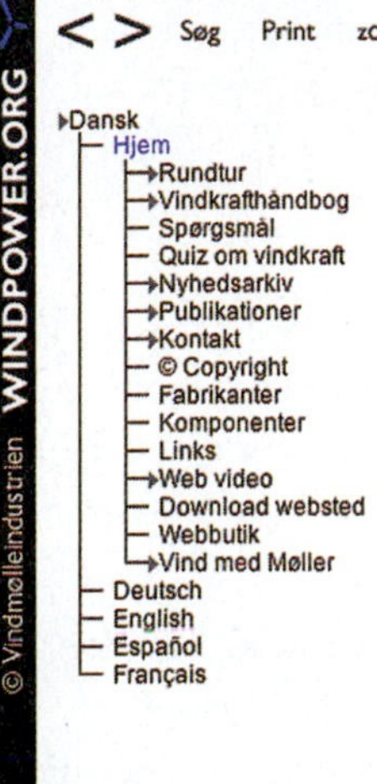

ENERGIFORLIG SKABER OPTIMISME
Det energipolitiske forlig indgået 29. marts 2004 er en anerkendelse af, at vindmølleindustrien er en vigtig velfærdsdynamo, som skaber beskæftigelse til mere end 20.000 danskere.
"Kontinuiteten på det strategisk vigtige danske vindkraftmarked er en forudsætning for, at vi fortsat kan beholde førertrøjen i den skærpede konkurrence på det globale marked," siger Vindmølle-industriens direktør Bjarne Lundager Jensen.
Læs pressemeddelelsen

ÅRSBERETNING 2003
Hent Vindmølleindustriens årsberetning 2003 som pdf.

VINDFORMATION
Download december 2003 og tidligere numre af kvartalsbladet.

STATEN KAN TJENE PÅ CO_2-KØB
»Køb af CO_2-kvoter baseret på vindkraftprojekter kan give nettogevinster til statskassen«, fortæller Vindmølleindustriens direktør Bjarne Lundager Jensen.
Se pressemeddelelsen og download rapporten

UD AF DET BLÅ er en 28 minutters Quick Time video med introduktion til vindkraft.

RUNDTUR I VINDKRAFTENS VERDEN er den mest besøgte del af dette websted. Den har mere end 100 sider om vind, placering af vindmøller, teknologi, elnet, miljø, økonomi og historie.

VIND MED MØLLER er et websted til skolebrug med opgaver og lærervejledning. Målgruppen er 5. klasse og opefter, men mange andre synes, dette er en sjov og hurtig måde at lære om vindkraft.

< > Search Print zoom

© Danish Wind Industry Association WINDPOWER.ORG

English
Home
Guided tour
Wind energy manual
FAQs
Wind energy quiz
News
Publications
Contact
Copyright
Manufacturers
Components
Links
Web video
Download web site
Web shop
Wind With Miller
Dansk
Deutsch
Español
Français

ENERGY AGREEMENT ENSURES CONTINUITY IN DENMARK
"The broad energy agreement reached 29. march 2004 shows, that there is a broad consensus in the Danish Parliament behind continued wind power development", says Bjarne Lundager Jensen, managing director of the Danish Wind Industry Association.
"Danish politicians recognize that the wind industry is an important engine for growth and welfare contributing billions of Euros to the Danish balance of payments as well as creating employment for more than 20.000 Danes."
Read press release

NEW SCIENCE PROJECT
Download a pdf with building instructions and experiments with a paper wind turbine - complete with a gearbox.

WIND VIDEO Out of the Blue is a 28 minute Quick Time video introducing wind energy.

THE GUIDED TOUR is the most visited section of this web site. It has more than 100 pages on wind, turbine siting, technology, electrical grid, environmental and economic aspects of wind energy and history.

WIND WITH MILLER is a web site for schools with class assignments and a teacher's guide. The target audience is 5th grade and up, but many others think this is a fun and fast way to learn about wind energy.

© Copyright 2004 Danish Wind Industry Association
Updated 29 March 2004
http://www.windpower.org/en/core.htm
Version 4.1

Preparing a Technical Description

To prepare a technical description, you need to identify the audience and task, determine the components, choose precise diction, design effective visuals, and select an appropriate format.

Audience's Task

Technical description should address the intended audience. The only way for you to make sure the description meets the needs of members of your audience is to analyze their purpose in reading the document and identify the questions they expect to have answered. You may find talking with representatives of the actual users very productive. At this stage, you should ask several questions:

What additional questions might be useful to anticipate?

- Why do users want or need the information? What is their task? In what ways will the information be important?
- Do they need information in order to understand a more detailed discussion that follows? Do they need to make a decision?
- Are users interested in a general overview or a detailed description?
- What details do the users need: Dimensions? Materials? Assembly? Function? Capabilities? Benefits?

Giving insufficient information leaves the audience with unanswered questions, but be equally wary of including unnecessary information; you may obscure facts you want to convey.

As you prepare a technical description, select information that responds to the audience's probable questions. The more removed the audience is from actually using the information in the description, the more general it can be. For example, the excerpt from the *Voyager Bulletin* at the beginning of this chapter is easy to read despite the inclusion of specific data; the readers of this status update report are generally interested nonexperts, not astrophysicists or aerospace engineers. Precisely identifying the audience also helps you decide on such crucial aspects of the description as components, diction, visuals, and format.

Components

Before you can describe something, you must separate it into parts or components because the description emphasizes the physical characteristics of each part. But people's concepts of "part" differ greatly. For instance, should a description of the moon Io be separated according to elements, geologic structures, or electromagnetic fields? Or consider the case of mechanical engineers asked to specify the number of parts in a simple house key. Their answers range from 1 to 27, with the mode (the number occurring most frequently) being 5. Their answers differ because they do not define *part* in the same way.

You can easily see that how you partition something depends on your purpose as the writer and on the background and task of the audience. Components can usually be separated into structural parts and functional parts.

- *Structural parts* comprise the physical aspects of the device, without regard to purpose. For example, a simple house key is made of a single piece of metal.
- *Functional parts* perform clearly defined tasks in the operation of the device. Although the key has a single structural part, it has multiple functional parts.

What other items have distinct structural and functional parts? What are the structural and functional parts of a fiberglass sailboat hull? A screwdriver? A car tailpipe?

How many parts does your house/apartment key have? What about your car key?

Applying your knowledge about the audience can help you decide whether one method of separating an object into its parts is more appropriate than another. Thinking about the audience and the purpose for writing the description also helps you decide whether you need to describe all the parts or only some of them.

Diction

The diction of a technical description should be precise, so that the information is verifiable. You can achieve this precision in three ways: choose the most specific terms appropriate for your audience, choose technically accurate terms, and consider the value of metaphor to convey descriptions.

Audience-appropriate Terms. Whether you select general or specific terms depends on the needs of your audience. Generally, nonexperts need accurate information, but they do not require extraordinary detail. Readers with more technical background need more technical details. For example, a general description of a lawnmower might appear in an advertising flyer from a chain store sent to all residents in an area. A more detailed description could be in a product brochure that sales reps could use to explain the mower's specifications to interested customers. Figure 15.11 presents two lawnmower descriptions that illustrate how characteristics can be described using general or specific diction, depending on audience needs.

Accurate Terms. A second way for a writer to ensure precision is to use the most accurate terms available. For example, many writers should differentiate more accurately between two- and three-dimensional objects. How often have you heard someone mistakenly refer to a ball as *round* instead of *spherical* or a box as a *square* rather than a *cube*? These geometric shapes — sphere/circle, cube/square, cone/pyramid/triangle — are commonly misused.

What other terms have you heard people misuse?

Not only is careless diction inaccurate, it also causes confusion for the audience. For example, if a three-dimensional object is described as triangular, how will the audience know if the solid form is really a cone or a pyramid? Figure 15.12 on page 565 reviews the terminology of geometric shapes. You can use these terms for figures, solids, and surfaces if your audience is familiar with them. If your audience is unlikely to know these terms, you may need to define the term or use a diagram.

FIGURE 15.11 Gaining Precision in Technical Descriptions

What other details might be important to engineers? To repair technicans? To consumers?

General Abstract Terms		Specific Concrete Terms
dependable mower	(specify brand)	Briggs and Stratton
powerful	(specify amount)	4 cycle, 3 1/2 HP
self-propelled	(specify type)	rear-wheel belt-to-chain drive
wide blade	(specify size)	21" blade
adjustable height	(specify variation)	7 positions, 1-3"
powerful, dependable, self-propelled mower with wide blade adjustable to cut different heights		Briggs and Stratton mower with 4-cycle, 3 1/2 HP engine; self-propelled by rear-wheel belt-to-chain drive; 21" blade; 7 cutting heights from 1-3"

Figurative Language. A third way to ensure precision is to consider whether figurative language such as metaphors, similes, and analogies would give readers a clear description. The example in Figure 15.13 on page 566 is taken from *Air & Space,* a publication read by aerospace engineers. The example presents a technical description of a microelectromechanical system (MEMS) that embeds both verbal and visual metaphors. The visual metaphor uses a playing card to illustrate the size of the MEMS in relation to the stealth aircraft that uses it. The language includes not only metaphors (e.g., "smart cars," "radar signatures"), but also similes (e.g., "like a spoiler," "flap-like surfaces," "as delicate as butterfly wings") and analogies (e.g., "sequins too small for a Barbie doll's cocktail dress").

Visuals

Precise visuals are as important in effective technical descriptions as is precise diction. Visuals enable the audience to form a mental image of the subject being described. Of course, all visuals should be labeled and titled and referred to in the text. Dimensions are usually more appropriately presented in visuals so the text is not cluttered or difficult to read.

In organizing a typical technical description of a mechanism, you could have an introductory section with a drawing or a photograph that shows the overall

FIGURE 15.12 **Geometric Shapes**[18]

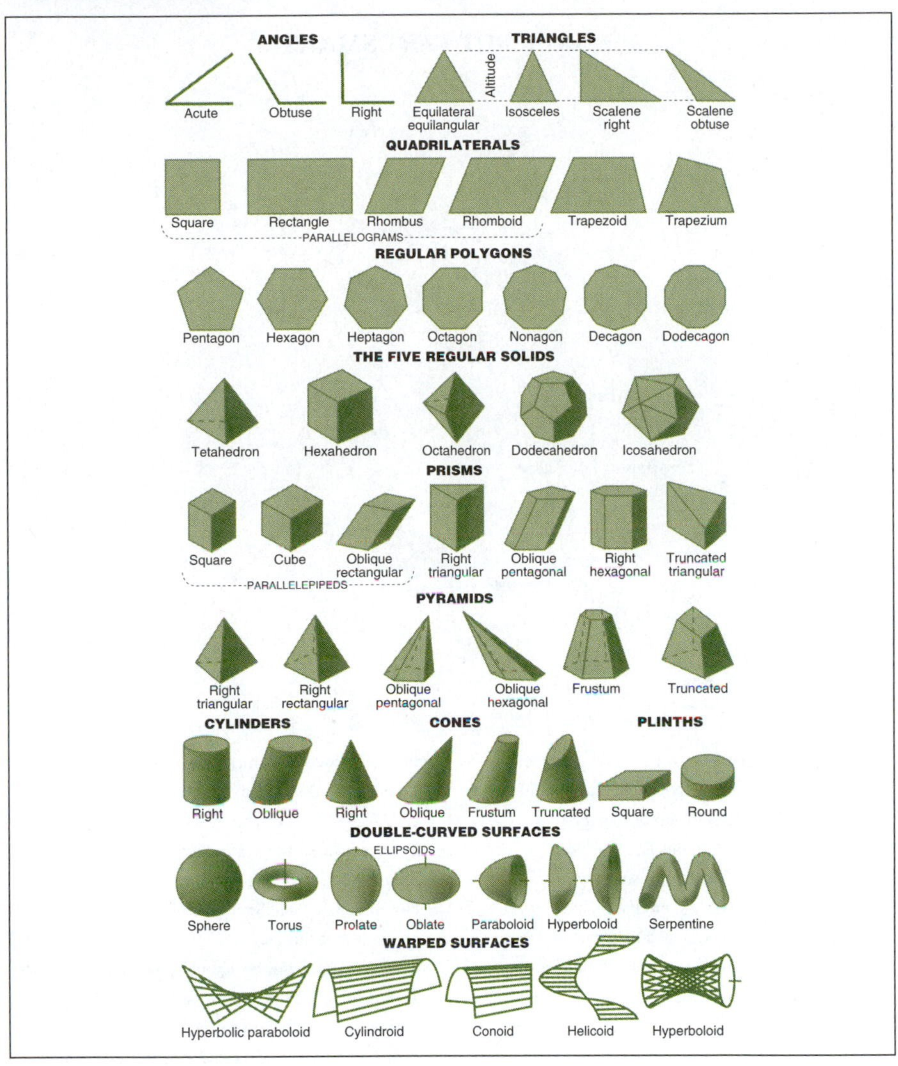

features. In organizing the rest of the description, you could use detailed views, such as phantom or cutaway views, to show the location of parts, as well as enlarged drawings, which could be placed adjacent to the text that describes them. Figure 15.14 on page 567 suggests that different types of visuals can be used to illustrate the exterior, the interior, and individual components as they relate to the whole. (Chapter 12 discusses visuals in more detail.)

FIGURE 15.13 Visual and Verbal Metaphor in Technical Text[19]

How effective is the visual metaphor of a playing card to illustrate the relative size of MEMS?

What specific examples of figurative language such as metaphors, similes, and analogies can you identify in this article?

How does the figurative language help readers understand the features of MEMS?

SMALL BUT *VERY* SMART

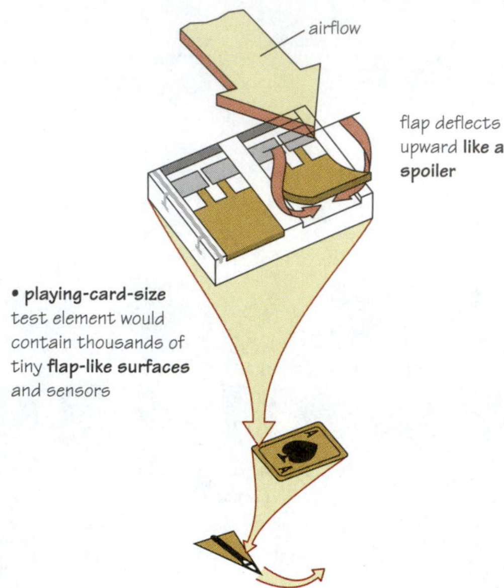

Smart is an engineering buzzword . . . ; there are smart cars, smart highways, smart bombs, smart TVs, and, if a group of University of California at Los Angeles and Cal Tech researchers can work it out, smart wings.

These scientists and engineers have developed a "microelectromechanical system" that, instant by nano-instant, alters the airflow over a wing to maintain laminar flow. Whenever microscopic sensors detect the changes in airflowthat foretell an incipient burble, minute tabs that function as tiny spoilers bend upward into the airflow to create counter-burbles that cancel out the boundary layer separation.

Microelectromechanical systems—let's henceforth revert to the official acronym MEMS—include devices (flaplets, in this case) so small that thousands of them can be built into microchips that also house the controlling sensors as well as the actuators that activate the surfaces themselves. (We're talking *way* tiny here: Think of sequins too small for a Barbie doll's cocktail dress. The UCLA/Cal Tech team foresees important uses for a version of the system implanted in the human body to detect and correct the adverse blood flowthat can lead to arterial clotting, among other things.)

If you have doubts about systems nearly as delicate as butterflywings surviving on an airliner in today's air transportation environment, it helps to understand that the research project is largely driven by the needs of future military aircraft. For stealthy airplanes, MEMS arrays that maintain laminar flow could also be used as substitutes for conventional movable control surfaces, which can create radar signatures when they're deflected. Deflect a bunch of MEMS on one wing or the other, say, and considerable lift asymmetry can be created, literally invisibly, rolling the airplane just the way an aileron would.

FIGURE 15.14 **Visuals for Technical Descriptions**

Purpose of Visual	Selected Visual to Use in Technical Descriptions
Visuals to give an overview	■ photographs ■ realistic drawings ■ topographic or contour maps
Visuals to describe interior components, to give an image of the way the parts fit together	■ phantom views (drawings that depict an exterior surface as transparent so the inside structure can be viewed) ■ schematics and wiring diagrams ■ cross-section maps
Visuals to describe individual parts in relation to the whole, to give an image of each individual component	■ exploded views (drawings that separate all the components and display them in the proper sequence and relationship for assembly) ■ cutaway views (drawings that slice a section out to show a full or partial cross-section) ■ blueprints ■ photographs or drawings of individual parts
Visuals to show patterns	■ photographs ■ videos ■ realistic drawings ■ pairs or groups of photographs or drawings

What kinds of visuals have you seen used in technical descriptions?

Some concepts and mechanisms are extraordinarily difficult to explain without the use of visuals. One example is the subject of the recent book *Self-Organization in Biological Systems,* which explores "diverse pattern formation processes in the physical and biological world."[20] What kinds of patterns? Zebra stripes. Stamen clusters. Designs on butterfly wings. The visual images provide explicit descriptions, as shown in the photos below and on the next page.

William L. Jeffries © 2002

The primary author of *Self-Organization in Biological Systems,* Scott Camazine, describes self-organization as "the various mechanisms by which pattern, structure and order emerge spontaneously in complex systems . . . the pattern of sand ripples in a dune, the coordinated movements of flocks of birds or schools of fish, the intricate earthen nests of termites, the patterns on seashells, the whorls of our fingerprints, the colorful patterns of fish and even the spatial pattern of stars in a spiral galaxy."[21]

William L. Jeffries © 2002

WEBLINK

Self-organization first catches our attention because of the amazing visual patterns. Then it captures our attention because of the mathematical "rules" that can be followed to simulate these patterns. For a link to an article in *Natural History* about this way of describing patterns and for extraordinary still and animated images that illustrate the principles, go to http://www.english.wadsworth.com/burnett6e.

CLICK ON WEBLINK

CLICK on Chapter 15/self-organization

Organization

When preparing a technical description, you have to make decisions about the sequence of information. Writers conventionally organize technical descriptions in spatial order to give the audience a clear view of appearance and structure. Occasionally, writers use chronological order, describing the components in order of assembly, or use priority order, describing the components in order of importance. Figure 15.15 outlines a conventional sequence of information you can use to plan a detailed technical description.

FIGURE 15.15 **Planning a Technical Description**

Title

1.0 Define the object (or substance, mechanism, organism, system, or location) in the introduction.

1.1 Define the object. Identify whose perspective the definition is from.

1.2 Identify the purpose of the object. Indicate the importance or impact of this purpose.

1.3 Describe the characteristics of the whole. Indicate acceptable tolerances or specifications.

1.4 Present a visual that provides an overall view of the object.

1.5 Identify the parts of the object.

2.0 Present a part-by-part description arranged in order of the parts' assembly, location, or importance.

2.1 Describe part one.

2.1.1 Define the part.

2.1.2 Identify the purpose of the part.

2.1.3 Describe the general appearance of the part (including a visual if useful).

2.1.4 Describe the characteristics of the part. Indicate acceptable tolerances or specifications.

2.1.4.1 Identify the general shape and dimensions.

2.1.4.2 Identify the material type and characteristics (color, flammability, optical properties, solubility, density, conductivity, magnetism, and so on).

2.1.4.3 Identify the surface treatment, texture.

2.1.4.4 Identify the weight.

2.1.4.5 Identify the method of manufacture.

2.1.4.6 Identify the subparts of part one.

2.1.5 Describe its attachment to other parts.

2.2 Describe part two.

. . . and so on

3.0 Conclude the description.

3.1 Explain how the parts fit together.

3.2 Explain how the parts function together. Explain what criteria are used to establish effectiveness.

What visuals would be appropriate for various sections of a technical description?

Technical descriptions should have a title if they are printed as a separate document or a section heading if they are incorporated into one section of another document. The introductory section usually begins with a definition suitable for the intended audience. The definition can include or be followed by a statement of the purpose or function of the document, as in this example:

> As the new owner of a wood-burning stove, you should be familiar with its structure and components. This information will help you safely maintain your stove as a supplemental source of home heat.

The introductory material presents an abbreviated version of the description. It includes characteristics of the whole: overall shape and major dimensions, primary color and texture, and any distinctive aspects. A photograph or realistic drawing often supplements this overall description. The final part of the introductory section partitions the whole into its major parts, in the order they will be described. This partition can be illustrated with an exploded or cutaway view.

To appeal to a particular audience, you may incorporate into the introductory material elements that increase audience interest and background knowledge but do not add substantively to the technical content. Keep in mind that experts are usually annoyed by the inclusion of what they consider extraneous information. However, if you are writing for or speaking to a general or nonexpert audience, consider some of these elements that may add interest or appeal to the introductory section:

- *Background information:* What is the history? What are current developments?
- *Parts–whole relationships:* Where does the object fit in relation to similar ones?
- *Qualitative distinctions:* What separates it from similar objects?

The body of a technical description involves a part-by-part description arranged in order of location, assembly, or importance. Each section of the body follows the same format. Initially, the part, and sometimes its purpose, is defined. Then a description of the general appearance of the part, including shape, major dimensions, and material, follows, often accompanied by a visual presenting detailed dimensions. The outline in Figure 15.10 identifies additional characteristics that are relevant for some audiences. Specifics are added according to the needs of the audience.

An architect designing a passive solar house would want information about surface treatments, optical and insulating properties, and weights of specially treated glass. An interior designer would be more concerned with color and texture. Both would be interested in subparts and methods of attachment to other parts.

The conclusion explains how the parts fit and function together. Just as you can stimulate reader interest in the introduction, you can also create a more lively conclusion by including some of these elements:

- *Applications:* How is it used?
- *Anecdotes or brief narratives:* Who uses it?
- *Advantages/disadvantages:* What are the benefits and/or problems?

Often a technical description does not have a concluding section, but simply ends when the last part has been fully described.

Individual and Collaborative Assignments

1. **Identify parts for analysis.** (a) Identify ways to describe the following subjects. Each can be separated into a number of different parts or subsystems. (b) Add one subject from your own profession to each category and identify its parts or subsystems, too.

   An organism such as a wolf can be separated in a number of ways. Some ways include by cellular composition; by categories of fluids, such as blood; by categories of systems, such as muscles; by categories of physical functions, such as reproduction and respiration; and by categories of ecological functions, such as controlling the elk population.

Subjects	Examples	Parts or systems
Objects	■ golf ball ■ wood screw	
Mechanisms	■ carbon monoxide detector ■ kidney	
Substances	■ granola ■ ocean water	
Systems	■ immune system ■ workflow in office	
Organisms	■ decathlon athlete ■ experimental variety of corn	
Locations	■ site of a new building ■ protected wetland	

2. **Distinguish between structural and functional parts.** Identify the structural and functional parts for one column of the following items. Both word lists and diagrams will help you identify the distinctions.

AAA battery	comb	light bulb	pocket lighter
ballpoint pen	field daisy	magnet	pocket knife
baseball	flashlight	oxygen tank	scissors
candle	hand saw	pinecone	zipper

3. **Choose appropriate diction for specificity.** Some descriptions might be inappropriate because they're negative or vague.

   (a) Under what circumstances might the following descriptions be inappropriate? Suggest possible alternatives that would provide more detailed and accurate descriptions.

   - big equipment shed
   - cheap replacement part
   - cold weather complicated step
   - easy-to-assemble desk
   - fast microprocessor
   - fast photocopier
   - fat patient
   - hard surface
   - recent decision
   - sharp angle

   (b) What other general description terms can you think of that might be inappropriate? In what circumstances?

4. **Consider audience needs.** Imagine that you work for a company that manufactures modular houses and sells directly to individual customers. Several publications are being prepared, and you have been assigned responsibility for creating descriptions of the modular houses. How would you identify the parts of the modular houses for the following audiences?

   - government agency with information about fuel-efficient construction methods
   - production line manager
   - architect
   - lumber wholesaler
   - municipal wiring and/or plumbing inspector
   - prospective homeowner

5. **Choose visuals for certain descriptions.** Several kinds of visuals are described in this chapter. Which would be appropriate to incorporate into a description of each of the following items? What would make one kind of

visual more appropriate than another kind for each item — considering, for example, context, purpose, and audience?

- apple, from bud to harvest
- computer mouse
- condominium complex
- crankshaft
- espresso coffee maker
- gas well
- human leg
- landfill
- mints: catnip, spearmint, peppermint, applemint
- paint sprayer
- reclining chair or geriatric chair
- septic tank and leach field
- silicon chip
- snowblower
- sprocket gears (as on a bicycle)

6. Analyze technical descriptions.

(a) Bring two examples of technical descriptions to class. If possible, locate one description in a U.S. publication and the other description from a publication outside the United States.

(b) Work with classmates in a small group to conduct a preliminary analysis of two of the technical descriptions your group brought in, considering at least these points:

- *Context.* What assumptions seem to have been made about the context in which the descriptions will be used?
- *Audience.* For what audience is each intended? Has the writer accurately analyzed the audience(s)?
- *Visuals.* Are the visuals appropriate and helpful? Are additional visuals needed? Do the visuals differ in descriptions from different cultures?
- *Format.* Does the format make the descriptions easy to read? Does the form differ in descriptions from different cultures?

(c) Based on your preliminary analysis, create an evaluation rubric that shows how you've incorporated the criteria discussed above as well as other criteria that you and your classmates have determined are helpful in analyzing and assessing technical descriptions.

(d) Test and refine the rubric by using it to conduct a complete analysis and evaluation of at least two other descriptions you and your classmates brought in.

(e) Give a brief oral presentation to the class about the criteria you selected and the assessment of one of your group's descriptions.

7. **Select two visuals.** Writers and designers frequently decide to use more than one visual in combination to convey information to readers. The figure below[22] shows a cross section of an eye and then shows an enlargement of a segment of that cross section. The cross section shows the basic structure; the enlargement shows how these structures work. The caption explains to readers how they can interpret the drawings.

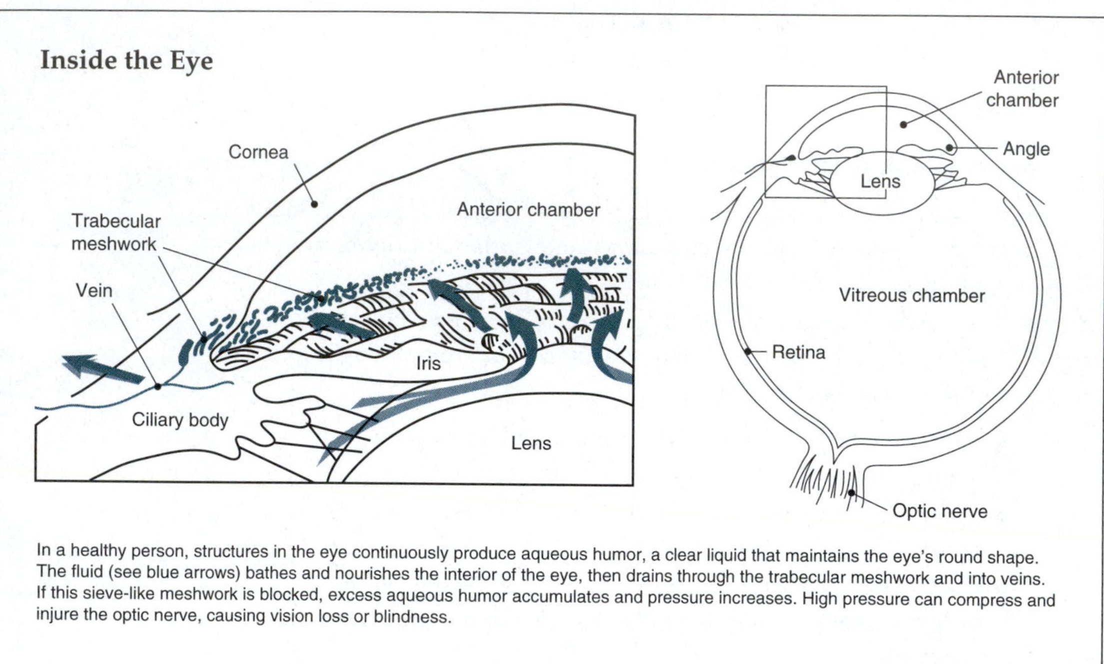

In a healthy person, structures in the eye continuously produce aqueous humor, a clear liquid that maintains the eye's round shape. The fluid (see blue arrows) bathes and nourishes the interior of the eye, then drains through the trabecular meshwork and into veins. If this sieve-like meshwork is blocked, excess aqueous humor accumulates and pressure increases. High pressure can compress and injure the optic nerve, causing vision loss or blindness.

Select a concept or mechanism that is important in your discipline. Locate two figures from the Web that work together to describe the concept or mechanism (although you may not have found them together). Write a clear, succinct description that can serve as a caption for the two visuals. (Make sure to provide appropriate citations.)

8. Write a technical description.

(a) Select an object, mechanism, substance, organism, or location from the following lists, or choose another subject that relates to your field:

Objects	*Mechanisms*	*Substances*
electrical cable	cider press	photographic developer
drill bit	pool filter	acetylene
shotgun shells	spinning reel	baking powder
contact lenses	spinning wheel	baby food
mallet	camera lens	blood sample
gable roof	combination lock	measles vaccine
computer chip	solar panel	yogurt
polarizing filter	smoke alarm	plant fertilizer
photographic film	semicircular canals	effluent
golf club	transit	cough medicine

Systems	*Organisms*	*Locations*
respiratory system	termites	a harbor mooring
photovoltaic system	dolphins	cross section of well site
braking system	tapeworms	R&D section of a plant
electronic auto inspection system	algae	forest marked for logging
	chickens	layout for vegetable garden
planetary system	yeast	geologic fault
HVAC system	mosquitoes	archaeological excavation
irrigation system	foxgloves	traffic cloverleaf pattern
photocopying machine	pumas	underground storage tank
scrubber for contaminants	protozoa	runway grid for airport

(b) Modifying the format outlined in Figure 15.15, write a description that appropriately considers context, audience, and purpose of your description.

(c) After writing your description, work in a small group to compare and contrast it with those written by other students. Examine the organization, selection of details, adjustment for audience, and use of visuals. Explain your reasons for the choices you made in designing and developing your description.

Chapter 15 Endnotes

1 Galileo, G. (1960). The sidereal messenger. In F. R. Moulton & J. J. Schifferes (Eds.), *The autobiography of science* (2nd ed., pp. 65–76). Garden City, NY: Doubleday.

2 NASA. (n.d.). PIA00600: Family portrait of Jupiter's great reds and the Galilean satellites. *Planetary photojournal.* Retrieved November 15, 2003, from http://photojournal.jpl.nasa.gov/catalog/PIA00600

3 NASA-Jet Propulsion Laboratory. (1979, February). Mission status report. *Voyager Bulletin, 36*(23), 3.

4 Science@NASA. (n.d.) Io's alien vocanoes. Retrieved November 16, 2003, from http://science.nasa.gov/newhome/headlines/ast04oct99_1.htm

5 NASA. (n.d.). *Galileo: Journey to Jupiter.* Retrieved November 15, 2003, from http://galileo.jpl.nasa.gov/

6 NASA. (2003, September 17). *Galileo end of mission status.* Retrieved November 15, 2003, from http://www.jpl.nasa.gov/galileo/news/release/press030921.html

7 Dombroski, P. (1995). Can ethics be technologized? Lessons from Challenger, philosophy, and rhetoric. *Technical Communication Quarterly, 38*(3), 146–150.

8 Parkman, R. (1997). Management of the newborn. In J. W. Graef & T. E. Cone, Jr. (Eds.), *Manual of pediatric therapeutics* (pp. 99–100). Boston: Little, Brown and Company. Reprinted with permission of Little, Brown and Company.

9 Stanley PROTO. (n.d.). *Inspection mirror.* Retrieved November 15, 2003, from http://www.stanleyproto.com/default.asp?CATETORY=INSPECTION+MIRRORS&TYPE=PRODUCT&PARTNUMBER=2374&Sdesc=1%2D1%2F4%26quot%3B+Circular+Inspection+Mirror

10 Cocozziello, D. Circular inspection mirror. *Technical and Scientific Writing,* 42.225. University of Massachusetts at Lowell.

11 *The common gateway interface.* (n.d.). National Center for Supercomputing Applications at the University of Illinois at Urbana-Champaign. Retrieved November 15, 2003, from http://hoohoo.ncsa.uiuc.edu/cgi/

12 *Overview.* (n.d.). Retrieved November 15, 2003, from the University of Illinois at Urbana-Champaign, National Center for Supercomputing Applications Web site: http://hoohoo.ncsa.uiuc.edu/cgi/intro/html

13 *How do I get information from the server?* (n.d.). Retrieved November 15, 2003, from the University of Illinois at Urbana-Champaign, National Center for Supercomputing Applications Web site: http://hoohoo.ncsa.uiuc.edu/cgi/primer.html

14 Ruud Air Conditioning Division. (2002). *Achiever 90 plus modulating gas furnace with contour comfort control* (No M22-6003). [Brochure]. Fort Smith, AR: Ruud.

15 Danish Wind Industry Association. (2003). Home page. Retrieved November 15, 2003, from http://www.windpower.org/index.htm

16 Danish Wind Industry Association. (2003). Contents in Danish. Retrieved November 15, 2003, from http://www.windpower.org/da/core.htm

17 Danish Wind Industry Association. (2003). Contents in English. Retrieved November 15, 2003, from http://www.windpower.org/en/core.htm

18 French, T. E., & Vierck, C. J. (1970). *Graphic science and design* (p. 79). New York: McGraw-Hill. Reprinted with permission of McGraw-Hill Book Company.

19 *Air & Space.* (1995, June/July). p. 35.

20 *About the book.* (n.d.). (n.a.). Retrieved November 15, 2003, from http://www.scottcamazine.com/personal/selforganization/about.htm

21 Camazine, S. (n.d.). *Self-organization in biological systems.* Retrieved November 15, 2003, from http://www.scottcamazine.com/personal/research/index.htm

22 Illustration by Harriet Greenfield, in Dryer, E. B. (1994, October). Preserving eyesight with foresight. *Harvard Health Letter,* 19(12), 4–6. Copyright © 1994, President and Fellows of Harvard College. Reprinted by permission.

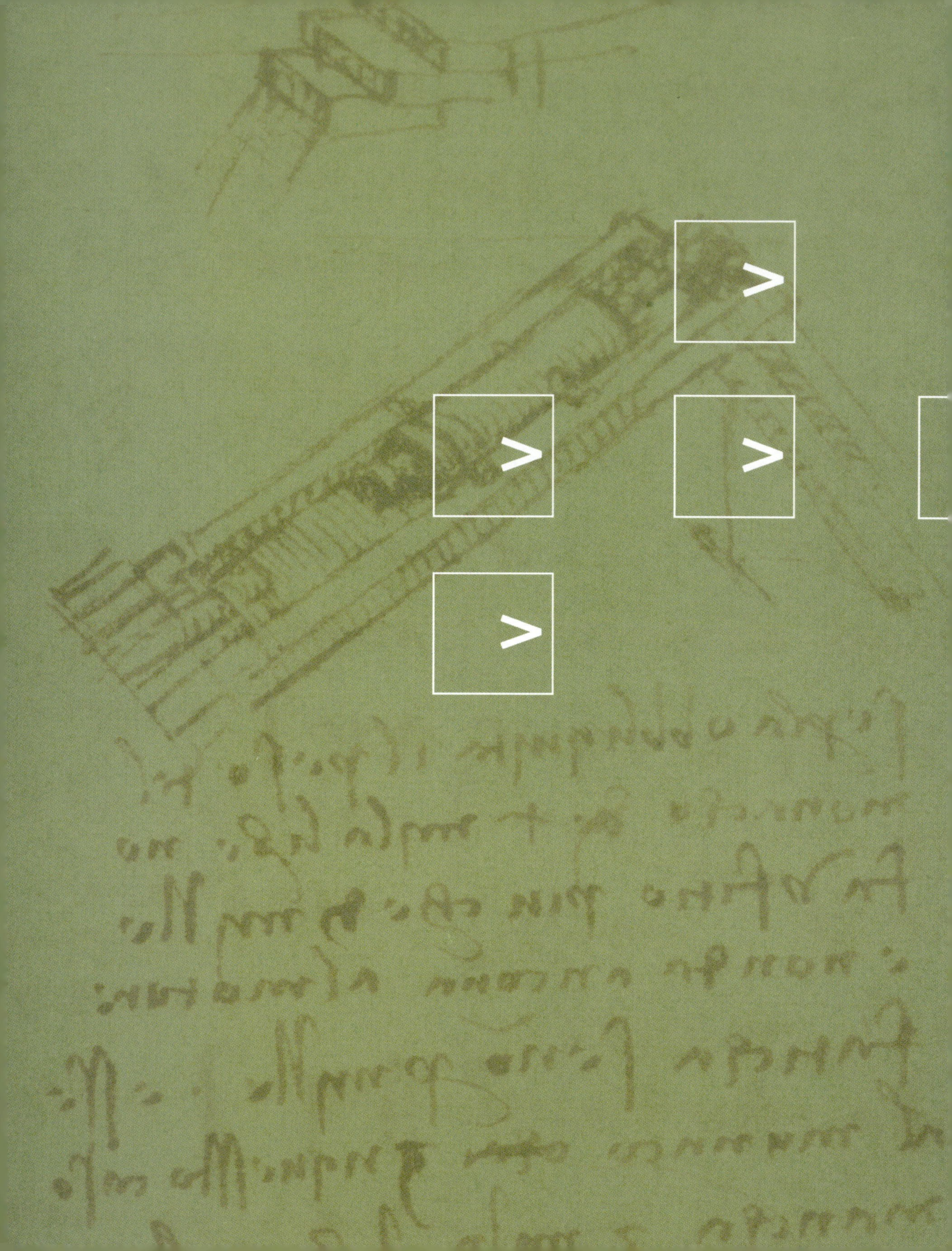

CHAPTER 16

Creating Process Explanations

Objectives and Outcomes

This chapter will help you accomplish these outcomes:

- Understand that process explanations present an overview of sequential actions in chronological order
- Use process explanations as part of larger documents, including manuals, orientation and training materials, marketing and promotional materials, and public information
- Use a conventional sequence of technical description, process explanation, and benefits or advantages that an audience can use
- Prepare effective process explanations by following these steps:
  - Identify your audience and its reasons for needing the information
  - List the steps of the action
  - Choose visuals to illustrate the sequence: flowcharts, timelines, schedules, drawings showing each element in a process, time-lapse photographs, drawings with overlays of changes, drawings showing the final product, and sequential drawings
  - Choose active or passive voice based on audience and purpose
  - Understand and use the appropriate organization and format

Process explanations play an important role in technical communication by providing information about the sequence of steps in any action, from blood donation to operation of a jet engine. Generally, *process explanations* (also called process descriptions) provide an overview or background, regardless of the audience's specific tasks. They are often embedded in a longer discussion that has already presented a definition and general description, as in this example taken from a U.S. Department of Energy (DOE) Web site about the Wind Energy Program:

> So how do wind turbines make electricity? Simply stated, a wind turbine works the opposite of a fan. Instead of using electricity to make wind, like a fan, wind turbines use wind to make electricity. The wind turns the blades, which spin a shaft, which connects to a generator and makes electricity. Utility-scale turbines range in size from 50 to 750 kilowatts. Single small turbines, below 50 kilowatts, are used for homes, telecommunications dishes, or water pumping.[1]

What works in process explanations? Analogies the audience can understand, statement of purpose, explanation of actions, and relation of equipment to application. All of these strategies in the short DOE process help explain how wind turbines make electricity.

Sometimes process explanations don't need to be any longer or more complicated than the example above, but workplace professionals usually need more detailed, technical information. For example, managers often read process explanations in marketing brochures to help them make purchasing decisions. Supervisors often read process explanations, like those in many manuals, to gain an understanding of a process they're responsible for but don't actually do themselves. Technicians and operators are usually encouraged to read a process explanation before following the directions to actually conduct a process. General audiences find that process explanations satisfy their curiosity about many things — how wine is made, how hurricanes are tracked, how oil wells are drilled.

WEBLINK

For a link to a terrific Web site, howstuffworks.com, go to **http://english.wadsworth.com/burnett6e.** The site is frequently updated and has very interesting process explanations.
CLICK ON WEBLINK
CLICK on Chapter 16/howstuffworks

Defining Processes

Process explanations explain sequential actions to members of an audience who need enough details to understand an action or process, but not enough to necessarily enable them to complete it. The following example of one step in a mechanical inspection process illustrates the difference between a process explanation and directions (which you'll read more about in Chapter 21). The process explanation identifies the general nature of the task; it is valuable precisely because it provides an overview rather than focusing on the details. In contrast, this step of the directions that enables a mechanical inspector to complete the task is very specific:

Process Explanation

A mechanical inspector initially ensures that the labels on the packages are correct.

Directions

Ensure that the computer-printed label contains all of the following information:

- customer contract number
- contract annex and line item
- part number
- nomenclature
- NSN (national stock number)
- quantity
- date packaged
- serial number (if applicable)

Accurate, accessible processes are a large part of what makes an organization function safely and legally. When the processes are inaccurate or inaccessible, problems arise. Most organizations work diligently to keep processes accurate and up to date; in fact, many organizations have built-in reviews to assure that all processes are current. However, some organizations have flawed processes — sometimes by accident, sometimes on purpose — that put people at risk. When an organization ignores such problems, some people believe they have an ethical, professional responsibility to make the problem public. These whistleblowers, now protected by law, make the workplace safer for everyone, but usually at great cost to themselves. The ethics sidebar on the next page addresses the risks and benefits of whistleblowing.

Visual inspection is a critical part of quality control/quality assurance (QC/QA). Not only do supervisors and managers need process explanations that provide an overview of the requirements and parameters of the inspection process, but the inspectors themselves need instructions that provide the specific steps required to implement rigorous inspection.

ETHICS SIDEBAR

Whistleblowers: Ethical Choices and Consequences

Would you be able to face the potential scorn and repercussions of being a whistleblower? Would you be compromising your personal ethics to keep information secret you felt the public should know? How do you make such a difficult ethical decision?

How would you write a process explanation for a process with potentially harmful aspects?

Dealing with the consequences of an ethical conflict can be difficult, especially when the conflict arises because of a difference between an employee's personal values and an organization's goals. Some technical professionals, faced with workplace situations they find unethical, decide their only viable option is to reveal corporate secrets to the public. These professionals, referred to as whistleblowers, jeopardize their career and personal safety because they believe the public's right to know supersedes corporate obligations.

For ethical, legal, or financial reasons, whistleblowers publicly expose a company's internal secrets. Movies like *The China Syndrome, Silkwood, The Insider,* and *Erin Brockovich* dramatically indicate the extreme repercussions some whistleblowers have faced. In these movies and in the actual situations on which these movies were based, the whistleblowers lost their jobs, found their careers finished, and faced personal assaults. While these movies dramatized extreme examples, many whistleblowers deal with less dramatic but still very important issues. Beyond the probable loss of a job, many whistleblowers have a difficult time getting hired at other companies within the same field and often must change careers.

So, faced with these negative consequences, why would a technical professional choose to be a whistleblower? Legal regulations provide one answer: a technical professional has a legal obligation to inform the public of potential harm. Researcher Carolyn Rude refers to this legal obligation as the "duty of due care":

> Technical communicators could be negligent, legally and ethically, if they knew that a product being documented could be hazardous, knew of the responsibility to provide clear instructions and adequate warnings but did not make an effort to do so, or to investigate hazards in the use of the product. (p. 179)[2]

Legal requirements are not the only factor behind the decision to become a whistleblower. Technical professionals may also respond from a sense of responsibility to the community. Reporter Todd Crowell of the online news service *Asia Now* believes whistleblowers "can be the catalysts for needed change" within a company.[3] Crowell believes that a whistleblower could have prevented Japan's worst nuclear power plant accident. Workers at the plant used an unauthorized operations manual that sacrificed safety for expedience; using this manual, employees accidentally set off a uranium chain reaction in 1999 that killed one worker and exposed hundreds of citizens to dangerous levels of radiation. Crowell believes the accident could have been avoided if someone had been willing to expose the improper procedures being used.

Making the decision to reveal company secrets is not easy, and it does not come without consequences. However, the consequences of not revealing information can be equally troubling.

To read more about whistleblowers and the issues involved, go to http://english.wadsworth.com/burnett6e for relevant links.

CLICK ON WEBLINK

CLICK on Chapter 16/whistleblowers

WEBLINK

Using Process Explanations

Process explanations often appear in the same kinds of documents as technical descriptions. You can decide whether to include a process explanation and, if so, what kind of details to incorporate by examining the purpose and task of your document. Will the process explanation help accomplish your purpose? Will it help the reader understand the process? The following discussion identifies and illustrates some of the common applications of process explanations: manuals, reports, orientation and training materials, marketing and promotional materials, and public information and education.

Reports

Reports — whether print documents or online help systems — frequently provide the audience with background information for understanding critical technical processes. The following sequence of information is typical, especially in a report's introductory section:

- *Technical description* — what a mechanism is
- *Process explanation* — how it works
- *Benefits or advantages* — why it's useful

This sequence is a particularly important part of documentation accompanying equipment that may be unfamiliar to either the user or manager.

The example in Figure 16.1 is from a type of short, regularly published report: a U.S. Geological Survey Fact Sheet. This particular fact sheet describes the testing process used by the Cone Penetration Testing (CPT) truck, which is used in geologic-hazard, hydrologic, and environmental studies. As a rapid and cost-effective approach, CPT is particularly useful in urban environments.[4] This USGS fact sheet balances text about the overview, process, and benefits with a simple and effective figure, a photograph, a sounding chart, and a diagram illustrating the relationship among the equipment, process, and output.

Task Manuals

One of the most frequent uses of process explanations is in task manuals. While the primary purpose of such manuals is to provide clear, step-by-step instructions to complete a task and to include cautions to avoid problems, users often

FIGURE 16.1 **Process Explanation Combining Textual and Visual Information**[5]

Accessible

- Information is chunked in relatively short paragraphs.

Comprehensible

- Critical technical vocabulary is defined or explained.

Usable

- A specific title identifies the focus.
- Headings identify the categories of information.
- Benefits are clearly identified, making decisions about use much easier.

Subsurface Exploration with the Cone Penetration Testing (CPT) Truck

The U.S. Geological Survey Cone Penetration Testing (CPT) truck is a fast and inexpensive way to conduct shallow subsurface exploration. Detailed data are available immediately, permitting on-the-fly mapping of stratigraphy and other subsurface features. CPT is a useful tool in geologic-hazard, hydrologic, and environmental studies. This rapid and cost-effective approach is particularly advantageous in urban environments because no drill spoils are produced.

Overview. Cone penetration testing (CPT) permits rapid exploration of shallow (less than 30 meters) subsurface conditions while minimizing retrieval of subsurface materials, an inconvenient and occasionally expensive byproduct of conventional drilling. CPT uses sensors that are pushed into the ground to infer the properties of both soils and pore fluids. Known as direct-push technology, this method can map the vertical and lateral extent of stratigraphic layers as well as the distribution of subsurface contaminants. Standard engineering correlations allow the geotechnical properties of stratigraphic layers to be inferred.

Process. A CPT sounding is made by pushing a small probe into the ground. Typically, a 3.6-centimeter-diameter probe (cone) is pushed into the ground to depths ranging from 15 to 30 meters. The cone is advanced downward at a constant velocity of 2 centimeters per second, using hydraulic rams that apply the full 23-ton weight of the CPT truck to push the probe rods to depth, as shown in Figure 1. In typical CPT soundings, the resistance to penetration is measured. Continuous measurements are made of the resistance to penetration of the tip and the frictional sliding resistance of the sleeve of the cone. The penetration resistance, which is digitized at 5-centimeter depth intervals, permits detailed inferences about stratigraphy and lithology. Soil type is inferred from a chart that compares these two measurements with the known physical properties of various soils.

Benefits. CPT is a much more rapid and cost effective approach than conventional drilling for shallow subsurface exploration. Typically, four to five 15- to 30-meter-deep soundings per day can be accomplished, in contrast to one or two per day with conventional drilling and sampling. Soundings also have the great advantage of not producing any drill cuttings, spoils, or fluids. This aspect is particularly advantageous where subsurface contaminants are present or suspected. Data are automatically logged onto a rugged field computer and are ready for immediate viewing and analysis in the field. CPT is a reliable and efficient method for stratigraphic profiling and obtaining soil-engineering parameters for geotechnical design, as well as being widely accepted and encouraged by regulators as an effective environmental-investigation technology.

FIGURE 16.1 **Process Explanation Combining Textual and Visual Information[5] (continued)**

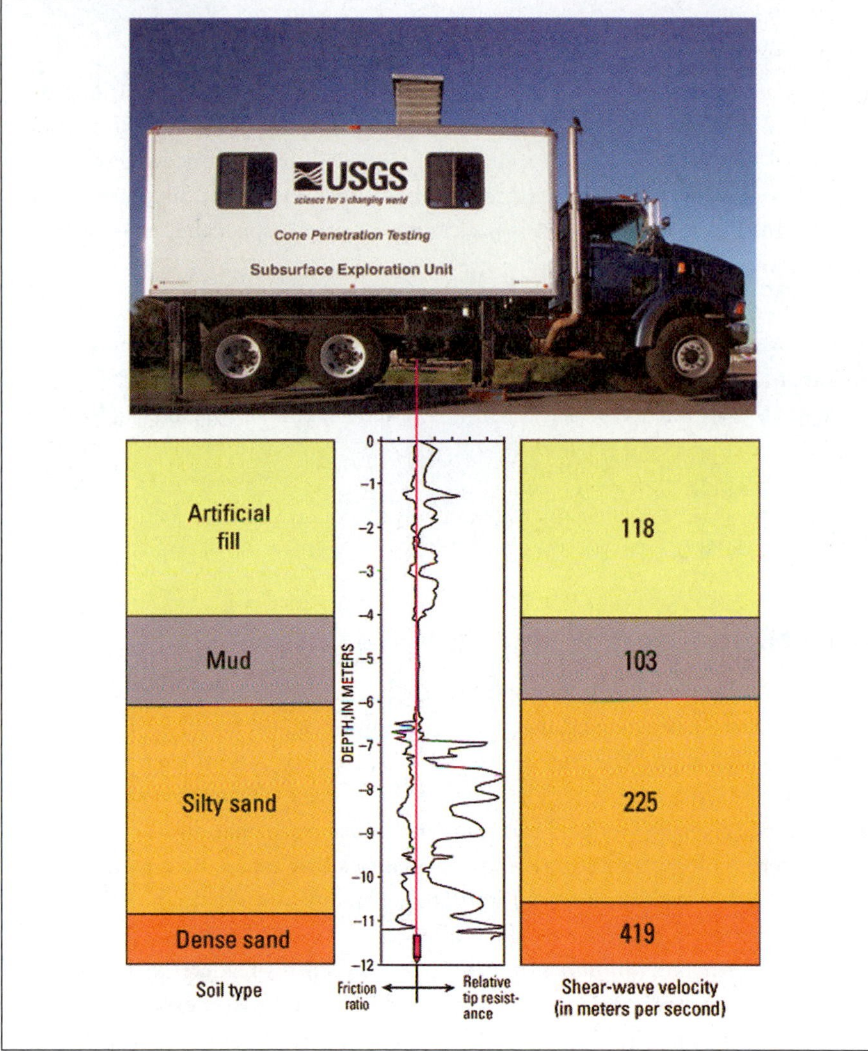

Accessible

- Color is used effectively in the accompanying figure to identify the probe and to differentiate the soil types of comparable shear-wave velocities.

Comprehensible

- The accompanying figure effectively incorporates a photograph, sounding chart, and diagram to show how the equipment, process, and output are related.

complete complex tasks more accurately and more cooperatively if they understand the overall process. A process explanation preceding step-by-step instructions is worth the time to prepare and the space to present.

Flowcharts are an effective way to present such overviews, as Wendy Phillips, a technical communicator at SYNERGEX.com, explains:

> I consider flowcharting processes very important to the documentation of any product. If you write programmer reference guides, you need to flowchart both

> how the software works and how it fits together with other software tools. For end-user documentation, you need to flowchart the process of how the user uses the software. . . . Creating a flowchart will clarify your thoughts, help you get organized, and create a deliverable you can check with your SMEs (to be sure you've grasped the big picture about how the product works).[5]

Software is not the only product that benefits from process explanations, as Figure 16.2 illustrates. Rather than a textual explanation, Figure 16.2 presents a Web-based flowchart — a single example of a quality control process description used by companies that are ISO (International Organization for Standardization) certified. In this case, the process is packaging, a critical aspect that can make or break a company.

WEBLINK

To access a basic tutorial about flowcharting, go to http://english.wadsworth.com/burnett6e. The tutorial reviews types, tips, symbols, and suggestions for analysis.

CLICK ON WEBLINK

CLICK on Chapter 16/flowcharting tutorial

Orientation and Training Materials

Although managers frequently appreciate simple, straightforward process explanations, students may need more detailed information because they are often expected to understand the reasons behind a sequence of actions even when the information is just a summary. The illustrated summary of a natural process in Figure 16.3 is given to students in a seminar about bog formation; the article contains terminology, definitions, and explanations that make the material inappropriate for readers interested in more general information, which is typical for most orientation and training materials.

The writer makes a number of assumptions about his readers' background knowledge but still adheres to the format of an effective process explanation. He begins with a broad definition of a bog before summarizing the sequence of its development; however, because he assumes that readers have some prior knowledge, he does not define all technical terms. The writer uses chronology to explain the development but also orients readers with precise spatial references.

Marketing and Promotional Materials

Some actions are far easier to delineate than bog development. For instance, Figure 16.4 presents an example written primarily for professionals; it explains the operation of a thermal inkjet cartridge for computer printers. This example provides a good illustration of the way in which process explanations are often

FIGURE 16.2 Visual Overview of a Process[6]

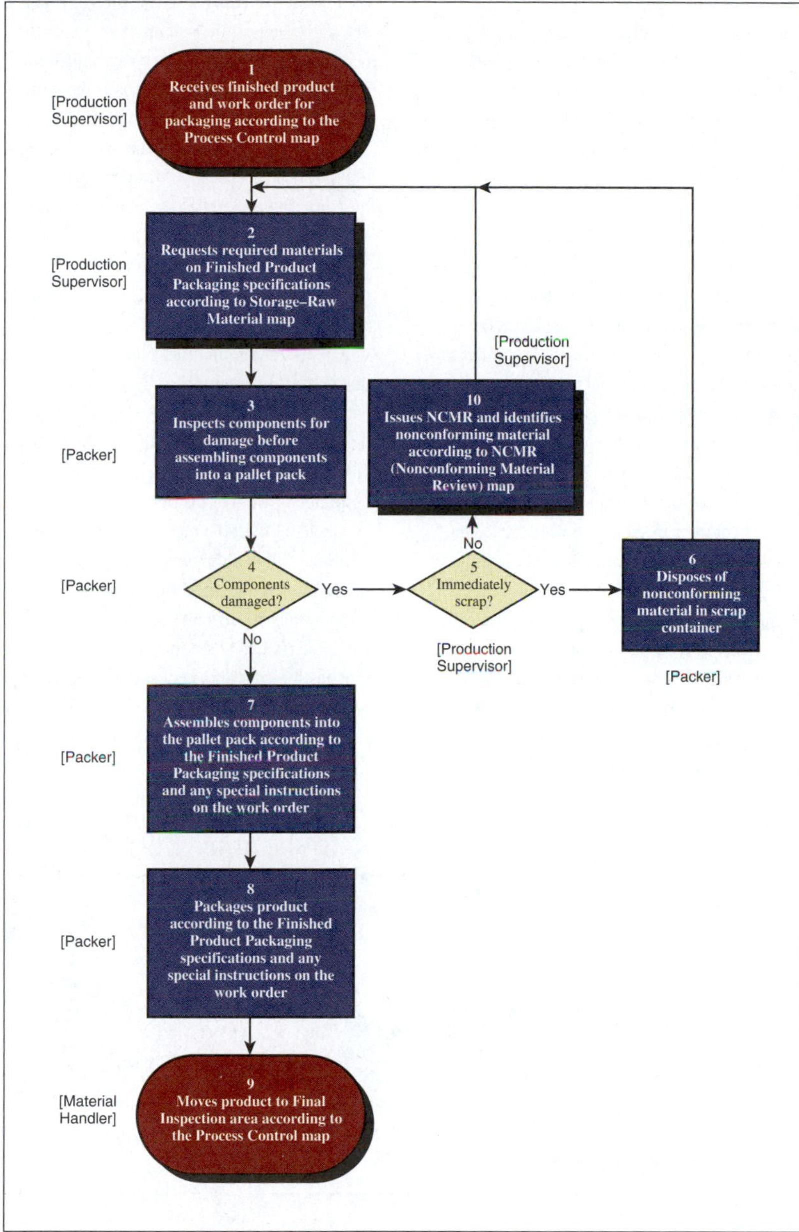

Accessible

- *Conventional flowchart shapes and bright colors differentiate the steps.*
- *The figure-ground contrast makes the text easy to read.*

Comprehensible

- *The flowchart templates — for example, a rectangle is a process step; a diamond is a decision box — are recognizable by most professionals.*
- *Arrows remove any ambiguity about the flow of the process.*
- *Identifying the people associated with each step makes individual responsibilities clear.*
- *The flowchart defines the overall scope of this part of the quality control process.*

Usable

- *The four shadowed boxes are linked to other parts of the process, so clicking on those boxes moves users to a related flowchart.*
- *The use of a flowchart and active links assumes the audience has specialized prior knowledge.*

FIGURE 16.3 Illustrated Summary of a Natural Process[7]

Accessible

- *The two-column layout creates short lines that make the information accessible.*
- *The elements in the pen-and-ink drawing are legible.*

Comprehensible

- *Explanations help readers understand the origin of common terms such as "quaking bog" and phenomena such as the high level of acidity of bog water.*
- *The chronology of the process is clearly signaled to readers by phrases such as "succession begins" and "growth proceeds."*

Usable

- *The drawing illustrates each of the main elements identified in the text.*

As you read the article about bog development, identify the technical terms that are not defined. Explain whether the lack of definitions hinders your understanding.

The characteristic bog develops over several thousand years in a relatively deep glacial depression, a "kettle," which is either poorly drained or has no outlet. Water is gained through precipitation alone and is lost by evapotranspiration. As flow through the pond in the depression is sluggish or nonexistent, there is no source for minerals, and the bog water is deficient in nearly all major plant and animal nutrients, especially nitrogen. Certain plants, particularly Sphagnum moss, may dominate the lower levels of vegetation, crowding out less well-adapted species. The Sphagnum also withdraws nutrients from the water, replacing them with acids. Bog water is, thus, acidic—especially beneath the Sphagnum-dominated zones.

Bog succession begins as horizontal growth over the surface of the water, since the bottom is generally too deep near the shore to allow plants to root as in a typical marsh or swamp. This lateral growth forms a dense mat of intertwining stems that supports all further growth. This mat characteristically grows out over the water, closing in on the pond center from all sides and eventually covering all open water. As there is still water beneath the mat, the mat is essentially floating; though it will generally support the weight of a person, the person gets the feeling that the earth is trembling. From this phenomena stems the term "quaking bog."

Growth proceeds upon the mat as vegetation builds up vertically, the accumulating mass forcing underlying vegetation downwards and below the static water level. This plant matter decays slowly, if at all, because the acidity of the water coupled with its coldness (it is, after all, well insulated from the warmer air above) inhibits bacterial action that causes decomposition. Thus, the basin becomes filled with partially decayed vegetation, and the mat eventually supports trees, which grow first over the landward, more "grounded" parts of the bog. Trees will advance out over the mat as the depression becomes filled, ultimately closing over the original open bog altogether. At this stage, the old bog may be difficult to recognize, though for some time to come, the acidity of the soil dictates which plants may survive there and which may not. The accompanying figure shows the cross section of a typical bog.

FIGURE 1
Cross Section of a Typical Bog

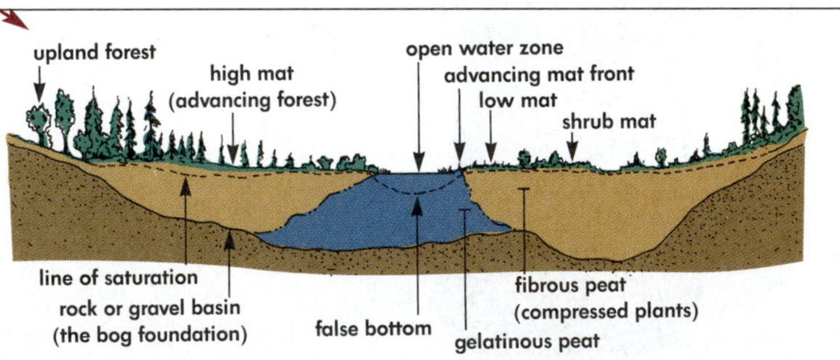

embedded in longer pieces of writing, such as marketing and promotional materials.

The process explanation makes more sense in the context of complete information. In this case, the first five paragraphs provide an overview that

FIGURE 16.4 **Process Explanation of How a Thermal Inkjet Cartridge Works**[8]

THERMAL INKJET REVIEW, OR HOW DO DOTS GET FROM THE PEN TO THE PAGE?

In its simplest form, an inkjet device consists of a tiny resistor aligned directly below an exit orifice. Ink is allowed to flow into the resistor area, and when the resistor is heated, the ink on the resistor essentially boils and forces a tiny droplet of ink out of the aligned orifice. This is called firing the nozzle.

A cross-sectional view of a single inkjet nozzle is shown in Fig. 1. On the foor of the firing chamber is a resistor. This resistor is patterned onto a silicon substrate using conventional thin-film farication procedures. Leads are connected to the resistor through the thin-film substrate. These leads ultimately travel out to the flexible circuit on the body of the print cartridge, through which a voltage can be applied across the resistor. The resistor is the heart of the thermal inkjet device, and the size of the resistor is the primary factor governing the volume of the ejected droplets.

FIG. 1.
An exploded cross-sectional view of a singleinkjet nozzle

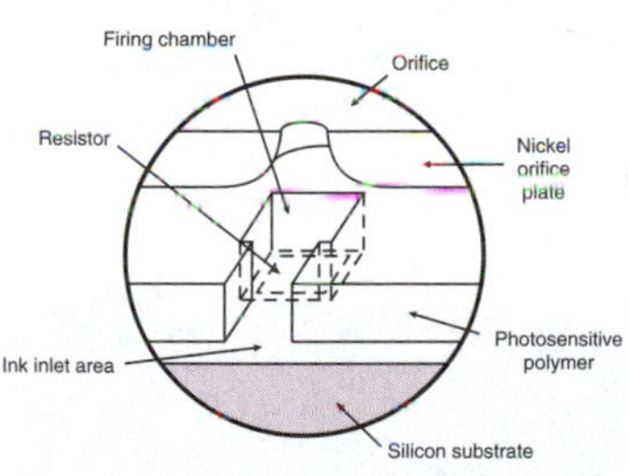

The walls of the firing chamber are made up of a photosensitive polymer. This polymer serves to define the walls of the firing chamber and determines the spacing between the resistor surface and the orifice. The thickness of this photosensitive barrier and the dimensions of the firing chamber are critical to the production of a well-formed droplet.

The photosensitive polymer also defines the dimensions of the inlet area to the firing chamber. Ink enters into the firing chamber through this inlet area. Like the barrier thickness, the inlet dimensions greatly affect the characteristics of the ejected droplet.

Finally, a gold-plated nickel orifice plate sits on top of the barrier. An orifice is formed in this plate directly above the firing chamber. This orifice hole is formed using an electro-forming process. The diameter of the orifice has a direct bearing on the volume and velocity of the ejected droplets.

To fire a drop, a voltage pulse is applied across the resistor. This pulse is typically very short, on the order of 2 to 5 microseconds in duration. The voltage pulse causes the resistor to heat up, temporarily bringing the resistor surface to temperatures up to 400°C. Heat from the resistor causes ink at the resistor surface to su-perheat and form a vapor bubble. Formation of this vapor bubble is a fast and powerful event, and expansion of the bubble forces some of the ink in the firing chamber out of the orifice at velocities of typically 10 meters per second.

By the time a droplet is ejected, the resistor has cooled down and the vapor bubble has collapsed. Through capillary forces, more ink flows into the firing chamber through the inlet area, thus readying the system for the firing of another droplet. The frequency at which the printhead can repeatedly fire droplets is determined by several factors including the inlet dimensions, the barrier thickness, and the fluid properties of the ink.

The device described above is essentially a droplet generator. The device designer has a fair amount of control over the characteristics of the ejected droplets. For example, the volume of the ejected droplet can be controlled by changing the size of the resistor—bigger resistors give droplets of larger volumes. In addition, the diameter of the orifice can be used to control droplet volumes. Droplet velocity is also controlled primarily by the diameter of the orifice.

The frequency at which droplets are ejected can be controlled by altering the size and shape of the barrier and by changing the rheological properties of the ink. . . .

Droplet characteristics, as they relate to print quality on the media surface, can be optimized through careful control of orifice profiles and resistor/orifice alignment. . . . Ink properties such as surface tension, viscosity, and thermal stability all play important roles in the production of useful droplets.

Accessible

- *The enlarged cross-sectional view lets readers see insider the nozzle.*
- *Serif font is conventional for documents that exceed one or two pages.*

Comprehensible

- *The initial paragraphs provide an overview that defines this type of printer, illustrates its critical parts, and explains their purpose and importance.*
- *The explanation of the process of firing a drop of ink is embedded in the overall document.*
- *Clear transitions help readers understand the chronology of the process.*
- *The concluding paragraphs discuss the effect that the process has on the product.*

Usable

- *Relatively short paragraphs increase ease of moving through the text.*
- *The figure is placed in the text following the text reference.*

defines, describes, and illustrates the printer, identifying the critical components that readers need to know in order to understand how the process of firing a drop of ink works. The overall definition and description are followed by two paragraphs that identify the sequence involved in firing a drop of ink. The concluding three paragraphs deal with ways in which the process can be controlled to produce high-quality printing.

Public Information and Education

Readers of general-interest publications such as daily newspapers or Web sites are often interested in technical information, but they may not have the experience or expertise to understand complex explanations. Instead, they need simple and appealing explanations. Depending on the subject and the audience, these explanations can be largely visual or largely textual.

Figure 16.5 is a primarily visual presentation that explains how acid rain is formed. Although the term *acid rain* is familiar to most people, the complex natural process by which it is formed is not. Figure 16.5 simplifies the process with a drawing that includes the major elements of the cycle of acid rain formation and provides an explanation appropriate for its intended audience.

FIGURE 16.5 **A Primarily Visual Process Explanation**[9]

Accessible

- *Readers will relate to the easily identified elements such as cars and exhaust, factories and smoke stacks.*

Comprehensible

- *Each major step in the process is identified, but few details are provided.*
- *Clear causal relationships are identified.*
- *The visual elements reinforce the information in the text.*
- *Active voice is appropriate here.*

Usable

- *Users typically enter the figure at the upper left corner and follow the arrows to move through the process.*

Explain whether simplified process explanations such as the one about acid rain in Figure 16.5 should be avoided. Do they mislead readers by omitting key information?

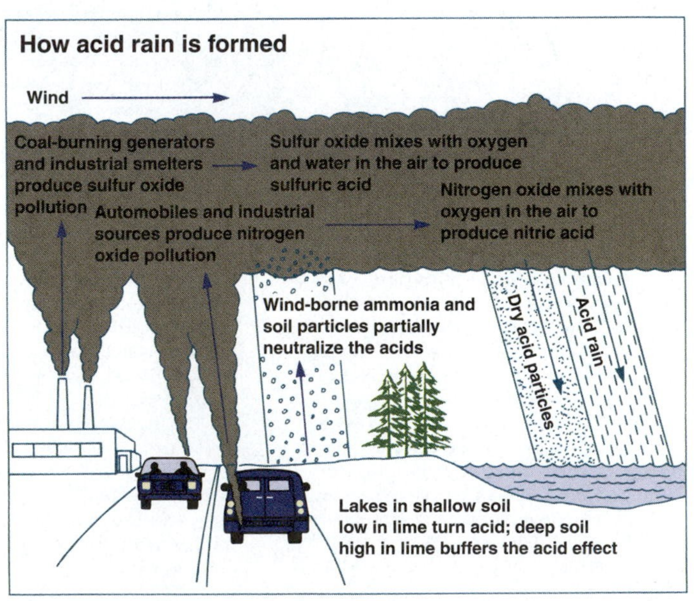
How acid rain is formed
Wind
Coal-burning generators and industrial smelters produce sulfur oxide pollution
Sulfur oxide mixes with oxygen and water in the air to produce sulfuric acid
Automobiles and industrial sources produce nitrogen oxide pollution
Nitrogen oxide mixes with oxygen in the air to produce nitric acid
Wind-borne ammonia and soil particles partially neutralize the acids
Dry acid particles
Acid rain
Lakes in shallow soil low in lime turn acid; deep soil high in lime buffers the acid effect

To read more about the entire printing process, go to http://english.wadsworth.com/burnett6e for relevant links.

CLICK ON WEBLINK

CLICK on Chapter 16/printing process

WWW WEBLINK

Preparing Processes

In preparing a process explanation, you need to consider the audience and purpose, identify the steps in the process, select or design visuals, and organize information.

Audience and Purpose

Identifying members of your audience and their purpose for reading your document or listening to your oral presentation will help you to prepare a process explanation. Most professionals are initially interested in an overall explanation of an action rather than the precise details necessary to complete that action. You need to ask not only who is going to use the process explanation and why, but you also need to know the circumstances under which they're likely to need such an explanation.

Identification of Steps

An essential part of preparing process explanations involves listing the steps of the action. If the time needed for each step or the time between steps is important, it too should be recorded. The sequence of steps forms the basis for the process explanation and also aids in designing visuals. For example, a common procedure in a hospital pediatrics unit is setting up a croup tent, which aids infants and children in breathing by surrounding them with moist, oxygen-enriched air. Both parents and young patients demonstrate anxiety about the tent, but parents seem to calm down (and thus the children relax) after they learn about what the croup tent does, how it is set up, and how it operates. The brief outline in the following example identifies the basic steps you could use to prepare a process explanation.

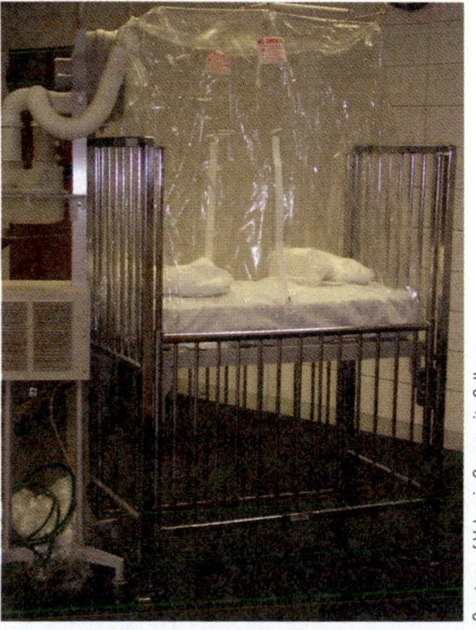
Courtesy of Valencia Community College

1. Defining a croup tent
2. Setting up a croup tent
   a. Metal frame attached to crib
   b. Canopy placed over the frame
   c. Water bottle filled and attached to frame
   d. Ice placed in chamber
   e. Valve inserted and hoses attached

3. Operating a croup tent
    a. Oxygen flow turned on to prescribed level
    b. Oxygen forced through water
    c. Oxygen enters tubing and passes through ice
    d. Moist, cooled oxygen enters tent

Visuals

You can choose from several types of visuals to illustrate the overall sequence of a process. Most common are flowcharts that give a visual overview in the same way that the introductory section of the text defines the action and identifies the major steps. Other visuals that provide an overview of the entire sequence are timelines and schedules.

A drawing can effectively show each element in a process, as in the cross section of a bog accompanying Figure 16.3. More common, though, are step-by-step changes that can be illustrated in a variety of ways: time-lapse photographs, drawings with overlays of changes, and drawings showing the final product. One other way is to use sequential drawings, as in Figure 16.6, which shows eight steps in the deployment of a satellite, from launch through to the final operational position. In this example, the steps in the action view are accompanied by very brief captions that are supplemental rather than essential; the drawing can stand alone.

FIGURE 16.6 **Sequential Drawing Showing a Process: Satellite Deployment**[10]

Accessible
- *The curved band shows the path of deployment.*

Comprehensible
- *The very brief captions for each stage of deployment are in parallel structure.*
- *The drawings are much easier to understand quickly than comparable textual explanations.*

Usable
- *Locating any particular stage of deployment quickly would be easy.*

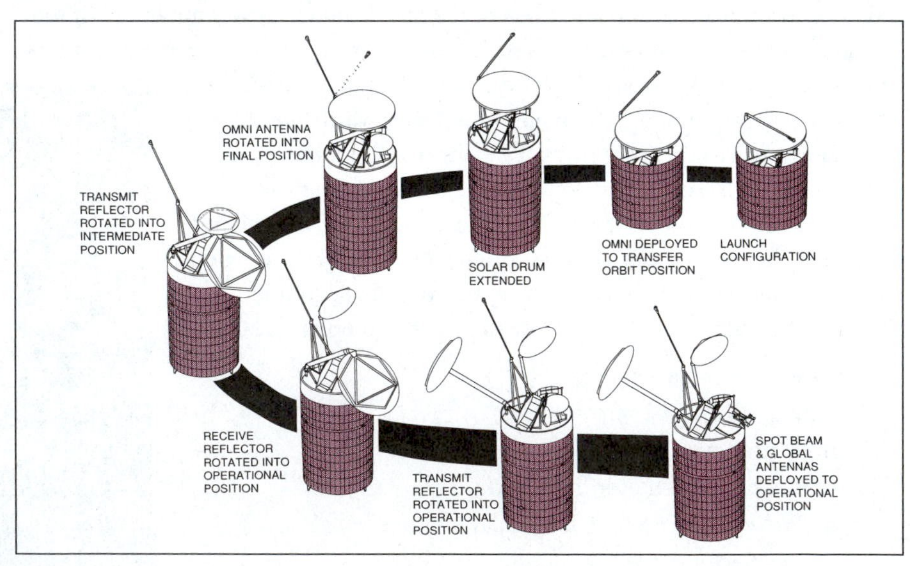

Process explanations can be presented in a number of ways, depending on the audience. Go to http://english.wadsworth.com/burnett6e for links to three different ways to display identical information — in this case the process of peer-review, publication, and discussion in the online scientific journal *Atmospheric Chemistry and Physics (ACP)* and its discussion forum Atmospheric *Chemistry and Physics Discussions (ACPD).*

CLICK ON WEBLINK

CLICK on Chapter 16/ACP and ACPD

WEBLINK

Diction

The audience and purpose of your process explanation affect your diction, or the language you use. One of your most important decisions is whether to select active or passive voice. If you want to emphasize the operator, the doer of the action, or the activating agent, use active voice. However, if you want to emphasize the recipient of the action or if the person doing the action is insignificant or unimportant, use passive voice. Figure 16.7 summarizes these key

FIGURE 16.7 **Active or Passive Voice in Process Explanations**

Active Voice	Passive Vloce
Use active voice when the action involves a person and you want to emphasize the operator or doer of the action. EXAMPLE: *Dr. Hunt attended* a seminar to learn about new techniques for treating kidney disease.	**Use passive voice** when the action involves a person and you want to emphasize the recipient of the action *or* when the person doing the action is insignificant or unimportant. EXAMPLE: *The marathon runner was treated* for dehydration by the doctor on duty in the emergency room.
Use active voice when the action does not involve a person and you want to emphasize the activating agent. EXAMPLE: *Torrential rains weakened* the dam.	**Use passive voice** when the action does not involve a person and you want to emphasize the recipient of the action. EXAMPLE: *The machine was activated* by the automatic timer.

factors in making a decision and provides examples. (See the Usage Handbook on the Web site for more details about active and passive voice.)

Organization and Format

Because process explanations are chronological, writers often use headings to distinguish their major steps. Section headings and subheadings help readers by signaling the movement from one part of the process to the next. Processes follow a format, summarized in Figure 16.8, that can be varied according to audience

FIGURE 16.8 **Planning a Process Explanation**

Title

1.0 Identify the process in an introduction.
- 1.1 Define the process.
- 1.2 Identify purpose or goal.
- 1.3 *Optional:* Identify the intended audience and the purpose.
- 1.4 *Optional:* Explain background needed by the intended audience.
- 1.5 *Optional:* Identify the relevant parts, materials, equipment, ingredients, and so on.
- 1.6 *Optional:* Design a flowchart or other visual to provide an overview.
- 1.7 Enumerate the major steps.

2.0 Present a step-by-step explanation of the process in chronological order, including cause-and-effect explanations as necessary.
- 2.1 Explain step one of the process.
  - 2.1.1 Define step one.
  - 2.1.2 Present the purpose or goal of step one.
  - 2.1.3 *Optional:* Identify the necessary parts, materials, equipment, or ingredients for step one.
  - 2.1.4 *Optional:* Illustrate step one.
  - 2.1.5 Present chronological details of step one, including any substeps.
  - 2.1.6 *Optional:* Present chronological details of the substeps if they're relevant to the audience.
  - 2.1.7 *Optional:* Explain the theory or principle of this step's operation or function if it's relevant to the audience.
  - 2.1.8 Explain how this step relates to the next step.
- 2.2 Explain step two . . .

3.0 *Optional:* Present a conclusion if it will increase the reader's understanding of the sequence, its theory, or applications.
- 3.1 Summarize the major steps only if the action is long and/or complex.
- 3.2 Discuss the theory or principle of operation if it has not already been incorporated in the primary discussion.
- 3.3 Explain applications if they're not self-evident.

needs. The less expert the audience, the less complex the information should be. However, less-informed nonexperts often need very careful explanations, which may take up more space than the highly technical explanations appropriate for experts. You will notice that a number of the steps are optional; they can be included or not, depending on the audience's needs.

Examining a Sample Process Explanation: Develping Low-Cost Roofing Materials

When Nilsa Cristina Zacarias was the Coordinator of the Research Division of the Appropriate Technology Center (CTA: Centro de Tecnologia Apropiada) in Asuncion, Paraguay, she wrote a process explanation as part of a final report for a German agency. This agency had funded a pilot project to develop building materials that low-income people in rural areas of a developing country could make and install themselves.

The logging industry in Paraguay's forests produces a great deal of sawdust, which is considered waste. One of Zacarias's colleagues, a researcher in CTA's laboratory, was supported by the International Development Research Center in Canada to develop an ecologically sound, low-cost material that could be used to make roofing material. Zacarias field-tested this process to determine if it could be used in an actual community. Figure 16.9 shows one of her early drafts of this process explanation, with comments about things that she wants to change in her revision.[11]

Zacarias knew that her readers weren't going to complete the process themselves, but she wanted them to understand it. Because her process explanation was part of a much longer report, she wanted to keep this section brief and easy to read, while emphasizing the economic and ecological goals expected by the funding agency. She decided that using a modified outline structure would make the process easy for readers to skim quickly (Figure 16.10), and that passive voice would be appropriate since her emphasis was on the process rather than the people completing the process. She also knew that including a figure would help readers visualize the way sawdust plates were attached to a roof. Although her explanation of building sawdust plates has few details, it follows the basic structure of more elaborated process explanations.

Construction in rural areas of developing countries depends on low-cost, readily available materials. One solution creates building materials from local resources. For example, sawdust, a byproduct of logging, is mixed with cement. The mixture is poured into wooden molds and dried for use as plates in roof construction.

Courtesy of Centro de Technologia Apropiada

FIGURE 16.9 **Preliminary Draft of Process Explanation**

BUILDING WITH SAWDUST PLATES

Add an introductory paragraph.

Three major steps are necessary to build with sawdust plates: 1. treating the sawdust, 2. fabricating the plates, and 3. placing the plates.

List steps to help them stand out.

Make wording of headings and steps parallel.

1. **Treatment** ing the sawdust

Sawdust is submerged into boiling water treated with 2% ferric sulfate for 10 minutes. Then the sawdust is passed through a strainer and rinsed with cold water. Finally it is placed on a clean surface to dry.

2. **Fabrication** ing the plates

First the sawdust and the cement are mixed with 1:7 proportion (cement: sawdust). Water is added until the mixture gets a determined level of plasticity. Second, this mixture is poured into wood molds and floated for getting a flat surface with a wood or steel float. This procedure is fast, but it has to be conducted with a smooth pressure. After fabrication, the plates could be lifted and put into storage for the drying process.

How long?

3. **Placement** ing the plates

How the plates are placed over the roof and fastened with bolts to the wooden structure. The plates are painted with a white, waterproof paint to protect the surface against the rain and sun.

Separate each major step to make the process easier to read.

FIGURE 16.10 **Revision of Process Explanation Using an Outline**

BUILDING WITH SAWDUST PLATES

Fabricated sawdust plates are an inexpensive alternative building material for low-cost roof construction. These plates are fabricated by mixing sawdust (waste from the logging industry) with cement and water. After the plates are fabricated, they can be used to insulate abnormally high roof temperatures from the interior rooms.

Three major steps are necessary to build with sawdust plates:

1. treating the sawdust.
2. fabricating the plates, and
3. placing the plates.

1. Treating the sawdust

- To avoid having the sawdust rot it is submerged into boiling water treated with 2% ferric sulfate for 10 minutes.
- Then the sawdust is passed through a strainer and rinsed with cold water.
- Finally it is placed on a clean surface to dry.

2. Fabricating the plates

- First the sawdust and the cement are mixed with 1:7 proportion (cement:sawdust). Water is added until the mixture gets a determined level of plasticity.
- Second, this mixture is poured into wood molds and floated for getting a flat surface with a wood or steel float. This procedure is fast, but it has to be conducted with a smooth pressure.
- After 12 days of fabrication the plates can be lifted and put into storage for the drying process.

3. Placing the plates

- The illustration shows how the plates are placed over the roof and fastened with bolts to the wooden structure.
- The plates are painted with a white, waterproof paint to protect the surface against the rain and sun.

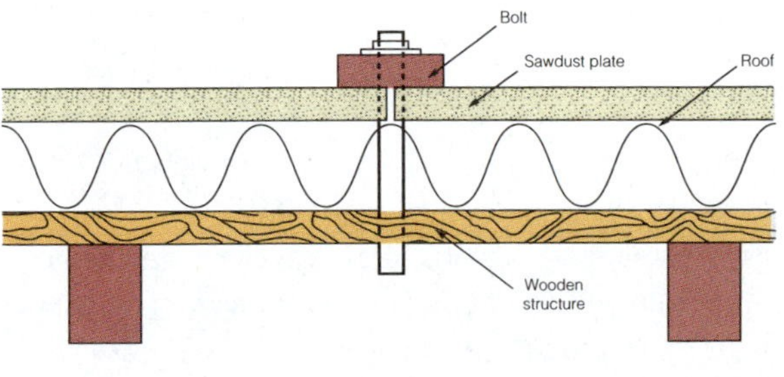

Accessible
- *Listing steps makes them easier to read.*

Comprehensible
- *Rationale helps readers understand the process.*
- *Specifying time helps readers better understand the fabrication process.*
- *Adding an illustration helps readers understand how the plates are attached to the existing roof.*

Comprehensible
- *Introductory paragraph situates the process for readers.*
- *Text is more coherent when steps and section headings are parallel.*
- *Clarifying substance for each proportion reduced ambiguity.*
- *Identifying the substeps in the process makes the simplicity of the process immediately apparent.*

Usable
- *Outline structure makes information easy to locate.*

PROCESSES INVOLVING SOCKEYE SALMON

The processes involved in catching, preserving, and cooking salmon have a long tradition in Native culture in Alaska. Sockeye salmon, which belongs to the family Salmonidae and is one of the seven species of Pacific salmonids in the genus *Oncorhynchus,* is a critical part of subsistence and commercial fishing in Alaska.[12]

Subsistence, an essential source of employment and sustenance for people in rural Alaska, enables people to feed and clothe their families. Even though wild food supplies one-third of the calories required by rural Alaskans (who make up 20 percent of the population of the state), subsistence accounts for only two percent of the combined fish and game harvest in Alaska.[13]

Traditionally, sockeye salmon has been central to the life of the Tlingit Indians in Alaska, which accounts for their long history in the development of fishing equipment and techniques[14] as well as preservation and preparation processes.

The processes continue, not only as part of the Tlingit heritage but as part of contemporary life. However, as people move to towns and cities, some of the traditional processes are modified. The following poem, "How to Make Good Baked Salmon from the River," by Tlingit elder, Nora Marks Dauenhauer, captures and honors the details of the traditional technical process so that it will not be forgotten or lost.

The poem also exemplifies the idea that process explanations are a critical part of our everyday lives and reminds us that processes evolve.

Sockeye salmon are *anadromous,* which means they migrate from the ocean to spawn in fresh water. During migration, sockeye typically have bluish back and silver sides. Then during spawning, the adults typically turn bright red, with a green head. The name "sockeye" is most likely a corruption of the Indian word "sukkai."[15]

Nora Marks Dauenhauer

How to Make Good Baked Salmon from the River[16]

for Simon Ortiz, and for all our friends and relatives who love it

It's best made in dryfish camp
on a beach by a fish stream
on sticks over an open fire,
or during fishing
or during cannery season.

In this case, we'll make it in the city,
baked in an electric oven on a black
fry pan.

INGREDIENTS
Bar-b-q sticks of alder wood.
In this case the oven will do.
Salmon: River salmon,
current supermarket cost
$4.99 a pound.
In this case, salmon poached from river.
Seal oil or hooligan oil.
In this case, butter or Wesson oil,
if available.

WEBLINK

For a close-up visual story of the largest sockeye salmon run in the world, in the Adam's River Run, in British Columbia, go to www.english.wadsworth.com/burnett6e for a link to this remarkable natural process.

CLICK ON WEBLINK

CLICK on Chapter 16/salmon run

DIRECTIONS
To butcher, split head up the jaw.
Cut through.
Remove gills. Split from throat down
the belly.
Gut, but make sure you toss all to
the seagulls
and the ravens, because they're your kin,
and make sure you speak to them
while you're feeding them.
Then split down along the backbone
and through the skin.
Enjoy how nice it looks when it's split.

Push stake through flesh and skin
like pushing a needle through cloth,
so that it hangs on stakes
while cooking over fire made from
alder wood.

Then sit around
and watch the slime on the salmon
begin to dry out. Notice how red the
flesh is,
and how silvery the skin looks.
Watch and listen
how the grease crackles, and smell
its delicious
aroma drifting around on a breeze.

Mash some fresh berries to go along
for dessert.
Pour seal oil in with a little water.
Set aside.

In this case, put the poached salmon
in a fry pan.
Smell how good it smells while it's
cooking,
because it's soooooooooooo important.

Cut up an onion. Put in a small dish.
Notice
how nice this smells too,
and how good it will taste.
Cook a pot of rice to go along with
salmon.
Find some soy sauce to put on rice,
or maybe borrow some.

In this case, think about how nice the
berries
would have been after the salmon,
but open a can
of fruit cocktail instead.

Then go out by the cool stream
and get some skunk cabbage,
because it's biodegradable,
to serve the salmon from.
Before you take back the skunk cabbage,
you can make a cup out of one
to drink from the cool stream.

In this case, plastic forks,
paper plates and cups will do,
and drink cool water from the faucet.

TO SERVE
After smelling smoke and fish and
watching
the cooking, smelling the skunk cabbage
and the berries mixed with seal oil,
when the salmon is done,
put salmon on stakes on the skunk
cabbage
and pour some seal oil over it
and watch the oil run
into the nice cooked flaky flesh
which has now turned pink.

Shoo mosquitoes off the salmon,
and shoo the ravens away,
but don't insult them, because
mosquitoes
are known to be the ashes of the
cannibal giant,
and Raven is known to take off
with just about anything.

In this case, dish out on paper plates
from fry pan. Serve to all relatives
and friends
you have invited to the bar-b-q
and those who love it.

And think how good it is
that we have good spirits
that still bring salmon and oil.

TO EAT
Everyone knows that you can eat
just about every part of the salmon,
so I don't have to tell you
that you start from the head,
because it's everyone's favorite.
You take it apart,
bone by bone,
but be sure you don't miss
the eyes,
the cheeks,
the nose,
and the very best part—
the jawbone.

You start on the mandible
with a glottalized alveolar fricative
action
as expressed in the Tlingit verb *als'óos.'*

Chew on the tasty, crispy skins
before you start on the bones.
Eiiiiiiii!!!!!!
How delicious.

Then you start on the body
by sucking on the fins
with the same action.
Include the crispy skins, and then
the meat with grease oozing all over it.

Have some cool water from the stream
with the salmon.

In this case,
water from the faucet will do.
Enjoy how the water tastes sweeter
with salmon.

When done, toss the bones to
the ravens
and seagulls, and mosquitoes,
but don't throw them in the salmon
stream
because the salmon have spirits
and don't like to see the remains
of their kin thrown in by us
among them in the stream.

In this case, put bones in plastic bag
to put in dumpster.

Now settle back to a story-telling
session
while someone feeds the fire.

In this case,
small talk and jokes with friends will do
while you drink beer.
If you shouldn't drink beer,
tea or coffee will do nicely.

Gunalchéesh for coming to my bar-b-q.

Who are the audiences for this poem? How does their prior knowledge affect their understanding? Their response?

The poem explains a traditional process for preparing salmon. How could this process — part of an oral tradition — be preserved in a recipe? How would a recipe change the process and the product?

What tribal knowledge is embedded in the poem? In what ways is culture always part of a process explanation?

How does tribal knowledge accommodate itself to change? How does the poem show such changes?

1. **Evaluate a process explanation.**
   (a) Read the following explanation about the operating process of a hydrogen/oxygen torch ignitor.
   (b) Identify the probable audience(s).
   (c) Evaluate all aspects of the process explanation, including the effectiveness of the organization, the visual, and the language choices.
   (d) Create a rubric for assessing this process explanation.

HYDROGEN/OXYGEN TORCH IGNITOR[17]
This reliable device can be used to ignite a variety of fuels.
Lewis Research Center, Cleveland, Ohio

The figure illustrates a hydrogen/oxygen torch ignitor that is reliable and simple to operate. This device is the latest in a series of such devices that have been used for more than 20 years to ignite a variety of fuel/oxidizer mixtures in research rocket engines. The device can also be used as a general-purpose ignitor in other applications, or as a hydrogen/oxygen torch.

The operation of this device is straightforward. Hydrogen and oxygen flow through separate ports into a combustion chamber in the device, where they are ignited by use of a surface-gap spark plug. The hot gases flow from this combustion chamber, through an injector tube, into the larger combustion chamber that contains the fuel-oxidizer mixture to be ignited.

The pressures and flows of hydrogen and oxygen are adjusted to obtain a pressure of about 135 psig (gauge pressure of 0.93 MPa) in the combustion chamber during operation. The pressures and flows are also adjusted for an oxidizer/fuel ratio of 40 to obtain a combustion temperature of 2,050 K, which is low enough that there is no need to cool the combustion chamber if the operating time is short enough.

Some of the flow of hydrogen is diverted to the annular space surrounding the injector tube to cool the injector tube. The rate of this cooling flow is chosen so that when it mixes with the hot gases at the outlet of the injector tube, the resulting oxidizer/fuel ratio is 5. The resulting flame at the outlet is about 12 in. (about 30 cm) long and its temperature is about 3,100 K.

The Hydrogen/Oxygen Torch Ignitor can be used as a general-purpose ignitor or as a hydrogen/oxygen torch.

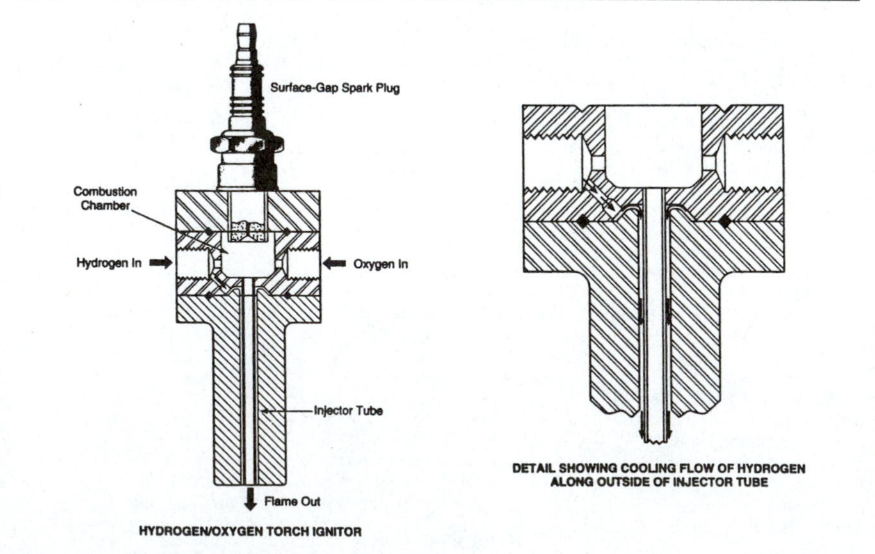

2. **Write a process description.** Select one of the topics below or choose one in your field or discipline. Identify the intended audience and specify the purpose of the process explanation. Analyze the audience to determine ways to increase (a) accessibility (for example, headings and subheadings, labeled steps), (b) comprehensibility (for example, content complexity; active or passive voice; first, second, or third person), (c) usability (for example, clear definitions, relevant visuals).

   birth of a calf or foal
   breeding of genetically pure lab animals
   creation of a silicon crystal
   depreciation
   design of a dietary program
   development of genetically modified organisms
   development of hybrid varieties of corn
   development of a hydrogen engine
   energy audit of a house
   extrusion of polymer parts
   fabrication of metal optics
   formation of a weather front
   formation of kidney stones
   formation of plaque on teeth
   grafting of plants
   installation of a wood-burning stove
   inventory control
   maintenance of a spinning reel
   manufacture of carbon fibers
   operation of a laser
   operation of a rotary engine
   operation of a septic tank or leach field
   operation of a transit
   packing a parachute
   refinishing a piece of furniture
   regeneration of tails in lizards
   self-regulated pain control (e.g., biofeedback)
   test/inspection procedure
   thermal aging process
   welding or brazing

3. **Design a visual to depict a process.**
   (a) Consider one of the topics in the preceding list. Identify the intended audience and specify the purpose of the process that you can present in a visual (the way Figure 16.5 explains acid rain). Design the visual so that it is both accurate and appealing.
   (b) Write a direct, brief caption that explains and supports the visual.

4. **Evaluate a Web-based process explanation.**
   (a) Visit the following Web site: www.howstuffworks.com. Choose a procedure to evaluate.
   (b) Identify the definitions and descriptions that are embedded as part of the process explanation.
   (c) Evaluate the process explanation by considering whether the information is effectively explained and organized (including whether the site is navigable), how visuals are used, and what language choices are effective (or ineffective).

(d) Create a rubric to identify the criteria you'll use for evaluating the process explanations.

5. **Analyze a brief process explanation.** What features of the following sentence qualify it as a process explanation?

   Routine maintenance involves changing the filter when it's dirty and making sure the heat exchanger and smoke pipe are clear of accumulated creosote.

6. **Design a brochure to explain a process.** The Equal Economic Opportunity Commission's (EEOC) enforcement guidance on harassment by supervisors recommends questions to be asked during the process of investigating a harassment complaint.[18] However, this list of questions doesn't by itself explain the process of an investigation or encourage any complainant to come forward. Search the Web to check the EEOC enforcement guidelines on harassment. Create a brochure that provides a process explanation for employees about harassment by supervisors, including the following questions. The brochure should be accessible, comprehensible, and usable.

   For the complainant — who, what, when, where, and how:

   - Who committed the alleged harassment?
   - What actually occurred or what was said? When did it occur? Is it ongoing?
   - Where did it occur? How often did it occur? How did it affect you?

   Determining employee reaction:

   - What response did you make when the incident(s) occurred or afterward?
   - Effect of harassment: Has your job been affected in any way by the harasser's actions?

   Collecting relevant information from other sources:

   - Was anyone present when the harassment occurred? Did you tell anyone about it?
   - Did anyone see you immediately after episodes of harassment?

   Determining a pattern of harassment:

   - Do you know whether anyone complained about harassment by that person?
   - Are there any notes, physical evidence, or other documentation regarding the incident(s)?
   - How would you like to see the situation resolved?
   - Do you have any other relevant information?

   Getting input from the alleged harasser:

   - What is your response to the allegations?
   - If the harasser claims that the allegations are false, ask why the complainant might lie.

- Are there any persons who have relevant information?
- Are there any notes, physical evidence, or other documentation regarding the incident(s)?
- Do you have any other relevant information?

For third parties:

- What did you see or hear? When did this occur? Describe the alleged harasser's behavior toward the complainant and toward others in the workplace.
- What did the complainant tell you?
- When did she or he tell you this?
- Do you have any other relevant information?
- Are there other persons who have relevant information?

The EEOC lists some factors to consider when determining credibility:

- Is the testimony believable on its face? Does it make sense?
- Did the person seem to be telling the truth or lying?
- Did the person have a reason to lie?
- Is there witness testimony (eyewitnesses, people who saw the person soon after the alleged incidents, or people who discussed the incidents with the complainant around the time the incidents occurred) or physical evidence (written documentation) that corroborates the complainant's testimony?
- Does the alleged harasser have a history of harassing behavior?

Chapter 16 Endnotes

1 U.S. Department of Energy, Wind Energy Program. (n.d.). *Find out about how the turbine works.* Retrieved November 16, 2003 from http://www.eere.energy.gov/wind/feature.html

2 Rude, C. (1994). Meaning publications according to legal and ethical standards. In O. J. Allen & L. H. Deming (Eds.), *Publications management: Essays for professional communicators* (pp. 171–187). Amityville, NY. Baywood.

3 Crowell, T. After Toramura: Ratting on Japan's Nuclear Sloppiness. *Asia Now.* 13 October 1999. Retrieved 2001, from http://cnn.com/ASIANOW/asiaweek/intelligence/9910/13

4 Modified from Noce, T. E., & Holzer, T. L. *U.S. Geological survey* (S. L. Scott, Graphics) (Fact Sheet 028-03). Retrieved October 25, 2003, from http://geopubs.wr.usgs.gov/fact-sheet/fs028-03/

5 Phillips, W. (1998, February 3). Flow charting and technical writers. Message posted to TECHWR-L electronic mailing list, archived at http://www.raycomm.com/techwhirl/archives/9802/techwhirl-9802-00133.html

6 ISO 9000 Maps. (n.d.). *Packaging map: ISO 9002.* Retrieved November 25, 2003, from http://Elsmar.com/9000maps/packagng.htm

7 Oltsch, F. M. (n.d.). *Bog development.* Becket, MA.

8 Shields, J. P. (1992, August). Thermal inkjet preview, or how do dots get from the pen to the page? *Hewlett-Packard Journal,* 67. © 1992 Hewlett-Packard Company. Reproduced with permission.

9 How acid rain is formed. *Lowell Sun.*

10 Slafer, L. I., & Seidenstucker, V. L. (1991). INTELSAT VT: Communications subsystem design. *COMSAT Technical Review, 21* (Spring), 61.

11 Zecarias, N. C., Iowa State University.

12 Office of Protected Resources, NOAA Fisheries, National Marine Fisheries Services. (2003). *Sockeye salmon.* Retrieved on November 18, 2003 from http://www.nmfs.noaa.gov/prot_res/species/fish/sockeye_salmon.html

13 Alaska Seafood Marketing Institute. (2002). *Alaska Federation of Natives: Subsistence facts.* Retrieved on November 18, 2003 from http://www.alaskaseafood.org/aboutus/salmonwk2003.htm

14 Alaska Fishing on the Web. (n.d.) *Fishing and the Tlingit Indian.* Retrieved on November 18, 2003 from http://www.alaskafishingontheweb.com/tlingit_fishing/

15 Office of Protected Resources, NOAA Fisheries, National Marine Fisheries Services. (2003). *Sockeye salmon.* Retrieved on November 18, 2003 from http://www.nmfs.noaa.gov/prot_res/species/fish/sockeye_salmon.html

16 Dauenhauer, N. M. (1998). How to make good baked salmon from the river. *First Fish–First People: Salmon Tales of the North Pacific Rim.* Seattle, WA: One Reel/University of Washington Press.

17 *NASA Tech Briefs.* (1995, December). 58.

18 Long, S. E., & Leonard, C. G. (1999). The changing face of sexual harassment. *HR Focus,* (Oct), p. S1.

PART V

Preparing Professional Communication

CHAPTER 17

Engaging in Oral Communication

Objectives and Outcomes

This chapter will help you accomplish these outcomes:

- Communicate ideas and plans effectively in front of an audience
- Focus on purposes for your presentation and determine information to include
- Engage listeners by organizing information in various ways and by using notes or outlines to help you keep track of what you're doing
- Design appropriate visuals and handouts to accompany your presentation
- Create a professional image based on appearance and demeanor, vocal characteristics, and the way you handle questions from the audience
- Be a good, active listener
- Evaluate presentations fairly and thoroughly

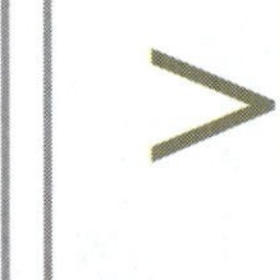

Oral presentations play an important part in professional communication. A survey of more than 700 managers rated "the ability to communicate ideas and plans effectively in front of an audience" as the most important career skill.[1] Twenty-five percent of these managers said they gave presentations at least once a week. Their views confirm that professional reputations (as well as promotions and raises) are positively related to effectiveness in making oral presentations.

You can begin to develop oral presentation skills by learning about different types of presentations and identifying your audience and purpose. The needs of the audience should influence the organization of the information. You also need to consider the actual presentation — your appearance and your voice. Finally, criteria for evaluating presentations will provide a way to assess them in the way that listeners do.

Types of Presentations

Throughout your career, you will give presentations in a variety of settings: meetings, seminars, academic and industrial classes, community meetings, and professional conferences. The different settings will affect the audience's expectations and thus your approach. Think of presentations as existing on a continuum from informal to formal, with class presentations as a unique variation.

A day-to-day working relationship with a group usually gives the members prior knowledge about your topic; thus, you can easily reduce the amount of detailed background information you present.

People often speak about the same subject to different audiences in different settings. For example, Lorraine Higgins, a member of a county extension agricultural research group, is investigating the impact of daminozide (commonly known as alar), a pesticide that has been used on some of the apple crop in the United States. Lorraine has sorted out the fiction from the facts of the daminozide controversy in an effort to inform interested people. In disseminating her information, Lorraine will speak to a number of groups: coworkers, farmers, nutritionists, counselors for human service agencies, parents, newspaper food editors, and Environmental Protection Agency (EPA) officials. Some of her presentations will be informal; others will be formal.

© Yellow Dog Productions/Getty Images

Informal Presentations

One audience for an informal presentation consists of professional peers in your own organization or your immediate subordinates or supervisors. You can usually be relatively informal with this audience because these people are probably already familiar with you and your work. Informal presentations often take place in weekly departmental meetings, when you are asked, for example, to describe how you're solving an inventory control problem or to explain progress you've made designing a circuit board for a customer. Informal meetings, which generally include discussion, are often an extension of day-to-day professional activities and conversations.

Another type of informal presentation occurs when a group of people wants to learn new information that you provide, followed by extensive discussion. Your presentation is a springboard for exploring an idea or issue. Such informal presentations might be given to community groups or special-interest groups.

Whether your informal presentation is for an internal or an external audience, do not think of it as an impromptu presentation that doesn't require preparation. Informal simply means that the people know you, and you already have a working relationship with them. It can also mean that the information you're providing will act as the stimulus for a discussion. Even though the presentation is informal, you still need to prepare so that your information is logically organized and your explanations are clear. You may use visuals or handouts. Planning ahead helps ensure that you present information efficiently.

When Lorraine Higgins talks to her coworkers about daminozide, she'll be informal, summarizing her reexamination of the available data. Lorraine will also be informal when she talks to members of the local parents' association, knowing that they not only want facts but also want to discuss possible ways daminozide might affect their children. For example, she can tell them that daminozide does penetrate the skin of apples so peeling the fruit will not remove it, but she can also tell them that it's used on less than five percent of the apple crops and that less than one percent of apple juice for infants has shown traces of daminozide. She also will speak informally to a group of food editors from regional newspapers who invite her to a roundtable discussion about conventionally grown versus organic produce.

What types of informal presentations are you likely to give as part of your professional work?

Formal Presentations

The audience for a formal presentation may not be familiar with either you or your work. As a result, formal presentations usually take more time and effort to prepare because you need to provide more background information and adjust the material to the audience's needs. For example, in a formal presentation you might justify departmental reorganization to corporate executives, introduce your company's manufacturing and inspection capabilities to international customers, or explain research findings at a professional society conference.

Sometimes the distinctions between informal and formal presentations are blurred. A project progress report about the development of a new aircraft engine might be submitted informally to other project managers in the same group; however, when the report is given to the customers who are officials from a foreign government, the presentation will certainly be formal.

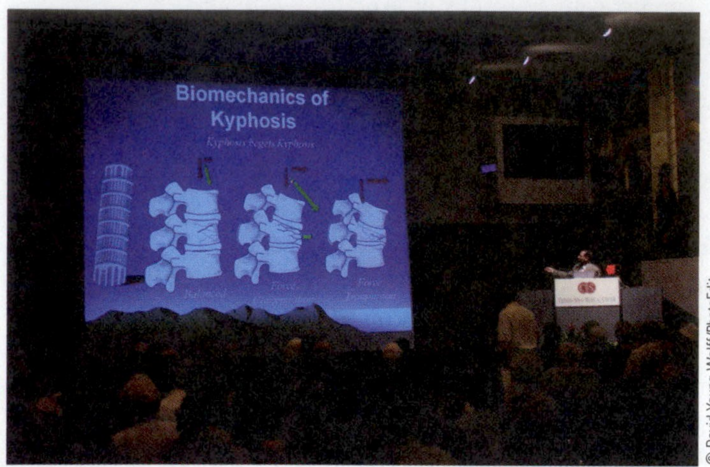

© David Young-Wolff/PhotoEdit

A formal presentation is usually to a group unfamiliar with your day-to-day work and without the prior knowledge that your coworkers have. For such audiences, you need to provide background information that contextualizes your discussion.

Both informal and formal presentations may be effectively used in lengthy sessions. For example, a day-long seminar for independent farmers interested in the uses of marginal land might begin with a formal presentation about techniques to extend land use that includes research by the local university and results from experimental farms. This session could be followed by an informal brainstorming session to exchange ideas about ways to practically and profitably use marginal lands. Other formal sessions could deal with financial implications, ecological impact, and government support. Informal sessions would give participants a chance to express their own views and talk with other farmers who have similar concerns and problems.

What types of formal presentations are you likely to give as part of your professional work?

Lorraine Higgins needs to prepare formal presentations for three very different audiences. When she is asked to appear on a panel with two other experts at the fall conference for the region's nutritionists and human service agencies, she prepares a 15-minute talk. She considers illustrating her presentation with PowerPoint slides but decides to use transparencies instead. She leaves plenty of time for questions and also hands out information sheets that the audience can pass on to their clients. When Lorraine is invited to address the annual meeting of Eastern Apple Growers, she prepares a formal 25-minute presentation, with PowerPoint to reinforce her points and handouts for later reference. When she prepares to testify at an EPA hearing, she writes a formal statement and also prepares responses to questions that she expects to be asked.

Class Presentations

Class presentations are a valuable opportunity to strengthen your oral presentation skills. The class audience provides reactions from a variety of academic and professional interests. In some situations the class can role-play a particular audience. For example, you might have developed an information session for summer interns in mechanical engineering. You could ask the class audience to imagine they are third-year mechanical engineering majors; their role-playing will give you a sense of how the presentation would be received by the actual intended audience.

Class presentations have another benefit that you seldom receive in the professional world. Members of the audience can give you honest, helpful criticism as well as identify areas in which you are particularly effective. Such feedback gives you the chance to maintain and develop your strengths while improving weak areas.

Preparing a Professional Presentation

To give an effective presentation, you must prepare carefully. Steps in preparation identifying the audience, focusing on the purpose for your presentation, and determining the kind of research you need to conduct.

Audience

The more you know about the listeners you're speaking to, the more likely you are to tailor the presentation to their needs and interests and to focus on information that is relevant to them. You must make adjustments for specific audiences in the complexity of your content, vocabulary, and amount of detail.

The four audiences in Figure 17.1 — professional peers, professional nonexperts, international audiences, and general audiences — are ones you can expect to address at various points in your career.

Who are you likely to speak with as part of your professional work?

FIGURE 17.1 **Characteristics of Audiences**

Possible Audiences	Characteristics
Professional peers	▪ Assume you have a high degree of technical expertise ▪ Expect support that substantiates points ▪ Typically ask more difficult questions than any other audience
Nonexpert professionals	▪ Include people in decision-making positions who may have limited technical experience, such as corporate officers, government officials, and military leaders ▪ Want technical information but do not need abundant peripheral technical details ▪ Listen especially carefully to conclusions and recommendations
International audiences	▪ Appreciate your awareness of and respect for their culture ▪ May interpret verbal and nonverbal parts of the presentation differently than you intend ▪ Appreciate well-designed visuals to help define terms and clarify the sequence of processes
General audiences	▪ May have multiple agendas, but typically come together for a common purpose ▪ Appreciate a clear statement of purpose, defined terms, useful analogies, interesting examples, effective visuals, and clear transitions ▪ Respond well to becoming involved

- *Professional peers* understand your field's jargon and can draw on relevant background knowledge to understand your points and follow complex ideas. Your presentations to them should be carefully organized, thoroughly documented, and supported by evidence.
- *Nonexpert professionals* have expectations similar to those of professional peers but are less comfortable with technical jargon and are less familiar with current theory and practice. They understand conclusions and recommendations more easily if you precisely define terms, clearly identify benefits as well as problems, employ visuals to stress key information, and provide a logical, easy-to-follow sequence of points.
- Addressing *international audiences* requires that you be familiar with the customs of those you are addressing, both for courtesy and effectiveness.
- *General audiences* have wide-ranging needs and interests. Such audiences might include a community group with people of varying backgrounds: auto mechanics, supermarket managers, biologists, kindergarten teachers, civil engineers, hair stylists, secretaries, dentists, highway department workers, and so on.

Purposes of Presentations

As with so many types of technical communication, once you identify the audience, you can establish the purpose. Knowing the purposes of oral presentations helps you define primary and secondary goals for your own presentations. The nature of a presentation, whether informal or formal, influences the tone you establish and your audience's expectations. Oral presentations in business and industry have several purposes that may occur separately or in combination.

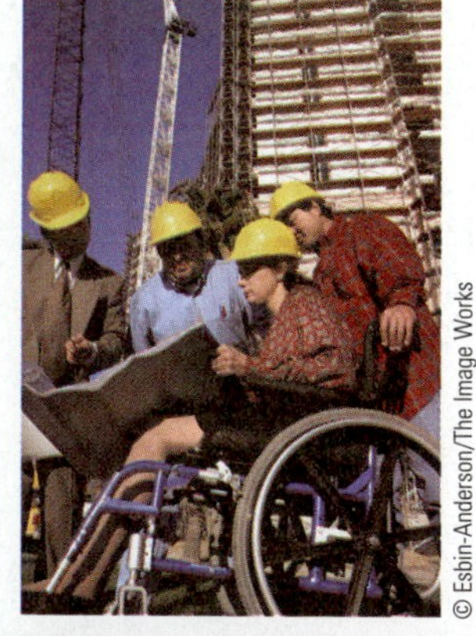

© Esbin-Anderson/The Image Works

- *Informative presentations* give your audience verifiable information, usually for decision making or background. For example, a project supervisor could present a status update report to other project supervisors in the same division of the company or explain changes that have been made to project plans.

© Jeff Greenberg/PhotoEdit

- *Persuasive presentations* attempt to convince your audience about the advantages of accepting a particular proposal or position. You establish common ground between your own ideas and the ideas of the audience. For example, an engineer could present an oral proposal to the division manager suggesting the restructuring of incoming inspection procedures.

- *Demonstrations* show your audience how something is done, define and describe a process as it happens, and educate the audience. For example, a company safety officer could demonstrate to summer interns how each piece of equipment in the department operates.

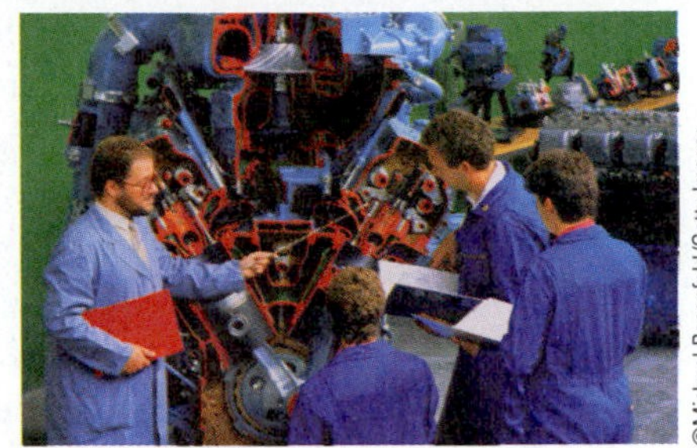
©Michael Rosenfeld/Getty Images

- *Training sessions* teach your audience how to do something, giving them the opportunity for hands-on experience and practice. As a trainer, you have a double task: to teach concepts and techniques to the entire group and also to provide individual assistance. For example, a lab manager could train operators and technicians to follow procedures for safeguarding against contamination of samples.

© Robert Essel NYC/CORBIS

What persuasive element is part of each of the following presentations?

- *Training to use software to monitor inventory*
- *Decision to change production procedures*
- *Explanation of control panel on ovens used to heat-treat metal parts*
- *Demonstration of computer numerical control (CNC) machines*

To accomplish your stated purpose in a presentation, you need specific information; obtaining it may require research or at least the coordination of information you already have. Generally, the research will fall into two categories: information you need in order to make the presentation and additional information the audience might ask about. The same primary and secondary research you do for a written report is useful for an oral presentation.

Organizing a Professional Presentation

No matter what your audience or purpose, you can make your presentation more professional by engaging the listeners, by organizing information in various ways, and by using notes or outlines to help you keep track of what you're doing.

Engaging the Listeners

Your listeners need a reason to pay attention. Some members of your audience will probably be receptive because they already know that they're interested in the information you have to present. However, others won't be sure you have anything important for them; you'll need a little entertainment value to get them to listen. Another group doesn't really want to be there; they'll need to be motivated to even pay attention. And probably a few members of your audience won't ever be engaged with you no matter what you do.)

So how can you engage your audience — to inform, entertain, and motivate the individuals listening to your presentation?[2] You have several options for increasing audience engagement. These four clusters of strategies won't ensure a perfect presentation, but they will help you get a good start.

What are some of the best things you have seen or heard people do in oral presentations to keep your interest and ensure your understanding?

Strategy Cluster 1: Create an audience-centered atmosphere.

- If you're making a presentation to a group outside your own organization, find out something about the audience beforehand so that you can relate your opening comments to the group.
- Show that you're glad to be giving the presentation and are interested in the audience as well as the topic.
- Use questions creatively:
  - Imagine the questions that the audience will ask and incorporate responses into your presentation.
  - Build in rhetorical questions that members of the audience can answer in their minds and relate to your presentation.
  - Provide the audience with a list of intriguing, problem-oriented questions that extend the ideas in your presentation.

Strategy Cluster 2: Encourage active involvement.

- Involve everyone in an activity such as solving a problem related to your topic, coming up with ways to approach an ethical dilemma, or viewing a brief videotape together and monitoring events on it.
- Get audience volunteers to help with a demonstration.
- Encourage audience members to answer specific questions you ask during the presentation, and then adjust your presentation to consider those responses.

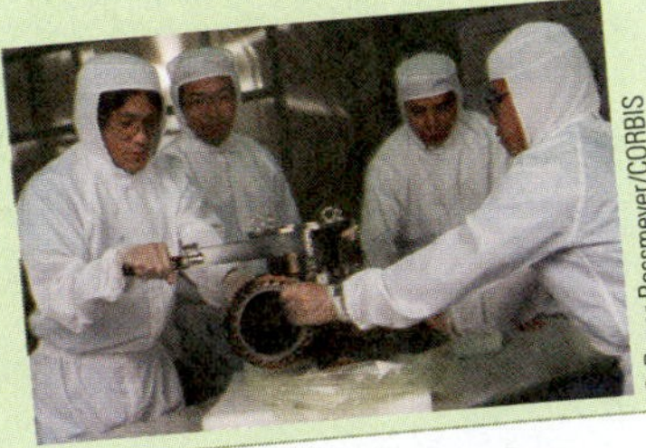

Strategy Cluster 3: Make what you say easy to listen to (and remember).

- Make some of your points with brief stories — not rambling stories with only the most tenuous relevance but, instead, tightly crafted anecdotes (what *Time* magazine writers call "nuggets"). People tend to remember information presented as part of a narrative long after they have forgotten separate pieces of information.
- Remember the limits of short-term memory, that is, the number of chunks of information we can hold in our mind at once. Typically, short-term memory holds a maximum of five to seven chunks of information, a number that decreases rapidly as the complexity of the information increases. It's also a number that decreases when information is presented orally. You're better off saying, "Let me illustrate the three main stages in this complex process, each of which has five steps" (which is well within short-term memory limits) than saying, "Let me move through the fifteen steps of this complex process."
- Connect new information to information the audience already knows. Too much new information all at once, unconnected to prior knowledge, ensures that your audience will remember almost none of the new information.

Strategy Cluster 4: Vary the pacing and structure of your presentation.

- Avoid cramming too much information into the time allotted for your presentation. Your audience will understand and recall what you've said far better if you avoid being cryptic or verbally dense.
- Intersperse difficult material with easier material.
- Schedule lengthy presentations in facilities that allow you to use a variety of activities to change the pace and encourage audience participation. For example, in a multihour session, follow your formal presentation with a break-out session, perhaps using a case situation or simulation activity that encourages members of the audience to work together in small groups to develop creative solutions related to the issues identified in your formal presentation.

Organizing for the Listeners

The way you organize your presentation determines its success. Your listeners must be able to follow your ideas easily. The same patterns that you use in writing — chronological, spatial, priority, comparison and contrast, and cause–effect — can and should help you organize oral presentations. For example, if you are giving a status update, chronological order can summarize the weekly or monthly progress and descending order can identify the most significant gains or losses. A marketing presentation can apply descending order to identify the features of your product and use contrast to identify significant differences between your product and your competitor's.

Additional conventions of organization beyond the standard patterns can make your presentation more appealing and understandable. Far more than in written documents, which audiences can reread, oral presentations need strong, obvious guideposts and transitions to indicate the organization of and movement through the presentation. Figure 17.2 summarizes structural and organizational strategies that make presentations easier for listeners to comprehend.

Using Note Cards or Outlines

Unless you are a very lucky, gifted, or experienced speaker, you will not be able to give an effective presentation beyond a few minutes without referring to notes or an outline. These notes, written on index cards or sheets of paper, should contain the main points of the presentation and specific facts, details, or statistics that you can refer to during your presentation. An outline will be far more useful to you than a complete text of your presentation.

If you write out the entire presentation, you might be tempted to memorize the material and then recite it as if giving a speech. An audience is interested in hearing you talk to them and with them, not read to them. Even if you prepare the complete text of the speech (perhaps for publication in an in-house newsletter or in conference proceedings), your presentation will be far more appealing if you do not read the paper.

An outline or note cards should also include key quotations and statistical information you'll need during a presentation. Even an informal presentation is strengthened by specific facts; you'll find them easier to remember and locate if you put them in your notes.

A topic outline is one of the most useful ways to organize main and subordinate ideas. Figure 17.3 shows the topic outline put together by a multidisciplinary engineering design team for a presentation to a client, Cottrell Technologies. One of the team members, Ned Hitch, was the liaison between his team and Danielle Conner, a project manager at Cottrell Technologies who had asked the team to design a digital controller system.

The design team had completed most of its work on the system, but as they were considering input interface devices, they discovered two options and could

FIGURE 17.2 **Ways to Organize Information**

Strategy	Suggestions and Examples
Use Purpose/Audience Statements	Draft a one-sentence purpose/audience statement to use in the introduction of your presentation. *This presentation introduces computer-aided design (CAD) to apprentice drafters by demonstrating the CAD software and beginning the training with a simulation program.*
Establish Organization	Use an appropriate organization and tell the audience how you're organizing the information. Help listeners differentiate main points and subordinate points. *I will give a time line of the steps to eliminate the asbestos problem.*
Preview Information	Preview what the presentation covers. *First I'll summarize design changes in this component. Then I will identify problems these changes have caused and explain the steps we've taken to eliminate them.*
Use Transitions	Include clear transitions to mark movement from one topic to the next or to indicate a shift in perspective. *Now that I've identified the few disadvantages of the continuous assembly line, I will discuss the significant benefits this change in the procedure could bring.*
Include Summaries	Periodically summarize what you've covered so far. *So far I've explained three common methods to eliminate exposed asbestos areas in your facility.*
Embed Verbal Signals	Signal particularly important points. Follow up with an example for each point. *The most severe health problems occur in two high-risk groups: infants and elderly. Let me compare infant mortality in this country with infant mortality in other industrialized nations. [specific example] . . . Now I will show that our country doesn't fare any better when we compare our geriatric care with that in other industrialized nations.*
Draw a Conclusion	Provide a conclusion that reviews the major points and indicates preferred action of the audience. *As you can see on the PowerPoint slide, we have covered five key points dealing with possible contamination of local wells. With this information, you should be able to justify your request for more frequent state inspection.*

FIGURE 17.3 **Outline for Oral Presentation**

Beginning each part of the outline with a verb helps Ned remember that he is engaged in a series of actions.

Using a numbering system helps organize main and subordinate points.

Boldfacing key terms helps Ned remember to stress these points.

Analysis of Input Interfaces for the Techni-Veyor Controller

1.0 Introduce the project
 1.1 The goals and objectives involve company and customer needs.
 1.2 Today's situation involves the need for a controller.

2.0 Identify the available options
 2.1 A bar code interface is easy to operate.
 2.2 A telephone-style keypad interface is inexpensive.

3.0 Specify the selection criteria
 3.1 The controller must be **compatible** with the existing system.
 3.2 The **cost** of the interface and system modifications is critical.
 3.3 The relative **ease of operation** is also important.

4.0 Present unit cost

5.0 Summarize conclusion
 5.1 Both systems are compatible though the bar code system is faster to install.
 5.2 The keypad is much less expensive.
 5.3 Both interfaces are easy to operate.

6.0 Make recommendation

not decide which one Cottrell Technologies would prefer. Ned had overseen the analysis of the two input interface devices. To explain the team's recommendation, Ned drew on the technical and communication skills of his teammates. Although he was identified as the principal author of the team's report and would present the team's recommendation, the work represented a collaborative effort.

Ned scheduled an hour-long meeting with Danielle Conner. His presentation would take 25 minutes, leaving him enough time to answer questions from the audience. Danielle Conner approved all major decisions in the project, though she worked with a group of technical experts who took responsibility for specific problems. All of these technical experts planned to attend Ned's presentation. (See the complete report on which this presentation is based in Chapter 20, Figure 20.10.)

Preparing Materials for a Professional Presentation

As you organize a presentation, you also need to think about the visuals and handouts that will support and clarify your information.

Visuals

Visuals are extremely valuable during an oral presentation. Because most people's visual memories are stronger than their auditory (hearing) memories, visual aids are particularly useful for presenting facts and statistics. In fact, visuals that illustrate or reinforce your information can increase most people's retention by some at least 10 percent and perhaps as much as 50 percent depending on the way the visuals are integrated with the verbal information.[3] Show rather than just tell whenever possible.

Content Complexity. You can use visuals to provide support for complex information as well as to preview and review key points. For example, if your presentation contains a great deal of numeric data, your audience members will understand and remember the information better by seeing as well as hearing (reinforcing their short-term memory). Showing an outline or diagram of a complex argument can help the audience understand and follow it. Also, since most people don't retain all of what they hear, even immediately after hearing it, visuals can reinforce your points.

WEBLINK

For an overview from OSHA about flip charts, transparencies, posters, slides, and videotapes, go to http://english.wadsworth.com/burnett6e. This site links you to some traditional information about the value of visuals in oral presentations and then reviews the uses, limitations, and applications of five types of visuals.
CLICK ON WEBLINK
CLICK on Chapter 17/OSHA

When you are deciding what types of visuals to select or design, remember that they are separated into these categories: charts, diagrams, schematics, graphs, tables, maps, drawings, and photographs. You can use visuals for the following purposes:

Organize Information for Listeners

- Preview and review main points.
- Differentiate main and subordinate points.

Support the Development of Information

- Illustrate or exemplify points.
- Display complex data or information.

Encourage Attention and Engagement

- Provide sensory variety to stimulate interest.
- Vary the pace of the presentation.
- Provide humor.

Keep your visuals simple. If you try to squeeze too much onto one transparency or slide, it loses its effectiveness. But simple doesn't mean that your visuals can't have a strong and appealing visual identity. Maintaining this identity will be easier if you establish consistent specifications for all your visuals (e.g., consistent font style and size, consistent use of color, consistent spacing, and so on).

In most cases, each visual in a presentation should support only one or two of your ideas. For example, the graph in Figure 17.4 would be difficult for an audience to read. Instead, the presenter could prepare four separate graphs, as in Figure 17.5, each clearly showing the changes. If direct comparison is necessary, the presenter could follow the separate graphs with a composite graph. Consider the limits of short-term memory (the ability to hold five to seven chunks of information in your mind at the same time) when you're designing visuals.

Types of Visuals. Visual information can be presented in many ways, depending on the audience, purpose of the visuals, art or graphic design capabilities (your own talent or that of the organization's graphic artists), finances, physical limitations of presentation room, time available, equipment available, your familiarity with each format, and so on. Each type of visual presents both benefits and problems:

- chalkboards or white boards
- flip charts (large pads of newsprint for writing or drawing)
- prepared posters, charts, tables, diagrams, maps, or photos
- slides

FIGURE 17.4 Complex Graph Inappropriate for Oral Presentation

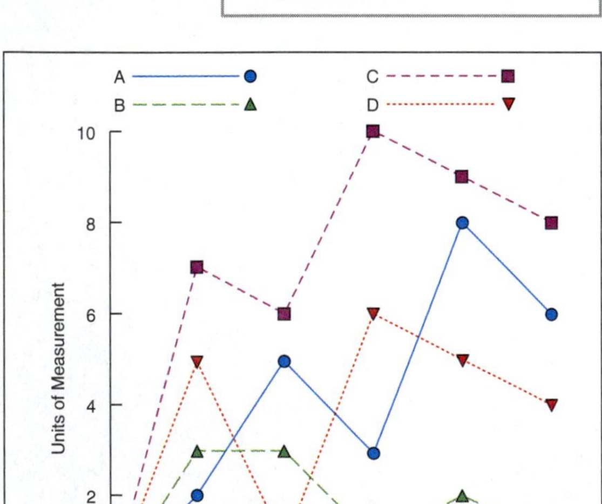

FIGURE 17.5 Series of Simple Graphs Appropriate for Oral Presentation

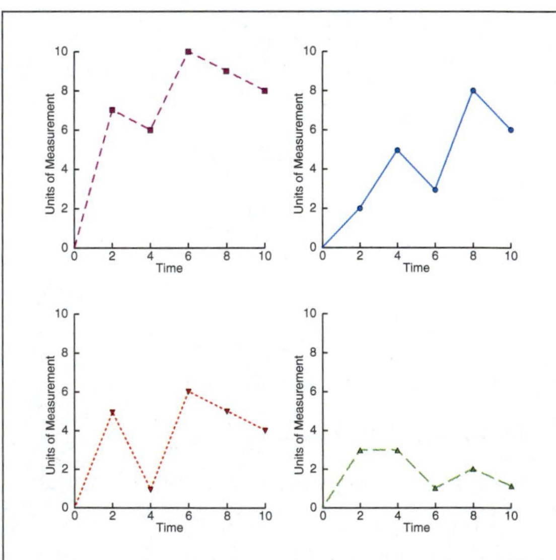

- presentation software slides (e.g., PowerPoint)
- transparencies
- videotapes, laser disks, DVDs, films
- physical models
- demonstrations

All take time to prepare, ranging from a few minutes for arranging for delivery of a flip chart to several months for filming, developing, and editing a training film. All involve expenses, ranging from pennies for a piece of chalk or felt-tip marker to thousands of dollars for elaborate models or videos. Ease of use varies also, from virtually error-proof overhead transparencies to complicated demonstrations.

Audience size will influence your choice of visuals. Charts work well with small audiences; transparencies and demonstrations are fine for average-size audiences; presentation software slides and films work well for large audiences. For some audiences, distribute paper copies of your visuals, which should have plenty of white space for notes.

Size of Visuals. The size of visuals is important. Few things are more frustrating and annoying to an audience than visuals they cannot see. A good rule of thumb for using flip charts and posters displayed on easels is to allow a minimum of 1 inch of letter height for every 10 feet of audience. So if 30 feet exist between the flip chart and the last row of the audience, lettering should be at least 3 inches high. Clearly, this guideline becomes impractical in a large room, where a presenter may switch from charts to overhead transparencies and slides, which can be projected even in a large auditorium.

Color. Not only are complexity, type, and size important in presentation visuals, but you also need to consider color. Graphic designer Jan White has developed a series of suggestions about ways to use color effectively.[4]

1. Match your use of color in slides or transparencies to the physical environment.
    - In a dark room, use slides or transparencies with a dark background and light, bright drawings and text.
    - In a light room, use slides or transparencies with a light background and dark drawings and text.

Light on Dark

Dark on Light

2. Use color to emphasize an important point. You can highlight segments of a pie graph, one bar on a graph, or one row of figures on a table.

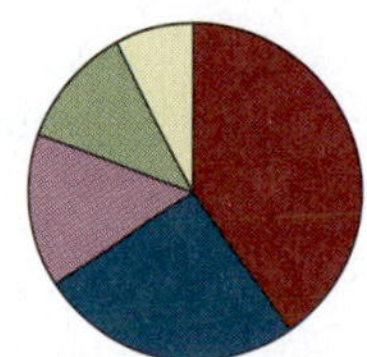

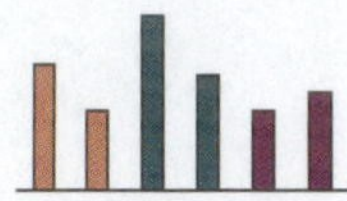

3. Use color to establish the relative importance of a series of points, for example, putting the most important information in the brightest color.

given information given information
given information given information
new information given information
given information new information
new information given information
given information given information

4. Use color to establish a pattern that the audience will recognize, for example, using a particular color to signal new information or information of a particular type (such as estimates for the upcoming fiscal year).

5. Use color to show a progression through a series of slides or transparencies, moving from dark to light or from dull to bright, so that the audience can see that the movement of your ideas corresponds to the changes in color.

PowerPoint for Oral Presentations. PowerPoint is a powerful and pervasive tool, and slide shows that can be created in PowerPoint now accompany many professional oral presentations. Despite its ubiquity, PowerPoint is easy to misuse because of three categories of problems: how you conceive of the slides during planning and drafting, what content you choose to put on the slides, and how you use the slides during your presentation. However, if you use PowerPoint rhetorically, responding to the questions posed in Figure 1.2 in Chapter 1, most of the problems associated with PowerPoint will be minimized or disappear.

- *Concept.* What is the role of PowerPoint as you plan and draft? If you think of PowerPoint as a template to fill in, with each slide as a miniature hierarchy of title and bullets, you're likely to produce a poor presentation and a boring series of slides. Why? Your thinking will be constrained by the form. Don't let your presentation be limited by the worst features of your tools. Instead, use the best and strongest features of tools. In planning a presentation, you need to respond to the context, the needs of your audience, and the purpose. Use PowerPoint to help you make key information more accessible, comprehensible, and usable.
- *Content.* What you choose to put on your slides and how that information is presented affects whether it is accessible, comprehensible, and usable. What can cause problems with accessibility?

> *What items on this list are also problems in poorly designed transparencies? What additional problems could occur only with PowerPoint?*

- ☐ Too many bullets
- ☐ Inappropriate font for electronic display
- ☐ Too small a font for the size of the room and audience
- ☐ Poor resolution of visuals
- ☐ Inappropriate and unnecessary animation
- ☐ Complex data displays
- ☐ Distracting background color and/or texture
- ☐ Low figure-ground contrast
- ☐ Inappropriate color palette for room lighting
- ☐ Distracting transitions between slides

What can cause problems with comprehensibility?

- □ Too much or too little information on a slide
- □ Lack of coherence between what you're saying aloud and what's displayed on the slide
- □ Insufficient use of PowerPoint's capability to animate certain kinds of complex relationships
- □ Using PowerPoint as a template for a written report (thinking the slides provide a chunking device)
- □ Clip art, especially badly done clip art
- □ Multiple levels of hierarchy
- □ Omitted or inadequate labels
- □ Information that needs to be carefully studied displayed for only a short period during a presentation

And, finally, what can cause problems with usability?

- □ Information that attempts to be a complete record of the presentation
- □ Information that is so distracting or riveting that the audience pays little attention to the speaker
- □ Information that doesn't acknowledge the audience's prior knowledge
- □ Information so cryptic that after the presentation it makes little or no sense if the audience if reviewing the slides

- *Presentation.* What is the role of PowerPoint during your presentation? If you simply read your slides, your presentation is doomed. Why would anyone want to sit in the audience to have you read what they can read for themselves? The information on your slides can appropriately fulfill any of several important functions: preview, explicate, illustrate, elaborate, contrast productively, summarize, or review aspects of your presentation. However, you need to remember that the slides are not the presentation. With few exceptions, what you *say* is the presentation. The PowerPoint slides are support.

Visuals Accompanying the Presentation to Cottrell Technologies. Ned Hitch planned to offer the recommendation made by the multidisciplinary engineering design team during his presentation at Cottrell Technologies. He requested a large conference room with computer projection systems for his laptop and a screen. He asked Danielle Conner how many people would attend the meeting and prepared enough handouts (with extra copies) for all the participants. He also had a copy of the recommendation report printed on high-quality paper to leave with Danielle at the end of the presentation.

Ned worked with his teammates to prepare the visuals to support his presentation. He wanted the PowerPoint presentation to reflect the content and structure of the written report but not contain the same amount of detail. Ned rehearsed the details he planned to share with the audience for each overhead transparency. He did not plan to discuss every detail in the report during his

FIGURE 17.6 **Draft of Transparencies for Cottrell Technologies Presentation**

The title slide identifies both the subject of the presentation and the speaker(s). The title slide also displays the basic design features that are repeated in all slides. Design consistency helps make your presentations accessible, coherent and professional. Choose a design that complements the ideas you are presenting. These design from the initial slide is carried over to all subsequent slides.

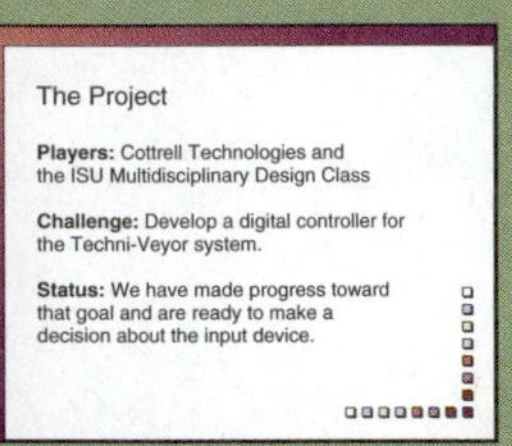

An audience expects an overview of the presentation, which may include identifying who and what as well as the expected outcomes of the presentation.

The presenter should not simply read this slide (or any other one). The slides can introduce, reinforce, and summarize information, but they are not the text of the presentation.

Goal and Objective

- To select an input device that meets your company's goals.
- To satisfy current and future customers with an affordable, state-of-the-art controller system.

You have lots of ways to make a presentation comprehensible. Audiences want to know why they're listening to a presentation and what's at stake, so goals and objectives should be presented in terms of the audience's concerns. Using bulleted lists keeps parallel ideas visually distinct so that the audience can quickly see the major points.

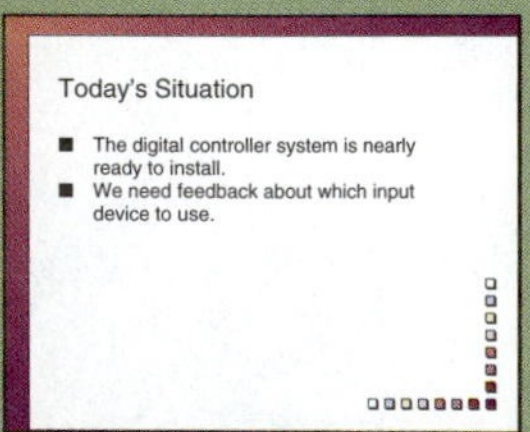

After identifying the goals and objectives, you can identify the factors, such as the status quo, that the audience needs to know. The bulleted list simply suggests the scope and sequence of points, not all the details; you present the details orally.

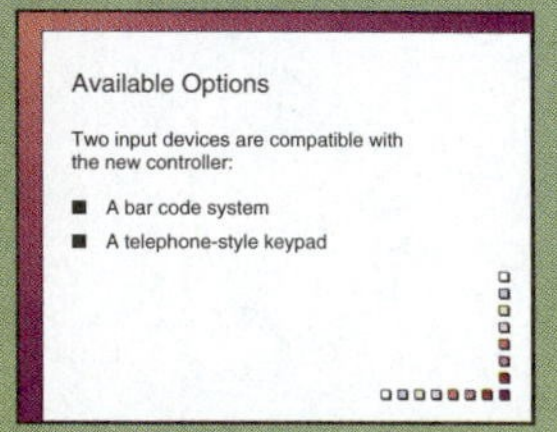

The options for audience members to consider should be clearly identified, so they can understand the available alternatives. You can use slides like this one to identify the options that you elaborate in your presentation. What are these options? What do they involve? Why are they the most appropriate options?

presentation, although he was familiar with those details should he need to use them during the question-and-answer session after the presentation.

The team decided that in addition to the PowerPoint presentation, Ned would show the input interfaces being considered. As he described the devices, he would indicate the features on actual models. Figure 17.6 shows the penultimate draft of the nine simple transparencies that Ned and his teammates designed using PowerPoint. They selected a simple, engaging, and professional template that formed the background for all the transparencies.

What is your response to this sequence of PowerPoint slides? They are the penultimate draft, not the final draft. Would you recommend revisions, or would you leave them as they are?

FIGURE 17.6 **Draft of Transparencies for Cottrell Technologies Presentation (continued)**

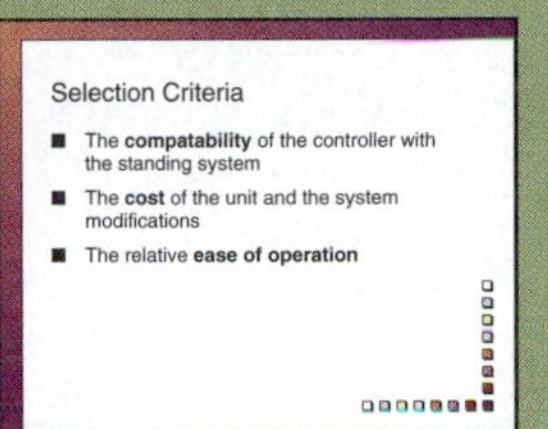

After the options are identified, audience members need to know the criteria for decision making. You can identify the criteria succinctly on a slide and then explain them during the presentation. You can boldface key words to emphasize the criteria, which will be the focus of your comments.

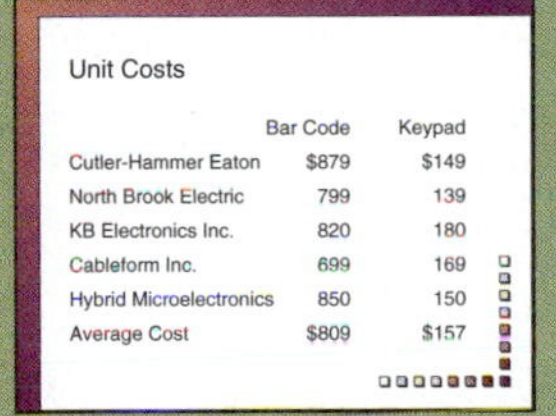

Unit Costs

	Bar Code	Keypad
Cutler-Hammer Eaton	$879	$149
North Brook Electric	799	139
KB Electronics Inc.	820	180
Cableform Inc.	699	169
Hybrid Microelectronics	850	150
Average Cost	$809	$157

Visuals help audience members grasp ideas quickly. For example, a table can help organize complex data so the audience can compare the alternatives according to cost.

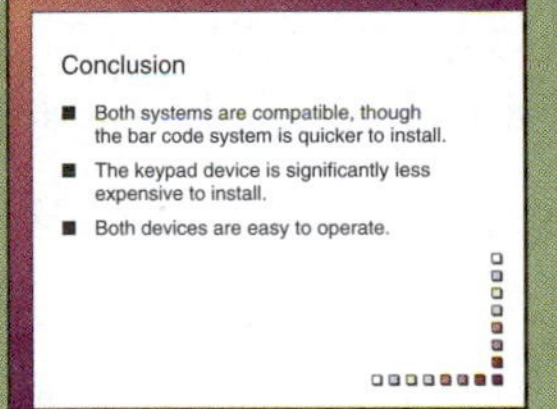

After presenting all the information that is important to audience members, you can remind them about the key ideas that you have discussed. You can make ideas more usable by briefly reviewing them, reminding the audience about what was most significant.

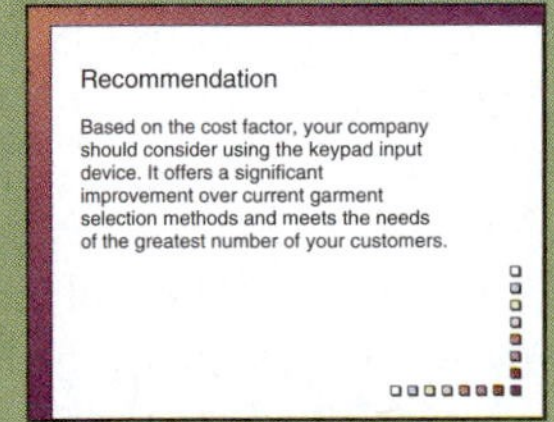

After presenting the audience with important information, you are ready to make recommendations. Audience members want to know what to do next, so they expect you to outline a clear plan, whether it is an action they must take or a decision that they must make.

WEBLINK

PowerPoint has come under intense scrutiny and criticism as a tool that is carelessly and sloppily used, too often substituting careful analysis for glitzy animation and dense text that mask poor thinking and weak writing. Go to http://english.wadsworth.com/burnett6e for a link to a site that identifies and explains many of PowerPoint's weaknesses and provides an ongoing dialogue about these problems.

CLICK ON WEBLINK

CLICK on Chapter 17/PPT criticism

WEBLINK

Despite the aggressive criticism of PowerPoint, many people recognize it as a valuable tool that can be used effectively. Go to http://english.wadsworth.com/burnett6e for a link to a site that responds to the criticism of PowerPoint and argues that its effectiveness is largely determined by the users, not by inherent factors in the tool.

CLICK ON WEBLINK

CLICK on Chapter 17/PPT tool

Handouts

Handouts can be a valuable asset for your presentation:

- They give your audience a convenient place to take notes.
- They provide a close-up view of essential information the audience will need to follow your presentation or complete an activity.
- They provide copies of primary visuals for immediate study and for future reference.
- They explain complex terminology.
- They summarize key points.
- They provide reading lists or bibliography references.
- They give the audience something to refer to after the presentation.

In considering whether to utilize handouts, you must resolve certain questions that are discussed below.

How Should Handouts Be Packaged? Handout material should be collated and bound, if only with a single staple. Loose pages create the impression of inadequate preparation and cause confusion; someone is always missing page 4 while someone else needs page 7. The packet should have a cover page that gives the title of your presentation, your name and organization, the meeting, the date, and the location. Sometimes you may also include an address, mail stop or E-mail, and telephone number. This cover information enables the packet to be filed and also provides a way to contact you later for questions or discussion. If you are making an important presentation, consider putting the packet in a folder with the organization's logo.

When Should You Distribute Handouts? Depending on their purpose, handout packets can be distributed at the beginning or end of a presentation. If the packet is needed for reference or note taking during the presentation, the packet should be distributed at the beginning of the session. You generally can keep people from riffling through the pages by saying, "Periodically during my presentation I'll refer to specific sections of your packet. I'll tell you what pages you need to refer to as they come up." If the packet contains sections that "give away" your presentation, or if it is simply a review of the oral and visual information, tell people that a summary will be available after the presentation for those who need it for reference. If you do not plan to refer to the packet during the presentation, it is certainly not necessary to distribute it at the beginning.

How do you prefer to use handouts during a presentation — as a place to take notes and confirm details or simply as a post-presentation reference?

How Much Detail Should Handouts Include? Handouts usually do more than duplicate the presentation; they highlight or outline key points and provide details of factual or statistical information that might be difficult to remember. Handout material is similar to visual material in that it supports the presentation. Particularly when the packet is distributed at the beginning of the presentation, the handout material referred to should illustrate the points in the presentation.

Visual material (e.g., transparencies or PowerPoint slides) may be reproduced so that the audience can refer to them quickly. Reproducing visuals also safeguards against breakdowns of audiovisual equipment and eliminates issues of poor visibility from a distance.

A handout should exactly reproduce the presentation only for formal conferences where someone presents a prepared paper. In that situation, copies of the paper are usually available after the presentation and are often published in a collection of conference papers.

How Should You Refer to Handouts during the Presentation? If you distribute the packet at the beginning of the presentation, tell your audience that you will refer to the appropriate pages as you go along. Each page should have a heading and a page number for easy reference. In a long packet with several sections, you will find references faster if you insert a sheet of colored paper between each section. Then you can say, "Turn to the page immediately after the blue page divider." Audiences seem to find this much easier than turning to page 9.

After you refer to a page in the packet, give the audience a few moments to locate the proper place. This slight pause eliminates most of the distraction caused by paper rustling.

Carefully assess what you want the audience to do during and after the presentation. Having them engaged in reading and note taking may encourage attention, but only if the activities are necessary and enhance your presentation.

What Is the Real Value of Your Handouts? Take the time and effort to prepare a packet only if you refer to it directly during the presentation or if people in the audience would need to refer to it afterward. If the packet is never going to be used, spend your time elsewhere to ensure the presentation's success. If the packet will be used, make the information accessible and put it in a clear context. For example, a list of key words and phrases might be meaningful during or immediately after a presentation, but a week or two later, most people will have forgotten the context. The context defines the terms, identifies the situation to which they apply, or explains how they were interpreted in your presentation.

Handouts Accompanying the Presentation to Cottrell Technologies. Ned Hitch worked with his teammates to prepare the handouts to support the presentation. They decided to use a feature of the PowerPoint software that enabled them to print thumbnail versions of their transparencies. They wanted the handout to include the transparencies Ned showed, along with space for notes so audience members could record their ideas and questions during the presentation. Once Ned and his teammates printed enough copies of the handout sheets, they stapled them into packets to distribute at the beginning of the presentation.

Poster Displays

Scientists, engineers, and other technical professionals in a variety of fields from agriculture to zoology use poster displays to present and share preliminary research findings and to communicate new ideas about technology to their colleagues. You may prepare poster displays for different types of professional conferences or seminars, any gathering that attracts colleagues from your field and asks them to present and share ideas.

Although the type of meeting or conference at which a poster is displayed dictates many of its characteristics, all posters strive to be accessible, comprehensible, and usable. The accessibility is particularly important: An effective poster is aesthetically appealing and easily legible. As you create your poster, use the following questions about purpose, design, transportation, and interaction to guide the preparation. Make a point to address the following questions and considerations:

What's the purpose? What's the message?	■ Decide the story you want to convey in the poster. Identify critical incidents or major points that tell the story. Determine a sequence that effectively tells the story. ■ Identify the primary audience and determine the likely level of prior knowledge. Select content complexity that will be appropriate. ■ Consider the verbal/visual balance that most effectively tells the story. Do you want more text, more images, or a balance? ■ Write succinct text in short paragraphs and bulleted lists to tell the story. Present equivalent text information in grammatically parallel structures. Label each chunk of text to help the audience understand the story. ■ Create easy-to-interpret, well-labeled images (such as photographs, lists, and graphs) to tell the story. Present equivalent images in visually parallel forms. Label each image to help the audience understand the story.
What will contribute to an effective, appealing design?	■ Use the design specifications specified by the conference. The amount of information presented and the poster's design are both affected by poster dimension restrictions (typical poster sizes are 2 to 3 feet wide and 3 to 6 feet long). ■ Conduct audience testing to determine the size, shape, texture, and color of the verbal and visual elements that attract favorable attention. Assess whether the poster's main message can be absorbed quickly. ■ Determine the budget for materials and preparation, including poster paper or card stock, photographs, enlargements, laser printing, lamination. ■ Check that all labels and text are (a) easy to identify, (b) short enough to read quickly, (c) interesting enough to hold attention, and (d) substantive enough to be useful. Test whether the print is large enough to read at a distance (1 inch height per 10 feet of distance). ■ Ensure that the design enables the audience to discern the logical flow of the information.
How will you transport and display of your poster?	■ Decide how you will transport your poster to the conference site. Roll it? Cut it into sections? Fold or hinge it? Determine how you will protect your poster during transportation. Mailing tube or a mailing box? ■ Find out what display options you have, for example, table, easel, wall, plastic- or felt-backed panel. Identify the supplies (e.g., Velcro, pins, tacks, tape) needed to set up your poster. Will conference organizers provide these supplies? ■ If your poster is in segments, make a sketch or photograph of your completed poster to use as a guide during setup. Bring a measuring tape or straightedge to position segments accurately. ■ Ask about how much time has been allotted — 10 minutes? 30 minutes? — to set up before the poster session begins.
How will you interact with people viewing your poster?	■ Determine whether you need to be present at an assigned time during the poster session to briefly present your poster or to answer questions. ■ Determine whether poster presenters are expected to bring an abstract or a fully developed paper. Find out how many copies you should bring. Will a table be available for the abstract or paper as well as your business cards? ■ Prepare a 30-second, 2-minute, and 5-minute explanation of your work. Always start with the 30-second version. If someone wants to know more, you'll be prepared with additional information.

Tara Barrett Tarnowski, now a graduate student in plant pathology at the University of Georgia, created this poster as part of her senior honors project. She imagined her audience would be graduate students, professors, and scientists in plant pathology. The audience would understand basic principles of plant pathology but have limited knowledge of the apple disease complex of her research.

Tara took twelve months to collect, analyze, and interpret her data. She took three weeks to create the poster itself. In commenting about the process of creating her poster, she said, "With limited space on my research poster, I needed to decide what was most important to convey to the audience. The poster gave me a chance to creatively combine images and text in a way that is easy to understand and shows my research in a compelling way."

The work represented in this poster has been turned into a manuscript, published in the online, peer-reviewed journal *Plant Health Progress.*

Accessible

- *The typeface is Comic Sans MS, with the title in 60-point type, section labels in 44 point, and text in 28 point — legible from a typical viewing distance.*
- *The visuals and their labels are sufficiently large to be legible.*
- *The white space between the boxed chunks and between items in the lists makes the information more legible.*
- *High figure-ground contrast between text and images and their backgrounds increases legibility.*

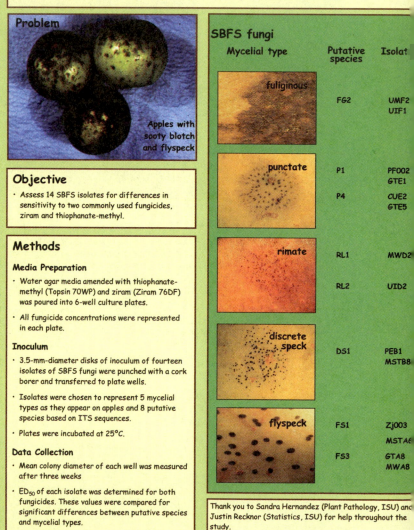

Fungicide sensitivity differs among newly discovered fun

Tara Barrett, Jean Batzer, Mark Gleason, and Phillip Dixon ~ Departmen

Introduction

Sooty blotch and flyspeck (SBFS) is a complex of several species of fungi resulting in the spotting and smudging of apple cuticles. The disease complex is economically significant as it decreases the market value of apples.

Several previously unknown members of the disease complex were found in a recent survey of nine orchards in the Midwest (IA, IL, MO, WI). The fungicide sensitivity of these fungi has not been characterized.

Objective

- Assess 14 SBFS isolates for differences in sensitivity to two commonly used fungicides, ziram and thiophanate-methyl.

Methods

Media Preparation

- Water agar media amended with thiophanate-methyl (Topsin 70WP) and ziram (Ziram 76DF) was poured into 6-well culture plates.
- All fungicide concentrations were represented in each plate.

Inoculum

- 3.5-mm-diameter disks of inoculum of fourteen isolates of SBFS fungi were punched with a cork borer and transferred to plate wells.
- Isolates were chosen to represent 5 mycelial types as they appear on apples and 8 putative species based on ITS sequences.
- Plates were incubated at 25°C.

Data Collection

- Mean colony diameter of each well was measured after three weeks
- ED_{50} of each isolate was determined for both fungicides. These values were compared for significant differences between putative species and mycelial types.

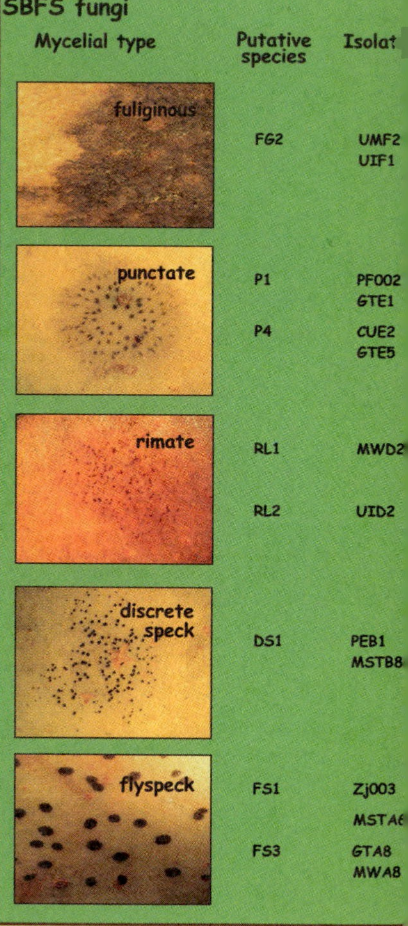

SBFS fungi

Mycelial type	Putative species	Isolat
fuliginous	FG2	UMF2 UIF1
punctate	P1	PF002 GTE1
	P4	CUE2 GTE5
rimate	RL1	MWD2
	RL2	UID2
discrete speck	DS1	PEB1 MSTB8
flyspeck	FS1	Zj003 MSTA6
	FS3	GTA8 MWA8

Thank you to Sandra Hernandez (Plant Pathology, ISU) anc Justin Recknor (Statistics, ISU) for help throughout the study.

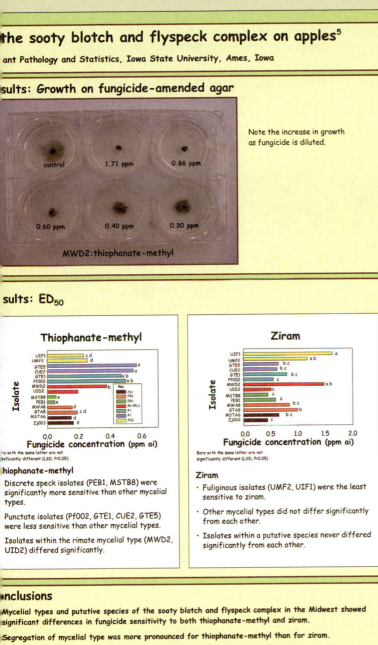

the sooty blotch and flyspeck complex on apples[5]

ant Pathology and Statistics, Iowa State University, Ames, Iowa

sults: Growth on fungicide-amended agar

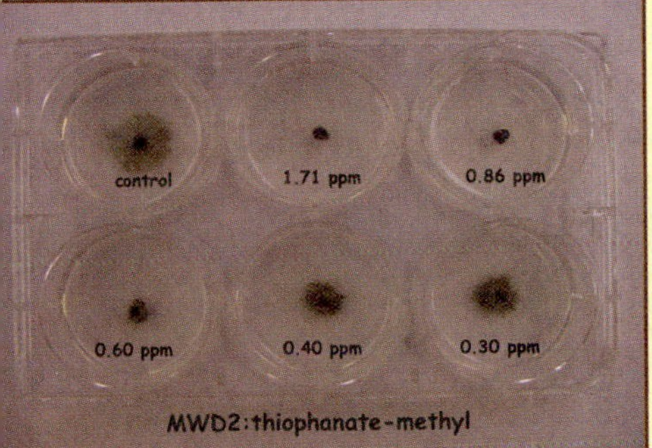

Note the increase in growth as fungicide is diluted.

sults: ED_{50}

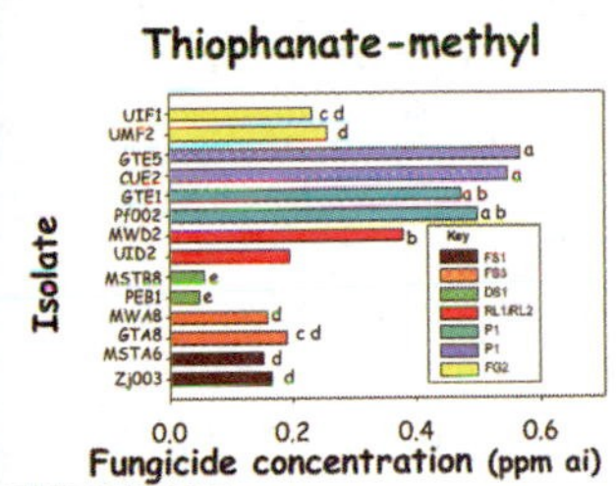

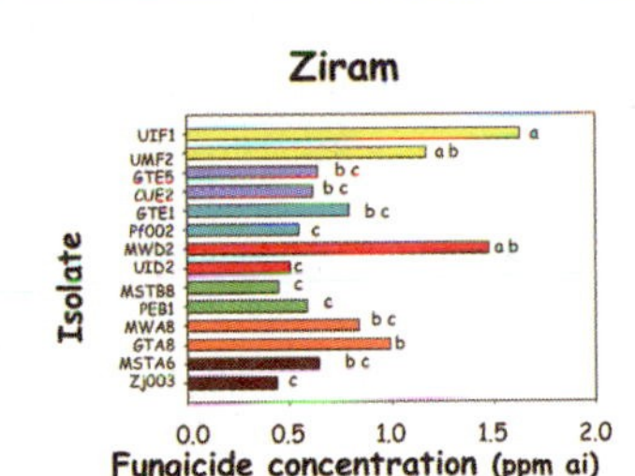

rs with the same letter are not ...nificantly different (LSD, P<0.05)

Bars with the same letter are not significantly different (LSD, P<0.05)

hiophanate-methyl

Discrete speck isolates (PEB1, MSTB8) were significantly more sensitive than other mycelial types.

Punctate isolates (Pf002, GTE1, CUE2, GTE5) were less sensitive than other mycelial types.

Isolates within the rimate mycelial type (MWD2, UID2) differed significantly.

Ziram

- Fuliginous isolates (UMF2, UIF1) were the least sensitive to ziram.
- Other mycelial types did not differ significantly from each other.
- Isolates within a putative species never differed significantly from each other.

nclusions

Mycelial types and putative species of the sooty blotch and flyspeck complex in the Midwest showed significant differences in fungicide sensitivity to both thiophanate-methyl and ziram.

Segregation of mycelial type was more pronounced for thiophanate-methyl than for ziram.

With the exception of the rimate isolates (RL1, RL2), isolates within both a putative species and mycelial type did not differ significantly in their fungicide sensitivity.

Comprehensible

- *Clear labeling of the types of isolates helps audience understanding.*
- *Additional information — for example, the suggestion next to the photos of the culture plates to "note the increase in growth as fungicide is diluted" — increases the speed and accuracy of audience interpretation.*
- *Photographs increase understanding of ways in which different isolates affect apple fruit appearance.*
- *Fungicide concentrations superimposed onto culture plate photo enables audience to conceptualize results.*

Usable

- *Sufficient detail enables the audience to assess appropriateness of methodology and consider replication of the results.*
- *Graphs visually display differences in fungicide sensitivity and allow audience to form conclusions regarding the results of the study.*
- *Conclusions address both results of study and implications, which enables audience to consider ways to respond to these diseases.*

Tara Barrett Tarnowski at work in a plant pathology lab

Courtesy of Tara Barrett Tarnowski

Presenting Yourself

What aspects of making an oral presentation make you the most anxious? The most excited?

Making an interesting, informative, well-organized presentation in a confident, appealing manner does not happen automatically. It takes the careful preparation just discussed as well as control of your body and voice, the ability to respond effectively to questions, and lots of practice.

Part of the manner in which you present yourself is based on the *ethos* you present to the audience, that is, the image of your personality and character. The ethics sidebar below explores the historical origins of ethos and its contemporary importance in oral presentations.

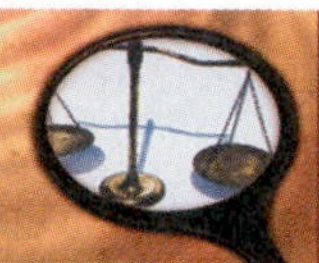

ETHICS SIDEBAR

Speaking Well: Ethics in Oral Presentations

Classical Greek and Roman texts about rhetoric were primarily concerned with oral communication; more than 2,000 years ago, in the days of scholars such as Plato, Aristotle, and Cicero, communicators were highly valued for their ability to present and defend arguments orally, both in law courts and in public, political meetings. Education for these ancient orators centered primarily on developing the orator's ability to persuade an audience.

This training had at its foundation a recognition of the value of the orator's personal integrity and character in influencing the audience. These characteristics of personality and character, referred to as *ethos,* were important components in building an audience's trust and respect for the orator. Quintilian, a classical Roman scholar, stated the connection between ethos and oration succinctly (despite being sexist): an orator is a "good man skilled in speaking."

Contemporary discussions about ethics and professional communication often focus on the importance of ethos. Researcher Stuart Brown, for example, claims that establishing a respectable ethos is the best way for technical professionals to produce ethical communication. Instead of viewing ethics as rules that you apply, much like applying spell checkers in word processing programs to catch spelling violations, you can view ethics as a reflection of a writer's personal integrity — that is, a reflection of the person's ethos.[6]

The connection between ethics and ethos is especially important for presenters, as the classical scholars recognized, because spoken words are so closely linked to the speaker. Researcher Laura Gurak, a specialist in technical presentations, considers the ethos of the presenter important because the presenter is in a position of "power, authority, and responsibility" to the audience. Historical examples such as Adolf Hitler clearly demonstrate the extreme consequences of speakers using their position to advocate unethical causes. Because a presenter can be so influential, distinguishing between a projection of ethos (what the audience thinks the speaker is like) and a speaker's true ethos (what the speaker is actually like) is very important. Matching our public and private ethos allows us to speak with more sincerity and make more believable speeches. When public and private ethos differ, however, deception and insincerity are more likely and our speeches are less believable. "As a presenter," Gurak explains, "you are responsible for not only structuring the most effective presentation, but you are also charged with the job of knowing how you feel about the topic and of being the most honest and sincere presenter that you can be."[7]

What are some ways to avoid presenting deceptive information?

Why do we expect presentations to be a reflection of a speaker's ethics? Are you willing to publicly present your beliefs and ideas? Would you state ideas during a public presentation that you don't actually believe? If you were asked to present a position on an issue that you disagree with, how would you respond?

Professional Appearance

Your professional image depends in part on your demeanor during the presentation. Your material can be accurate and interesting, and your voice can project to the back of the room, but if your behavior or appearance detracts from the presentation, its effectiveness will be diminished. Some people do strange things when they are in front of a group, partly from nervousness, partly because they do not consider how the audience will react.

What is especially distracting to you when listening to an oral presentation?

Most distracting behaviors disappear when a speaker relaxes in front of an audience. To discover your own distracting behaviors, arrange for a presentation to be videotaped. View the tape and note the areas that you need to improve, then practice to eliminate the problems. Work with groups such as Toastmasters, a professional organization with chapters throughout the country, for improving workplace presentations. Such a group often helps because you get immediate, constructive feedback from other people who also make professional presentations. Also consider using your company's training and development staff. They often are willing to arrange for practice sessions for an important presentation you have to give and may provide a staff member to help you polish your delivery. Figure 17.7 identifies some suggestions that will help you give a professional presentation.

WEBLINK

Toastmasters International is a nonprofit organization founded in 1924 to help people give better presentations, develop leadership and management skills, work more productively with colleagues, and offer and accept productive criticism. For a link to Toastmasters, go to http://english.wadsworth.com/burnett6e and then access the About Toastmasters menu.

CLICK ON WEBLINK

CLICK on Chapter 17/Toastmasters

FIGURE 17.7 Suggestions for Professional Appearance During a Presentation

Goal	Suggestions to Eliminate Potential Problems
Wear Appropriate Clothing	■ Wear clothes that make you feel good and that you don't have to adjust. ■ Avoid clothes and jewelry that detract from the presentation. ■ Check and adjust your clothing and hair *before* getting up to speak.
Handle Notes Comfortably	■ Number the cards or pages; this is particularly useful if you should accidentally drop them. ■ If there is no podium, have a firm backing for your notes so that the sheets don't bend and shake.
Make Eye Contact	■ Look directly at North American audiences to establish a rapport and also give you the appearance of confidence. People are unlikely to let their attention wander if you look, even briefly, directly at them. ■ Don't stare over the heads of the audience; it makes them uncomfortable and may cause them to question your credibility.
Handle Mistakes Smoothly	■ If you make a noticeable error, simply apologize and continue with the presentation. ■ Don't scrunch your lips, move your tongue inside your cheek, squint your eyes, or giggle.
Relax Your Hands	■ Use your hands to hold your notes. ■ If you're not holding notes, rest your hands on the podium or lectern. ■ Try one hand casually in your trouser or skirt pocket. ■ Relax your hands by your side. ■ Gesture naturally, as you normally would in a discussion. Hold the pointer for the flip chart or overhead projector. ■ Don't shove both hands in your pockets, clutch the podium, or gesture wildly. This will distract your audience.
Relax Your Feet	■ Stand on both feet during your presentation. ■ Wear shoes that are comfortable (and keep them on your feet). ■ Don't rock back and forth.
Move Naturally	■ Do a few relaxation moves before you get up to speak. ■ While you're speaking, avoid nervous actions such as knuckle cracking, shoulder shrugs, stretches, knee bends. ■ Avoid standing ramrod straight without moving. ■ Try to move naturally and comfortably by focusing on what you are saying, not on how you look.
Use the Podium Comfortably	■ Don't lean on the podium. Don't clutch the podium. ■ If you're tall, don't lounge on the podium. ■ If you're short, arrange to have a riser to stand on, or just move to the side of the podium and do not use it at all.

Vocal Characteristics

Your "voice print" represents a unique combination of vocal characteristics: volume, articulation and pronunciation, rate, and pitch. Everyone has a distinctive voice that can be used effectively in an oral presentation. Vocal characteristics are so closely intertwined that the improvement of one characteristic nearly always improves the others.

A speaker has to be heard. The speaker's voice needs sufficient volume to project throughout the entire room. If possible, practice your presentation in a room the same size as the one you'll make the presentation in. Speak in a voice that can be clearly heard by a person sitting in the back of the room. If you are speaking in a very large room, arrange to have a microphone available and practice with it before you speak.

Your audience will react more positively if you clearly articulate each word. Pronounce every part of each word, without dropping, adding, or slurring letters or syllables. Your presentation will go more smoothly if you are confident of the pronunciation of every word in your presentation; mispronouncing or stumbling over words will make you uncomfortable and may lower your credibility with the audience.

The rate at which you speak affects how the audience reacts and how well they can listen. The average speaking speed is approximately 150 words per minute. Variations in your rate of speech — your pacing — give you some control over the audience's attention. Rapid delivery demands the audience's full attention, and slow delivery allows carefully placed emphasis. Occasional pauses do not detract from a presentation but rather permit the audience (and you) time to collect thoughts.

Use your voice as a tool. Its pitch or tone, its highness or lowness, helps determine your credibility and appeal. Pitch is controlled by muscle tension, so relaxing helps your voice sound natural (and usually prevents the nervous squeak caused by tense muscles). An unvarying pitch creates a monotone, which is boring and difficult to listen to for very long. Your control of pitch allows you to inflect or emphasize important words and thus your speech.

What are your strongest vocal characteristics? Those that need improvement?

Handling Questions

Members of your audience almost always ask questions, so you need to plan not only how but also when to respond. Whether you permit questions during the presentation depends on the nature of the presentation, your reaction to being interrupted, and the size of the audience. If you prefer not to be interrupted, ask people to hold their questions until the question-and-answer session at the end. Questions are usually held until the end in formal presentations, particularly with large audiences. However, if the session is small and informal and you do not mind interruptions (and can easily get back on track), tell the audience you will respond to questions as they arise.

You need to consider problems that might result from questions. For example, if you don't know the answer to a question, don't panic. A presenter is not expected to know everything. An experienced, confident presenter will say, without embarrassment, "I'm sorry, I don't know, but if you'd like me to find out, contact me after the presentation."

A person may disagree with you, in which case you will at least know people are listening. In some cases, you may simply acknowledge the specific area of disagreement, noting that multiple views or interpretations are possible. In other situations, you may want to focus on the likely cause of the disagreement or restate your case, including a response to the person's disagreement.

> I understand your reluctance to commit to the expense of retooling an entire production line, but the market research shows first-quarter sales of this product will cover all equipment and tooling.

If one person monopolizes the questioning, you can ask if other people have questions, and then call on another person. If no one except the monopolizer has a question, suggest that the person meet you after the presentation to continue the discussion and end the question-and-answer session.

Occasionally, someone will ask a seemingly stupid question or perhaps a question about something you have carefully explained. Act as if the question is legitimate. You can't tell what provoked the person to ask the question. Keep a straight face and give a straight answer.

Other times a person will make a statement rather than asking a question. In such situations, you can express interest in the person's view and then ask if the person has a question.

If you or the audience can't hear the question, tell the questioner, and ask him to repeat it. If parts of the audience can't hear a question, you should repeat it so that everyone can hear. Repeating the question has the additional benefit of affirming that you understand what was asked.

When a question is asked that has nothing to do with the topic or that you don't understand, you can ask the questioner to rephrase the question; it may just be poorly worded. If you ask the person to repeat the question, you might get a repetition of the very same one you didn't understand. If the rephrasing doesn't help, you might not be able to provide an answer. Finally, if no one has a question, thank the audience and conclude the presentation.

Evaluating Presentations

To be a good critic, you have to be a good listener, an active listener. You also need to know what to look for and listen to in the presentations you are asked to evaluate.

Active Listening

You can't accurately or fairly evaluate any presentation that you haven't listened to closely. Most good listeners have developed specific techniques that help them listen more effectively — staying attentive, following the speaker's presentation, understanding the content — by engaging internally in a number of activities that increase their listening comprehension. As you listen to a presentation, try to practice the behaviors identified in Figure 17.8, which focus on determining purpose, identifying organization, distinguishing critical elements, monitoring reactions, and making connections.

As a speaker, you must encourage the audience to listen to your speech so that members recognize the purpose of your presentation, recall significant points,

How can you try to engage an audience that is not paying attention?

FIGURE 17.8 **Strategies to Promote Active Listening**[8]

Strategies	Activities
Determine Purpose	▪ Determine your own purpose for listening. ▪ Identify the speaker's purpose.
Identify Organization	▪ Identify and follow the speaker's plan of organization. ▪ Accurately identify the speaker's main points and ideas. ▪ Keep track of main points by note taking, using the outline in handout, or mental recapitulation. ▪ Accurately note the speaker's supporting details and examples. ▪ Note transitional words and phrases.
Distinguish Critical Elements	▪ Distinguish between old and new material. ▪ Distinguish between relevant and irrelevant material. ▪ Distinguish between facts and opinions.
Monitor Your Own Reactions	▪ Note possible speaker bias and emotional appeals. ▪ Note your own bias and reaction to appeals. ▪ Recognize speaker inferences. ▪ Delay criticism until the speaker is finished. ▪ Ask questions (mentally or on paper) as the talk proceeds.
Make Connections	▪ Predict outcome of presentation. ▪ Summarize and paraphrase the speaker's main points (mentally or on paper) after the presentation. ▪ Anticipate possible impact of speaker's remarks. ▪ Draw conclusions from the presentation. ▪ Relate the speaker's ideas to your own.

and critically evaluate the content. If research done on the listening patterns of college students is any indication, you have a challenging job. One widely cited study reported that after 15 minutes, 10 percent of the audience was inattentive; after 18 minutes, 33 percent; and after 35 minutes 100 percent.[9] If you develop good listening techniques, you should be able to use this knowledge to improve your skill as a presenter.

Assessing Presentation Skills

Oral presentations are a regular part of academic and professional life. Most of the time you will attend a presentation to acquire information. Sometimes, however, a classmate or colleague will ask you to assess the presentation he or she has given. The questions in Figure 17.9, though not exhaustive, will help you with such an evaluation.

What are your presentation strengths? What areas need improvement?

You can also use these evaluation questions to assess your final rehearsal of a presentation you're going to give. Make an audiotape or a videotape of your presentation, then evaluate this final rehearsal. Listen to or watch the tape critically, and note the places where you can strengthen the presentation. Play the rehearsal tape at least twice because you'll pick up different things each time. Once you have a list of things to change, you will probably need to rework the presentation. People who give important professional presentations often go through two or three tape-and-review cycles before they are satisfied with their presentation.

WEBLINK

For a link to checklists to assess and evaluate oral presentations, go to http://english.wadsworth.com/burnett6e.
CLICK ON WEBLINK
CLICK on Chapter 17/checklist

FIGURE 17.9 Questions to Guide Evaluation

Categories	Critical Questions	Yes	No
Physical Environment	■ Is the presentation area set up before the presenter begins?	☐	☐
	■ Is the presentation area arranged so that everyone in the audience has a clear view?	☐	☐
Content	■ Is the information accurate and verifiable?	☐	☐
	■ Is the information adapted to the audience?	☐	☐
	■ Are the details and examples relevant and appropriate?	☐	☐
Professional Demeanor	■ Does the presenter appear well-prepared?	☐	☐
	■ Does the presenter appear poised and confident?	☐	☐
	■ Does the presenter handle notes unobtrusively?	☐	☐
	■ Does the presenter adhere to the time limits?	☐	☐
	■ Does the presenter provide helpful handout material?	☐	☐
	■ Is the presenter dressed appropriately?	☐	☐
Organization for the Audience	■ Does the presenter identify the presentation's purpose?	☐	☐
	■ Does the presenter make the opening interesting?	☐	☐
	■ Does the presenter preview and review periodically?	☐	☐
	■ Are the main points in the presentation clearly identified?	☐	☐
	■ Is the information logically organized?	☐	☐
	■ Does the presenter provide transitions between ideas?	☐	☐
	■ Does the presenter use effective examples?	☐	☐
	■ Does the presenter provide an appropriate conclusion?	☐	☐
Visual Support	■ Are the visuals used in the presentation large enough to be seen by everyone in the audience?	☐	☐
	■ Are the visuals appropriate in type/style and professional in appearance?	☐	☐
	■ In a demonstration, does the presenter complete the actions and speak simultaneously?	☐	☐
Presentation Style	■ Is the presenter's voice pleasant and professional?	☐	☐
	■ Is the presenter's body language appropriate?	☐	☐
	■ Is the presenter's voice loud enough for the size of the room?	☐	☐
	■ Is the presenter's inflection varied? pronunciation correct? pacing appropriate?	☐	☐
	■ Does the presenter make eye contact with the audience?	☐	☐
	■ Does the presenter respond directly to audience questions?	☐	☐

Individual and Collaborative Assignments

1. **Design a plan for an effective oral presentation.** Thelma Michelson is a staff chemist doing successful R&D work a year after her graduation. She has returned to college to take an evening course in technical communication to upgrade her professional skills. The only assignment she is apprehensive about is an oral presentation near the end of the term. Throughout the semester, she has used various aspects of her research at work for class assignments. She has often been able to use completed class assignments for writing she is required to do as part of her job.

   One paper describing an innovative analytical process is particularly important to her because when she showed it to the project manager, he added the names of the other project members and submitted it for presentation at the upcoming American Chemical Society conference. Thelma was ecstatic when the paper was accepted for presentation at the conference and publication in the conference proceedings. Now she is panic-stricken — her project manager has suggested that she present the findings because she was instrumental in the process development.

   Thelma has a list of reasons why she is the wrong person to represent her company: She'd turn red, her voice would crack or disappear, she'd forget her material, she'd forget to move to the next screen on the computer projection system, she'd forget to refer to the handouts, she'd be unable to answer questions. She is very anxious that her nervousness about talking in front of a group and her inexperience as a presenter will detract from her credibility as a chemist. Her manager has insisted: Thelma will make the presentation.

   She has five weeks to prepare for a full dress rehearsal of her presentation for the upper management of her company. She believes if she can do well in this rehearsal, she will do well in the actual presentation. How should she prepare? Design a practical and effective plan for Thelma to follow. Be very specific. She will need a plan that identifies what she should do and when she should do it. Build in ways that will give her constructive feedback as well as increase her confidence.

2. **Prepare an informative presentation.** Educate your audience about a product, practice, or procedure. Consider these topics:
   - Start-up of a new manufacturing facility
   - Alternatives to traditional burial
   - Expectations in workplace culture in [select country you're interested in]
   - Career options for social workers
   - Polymer film used to cover food products
   - Building of a microchip

3. **Prepare a presentation to demonstrate how to use equipment or complete a process.** Choose a process or equipment with which most of your audience is unfamiliar. Consider these topics:
   - Mountaineering equipment
   - Sudden infant death syndrome (SIDS) monitor
   - Film or tape splicing
   - Quilting
   - Cleaning oil-soaked water birds after an oil spill
   - Restringing a tennis racquet
4. **Prepare a persuasive presentation.** Try to persuade your listeners to accept your position. Consider these topics:
   - Advocating tool cribs or individual tool boxes for a machining company
   - Assuming all information available on the Internet is in the public domain
   - Lowering industrial emissions standards
   - Providing medical care in hospices for terminal patients
   - Installing filtering software for corporate Internet users
   - Mandating smallpox vaccinations for all military personnel
5. **Prepare a training session that teaches your audience to complete a task.** Consider these topics:
   - Successfully going through airport security
   - Polishing an optical surface
   - Caning a chair
   - Air layering a plant
   - Inspecting a machined part
   - Creating a Web site
6. **Utilize visuals effectively in a presentation.** Revise one of the presentations in Assignments 2–5 to utilize PowerPoint or another visually based presentation system.
7. **Conduct a survey.** Go to http://english.wadsworth.burnett6e for a link to a survey to examine. Click on Chapter 17/oral communication survey. Modify it to use in a local survey for identifying the kinds of oral communication that are a critical part of workplace activities in an organization. Each group in the class should focus on a different local organization.

   (a) Work in a small group to select an organization whose oral communication you'll explore, and then identify your goals for that investigation. Contact the organization to enlist its cooperation.

(b) Work together to modify the survey.
(c) Contact the organization to arrange to administer the survey to ten employees.
(d) Compile the results and, in your small group, interpret these results.
(e) Prepare an oral presentation to the class in which you present your purpose, methods, results, conclusions, and recommendations.

8. **Critique a PowerPoint presentation.** To apply the principles and guidelines you've learned, go to **www.osti.gov/govtec.html** for links to 12 slides and text of a presentation about the Digital National Library of Energy Science and Technology. Work in a small group to critique this presentation. Compare your critique with that of other groups in the class.

Chapter 17 Endnotes

1 Critical link between presentation skills, upward mobility. (1991). *Supervision, 52*(24), 6.

2 Wickman, F. (1992). Getting them to "buy in" to your message. *Supervisory Management.*

3 A number of Web sites report similar information about the value of visuals, most depending on two widely cited but old studies: These following sites are useful, in part because they cite their sources. (1) Epson America. (2004). The numbers on why you need visuals. Retrieved January 23, 2004, from http://www.presentersonline.com/basics/visuals/needvisuals.shtml (2) Media Services. (1999). Why use visuals in the first place? Pacific Lutheran University. Retrieved January 23, 2004, from http://www.plu.edu/~media/why_visuals.html (3) Nordgren, L. (1998). Why use visuals? Designing and developing multimedia productions. Retrieved January 23, 2004, from http://www.plu.edu/~libr/workshops/multimedia/why.html

4 Modified from White, J. V. (1990). *Color for the electronic age.* New York: Watson-Guptill Publications.

5 Tarnowski, T. B., Batzer, J., Gleason, M., & Dixon, P. (2003). Fungicide sensitivity differs among newly discovered fungi in the sooty blotch and flyspeck complex on apples. Iowa State University, Ames, Iowa.

6 Brown, S. (1994). Rhetoric, ethical codes, and the revival of *ethos* in publication management. In O. J. Allen & L. Deming (Eds.), *Publications management: Essays for professional communicators* (189–200). Amityville, NY: Baywood.

7 Guark, L. (2000). *Oral presentations for technical communication.* Boston: Allyn and Bacon.

8 Adapted from Devine, T. (1982). *Listening skills schoolwide: Activities and programs.* Urbana, IL: National Council of Teachers of English.

9 Bauer, L. (1983). Best uses of the lecture. Project PHYSNET. Michigan State University. Retrieved January 21, 2004, from http://35.8.247.219/home/modules/pdf_modules/m83.pdf

CHAPTER 18

Preparing Correspondence

Objectives and Outcomes

This chapter will help you accomplish these outcomes:

- Understand that correspondence, an important kind of technical communication, does not exist in a vacuum, nor does it have rigid format prescriptions
- Compose correspondence using appropriate planning strategies, selecting appropriate content, and developing an appropriate organization so the document is legible, readable, and usable
- Present good news in direct (descending) order. Present bad news in indirect (ascending) order
- Positively influence readers' perceptions of your professional competence by using direct language, adopting a *you*-attitude rather than an *I*- or *we*-attitude, focusing on readers rather than yourself, and avoiding exclusionary language
- Understand and respond to the factors in the rhetorical situation that affect the composition and interpretation of correspondence

> How much correspondence do you typically receive each week — total number of e-mails, memos, and letters? How much of it is important to you?

Correspondence includes all types of e-mail, memos, and letters. Unlike either face-to-face conferences or telephone conversations, correspondence documents interactions, which can be referred to and checked.

This chapter includes guidelines and examples to help you prepare correspondence. You'll begin by reading about general characteristics of correspondence. Then you'll read about composing correspondence: following guidelines, using the appropriate attitude and tone, and responding to audience, organization, format, and visual displays. Finally, you'll read a case that illustrates the *domino effect* of correspondence, one letter or memo generating a chain reaction of correspondence.

Characterizing Correspondence

Accurate and unambiguous correspondence — e-mail, memos, and letters — is the everyday communication, both internal and external, that moves workplace activities forward. Short e-mail messages are effective for relatively straightforward communication, for example, inquiring about costs, deliveries, or specifications; responding to a query; or reminding colleagues about a meeting. For complex interactions, written correspondence, whether paper or electronic, is appropriate; you can revise phrasing until it is precise and unambiguous, clarify complex issues, and archive a permanent record.

Correspondence can be distinguished from other genres of technical communication in several ways: audience, composing and revising, datedness, and conventions.

- **Audience.** Although notable exceptions exist, correspondence typically addresses one person or an identified group of people — for example, coworkers, managers, customers and clients, suppliers, and the press. E-mail can be used for internal or external readers, memos for people or specific groups internal to the organization, and letters for people external to the organization. Many people do not consider most correspondence confidential, so the e-mail, memos, and letters that you write to one person may be read by others.

> How can you respond to both the time demands of correspondence (usually it's needed right away) and the need to create professional, high-quality documents?

- **Composing and Revising.** Because you are usually familiar with your audience's expectations and your content (queries from you or information for them), correspondence is often written fairly quickly and may not undergo as many revisions as other technical documents. However, dashing off a message and sending it before you reconsider the content, clarity, and conventions can lead to problems. Always reread correspondence before sending it.
- **Datedness.** Because correspondence usually responds to a current situation, the information in most e-mail messages, memos, and letters needs to be updated more frequently than other technical documents.

- **Conventions.** Because correspondence is often prepared quickly, writers sometimes pay insufficient attention to conventions such as mechanics, grammar, diction, and sequence of information. Inattention to such conventions might make the audience think either that accuracy and details don't matter to the writer or that the readers aren't important enough to merit attention to conventions.

Correspondence is similar to other genres of technical communication in three important ways. First, your company or organization owns all work-related correspondence you write. Just because you write it doesn't means it's yours. More specifically, anything you write at work can be used in court proceedings, either for or against the organization. Second, correspondence should follow Grice's overarching principle that information should be timely as well as purposeful and also follow his four maxims (as detailed in Chapter 1): be accurate and verifiable, be as informative as necessary, be relevant, be "perspicuous" (so avoid obscurity and ambiguity, be brief and well organized). Third, correspondence should consider factors affecting accessibility, comprehensibility, and usability that are critical for all technical documents (presented in Figure 1.2 in Chapter 1).

Who should be responsible for each of the following elements of correspondence — the originator or the transcriber (secretary)?
- *timely response*
- *factual accuracy*
- *grammatical precision*
- *typographical correctness*
- *clear, tactful, and understandable wording*
- *relevant, usable information*
- *print quality*

Delivering Correspondence

Paper correspondence has traditionally been sent by the U.S. Postal Service (USPS), usually arriving in one to five business days, depending on the method of delivery selected. For an additional fee, the USPS can also track an individual envelope or package. Many organizations send very important documents such as contracts, other legal documents, one-of-a-kind documents, and time-sensitive documents by overnight delivery such as UPS and FedEx, which also tracks each individual envelope or package.

Common electronic modes of delivery include faxes, various kinds of asynchronous communication, and synchronous communication. As more organizations develop policies for accepting electronic signatures, electronic correspondence for legal documents will become more accepted.

E-mail is the most common method of asynchronous correspondence. However, electronic mailing lists and electronic bulletin boards are also widely used. Lists are grouped e-mail addresses that allow you to send one message to multiple recipients — for example, all employees, all team members, or all members of a professional organization — without typing in every individual address. Bulletin boards allow users to post messages to large audiences, but instead of receiving an e-mail with the message, users must log on to the board to see messages and responses. Bulletin boards are often used by organizational intranets.

E-mail has changed the nature of workplace correspondence. Go to http://english.wadsworth.com/burnett6e for links about e-mail trends in the workplace, specifically the ways in which "e-mail affects productivity, communication, and quality of life at work." The survey was conducted by Vault.com, an online career information company that surveyed over 1,000 employees to determine the ways in which e-mail affects workplace life.

CLICK ON WEBLINK

CLICK on Chapter 18/email survey

Synchronous correspondence such as instant messaging is increasingly popular in workplaces, in some instances replacing telephone calls. Instant messaging allows individuals and team members to instantly send and receive informal messages.

All electronic messages, whether asynchronous or synchronous, should be as correct and informative as any other type of correspondence you might write. The following *netiquette* (a contraction of "network" and "etiquette") guidelines should apply to all of your electronic correspondence.

Headings

- Check the *To* line of your e-mail message to confirm that you are responding to the appropriate person or persons (and not to an entire list if you don't intend to).
- Include a brief descriptive note in the *Subject* field to aid reading, storing, and searching at the other end. "RE: RE: FWD: RE: Phone Call" is not a descriptive subject line.

Content

- Cover only one topic per message to make replying, forwarding, or organizing archived messages easier.
- Keep the message brief and on topic.
- Indicate the content of the original message when replying by quoting pertinent portions or by summarizing the subject. You do not need to copy the entire message.
- Do not respond immediately to a message that upsets you; in all cases, avoid *flaming,* which is an unprofessional, emotional, and usually rude electronic response. If you would be unprofessional to say the words over the phone or face-to-face, avoid sending them in an electronic message.

Audiences

- Write as if the whole world will read your message, because messages can be easily and accidentally forwarded.
- Confirm that the recipient actually received an important message by asking for acknowledgment.

Conventions

- Begin with an appropriate salutation such as the person's name. Do not begin a workplace e-mail with "Hey." (Sometimes in a rapid-fire series of very short e-mails between a small group of people, salutations may be omitted after the initial identities have been established.)
- Spell and use words with care.
- Avoid emoticons and cutsey, abbreviated spellings in professional correspondence, such as :-) or "c u" for "see you."
- Avoid using all caps. This is considered SHOUTING. Use upper- and lowercase text.

An innovative letter format omits the salutation and complimentary closing and incorporates the addressee's name in the first sentence. Can you find any reasons why an organization (or individual) should adopt or reject this style?

Keep in mind that all electronic messages can be forwarded, printed, or permanently stored. E-mail can also be misdirected, even when you are careful, so do not send an electronic message that you would not want everybody to read. This caution is reinforced in the ethics sidebar below.

What's the most annoying thing about some of the e-mail you receive?

ETHICS SIDEBAR

To Delete or Not to Delete: E-Mail and Ethics

Consider the following actual e-mail exchange:

To: David Smith
From: Laura
Message: Hi, David, Please destroy the evidence on the [name of case] litigation you and I talked about today. Thanks, Laura.

\- - - - - - -

To: Laura
From: David
Message: Hi, Laura, Acknowledged your message and taken care of. Aloha, David.

Consider another e-mail sent to a coworker:

Message: Did you see what Dr. [name omitted] did today? If that patient survives, it will be a miracle.

These e-mails were recovered from computer hard drives by Electronic Evidence Discovery, one of a growing number company specializing in data and e-mail retrieval.[1] Apart from being embarrassing, e-mails like these are often used in lawsuits. For example, in the past ten years, several cases have revolved around e-mail:

- Nissan Motor Corporation fired two employees for receiving sexually suggestive e-mails. The court upheld the decision, accepting Nissan's argument that the company owned anything sent or received by their network.

- Microsoft Corporation was found guilty of unfair business practices, in part on the basis of e-mails sent by CEO Bill Gates and other corporate officers.
- A damaging e-mail was recovered from pharmaceutical company employees proving that the company knew about the dangerous side effects of the diet drug Phen-Fen. This discovery led to a quick settlement.[2]

Do you view e-mail as private?
How do you feel about a company having the right to read an employee's e-mail?
Should you be responsible for e-mail you receive?
What are some ways to write e-mail that deals with sensitive issues?

Each of these court cases deals with ethical issues concerning e-mail: Is e-mail private? Who is responsible for the content of an e-mail message — the sender or the receiver? Who owns an e-mail message — the employee or the employer? The growing field of computer forensics reminds us that e-mail messages are accessible to others, even after they have been deleted.

Computer forensics engineers are experts at recovering active and deleted data from a variety of electronic sources, including storage systems, operating systems, password-protected and encrypted areas, inactive file areas, databases, calendar software, and proprietary software programs.[3] These engineers can also provide a variety of analyses of the recovered data, including reconstructing deleted files, proving file tampering, and recreating user activity of a certain drive.[4] Many of these specialists are being called upon by attorneys to search for evidence for or against parties in lawsuits and are then being called as expert witnesses in these cases.

The best advice for using e-mail? Read and follow your company's e-mail usage policy very carefully. A 2003 survey by the American Management Association, the ePolicy Institute, and Clearswift found that 75% of the respondents worked at a company that had written e-mail policies and just 48% were educated about these policies.[5] Formal retention and deletion policies are also on the rise due to new SEC regulations and provide useful guidelines for archiving and retaining electronic files for future usage.

Bottom line? Never write anything in an e-mail that you wouldn't be willing to say to your boss or in a court of law — you never know when you might have to do just that.

Composing E-Mail Messages, Memos, and Letters

The following guidelines should help you compose effective correspondence:

- Include a descriptive subject line if appropriate.
- Address your audience directly.
- State objectives or ask questions initially; follow with explanatory material.
- Organize material in direct (descending) order if you anticipate a neutral or positive response.
- Organize material in indirect (ascending) order if you anticipate a negative response.
- Enumerate or bullet items for clarity.

- Be specific about the action (if any) that you want the reader to take.
- End with a friendly comment.

When composing e-mail, memos, and letters you should consider these factors to make your document more comprehensible and usable:

- Attitude and tone
- Organization of information
- Format

Attitude and Tone

Some people will know you only through your correspondence and will form an opinion of your competence from the content and style of your e-mail messages, memos, and letters. Although this may not seem fair, it is realistic. You can create a professional image in your correspondence as well as make your correspondence easy to understand and use by replacing trite, outdated expressions, involving the reader with a *you*-attitude, and avoiding exclusionary language.

What is ethos?

What contributes to ethos?

What affects the ethos of your correspondence?

How does ethos affect your correspondence?

Simple, Direct Language. One way you can create a positive, sincere approach is to use simple, direct language. Your readers are as busy as you are. They don't have time to wade through the formal, pompous phrasing some people employ in a misguided attempt to enhance their image of competence.

Few people begin a telephone conversation or open a business meeting by saying, "This serves as notification of our agreement, as per our conversation of 10 October, that all orders processed on and following 12 October will meet the rev. 2 specifications." Why not just say, "Beginning on October 12, all orders will meet specifications of revision 2"? To clarify your correspondence, avoid trite phrases:

Trite Phrases	***Possible Revisions***
attached please find	attached enclosed
in reference to said specification	in reference to the specification
re your claim	regarding your claim about your claim
under separate cover	separately sent by UPS (or FedEx or DHL)
each and every one of you	each employee all team members
If I can be of any further service to you, please do not hesitate to let me know.	You can reach me at 555-3941 if you have any questions or comments.

***You*-Attitude.** Substitute the *you*-attitude for the *I*- or *we*-attitude whenever possible. Focus on and emphasize the readers rather than yourself as the writer:

> What is an alternative you-attitude version for each of the I- or we-examples here?

I- *or* we-*attitude*	***You-attitude***
I appreciate your hard work on this project.	Your hard work on this project has been valuable.
We need your crew to follow the new procedures.	Your crew will benefit from following the new procedures.
Our department is backlogged, so your order will be delayed a week.	Your order will be shipped on March 17, one week later than originally scheduled.

Your writing should be sincere; using the *you*-attitude helps achieve this tone by implying that you recognize readers' perspectives. However, overuse of the *you*-attitude results in writing that sounds insincere; be aware not only of the language in your writing but also of the image it projects.

Exclusionary or Biased Language. Exclusionary language (for example, assuming all dieticians are women or engineers are men) and biased language (for example, assuming certain ethnic groups are lazy or people with disabilities are incapable) cause problems. Such statements can simply be inaccurate; technical professionals, who are seldom limited to a single gender in any particular field, pride themselves on precise, factual presentations. Moreover, use of exclusionary language reflects badly on you as an individual, reflects a negative corporate image, and has resulted in legal battles because corporate language is often assumed to reflect corporate policies.

WEBLINK

Avoiding exclusionary language is an international concern in most countries where English is the primary language. Go to http://english.wadsworth.com/burnett6e for links to representative sites in other countries that encourage inclusive language.

CLICK ON WEBLINK

CLICK on Chapter 18/international inclusive language

To avoid exclusionary language, always consider what assumptions your word choice suggests and how your language might be interpreted. Be especially careful to avoid words that are biased in terms of gender, race, ethnicity, marital status, sexual orientation, or religion. For example, inviting "Congressmen and their wives" to a fundraiser assumes that members of Congress are all (1) male, (2) heterosexual and (3) married, thus excluding all others. Even changing the invitation to say "Congressional representatives and their partners" still assumes all have a partner. A nonexclusionary (that is, inclusive) wording could be "Congressional representatives and guests."

If you address people as individuals rather than as members of different groups, you are more likely to avoid exclusionary or biased language that might offend someone. The following list is a representative of the kind of exclusionary or biased language that can easily be changed.

Avoid Exclusionary Language	Choose Inclusive Language
Job titles	
newsman	reporter, journalist
fireman	firefighter
postman	postal worker, mail carrier
policemen, policewoman, G-men	police officer, FBI agent
steward, stewardess	flight attendant
References to "man"	
man, mankind	humanity, human race, human beings, people
layman, common man, man on the street	lay person, people in general, average person, ordinary people
man and wife	man and woman, husband and wife
man's achievements	human achievements, social achievements
brotherhood	kinship, unity, collegiality, community
man-made	artificial, synthetic, machine-made, manufactured
man-hours	employee hours, work hours
manning	staffing, working, covering, running
Roles	
chairman	chair, coordinator, supervisor, facilitator
businessman	business executive
Descriptions related to disability[6]	
the disabled	people with disabilities
suffering from multiple sclerosis	people who have multiple sclerosis
patient management, case management	care coordination, supportive services, resource coordination, assistance
community support needs of individuals	responsibilities of communities for inclusion and support

What additional categories can you add to this list?

What additional exclusionary terms and their inclusive counterparts can you add?

WEBLINK

Exclusionary language is not just about sexist language. It also deals with other kinds of exclusions in the workplace, for example, disability, race, sexual orientation, and age. Go to http://english.wadsworth.com/burnett6e for links to sites about avoiding other kinds of language exclusion.

CLICK ON WEBLINK

CLICK on Chapter 18/avoiding exclusion

Organization of Information

The content in all correspondence should be organized so that it can be read easily and quickly. You can effectively organize correspondence to fulfill these common functions:

- disseminating information
- making requests or inquiries
- responding to requests or inquiries

The following suggestions are not unbreakable rules but, rather, guidelines that can be modified to suit specific situations.

Disseminating Information. Information can be disseminated through a variety of documents: directives, policy statements, announcements, and press releases. The standard 5 *W*s plus *H* formula used in journalism — who? what? when? where? why? how? — forms the basis for organizing information for readers within your own organization, other organizations, and the general public. Internal information can be dispensed through memos, posted on bulletin boards, or delivered via a paper or electronic distribution list. These postings convey information about such diverse subjects as changes in insurance coverage, holiday party plans, and quarterly quotas.

Correspondence with other organizations may be addressed to a particular person or department. For example, a company might announce to customers dates for the summer shutdown or changes in the price structure. Press releases within an industry or for the general public announce such things as new products, research successes, and outstanding employee achievements. The organization of correspondence that disseminates such information varies, depending on the receptivity of the audience:

Receptive or Neutral Audience

1. Identify the *who* (including individual and company).
2. Identify the *what.*

Negative or Neutral Audience

1. State the relevant background (such as the problem or situation that necessitates the change).
2. Identify the *who* and *what.*

3. Include details of *when, where, why, how.*
4. Explain the impact.
5. Provide a name and number for people to contact for additional information. (optional)

3. Include details of *when, where why, how.*
4. Explain the impact.
5. Provide a name and number for people to contact for additional information.

Making Requests or Inquiries. People making routine requests and inquiries usually seek information; however, they may also ask for actual samples of a product or a specific action. Requests or inquiries usually contain the following elements, with amount of detail dictated by the situation:

1. State the reason for the request.
2. State the request directly.
3. Explain the benefit of the information, sample, or action.
4. Ensure confidentiality, if warranted.
5. Identify when the information, sample, or action is needed.
6. Thank the recipient.

Responding to Requests or Inquiries. Some routine requests can be handled by sending prepared information, such as a price list, catalog, spec sheet, or brochure. Other routine requests can be handled with a form letter that is personalized by using the inquirer's name and address. One of the many benefits of word processing is that it enables a company to generate original copies of form letters.

When a request or inquiry requires an original response, the letter or memo can be either positive or negative. A positive, nonroutine response not only answers the question(s) but also offers additional information if appropriate and builds good will. Letters that deny a request, claim, or order usually follow a sequence that explains the negative response while maintaining good will:

Positive Responses

1. Acknowledge the request or inquiry.
2. Say "yes."
3. Include the information or identify an accessible source for the information.
4. Offer additional helpful suggestions, if appropriate.
5. Build goodwill.
6. Conclude in a friendly manner.

Negative Responses

1. Acknowledge the request or inquiry.
2. Explain briefly what makes a refusal necessary.
3. Say "no" directly to avoid misunderstanding.
4. Offer an alternative.
5. Build goodwill.
6. Conclude in a friendly manner.

Format

Literally hundreds of minor variations exist in the formats used for correspondence, and with the widespread use of computers, conventions are changing rapidly. You need to be familiar with formats and conventions for two reasons. First, many document originators use word processing to enter their own drafts directly into a computer. Second, when you sign a letter or initial a memo, you are acknowledging that it meets your standards of format as well as content.

WEBLINK

Correspondence often responds to predictable situations, leading some people to believe that it can be formulaic. As a result, some companies advertise memo and letter templates. The problem is that such templates ignore rhetorical elements that make all but the most routine correspondence distinctive. Go to http://english.wadsworth.com/burnett6e for links to sites that advertise or provide such templates. On the surface, using templates might be initially appealing. What do you see as the benefits of templates? What do you see as problems? What aspects identified in Figure 1.2 (in Chapter 1) do templates ignore?

CLICK ON WEBLINK

CLICK on Chapter 18/letter templates

In deciding which format to use, search your organization's files for examples of previous correspondence that are logically arranged and that meet the approval of your supervisor. Also, many organizations have a style guide that specifies preferred formats. If these avenues are closed to you, consult any good technical or business communication handbook. Composing your own letters lends a personal tone that can never be achieved with a form letter. However, if you are rushed or encounter a mental block, you can get ideas from books of form letters designed to meet hundreds of situations you might encounter in the workplace. The Tele-Robics examples that follow include annotations about format.

Domino Effect of Correspondence

No piece of correspondence exists in a vacuum; a single message can trigger a series of e-mail messages, memos, and letters, as the organizational chain reactions that occurred at Tele-Robics illustrates. As you read each e-mail message, memo, and letter in that case study, think of the workplace environments as well as the specific purposes and audiences. The correspondence is presented chronologically so that you can see the evolving story and recognize that people at Tele-Robics were writing in response to the situation in which they were involved. For example, John Hernandez did not say, "How can I write a response letter to a customer?" Instead, he said, "How can I best respond to James Crocker's concerns?"

As you read the sequence of e-mail messages, memos, and letters, you see that this correspondence produced the reactions that the various professionals involved had hoped they would. In other words, the readers seemed to interpret the messages the way the writers intended. Tele-Robics resolved the technical problems, and corporate accounts with customers were maintained.

The examples in this series can serve as guidelines as you compose your own correspondence. While correspondence does not have rigid prescriptions, it should follow the conventions in format, organization, and tone that are illustrated by the following samples. Generally, if your correspondence has the

FIGURE 18.1 **Domino Effect of Correspondence: The Tele-Robics Case**

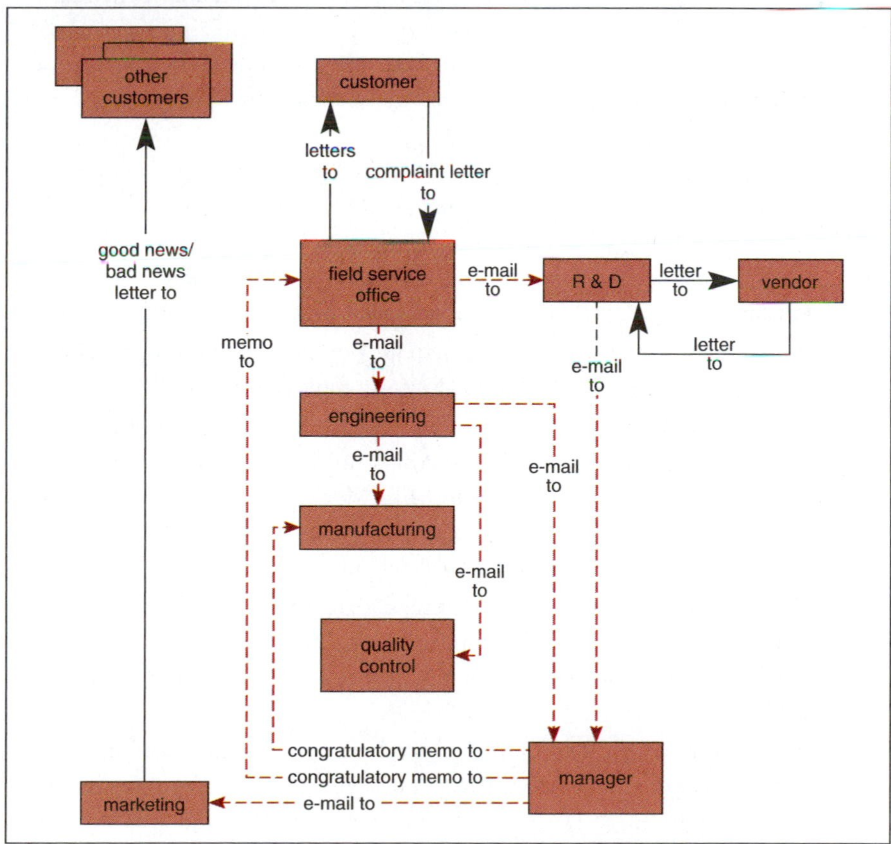

effect you desired on your intended audience, it will be effective. Each example is annotated to point out the tone, style, and format used in each message.

In the Tele-Robics case, the letters and e-mail messages involve a number of people, both within and outside the company. The case deals with one of Tele-Robics' products, the Lightening IV Illuminator. This chain reaction starts when an important Tele-Robics customer, James W. Crocker at Crocker Computers, Inc., voices a complaint in a letter. The chronology of the correspondence will help you preview the situation.

1. Mr. Crocker, a major customer, writes a formal letter to John Hernandez, the field service manager for Tele-Robics, expressing disappointment with equipment and service that his company received. He wants the costly problem to be resolved immediately (Figure 18.2).
2. John Hernandez, Tele-Robics' Field Service Manager, writes a letter to Crocker, acknowledging his complaint, identifying the cause of the problem, and explaining how and when it will be resolved (Figure 18.3). Hernandez realizes the company's credibility is damaged, not because the problem exists but because it has taken so long to correct.
3. John Hernandez also sends two e-mail messages: one to Sarah Bell, Tele-Robics' manager of R&D, who can study how the complaint affects the development of similar products, and the other to Tom Watt, manager of engineering, the group that developed the specifications for the illuminator (Figure 18.4 and Figure 18.5).
4. Based on group's research findings, Sarah Bell writes a letter to Chuck Taylor at Optical Coatings, Inc., one of Tele-Robics' prime vendors, requesting a change and a revised cost opinion (Figure 18.6). She also sends an e-mail message to Brian Bartlett, division manager for Tele-Robics, to alert him of a change that might affect his new product line (Figure 18.8).
5. Chuck Taylor at Optical Coatings responds to Sarah Bell's inquiry with exact figures and requests additional information (Figure 18.7).
6. Tom Watt, who earlier received the message from John Hernandez about the illuminator problem, sends two e-mail messages: one accompanying an engineering change order (ECO) to Mike Maxwell in manufacturing, with a copy to quality control; the other to Brian Bartlett, his manager, explaining his recent actions (Figure 18.9 and Figure 18.10).
7. Brian Bartlett later writes an e-mail message to Shawna Simons, marketing manager, updating her on the problem and asking that the marketing representatives contact customers with the problem equipment (Figure 18.11).
8. Shawna Simons writes an individualized letter to customers who have bought a Lightening IV Illuminator. She sends a blind copy of the letter to Brian Bartlett — that is, she sends him a copy, but the original letter does

What circumstances might warrant a blind copy of a letter or memo?

FIGURE 18.2 **Complaint Letter from Customer**

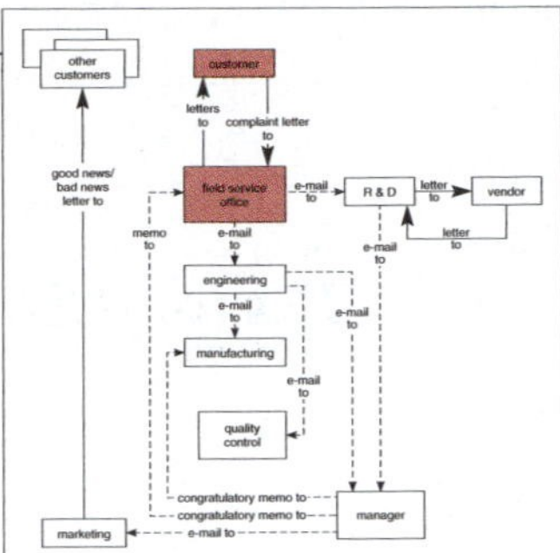

Crocker Computers, Inc.
1932 Sylvan Avenue
Granbury, Texas 78048

January 21, 20—

Mr. John Hernandez
Manager, Field Service
Tele-Robics, Inc.
799 Warren Avenue
Greenville, TX 75933

Dear Mr. Hernandez:

Our recently purchased Lightening IV Illuminator is a disappointment. Instead of increasing our productivity, it has cost Crocker Computers, Inc., thousands of dollars in downtime.

Your field service office has sent three different technicians in the past week to service this new illuminator, with no success. Either your technicians do not know how to repair the Lightening IV Illuminator, or your company cannot manufacture this illuminator to meet our specs.

I expect that you will correct this problem immediately.

Sincerely,

James W. Crocker

James W. Crocker

JWC:cd

- ***Professional practice.*** *Crocker chooses a formal salutation, which suggests that he doesn't know Hernandez very well.*
- ***Audience.*** *Crocker assumes a professional tone as he approaches the problem, effectively stating his complaint early and explicitly.*
- ***Action.*** Crocker *directly asks that the problem be corrected.*
- ***Information.*** *Crocker reviews the history of the problem, explaining steps that have been taken to address the problem and the lack of success thus far. The explanation should be entirely comprehensible to Mr. Hernandez.*
- ***Format.*** *Crocker chooses a usable format: a conventional* standard block style that *begins all lines at the left margin. The reference* initials *indicate that James W. Crocker (JWC) wrote or dictated the letter, and someone with the initials C. D. typed it.*

What are the advantages and disadvantages of Crocker writing a formal letter rather than simply calling John Hernandez on the phone or sending e-mail?

FIGURE 18.3 Response Letter to Customer Complaint Letter

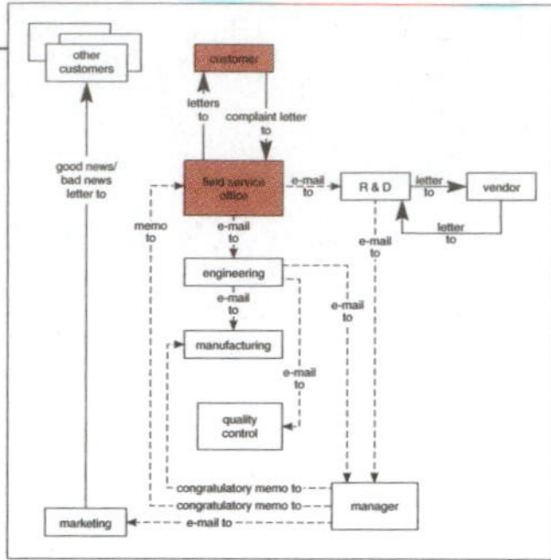

■ ***Action.*** *Once Hernandez identifies the cause of the problem, he explains how and when it will be resolved. He offers a solution that Crocker might reasonably have expected three service calls ago, which calls into question the competence of the service technicians and the supervision of those technicians.*

■ ***Audience.*** *Based on Hernandez's letter, Crocker reasonably expects a response within one week. He can use this information if the promised response is not satisfactory.*

TELE-ROBICS, INC.
799 Warren Avenue Greenville, TX 75933

January 23, 20—

Mr. James W. Crocker
Crocker Computers, Inc.
1932 Sylvan Avenue
Granbury, TX 76048

■ ***Professional practice.*** *Hernandez responds to Crocker immediately in a conventional and accessible letter. Hernandez adopts Crocker's formal tone by addressing his letter to "Mr. Crocker," not "Jim."*

Dear Mr. Crocker:

Our field service representatives have filed reports detailing the energy failure of the Lightening IV Illuminator reported by your company. Based on this information, I have concluded that your illuminator contains lenses that have defective coatings.

The engineering staff here at Tele-Robics, Inc., assures me that your energy spec can easily be met once the original lenses are replaced. Our manufacturing facility has already sent a new illuminator to you. Your field service representative will contact you to install the replacement this week.

The replacement illuminator will meet your specs, enabling you to increase productivity.

Sincerely,

John Hernandez

John Hernandez
Manager, Field Service

JH:jn

■ ***Information.*** *Hernandez's explanations should be comprehensible to Crocker; the language is straightforward. Hernandez could apologize but does not. Instead, he acknowledges the complaint.*

■ ***Format.*** *This letter uses a* conventional *standard block style, which means that all the lines start at the left margin. Formal letters require a handwritten signature above the typed name and title.*

> *Would Hernandez look more responsible and more professional if he apologized to Crocker for the delay and expense?*

> *How does Crocker's letter stimulate Tele-Robics' attention to a problem that extends beyond a single customer?*

not indicate that she has sent a copy to anyone (Figure 18.12). This "blind copy" procedure is used when, for whatever reason, it is not diplomatic for the receiver of the correspondence to know that a copy has been sent to someone else.

9. Brian Bartlett writes memos to Mike Maxwell and John Hernandez. He thanks them and their coworkers and crews for their quick and appropriate response to the problem (Figure 18.13 and Figure 18.14).

FIGURE 18.4 **E-Mail Message to Tele-Robics's Manager of R&D**

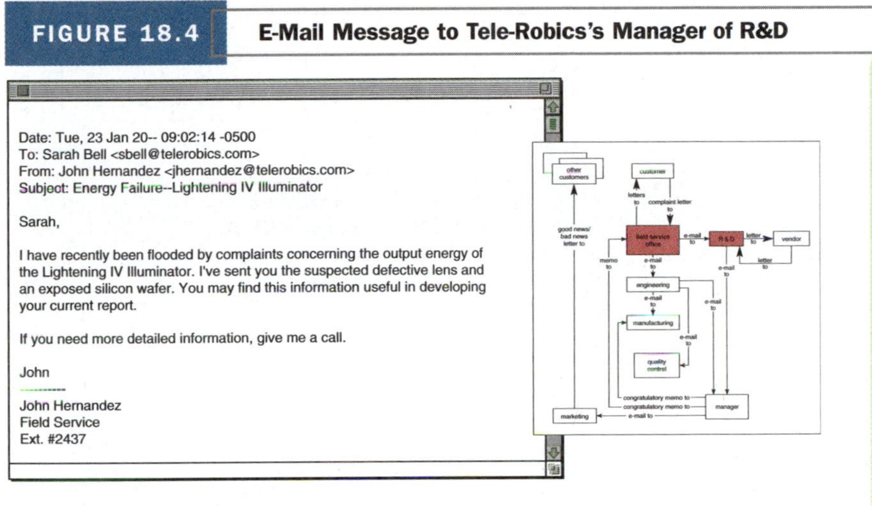

Date: Tue, 23 Jan 20-- 09:02:14 -0500
To: Sarah Bell <sbell@telerobics.com>
From: John Hernandez <jhernandez@telerobics.com>
Subject: Energy Failure--Lightening IV Illuminator

Sarah,

I have recently been flooded by complaints concerning the output energy of the Lightening IV Illuminator. I've sent you the suspected defective lens and an exposed silicon wafer. You may find this information useful in developing your current report.

If you need more detailed information, give me a call.

John

John Hernandez
Field Service
Ext. #2437

FIGURE 18.5 **E-Mail Message to Tele-Robics's Manager of Engineering**

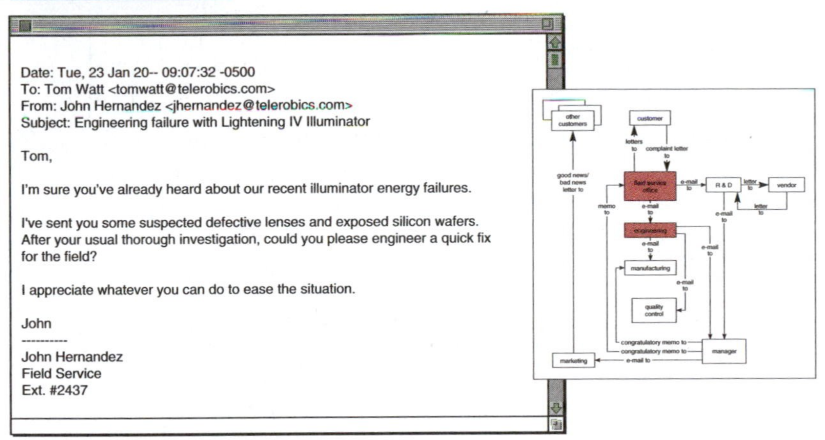

Date: Tue, 23 Jan 20-- 09:07:32 -0500
To: Tom Watt <tomwatt@telerobics.com>
From: John Hernandez <jhernandez@telerobics.com>
Subject: Engineering failure with Lightening IV Illuminator

Tom,

I'm sure you've already heard about our recent illuminator energy failures.

I've sent you some suspected defective lenses and exposed silicon wafers. After your usual thorough investigation, could you please engineer a quick fix for the field?

I appreciate whatever you can do to ease the situation.

John

John Hernandez
Field Service
Ext. #2437

Hernandez doesn't attribute blame for the problem either to R&D or to engineering. How does his approach contribute a positive working environment?

- ***Professional practice.*** *These are brief internal messages to colleagues who know each other well, so the tone is informal and first names are appropriate.*
- ***Format.*** *Effective e-mail messages contain a specific subject line and an appropriate salutation. A "signature block" at the end of the message includes the writer's title and contact information.*
- ***Information.*** *The clear, direct messages provide essential information. Hernandez informs Bell about the recent complaints so that her R&D group can make appropriate changes in their current project, a new version of the illuminator. Hernandez also informs Tom Watt so that his engineering group can adjust the specifications for the illuminator.*
- ***Audience.*** *The phrasing indicates that he understands the prior knowledge of this audiences about both the context and the technical details and that he respects their professional*

FIGURE 18.6 Letter to Vendor

■ ***Audience.*** *Bell writes a letter to Charles Taylor at Optical Coatings, Inc., one of Tele-Robic's prime vendors. Bell informs Taylor about a design change that affects an order. Even though Bell is on a first-name basis with the vendor, her letter is clear and precise, not chatty.*

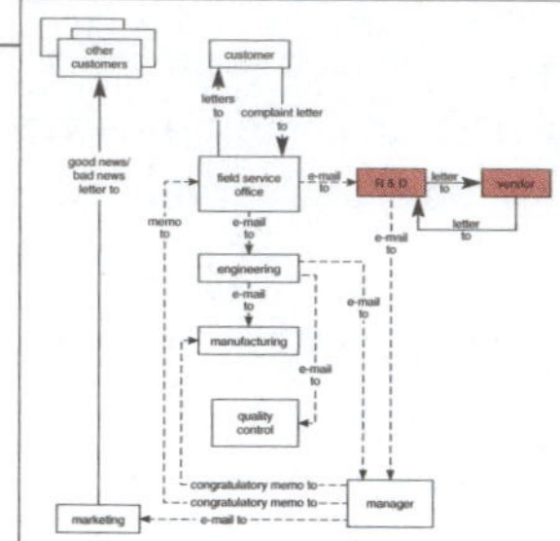

TELE-ROBICS, INC.
799 Warren Avenue Greenville, TX 75933

January 28, 20—

■ ***Action.*** *When Sarah Bell receives the e-mail message from John Hernandez, she and her team study how the complaint affects the development of similar products.*

Charles D. Taylor
Optical Coatings, Inc.
536 Stevens Street
Cypress, TX 77429

Dear Chuck:

SUBJECT: Revision of spec for lens coating on Lightening IV Illuminator

The engineering staff at Tele-Robics has informed me that the lens coatings that Optical Coatings did for the Lightening IV Illuminator are not adequate for long life. The coating degenerates with the lens use, allowing longer exposure times because of lowered energy output.

This, of course, affects our work in the Lightening V. As a result, we're changing the specs for the coatings, which should eliminate any problem.

Hold up on coating all Tele-Robics lenses you currently have in-house.

Coat the enclosed lens to the following revised spec:

narrow band coating 405 nm 1/4 mag fluoride

How does this change affect the price and delivery schedule for the 25 lenses you've already quoted?

Please let me know when you'll have the lens with the new coating ready for testing.

Sincerely,

Sarah Bell

Sarah Bell
Manager, Research and Development

SB:mj

■ ***Information.*** *Bell provides the necessary technical information and asks a specific question about price and delivery. (See Figure 18.8 for her second piece of correspondence.)*

FIGURE 18.7 **Response from Vendor**

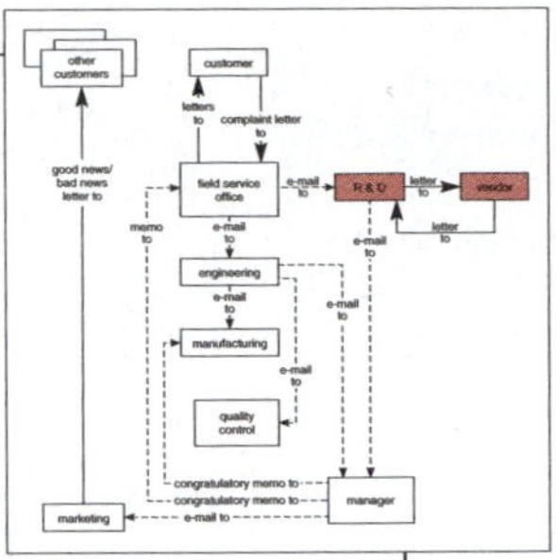

OC
Optical Coating, Inc.

1701 Industrial Park
Titusville, FL 32780

January 30, 20—

Sarah Bell
Manager, Research and Development
TELE-ROBICS, INC.
799 Warren Avenue
Greenville, TX 75933

- ***Audience.*** *Within a few days, Bell receives a reply from Charles Taylor at Optical Coatings. He responds to her inquiry and provides specific information.*

Dear Sarah:

We will have no problems in coating the lenses to your new specs.

We currently have 10 lenses already coated to the old specs and another 15 lenses that have not yet been coated. Would you like us to strip the coating and recoat the lenses to the new specs? There is no change in the per piece price for the new coating.

- ***Action.*** *Taylor also asks Bell about stripping and recoating 10 lenses already coated to the old specs, maintaining good business relations with her company by adding a low charge for this extra work.*

	per piece x	# pieces =	net $
Coat experimental lens	$33.00	1	$ 33.00
Strip lens coating	$ 6.00	10	$ 60.00
Coat lens to new spec	$27.00	15	$405.00

The coating on the one lens you sent will be ready on Tuesday morning. As soon as you contact me, we can begin the new coatings on the other 25 lenses.

Sincerely,

Charles D. Taylor

Charles D. Taylor

CDT:lp

- ***Professional practice.*** *He signs the letter formally, indicating that Bell may have misjudged how well she knows him.*

FIGURE 18.8 E-Mail Message

- ***Audience.*** *The second piece of correspondence Bell writes is an e-mail message to Brian Bartlett, Division Manager for Tele-Robics.*
- ***Information.*** *She informs Bartlett about the work her group is doing to solve the problem of the energy failure, both with the current and the new models of the illuminator.*

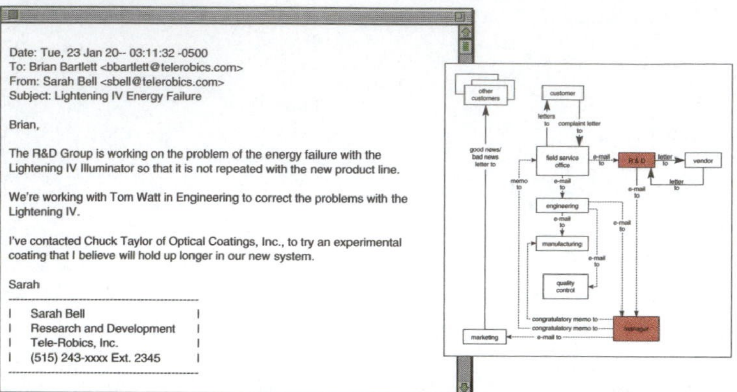

Date: Tue, 23 Jan 20-- 03:11:32 -0500
To: Brian Bartlett <bbartlett@telerobics.com>
From: Sarah Bell <sbell@telerobics.com>
Subject: Lightening IV Energy Failure

Brian,

The R&D Group is working on the problem of the energy failure with the Lightening IV Illuminator so that it is not repeated with the new product line.

We're working with Tom Watt in Engineering to correct the problems with the Lightening IV.

I've contacted Chuck Taylor of Optical Coatings, Inc., to try an experimental coating that I believe will hold up longer in our new system.

Sarah

Sarah Bell
Research and Development
Tele-Robics, Inc.
(515) 243-xxxx Ext. 2345

FIGURE 18.9 E-Mail Message

- ***Audience.*** *Tom Watt received an e-mail message from John Hernandez (Figure 18.5).*
- ***Professional practice.*** *Notice that Watt also sends a copy of the message and ECO to Quality Control so they'll know about the changes in specifications.*

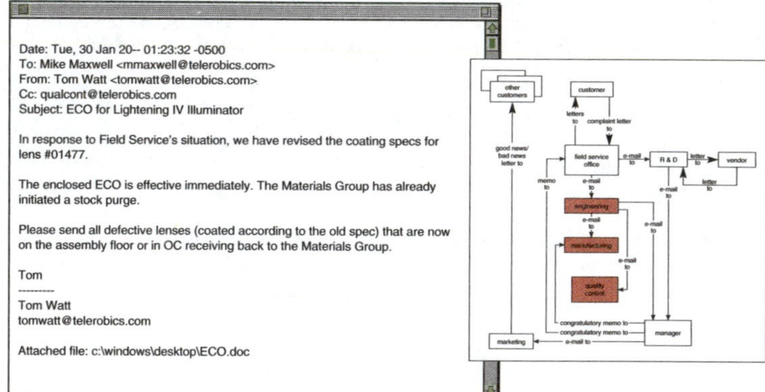

Date: Tue, 30 Jan 20-- 01:23:32 -0500
To: Mike Maxwell <mmaxwell@telerobics.com>
From: Tom Watt <tomwatt@telerobics.com>
Cc: qualcont@telerobics.com
Subject: ECO for Lightening IV Illuminator

In response to Field Service's situation, we have revised the coating specs for lens #01477.

The enclosed ECO is effective immediately. The Materials Group has already initiated a stock purge.

Please send all defective lenses (coated according to the old spec) that are now on the assembly floor or in OC receiving back to the Materials Group.

Tom

Tom Watt
tomwatt@telerobics.com

Attached file: c:\windows\desktop\ECO.doc

- ***Information.*** *This e-mail message prompted people in the engineering department to locate the problem, necessitating that an ECO, accompanied by an e-mail message, be sent to Mike Maxwell in manufacturing.*

FIGURE 18.10 E-Mail Message

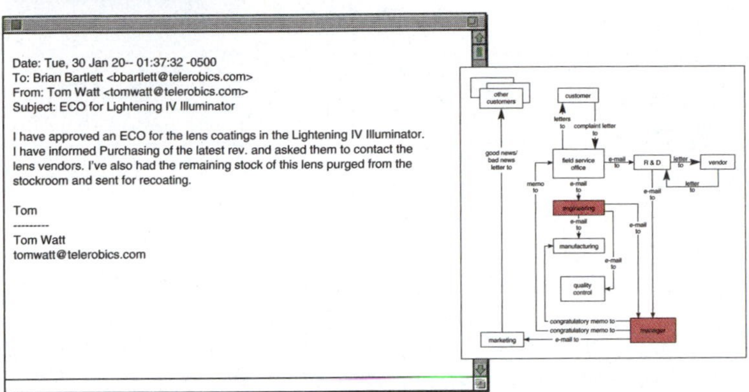

Date: Tue, 30 Jan 20-- 01:37:32 -0500
To: Brian Bartlett <bbartlett@telerobics.com>
From: Tom Watt <tomwatt@telerobics.com>
Subject: ECO for Lightening IV Illuminator

I have approved an ECO for the lens coatings in the Lightening IV Illuminator. I have informed Purchasing of the latest rev. and asked them to contact the lens vendors. I've also had the remaining stock of this lens purged from the stockroom and sent for recoating.

Tom

Tom Watt
tomwatt@telerobics.com

- ***Audience.*** *Tom Watt also sends an e-mail message to Brian Bartlett to let his manager know the major steps he's taken to resolve the problem.*
- ***Information.*** *This is the second e-mail message Bartlett has received; the department managers believe he should be informed about problems and their solutions to these problems.*

FIGURE 18.11 E-Mail Message

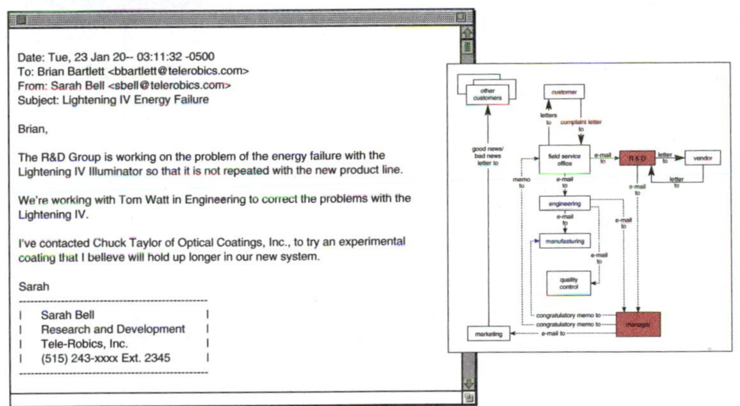

Date: Tue, 23 Jan 20-- 03:11:32 -0500
To: Brian Bartlett <bbartlett@telerobics.com>
From: Sarah Bell <sbell@telerobics.com>
Subject: Lightening IV Energy Failure

Brian,

The R&D Group is working on the problem of the energy failure with the Lightening IV Illuminator so that it is not repeated with the new product line.

We're working with Tom Watt in Engineering to correct the problems with the Lightening IV.

I've contacted Chuck Taylor of Optical Coatings, Inc., to try an experimental coating that I believe will hold up longer in our new system.

Sarah

Sarah Bell
Research and Development
Tele-Robics, Inc.
(515) 243-xxxx Ext. 2345

- ***Audience.*** *Brian Bartlett examines the information he receives from John Hernandez in field service and from Mike Maxwell in manufacturing. He gathers some additional information beyond what Hernandez or Maxwell would be expected to provide in short update memos, and then writes an e-mail message to Shawna Simons, marketing manager for Tele-Robics.*

- ***Action.*** *Bartlett gives Simons an update of the problem and asks that the marketing representatives contact customers with the problem equipment, informing them that Tele-Robics will retrofit their illuminator lenses at no charge.*

FIGURE 18.12 **Letter to Customer**

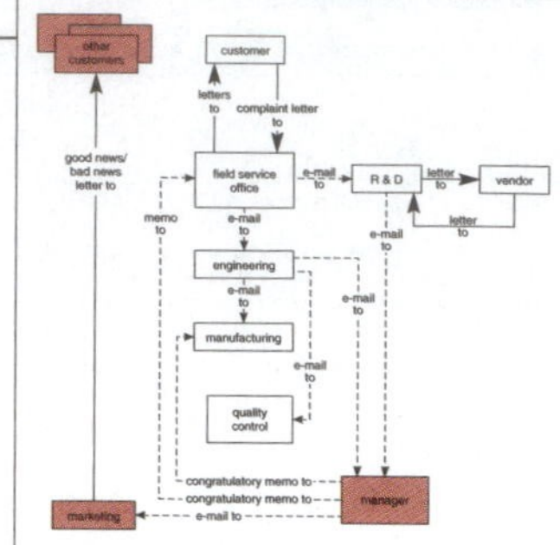

TELE-ROBICS, INC.
799 Warren Avenue Greenville, TX 75933

February 7, 20—

Leigh Ward, Buyer
Omnitech, Inc.
17 Industrial Park
Scranton, PA 18505

■ ***Audience.*** *The letter (here addressed to Leigh Ward, a buyer for Omnitech) shows that Shawna Simons writes to customers who have bought a Lightening IV Illuminator.*

Dear Ms. Ward:

SUBJECT: Lightening IV Illuminator Modification

Your Lightening IV Illuminator is designed for continual use. Recently, we became aware of less than 100% operation with some of these systems. Specifically, the original lens coatings—after extended use—gradually permit higher than normal exposure times due to lower energy output.

Your field service representative will contact you this week to arrange a time to replace the original lenses, as part of the equipment's warranty. Down time will not exceed one hour. This short interruption in your production schedule will assure continued operation of your Lightening IV Illuminator.

Sincerely,

Shawna Simons

Shawna Simons
Marketing Manager

SS:rg

■ ***Action.*** *Each customer will receive an individualized letter, signed by the marketing representative who handles the account. Simons explains that the retrofit is covered under the equipment warranty and will cause little interruption in the production schedule.*

■ ***Professional practice.*** *Simons sends a blind copy of the letter to Brian Bartlett (that is, on her letter to Ms. Ward she doesn't include the copy notation that she is sending a copy to Brian Bartlett).*

FIGURE 18.13 Memo to Employee

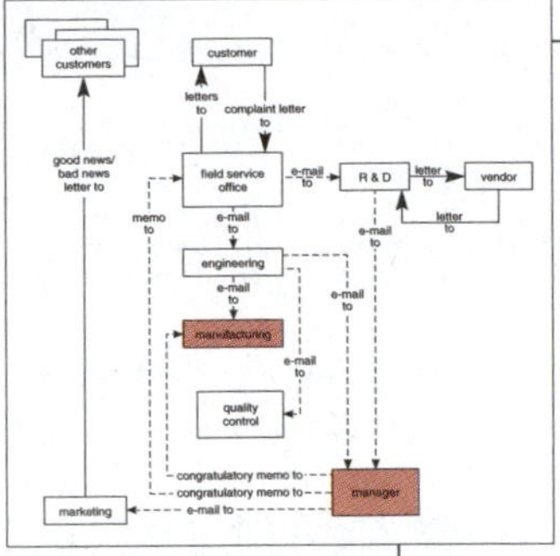

TELE-ROBICS, INC. MEMO

February 12, 20—

TO: Mike Maxwell, Manufacturing Manager
FROM: Brian Bartlett, Division Manager *BB*
SUBJECT: Lightening IV Illuminator

You and your manufacturing crew did a great job in handling the changes in the assembly of the Lightening IV Illuminator. The long hours you all put into rebuilding the Lightening IV's eliminated shipment of defective illuminators. As always, the manufacturing group gets the job done!

BB:ft

- ***Professional practice.*** *Once Brian Bartlett is assured that the problem is resolved and that customers will be contacted, he takes the time to write a memo to Mike Maxwell, thanking him for his role in resolving the illuminator problem and giving additional thanks to his crew in manufacturing.*

- ***Format.*** *The "to/from/subject" heading is so widely accepted that many companies have memo forms printed to eliminate having to write the headings. The signature or initials of the document originator indicate the person has read and approved the correspondence after it was input.*

FIGURE 18.14 Memo to Employee

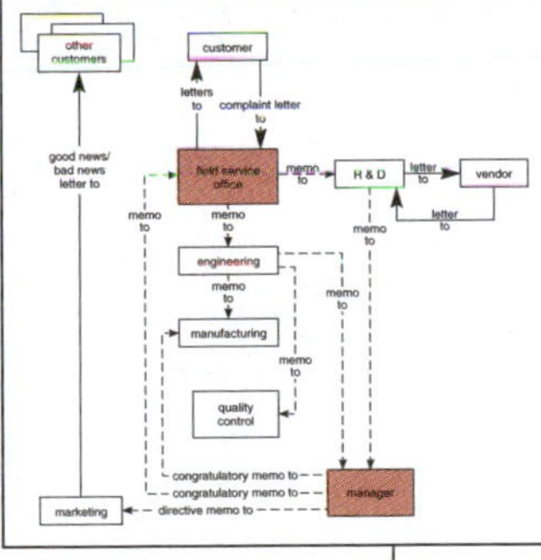

TELE-ROBICS, INC. MEMO

February 12, 20—

TO: John Hernandez, Field Service
FROM: Brian Bartlett, Division Manager *BB*
SUBJECT: Lightening IV Illuminator Servicing

Congratulations to you and your group on the way you handled the coating failure on the Lightening IV. Due to your quick response to a customer, the manufacturing plant was able to avert further shipment of defective illuminators. Good job!

BB:ft

- ***Audience.*** *Bartlett also sends a memo to John Hernandez, thanking him and his group for their quick and appropriate response to the problem. Managers who take the time to thank employees for doing a good job generally have smooth-running, productive organizations.*

> *What's your opinion about whether a manager or supervisor should write a letter to an employee expressing thanks or acknowledging excellence for doing something that is part of the employee's regular job?*

> ***Rhetorical problem-solving.*** *After reading the series of letters and memos in Figures 18.2–18.14, you can see that correspondence creates a complex network. Solving the technical problem with the Lightening IV Illuminator required a chain of correspondence that articulated the problem, asked questions, and explained decisions. The writers not only considered content, purpose, audience, organization, language, and format, but they were also sensitive to interpersonal and political factors.*

WORLD CUP CORRESPONDENCE

What kind of correspondence is involved in getting to a World Cup soccer match? Lee Tesdell spent more than a year in correspondence to get tickets for the 2002 World Cup, which was held in Korea and Japan.

Soccer (association football) is the world's only universal sport. The world championship of soccer, the World Cup, takes place every four years. It first was held in 1930 in Uruguay. National teams made up of the best players in each country compete for the title of best soccer team in the world.

The ticket application process begins long before the World Cup matches themselves. Since soccer is so popular around the world, fans from all countries want to attend games at the World Cup. In order to make this possible, Federation Internationale de Football Association (FIFA), soccer's world governing body, tries to make the process fair and open to all people.

The following correspondence shows the steps that Lee Tesdell followed to obtain tickets to take his two sons to the 2002 World Cup in Daegu, Korea. While a significant portion of the correspondence was electronic, the actual paper tickets for three games he and his sons eventually attended were elaborately designed.

In the next World Cup, to be held in 2006 in Germany, a total of 203 countries will compete. Only 32 teams will progress successfully through the qualification schedule to the 2006 World Cup championships, a month-long series of games that culminate in a world championship final game. The 2006 World Cup information is online, although there are no tickets available yet: http://fifaworldcup.yahoo.com/06/en/loc/f/index.html.

As you read these six representative stages of the process, try to identify elements that suggest this correspondence is for a worldwide audience rather than a North American audience.

1. *13 April 2001*
Electronic Correspondence: Online research about venue-specific tickets (VST) on the fifa-tickets.com Web site

I began my Internet search for World Cup tickets at www.fifa.com, which provided two kinds of tickets:

1 | FIFA Information about World Cup tickets

Venue Specific Ticket Series (VST Series)

The Conventional Approach to FIFA World Cup™ Ticketing

For previous FIFA World Cups™, the tickets made available by FIFA and the Local Organising Committees to the general public were for a number of matches to be played at specific venues, the so-called Venue Specific Ticket Series.

For 2002, VSTs will be sold in series consisting of tickets to all the matches that are to be played in a given venue, with the exception of the Opening Match, the Semifinals and the Final.

There is one VST Series per venue. The price of each VST Series is equal to the sum of the face price(s) of the tickets contained in the VST Series. The number of matches included in each VST Series and the corresponding prices in US Dollars are shown below:

	VENUE	VST SERIES CODE	COST PER VST SERIES IN US DOLLARS			NUMBER OF MATCHES			
			CATEGORY 1	CATEGORY 2	CATEGORY 3	GM	R16	QF	3RD
KOREA	BUSAN	BSN	$450	$300	$180	3			
	DAEJEON	DJN	$525	$375	$220	2	1		
	SUWON	SWN	$675	$475	$280	3	1		
	SEOUL	SEL	$150	$100	$60	1			
	DAEGU	DGU	$675	$475	$280	3			1
	ULSAN	USN	$600	$400	$245	2		1	
	INCHEON	INC	$450	$300	$180	3			
	JEONJU	JNJ	$525	$375	$220	2	1		
	GWANGJU	GWJ	$600	$400	$245	2		1	
	SEOGWIPO	SOG	$525	$375	$220	2	1		
JAPAN	SAPPORO	SAP	$450	$300	$180	3			
	MIYAGI	MIY	$525	$375	$220	2	1		
	NIIGATA	NII	$525	$375	$220	2	1		
	IBARAKI	IBA	$450	$300	$180	3			
	SAITAMA	SAI	$450	$300	$180	3			
	YOKOHAMA	YOK	$450	$300	$180	3			
	SHIZUOKA	SHI	$600	$400	$245	2		1	
	OSAKA	OSA	$600	$400	$245	2		1	
	KOBE	KOB	$525	$375	$220	2	1		
	OITA	OIT	$525	$375	$220	2	1		

TOURNAMENT STAGES

- Group Matches (GM)
- Round of 16 (R16)
- Quarter-finals (QF)
- 3rd Place Play-off (3RD)

Request Number:	**0030939**
Date:	**Thursday, August 16, 2001**
Status:	**APPLICATION RECEIVED**
Total Request Cost:	**$840**
Charged Amount:	

Name				
Requested Ticket Series	**Team/Venue**	**Requested Category**	**Requested Price**	**Status**
Allocated Ticket Series	**Team/Venue**	**Allocated Category**	**Allocated Price**	
Lee TESDELL				
DGU	DAEGU	Category 3	$280	Accepted
DGU	**DAEGU**	**Category 3**	**$280**	
Omar TESDELL				
DGU	DAEGU	Category 3	$280	Accepted
DGU	**DAEGU**	**Category 3**	**$280**	
Ramsey TESDELL				
DGU	DAEGU	Category 3	$280	Accepted
DGU	**DAEGU**	**Category 3**	**$280**	

<< Back to Login Screen **Continue (New Request) >>**

2 | Notice that Online Application was Received

VST (venue-specific tickets) and TST (team-specific tickets), as the screen capture shows. In the first case, a fan could purchase tickets to a match in one of the Korean or Japanese venues, but only in a specific stadium. With ten stadia in each country, a total of 20 venues existed for the 64 matches. In the second case, a fan could buy tickets for games for only one team, say the French national team, wherever that team might be playing in either Korea or Japan. I wanted to purchase VST for my two sons and me.

2. 16 August 2001
Electronic Correspondence: Online application submitted and confirmation of ticket order received

I first tried to get tickets at the stadium in northern Japan at Sapporo, Hokkaido, but none were available. I found that the category 3 tickets — the least expensive — were available only at Daegu, Korea, which was why I applied for tickets to games in Daegu. I submitted an online application for three sets of category 3 tickets. I wasn't sure we actually had the tickets reserved, but I did receive electronic notification that our application had been received.

3. 04 September 2001
Paper Correspondence: Received ticket allocation confirmation. Credit card charged for amount of tickets

Finally! I had official confirmation in a letter and an invoice that we had our World Cup tickets for matches in Daegu, Korea. My credit card had been charged for the total amount of the tickets in dollars — $840. At the time I did not realize that I was getting tickets to four matches, since the third-place match was going to be held in Daegu. We eventually sold our tickets for the third-place match to an American on E-bay, since we could not stay in Korea long enough to see this match.

3 | Ticket Allocation Letter and Sales Invoice

PO Box 2002
Cheadle Hulme SK8 7RR, England
Tel: +44 (0)870 123 2002
Fax: +44 (0)870 124 2002
Email: sales@fifa-tickets.com

Lee TESDELL
300 NW 158th Avenue
Slater
IOWA
50244
United States of America

4 September 2001

Dear **Lee TESDELL**

Ref: Intermediate Sales Phase - Your Application **LEETES44** Request **30939**

Thank you for applying to the 2002 FIFA World Cup™ Ticketing Bureau.

We are pleased to inform you that you have been allocated the ticket series identified in the table below as "Allocated Ticket Series". If you answered "yes" to the lower category and/or fewer matches question in the Ticket Application, and your ticket series was not available, you will have been allocated the next available ticket series for your team/venue.

Details of the results of your application follow:

Requests Accepted

Name	Requested Ticket Series			Allocated Ticket Series		Amount Due
	Team/Venue	Type	Category	Type	Category	
Lee TESDELL	**DAEGU**		**CAT3**		**CAT3**	**US$280**
Omar TESDELL	**DAEGU**		**CAT3**		**CAT3**	**US$280**
Ramsey TESDELL	**DAEGU**		**CAT3**		**CAT3**	**US$280**
Total to be charged in US Dollars						**US$840**

Your credit card, as detailed in your application, has been charged 840 US dollars and your application for the tickets identified as "Allocated Ticket Series" is successful. The attached invoice provides details of the allocated ticket series and the total amount that was charged to your credit card.

You will receive a confirmation letter in approximately 6 to 8 weeks. This communication will include, amongst other things, an Official FIFA Ticket Certificate per person which serves as "proof of purchase", and details of how to contact us regarding change of address and ticket delivery.

Your application has been made pursuant to the provisions and procedures set forth in the 2002 FIFA World Cup™ Ticket Application Form, including the Ticket Terms and Conditions set forth therein. Any application which, in the sole discretion of FIFA and/or the 2002 FIFA World Cup™ Ticketing Bureau, is not in compliance with these provisions and procedures can be rejected, and/or the tickets issued can be cancelled. Please refer to the 2002 FIFA World Cup™ Ticket Application Form for details.

If your selected application has not reached the limit of 4 ticket series per household (address) you may apply for more ticket series. Please remember that you must provide the personal details of each

Sales Invoice

To: **Lee TESDELL**
300 NW 158th Avenue
Slater
IOWA
50244
United States of America

Invoice No. WCTB009783
Date: 4 September 2001
Our Ref: LEETES44 / 030939

DESCRIPTION

NAME	SERIES	PRICE
Lee TESDELL	DAEGU CAT3	US$280
Omar TESDELL	DAEGU CAT3	US$280
Ramsey TESDELL	DAEGU CAT3	US$280
	GRAND TOTAL	**US$840**

2002 FIFA World Cup™ Ticketing Bureau is a trading name of SEAMOS Marketing AG, a Swiss Company with limited liability.

FOR INFORMATION ONLY

The credit card, as detailed in your application, is being charged.

4. 26 October 2001

Paper Correspondence: Official paper ticket certificate with notification that we would receive actual tickets six weeks before World Cup

We knew we had World Cup tickets now, but we didn't have the actual paper tickets. For security reasons, the 2002 FIFA World Cup Ticketing Bureau initially sent only ticket certificates. These certificates are very ornate documents incorporating the official 2002 World Cup logo and the three official languages: Korean, Japanese, and English.

5. 05 April 2002 to 01 June 2002

Electronic correspondence: E-mail about home stay

We arranged our home stay through the Korean Committee, which matched us with a family that had volunteered to host visitors to the World Cup. Before making the reservations, I calculated the cost of the rooms based on the current exchange rate. The family e-mailed us promptly, and we exchanged several e-mails in English prior to arriving in Korea, where we finally met them. We paid a homestay fee of 500,500 Korean won (US $391.23) on May 14, 2002, for the three of us for seven nights.

PO Box 2002
Cheadle Hulme SK8 7RR, England
Fax: +44 (0)870 124 2002
Email: sales@fifa-tickets.com

Lee Tesdell
300 Nw 158th Avenue
Slater
Iowa
50244
United States Of America

26th October 2001

Dear Lee Tesdell,

It gives us great pleasure to enclose the FIFA Official Ticket Certificate(s) issued by FIFA to confirm the 2002 FIFA World Cup™ tickets you have purchased.

For security reasons, you will not receive actual Admission Tickets to the 2002 FIFA World Cup Korea/Japan™ until approximately six weeks prior to the commencement of the event. Admission Tickets for the first round matches for Team Specific Ticket Series that include later round match tickets will be accompanied by instructions regarding later round Admission Tickets. Note that the enclosed Ticket Certificates cannot be used in place of actual match tickets to gain admission to the stadia and will not be exchanged for Admission Tickets. If you do not receive your match tickets by 1 May 2002, please contact us immediately.

Please be aware that the address we have used to send you this communication is the address on record that will be used next year to send you your actual Admission Tickets. Should you change address between now and then please inform us by e-mail, letter or fax at the above address or number. Please note we can only send the Admission Tickets to the Main Applicant indicated on the Application form.

10539202

Official Partners of the 2002 FIFA World Cup Korea/Japan™

adidas AVAYA Budweiser Coca-Cola FUJI XEROX FUJIFILM Gillette HYUNDAI JVC KT NTT MasterCard McDonald's PHILIPS TOSHIBA

2002 FIFA World Cup™ Ticketing Bureau is a trading name of SEAMOS Marketing AG, a Swiss Company with limited liability.

...ation has an ...u, him or her ...bmit a "Request to ...002 FIFA World ...'A may consent to ...hat are outlined in ...structions and ...ase note that for ...nd submit a ...subject to the 2002 ...er requirements set

... of tickets that you ...at you can attend ... to follow. The ... World Cup ...ound matches, will

... please do not ...ve.

... exciting and enjoyable time

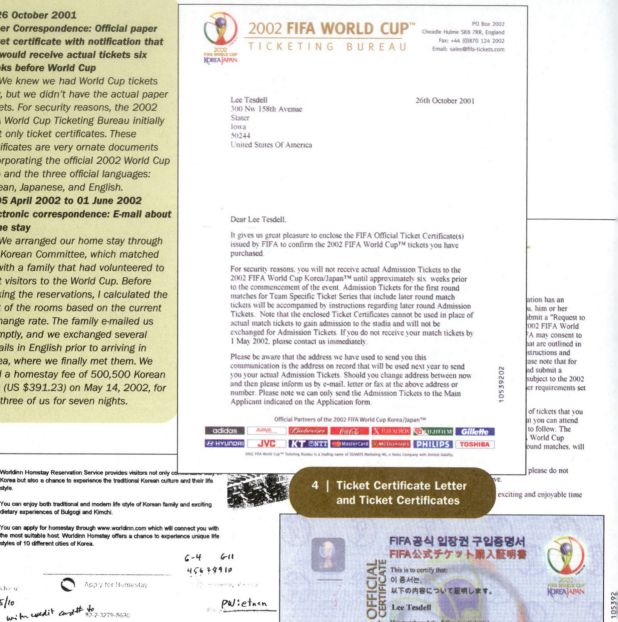

4 | Ticket Certificate Letter and Ticket Certificates

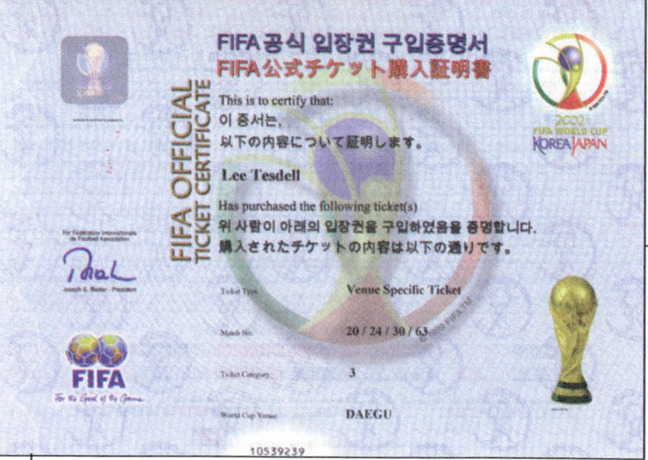

105392

Worldinn Homestay Reservation Service provides visitors not only com... stay in Korea but also a chance to experience the traditional Korean culture and their life style.

You can enjoy both traditional and modern life style of Korean family and exciting dietary experiences of Bulgogi and Kimchi.

You can apply for homestay through www.worldinn.com which will connect you with the most suitable host. Worldinn Homstay offers a chance to experience unique life styles of 10 different cities of Korea.

Apply for Homestay

2002 6-4 6-11 45678910

confirmed on 5/10 ... by 5/17 with credit card# to ... 82-2-3279-5690

PW:etnen

30K = 23.38
50K = 38.9 USD

- Reservation can be made through internet(www.worldinn.com) or call center(82-2-3279-5690, 82-2-555-5555). You need to make reservation at least 2 weeks prior to your preferred check-in date.
- Worldinn will select an appropriate homestay for you and notify you via e-mail within one week. (If worldinn could not find an appropriate homestay, we will e-mail to you.)
- Payment must be made within two week from your reservation. Your reservation will be automatically cancelled unless the payment is made.
- Payment for homestay can vary and average lodging fee is between 30,000won~50,000won(KRW) per person per night. (The reservation service charge is 10% of homestay fee.)
- Some homestays are free of charge. In this case, you only need to pay reservation service charge of 5,000 KRW per person.
- The inf... host. Type your name and your address...

5 | Electronic Correspondence Reserving Homestay

6 | Letter Accompanying Tickets and Tickets

6. April 2002

Paper Correspondence: letter and paper ticket receipt

At the sixth stage, we received a letter from the 2002 FIFA World Cup Ticketing Bureau as well as our tickets for each game, ones we would present at the gate in Daegu Stadium. The tickets themselves are complex documents that include the official FIFA World Cup logo. The text is in the two host country languages — Japanese and Korean —and, in addition, English. Each game (match) has a number. Our matches were 20 (Denmark v. Senegal), 24 (South Africa v. Slovenia), and 30 (USA v. Korean Republic).

2002 FIFA WORLD CUP™
TICKETING BUREAU

2002 FIFA WORLD CUP KOREA JAPAN

PO Box
Cheadle Hulme SK8 7RR, En
Tel. +44 (0) 870 123
Fax: +44 (0)870 124
Email: sales@fifa-ticket

Lee Tesdell
300 Nw 158th Avenue
Slater
Iowa
50244
UNITED STATES OF AMERICA

Delivery Seq No 20378
No.Tickets/Pack 1
No.Smartcards/Pack 0

April 2002

Dear Lee Tesdell

It gives us great pleasure to enclose your tickets for the 2002 FIFA World Cup Korea/Japan™. If you have purchased tickets that are conditional upon your team's performance during the First Round Group Matches, you will also find enclosed your FIFA SmartCard with instructions on how to use it. There are essentially three types of ticket products for which FIFA SmartCard may be used; Fixed or Conditional Team Specific Ticket Series 4, 5, 6 or 7 or individual Conditional Team Specific Tickets for those who purchased Second Round Tickets from their National Association.

May we take this opportunity to wish you an exciting and rewarding 2002 FIFA World Cup™ experience.

2002 FIFA World Cup™ Ticketing Bureau
For and on Behalf of

f Executive Officer
Football Association

Lee Tesdell, PhD., is a full-time professor at Minnesota State University, Mankato, a soccer referee, and part-time farmer.

Courtesy of Lee Tesdell

Individual and Collaborative Assignments

1. **Write a response to a memo.** Assume the role of Carole Marcotte, general manager, and compose a response to the following memo. Your memo should be addressed to Richard Curtis in the marketing department.

October 3, 20—

TO	Carole Marcotte, General Manager
FROM	Richard Curtis, Marketing
SUBJECT	Major order for 024BL6 carrier

We have a possibility of receiving an order for 200,000 nibs for the 024BL6 carrier, but it must be delivered before July 31. Do you think we have the production capability to accept such an order?

Here are some data to help you come to a decision: Last year in January we produced 150,000, in February 155,000, in March 145,000, in April 140,000, in May 143,000, in June 147,000, in July 90,000 (plant closed two weeks for vacations), in August 120,000, in September 160,000, in October 25,000 (strike), in November 120,000, and in December 153,000.

We have been very successful with this product. Some of our largest sales have been to Fortune 500 companies, although last month we shipped to many less well known organizations. Lagassee, Inc., bought 10,000. Demonsthanes bought 16,000, P&C bought 22,000, Clairbridge Brothers bought 17,500, and Harpswell bought 8,250. We must not commit ourselves to one large order to the exclusion of these others.

One of my main concerns about accepting so large an order is the question of material supplies. Before coming to a decision, you might want to contact some of our more reliable vendors. Mr. Edwin Mundale of LaSol Iron Works, at 1125 Industrial Avenue in Tyngsboro, Missouri, is one suggestion. Another is Ms. Ramona Fitzdurwood who is the liaison officer for Abruters, Inc. Their plant is on Liatris Road in Pobgardner, Kansas.

Let me have your advice within two weeks. I hate to turn down such a plum, but I don't want to risk our reputation for reliability, either.

2. **Write two different types of letters or memos.** Incorporate the following facts into two different letters or memos. Use priority order in the first and

chronological order in the second. Determine which you prefer and justify your preference in a memo to your instructor.

- Our production of component Z43X has dropped.
- A meeting of production supervisors will be held.
- Two of our best customers have complained about late shipments.
- Supervisors should bring to the meeting figures on costs, equipment, personnel, and so on.
- One customer has canceled his standing order and is buying his Z43Xs from our competitor.
- The meeting will be held on March 4 at 2:30 p.m.
- Last month four machines had excessive down time.
- The supervisors are to meet in Conference Room G in the administrative wing.
- Two months ago, production of component AB22X also plummeted.

3. **Write memos or letters to respond to a situation.** Early this morning you found out that part of an already-delayed order for a major account (Bit-Byte Corporation) will not meet the new deadline and that the customer has serious complaints about some of the delivered portion of the order.

   Specifically, the production schedule for system XY-9 has been set back because of manufacturing backlogs, equipment malfunction, and duplicate scheduling. Your supervisor is busy sorting out another complex problem and asks you to write a detailed memo explaining how you have decided to handle the problems with manufacturing and with the customer.

   The customer's serious complaints are that the equipment that was delivered is missing an I/O board for the CPU, has the wrong character set for the printer, and has broken prongs on the disk-drive plug. Beyond these problems, the shipment was five weeks late because it was originally delivered to the wrong customer. Bit-Byte Corporation demands an explanation and a guaranteed solution to the problems.

   Help! You're due in Maynard for an important lunch meeting, but you must write a memo to your supervisor and then a letter to the customer before you leave. To help you plan, answer the following questions:

   - Can you send both the same information?
   - What should you say to each?
   - What tone should you take in each?
   - How much of an explanation should you include?
   - Should you make excuses?
   - Should you make promises?

- What should you focus on or emphasize?
- Could you respond to either of these people with an e-mail? Why or why not?

Now, write the message to your supervisor and the letter to the customer based on your assessment of the situation.

4. **Analyze a piece of correspondence.**
   (a) Locate a piece of business correspondence, either something that was addressed to you or that you have access to.
   (b) Using the criteria you learned from this chapter and the existing criteria you have established for effective technical communication, create a rubric to assess the document.
   (c) Write a brief report for your instructor in which you include the original document and your rubric.

Chapter 18 Endnotes

1 Electronic Evidence Discovery. (2004). *Home page.* Retrieved October 5, 2003, from http://www.eedinc.com/

2 Nimsger, K. M. & Lange, M. C. S. (2002, May 10). Computer forensics experts play crucial role [Electronic version]. *Lawyers Weekly,* 14. Retrieved October 5, 2003, from http://www.krollontract.com/LawLibrary/Articles/lawyersweekly.pdf

3 Nimsger, K. M. & Lange, M. C. S. (2002, May). *Examining the data: A beginner's guide to computer-based evidence.* Kroll Ontrack Law Library. Retrieved October 5, 2003, from http://www.krollontrack.com/LawLibrary/Articles/securityproducts.pdf

4 Nimsger, K. M. & Lange, M. C. S. (2002, May). *Examining the data: A beginner's guide to computer-based evidence.* Kroll Ontrack Law Library. Retrieved October 5, 2003, from http://www.krollontrack.com/LawLibrary/Articles/securityproducts.pdf

5 American Management Association. (2003). *2003 e-mail rules, policies, and practices survey.* Retrieved October 5, 2003, from http://www.amanet.org/research/pdfs/Email_Policies_Practices.pdf

6 APA, Committee on Disability Issues in Psychology. (1992, April). *Guidelines for nonhandicapping language in APA journals.* Retrieved November 25, 2003, from http://www.apaastyle.org/disabilities.html

CHAPTER 19

Preparing Proposals

Objectives and Outcomes

This chapter will help you accomplish these outcomes:

- Identify types of proposals
- Locate requests for proposals (RFPs)
- Identify the appropriate means of persuasion for the proposal's audience
- Understand the different necessary parts of each type of proposal
- Prepare, organize, and write a proposal

Proposals are used in a variety of situations, as the following list suggests:

- The EPA is willing to support state and local microbiological testing and monitoring of coastal recreational waters near publicly accessible beaches in order to raise awareness of exposure risks to disease-causing microorganisms.
- The Kellogg Foundation is willing to fund plans for research into new sustainable and ecofriendly food systems.
- Media Designs' warehouse for storing damaged and returned goods is overflowing. The warehouse manager meets with a business consultant to discuss the problem and identify the goal that she has in mind.
- A computer network engineer believes that creating a company intranet with calendar and project management features would help cross-functional teams across regional offices collaborate more effectively. The chief information officer and chief operating officer are willing to discuss the matter.

All of these cases result in the development and submission of proposals. The EPA and the Kellogg Foundation solicit input through a published request for proposal (RFP). The consultant submits to Media Designs a proposal to develop a plan for managing its warehouse full of damaged and returned goods. The computer network engineer writes an unsolicited proposal to assess the potential benefits of implementing a company-wide intranet.

In each situation, the readers — the decision makers — have certain expectations about the proposal they read. They typically expect some or all of the following information to be included:

- ***Situation.*** Provide a definition of the problem or opportunity, including information that situates it in the organization.
- ***Plan.*** Present a plan for resolving the problem or addressing the opportunity.
- ***Benefits.*** Explain probable benefits that will result from adopting the plan.
- ***Approach.*** Outline methods for implementing the plan, including management plans, schedule, and costs.
- ***Evaluation.*** Identify an evaluation strategy for determining whether the proposed plan works.
- ***Qualifications.*** Establish your qualifications for submitting the proposal and implementing the plan.

Sometimes this information is presented in sections that correspond to the six categories; other times, the information is combined into fewer sections. The selection and focus of the information depend on the purpose and audience for the proposal.

Preparing and organizing proposals differ in important ways from preparing and organizing reports (which you will read about in Chapter 20). Reports are about information the writer already knows and things that have already happened; they present an "answer." Proposals are about things that the writer wants to happen and offer ways to do those things; they suggest approaches to discover an "answer."[1]

WEBLINK

How much money is given away each year in the United States? By philanthropic foundations? By corporations? By community foundations? Go to www.english.wadsworth.com/burnett6e for a link to learn about the top 100 philanthropic foundations, corporations, and community foundations in the United States. You can rearrange each list by asset size and by total giving.

CLICK ON WEBLINK

CLICK on Chapter 19/reasons for rejection

Characterizing Proposals

A form of persuasive writing, *proposals* attempt to convince an audience that a proposed plan or set of objectives responds to a problem or addresses a problem while being workable, manageable, logically organized, and cost efficient. As the list at the beginning of this chapter indicates, proposals range from responses to formal requests for proposals (RFPs) to unsolicited proposals for solving perceived problems. Proposals can have one of several purposes:

Which kind of proposals are you likely to need to write or contribute to in your professional work?

- ***Solve a problem*** — The problems that proposals try to solve vary from designing or manufacturing a mechanism to modifying a process to establishing a new business procedure.
- ***Investigate a subject*** — An important part of technical communication in business and industry as well as in government and research institutions is generating investigative proposals for a variety of projects.
- ***Sell a product or service*** — Sales proposals offer a service or product, providing the potential customer with information needed to make a decision.

In this section, you'll read about differences between solicited and unsolicited proposals as well as about sources of RFPs.

Types of Proposals

A *solicited proposal* is written in response to an RFP. When an organization turns to an outside vendor, consultant, or researcher to address a problem, it issues an RFP that identifies all of the specifications the proposal must fulfill in order to be accepted. The requirements and restrictions in RFPs vary according to the needs of the issuing organization. Since an RFP provides the specifications for the proposal, a successful proposal must adheres to its guidelines. Some RFPs even indicate a point value or a weight assigned to each section, so a proposal writer can emphasize the most important sections.

When the need that stimulated the RFP is for a product or service, the solicited proposal may also be called an invitation to bid, a bid request, a purchase request, an invitation for proposal, or a request for quotation (RFQ). RFPs, RFQs, and similar requests are usually external documents, directed to and responded to by people outside an organization.

> *Under what circumstances would you submit an unsolicited proposal? Would you be inclined to discuss your plans with your supervisor before you submitted it?*

Sometimes a problem exists but no one issues an RFP. A person who identifies the problem and has the skill or experience and initiative to solve it may submit an unsolicited proposal. This is written in response not to an RFP but to a perceived need. Unsolicited proposals are often internal documents, responses to a perceived need by someone in an organization.

WEBLINK

A quick Web search will identify more than ten million sites that have something to do with proposals. The WEBLINKS in this chapter provide a more focused approach by suggesting specific sites for guidelines, funding organizations for different purposes, cautions, and so on. Two very useful online resources, the Grantsmanship Center and the Foundation Center, provide easily accessible information, including funding sources, an archive of articles, and useful tutorials. For links, go to www.english.wadsworth.com/burnett.6e.
CLICK ON WEBLINK
CLICK on Chapter 19/TGCI and TFC

Sources of RFPs

RFPs are not always sent to every organization or company that might want to respond, so an interested person might be unaware of a particular RFP. However, because approved proposals are a potential source of income for business and of funding for human services organizations, an enterprising individual should seek out RFPs pertinent to corporate or organizational objectives. The acceptance of a proposal means the award of a contract or a financial grant. RFPs are regularly issued by several types of organizations:

- research and nonprofit foundations
- educational institutions
- government agencies
- private business and industry

You can locate funding agencies and RFPs from a variety of sources, including university grant offices, which list available RFPs as well as assist in preparing the proposals.

Public, university, and corporate libraries have resources that can assist you. For example, the *Annual Register of Grant Support* provides listings of grant programs and requirements for applications. Another useful reference is the *Foundation Grants Index,* published by the Foundation Center. The Foundation Center's comprehensive Web site has numerous articles about writing successful grants. One of the most comprehensive collection of resources is available through the Grantsmanship Center. The *Granstmanship Center Magazine* publishes articles about grant writing and advice for nonprofits, which are archived on their Web site. This Web site also includes sources of funding for nonprofit organizations, a schedule of the Grantsmanship Center's training programs for grant writing, and a list of publications about proposal writing. Other excellent resources that identify specific grant and contract announcements include these publications, available in libraries or university grant offices:

- print and electronic versions of the *Catalog of Federal Domestic Assistance*
- *Federal Grants and Contracts Weekly,* available in a subscription-based e-mail newsletter
- *Federal Register*

Figures 19.1 and 19.2 reproduce RFPs from the *Federal Register* (which publishes synopses of proposed federal contract actions that exceed $25,000) and from businesses. Just these three announcements illustrate the tremendous range of possibilities.

An RFP from the National Center for Environmental Research (NCER), which is part of the EPA invites proposals to conduct "research on algal species whose populations may cause or result in deleterious effects on ecosystems and human health."[2] Figure 19.1 shows the announcement that was published in the *Federal Register.*

An RFP from another government agency, the Centers for Disease Control (CDC), which is part of the U.S. Department of Health and Human Services (HHS), invites proposals to "evaluate strategies to reduce the number of residential fire-related injuries and fatalities in high-risk communities."[3] Figure 19.2 shows the announcement that was published in the *Federal Register.*

FIGURE 19.1 Request for Proposal: NCER Research on Algal Blooms[4]

The initial description by NCER provides a number of specific details:

- *Duration: 3-year research projects*
- *Approach: multi-disciplinary regional studies*
- *Focus: causes, detection, effects, mitigation, and control*
- *Location: U.S. coastal waters (including estuaries and Great Lakes)*

Program Title: Ecology and Oceanography of Harmful Algal Blooms Program
Agency: U.S. Environmental Protection Agency (EPA)
Sub-Agency: National Center for Environmental Research (NCER)
Program URL: http://www.tgci.com/fedrgtxt/03-27674.txt

People interested in further information about this opportunity can simply click on the highlighted link to read the more detailed RFP.

Summary

The purpose of this notice is to advise the public that the participating agencies are soliciting individual research proposals of up to 3 years duration, and depending on appropriations, multi-disciplinary regional studies of 3 to 5 years duration for the Ecology and Oceanography of Harmful Algal Blooms (ECOHAB) program. This program provides support for research on algal species whose populations may cause or result in deleterious effects on ecosystems and human health. Studies of the causes of such blooms, their detection, effects, mitigation, and control in U.S. coastal waters (including estuaries and Great Lakes) are solicited.

FIGURE 19.2 Request for Proposal: Reducing Residential Fire-Related Injuries and Fatalities[5]

The CDC clearly identifies the purpose of this project and specifies that the outcomes must be measurable and aligned with existing performance goals.

Calling attention to the performance goals by putting them in a numbered list increases the likelihood that readers will interpret them as important.

This preliminary description concentrates on the objective and provides sufficient background information to make clear the probable directions the proposals should take.

Program Title: Community Trial to Test the Effectiveness of the Smoke Alarm Installation and Fire Safety Education Program
Agency: U.S. Department of Health and Human Services (HHS)
Sub-Agency: Centers for Disease Control (CDC)
Program URL: http://www.tgci.com/fedrgtxt/03-29634.txt

Summary

The purpose of this program is to evaluate strategies to reduce the number of residential fire-related injuries and fatalities in high-risk communities. This program addresses the "Healthy People 2010" focus area of Injury and Violence Prevention. Measurable outcomes of the program will be in alignment with one or more of the following performance goals for the National Center for Injury Prevention and Control:

1. Increase the capacity of injury prevention and control programs to address the prevention of injuries and violence.
2. Monitor and detect fatal and non-fatal injuries.
3. Conduct a targeted program of research to reduce injury-related death and disability.

The objective of this cooperative agreement is to rigorously evaluate strategies to reduce the number of residential fire-related injuries and fatalities in high-risk communities. Smoke alarms have proven effective in reducing the fire death and injury toll. Research shows that functioning smoke alarms are more likely to be present in a home when a fire safety program provides and installs them, rather than simply providing vouchers and/or discounts to individuals to obtain alarms that require resident installation. There are CDC

An RFP from We Energies, a principal subsidiary of Wisconsin Energy Corporation, invites proposals to provide "up to 200 megawatts of wind generation. Contracts resulting from proposals will be for a term of twenty (20) years and for a minimum of 20 megawatts."[6] Figure 19.3 shows the Web announcement.

FIGURE 19.3 **Request for Proposal: 2003-2004 Wind Energy**[7]

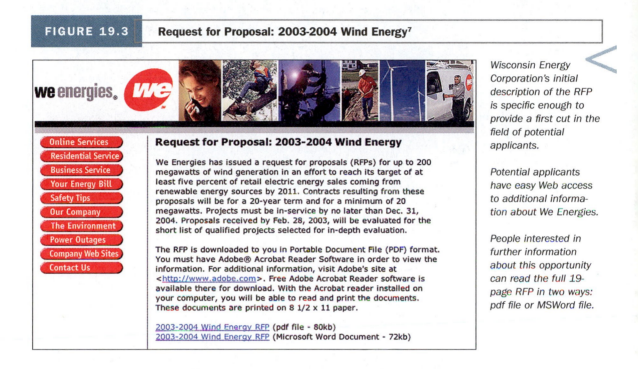

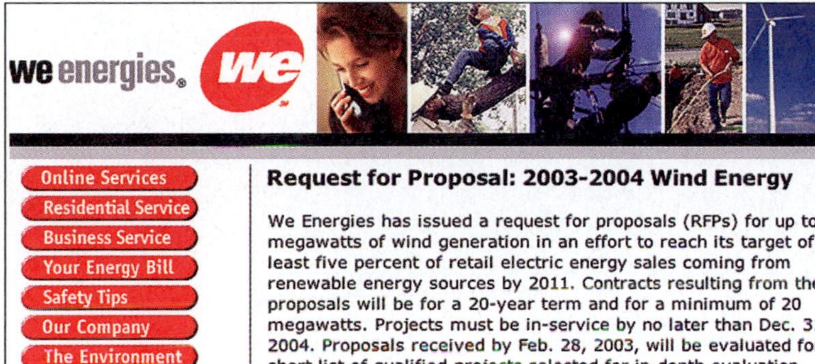

Request for Proposal: 2003-2004 Wind Energy

We Energies has issued a request for proposals (RFPs) for up to 200 megawatts of wind generation in an effort to reach its target of at least five percent of retail electric energy sales coming from renewable energy sources by 2011. Contracts resulting from these proposals will be for a 20-year term and for a minimum of 20 megawatts. Projects must be in-service by no later than Dec. 31, 2004. Proposals received by Feb. 28, 2003, will be evaluated for the short list of qualified projects selected for in-depth evaluation.

The RFP is downloaded to you in Portable Document File (PDF) format. You must have Adobe® Acrobat Reader Software in order to view the information. For additional information, visit Adobe's site at <http://www.adobe.com>. Free Adobe Acrobat Reader software is available there for download. With the Acrobat reader installed on your computer, you will be able to read and print the documents. These documents are printed on 8 1/2 x 11 paper.

2003-2004 Wind Energy RFP (pdf file - 80kb)
2003-2004 Wind Energy RFP (Microsoft Word Document - 72kb)

Wisconsin Energy Corporation's initial description of the RFP is specific enough to provide a first cut in the field of potential applicants.

Potential applicants have easy Web access to additional information about We Energies.

People interested in further information about this opportunity can read the full 19-page RFP in two ways: pdf file or MSWord file.

WEBLINK

Finding the best match between the needs of an organization and the opportunities offered by a funding agency is often difficult. Here are two useful government sites: (1) Grants.gov describes itself as a site that "allows organizations to electronically find and apply for competitive grant opportunities from all Federal grant-making agencies." (2) FedBizOpps.gov describes itself as "the single government point-of-entry (GPE) for Federal government procurement opportunities over $25,000." For links that will let you explore the way these sites work, go to **www.english.wadsworth.com/burnett6e**.

CLICK ON WEBLINK

CLICK on Chapter 19/gov grant opportunities

You can also identify possible sources of funding through computerized databases that use a keyword system to help you locate grant announcements for research in specific areas. Most university grant offices offer assistance in computerized searches.

WEBLINK

For local projects, the most appropriate source of support may be community foundations, which are nonprofit, tax-exempt, publicly supported grant-making organizations. They are not only a potential source of financial support, but they are also a good place to begin networking to learn about other funding possibilities and about possible collaborations with other nonprofits in your geographic area. For more information about grants for nonprofit organizations, go to **www.english.wadsworth.com/burnett6e**, where you will find links to search for community foundations by state as well as a link to literature about funding for nonprofit organizations.

CLICK ON WEBLINK

CLICK on Chapter 19/nonprofits

Using Persuasion in Proposals

To write effective proposals, you need to understand persuasive techniques. Effective use of such techniques helps convince the nonprofit foundation, educational institution, government agency, or private business to approve or accept your proposal. Essentially, you're establishing agreement about the situation (that is, the problems or opportunities that need to be addressed), posing a plan to address them, and identifying benefits that will accrue if your plan is adopted. The process of preparing a proposal is like a debate, with you imagining, anticipating, and responding to potential arguments that the reader may raise.

What are the ethics of using what are popularly called propaganda techniques (fallacious appeal to authority, appeal to pity, appeal to emotion, question begging, slanting, etc.) in preparing a proposal?

Traditionally, persuasion has been identified with appeals to emotion and, therefore, has sometimes been viewed as inappropriate for technical communication. The actual practice of technical communication, however, always includes a persuasive element. Persuasion doesn't mean manipulation; rather, it means applying credible, logical arguments to convince readers that the writer's view is appropriate. Audience members who develop positive attitudes toward the subject of a proposal are more likely to accept that proposal.

Technical communicators have a responsibility to produce proposals that are ethical, credible, and logical. These factors have been identified as critical aspects of persuasion since Aristotle described them in his *Rhetoric* in the fourth century BCE:[8]

- concerns of the audience
- credibility of the speaker or writer
- logic of the message

Clearly, proposals have strong persuasive elements. In what ways are other kinds of technical documents also persuasive?

How can you increase the likelihood of acceptance? Go to www.english.wadsworth.com/burnett6e for links to sites that provide tips for getting it right! After you've read a few of these tip lists, you see common patterns of what's expected.

CLICK ON WEBLINK

CLICK on Chapter 19/tips

WEBLINK

Audience Concerns

The audience must have some sense of the problem or opportunity in order to agree with the changes advanced by your proposal. Part of your job is to identify this problem or opportunity. You can identify it by conducting a careful audience analysis and assessing receptivity and resistance. Once the problem or opportunity has been identified, you can take time to think about how you will respond.

The more sweeping the changes you advocate, the greater the readers' concerns are likely to be and the more completely the proposal must establish the need and then substantiate the validity of the changes. This is one of the single most important aspects of preparing a proposal:

> Establish that the problem or opportunity exists and then clearly show how your plan addresses it.

How can you explicitly establish a relationship between the problem or opportunity and your plan?

Understanding why readers react the way they do to new material in a proposal may help you write more persuasively. Theorists in psychology and communication believe that people reject or at least devalue information that conflicts with their existing beliefs. As a writer and presenter, you will find that this principle, called *cognitive dissonance,* has important, practical applications. The more readers and listeners adhere to their existing beliefs, the less likely they are to accept proposed changes.

You can negate the effects of cognitive dissonance in your proposal by suggesting why the change will be beneficial. After you identify elements that may cause anxiety in your audience, offer a solution that dissipates the anxiety. Explain how the solution meets the readers' needs and thus results in a decision that benefits the organization as well as the decision maker.

Persuaders' Credibility

After you have established the problem or opportunity, you must develop your credibility. If readers believe that you are reasonable, honorable, and display goodwill, then you have credibility. You gain credibility by demonstrating a variety of qualities: technical expertise, favorable reputation, organizational status, values similar to those of the audience, and characteristics in your writing that show you to be understanding, well informed, carefully organized, articulate, trustworthy, and fair minded.

In many situations, credibility is the single most important factor in awarding contracts. For example, some firms choose a vendor on the basis of reliability for delivery and reputation for integrity, even before they consider cost (W. L. Jeffries, personal communication, 2004). Similarly, your credibility may be the deciding factor in your proposal's approval.

The more far-reaching or expensive the changes that you have proposed, the more credibility you must have if the proposal is to be seriously considered. In fact, credibility is so important that gaining the support of a person with higher credibility than yours increases the likelihood that a proposal will be accepted. Although this is a political issue, you should be aware that it may affect the acceptance or rejection of your ideas.

Logic of Message

After readers' needs and your credibility are established, you should make sure the proposal is logical. First, your document must be based on sound assumptions. Then you should build a reasonable case that explains and relates audience needs to your proposed plan, supporting each point with valid, reliable evidence. In addition to presenting a logical case, you also need to acknowledge and respond to opposing views.

You can develop a sound argument and organize your ideas by reasoning, either inductively or deductively.

Induction. *Induction* is reasoning from the particular to the general. You reach a conclusion about all of the members of a group after examining representative examples (for example, the Gallup poll or the Neilsen ratings). For example, imagine your proposal about employee health seminars depends in part on the audience accepting your argument that ingestion of any alcohol during pregnancy can cause fetal damage. Your goal is to reduce the audience's uncertainty about your argument. Is the audience justified in some uncertainty? Of course. You cannot examine all instances of ingestion of alcohol during pregnancy. Instead, you need evidence that your audience will see as accurate, current, and representative. If you don't present such evidence, your inductive argument will fail because your evidence will not be seen as reflecting the whole.

Deduction. *Deduction* is reasoning from the general to the specific. Traditionally, this reasoning takes the form of a *syllogism,* a three-part argument:

Major premise — general statement about an entire group
Minor premise — statement about an individual within the group
Logical conclusion — conclusion about the individual

For example, imagine your proposal depends in part on the audience accepting your argument that the Toyota RAV (your choice for a fleet vehicle) is safer in accidents than larger SUVs. Here's a syllogism:

In an accident, smaller SUVs are safer than larger SUVs. (*major premise: all A are B*)
The Toyota RAV is a small SUV. (*minor premise: C is an example of A*)
In an accident, the Toyota RAV will be safer than a larger SUV. (*logical conclusion: therefore, all C are B*)

An audience will accept the logical conclusion of your syllogism if your major premise and your minor premise are both accurate. Beware of false syllogisms, in which the logic does not carry over (All redheads have pale skin. Marcia has pale skin. Therefore, she is a redhead).

A variation of a syllogism called an enthymeme is especially useful in creating effective arguments. An *enthymeme* is actually a syllogism with a few differences: The major premise is unacknowledged or assumed, it can address probabilities rather than verifiable information, and it can employ ethical and emotional arguments as well as logical arguments.[9] Here's an enthymeme:

Because the Toyota RAV is a small SUV, in an accident it will be safer than a larger SUV.

What is the unstated premise in this enthymeme?

Think of an enthymeme as a claim with evidence. It often includes a cue word about the relationship between the claim and the evidence: *because, for, hence, since, so, therefore, thus.*

Developing a strong proposal that persuades your audience is easier if the readers or listeners believe you have advanced a solid plan to meet identified problems or opportunities. Your analysis and response to audience needs, awareness of your own credibility, and attention to logic will go a long way in building a case that wins approval.

WEBLINK

While lots of advice is available about writing proposals, you can be discriminating about your sources. Go to www.english.wadsworth.com/burnett6e for links to several sites that offer practical suggestions for developing and preparing proposals.
CLICK ON WEBLINK
CLICK on Chapter 19/help with proposals

Many of the proposals you write will be *solicited* — that is, written in response to RFPs. One way to become a more successful proposal writer is to temporarily shift your perspective to consider the concerns of people who write RFPs, specifically the RFP you're addressing. Why? This lets you consider the problem from the perspective of the person who will decide if your proposal merits attention and will get accepted. If you're writing an *unsolicited* proposal, one that isn't in response to an RFP but is based on your personal observation and initiative, you will write a better proposal if you imagine (and perhaps even actually outline) a fictional RFP that defines the problem or opportunity you're addressing.

Writing an RFP

How do you go about writing a useful RFP, actual or fictional? It takes more time, thought, and effort than many people realize. A carelessly researched and written RFP typically gets inadequate proposals in return. The preparation of an effective RFP typically includes several steps, although the process is not at all linear nor prescriptive.

Identify the Problem or Opportunity. If you can't identify the broad problem or opportunity for change, you're not likely to be able to describe what you want vendors, consultants, or researchers to do. Who cares about this problem or opportunity? Why is it important?

Provide Background Information. This information is necessary, maybe even critical, for the vendors, consultants, or researchers responding to the RFP. The more specific you are about the situation, the more helpful you'll be. Your information will probably include boilerplate text describing your organization, but you should also include information related to the specific problem or opportunity.

Define the Desired Outcome. An outcome is not the same thing as a specific product or service that might be proposed to address a problem or opportunity. Consider how the outcome can be described. What do you think you want to happen? Consider the ramifications of implementing this outcome. In considering these ramifications, draw on the opinions and expertise of colleagues who have a vested interest in the situation as well as those not directly involved but who may have useful insights or opinions. Don't describe the outcome so strictly that you eliminate the creativity of the vendors, consultants, or researchers.

Specify the Product or Service You Need. While you should clearly convey the product or service you need, be flexible enough to realize that once you begin in-depth exploration, your perspective about needs and preliminary ideas about possible solutions might change. How much might things change once you begin exploration? Management expert Judith Harkham Semas tells the story about a dramatic and unexpected change in perspective affected an entire project:

> Holly Townsend, president and chief executive officer of Association Publishers, Inc., Bethesda, Maryland, an association publication research firm, was once asked to conduct an editorial research study for a large health care association.
>
> "When we initially bid on the project, it was described in the RFP as a standard, four-page readership study aimed at providing data for a graphic redesign of their magazine," Townsend explains. In the process of creating the survey instrument, she met with several of the publication's staff members. During these conversations, "staff made it clear that their larger concerns had to do with editorial content rather than design. They felt certain that their take on what the magazine should be differed from what others in the organization wanted.
>
> "As they aired these concerns, it became obvious that their real need was for a comprehensive publications audit," she recalls. "This required a series of studies that included focus groups and mailed studies, as well as several meetings with staff and leadership to address a wide range of concerns that went well beyond a visual face-lift."
>
> What Townsend had expected to be a $10,000 study ended up as a vastly more complex and expensive project, with a price tag exceeding $30,000. "Luckily, our client realized that they had underestimated their needs just about the time that we did; they were willing to revisit the RFP and fund the larger project." But the situation could have been a disaster. Had Townsend delivered exactly what the association initially asked for, and no more, the group's needs would not have been met. At $10,000, that's an expensive lesson.
>
> "That experience taught me how important it is for us to thoroughly interview prospective clients about their research goals prior to responding to RFPs. [By doing so], we can determine exactly what problems we are being hired to address. It doesn't hurt to understand who at the association is behind the decision to do the research, what their goals are, and how they intend to use the data. In most cases, many people within the organization use research studies; each may have slightly different goals. The challenge is to be sure that everyone's needs are being met."[10]

Require Detailed Information about the Organization and Personnel. You need detailed information about the background of the vendors, consultants, or researchers so that you can make a fair comparison among those who respond. Depending on what the RFP is for, you should consider which of the following information is relevant:

- ***Personnel***
  - Capabilities of key players and staff, including longevity and turnover, education, professional background, and relevant experience
  - Experience and success with other organizations of similar size and scope

- ***Available equipment***
  - Equipment for both production and testing
  - Condition and maintenance
- ***Work history***
  - Flexibility or limitations in scheduling
  - Ability to handle possible but unexpected problems beyond their control (e.g., transportation problems, labor disputes, limitations in raw materials, severe weather such as hurricanes) and pinpointing possible workarounds
  - Specific products and services they have produced for others or projects they have completed; you want a history, a track record
  - Recommendations from previous projects[11]

Provide Process Information. You need to provide essential details about the entire proposal process so that potential vendors, consultants, and researchers can decide whether they have the capability to fulfill your expectations. This information also enables them to submit a proposal that meets your expectations. Specifically, you need to provide information about these factors:[12]

- ***Dates*** — due date for proposal, response time for acceptances/rejections of the proposal, timetable for the project
- ***Project details to be considered***
  - Confidentiality and compliance expectations
  - Support requirements — oral presentations, written reports, staff briefings, etc.
  - Travel requirements — frequency, locations, schedule
- ***Proposal submitted for consideration***
  - Organization of proposal — required section headings, sequence of information
  - Document design specifications — length, font style/size, headings, margins, line spacing
  - Budget — products and services you want priced
- ***Criteria for evaluating the proposals***

Establish Criteria for Selection. A detailed RFP often identifies the selection criteria and then defines and explains each category. Clear, measurable criteria let you more easily assess and compare the proposals that are submitted. Some RFPs include a grid or matrix for assessing each proposal.[13]

Selection Criteria	Assessment (1–10)	Weight	Score
Understanding requirements		× 2.0	=
Technical capability		× 2.0	=
Analysis of problem		× 2.0	=
Proposed solution		× 3.0	=
Project management		× 1.5	=
Likelihood of compliance		× 1.5	=
Budget and budget narrative		× 3.0	=
Other factors		× 1.0	=
Total			

If you provide criteria for evaluation, people submitting proposals are likely to use them for both self-assessment and peer-assessment as they prepare their documents. As a result, you're likely to receive higher-quality proposals.

Managing the Proposal Process

Some organizations have an RFP coordinator or a manager who approves the RFPs to assure organizational consistency. This person also may act as the point person to respond to general questions from vendors, consultants, or researchers who are interested in submitting a proposal. He or she may coordinate the schedule for submission of proposals, evaluation of these proposals, and the responses about acceptance or rejection to those who submitted proposals. Sometimes debates involving complex decisions can take many years to make and involves many segments of our society, as the ethics sidebar below shows.

ETHICS SIDEBAR

The Case of Nuclear Waste Disposal[14]

Since the 1950s, nuclear power plants in the United States have produced more than 80,000 tons of radioactive waste, the most dangerous substance known to humans. The question of how to safely dispose of this waste, which remains toxic for hundreds of thousands of years, has proven daunting. From an engineering standpoint, the disposal issue presents unique design challenges. From a public communication standpoint, the issue challenges technical professionals to translate scientific information about health hazards into language that the average person can understand. Done effectively, risk communication not only informs the public but also involves the public in decisions that could impact the safety of generations to come.

In the debate over nuclear waste disposal, many interested parties (such as scientists, government agencies, and nonprofit organizations) have weighed in on the risks of various proposals. Most have focused on a proposal to store the waste in a chamber 1,000 feet underground at Yucca Mountain, Nevada. At issue is not only the relative safety of the site, which sits on a major fault line, but also of a proposal to transport the waste by highway, rail, and barge in massive, vault-like casks. Opponents of the transportation plan, including many environmental groups, claim the risk to public health is simply too great. According to some estimates, as many as 310 accidents will occur on the nation's highways, some severe enough to cause leakage of deadly radiation. To opponents, this possibility of leakage means the plan is unsafe. Yet proponents of the proposal, including the Department of Energy, do not see the possibility of accidents as a risk to public health. While they acknowledge that the projected 105,000 truck shipments over four decades will not all be accident-free, they claim that the casks are built to withstand accidents and that the likelihood of leakage is slim to none. Thus, proponents argue, the transportation proposal is safe.

Given these conflicting assessments, who should we believe? Who is providing a more accurate assessment of the risks? As this case illustrates, risk communication can involve as much opinion as fact. Those assessing the risks may also be attempting to persuade the public to support their positions. Is this ethical? Where is the room for disagreement? Do differing opinions mean that someone is distorting the facts?

What other issues have you heard about involving discussion of risks?

What stake did the communicators have the issue?

Did the communicators adhere to the above guidelines?

One way for the public to evaluate risk communication and for technical professionals to ensure the ethical integrity of their communications is to use the following guidelines published by the U.S. EPA:

- Accept the public as a legitimate partner in the decision-making process.
- Be honest and open. Speculate only with the utmost caution and be willing to admit mistakes.
- Use simple, non-technical language.
- Obtain feedback. Make sure that the message was understood.

Preparing Proposals

Preparing a good proposal is easier if you know why so many are turned down or rejected. The reasons for rejection have stayed much the same for more than 40 years. In 1960, Dr. Ernest Allen, then Chief of the Division of Research Grants at the National Institutes of Health (NIH), published an article in *Science,* a journal for scientists, that compiled a list of reasons why more than 600 NIH proposals were rejected. In the early- to mid-1980s, the NIH again published a list of the ten most common reasons that proposals submitted to them are rejected. Nothing in the 20 years since then has diminished the accuracy of this list. These reasons can be easily adapted to other kinds of proposals.[15]

- Lack of new or original ideas
- Diffuse, superficial, or unfocused research plan
- Lack of knowledge of published relevant work
- Lack of experience in the essential methodology
- Uncertainty concerning the future direction
- Questionable reasoning in the experimental approach
- Absence of an acceptable scientific rationale
- Unrealistically large amount of work
- Insufficient experimental detail
- Uncritical approach

Referring to these reasons and making sure none apply to your proposals may increase the chances that your document will be approved.

WEBLINK

Many more proposals are rejected than accepted. The reasons have been collected and classified. For links to U.S. and Canadian sites that offer reasons for rejections, go to **www.english.wadsworth.com/burnett6e**.

CLICK ON WEBLINK

CLICK on Chapter 19/reasons for rejection

Preparing a proposal is easier if you are familiar with the preparation process and know some of the guidelines that professionals find especially helpful. Like any writing task, preparing a proposal benefits from your awareness of the writing process — managing, planning, drafting, evaluating, and revising. Although the steps are listed as if the process is linear, the process is in fact recursive — you repeat some steps several times at various stages of the process.

Of the five steps described for preparing a proposal, which ones are you skillful at doing?

Which ones do you need to strengthen?

Planning

Preparing proposals will generally go more smoothly if you plan the project. The following guidelines should help:

- Be aware of deadlines. If possible, submit the proposal early.
- Establish an achievable schedule for completing the proposal.
- Know the review and evaluation procedure that will be used to assess the document.
- Analyze the background knowledge and experience of the intended readers/decision makers.

No matter how good the proposal you prepare, if it misses the deadline, it won't be considered. Many proposals for external funding have absolute deadlines: If a proposal is submitted after a certain time on a particular date, it won't be considered. Internal proposals usually don't have deadlines that are quite as rigid, but still need to be completed on time.

You also need to know the procedure that will be used for reviewing your proposal. Will the decision makers be experts about the subject? Will the reviewers have a very limited amount of time to evaluate the proposals? You can adjust things such as the amount and kind of details you choose to include in the document or its structure and design once you know the evaluators' background and the circumstances in which they'll be reading your document.

Once you have the project schedule established, you need to plan a detailed review of what's expected by studying the RFP very carefully. The following guidelines should help your preliminary planning:

- Read and reread the RFP. Characterize the organization (e.g., read the mission statement), analyze the situation.
- Identify and substantiate the problems or opportunities you are addressing. Include the hot buttons associated with those problems or opportunities.
- If at all possible, meet with the key people involved to discuss the problem or opportunity.
- Propose a plan that responds to the problems or addresses the opportunities.
- Organize the plan in an outline or a flowchart to help you create schedules and budgets.
- Know the evaluation criteria that will be used to determine acceptance or rejection.
- Analyze probable competition.
- Consult with colleagues to receive feedback about the plan.
- Create a manageable budget for implementing the proposal.

Note the specifications that are identified in the RFP, paying special attention to the proposal structure that is expected and the evaluation criteria that will be used. Keep track of the names and numbers of people to contact to discuss your plan — and call them. They very often have extremely useful information that will help you shape the proposal.

You may need to spend time documenting the problem or opportunity. Even if you know the problem or opportunity exists, you should take the time to find supporting evidence. After you have defined and substantiated the problem or opportunity, you can start to rough out a plan. As you develop this plan, make sure that it fits the evaluation criteria established in the RFP. For example, is low cost a prime consideration, or is ease of implementation more important? After your plan is written, you should take a little time to imagine what competitive proposals may suggest. How is your plan better?

Drafting

When you're ready to begin drafting the proposal, you need to turn again to the RFP. Follow the recommended structure or sequence of information. In general, these guidelines may help you during drafting:

- If an RFP exists, follow it exactly. If no recommended format exists, use the generic one presented in this chapter. If you use the language of the RFP, you demonstrate to the audience that you understand the situation.
- Establish a clear link between the problem or opportunity you have identified and substantiated and the plan that you are proposing to solve the problem or respond to the opportunity.
- Provide information about the implementation of your plan: Who? When? How? Where? How much?
- Anticipate and address potential objections.
- Support your generalizations with specific details and examples. Cite your sources. Use visuals and tables to support or make points when possible.
- Use a *you*-attitude when possible and appropriate.

While many RFPs mandate a specific format, others do not. If no structure is recommended, first inquire if one is typically used in that particular business or organization or discipline. If no standard exists, then you can use the generic structure for a proposal that is presented in this chapter.

You can't assume that what you see as an obvious relationship between the problem and your proposed solution is obvious to everyone else. You need to make the connections clear. You also need to explain how your plan will be implemented: Who is involved? What is the schedule? What facilities will be needed? How much will the plan cost?

You can help readers accept your plan by anticipating their objections. What will they find confusing? What will they criticize? If you address potential objections and support your explanations with specific details and examples, you are more likely to convince readers. Another way you can help your cause is to adopt a *you*-attitude, which emphasizes the attitudes and interests of readers. The emphasis moves from *I* to *you.* Such an attitude requires that you identify with readers and understand their perspectives.

Budgeting

Budgets are usually a particularly important feature of proposals. A budget identifies projected expenses (and, in some cases, income), sometime in a list and sometimes in a fully developed spreadsheet. Typical budget items include direct costs for items such as personnel (percentage of salary plus benefits), equipment, travel, office supplies, and postage. If the budget extends beyond one year, percentages for inflation are often included. You also need to identify what are called indirect costs (overhead expenses), which include a percentage of the expense of operating the facilities in which the work will be done, using the services of staff, and using specialized equipment. This is a percentage that is negotiated between the funding agency and your organization. You also need to identify what are called in-kind contributions from your organization, which are nonmonetary contributions such as time from support staff (e.g., secretaries) or experts (e.g., statisticians), use of specialized facilities (e.g., a wind tunnel or a greenhouse), or complimentary attendance at a particular event (e.g., a professional conference). In constructing your budget, you need to carefully read the RFP because many funding agencies and organizations explicitly exclude certain categories of expenses.

Most proposals also require a budget narrative, which explains each item in your budget, linking each one to the implementation and evaluation of your plan. Unless the RFP specifies a specific format for the budget narrative, you can present it in one of three ways:

1. Add a column to the budget summary or spreadsheet called "Budget Narrative" and provide an explanatory sentence for each line item.
2. Add a footnote reference to each line item and list the explanatory sentences in footnotes directly following the budget itself.
3. Provide a separate subsection in which you explain the rationale for each category (and line items as necessary) in short, coherent paragraphs.

Here's one additional piece of information: Funders often begin reading proposals by turning to the budget and budget narrative even though they are near the end of the actual document. These readers use the budget to get a sense of the focus and scope of your proposal, which makes the accuracy, detail, and narrative explanation of your budget critical.

Evaluating

Once the draft is done, you need to evaluate it, trying to view it the same way as the intended audience. The following guidelines may be useful at this stage:

- Determine if RFP/RFQ directions have been followed.
- Determine if the draft meets or exceeds the criteria for evaluation.
- Examine the accuracy of technical content.
- Study the feasibility of the plan.
- Review acceptability of cost.
- Solicit reviews of the draft from colleagues.

This evaluation is the document-testing phase. One of the most difficult things will be building time enough into the process to actually review and evaluate your proposal. This is also a time to assess whether your document is persuasive. Ask whether you have understood and responded to the audience's needs, established your own credibility, and built a logical case. If the RFP included a rating scale or matrix, use it to assess where you need to revise.

Revising

Your own careful evaluation of the draft as well as feedback from other reviewers will give you ideas for revising the proposal. As you make this close-to-final pass through the document, check the following elements:

- Add, modify, or delete information to meet RFP evaluation criteria.
- Make sure that the argument is coherent, especially the connection between the problem/opportunity and the plan.
- Check that the document design conforms to the design specifications in the RFP and, within those constraints, is visually appealing and consistent.

The revision will give you another chance to check that the relationship between the problem and the plan is clearly established and that you have responded to the evaluation criteria in the RFP. You also need to check the various levels of edit (see Chapter 8) to catch any errors and inconsistencies that have eluded you and other reviewers. One easy place to slip up is having inconsistent wording or order in the table of contents, the summary or abstract, and the text of the document itself. Finally, you need to ensure that the design of the document makes all the information accessible.

Organizing and Submiting Proposals

Proposals can be organized in a variety of ways and presented in a variety of formats, ranging from an informal, one-page memo to a business letter to an extensive formal document of perhaps several hundred pages.

Sequencing Information

If you are not required to present your proposal in a specific sequence, the following one works in many situations.

- The ***introduction*** describes the situation by defining and substantiating the problem or opportunity. Unless the problem or opportunity is clearly established, the plan will have no context; the members of the audience won't be able to assess whether the plan is appropriate if they don't know what it's intended to address. This section gives the background of the problem or opportunity in order to substantiate that you understand the situation. Sometimes you establish familiarity with the situation by summarizing previous attempts to deal with the problem or opportunity and reviewing current work that would make your plan acceptable.

- The ***plan*** gives the audience the details about the specific ways in which the problem or opportunity will be addressed — that is, it identifies objectives to be met and outcomes that are expected.
- The ***benefits*** section explains probable tangible and intangible benefits that should result from adopting the plan.
- The overall ***approach*** explains the methodology that will be used to implement the plan, the management of the plan, the budget (including specific information about all direct and indirect costs), and the schedule, which identifies when things will happen (the implementation, the milestones, the evaluation). Sometimes these are separate sections.
- The ***evaluation*** of the plan identifies ways to determine whether your proposed plan accomplishes its objectives and delivers what is promised.
- The ***qualifications*** of the organization usually describe the proposers' capabilities and the structure of the proposers' organization, if any. It may include resumes of key personnel who will implement the plan. The qualifications section usually begins as boilerplate — "canned," prewritten material that is used to save time when preparing this standard section of a proposal. This boilerplate is tailored so that the discussion of the qualifications can be adapted to this specific situation and, thus, be more persuasive.

Figure 19.4 identifies components that are typically found in proposals. However, only the most formal proposals contain all of them; less formal proposals eliminate several components; informal ones eliminate even more. Also, a formal proposal contains more front matter, and it usually presents background information, organizational capabilities, and a summary, which are seldom included in less formal proposals. In contrast to the complexity of a formal proposal, informal proposals contain only essential material: an introductory statement, the technical solution, and a schedule. In most proposals, these components are easily identifiable, but they are often combined and reordered to meet the perceived needs of the audience.

WEBLINK

Proposals are typically complex documents that have multiple collaborators working together over an extended period. Using a checklist is one way to ensure that all the necessary components have been completed. For a link to additional checklists, go to www.english.wadsworth.com/burnett6e. You may be surprised at the variety of checklists; the items vary depending on the purpose of the list, the type of proposal, and the organization.

CLICK ON WEBLINK

CLICK on Chapter 19/checklists

The three general types of proposals mentioned earlier in the chapter for solving a problem, investigating a subject, and selling a product or service have similar components, but their content varies in their introductions and their

FIGURE 19.4 **Components of Proposals**

	Formal	Less Formal	Informal
Front Matter			
Title Page	•	•	
Table of Contents	•		
Abstract/Executive Summary	•		
Introduction			
Establish current situation	•	•	•
Define the problem	•	•	•
Provide evidence to substantiate problem	•	•	
Provide relevant background	•		
Plan			
Define proposed plan	•	•	•
Present objective of the plan	•	•	•
Discuss plan in relation to established criteria, standards, or specifications	•	•	
Benefits			
Explain probable benefits accruing from plan	•	•	•
Approach (may be separate sections)			
Methodology: Relate the plan's objective to specific tasks for achieving it	•	•	•
Schedule: Identify time for planning, implementation, and evaluation	•	•	•
Costs/Budget: Present direct and indirect costs for personnel, materials and supplies, and support services	•	•	
Management: Explain the administration, including organization and personnel	•		
Evaluation			
Assess progress in meeting plan objective	•		
Qualifications			
Consider capabilities of personnel and appropriateness of organization's experience, facilities, resources	•	•	
End Matter			
Appendix: Include supporting information useful to readers that would interrupt flow of the proposal's persuasive argument	•		
Sources Cited: Include references	•		

technical solutions. Other sections, such as management, budget, and schedule, are less influenced by the type of proposal.

WEBLINK

For more information about proposal budgets and narratives, go to www.english.wadsworth.com/burnett6e for a link to a short online tutorial.
CLICK ON WEBLINK
CLICK on Chapter 19/budget

Analyzing Parts of a Proposal

This section provides examples of sections of a proposal so that you can see examples of the ways in which one might be developed. All of these examples are taken from a proposal for a three-year genome/phenome research project that was submitted to the Australian Research Council (ARC). The 45-page proposal includes five main sections and 32 subsections:

Part A — Administrative Summary
(Pages 1–4 = 4 pages)

- A1 Organisation to Administer Grant
- A2 Participant Summary
- A3 Support Being Applied For
- A4 Project Title
- A5 Project Summary
- A6 Classifications and Other Statistical Information
- A7 Additional Details
- A8 Research Students
- A9 Certification

Part B — Personnel
(This section is repeated three times, once for each researcher.
Pages 5–28 = 23 pages)

- B1 Person Number
- B2 Abbreviated Details
- B3 Postal Address
- B4 Memberships
- B5 Current Holder of an ARC Fellowship
- B6 Affiliations
- B7 Qualifications
- B8 Academic, Research, Professional and Industrial Experience
- B9 Additional Fellowship Details
- B10 Research Record Relative to Opportunities

Part C — Project Cost
(Page 29–33 = 5 pages)

- C1 Budget Details
- C2 Justification of Funding Requested from the ARC
- C3 Details of Non-ARC Contributions

Part D — Research Support
(Pages 34–35 = 2 pages)

- D1 Research Support of All Participants
- D2 Reports on ARC Grants

Part E — Project Description
(Pages 36–45 = 10 pages)

- E1 Project Title
- E2 Aims and Background
- E3 Significance and Innovation
- E4 Approach
- E5 National Benefit
- E6 Communication of Results
- E7 Description of Personnel
- E8 References

Most of Parts A and B of the proposal are required forms to be completed, with a few critical exceptions. Figure 19.5 presents the first page of the proposal. The top two-thirds identifies the institution before the individuals, specifies the type of research project, and indicates the length of the project. The bottom third provides three critical pieces of information that shape the way the rest of the proposal is read: research area, title, and project summary.

FIGURE 19.5 **Excerpt from Part A of the ARC Proposal**[16]

PART A—ADMINISTRATIVE SUMMARY

A1 ORGANISATION TO ADMINISTER GRANT

The University of Queensland

A2 PARTICIPANT SUMMARY

Chief Investigators (CI), Partner Investigators (PI) and ARC Research Fellows (APF, ARF/QEII or APD). Participant details are provided in Part B.

Person number	Family name	Initials	Organisation	Role	ECR
1	Trott	DJ	The University of Queensland	CI	☒
2	Bulach	DM	Monash University	CI	☒
3	Cordwell	SJ	Macquarie University	CI	☒

A3 SUPPORT BEING APPLIED FOR

A3.1 Type

		Number sought
×	Research Grant (personnel and project costs other than Fellowship salaries)	
	Australian Postdoctoral Fellowship (APD)	0
	Australian Research Fellowship/Queen Elizabeth II Fellowship (ARF/QEII)	0
	Australian Professorial Fellowship (APF)	0

A3.2 Years that support is being sought from the ARC

2003 ☒ 2004 ☒ 2005 ☒ 2006 ☒ 2007 ☒

A3.3 Priority Area

Genome/Phenome Research

A4 PROJECT TITLE

Global and functional analysis of intestinal spirochaete (*Brachyspira pilosicoli*) outer membrane proteins expressed during infection

A5 PROJECT SUMMARY

The intestinal spirochaete *Brachyspira pilosicoli* is an important cause of diarrhoea in production animals. The bacteria attach to epithelial cells by a unique yet uncharacterised mechanism, forming a dense, hair-like covering on the surface of the colon. We will analyse the *B. pilosicoli* outer membrane proteome to identify and characterise unique, surface exposed proteins expressed during infection and determine their role in attachment. Our results will generate insight at the molecular level into the intimate relationship between parasite and host. This will provide new perspectives on intestinal spirochaete evolution, host adaptation and future novel intervention strategies for controlling *Brachyspira* infections.

The sans serif font is accessible. It is more common in professional documents outside the United States than within the United States.

The clear sections headings make information easy to locate.

The use of forms makes reading and assessing large numbers of proposals easy for the reviewers.

Inter-institutional collaboration is immediately obvious.

The title includes technical terms only familiar to experts as well as terms with specialized meanings (such as "global" and "functional").

The specific focus of this investigation is clearly identified in the priority area box.

The project summary includes technical terms, but the information is comprehensible to educated nonexperts. The summary clearly presents the problem, identifies the unknown information, describes the process to be used, and suggests the importance of this new information.

Each chief investigator (or CI in Australia; called principal investigator or PI in the United States) describes his contributions to the field, thus establishing his credentials. The paragraphs for each CI are followed by each one's list of publications and presentations and by an additional paragraph for each describing "[o]ther evidence of impact and contributions to the field." Figure 19.6 shows the beginning of one of the statement about disciplinary contributions, written with the straightforward use of first person, which is entirely appropriate.

In this proposal, the budget precedes the project description, so readers' perceptions of the project will be shaped by the budget categories as well as the amount of the budget before they ever read any details about the project itself. The excerpt of the budget in Figure 19.7 is a straightforward presentation of the expenses. This figure shows the budget for year three, but the categories are identical for the other two years. In this proposal, the budget narrative follows the budget in a series of paragraphs that justify and explain the proposed expenses.

Part E is the heart of the proposal — the description of what will happen and why it matters. At the beginning of this description (Figure 19.8), readers need to understand the background that will help the rest of the proposal make sense. The researchers condense the research literature and use those summaries as part of their background explanation. This section prepares readers for the detailed

FIGURE 19.6 **Excerpt from Part A of the ARC Proposal[17]**

CIs are not asked to provide their entire background in textual explanations, but, instead, to highlight the most relevant aspects that relate to this project.

First-person is appropriate because each CI is asked to describe himself. Third-person would be awkward and stilted.

The hierarchy is made clear in several ways: the labeling system (e.g., B10 and B10.1), the style of the font (all-cap vs. upper- and lowercase), the size of the font, and the indentation of the subordinate category.

PERSON NUMBER: 1 **DarrenTrott** **GAMS ID** G69937

B10 RESEARCH RECORD RELATIVE TO OPPORTUNITIES

B10.1 A statement on your most significant contributions to this research field

I am a veterinarian with 12 years experience researching zoonotic and production animal bacterial diseases, focusing on population genetics and molecular epidemiology. Since 1994, my principle research interest has been the classification, pathogenicity and epidemiology of anaerobic intestinal spirochaetes. I completed my Ph.D in this field in 1998 in the laboratory of Professor David Hampson. During this highly productive

FIGURE 19.7 Excerpt from Part C (Budget) of the ARC Proposal[18]

C1 BUDGET DETAILS

C1.3 YEAR 2005

Column 1	COSTING			
	2	3	4	5
Source of funds	ARC	University	Other	Total
DIRECT COSTS				
Personnel (Salaries + On-costs)				
CI (Trott) @ 0.2 FTE + 26% on-costs	0	14868	0	14868
CI (Bulach) @ 0.2 FTE + 26% on-costs	0	13686	0	13686
CI (Cordwell) @ 0.1 FTE + 26% on-costs	0	7560	0	7560
Research Associate @ FT + 26% on-costs	63518	0	0	63518
Ph.D Scholarship	22771	0	0	22771
Total Personnel (a)	86289	36114	0	122403
Teaching Relief				
Total Teaching Relief (b)	0	0	0	0
Equipment				
Total Equipment (c)	0	0	0	0
Maintenance				
Expend[illegible]	[illegible]		[illegible]	[illegible]
Animals				
Proteon				
Travel				
Airfares				
Other				
Video C				
[illegible] (f)		0	[illegible]	[illegible]
TOTAL DIRECT COSTS (g)	132989	36114	0	169103
INDIRECT COSTS				
CIs, PIs and any researcher Level A or above x multiplier				
CIs and Professional Research Staff x 1.25 (AVCC multiplier)		124540	0	24540
TOTAL INDIRECT COSTS (h)		124540	0	124540
TOTAL COSTS (i)	132989	160654	0	293643

C2 JUSTIFICATION OF FUNDING REQUESTED FROM THE ARC

Maintenance: Expendable research materials are an essential component of this project. This budget item will cover items such as bacterial culture medium, cell culture medium ($3,500 p.a); chemicals and reagents ($3,000 p.a.); molecular biology reagents and kits eg. resins for protein and DNA purification, DNA modifying enzymes and sequencing costs ($9,500 p.a.); plasticware eg. tips and tubes ($4,000 p.a.) and electron microscopy materials eg. immuno-gold labelling reagents, grids and chemicals ($2,000 p.a.). The total budget requested for consumables is $22,000 p.a.

The source of each category of funds is labeled.

Direct and indirect costs are separated.

Each separate item within each category has its own line in the budget.

Personnel costs are usually calculated with the salary plus benefits. Each subcategory is totaled.

Many of the categories in the budget have a short corresponding paragraph in the budget narrative, which justifies that particular expense, as in this explanation of the maintenance expenses. narrative.

Specific amounts are presented in the budget narrative so readers do not have to flip back and forth between the budget and the budget narrative.

The line item for "Maintenance" C1 BUDGET DETAILS is explained and justified in the narrative in Section C2.

FIGURE 19.8 **Excerpt from Part E (Project Description) of the ARC Proposal: Background[19]**

The problem is identified as international, accepted as "a significant cause of colitis in all major pig-producing countries, including Australia."

Confirmation that other researchers in the UK also agree with the cause of the problem strengthens the argument.

The text and table reinforce each other, with the text referring to "seven distinct species of Brachyspira*" and the table elaborating the hosts, diseases, and polar attachment for each of the seven species.*

*The seriousness of the problem is reinforced by the explanation that "*B. pilosicoli *is not just a pathogen of pigs." The pathogen occurs in a range of animals as well as in human communities.*

E2 AIMS AND BACKGROUND

***Brachyspira pilosicoli:* an emerging cause of colitis in many host species.**
Research in the last ten years has clarified the taxonomy, epidemiology and pathogenic capability of spirochaetes that inhabit the large intestine of monogastric animals. Seven distinct species of *Brachyspira* are recognised and the genus contains both pathogenic and non-pathogenic representatives (Table 1). *Brachyspira* pilosicoli was confirmed in 1996 as the agent of porcine intestinal spirochaetosis (PIS), a colitis of growing pigs that results in diarrhoea and failure to gain weight (T18). PIS is a significant cause of colitis in all major pig-producing countries, including Australia (30). In the United Kingdom, a comprehensive eight-year survey of piggeries with colitis consistently identified *B. pilosicoli* as the major causative agent (33).

Table 1. Host species and disease associations of the recognised species of *Brachyspira*

***Brachyspira* species**	**Known hosts**	**Disease association**	**Polar attachment***
B. hyodysenteriae	Pigs Rheas	Swine dysentery Necrotising typhlitis	No
B. intermedia	Pigs Poultry	Unknown Wet litter/poor production	No
B. innocens	Pigs	Non-pathogenic	No
B. murdochii	Pigs, rats	Non-pathogenic	No
B. alvinipulli	Poultry	Wet litter/poor production	No
B. pilosicoli	Pigs, humans, dogs, avian species, non-human primates	Intestinal spirochaetosis	Yes
B. aalborgi	Humans, non-human primates	Intestinal spirochaetosis	Yes

*defined as attachment of the spirochaete by one cell end to the colonic epithelium.

B. pilosicoli is not just a pathogen of pigs. It has established a niche in a wide range of monogastric hosts. B. pilosicoli is now recognised as a cause of wet litter and decreased egg production in poultry (29, T2). In addition, *B. pilosicoli* has been isolated from dogs (T7), game birds (T21) and water birds (T8). *B. pilosicoli* also commonly colonises humans. High rates of *B. pilosicoli* carriage (,30%) have been demonstrated in individuals in developing communities (including Australian Aborigines) (14, T4), AIDS patients (11) and homosexual males (T9). Rates of carriage amongst other groups of individuals are extremely low (,1.5%). In studies conducted in Papua New Guinea, the same strain of *B. pilosicoli* was shown to colonise individuals for up to six weeks and cross-species transmission was demonstrated to occur between dogs and humans (T4, T11).

explanation of the plan, using microscopy to illustrate the points (Figure 19.9). After describing recent advances in the global analysis of outer membrane proteins using proteomics, the researchers present their hypotheses, speculate on the significance of their work, and then provide details about the approach they'll take. In what is the most technical part of the proposal, they identify four specific objectives (Figure 19.10).

To read the entire proposal, including some revision decisions the researchers made, go to www.english.wadsworth.com/burnett for a link to the document.
CLICK ON WEBLINK
CLICK on Chapter 19/*Brachyspira pilosicoli*

WEBLINK

FIGURE 19.9 **Excerpt from Part E (Project Description) of the ARC Proposal: Photos**[20]

The photos — good examples of microscopy — are clearly labeled, referred to in the accompanying text, and help readers understand the textual explanation.

The use of metaphor ("resembles a hairy covering on the surface of the colon"), an etymological explanation (of pilosicoli), and an analogy ("cap-like") all help readers better understand the information.

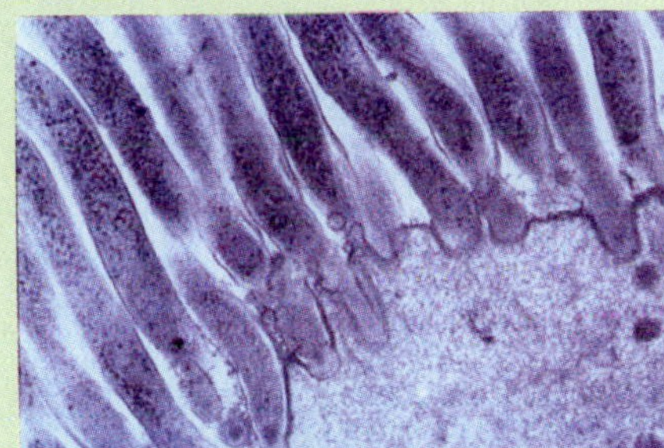

Figure 1: Transmission **(A)** and scanning **(B)** electron micrographs of the intimate attachment of *Brachyspira pilosicoli* to the colonic epithelium Deep indentation of the spirochaete tip into the host cell plasma membrane is shown (Figure 1A arrow).

The attachment mechanism of *B. pilosicoli.* The unique polar attachment of *B. pilosicoli* to the colonic epithelium has been described in animal models of intestinal spirochaetosis (18, T16). Large numbers of spirochaetes invaginate into the mature columnar epithelium by one cell end without penetration of the host cell membrane and resemble a hairy covering on the surface of the colon (*pilosicoli* translates to "of a hairy colon") (T18). The attachment results in a cap-like elevation of the underlying parasitised cell and effacement of its microvilli (18). Compared to non-attaching *Brachyspira*, *B. pilosicoli* cells possess an electron-lucent lattice-like structure at the pointed end of the cell. Transmission electron micrographs of attached *B. pilosicoli* cells show deep indentations at the point of polar attachment (Figure 1A arrow). An 8-12 nm gap, filled with electron-dense fibrillar material is present along the Hampson. During this highly

FIGURE 19.10 **Excerpt from Part E (Project Description) of the ARC Proposal: Process[21]**

The opportunity the proposed research provides is clearly stated in the first sentence.

The relevant aspects of the CIs' professional expertise and experience are highlighted.

The specific process is presented in a way that is comprehensible to both experts and interested nonexperts. Technical terminology is used only when essential for accuracy.

Hypotheses and aims of the current project. The virulence mechanisms of the pathogenic intestinal spirochaetes are yet to be characterised. Determining the role of OM proteins expressed during pathogenesis is essential to facilitate the study of parasite and host interactions at a molecular level. There is a number of reasons why the identi...

The project will bring together a veterinarian with extensive experience in spirochaete biology, pathogenesis, animal models and ultrastructure; a protein specialist with a microbiology background who is one of the world's leading proteomics experts; and a molecular biologist with extensive experience in spirochaete genetics and expression systems. We will use a four-step process to identify unique, surface exposed *B. pilosicoli* proteins and investigate their roles in pathogenesis, with a priority placed on the attachment process:

1) Isolate purified OM from *B. pilosicoli* grown in media that simulates *in vivo* conditions.
2) Undertake comparative analysis of *Brachyspira* OM proteomes to identify unique *B. pilosicoli* proteins that are candidates for involvement in pathogenesis.
3) Identify, clone, and express the genes encoding these candidate proteins in order to develop protein-specific reagents.
4) Use these protein-specific reagents and constructed knock-out mutant strains to investigate the role of candidate proteins in pathogenesis.

Submitting a Proposal

The type of proposals and its audience influences its delivery. Proposals have traditionally been paper documents, carefully crafted to respond to the situation described in the RFP or to observed problems or opportunities. Recently, some funding agencies and foundations have broadened their process to encourage online submission and oral presentation.

Electronic submission of proposals is fast becoming the norm in North American and parts of the European Union for a number of reasons:

- "***Money.*** Completed survey questionnaires, import/export declarations and application forms can be sent over the Internet at no additional cost. Postage or courier fee is no longer required.
- ***Time.*** The delivery time can be reduced significantly, thus expediting the processing of applications for services.
- ***Space.*** Duplicate copies of completed survey questionnaires, import/export declarations and application forms (in softcopy format) can be saved

electronically with minimal effort. Additional space to retain hardcopies for cross-referencing and documentation is no longer required.

- ***Flexibility.*** Completed survey questionnaires, import/export declarations and application forms can be submitted electronically, on an anytime and anywhere basis.
- ***Data Protection.*** Important data can be better protected by electronically encrypting the completed survey questionnaires, import/export declarations and application forms, thus ensuring that only authorized [officials] can read the contents."[22]

© Fisher/Thatcher/Getty Images

In creating a strong oral proposal, a team is able to use strategies too often used only in presentations, such as storyboarding, which is a powerful planning tool for all kinds of written and oral presentations. A strong proposal can be made even stronger by a confident, coherent oral presentation. Why? In a well-designed, well-rehearsed presentation, the crucial information is reduced to key points (respecting limits of the audience's short-term memory) so the features of the proposal stand out. The presenters of an oral proposal, typically the key players on the project team, are involved from the start, so they are comfortable making the presentation.[23] As a result of the involvement of the key players, people at the funding organization get to see the team they're be working with in action and can determine if the fit is right.[24]

© Left Lane Productions/CORBIS

A number of corporations hire presentation coaches like psychologist Carol Fleming to prepare "engineering project managers and teams to make final oral proposal presentations. These presentations [are] typically for contracts involving billion dollar commitments." One manager who hired Fleming commented that she "was extremely successful in orchestrating the dynamics and strategy of the presentations as well as grooming and coaching the team members in their individual areas of the presentations. . . . [One of her] many strengths is the ability to focus on the sometimes hidden needs of the customer and to adjust her approach to meet both organizational goals and the specific needs of the individual."[25]

More and more agencies and foundations are encouraging online submission of proposals. Go to **www.english.wadsworth.com/burnett6e** for a link to online submission of government proposals.

CLICK ON WEBLINK

CLICK on Chapter 19/online submission

The "shift to oral presentations as part of the procurement process has forced many companies to rethink their proposal efforts."[26] Go to **www.english.wadsworth.com/burnett6e** for a link to brief discussion about oral proposal presentations.

CLICK ON WEBLINK

CLICK on Chapter 19/oral submission

Examining a Sample Proposal

Proposals are usually intended for management personnel or external readers and may range from a few to several hundred pages. Figure 19.11 is an unsolicited internal proposal that offers a plan to merge two departments in a company. As you read this problem-solving proposal, you should consider ways in which it could be improved.

Responses to FAQs about writing proposals is a good way to review basic considerations in researching and preparing a proposal. For links to FAQs and straightforward responses, go to **www.english.wadsworth.com/burnett6e**.

CLICK ON WEBLINK

CLICK on Chapter 19/FAQs

FIGURE 19.11 **Unsolicited Internal Proposal**[27]

PROPOSED MERGER:
MECHANICAL ASSEMBLY WITH FINAL ASSEMBLY

Submitted to
Brian Donnelly
Manager, Production
Cybertronics, Inc.

Prepared by
Craig Wilder
Supervisor, Mechanical Assembly
Cybertronics, Inc.
October 27, 20—

PROPOSAL ABSTRACT

The cost and inconvenience of moving an entire department must always be weighed against the long-term benefits to be derived from such a move. Both the company and the employee could benefit from moving the Mechanical Assembly line from its current building into the Final Assembly area. Valuable work space would be better utilized, opening up the much needed space for new products scheduled for production. The move would also reduce handling and shipping costs, cross-train employees, and reduce overhead by 50 percent.

FIGURE 19.11 **Unsolicited Internal Proposal (continued)**

Proposed Merger: Mechanical Assembly with Final Assembly

The Situation at Cybertronics

Understanding the proposed merger requires a review of recent sales history at Cybertronics, a clear statement of the space problem, and an explanation of the differences between the Mechanical Assembly and Final Assembly areas.

How effective is the writer in describing the situation, including defining and substantiating that problem?

History

For the past seven years, Cybertronics has been a leader in the ink jet printer industry. This success is largely because of the company's model 800 series printers. This product line has been the heart of the company's revenue growth and rapid expansion. Five years ago Cybertronics was shipping 300 units per day of 800 series products alone. Today, however, Cybertronics is shipping only 75 series 800 printers per day. The forecast ramps down to 50 per day in three months, and the backlog for sales is soft for the next fiscal quarter. Cybertronics is shifting away from the series 800 products and emphasizing the new product lines.

Statement of Problem

Two new products going into production need the space, tools, and equipment now used for older products. Specifically, the Mechanical Assembly and Final Assembly areas need more space.

The current locations of the Mechanical Assembly and Final Assembly areas do not make efficient use of space. Space is being wasted in both areas, as Table 1 shows. Combining the Mechanical Assembly area and the Final Assembly area would free approximately 14,500 square feet.

Displaying square footage in a table emphasizes the inefficient use of space in both assembly areas.

TABLE 1

Allocation of Space in Mechanical Assembly and Final Assembly

AREA	SPACE Available	Used	Unused
Mechanical Assembly	14,500 sq ft	8,000 sq ft	6,500 sq ft
Final Assembly	13,000 sq ft	7,000 sq ft	6,000 sq ft

FIGURE 19.11 **Unsolicited Internal Proposal (continued)**

Background
At present, the two assembly areas are operated in different ways. The Mechanical Assembly area has a conveyor that is automatically timed to advance one unit every $3^1/_2$ minutes. The assembly line has nine different positions, each with a certain job on mechanisms as they pass through that station. After a mechanism reaches the end of the line, it is a finished unit. This "progressive assembly line"—one mechanism every $3^1/_2$ minutes—is a steady flow that is easily controlled and very predictable. It is efficient for two reasons:

1. The mechanism moves to the worker rather than the other way around.
2. Each worker becomes specialized by repeating the same job.

Workers are rotated from job to job on a weekly basis to avoid boredom and excessive repetition.

In Final Assembly, on the other hand, mechanisms are unpackaged and put on a long conveyor line, about 20 at a time. When workers select (or are assigned) a task, they take the necessary tools, parts, and hardware and walk up and down the line doing their job. Each worker goes along at an individual rate; when all the jobs are done, the line is emptied and the process begins again. The process offers no incentive to work quickly because the mechanisms will stay on the line until everyone is done. A lot of time is wasted; a lot of effort is spent walking around and looking for tools and parts. The units are not finished one at a time but rather in batches of 20. This uneven flow makes predicting production speed nearly impossible; management has little control over the speed of production in the assembly area.

Proposed Plan to Merge
Mechanical Assembly and Final Assembly
The proposed plan has specific goals and a clear way to reallocate available space.

Goals
The proposed plan has two goals:

1. Make more efficient use of space in manufacturing.
2. Provide space for the new products going into production.

Reallocation of Space
Both the Mechanical Assembly and Final Assembly areas could fitinto 13,000 square feet because some of the area now needed would not have to be used if the departments were combined. Most of this space saving comes from two areas:

1. Packing and shipping in Mechanical Assembly would be combined with receiving and unpackaging in Final Assembly.
2. Final Assembly could eliminate their entire storage location that is now being used to hold spare parts to fix defective mechanisms.

2-

Using numbered lists to present reasons for progressive assembly (here) and plan goals (below) helps readers understand and remember these critical parts of the proposal.

After reading about the situation at Cybertronics, what questions come to mind that you hope the writer considered in shaping a plan?

FIGURE 19.11 **Unsolicited Internal Proposal (continued)**

> What additional information would you need to know in order to accept this proposal?

> The writer could increase the readers' retention of the information by creating new subtitles that identify each actual space-saving measure — for example, "Reducing the Packing Area" rather than "Space Saving Measure 1."

IMPLEMENTING THE PLAN

The 13,000 square feet could be saved by implementing the three following space-saving measures.

Space Saving Measure 1

The packing area could be nearly eliminated. At present we use about 1,500 square feet for packing mechanisms, holding them in a loading area while waiting for pickup. Final Assembly uses more than 500 square feet to receive and unpack mechanisms received from stock. Nearly all of the 2,000 square feet now allocated to packing, shipping, and receiving would be eliminated. A small area—less than 500 square feet—would still be needed for packing and shipping overseas and for occasional excess being moved into stock. In addition to the space savings, the overhead would be reduced.

Space Saving Measure 2

The storage space in Final Assembly that is now used to store spare parts to repair defective mechanisms would be totally eliminated. If there were one department containing both the mechanism build and the final build, then all the mechanism parts would already be in the department.

Space Saving Measure 3

To further reduce the materials handling requirements and inefficient use of space, we could join the two assembly-line conveyors and make them into one continuous flow of production.

BENEFITS OF IMPLEMENTING THE PROPOSED PLAN

Implementing the proposed plan would result in a number of benefits for Cybertronics.

Reduction in Supervision

If the two areas were combined, supervision overhead would be cut in half because there would be only one line and a smaller total work force; one supervisor could do an effective job.

Cross-Training

Cross-training requirements would also be an advantage. According to a manual, *Modern Supervisory Techniques,* when people become bored with their jobs, they are unmotivated. However, cross-training on new jobs sparks interest and, therefore, increases motivation. Additionally, the National Foreman's Institute's manual, the *NFI Standard Manual for Supervisors,* says that when employees understand the "big picture" of how their jobs affect the product and the company, they will have more pride in

-3-

FIGURE 19.11 Unsolicited Internal Proposal (continued)

their work, improve the quality of their work, and increase their motivation. Combining the two assembly areas would provide many opportunities for cross-training, so employees could understand the entire assembly operation.

Summary of Benefit

The combination of Mechanical Assembly and Final Assembly into a single area that uses progressive assembly has distinct benefits

1. It frees space for new product line.
2. It increases productivity in Mechanical and Final Assembly.
3. It increases employee motivation.
4. It increases management control of production.
5. It lowers overhead and management costs.

-4-

Do these benefits seem convincing? Can you recommend any strategies to strengthen the benefits section?

The writer needs to revise the entire benefits section so that the discussion of the benefits (above) corresponds exactly with the summary presented here.

Individual and Collaborative Assignments

1. **Revise a proposal.** Reread the proposal in Figure 19.11. Based on what you have learned in this chapter, work with a small group to revise this proposal. (NOTE: You will find an electronic version of the proposal in the Chapter 19 section of the companion Web site at www.english.wadsworth.com/burnett6e.)
2. **Identify a problem and propose a solution.** Work with a group of classmates who are in the same academic field or department.
   (a) Identify a problem in your field or department that you all agree with. Record the process or steps your group uses to identify the problem.
   (b) Propose a solution that you all agree upon. Record the process or steps the group uses to evaluate the proposed solution.
3. **Outline a proposal.** You need to persuade an engineering committee in your company to support and fund leaves of absence for employees to teach technical courses at City College. These four-month leaves will enable the college to offer courses in state-of-the-art technology. Outline the proposal you could present to have the plan approved.
4. **Prepare an RFP.** Assume you represent a private foundation whose concern is to increase scholarship in and public support for foreign language capability, particularly for technical experts working for multinational companies. Prepare an RFP and decide to whom you will send it.
5. **Write a proposal.** In a small group, identify a problem or opportunity in your school, community, or workplace. Develop a proposal that addresses the problem. Be sure that you speak with the people directly involved to learn more about the history of the problem, the solutions that have already been attempted, the constraints that must be considered, and so on. Direct your proposal to the decision maker(s) involved.
6. **Analyze resources for proposal writers.**
   (a) Visit the following Web site: www.kn.pacbell.com/wired/grants/. What kind of information does this site provide for writers of proposals? Analyze the rhetorical situation (audience, purpose, context) of this site, and determine whether you think the site is effectively designed in terms of the rhetorical situation. For whom would this site be particularly useful?
   (b) Conduct a separate search for additional Web sites that might be useful for people researching and preparing proposals. Work with a small group to combine your site recommendations into an online reference list for proposal writers. Include features that would make it usable for a broad range of users.

7. **Assess an RFP.** Visit the following Web site, **http://www.we-energies.com/windrfp.htm**, to download the full RFP that you read about briefly in Figure 19.3. Read the RFP and make particular note of the sections and their sequence. Then review the section of this chapter that discusses writing RFPs. Work with a small group to assess the effectiveness of this RFP. Write a collaborative memo to your instructor indicating which aspects of this RFP are well done and which could be strengthened.

Chapter 19 Endnotes

1 This differentiation of reports and proposals is discussed in more detail in a very useful book: Freed, R. C., Freed, S., & Romano, J. (2003). *Writing winning business proposals: Your guide to landing the client, making the sale and persuading the boss* (2nd ed). New York: McGraw-Hill.

2 U.S. Environmental Protection Agency, NCER. (n.d.). *Ecology and oceanography of harmful algal blooms program.* Retrieved December 8, 2003, from http://www.tgci.com/funding/fedResult.asp?thisId=3950

3 U.S. Department of Health and Human Services, CDC. (n.d.). *Community trial to test the effectiveness of the smoke alarm installation and fire safety education program.* Retrieved December 8, 2003, from http://www.tgci.com/funding/fedResult.asp?thisId=4011

4 U.S. Environmental Protection Agency, NCER. (n.d.). *Ecology and oceanography of harmful algal blooms program.* Retrieved December 8, 2003, from http://www.tgci.com/funding/fedResult.asp?thisId-3950

5 U.S. Department of Health and Human Services, CDC. (n.d.). *Community trial to test the effectiveness of the smoke alarm installation and fire safety education program.* Retrieved December 8, 2003, from http://www.tgci.com/funding/fedResult.asp?tisId=4011

6 We Energies. (n.d.). *Request for proposal: 2003–2004 wind energy.* Retrieved December 8, 2003, from http://www.we-energies.com/windrfp.htm

7 We Energies. (n.d.). *Request for proposal: 2003–2004 wind energy.* Retrieved December 8, 2003, from http://www.we-energies.com/windrfp.htm

8 Discussion draws on these sources: (1) (n.d.). Aristotle's *Rhetoric:* A hypertextual resource compiled by Lee Honeycutt. Retrieved December 5, 2003, from http://www.public.iastate.edu/~honeyl/Rhetoric/ (2) Wilcox, R. (1997). Persuading your reader or listener. In *Communication at work* (pp. 284–292). Boston: Houghton Mifflin.

9 Discussion draws on this source: Jacobi, M. (1987). Using the enthymeme as a heuristic in professional writing courses. *Journal of Advanced Composition, 7.* Retrieved December 5, 2003, from http://jac.gsu.edu/jac/7/Articles/5.htm

10 Semas, J. H. (1999, September). Goof-proofing your RFPs. *Association Management, 51*(9), 41–46. Reprinted with permission.

11 Semas, J. H. (1999, September). Goof-proofing your RFPs. *Association Management, 51*(9), 41–46. Reprinted with permission.

12 Semas, J. H. (1999, September). Goof-proofing your RFPs. *Association Management, 51*(9), 41–46. Reprinted with permission.

13 Modified from Klariti Writing Services. (n.d.). *Golden rules before starting your proposal.* Retrieved December 5, 2003, from http://www.klariti.com/business-writing/Golden-Rules-before-starting-RFP-ITT-proposal.shtml

14 Discussion in the Ethics Sidebar draws on these sources: (1) Chalfant, R. (1998). Keys to communicating with the public. *New Steel,* May, 82. (2) Markiewicz, D. (2002). How to talk about risk: Your job is to inform, not disarm. *Industrial Safety and Hygiene News,* April, 20–21. (3) Sadik, P. (2002, June). *Radioactive roads and rails: hauling nuclear waste through our neighborhoods.* U.S. PIRG Education Fund. (4) U.S. Department of Energy, Office of Public Affairs. (n.d.). *Transportation of spent nuclear fuel.* Retrieved August 14, 2002, from www.ocrwm.doe.gov/wat/pdf/snf_trans.pdf

15 Discussion draws on these sources: (1) U.S. Department of Health and Human Services. (1983). *NIH peer review of research grant applications* (148–149). Washington, DC: U.S. Government Printing Office. (2) Thackrey, D. (n.d.). *University of Michigan proposal writer's guide.* Retrieved December 5, 2003, from http://www.research.umich.edu/proposals/pwg/pwgrejected.html

16 Trott, D., Bulach, D., & Cordwell, S. (2003). *Global and functional analysis of intestinal spirochaete* (Brachyspira pilosicoli) *outer membrane proteins expressed during infection.* Proposal submitted to the Australian Research Council.

17 Trott, D., Bulach, D., & Cordwell, S. (2003). *Global and functional analysis of intestinal spirochaete* (Brachyspira pilosicoli) *outer membrane proteins expressed during infection.* Proposal submitted to the Australian Research Council.

18 Trott, D., Bulach, D., & Cordwell, S. (2003). *Global and functional analysis of intestinal spirochaete* (Brachyspira pilosicoli) *outer membrane proteins expressed during infection.* Proposal submitted to the Australian Research Council.

19 Trott, D., Bulach, D., & Cordwell, S. (2003). *Global and functional analysis of intestinal spirochaete* (Brachyspira pilosicoli) *outer membrane proteins expressed during infection.* Proposal submitted to the Australian Research Council.

20 Trott, D., Bulach, D., & Cordwell, S. (2003). *Global and functional analysis of intestinal spirochaete* (Brachyspira pilosicoli) *outer membrane proteins expressed during infection.* Proposal submitted to the Australian Research Council.

21 Trott, D., Bulach, D., & Cordwell, S. (2003). *Global and functional analysis of intestinal spirochaete* (Brachyspira pilosicoli) *outer membrane proteins expressed during infection.* proposal submitted to the Australian Research Council.

22 *Benefits of electronic submission.* (n.d.). Retrieved December 5, 2003, from http://www.info.gov.hk/censtatd/eng/prod_serv/services/e_submission/e_sub_benefits.htm

23 Van Akkeren, P. *Proposal development information.* Retrieved December 5, 2003, from http://home.earthlink.net/~paulvana/proposal_info.htm#Oral%20Proposals

24 Pease, G. W. (1999, Fall). Persuasive oral proposal presentations. *Proposal Management,* 45–47. Retrieved December 5, 2003, from www.apmp.org/pdf/fall99/45oralproposals.pdf

25 Fleming, C. (n.d.). *The sound of your voice: What our clients say.* Retrieved December 5, 2003, from http://www.speechtraining.com/Pages/about_testimonials.html

26 Pease, Gregory W. (Fall 1999). Persuasive Oral Proposal Presentations. APMP, pp 45–47. Retrieved December 5, 2003, from www.apmp.org/pdf/fall99/45oralproposals.pdf

27 Wilder, C. Proposed merger: Mechanical assembly with final assembly. *Technical Writing,* 42.225, University of Lowell.

CHAPTER 20

Preparing Reports

Objectives and Outcomes

This chapter will help you accomplish these outcomes:

- Learn to analyze and manage these critical aspects of the reports you'll read and write: purposes, formality, audience, organization, and genres
- Understand and appropriately use various types of reports, such as research, task, activity, progress, minutes, and trip/conference
- Select elements of report formats that are useful to your intended readers
- Analyze and be able to recommend revisions to reports that others write

Reports account for a substantial amount of writing in business, industry, government, and nonprofits. In fact, unless writing is your primary job, preparing reports and correspondence will make up most of the writing you do. You'll be more successful as a professional if you can differentiate informal and formal reports, examine reports' purposes and characteristics, learn ways to plan and design them, and understand specific types and formats of informal and formal reports.

Planning Reports

Reports record and sometimes influence current operations, as well as information that may be needed in the future. How could you buy a computer printer without organized information about comparable models? Why would you finance a business trip or attendance at a conference if you never expected to be told what had been accomplished? How could you plan a production schedule if you didn't know the project's current status — how much was finished, how much was yet to be completed, what supplies were on hand, what personnel were available?

When you begin to plan a report, you need to ask yourself a series of questions to determine the purposes, assess the level of formality and identify the audiences, decide the most appropriate and effective organization, and select the genre:

- ***Purpose.*** What is the overall purpose of this report? What is my goal as writer?
- ***Formality.*** What's the appropriate tone? Approach?
- ***Audience.*** Who will read this report? What do they already know, and what do they want to know? What will they do with the information in the report?
- ***Organization.*** How should the report be organized? How should information be designed?
- ***Genre.*** Which information should be presented in writing, which in visuals, and which orally? What genres will be most appropriate?

Determine the Purposes

Reports generally have one of three (sometimes overlapping) purposes:

- to report information, including progress on projects as well as key aspects of daily activities, meetings, trips, and conferences
- to analyze information, including data for decision making and data from laboratory studies as well as field investigations
- to persuade readers to consider the analysis and accept any recommendations

> *Regardless of the purpose of a report, can one be written that has no argument? No persuasive element? No position?*

Regardless of the type of report you are preparing, always consider *why* it is needed: Someone needs information to stay informed, make a decision, or justify an action.

Writing a purpose statement gives you a clear focus for a report and provides a way for you to control content and organization. The readers benefit from a purpose statement that focuses on the intent of a report and explains how it fulfills their need(s) for certain information. The purpose, which should be placed at the beginning of a report, should follow a specific pattern: State the problem, identify questions and activities related to the problem, and explain the ways in which the report responds to the problem and to the related questions and activities.

Preparing a report often seems fairly straightforward — state the problem, present the information about it . . . and that's it. But the writers of reports often find themselves entangled in complex situations that require them to make decisions about the accuracy of the data they're using. The ethics sidebar below suggests some of the issues that workplace professionals need to consider.

ETHICS SIDEBAR

Hard Facts or Mere Guesses? Ethics and Statistics

Consider the following scenario:

> You are watching a television newscast, and you see a report on the dangers of anorexia nervosa (a type of eating disorder). To illustrate the scope of the problem, the reporter cites a statistic that 150,000 women die each year from anorexia. Though surprised by the high number, you accept the statistic at face value. After all, it seems to come from an "official" source. However, you have been misled. Far fewer deaths per year are attributed to anorexia.[1]

So why did the reporter offer such an exaggerated statistic? At one time, public health advocates estimated that 150,000 American women are anorexic. The statistic later "mutated" when some people began reporting 150,000 as the number of deaths per year. This mutated version then spread as more and more people repeated it — partly because it is such a shocking statistic — and few people questioned its accuracy.

Misleading statistics like this one are not uncommon. In all kinds of communication, statistics fill an important role, though they can easily be manipulated in an ethically questionable manner. Activists use statistics to draw attention to social problems. The media use statistics to make news stories more dramatic and compelling. Researchers use statistics to document findings and provide evidence for their conclusions. Corporations use statistics to promote their products and to boost their profits. However, according to sociologist Joel Best, sometimes people use inaccurate statistics in order to serve these purposes.

Can you think of statistics that you have heard? Asking yourself the three questions mentioned above, do you think those statistics were accurate or inaccurate? Why or why not?

Best describes two types of inaccuracies: mutant statistics, which become garbled through misunderstanding and miscommunication; and deliberate manipulations, or conscious efforts to use statistical information in misleading ways. Some examples of deliberate manipulations include selectively choosing which numbers to report or choosing words to convey a particular understanding of statistical information.[2]

So how can we deal with the problem of inaccurate statistics? According to Best, the answer is not to ignore all statistics, but to become better judges of the numbers we encounter. To serve this goal, Best offers three questions to ask whenever you encounter a new statistic:

- Who created the statistic? Does the figure come from activists, government officials, or corporations?
- Why was the statistic created? What motives might the authors have? How might these motives influence the presentation of statistical information in favor of either high or low numbers?
- How was the statistic created? Was it produced through a wild guess? Or did the authors use clear, verifiable measures?

Ultimately, no statistic is perfect. Each is the product of peoples' efforts, and sometimes those efforts are reasonable and objective. Other times they are not. As a technical professional, you may be involved in creating and presenting statistics, and thus you may have an opportunity to influence the quality of those statistics.

WEBLINK

Massaging data. Manipulating data. Suppressing outliers. Changing scales. For links to sites with discussing issues surrounding data manipulation, go to www.english.wadsworth.com/burnett6e.
CLICK ON WEBLINK
CLICK on Chapter 20/ethical data

Assess the Formality

Regardless of their purpose, reports can be described on a continuum from informal to formal. The primary differences have to do with the writer's familiarity with the audience, the amount of documentation in the report, and the report's tone.

Generally, informal reports can be appropriate if writers have regular contact with their readers because it's likely that they already know the background. This sidesteps the need for intensive documentation and detail. Regular contact also means that the tone can be informal; however, informal doesn't mean chatty and personal. Informal reports can be written as memos, letters, or short reports (without front matter or end matter).

In contrast, writers of formal reports probably don't have regular contact with their readers. Because the readers may not be familiar with the background, it needs to be included in formal reports, usually making the reports longer and more detailed. These reports need more documentation because the readers may not know the history or sources. Formal reports generally adhere closely to strict conventions of language. Most formal reports also have a number of standard sections, including front matter, such as an abstract and table of contents, and end matter, such as references and appendices.

Because of its extensive front matter, a formal report tends to repeat content, something that seldom occurs in an informal report. However, this is usually an advantage rather than a disadvantage because the multiple audiences for formal reports read various sections for different purposes. For example, managers focus on conclusions and recommendations so they can make informed decisions. Technical professionals focus on the accuracy and feasibility of the technical details. Individual readers seldom go through an entire formal report from beginning to end; rather, they read the sections relevant to their jobs.

Deciding whether a report should be informal or formal has less to do with the subject of the report than with your relationship with the readers and your assessment of their information needs. After you decide who your readers are and what they need to know, you can decide whether to write an informal or a formal report. For example, imagine you are writing a progress report about your investigation of several varieties of high-yield, disease-resistant wheat. You would probably write an informal report for the members of your research team. However, you would write a formal report to be submitted to the funding agency, the U.S. Department of Agriculture.

Identify Audiences

Knowledge about your report's audiences helps you clarify its purpose and organization. One of the most difficult aspects of preparing reports is tailoring your material for readers with a variety of backgrounds and needs.

When you begin planning a report, the information in Chapter 4 about audience analysis will help you to identify the organizational relationship between you and your readers. Are your readers on a different organizational level from you, or on the same organizational level but with a different technical specialization? Or are they in a different organization?

Once you have answered these questions, concentrate on how the readers will use the report. *Primary readers* fall into one of two categories: those who will use the recommendations for decision making and those who will be interested in the technical details, particularly the accuracy of data and the feasibility of the conclusions and recommendations, as they affect their work. *Secondary readers* are indirectly affected by the conclusions and recommendations of a document. For example, a report recommending reroofing and insulating a manufacturing plant would have the facilities manager as one of the primary readers. Department supervisors, whose work might be disrupted by the construction, would be secondary readers.

An audience's attitude toward the information in a report should influence the organization of that information. Should a recommendation come near the beginning of a report or near the end? In general, if the audience will be receptive, it can be presented and discussed near the beginning; if the audience is likely to be resistant, the evidence and discussion should be presented before the recommendation.

WEBLINK

What kind of guidelines do people in government agencies and businesses use to determine if a report is comprehensible and usable? For links to sites with useful criteria, go to **www.english.wadsworth.com/burnett6e**.
CLICK ON WEBLINK
CLICK on Chapter 20/guidelines

The report in Figure 20.1, opposing the use of foam-in-place (FIP) packaging, is written as a memo to a nonreceptive audience, in this case, a manager who has expressed enthusiasm for this packaging system. Because of his expressed support for FIP packaging, the manager is not likely to be initially receptive to the report's negative conclusion; therefore, the writer places his recommendation at the end of the report rather than its more typical place near the beginning. The writer hopes that by the time the manager has read the points opposing FIP packaging, the argument will have been strongly made. Once the manager understands the problems with FIP packaging, he may be receptive to the negative recommendation.

Organize the Information

Readers are generally neutral or positive because most are willing to form opinions based on currently available information rather than on previously held beliefs. Generally, the information in any report is organized so that it follows a sequence similar to this one:

1. ***Overview:*** This section states the purpose and/or problem that necessitates a report. (On a prepared form, the title and column heads may provide this information.) Sufficient information is provided for readers to understand the context of the report.
2. ***Background:*** This optional section presents information dealing with methods of investigation as well as materials and equipment used.

3. ***Recommendations:*** This section identifies any conclusions and/or recommendations (usually in priority order). Sometimes a summary of the recommendations is placed near the beginning to save busy readers time, with a more detailed discussion later in the report. Some informal reports do not need a separate section for recommendations.
4. ***Evidence:*** This section presents the results. Sometimes the results are summarized before being presented in detail. On a prepared form, this summary might be a final column showing net gain or good/fair/poor evaluation.
5. ***Discussion:*** This section explains or justifies the conclusions or recommendations on the basis of the supporting results. If the report does not contain recommendations, this section can review or summarize the major points. In an informal report, this section is often omitted.

As discussed in Chapter 10, you organize a document according to your understanding of your readers' attitudes. If they are generally receptive and positive, or at least neutral, your material should be organized with the most important information (problem statement and recommendations) first, followed by discussion and supporting details. If the readers have negative attitudes towards the subject of the document or might be opposed to its recommendations, the material should be organized with the recommendations at the end, to enable them to follow your reasoning.

You will find the patterns introduced in Chapter 10 particularly useful for responding to various situations and audiences. For some situations and audiences, only a straight *chronological* listing of actions or activities is required. For example, a report describing a new accounting procedure could easily list the revised steps. Occasionally, *spatial order* will help you describe physical characteristics of objects, mechanisms, organisms, or locations. *Cause and effect* is valuable for projections and estimates; *comparison and contrast* is useful for reports surveying the similarities and differences between various equipment or services. In every case, the organization you use should be determined by your purpose and your readers' task.

Figure 20.2 presents an example of a well-organized report prepared by the U.S. Army Corps of Engineers (USACE). The USACE is made up of more than 35,000 scientists and engineers, of which about 1.5% are military and 98.5% are civilian. These technical professionals, including biologists, biosystems engineers, environmental engineers, geologists, hydrologists, and natural resource managers, are responsible for providing a range of civil and military engineering services related to the environment:

- Planning, designing, building and operating water resources and other civil works projects (for example, navigation, flood control, environmental protection, disaster response, etc.)

FIGURE 20.1 Recommendation Report to Nonreceptive Audience[3]

*Because this is an informal report, the identifying information can be in the **heading**.*

*A **subject line** helps the nonreceptive reader focus, but it doesn't establish a position as it would for a receptive reader.*

*The **overview** introduces the circumstances necessitating the report.*

*The brief **background** section establishes the sources the writer checked and the criteria he used.*

*The **list** enumerating five separate points about the cost is easier for readers to comprehend than if the information had been presented in a single paragraph.*

Meredith Manufacturing Memo

TO Lan Wu

DATE 08 April 20—
FROM George Buchanan
DEPT Quality Assurance
EXT 555-5597
LOC/MAIL STOP NQ/509

SUBJECT Recommendation about using foam-in-place (FIP) packaging

We have been considering foam-in-place (FIP) packaging as an alternative that may resolve our packaging problems. Since a number of companies in the area have implemented FIP systems, we can benefit from their experiences. Many have found that FIP creates as many problems as it solves and have since turned to other packaging.

While investigating FIP, I have read FIP promotional material and discussed the product with representatives from three local FIP distributors. I have also spoken with quality assurance managers and packaging engineers at five local companies that have tried FIP.

FIP has many pitfalls that affect its use. This memo highlights information about material density, shipping, drop tests, costs, operator protection and health, and chemical disposal.

1. **Material Density.** Material density is very important in determining the "G" level or shock transmitted to a packaged product. The density of the foam in FIP is not uniform. Regardless of what the FIP suppliers advertise, the density varies from operator to operator.
2. **Shipping.** The effective static stress of FIP is from 0.04 to 0.2 pound per square inch (psi), which means a large contact surface area is required to cushion a dense product properly. Other cushioning materials such as polyethylene have an effective static stress range of 0.3 to 1.0 psi, which means less surface area is required. In general, a FIP package is larger than the equivalent expanded polystyrene or polyethylene package, based on the same fragility level. This means that shipping charges are greater for FIP packages.
3. **Drop Tests.** FIP obtains its cushioning ability from its foam structure. FIP suppliers in our area show cushioning curves for only the first drop test. Second and third drops are less favorable because the cell structure breaks down during the first drop. Other cushioning materials such as polyethylene do not break down and are much more consistent according to ASTM's standard 10-drop test. The probability that a packaged product will be dropped more than once during transportation is very high.
4. **Costs.** FIP has many hidden costs.
   a. Additional costs include continued maintenance and spare parts, gun cleaning solvents, nitrogen, protective clothing, and utilities.

FIGURE 20.1 **Recommendation Report to Nonreceptive Audience (continued)**

b. If the foam touches any part of a product, the product is damaged and must be reworked. Once the foam adheres to a surface, no chemical will remove it. If the person operating the FIP system is careful, most rework expenses can be avoided. However, many people working on the packaging line are the least skilled hourly workers; often they are not careful.

c. On an annual basis, the cost is substantial for plastic film to cover the product before the foam is molded around the product.

d. FIP chemicals are typically sold by weight. However, in every canister, some of that weight is not converted into foam. Chemicals that cannot be extracted are left in the bottom of the canisters. Additionally, a 15 to 20% weight loss occurs because of gasses that escape during chemical mixing.

e. The largest initial cost is the FIP equipment. The cost of providing ventilation equipment must also be included.

5. **Operator Protection.** Most FIP systems use an MDI form of isocyanate, which is an irritant to the human body. This means that FIP work areas must be well ventilated, operators must wear special clothing to protect their garments, and goggles are mandatory to prevent eye injuries.
6. **Operator Health.** Regardless of which FIP system—with operator safe-guards—is installed, the health of the operators is still a concern. All FIP systems use urea formaldehyde isocyanate chemicals, which according to medical experts can cause cancer. OSHA regulations clearly specify that overexposure to the chemicals results in irritation to the respiratory tract. (See attached OSHA regulations.)
7. **Chemical Disposal.** The biggest deterrent to using an FIP system is the fact that EPA regulations classify urea formaldehyde isocyanate as hazardous waste; any residual chemicals must be disposed of in accordance with local, state, and federal guidelines.

Because of the engineering, health, and environmental limitations of FIP systems, I recommend that we no longer consider installing such a system. Instead, we should investigate other alternatives for packaging our product.

George Buchanan

George Buchanan

GB/ml
Attachments

page 2

*The **design** of the memo — short paragraphs, enumerated list, bold, run-in headings — makes the information more accessible to the reader.*

*The **evidence** and **discussion** are combined in a seven-item list that establishes the reasons that FIP packaging is not the option that Meredith Manufacturing should choose.*

*The **recommendation** is at the end of the report because if it is earlier the reader is likely to be nonreceptive. After reasons for accepting the recommendation have been presented, the reader is likely to be more receptive.*

- Designing and managing the construction of U.S. military facilities
- Providing design and construction management support for the U.S. Department of Defense and other federal agencies.[4]

Every USACE project has a document trail dealing with the project itself (for example, planning reports, interim status reports, final recommendation reports for other agencies) as well as reports dealing with the interaction between the USACE and the communities in which their work is done. Typically, USACE reports are prepared for *sponsors* — that is, the municipality, county, state, conservation district, or other agency for which the work is done. The civil work that USACE does typically either provides background studies about some environmental aspect of the status quo for a sponsor or analyzes and recommends solutions to an environmental problem. In a problem-solving report, the USACE recommends a series of alternatives to solve a problem; it does not mandate that any recommendations be implemented.

Because of the wide range of audiences and purposes for USACE reports, the engineers and scientists need to ensure that their reports are accessible, comprehensible, and usable. They use a number of methods to help nonexpert audiences understand their reports:

- avoid acronyms
- provide definitions and explanations
- include high-quality photo images (accurately scaled and labeled)

The Hydrologic Engineering Branch of the Omaha District of the USACE is responsible for the hydrology and hydraulic design and maintenance of the Missouri River infrastructure, which includes both land and water resources.[5] The following USACE example is an excerpt from a technical report, the *Glovers Point Hydraulic Analysis,* written to the project sponsor, which is the advisory council of the Winnebago Indian Nation and the engineer they hired to oversee the project "to evaluate the feasibility of constructing an off-channel chute at the Glovers Point mitigation project site."[6] The purpose of this project is to increase the available habitat for wildlife and fish. Glovers Point is located on the Winnebago Reservation in northeast Nebraska at river mile 712 (that is, 712 miles from St. Louis, Missouri, where the river begins) on Glovers Point Bend on the Missouri River, approximately 10 miles north of Macy, Nebraska.

The *Glovers Point Hydraulic Analysis* is a 17-page recommendation report that was included in its entirely as Appendix A of a very lengthy report. *Glovers Point Hydraulic Analysis* was written primarily by Paul Boyd, a hydraulic engineer in the Hydrologic Engineering Branch of the Omaha District of USACE. The report includes seven main sections:

1.0 Introduction

2.0 Project Description

- ***2.1. Glovers Point location***
- ***2.2. Chute location***
- ***2.3. General conclusions***

3.0 Source Data

- ***3.1. Cross-sections***
- ***3.2 CRP Data***

4.0 HEC-RAS Analysis

- ***4.1. Model Construction and Calibration***
- ***4.2. Chute Geometry***
- ***4.3. Pilot chute connection***
- ***4.4. Site-specific model geometry***

5.0 Model Simulation Results

- ***5.1. Chute alternative selection***
- ***5.2. Velocity and sediment carrying capacity***
- ***5.3 Geometry selection***
- ***5.4. Chute crossing***
- ***5.5. Overbank flow***
- ***5.6. Riverbed degradation***
- ***5.7. Chute effects to navigation channel***

6.0 Summary

7.0 References

In this specialized use, "mitigation" means returning habitat to its natural condition — for example, returning prairie that has been used as farmland back into prairie or returning a river that has been diverted to its original flow.

The report in Figure 20.3 and all other USACE reports is designed so that each point raised in the project description (Section 2.0) is expanded in an individual subsection later in the report. You will be reading Section 1.0 and Section 2.0 from the report about the Glovers Point mitigation project and then Subsection 5.4, which is the elaboration of one sentence in Section 2.0 that deals specifically with alternatives in the "design of culvert flow pass that would allow for vehicular traffic and serve as a weir structure during high water flows." Put in lay terms, this means keeping the existing road open while limiting flood damage.

In mitigation projects like the one in Figure 20.3, computer models are often used to simulate water flow; the information from a series of computer simulations representing various conditions likely to actually occur helps the engineers and scientists recommend appropriate alternatives, ones that are environmentally sound, socially responsible, and fiscally feasible.

Courtesy of USACE

This photo shows the mitigation of a wildlife habitat on the Missouri River.

Select Genre

Reports vary not only in purpose but in genre as well. Information can be presented in prepared forms, memos, letters, or formal reports. The choice generally depends on a combination of factors: whether the document is routine or not, whether the audience is internal or external, and whether the tone should be informal or formal. Figure 20.2 identifies the most likely genre for various reports.

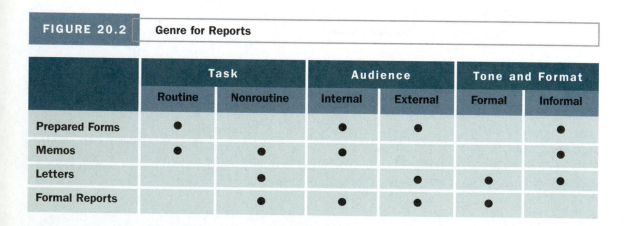

FIGURE 20.2 Genre for Reports

	Task		Audience		Tone and Format	
	Routine	Nonroutine	Internal	External	Formal	Informal
Prepared Forms	●		●	●		●
Memos	●	●	●			●
Letters		●		●	●	●
Formal Reports		●	●	●	●	

FIGURE 20.3 Recommendation Report with Effective Organization[7]

The report is in a conventional font that is sufficiently large to be read by most audiences.

Glovers Point
Hydraulic Analysis

Hydrologic Engineering Branch
USACE-NWO-ED-H

18 Feb 2003 – Original Draft
08 Apr 2003 – Revised Draft

US Army Corps
of Engineers
Omaha District

Do you know what these abbreviations mean?
HEC–RAS *= Hydrologic Engineering Center (USACE think tank) –River Analysis System (computer modeling system)*
CRP *= construction reference plane*
cfs *= cubic feet per second*
fps *= feet per second*
ft msl *= elevation in feet/mean sea level*

Section 1.0 Introduction identifies the completed task, its location, and the computer model used to simulate water flow. Such information orients readers to upcoming information.

Subsection 5.4 provides a more detailed discussion about two criteria for selecting a culvert system: (1) Can the culvert manage flow of water? (2) Can the culvert be topped with a road that provides vehicular access to the river?

Section 2.0 Project Description begins in Subsection 2.1 by identifying the location. Often these reports are sent to external reviewers who need to locate the place on a USACE map.

The hierarchy of the section heading and subheadings is designated in three ways: numerically, typographically, and spatially.

The same kind of information is provided about each alternative so they can be fairly compared. These alternatives are presented is descending order of effectiveness in solving the problem.

Sensitive economic, social, and political issues must be considered. Two examples: Regardless of the project, the river must remain navigable; regardless of the appeal of the land, the identity of private property owners and the precise locate of private property must be protected.

1.0 Introduction

A technical analysis was performed to evaluate the feasibility of constructing an off-channel chute at the Glovers Point mitigation project site. The numerical model HEC-RAS (v3.1, HEC 2002) was used to simulate water flow through the study area with the addition of the design features being considered. The model, when calibrated to measured conditions on the river, can estimate water surface elevations and flow rates necessary for proper evaluation of the proposed design alternatives.

2.0 Project Description

2.1. *Glovers Point location.* The study site is located on the Winnebago Indian Tribe Reservation in northeast Nebraska at river mile 712 (RM 712) on Glovers Point Bend, approximately 10 miles north of Macy, NE.

2.2. *Chute location.* Two chute locations, and a total of three alternatives, were considered. Alternative #1 would follow the general channel of a chute used in the past. The inlet to the chute would be located at RM 713.5 and the outlet located at RM 711.2. Following the former chute flow path, the length of flow through the off-channel chute would be approximately 10,700 feet. Along the flow path a culvert crossing was inserted into the model at 8773 ft from the chute outlet. This correlates with the location of an existing road crossing. Alternative #2 was considered as an alternative with a shorter flow length, starting at RM 712.64 and ending at the same location as Alternative #1, for a total flow length of 7,600 ft. HEC-RAS analysis was performed to evaluate the stability and flow velocities of multiple chute designs. The third option included both Alternative #1 and a backwater area just upstream of the outlet of the chute, considered as Alternative #3. Geometry for this option was not created as water stored in the backwater area would be considered ineffective flow area and would not cause any significant variation in model results from Alternative #1 alone.

FIGURE 20.3 **Recommendation Report with Effective Organization (continued)**

2.3. *General conclusions.* Simulations with HEC-RAS were used to evaluate the feasibility of multiple designs. The off-channel chute simulation results showed multiple scenarios in which the geometry of the structure would be sustainable for Alternative #1. The most ideal of these geometries would be the one that resulted in the lowest volume of excavated material while maintaining sufficient velocities to prevent significant future maintenance. None of the Alternative #2 pilot chutes produced results favorable enough to warrant further consideration. Alternative #3 would result in the same channel dynamics as Alternative #1 but with the addition of multiple acres of shallow water habitat. Depending on future water elevations at the site, Alternative #3 may incur additional maintenance. All alternatives included design of culvert flow pass that would allow for vehicular traffic and serve as a weir structure during high water flows. (See aerial photo below.)

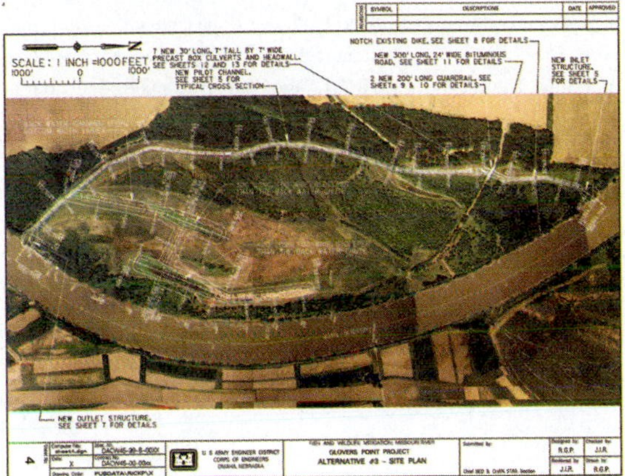

Based on modeling results, the long term sustainability of an off channel chute at the Glovers Point Mitigation Site is feasible. The Alternative #1 location is the only location that would have a significant self-sustaining life span.

The general conclusion in Subsection 2.3 presents the overall criterion and applies it to each alternative in descending order of effectiveness, paralleling the order in the preceding section.

The critical aspects of each alternative are that the culvert design allows for regular traffic and can serve as a small dam during high water flows.

The aerial photograph shows the project site with an overlay of the proposed excavation plan.

FIGURE 20.3 **Recommendation Report with Effective Organization (continued)**

Several pages later in the report, this point about chute crossing (highlighted here and in Section 2.3 only for our identification) is considered in greater detail.

Conditions in extraordinary circumstances (high flow rates) are explained. Specifically, the road would act as a weir for the high water.

5.4. *Chute crossing.* To allow continued access to the river, the chute designs included flow analysis for a set of box culverts inserted at the current crossing location with a roadway over the structures. The culvert system should be able to fully manage the CRP flow. At high flow rates, water would be expected to overtop the road surface and the crossing and road surface would serve as a weir. In addition, culvert flow length was limited to no less than 30 feet to allow for a 24 ft wide roadway across the culvert group. To accommodate these higher flows, the size and number of culvert boxes as well as the elevation and flow width of the road surface were examined.

Total culvert flow areas of 175 and 280 ft2 were considered. The 280 ft2 option allowed for greater flow at lower river flowrates and reduced the backwater effect seen at higher river flow rates. In addition, an elevation of 1057 ft msl at the top of the road surface allowed for best combination of weir flow and culvert flow. The effective flow width of the flow over the road surface was set at 150 ft. The design top of roadway is approximately eight feet lower than the existing roadway, and past the 150 ft width, the road will rise at a 10:1 slope to the existing road elevation. Figure 3 plots a cross section of the culvert and roadway. The culvert boxes were selected with a 7.0 ft by 7.0 ft inside dimension. The design sets the bottom elevation of the culvert 2.0 ft below the channel invert elevation. The analysis used a 7.0 ft by 5.0 ft size culvert. The 7.0 ft by 7.0 ft culverts were selected for construction to allow flexibility in matching projected degradation of the Missouri River channel bed. Table 2 summarizes the performance results for the selected pilot chute at various future widths and flowrates.

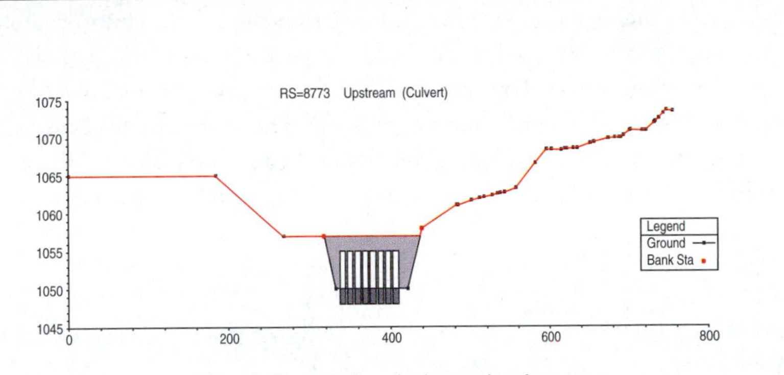

Figure 3. Cross-section of culvert and roadway.

The cross section of the culvert and roadway shown in Figure 3 is useful to readers because it provides some idea of how wide the river might be at flood stag

Table 2. Model performance of selected pilot chute

	32,000 cfs (CRP)				60,000 cfs			
	30 ft pilot	100 ft	150 ft	200 ft	30 ft pilot	100 ft	150 ft	200 ft
Flow rate in Chute (cfs)	309	746	1063	1216	1605	3096	3673	4223
Inlet velocity (fps)	0.35	0.84	1.22	1.41	0.77	1.51	1.80	2.09
Upstream channel velocity (fps)	1.89	1.62	1.61	1.38	3.50	2.79	2.33	2.07
Upstream culvert velocity (fps)	1.07	1.33	1.43	1.26	1.28	1.83	1.76	1.70
Inside culvert velocity (fps)	1.27	3.17	5.06	6.20	1.73	4.32	6.02	6.63
Roadway weir flowrate (cfs)	0	0	0	0	1119	1887	1987	2368
Water elev. at culvert inlet (ft msl)	1054.4	1054.4	1054.3	1054.3	1059.5	1059.4	1059.4	1059.4
Downstream culvert velocity (fps)	1.22	1.64	1.78	1.64	2.17	2.81	2.49	2.25
Downstream channel velocity (fps)	1.85	1.44	1.65	1.47	2.45	2.77	2.33	2.08

Table 2 is useful to readers because it lets them compare the way the culverts will perform with a standard water flow (for example, 32,000 cfs) and with a high water flow (e.g., 60,000 cfs). It provides information for accurate predictions.

Prepared Forms. An appropriate method for routine, informal communication for both internal and external readers is *prepared forms.* Accounts of day-to-day operations most frequently use prepared forms and memos, seldom formal reports. For example, inspection and evaluation reports are often done on prepared forms, whether the subject is a precision machined part or a patient admitted to a pediatric unit.

In what circumstances would you prefer to submit reports on a prepared form?

Forms can be a tremendous timesaver for both writers and readers. Forms assist writers by reminding them to include every essential item. And they assist readers by ordering information in the same sequence in each report of a series, so the readers can quickly locate whatever information they need. Unfortunately, forms are often poorly designed or outdated. As part of your work, you may need to design or redesign forms. Four areas of design are important to consider: content, layout, visual elements, and paper stock. The more carefully you adapt these four elements to the form's purpose and audience, the easier the form will be to complete and the more likely to provide you with useful information.

WEBLINK

Designing forms is a critical skill in some professions. For a link to useful information about designing effective forms, go to www.english.wadsworth.com/burnett6e.

CLICK ON WEBLINK

CLICK on Chapter 20/form design

Memos and Letters. Routine information can be easily presented in memos and letters. This ease accounts for the widespread use of form memos and letters, which were popular long before word processing made individualization possible. Often, though, memos and letters deal with nonroutine information, with memos going to internal readers (either an individual or a group) and letters going to external readers (a single individual, though copies may be sent to others). Both memos and letters range from informal to fairly formal. (Chapter 18 presents a detailed discussion of correspondence.) Headings and subheadings can ease reading. Data should be presented clearly and directly; tables, charts, and graphs are often used.

Reports. Communication that does not fit neatly on a prepared form and requires a more formal tone and format than a memo or letter is placed in a report. Reports can be intended for internal or external readers. Regardless of purpose, audience, or degree of formality, a report should be easy to read and visually appealing. As in memos and letters, headings and subheadings can ease reading. Data should be presented clearly and directly; tables, charts, and graphs are often used.

WEBLINK

Sometimes you want to see examples of reports, both good examples and bad ones, to get ideas of what you should do and what you should avoid. For links to a huge archive of technical reports, go to www.english.wadsworth.com/burnett6e.

CLICK ON WEBLINK

CLICK on Chapter 20/report archives

This section discusses the function and organization of the most frequently used reports:

- research reports and articles
- task reports (recommendation, justification, inspection, information, investigation)
- periodic activity reports (daily, weekly, monthly, quarterly)
- progress (interim, status) reports
- trip and conference reports
- meeting minutes

Research Reports and Articles

Research is a systematic investigation to establish verifiable information. It's about discovering new knowledge rather than applying existing knowledge. The discovery is sometime done in controlled experiments, sometimes by careful observation, sometimes by analysis of available artifacts or records. Research exists in virtually every discipline, from art and anesthesiology to zoology and zymurgy. Generally, research has several characteristics, regardless of the discipline:

- Research should be ***accessible,*** open to scrutiny by peers and available to the public.
- Research should be ***transparent,*** clear in its structure, process, and outcomes.
- Research should be ***transferable,*** useful beyond the specific research project, applicable in principles (if not specifics) to other researchers and research contexts.[8]

In Chapter 17, you read and viewed the poster that Tara Barrett Tarnowski prepared about her research. Figure 20.4 shows the beginning of the article that resulted from that research, which was published in the refereed online journal *Plant Management Network.* Tarnowski conducted the research, prepared the scientific poster, and coauthored this research article while she was an undergraduate in plant pathology. She is listed as the first author in both documents, indicating in this case that she is primarily responsible for conceptualizing and conducting the research.

If you were to read the entire article, you would learn that this is not the first article that Tarnowski and her coauthors (Jean Batzer, Mark Gleason, Sara Helland, and Phillip Dixon) have written. The literature review has 25 citations, including one with Tarnowski as first author, three with Batzer as first author, one with Gleason as first author, and one reference to Gleason's unpublished data.

Why should all research reports and articles present the rationale or justification for the investigation in the introduction?

FIGURE 20.4 Research Article Published in Refereed Online Journal[9]

Dates are important in scientific publications. Readers learn when the article was accepted as well as when it was published.

Key words that help readers locate the article in a search are included in the title.

The collaborators and their affiliations are listed, including identification of the corresponding author if readers have questions.

The complete citation is provided.

Virtually all published research articles and most unpublished research reports include an abstract.

The subject of the research — sooty blotch and flyspeck diseases — is defined (including clear citations) and then illustrated in Figure 1, which also appeared in the poster.

Accepted for publication 12 November 2003. Published 9 December 2003.

Sensitivity of Newly Identified Clades in the Sooty Blotch and Flyspeck Complex on Apple to Thiophanate-methyl and Ziram

Tara Tarnowski, Jean Batzer, Mark Gleason, and **Sara Helland,** Department of Plant Pathology, **Phillip Dixon,** Department of Statistics, Iowa State University, Ames, IA, 50011

Corresponding author: Mark Gleason. mgleason@iastate.edu

Tarnowski, T., Batzer, J., Gleason, M., Helland, S., and Dixon, P. 2003. Sensitivity of newly identified clades in the sooty blotch and flyspeck complex on apple to thiophanate-methyl and ziram. Online. Plant Health Progress doi:10.1094/PHP-2003-1209-01-RS.

Abstract

Sensitivity to the fungicides thiophanate-methyl and ziram was characterized for newly identified clades in the sooty blotch and flyspeck (SBFS) disease complex. Isolates obtained from apple orchards in Illinois, Iowa, Missouri, and Wisconsin were previously grouped into clades based on parsimony analysis of the internal transcriber spacer (ITS) and large subunit (LSU) regions of rDNA. Two isolates from each of eight clades were cultured on water agar amended with a range of fungicide concentrations. Radial growth of colonies was measured after incubation for 21 days at 25°C. Fungicide sensitivity (ED_{50}) did not differ significantly ($P = 0.05$) between isolates within a clade, but several clades differed significantly. There was a greater variability among clades for sensitivity to thiophanate-methyl than for ziram. Significant differences in fungicide sensitivity among, but not within, clades provide evidence that these clades differ physiologically. Understanding differences in fungicide sensitivity among SBFS clades could have practical implications for improved management of SBFS.

Introduction

Sooty blotch and flyspeck diseases (SBFS) are caused by a complex of saprophytic fungi that colonize the cuticle of pome fruits. Sooty blotch signs are variable, but generally appear as brown to black smudges on the apple peel, whereas flyspeck appears as groups of tiny brown or black spots (Fig. 1). The fungi blemish the fruit, downgrading its value from fresh-market to cull status, and thereby can cause severe economic losses for apple growers (24). The SBFS complex includes *Peltaster fructicola*, *Geastrumia polystigmatis*, and *Leptodontium elatius*, which cause sooty blotch, and *Zygophiala jamaicensis*, which causes flyspeck (16). Six mycelial types -- punctate, fuliginous, rimate, ramose, flyspeck, and discrete speck -- are currently used to describe the diverse morphology of members of the SBFS complex on apples (3,11).

Fig. 1. Sooty blotch and flyspeck signs on Golden Delicious apple from Pella, IA.

You would also learn that this article from *Plant Management Network* follows a number of conventions beyond those apparent on the initial screen. For example, at the end of the introduction, Tara and her coauthors present the rationale for their study of sooty blotch and flyspeck diseases (SBFS):

Current methods of control provide "promising but inconsistent results" in controlling SBFS.

To do a better job of controlling SBFS, people need "a better understanding of the environmental biology and fungicide sensitivity of members of the SBFS complex."

This study characterizes "the sensitivity of newly discovered SBFS fungi to the fungicides thiophanate-methyl and ziram." This characterization contributes to knowledge that will lead to better control against SBFS.

To prevent economic losses from SBFS, a protectant fungicide spray program is implemented in apple orchards throughout the eastern half of the United States, as well as in other humid areas of apple production worldwide. Use of weather-based disease-warning systems to reduce the frequency of fungicide sprays in orchards has provided promising but inconsistent results (12,13,21; Gleason, *unpublished data*). In both conventional and integrated pest management (IPM) approaches, management regimes have been applied without regard for the species of SBFS present in each orchard. More effective management may require a better understanding of the environmental biology and fungicide sensitivity of members of the SBFS complex.

Few fungicides are currently registered for control of SBFS. Among these, thiophanate-methyl and ziram are among the most affordable and widely used compounds in the U.S. The purpose of the present study was to characterize the sensitivity of newly discovered SBFS fungi to the fungicides thiophanate-methyl and ziram. A preliminary report has been published (2).[10]

Tara and her coauthors present their methods. While some of the same visuals that are on the poster are also used in the research article, the level of detail is considerably greater in the article:

The technical terminology and the specialized processes are clear signals that research articles are intended for experts.

The information is presented in sufficient detail to be usable by other plant pathologists.

The reference to the accompanying figure makes clear that all of the information is not in the text.

Stock suspensions of each fungicide were prepared in distilled water using 1:200 serial dilutions. The most concentrated suspension corresponded to the lowest labeled rate for the fungicide on apples (10). Fungicide stock suspensions were added to autoclaved 2% water agar (after cooling to approximately 40°C). Three ml of the amended media were added to each well in a six-well (well diameter = 3.5 cm) tissue culture plate (Costar #3516, Corning Inc., New York, NY).

Inoculum was prepared by streaking 30-day-old cultures onto potato dextrose agar (PDA) to achieve a uniform lawn of mycelium. After 17 days of incubation at 23°C, 3.5-mm-diameter fungal plugs were placed mycelium down onto the media in the six-well plates. The inoculated plates were incubated in the dark at 25°C (Fig. 2).

Methodological details are included in the caption.

Information about the numbers in figure is provided to reduce chances of misinterpretation.

Why do research reports and articles usually present methodological detail sufficient to enable another researcher to replicate the study?

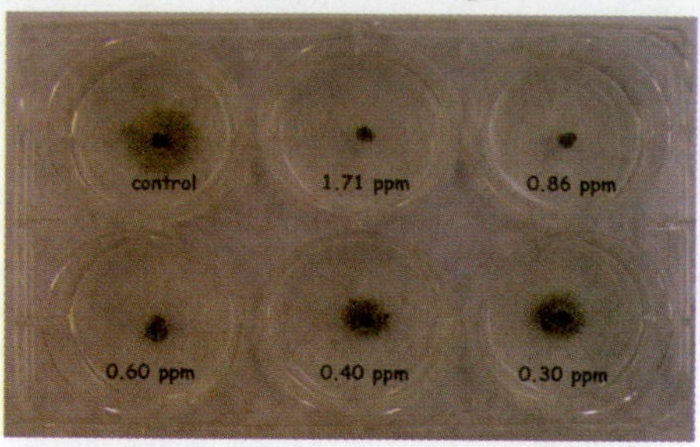

Fig. 2. Mycelial plugs were places on fungicide amended agar in wells of tissue culture plates. Isolated MWD2 is shown on thiophanate-methyl-amended agar. Fungicide concentrations (mg/ml) are shown below corresp... wells...

Why do research reports and articles usually present implications with caveats and cautions even if the results appear to be positive and encouraging?

Finally, Tara and her coauthors present their results and implications, again drawing on figures from the poster, but this time providing considerably more details in the discussion. The poster simply presents three technical conclusions while the research article discusses implications, mentioning, for example, the "potential to lead to more effective SBFS management strategies." This study's "discovery of sensitivity differences to commonly used fungicides in the SBFS complex on apples" provides information with two important implications: growers will be able to adjust their current "fungicide-based management strategies" and other researchers may develop "more effective SBFS management strategies."

Conclusions (from poster; see Chapter 17)

1. Mycelial types and putative species of the sooty blotch and flyspeck complex in the Midwest showed significant differences in fungicide sensitivity to both thiophanate-methyl and ziram.
2. Segregation of mycelial type was more pronounced for thiophanate-methyl than for ziram.
3. With exception of the rimate isolates (RL1, RL2), isolates within both a putative species and mycelial type did not differ significantly in their fungicide sensitivity.[12]

Implications (from the research article)

Our results have the potential to lead to more effective SBFS management strategies. Species composition within the SBFS complex varies among geographic locations. Furthermore, the SBFS assemblage within an orchard may be dominated by one or a few species. In a preliminary survey, the prevalence and incidence of SBFS mycelial types varied sharply among nine apple orchards in four states (IA, MO, IL, and WI) (4). *L. elatius*, a common SBFS species found in North Carolina, was not isolated from apples in the Midwest. In a North Carolina study, the prevalence and incidence of SBFS mycelial types differed among four orchards (22). Although in vitro fungicide sensitivity may not correlate precisely to SBFS control in the field (8), knowledge of the composition of the disease complex within a specific orchard and the fungicide sensitivities of the predominant clades could allow a grower or consultant to better select fungicides, rates, and spray timing to suppress SBFS.

The discovery of sensitivity differences to commonly used fungicides in the SBFS complex on apples indicates that members of these newly identified clades have significant functional and physiological differences. Thus, members of the SBFS complex may respond differently to various fungicide-based management strategies. With further understanding of fungicide sensitivity differences and ecological interactions within the SBFS complex, more effective SBFS management strategies may be developed.[13]

What about copyright and fair use of your documents? What are the guidelines? What's the law? What's a good idea? For links to useful explanations, go to **www.english.wadsworth.com/burnett6e**.
CLICK ON WEBLINK
CLICK on Chapter 20/copyright

WEBLINK

Task Reports

Technical professionals deal with many tasks that result in reports. Most common are recommendation or justification reports, inspection or examination reports, and information or investigation reports.

A *recommendation report* or *justification report* primarily presents or defends specific suggestions or solutions for a particular situation. Such a report might recommend purchasing particular equipment or might justify changes in procedures, personnel, or policies. For example, when a department manager wanted approval for changes in his department's procedures for storing solvents, he prepared a memo to his manager in which he explained the problems with the current procedures, recommended specific changes, and summarized reasons why the new procedures would be safer and encourage better record keeping. The reports in Figure 20.1 and 20.2 are both examples of recommendation reports.

An *inspection report* or *examination report* focuses on recording observable details, sometimes followed with recommendations. In some cases, an inspection report is completed on a prepared form, such as when the task involves the inspection of mechanical parts. In other cases, a report follows the sequence presented earlier in this chapter: overview, background, recommendations, evidence, and discussion. For example, imagine that a civil engineer filed a report with her supervisor, the director of public health, following her visit to the community's waste treatment plant. She gave an overview of the reason for her visit (to see if new regulations had been implemented), focusing on her observations of how well the workers were able to follow the new regulations using current equipment. As a result of her observations, she recommended equipment modifications to allow the new regulations to be more easily followed and offered evidence that her recommendations would not only improve plant efficiency but also be more cost effective. She concluded with a discussion about the overall benefits that implementation of the new regulations would have.

An investigation results in an *information* or *investigation report* that collects and evaluates information about some existing situation, but the writer need not always include a recommendation. For example, after a manufacturing engineer noticed a pattern of equipment malfunction following certain production runs, he reviewed the maintenance records for the quarter and compiled a report for his manager that identified the problem, provided a summary of the malfunctions, and suggested probable causes. However, he stopped short of making recommendations because his primary purpose was to provide his manager with information.

Periodic Activity Reports

Efficient organizations have developed reporting methods to keep track of ongoing activities within the organization. These reports, filed daily, weekly, monthly, or quarterly, are compiled by supervisors and managers to describe the work completed by their section or group. Typically, each engineer may compile weekly reports that an engineering manager uses for a monthly report; a section manager turns the monthly reports into a quarterly report. Sometimes information from these reports is used as the basis for projections that anticipate changes in project design, scheduling, or budgeting.

When the work is routine, daily or weekly periodic activity reports can be recorded on prepared forms. However, nonroutine activity reports require more than merely filling in the blanks. These reports, whether routine or not, should be factual and honest; exaggerating to make a particular time period look better than it really is is simply bad practice. When you are determining the content and format of a report, be particularly careful to include necessary information; omit information that is interesting but not critical. Organizations usually list information they expect in a periodic activity report:

1. ***Overview:*** Identify projects.
2. ***Activities:*** Specify project activities that are completed, in process, and planned.
3. ***Recommendations:*** Establish needed changes in scheduling, personnel, and budget.

Organizations also often have a required format (even if they don't have prepared forms) so you usually won't have to spend much time selecting the information to include or organizing it. In some situations, though, you may be given the criteria for the reports and then left on your own to organize and present the information in the activities section. Basically, you should organize in one of two ways, depending on your readers' expectations and needs.

- *Chronological order* is appropriate if readers expect a straightforward listing of the activities with no comment or evaluation.
- *Descending priority order* is appropriate if readers need the report to rank order the completed and in-process work by importance.

For example, in preparing a monthly report, the head of a software development department might use descending priority order, writing in great detail about the significant activities of the month and giving routine matters only cursory attention. But a hotel banquet manager could logically use chronological order for periodic activity reports because those reports must identify the daily use of the banquet facilities.

The recommendation section in a periodic activity report is used only if you need to suggest changes based on the activities. Evidence and discussion sections seldom appear in periodic activity reports.

Progress Reports

Unlike periodic activity reports that are prepared on a regular schedule regardless of current projects, progress reports summarize the progress, status, and projections related to a particular project. Progress reports (sometimes called status reports or interim reports) are part of almost all long-range projects and may also be expected in short-term projects.

Information about the progress of an activity answers a variety of reader questions:

- How is the project going?
- What has been accomplished during this phase of the project?
- How much time or effort, money, and so on did these tasks take?

Information about status answers additional reader questions:

- Where are we now?
- How do current activities relate to the overall project?
- How does this work affect other phases of the project?

Information about projections answers even more reader questions:

- Are we on schedule to meet our completion date?
- What plans need to be changed or altered?
- What will we do in the future?

A progress report generally follows this sequence of information:

1. ***Overview:*** Introduce the project.
2. ***Progress:*** Summarize the progress to date.
3. ***Recommendation:*** Identify major recommended schedule changes.
4. ***Evidence:*** Provide reasons for changes.
5. ***Discussion:*** Discuss the impact of the proposed changes.

The *overview* of a progress report orients readers to the project, identifying the purposes of both the project and the report. Additionally, it surveys the project and specifies the dates the report covers. This section of the report is crucial. Either specific times or number of tasks can be emphasized, depending on the desired focus. Some progress reports recommend changes necessitated by the progress or lack of progress. Less frequently, additional reports establish and discuss reasons for these changes.

Like periodic activity reports, progress reports can be organized in either chronological or priority order. A chronologically ordered progress report emphasizes the time period, listing dates and then specifying the tasks that were completed during that period. If you prefer to emphasize the tasks, place them before the time period in your report; these tasks can be arranged in chronological or priority order, depending on the readers' needs. The data should be clearly organized and presented in tables, graphs, and charts, which is more effective than placing the data in conventional paragraphs.

A progress report is particularly easy to prepare if the schedule with the original project proposal is structured so that the progress on each task can be monitored. Turn back to Chapter 7 to review the PERT chart in Figure 7.3 and the Gantt chart in Figure 7.4. If you prepare a PERT or Gantt chart (or some similar kind of project planning schedule), a substantial portion of your progress report can be updating this chart.

Meeting Minutes

The record of the proceedings of any deliberative group is usually called the *minutes,* or sometimes, particularly in legislative bodies, the *journal.* Minutes provide a record of the discussion and decisions that occur at meetings, serving as official (sometimes even legal) records. Beyond their function of official documentation, minutes provide a convenient review of the meeting for people who attended as well as for those who didn't.

Preparing the minutes for a meeting requires both organization and attention to long-established conventions. The agenda for a meeting can provide you with a basis from which you can organize the information. If the meeting was recorded, you can use the tape to fill in details you may have missed while taking notes. *Robert's Rules of Order* recommends the format and sequence of information for meeting minutes.

WEBLINK

Knowing *Robert's Rules of Order* is essential if you're going to be running or participating in formal meetings. For a link to this traditional source of information about meetings, go to www.english.wadsworth.com/burnett6e.
CLICK ON WEBLINK
CLICK on Chapter 20/Robert's Rules

Why make minutes a record of actions rather than a record of opinions that were voiced?

Unless the minutes are to be published, they should mainly record what was *done* at the meeting, not what was *said* by the members. The minutes should never reflect the secretary's opinion, favorable or otherwise, on anything said or done.

Trip and Conference Reports

Trip and conference reports are important for several reasons. They force the traveler to review and evaluate the activities of the trip or conference and distinguish the major accomplishments from those less important. They also enable the traveler to share activities and information with people who didn't make the trip or attend the conference. Of course, they also justify the time and expense of the trip or conference and can even be used to verify business trips for IRS inquiries and audits.

A trip or conference report is seldom completed on a prepared form. Instead, the information listed below is incorporated into a logically organized, clearly stated report:

Trip	*Conference*
purpose and date of trip	purpose and date of conference
primary task(s)	primary task(s)
personal role	personal role
people contacted	sessions attended
question(s) raised or resolved	information gained
conclusions	conclusions

As you can see, the information presented in trip and conference reports is very similar. Unlike periodic activity reports and progress reports, which frequently use chronological order, effective trip and conference reports employ priority order: You rank the tasks and information according to their value to your organization.

Formats of Reports

A report's audience and purpose influence not only its content but also the components you choose to include. Writing for a mixed audience requires you to select and organize the material so that it serves all categories of readers, whatever their different needs.

Prior knowledge of the audience plays an important role in choosing the overall sequence of information in a report. Recall that *inductive reasoning* moves from specific to general and *deductive reasoning* from general to specific. Receptive readers generally appreciate reports that are organized deductively, giving the conclusions first and then providing the substantiating details; most reports are arranged this way. Sometimes, however, a report might be better received if arranged inductively, giving specifics first and letting them lead to general conclusions and recommendations.

Complex reports are often prepared collaboratively. What can people do to ensure that the report is a coherent team document rather than a cobbled-together effort by individuals?

Most reports first state a position and then establish its validity. Thus, a deductively organized report usually uses a two-tiered approach:

1. Give an overview and then summarize the preferred recommendation or solution.
2. Cite evidence or support for the recommendation or solution in descending order of significance, dealing with both positive and negative points.

You might assume that reports have a specific format that everyone follows. However, few professionals agree as to what constitutes the best combination of components or the order of those components. This lack of consistency shows up in the dozens of style guides published by individual companies and professional associations. Further evidence for the disparity of report formats comes from NASA: "The preliminary findings of a NASA study revealed that (1) nearly one hundred components were used [in different technical reports], (2) there was an apparent lack of consistency in the terms used for the components, and (3) there was an apparent lack of consistency in the location of the components."[14]

This text employs a generic report format that incorporates the standard components of most reports; if your company prescribes its own format, use that instead. The components are first defined and then illustrated in sample formal reports reproduced at the end of the chapter.

Front Matter of a Report

The front matter in a report consists of the sections that come before the body of the document. The sections that are identified in Figure 20.5 are those typically used in reports. Although not every report includes every item, the more formal the report (and, thus, the more distant the audience) the more front matter is likely to be included. The less formal the report, the less likely it is to include front matter.

Some components of front matter require more discussion than the summary provided in Figure 20.5. Some issues need to be considered when you write a letter of transmittal and a table of contents for a report.

Whether to include a letter of transmittal depends on company policy. The author of the report can use a letter of transmittal to introduce the primary reader to the document. Using one of the standard letter formats (illustrated in Chapter 18), a letter of transmittal generally has three paragraphs. The first introduces the document's subject and purpose. The second usually focuses on one or two key points dealing with the document's preparation or content: problems, resources, additional work, conclusions, recommendations. The third paragraph is a courtesy that encourages the reader to contact the writer with any questions.

An effective table of contents uses subject headings, not section labels, to identify major sections and subsections of a report.

> *Section heading* 1.0 Introduction
>
> *Content heading* 1.0 Problems in Treating Postoperative Lung Congestion

FIGURE 20.5 **Purposes and Practices of Front Matter**

Letter of Transmittal

Purpose: Introduce the primary reader to the document.

Practice:

- Introduce the document's subject and purpose (¶1).*
- Focus on one or two key points dealing with the document's preparation or content: problems, resources, additional work, conclusions, recommendations (¶2).
- Encourage reader to contact the writer with any questions (¶3).
- See Figure 20.10a for a typical letter of transmittal.

Cover

Purpose:

- Secure the pages of the report.
- Create a professional image.

Practice:

- Match weight of cover to length of report (a brief report needs only a lightweight cover; a lengthy report requires a sturdy cover).
- Select cover that stays flat when report is open.

Title Page

Purpose: Identify title/subtitle, author(s) and organization, person and organization for whom report was prepared, date.

Practice:

- Sometimes include project identification and report numbers.
- See Figure 20.10b for a typical title page.

Table of Contents

Purpose: Identify major sections and subsections of the report.

Practice:

- Differentiate sections both typographically and spatially.
- See Figure 20.10c for a typical table of contents.

List of Tables and Figures

Purpose: Use as a table of contents for visuals in a report.

Practice: Omit if there are fewer than five tables or figure.

List of Appendixes

Purpose: Use as a table of contents for appendixes.

Practice: Omit if the report has only one or two appendixes; listing can be included as a major section in the table of contents.

Abstract or Executive Summary

Purpose: Summarize major points, findings, or recommendations of the report.

Practice:

- Do not use as introduction to the report but as independent, entirely condensed version—the report in miniature.
- See Figures 20.10d and 20.10e for a typical executive summary.

*Symbolizes paragraph.

The various sections can be differentiated both typographically and spatially. These are some of the common ways to differentiate levels of a document:

- ***Indentation:*** The more indented the heading, the more subordinate it is.
- ***Type size:*** The smaller the heading, the more subordinate it is.
- ***Type style:*** More aggressive type styles are used for titles and level-1 headings, less aggressive styles for level-2 and level-3 headings.

Figure 20.6 illustrates a numeric outline, a format often used by government agencies, companies that have government contracts, and some scientific organizations. The levels are indicated in three ways: numbers to signal the sections (as in the body of the report), typographic variation (the use of **BOLD SMALL CAPS** for level-1 headings), and indentation for each succeeding level of heading.

Figure 20.7 shows a more conventional table of contents. The levels are indicated in three ways: **BOLD ALL CAPS** for major headings, indentation to indicate level-3 subsections, and type style changes ***(bold italics).***

Body of a Report

The selection and sequence of components in the body of a formal report depend on audience and purpose. Usually some or most of the following components are included:

Part I

- statement of purpose or problem
- summary of findings
- summary of recommendations

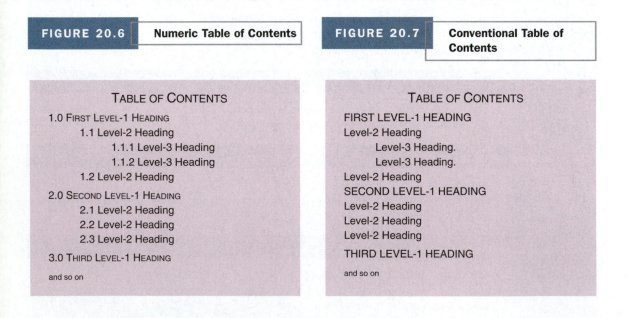

FIGURE 20.6 **Numeric Table of Contents**

TABLE OF CONTENTS

1.0 FIRST LEVEL-1 HEADING
    1.1 Level-2 Heading
        1.1.1 Level-3 Heading
        1.1.2 Level-3 Heading
    1.2 Level-2 Heading
2.0 SECOND LEVEL-1 HEADING
    2.1 Level-2 Heading
    2.2 Level-2 Heading
    2.3 Level-2 Heading
3.0 THIRD LEVEL-1 HEADING
and so on

FIGURE 20.7 **Conventional Table of Contents**

TABLE OF CONTENTS

FIRST LEVEL-1 HEADING
Level-2 Heading
    Level-3 Heading.
    Level-3 Heading.
Level-2 Heading
SECOND LEVEL-1 HEADING
Level-2 Heading
Level-2 Heading
Level-2 Heading
THIRD LEVEL-1 HEADING
and so on

Part II

- background to problem
- literature search of information relevant to problem
- approach, method, and materials (for reports of experiments, surveys)
- available options (solutions)
- results: collected data or findings
- discussion
- interpretation(s)
- conclusion(s)
- recommendation(s)

Separating the body into two major segments responds to the needs of the two very different kinds of primary readers, the decision makers and technical professionals. Part I of the body states the purpose or problem and then summarizes the findings and recommendations, immediately giving decision makers the information they need without making them sort through myriad technical details. Part II of the body provides details for technical professionals. The organizational patterns presented in Chapter 10 provide a variety of structures whose application depends on both purpose and content. The following list reviews these patterns and suggests appropriate uses:

- ***Chronological order*** — Chronology is used for explaining processes, such as setting up a new production line.
- ***Spatial order*** — Descriptions of objects or locations are best arranged spatially; that is, in the same sequence in which an operator or observer encounters the subjects described. For example, spatial order could be used in describing the physical facilities in a remodeled plant.
- ***Cause and effect*** — Problem-solving situations are easily explained this way, often arranged in descending order of importance. For example, suggestions of ways to increase the acceptance ratio of items passing through quality control could be explained using cause and effect.
- ***Comparison and/or contrast*** — Focusing on the similarities and differences, also often arranged in descending order, provides a reader with the basis for decision making. A report using comparison and contrast might advocate the lease of new equipment or a change in vendors.

End Matter of a Report

The end matter of a report comes after the body. It can be made up of any of the components in Figure 20.8. The formality of the report and the needs of the audience influence what is included. Very formal reports typically include a great deal of end matter; very informal reports usually include none.

FIGURE 20.8 Purposes and Practices of End Matter

Appendixes	
Purpose:	▪ Present useful information that might otherwise interrupt the flow of the report.
Practice:	▪ Include or omit items depending on purpose, audience, and situation: □ formulas used in calculations □ complex calculations of which only the results are used in report itself □ survey forms □ interview questions □ transcripts □ correspondence related to the document subject □ detailed figures from which selected information is taken for report □ references for further reading
Glossary, List of Symbols, List of Abbreviations	
Purpose:	▪ Define terms unfamiliar to the readers.
Practice:	▪ Signal defined terms in some consistent way (italics, asterisk).
Footnotes, Sources Cited, Works Cited, or References	
Purpose:	▪ Document the sources of all cited information.
Practice:	▪ Select format appropriate for discipline and profession. ▪ Provide internal (in-text) citations or footnotes for (1) direct quotations, (2) paraphrased information that's not common knowledge for the intended audience, (3) statistical and other factual information, and (4) visuals. ▪ List all sources referred to in the document in a sources cited, works cited, or references section.

Appendixes present supplementary information to increase readers' knowledge or understanding. What to include depends on the purpose and audience of the document (Figure 20.10). For example, primary readers who have a technical background will benefit from detailed technical data in the body of the document far more than will readers who rely on the writer's technical expertise to summarize pertinent information. For primary readers with backgrounds in business rather than a technical field, detailed data are appropriately placed in an appendix.

A ***glossary*** defines terms unfamiliar to readers (boldface the first occurrence of each term defined in the glossary). It is optional; you decide whether to include one according to the expertise of the audience and the complexity of the content.

Sometimes the glossary is placed with front matter if the primary reader is not likely to know the terms. If the report has only a few unfamiliar terms or concepts to define, information notes can be used instead of a glossary.

Information notes (usually numbered) are most convenient if placed at the bottom of the page on which the reference appears; occasionally, however, these notes are combined with the end notes.

The list of ***sources cited,*** or source notes, is the final section of the end matter. Complete citation information enables readers both to check where you obtained your information and to locate sources if they need to follow up a particular lead.

Design Elements

A writer of a formal report needs to be attentive to several aspects of design:

- headings and subheadings
- pagination
- figures and tables
- accessibility, comprehensibility, and usability

Headings and subheadings should match in typography, placement on page, and wording those listed in the table of contents. These headings and subheadings do more than provide visual breaks for the reader; they also act as cohesive devices, identifying the movement from one topic to the next.

By convention, most reports identify the first page of the body as page 1 of the report, continuing sequentially until the final page of the end matter. The front matter of the report is usually numbered with lowercase roman numerals (i, ii, iii, and so on).

Visual material (figures, tables, and so on) should generally be incorporated into the report, closely following their textual reference. Incorporating visuals makes more work for the writer in designing the layout of each page but benefits the reader, who can immediately refer to the appropriate figure or table without turning to an appendix.

Finally, effective design can be used to make information more accessible, comprehensible and usable to readers. Figure 20.9 shows an original paragraph from the USACE report that you read excerpts from earlier and then two alternative ways that the same information could be presented: (1) a bulleted list with each alternative italicized at the beginning of the opening sentence to provide a label; or (2) an embedded table with the grid lines suppressed that puts the three alternatives in one column and the explanations in a second column.

Each revision takes only slightly more space than the original, which is more than compensated by the increase in accessibility.

REPORTS, TRANSLATION, AND CULTURE

© Jerry Kobalenka/Getty Images

Consider the complexities of translating a technical report written in Ukrainian into English. This particular translation was used as background information in preparing a report in English about the management of the Ukraine Agricultural Policy Project by the Institute for Policy Reform and Iowa State University.

The Ukranian national symbol depicts the founding of the capital of Ukraine, Kyiv, in 482, by a Slavic prince, Kyy, and his younger brothers and sister.

The source text contained numerous instances of legislative, economic, and agricultural terminology. The project was 23 pages long in both the original and the translation (which is not very common; typically Ukrainian versions are longer than English versions).

The document was translated on a lexical level, but that wasn't sufficient to fairly represent the document or to meet the needs of the audience. Therefore, the translator also made the following changes:

- Added translator's notes that explain the need for structural and document design changes
- Altered the sequence of information at the beginning of the document
- Added headings and more white space
- Presented some textual information in tables
- Made some transitions more obvious

Why? Ukrainians tend to be "reader-responsible" writers — that is, they assume that readers bear a large responsibility for interpreting information they read. So, for example, transitions are rarely provided in the source text, and those transitions that are present are so subtle and idiosyncratic that they are almost invisible. The assumption is that Ukrainians readers will easily and naturally make the links that create coherence. In contrast, English readers expect transitions to help establish coherence. (See Chapter 2 for more detailed discussion about cultural expectations.)

© Sean Sprague/The Image Works

The Pochayiv Monastery in Ternopil, Ukraine, dates to 1527, founded by monks who fled the Tartar invasion in Kyiv in 1240. About 70 monks currently reside in the monastery, which exemplifies Western Ukranian Baroque architecture.

Courtesy of Oskana Hylva

Oskana Hylva was educated in Ukraine, the UK, and the United States. She has worked as a technical translator on a number of projects, teaches technical communication, and participates as a researcher on a university-wide communication-across-the-curriculum initiative.

ПРО ПІДСУМКИ ВИКОНАННЯ УКАЗУ ПР

від 3 грудня 1999 року
"Про невідкладні заходи щод
реформування аграрного сектор

В економіці України аграрний сектор залиша
ланкою, яка значною мірою визначає соціал
суспільства та продовольчу безпеку держави.

Тут зайнята п'ята частина працюючих у
діяльності, зосереджена майже чверть виробничих
частина продовольчих ресурсів та забезпечуєть
роздрібного товарообігу.

Існування, соціальне благополуччя і здоров'я
пов'язані із землею. Земельні ресурси, на використ
обсягу продовольчого фонду та 2/3 фонду това
вважається первинним фактором виробництва,
України. Частка земельних ресурсів у складі п
становить 40 відсотків (виробничі та оборотні засоб
40%).

Власне земельна площа (суша) України
(загальна – 60354,8 тис.га); її сільськогосподарськ
(41829,5 тис.га); частка ріллі в загальній площі с
сягає 79% (32669,9 тис.га).

Земельний фонд України характеризується н
потенціалу, в його структурі переважають землі з р
база землеробства країни розміщується на грунт
експертними оцінками, при раціональній структ
відповідному науковому та ресурсному забезпеченн
продуктів харчування на 145-150 млн. чоловік.

У державній власності знаходиться 50,9% з
приватній власності знаходиться 49,1% при наявно
власності та користувачів.

Ефективне використання наявного приро
потенціалу можливе лише за умов швидкого та р
сільського господарства на засадах приватної власно

On Immediate Measures to Accelerate Reforms in Agricultural Sector of the Economy

Presidential Decree N 1529/99 of December 3, 1999

Summary of the decree implementation

[Translator's note: Most headings and introductory transitions have been added and some facts have been put in the tabular format to enhance the readability of the document in English. Transitional, definitional, and explanatory information provided by the translator has been placed inside [xxxxx] to differentiate it from the translated text.]

Background

[This background section includes three sections that help orient readers about the overall issues addressed in this document. First, general demographic facts precede the problem statement. The problem is then situated historically, before a summary of relevant legislative acts and developments.]

Fast facts and problem statement

The total area of Ukraine is 60,354.8 thousand hectares (57,939.8 thousand hectares of landmass). [1 hectare equals 2.4711 acres]. Agricultural land constitutes 72.2 % of the landmass area or 41,829.5 thousand hectares. Seventy-nine percent (32,669.9 thousand hectares) of all agricultural land is arable.

Currently, with 21 million landowners and land users, 50.9% of the land is state owned and 49.1% is privately owned.

The agricultural sector is a critical aspect in Ukraine's economy and still considerably determines its social and economic well being. Specifically, the agricultural sector employs one-fifth of the Ukraine's workforce, possesses one-fourth of its manufacturing, is a major food producer, and supplies nearly three-fourths of the retail commodity circulation.

The social and material well being of the Ukrainian people are and have always been dependent on the land. Ninety-five percent of all food production and 66.7% of commodity manufacturing come from land. Land resources constitute 40% of all productive forces. It is no surprise, therefore, that agriculture is considered to be the primary factor of Ukraine's economy.

Ukraine's agricultural land has a significant bio-productive potential due to the

сільській місцевості - біля 11 млн. особистих підсобних госпо
свої земельні наділи.

У ході виконання Указу Президента України структ
сільськогосподарських підприємств на початку 2000 року зазн
Так, якщо за станом на 1 грудня 1999 року найбільш по
аграрних підприємств були колективні сільськогосподарс
(КСП) — 64%, то на початок квітня 2000 року формувань та
форми практично не залишилося (мал. 1).

Станом на 1.09.2000 проведено реорганізацію практи
колективних сільськогосподарських підприємств.

На їх базі, відповідно до вимог чинного законодавства
13723 нових сільськогосподарських підприємств, серед них з
селянських (фермерських) господарств (8%), 2840 прив
орендних) підприємств (20%), 6402 господарських товари
сільськогосподарських кооперативи (24%). Зазначені підприє
засадах приватної власності та оренди землі і майна (мал. 6 – 6

За період виконання Указу Президента України част
товариств у структурі аграрних підприємств збільшилас
сільськогосподарських кооперативів — з 2 до 24%, приватних
4 до 20% (мал. 2).

Найбільша кількість підприємств - 6261 (47,9%) засно
від 11 до 20 засновників мають 1341 підприємство (10,3%), в
707 підприємств (5,4%), від 101 до 500 - 1044 підприємст
до1000 - 318 (2,4%) і більше 1000 осіб - 56 підприємст
підприємств (3318) засновано однією особою, як правило за
працездатного віку і лише 2,7% - мають вік більше 60 років. С
засновників 227 {6,8%) жінок очолюють новостворені підприє

У новостворених підприємствах чисельність прац
2163,6 тис. осіб, з яких 9% складають спеціалісти - агроно
інженери, економісти, бухгалтери, ветлікарі та ін.

Земельні площі колишніх КСП зберегли єдиним
підприємств (56,9%), а 10378 (79,5%) - стали користува
майнового комплексу (або частини) колективних сільгосппідп

На початок липня 2000 року кількість фермерськи
Україні складала 37,1 тис. юридичних осіб (з 1.10.1999 р. зб
на 2000 господарств). Збільшення кількості фермерсь
зумовлена процесом реформування колективних сільс
підприємств, так як до 1999 року відмічалась певна тенден
кількості фермерських господарств.

Майже удвічі зросла частка землі, яка використовуєт
загальній площі сільськогосподарських угідь країни ф
займають 4,7% проти 2,7% минулого року. В цілому у корис
знаходиться 1982,8 тис.га сільськогосподарських угідь, з ни
Відбувається укрупнення фермерських господарств. В сер

collectively owned ones, specifically, 64%; at the beginning of April 2000, practically all of them disappeared (figure 1).

As of September 1, 2000, practically all 10,833 collectively owned agricultural enterprises were reformed into new-type agricultural enterprises. Specifically, 13,723 of the latter were registered. [The following table presents types, numbers, and percentages of agricultural enterprises in Ukraine.]

Type of agricultural enterprises	Number	Percentage (%)
farms	1,030	8
private leasing enterprises	2,840	20
agricultural partnerships	6,402	47
agricultural cooperatives	3,312	24

In Figure 6-6.1, the black section of the pie chart represents private enterprises based on leasing.

As a result of the Presidential Decree, significantly increased agricultural partnerships—from 14% to 47%; agricultural cooperatives —from 2%-24%; and private enterprises—from 4% to 20% (figure 2). [The following table presents numbers and percentage of enterprises owned by specific numbers of people.]

Number of owners	Number of enterprises	Percentage (%)
1	3,318	25.0
2 - 10	6,261	47.9
11 - 20	1,341	10.3
21 - 100	707	5.4
101 - 500	1,044	8.0
501 - 1000	318	2.4
More than 1000	56	0.4

In terms of demographics of single-individual owners, only 2.7% are at the retirement age and over [currently, in Ukraine the retirement age for men is 60 and for women 55], and only 6.8% (227) are women.

Out of 2163.6 thousand people employed in these enterprises, 9% are agronomists, livestock experts, engineers, economists, accountants, and veterinarians.

7,434 enterprises (56.9%) use the land that was once collectively owned by agricultural enterprises; 10,378 enterprises (79.5%) use property that was once collectively owned by agricultural enterprises.

Translator's notes in brackets explain the structural and document design changes.

The sequence of information at the beginning of the document has been changed to conform to Western expectations.

Headings and white space are more of a U.S. convention than a Ukranian convention.

Some of the transitions are more obvious in the English translation: currently, specifically, therefore.

Paragraphs of textual information have been converted into tables. For most U.S. readers, the tabular presentation is easier to comprehend.

The new tables — designed by the translator — needed to have column labels created. The two versions show the cultural difference in commas and decimal points in numerals: 5,4% in Ukraine = 5.4% in English

FIGURE 20.9 Redesign Information for Accessibility

ORIGINAL

2.2. *Chute location.* Two chute locations, and a total of three alternatives, were considered. Alternative #1 would follow the general channel of a chute used in the past. The inlet to the chute would be located at RM 713.5 and the outlet located at RM 711.2. Following the former chute flow path, the length of flow through the off-channel chute would be approximately 10,700 feet. Along the flow path a culvert crossing was inserted into the model at 8773 ft from the chute outlet. This correlates with the location of an existing road crossing. Alternative #2 was considered as an alternative with a shorter flow length, starting at RM 712.64 and ending at the same location as Alternative #1, for a total flow length of 7,600 ft. HEC-RAS analysis was performed to evaluate the stability and flow velocities of multiple chute designs. The third option included both Alternative #1 and a backwater area just upstream of the outlet of the chute, considered as Alternative #3. Geometry for this option was not created as water stored in the backwater area would be considered ineffective flow area and would not cause any significant variation in model results from Alternative #1 alone.

POSSIBLE REDESIGN

2.2. *Chute location.* Two chute locations, and a total of three alternatives, were considered.

- *Alternative #1* followed the general channel of a chute used in the past. The inlet to the chute would be located at RM 713.5 and the outlet located at RM 711.2. Following the former chute flow path, the length of flow through the off-channel chute would be approximately 10,700 feet. Along the flow path a culvert crossing was inserted into the model at 8773 ft from the chute outlet. This correlates with the location of an existing road crossing.
- *Alternative #2* had a shorter flow length, starting at RM 712.64 and ending at the same location as Alternative #1, for a total flow length of 7,600 ft. HEC-RAS analysis was performed to evaluate the stability and flow velocities of multiple chute designs.
- *Alternative #3* included both Alternative #1 and a backwater area just upstream of the outlet of the chute. Geometry for this option was not created as water stored in the backwater area would be considered ineffective flow area and would not cause any significant variation in model results from Alternative #1 alone.

POSSIBLE REDESIGN

2.2. *Chute location.* Two chute locations, and a total of three alternatives, were considered.

Alternative #1	This alternative followed the general channel of a chute used in the past. The inlet to the chute would be located at RM 713.5 and the outlet located at RM 711.2. Following the former chute flow path, the length of flow through the off-channel chute would be approximately 10,700 feet. Along the flow path a culvert crossing was inserted into the model at 8773 ft from the chute outlet. This correlates with the location of an existing road crossing.
Alternative #2	This alternative had a shorter flow length, starting at RM 712.64 and ending at the same location as Alternative #1, for a total flow length of 7,600 ft. HEC-RAS analysis was performed to evaluate the stability and flow velocities of multiple chute designs.
Alternative #3	This alternative included both Alternative #1 and a backwater area just upstream of the outlet of the chute. Geometry for this option was not created as water stored in the backwater area would be considered ineffective flow area and would not cause any significant variation in model results from Alternative #1 alone.

Why might the authors of the original USACE report not have used a design that makes the information more accessible?

Machine translation enjoys periodic popularity as a solution to international communication. For a link to explore issues related to machine translation, go to www.english.wadsworth.com/burnett6e.

CLICK ON WEBLINK

CLICK on Chapter 20/machine translation

WEBLINK

Examining a Sample Report

Examine the following technical report written by an eight-member team of engineering students. You'll notice that it's not very long. Workplace professionals are sometimes expected to produce very detailed and lengthy reports (the excerpt from the battery site report in Chapter 1 is an example from a series of reports that were typically longer than 30 pages). More frequently, however, you'll be expected to write relatively short, well-organized reports for experts who will be interested in the information you present.

Figures 20.10a–n present a collaborative recommendation report written by a student team in a multidisciplinary design class at the College of Engineering at Iowa State University. The class worked on a term-long project to develop a digital controller for Cottrell Technologies, Inc. During the project, the class and the company communicated frequently to ensure that the recommendations in the final report met the company's design, financial, and quality requirements.

Cottrell Technologies is a mid-sized production company that manufactures hardware for retail and industrial organizations. For example, they manufacture garment storage racks and conveyor systems for retail dry cleaners. The conveyor systems they have sold during the last two decades have failed to attract new customers. To become competitive in this market again, the company decided to manufacture a new conveyor system that includes state-of-the-art technology. After making arrangements with the university, they turned the project over to the multidisciplinary engineering design class and challenged the team of students to design a new system. It would be sold to new customers and marketed to past customers as a system upgrade.

The design team elected an electrical engineering major, Ned F. Hitch, to serve as the liaison with Cottrell Technologies. In this role, he was to present the team's recommendation for an input interface to be used with the controller system. The analytical report she coordinated was based on the class's investigation and comparison of input interface devices.

The design team members completed their preliminary design for the basic conveyor system and started work on the subsystems. Hitch took the lead during the input interface phase of the project, researching available technologies and analyzing them with her teammates. The team identified two interface devices that would work with the conveyor system.

The team was most impressed with the bar code scanning interface but noticed that it would be more expensive than the keypad units they were also considering. Unsure whether the price difference was an important factor for Cottrell Technologies, the team presented its findings in a recommendation report to the company.

Hitch coordinated both the draft and final versions of the report. Other team members contributed information for the analysis and reviewed the document. Several people on the team, including Hitch, were enrolled in a technical communication course at the same time as they were working on this project. As a team, they agreed that their report should reflect the high standards that their client would expect from workplace professionals.

The *penultimate,* or next-to-final, drafts of the executive summary (Figure 20.10d) and the body of the report (Figures 20.10f, h, k, m) have been edited by Hitch. In these penultimate drafts, she not only makes minor grammatical and mechanical corrections and runs the spell checker, but she also sees places that need foregrounding (previewing), additional clarifying information, or some reordering of information.

The final report begins with a letter of transmittal (Figure 20.10a), followed by a title page (Figure 20.10b), a table of contents (Figure 20.10c), an executive summary (Figure 20.11e), and the body of the report (Figures 20.10g, i, j, l, n). As the liaison between the team and the company, Hitch must make a formal presentation of the report's recommendations to Cottrell Technologies. (Also see Chapter 17, Figure 17.3 and Figure 17.6, to review the team's outline and overhead transparencies accompanying the presentation to Cottrell Technologies about their recommendations.)

In comparing the draft and final versions of this report, a number of revisions are obvious. The ability to assess the qualitative differences in these revisions helps distinguish skillful communicators from those less skillful.

- ***Identify the content and context.*** In preparing the final version of the report, the writers provide readers with more information right from the beginning. For example, they revise the executive summary to more accurately reflect the report's content and organization.
- ***Anticipate reader's needs.*** In preparing the final version, the writers anticipate readers' needs in a number of ways. For example, they present general information first, followed by specific information; they make sure to use terminology consistently; and they discuss the two alternatives in the same order for each criterion so that readers can make comparisons more easily.

FIGURE 20.10A **Final Version of Transmittal Letter Accompanying Recommendation Report**

Multidisciplinary Engineering Design
239 Zachry Engineering Hall
Iowa State University
Ames, IA 50011-1201

October 11, 20—

Ms. Danielle Conner
Cottrell Technologies, Inc.
1401 Highland Heights
Cedar Rapids, IA 52046

Dear Ms. Conner:

The multidisciplinary engineering design team at ISU is hard at work on the digital controller project we discussed at the beginning of the term. Before we may proceed much further, however, a decision must be made. We are considering two alternative input interfaces for the controller unit: a bar code scanner and a telephone-style keypad.

This report analyzes the two alternatives according to three criteria that you have suggested are important to Cottrell Technologies: compatibility with the existing Techni-Veyor, cost, and relative ease of operation. The input alternatives are described, analyzed, and compared based on these criteria; one is recommended for your approval.

Should you have any questions, please contact me. I look forward to hearing your opinion about our team's recommendation, so we may move forward with the next phase of the project.

Sincerely,

Ned F. Hitch

Ned F. Hitch
Multidisciplinary Engineering Design
515-555-3728
nhitch@iastate.edu

- ***Establish connections.*** In preparing the final version, the writers create clear links between ideas; thus, the information is more understandable. For example, they help readers by presenting parallel amounts of information about different interfaces in similar ways, so readers give these points similar consideration.

FIGURE 20.10B Final Version of Title Page for Recommendation Report

Analysis of Input Interfaces
for the
Cottrell Technologies, Inc.
Techni-Veyor Controller

Submitted to

Danielle Conner
Cottrell Technologies, Inc.
1401 Highland Heights
Cedar Rapids, Iowa 52046

Prepared by

Multidisciplinary Engineering Design Team
Iowa State University
Ames, IA 50011-1201

October 11, 20—

- ***Use effective design and visuals.*** In preparing the final version, the writers improve the design. For example, they more clearly chunk information, emphasize key points with design elements, use type choices and spacing consistently, and make information easier to locate by using elements such as headings and bullets.

FIGURE 20.10C **Final Version of Table of Contents for Recommendation Report**

TABLE OF CONTENTS

- ***Make the text accessible.*** In preparing the final version, the writers make the information more accessible by adding introductory paragraphs that preview upcoming information. For example, they add a preview paragraph in the Compatibility section to explain the importance of the criterion.

FIGURE 20.10D Draft of Executive Summary of Recommendation Report

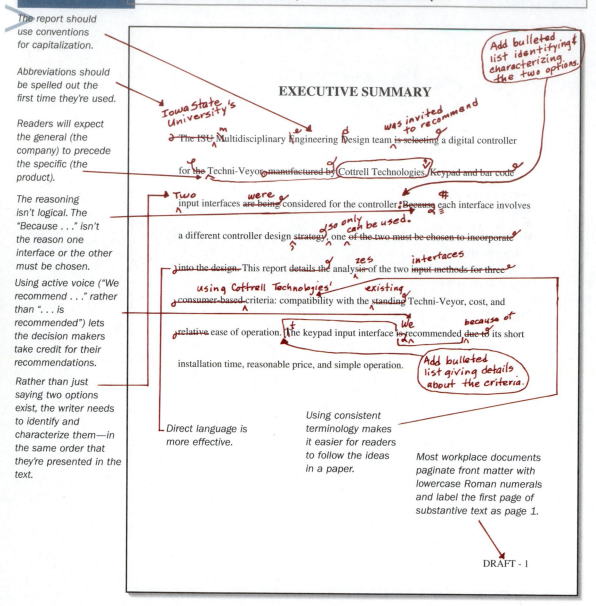

The report should use conventions for capitalization.

Abbreviations should be spelled out the first time they're used.

Readers will expect the general (the company) to precede the specific (the product).

The reasoning isn't logical. The "Because . . ." isn't the reason one interface or the other must be chosen.

Using active voice ("We recommend . . ." rather than ". . . is recommended") lets the decision makers take credit for their recommendations.

Rather than just saying two options exist, the writer needs to identify and characterize them—in the same order that they're presented in the text.

EXECUTIVE SUMMARY

The ISU Multidisciplinary Engineering Design team is selecting a digital controller for the Techni-Veyor manufactured by Cottrell Technologies. Keypad and bar code input interfaces are being considered for the controller. Because each interface involves a different controller design strategy, one of the two must be chosen to incorporate into the design. This report details the analysis of the two input methods for three consumer-based criteria: compatibility with the standing Techni-Veyor, cost, and relative ease of operation. The keypad input interface is recommended due to its short installation time, reasonable price, and simple operation.

DRAFT - 1

Direct language is more effective.

Using consistent terminology makes it easier for readers to follow the ideas in a paper.

Most workplace documents paginate front matter with lowercase Roman numerals and label the first page of substantive text as page 1.

- ***Reflect professional standards.*** In preparing the final version, the writers conform to conventions for style and usage. For example, they generally use active voice, select appropriate diction; adhere to conventions for capitalization and abbreviations, and eliminate problems with dangling modifiers and agreement.

FIGURE 20.10E **Final Version of Executive Summary of Recommendation Report**

EXECUTIVE SUMMARY

Iowa State University's multidisciplinary engineering design team was invited to recommend a digital controller for Cottrell Technologies' Techni-Veyor. Two input interfaces were considered for the controller:

- The **bar code interface** is similar to scanners used in video rental stores, automatically recording item when the scanning bar is passed over the appropriate code.
- The **keypad** interface also records the type of item, but its memory functions are performed manually rather than automatically with a scanner.

Each input interface involves a different controller design, so only one can be used. This report analyzes the two interfaces using Cottrell Technologies' criteria: compatibility with the existing Techni-Veyor systems, cost, and ease of operation.

- **Compatibility.** The bar code interface is not compatible with the systems currently used by Cottrell Technologies. The bar code interface could be installed only after modifying current systems—a process that could take several hours. The keypad interface is more readily compatible with the existing systems; installation would take less than an hour.
- **Cost.** Both bar code and keypad interfaces are widely available; however, bar code interfaces typically cost at least five times more than keypad interfaces.
- **Operation.** Either controller is an improvement over the current method for retrieving garments. Neither system would require lengthy training of operators. The bar code scanner has a slight input time advantage and reduces the possibility of operator error.

We recommend the keypad interface because of its short installation time, reasonable price, and simple operation.

page ii

The revised executive summary is easier to read because the two options are emphasized in several ways:

- *using each to signal each interface*
- *boldfacing the key terms*
- *indenting the short list*

Identifying the three criteria in a bulleted list helps readers anticipate both the organization of the report and the major point that will be made.

When you finish reading the draft and final versions of the report for Cottrell Technologies, you can use these guidelines to compare changes made in the final pages. You need experience in applying these guidelines, both to become more skillful in revising your own documents as well as to become a more constructive colleague when coworkers ask you to review their documents.

FIGURE 20.10F Draft of Recommendation Report

*The Use of **BOLD ALL CAPS** makes the long title and level-1 headings difficult to read. **Bold Small Caps** would be more effective.*

Readers can be distracted or misled by dangling modifiers.

Active voice is generally more effective unless the writer has a specific reason to use passive voice.

Readers are less confused if you use the same term consistently to refer to the same object ("interface" throughout rather than "system," "method," or "unit").

An additional level-2 heading would help readers.

Agreement errors can distract or confuse readers: "patron" is singular; "their" is plural.

ANALYSIS OF INPUT INTERFACES FOR THE COTTRELL TECHNOLOGIES, INC. TECHNI-VEYOR CONTROLLER

The ISU Multidisciplinary Engineering Design team has recommended a digital controller for Cottrell Technologies. The controller will be installed to work with Techni-Veyor, a dry cleaning conveyor system manufactured by Cottrell Technologies and used by cleaning companies nationwide. As part of the controller selection process, two popular input interfaces are being considered: a bar code method and a digital keypad method. This report analyzes both input methods and recommends one for your approval.

The two input interfaces are analyzed for three different criteria: compatibility with the current conveyor, cost, and relative ease of operation. Data is drawn from the most recent *Thomas Register* and from interviews with both Danielle Conner and a customer currently using the Techni-Veyor system.

ALTERNATIVE OVERVIEW

The two input interfaces in consideration are a bar code scanner and a telephone-style keypad. The bar code scanner is similar to those in many video rental stores. When a garment is brought in, it is assigned a bar code identification tag. The bar code is scanned into the computer's memory and assigned a patron's name. After the garment is cleaned, its bar code is scanned into the computer along with its position on the conveyor. Just as a video tape is assigned to a patron, each garment and respective location is assigned to a specific patron. For every patron, the computer retains their garment bar code and coded position on the conveyor. When the patron returns, their name is entered into the computer and their garment is automatically retrieved by the digital controller.

DRAFT - 2

FIGURE 20.10G **Final Version of Recommendation Report**

ANALYSIS OF INPUT INTERFACES FOR THE COTTRELL TECHNOLOGIES, INC. TECHNI-VEYOR CONTROLLER

The Iowa State University multidisciplinary engineering design team has recommended a digital controller for Cottrell Technologies. The controller will be installed to work with Techni-Veyor, a dry cleaning conveyor system manufactured by Cottrell Technologies and used by dry cleaning companies nationwide. As part of the controller selection process, our team considered two popular input interfaces: a bar code interface and a digital keypad interface. We analyzed both input interfaces and recommend one for your approval.

Our team used three criteria to analyze the two input interfaces: compatibility with the current conveyor, cost, and relative ease of operation. We took our data from the most recent *Thomas Register* and from interviews with both Danielle Conner and a Cottrell Technologies customer currently using the Techni-Veyor system.

This report presents the following information:

- an overview of the two alternatives: bar code interface and keypad interface
- an analysis of alternatives according to three criteria: compatibility, cost, and operation
- conclusions: cost and down time are critical in selecting appropriate option
- recommendation: keypad interface is better option

page 1

Using first person ("our team," "we") emphasizes the team's analysis and decision making. The people who made the recommendations are acknowledged. In addition, the audience of the report is identified.

Readers need information that identifies criteria and sources of data.

Foregrounding the sequence of sections in a preview paragraph lets readers know what's coming in the report.

FIGURE 20.10H Draft of Recommendation Report

An additional level-2 heading would help readers.

Agreement errors can distract or confuse readers: "patron" is singular; "their" is plural.

The spacing between the headings and the text must be adjusted. Here the spacing above and below each heading is the same, so the headings look as if they're floating equidistant between the chunks of text. The headings should be chunked with the appropriate text by putting more space above than below the heading.

A preview paragraph in the Compatibility section will convey the importance of this criterion.

Adding Manfield's title clarifies her position and contributes to the authority of her opinion.

"Therefore" is unnecessary because a cause-effect relationship doesn't exist.

Add level 2 heading — Keypad Interface

The keypad interface works like the bar code scanner, except [activating] its memory functions are [is] performed manually with a card file. When a garment is brought in, it is assigned a four-digit number that is recorded by hand with the patron's name in a card file. After the garment is cleaned, its position number on the conveyor is written on the appropriate patron's card and once again filed. When the customer returns, their card is taken from the file and their garment location number entered into the computer with a telephone-style keypad. The digital controller then turns the conveyor and retrieves the patron's garment.

ANALYSIS [of ALTERNATIVES]

The alternatives were analyzed for three criteria considered essential by Cottrell Technologies: the compatibility of the controller with the existing Techni-Veyor system, the cost of the unit and system modifications, and the relative ease of operation.

Present in a bulleted list.

Compatibility

Add ¶ about importance of compatibility.

Bar Code Input

The compatibility of the bar code system involves a lengthy installation period. [Add a reason] The Techni-Veyor rack is labeled with four-digit numbers above each slot. Installation of the bar code system requires labeling the entire rack with bar code decals. Depending on the size of the rack, this takes from a couple of hours to a day.

Charlotte Manfield [general manager] of Fairfield Cleaners in Ames, said a downtime could cost her considerable business. Ms. Manfield emphasized the importance of maintaining an efficient garment-retrieval system of some sort in the event of a mechanical problem, breakdown, or system upgrade. Therefore, some advance planning would help Ms. Manfield [(and other customers with similar large dry cleaning businesses)] to avoid delays and customer dissatisfaction due to [time delays caused by] the installation of a bar code controller system.

DRAFT - 3

FIGURE 20.10I **Final Version of Recommendation Report**

OVERVIEW OF ALTERNATIVES

The two input interfaces our team considered were a bar code scanner and a telephone-style keypad.

Bar Code Interface

The bar code scanner for use with the Techni-Veyor is similar to those used to control inventory in many video rental stores. When a garment is brought into a dry cleaners, it is assigned a bar code identification tag. The bar code is scanned into the computer's memory and assigned the patron's name. After the garment is cleaned, its bar code is scanned into the computer along with its numbered position on the conveyor. Just as a videotape is assigned to a patron, each garment and respective location is assigned to a specific patron. For every patron, the computer retains the garment bar code and coded position on the conveyor. When the patron returns, the name is entered into the computer, and the garment is automatically retrieved using the digital controller.

Keypad Interface

The keypad interface works like the bar code scanner, except activating its memory functions is performed manually with a card file. When a garment is brought in, it is assigned a four-digit number that is recorded by hand with the patron's name in a card file. After the garment is cleaned, its position number on the conveyor is written on the appropriate patron's card and once again filed. When the customer returns, the card is taken from the file and the garment location number entered into the computer with a telephone-style keypad. The digital controller then turns the conveyor and retrieves the patron's garment.

ANALYSIS OF ALTERNATIVES

The alternatives were analyzed for three criteria considered essential by Cottrell Technologies:

- the compatibility of the controller with the existing Techni-Veyor system,
- the cost of the unit and system modifications, and
- the relative ease of operation.

page 2

Headings and subheadings help readers preview and review the major chunks of information.

The reference to video stores reminds readers that they are probably already familiar with bar code scanners.

The operation of two different interfaces (the bar code scanner and keypad) is presented in similar ways so readers give the interfaces equal consideration.

The foregrounding of the analysis section of the report lets readers know what to anticipate.

FIGURE 20.10J Final Version of Recommendation Report

Readers need to know why each criterion is important.

The two alternatives should be discussed in the same order — first the bar code interface and then the keypad interface — for each criterion.

Compatibility

Because many customers already have a Techni-Veyor system in place, compatibility between the existing system and the new input controller is an important consideration. The new controller should be easily adaptable to customers' existing hardware. The bar code and keypad controllers differ in their compatibility with Techni-Veyor hardware.

Bar code interface. The bar code interface involves a lengthy installation period because of problems with compatibility. The Techni-Veyor rack is labeled with four-digit numbers above each garment slot. Installation of the bar code system requires labeling the entire rack with bar code decals. Depending on the size of the rack, this takes from a couple of hours to a day.

Charlotte Manfield, general manager of Fairfield Cleaners in Ames, said that the installation downtime could cost her considerable business. Ms. Manfield emphasized the importance of maintaining an efficient garment-retrieval system of some sort in the event of a mechanical problem, breakdown, or system upgrade. Some advance planning would help Ms. Manfield (and other customers with similar large dry cleaning businesses) to avoid delays and customer dissatisfaction due to time delays caused by the installation of a bar code controller system.

Keypad interface. The compatibility of the keypad interface with an existing Techni-Veyor is very good. The only installation time involves plugging the controller into the conveyor and calibrating it for the size of the rack. Installation time is no more than an hour, including learning to use the controller in the process of setting it up. The Techni-Veyor rack is already numbered for use with the keypad interface and requires only the patience of the operator in becoming familiar with the digital controller.

page 3

Individual and Collaborative Asignments

1. **Evaluate two reports.** Locate a report that has been approved and one that has been disapproved. Using what you have learned in this chapter and your own experience, develop a rubric to evaluate the effectiveness of reports. Review both reports and fill the rubric in accordingly. Discuss why you think one was approved and the other not.
2. **Write a series of reports.** Imagine that you would like your company to fund your travel to a professional conference in your field of study. Write the appropriate reports generated by such a trip: request for authorization to go, request for expense money, report of the trip, expense report, and a recommendation that your company adopt a procedure or product that you learned about.
3. **Conduct research for a report.** The local office of the federal Occupational Safety and Health Agency has requested a report of your plant's accommodations, or lack of them, for disabled workers. What reports would you need to obtain, and what would you need to write? (You are not preparing the actual documents; rather, you are explaining the probable source, type, audience, and context.)
4. **Revise a report.** Work with a small group to complete the following tasks based on excerpts from an analytical report below.
   (a) Read through the paragraphs and, with your group, decide a possible purpose and audience for the report from which these paragraphs are taken.
   (b) Decide the sequence of the paragraphs so that they form a logical, coherent discussion. Have reasons for your decisions.
   (c) Create headings and subheadings for this section of the report so that readers can easily follow the discussion. Have reasons for your decisions.
   (d) Compare and discuss your group's decisions and reasons with those of other groups in the class.[15]

> 1. Perhaps the most appealing aspect of developing this site is the availability of existing infrastructure. All major utilities are immediately available: electric, gas, water, sanitary sewer, storm sewer, telephone, and fire hydrant lines. All of these services are currently available on the site or are on the border of the site. All road surfaces are in moderately good-to-excellent shape, and Mortensen Road is being newly paved westbound from State Avenue. Mortensen Road, eastbound from the site, is already a 4-lane roadway leading to an exit ramp for Highway 30.
> 2. As noted in Perceptual Characteristics of the Site Analysis, the site is generally flat. More visual diversity would aid in the creation of a development concept and general excitement for what could be created. All other unique features are human-made structures that are intended to be torn down or relocated.

3. This site is located at the intersection of two major arterials, State Avenue and Mortensen Road. Additionally, reasonable and ready access is available to Highway 30 and Lincoln Way via Mortensen Road and State Avenue. With these three arterial roadways and access to Interstate 35, access to the site and the area is excellent. Appendix 5 shows an area map and traffic counts to give a feel for the area.
4. The close proximity to infrastructure is arguably the most compelling reason for development of this site. All current University facilities are easily relocated. The negative aspects of site development are readily overcome. Another compelling reason to develop is the need for affordable housing. With a proper combination of housing types, a site of this size and location could produce hundreds of safe, inexpensive, easily accessible, and well-planned homes.
5. This site has three major negative aspects of development. The first is its lack of visual diversity. Second, ownership of this property and the current lease status could create significant problems in terms of delays and stoppages. Finally, the existence of pollutants could cause limited but significant problems.
6. The residue of agricultural chemicals may or may not cause short-term problems. The primary concerns for ongoing pollution problems are the compost facility run by the university and noise pollution from the college dormitories and Highway 30.
7. This site is buildable for more than its physical characteristics. On the physical side, the topography is generally flat, and soils are favorable for or adequate for development. But other benefits immediately present themselves. It is a negative that there is a lack of visual diversity, but that also means there is less to do to prepare the site for construction. The vegetation is light to non-existent, wildlife is no factor, and other animals will relocate with the movement of the existing facilities. Existing facilities have already discussed moving or have made plans to relocate. The site is high above the flood way and has moderate-to-excellent drainage. Overall neighboring and existing land use is not an encumbrance to most if not all development plans.
8. This site has three major positive aspects of development. First is its excellent transportation access. Second is its favorable developmental characteristics. And third is its excellent access to existing infrastructure.
9. The University is the current owner of the site. Aside from the lengthy approval process for any sale involving a government agency of this type, there is also a 99-year lease that is currently only 17 years old. This lease is with the USDA. Potential delays should be anticipated when dealing with these agencies.

5. **Write a formal report.** Identify a problem that needs to be solved and prepare a formal report that supports the most appropriate solution. Subjects may come from your personal or community life, your academic work, or your professional work. The list of verbs in the left-hand column may help you decide how to approach a report. Some possible topics are listed in the right-hand column. Try matching the verbs with several of the topics to see how changing the action may change how you think about approaching the report.

Actions to Consider	***Topics for Inquiry***
analyze	agriculture on marginal lands
approve	artificial insemination
change	cable television

compare	capital expenditure for equipment
eliminate	corporate daycare facility
establish	corporate physical fitness program
initiate	drainage problem
install	energy alternatives
institute	food co-op
investigate	insulation alternatives
justify	machine or program X versus machine or program Y
organize	management system for department, warehouse
recommend	manufacturing, testing, marketing procedure
reorganize	over-the-counter cold remedies
synthesize	program to eliminate electrostatic discharge
verify	solar hot-water heater
	telemarketing program
	test and immunization program
	toxic waste dump in your community
	water quality in local lake, river, reservoir
	wetlands disruption

6. **Resolve a dilemma.** Imagine this scenario:

   You are working on an internal recommendation report at a company that manufactures plastics. The report recommends a new mixing process that will solve a product strength problem. In the process of writing the report, you discover that post-mix waste is being discarded into a local water basin. Although tests reveal that the current level of contaminants in the water basin is safe, you are concerned that accumulation over decades could have negative effects on the water's potability. The report, though, has no section designed to deal with waste products; the report was commissioned to deal with a single technical problem, strength.

   Work with a small group to discuss the following questions and decide on a group position that you then report to the class:

   - Do you include a section in the report about waste?
   - Are you responsible for looking at long-term environmental effects?
   - Would you include a section on waste disposal in the report described above?
   - What are the potential consequences if you do? If you don't?
   - What factors will influence your decision about including report sections that are not required (and may even place the organization in a bad light)?

Chapter 20 Endnotes

1 Statistics for the incidence and death rates for anorexia (and other eating disorders) are controversial. The example in the Ethics Sidebar is for the purpose of provoking discussion about ways in which statistics are derived and used. The statistics in this Ethics Sidebar can be derived from these references: (1) National Institute of Mental Health. (2003). Deaths from anorexia. *The numbers count: NIMH.* Retrieved January 26, 2004, from http://www.wrongdiagnosis.com/artic/the_numbers_count_nimh.htm and (2) O'Neill, B. (1996). Anorexia statistics. Retrieved January 26, 2004, from http://www.polisci.ucla.edu/faculty/boneill/anorexia.html

2 Best, J. (2001). *Damned lies and statistics.* Berkeley, CA: University of California Press.

3 Buchanan, G. (n.d.). FIP packaging. *Technical Writing, 42,* 225. University of Lowell.

4 US Army Corps of Engineers. (n.d.). *Our mission.* Retrieved December 13, 2003, from http://www.usace.army.mil/who.html

5 US Army Corps of Engineers. (n.d.). *Our mission.* Retrieved December 13, 2003, from http://www.usace.army.mil/who.html

6 US Army Corps of Engineers, Omaha District. (2003, April) *Glovers Point hydraulic analysis* (USACE-NEO-ED-H).

7 US Army Corps of Engineers, Omaha District. (2003, April). *Glovers Point hydraulic analysis* (USACE-NOW-ED-H).

8 The Centre for Research in Art & Design. (n.d.). *Definition of "research."* Retrieved December 13, 2003, from http://www2.rgu.ac.uk/criad/r2.htm

9 Tarnowski, T., Batzer, J., Gleason, M., Helland, S., & Dixon, P. (2003, December 9). Sensitivity of newly identified clades in the sooty blotch and flyspeck complex on apple to thiophanate-methyl and ziram. *Plant Health Progress,* doi:10.1094/PHP-2003-1209-01-RS. Retrieved December 13, 2003, from www.plantmanagementnetwork.org/pub/php/research/2003/apple/

10 Tarnowski, T., Batzer, J., Gleason, M., Helland, S., & Dixon, P. (2003, December 9). Sensitivity of newly identified clades in the sooty blotch and flyspeck complex on apple to thiophanate-methyl and ziram. *Plant Health Progress,* doi:10.1094/PHP-2003-1209-01-RS. Retrieved December 13, 2003, from www.plantmanagementnetwork.org/pub/php/research/2003/apple/

11 Tarnowski, T., Batzer, J., Gleason, M., Helland, S., & Dixon, P. (2003, December 9). Sensitivity of newly identified clades in the sooty blotch and flyspeck complex on apple to thiophanate-methyl and ziram. *Plant Health Progress,* doi:10.1094/PHP-2003-1209-01-RS. Retrieved December 13, 2003, from www.plantmanagementnetwork.org/pub/php/research/2003/apple/

12 Tarnowski, T. B., Batzer, J., Gleason, M., & Dixon, P. (2003). Fungicide sensitivity differs among newly discovered fungi in the sooty blotch and flyspeck complex on apples. Iowa State University, Ames, Iowa.

13 Tarnowski, T., Batzer, J., Gleason, M., Helland, S., & Dixon, P. (2003, December 9). Sensitivity of newly identified clades in the sooty blotch and flyspeck complex on apple to thiophanate-methyl and ziram. *Plant Health Progress,* doi:10.1094/PHP-2003-1209-01-RS. Retrieved December 13, 2003, from www.plantmanagementnetwork.org/pub/php/research/2003/apple/

14 Stohrer, F. F., & Pinelli, T. E., (1981, March). Marketing information: The technical report as a product. In *Technical writing: Past, present, and future* (NASA Technical Memorandum 81966 14).

15 Used with permission of Aaron Kurdle, Community and Regional Planning, Ames, IA: Iowa State University.

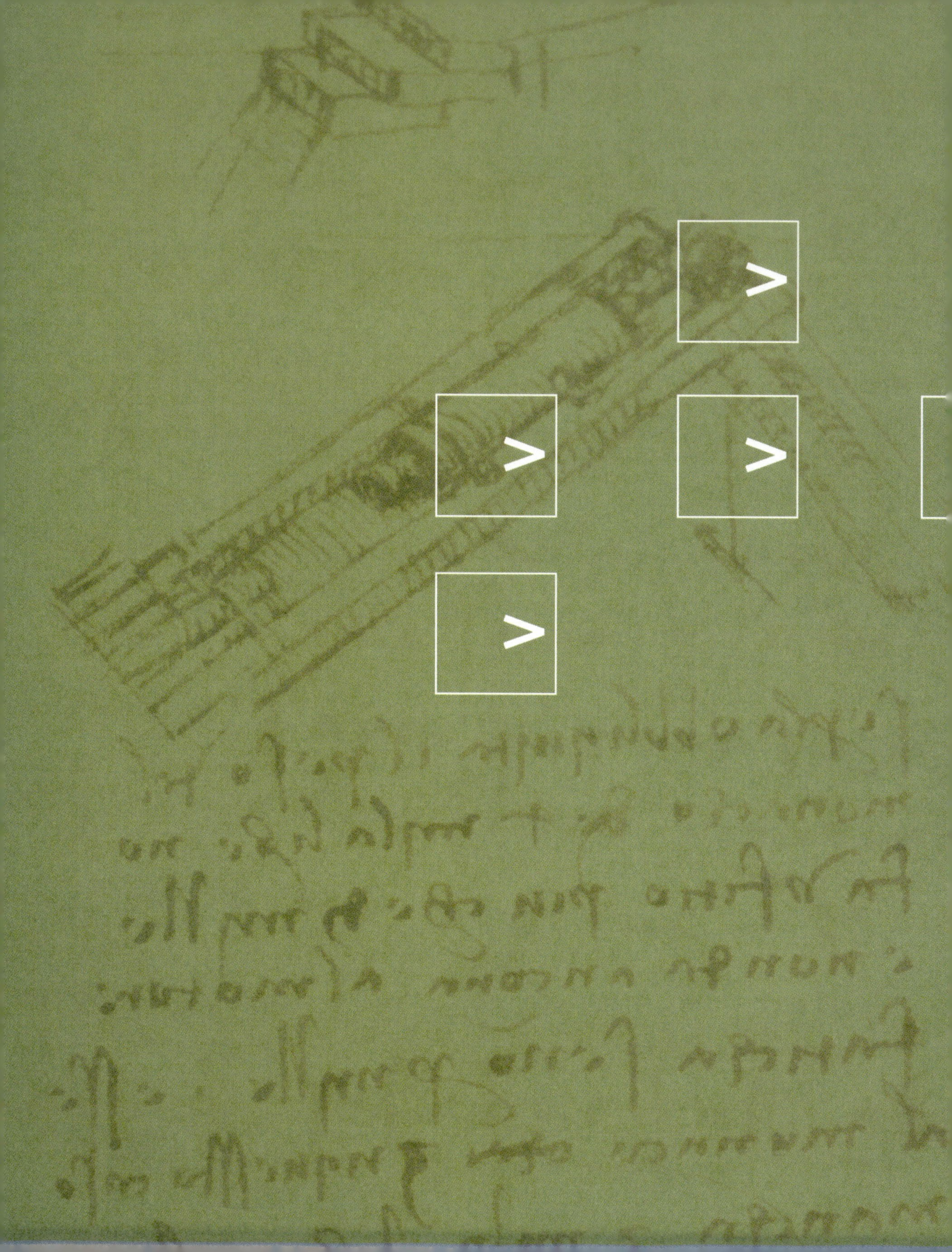

CHAPTER 21

Preparing Instructions and Manuals

Objectives and Outcomes

This chapter will help you accomplish these outcomes:

- Ensure instructions are usable by combating audience misunderstanding, using principles of adult learning, and confronting aliteracy
- Analyze task, audience, and genre
- Effectively use content elements of instructions
- Effectively use visual elements of instructions
- Provide necessary warnings and cautions

In this chapter, the term "instructions" refers to all kinds of instructional texts, including instruction sheets, electronic help systems, print manuals, training scripts, and tutorials.

Instructions are everywhere — from the arrow on the lid of a peanut butter jar indicating the way the lid twists off to multivolume manuals detailing every step in the maintenance of commercial aircraft to instructions for planting on seed packages to electronic help files provided with your word-processing software.

Lots of reasons exist for maintaining accurate and up-to-date instructions and for encouraging various groups of people to read and use them:[1]

- Personnel safety and performance
- Process/product safety and performance
- Summary or overview of process/product features
- Orientation for new team members as well as sales and marketing personnel
- Central location for documenting process/product specifications and modifications

Put more directly, instructions are an essential part of the process of creating products and of the products themselves, not a separate, valued-added component, regardless of the company or country. In fact, the need for effective instructions crosses national boundaries, as clearly shown in the following excerpt from the introduction to *Usable and Safe Operating Manuals for Consumer Good,* a guideline created in the SecureDoc project with the support of the Commission of the European Community:

> European Union legislation and regulations specify that a technical product is only complete when accompanied by an operating manual. Delivery or sale of a product without an operating manual or with an incomplete or deficient one infringes the law. In this case, consumers are entitled to assistance.[2]

Why do instructions have more and a broader range of users than most technical documents?

Issues such as audience comprehension, adult learning principles, and aliteracy strongly influence the way you define, design, and construct a wide range of instructional texts, oral presentations, and visuals. You can analyze and differentiate instructions by task, audience, and genre, in each case assessing whether the instructions are accessible, comprehensible, and usable.

Getting People's Attention

More than any other kind of technical document, instructions and manuals have an immediate purpose: to enable users to complete tasks. Three critical concerns — lack of audience understanding, adult learning, and aliteracy — can influence whether instructions are actually usable.

After reading the following two examples, identify at least five other situations in which understanding written, oral, or visual instructions is critical.

Increasing Users' Understanding

Is the problem of people not understanding instructions serious? Absolutely! Two examples are sufficient to suggest the breadth of the problem.

Jurors Misunderstand Judges' Instruction. Jurors frequently do not understand judges' instruction about legal rules or about their responsibilities in rendering a decision. Recommendations can be drawn from social science and linguistics about ways to improve instruction to juries:[3]

- Provide jury instruction earlier in a trial. One federal judge compares "the odd practice of waiting until the end of the trial to instruct the jury with 'telling jurors to watch a baseball game and decide who won without telling them what the rules are until the end of the game.'"[4]
- Begin each instruction with general information to help organize and contextualize it, such as "this instruction is in two parts," or "a person can become negligent in two ways"; provide lists and charts to help jurors understand and remember the rules.
- Make jury instruction case-specific. Jury instruction should contain the names of parties as well as actual issues and examples from the case.
- Simplify the language of jury instruction. Adjust the vocabulary to avoid legal jargon and uncommon words and phrases; substitute difficult words and phrases with commonly understood synonyms; avoid long, compound sentences and the use of double or triple negatives.

Why do you think that juries function more effectively if the members understand "from the very start what laws have allegedly been broken, the meaning of key terms, and the ways that witnesses' testimony is intended to relate to the charges," rather than waiting until the end of the trial to receive instruction about how to interpret what they've heard?[5]

Patients Misunderstand Physicians' Instructions. A study reported in the *Journal of General Internal Medicine* conducted at the Hospital of the University of Pennsylvania compared outpatients' "understanding of medication dosing instructions written in terms of daily frequency with patients' understanding of instructions specifying hourly intervals." The results are based on questionnaires completed by 500 patients presenting new and refill prescriptions to the hospital outpatient pharmacy. Prescriptions with dosing instructions specifying daily frequency, rather than hourly intervals, were far more easily understood.

	Total patients who completed survey questionnaire	**Patients who misunderstood dosage instructions**
Prescriptions specifying dosing instructions in hourly intervals (e.g., q6h)	71	55 (77%)
Prescriptions with dosing instructions specifying daily frequency (e.g., qid)	429	4 (0.99%)

According to the researchers, "this difference remained when patient subgroups were evaluated by education level, new versus refill prescriptions, and analgesic versus nonanalgesic medications." The researchers concluded that "the intended dosing regimen is frequently misunderstood when the physician writes outpatient

Why do you think people respond better to instructions that specify the number of times per day rather than hourly intervals?

prescriptions in hourly intervals. To promote optimal patient compliance, the outpatient prescription label should state the number of times a day a medication is to be taken."[6]

Using Principles of Adult Learning

In the workplace, the audience for instructions is usually adults who have limited time (and, probably, limited patience). Designing instructions that appeal to their particular needs makes good sense. More than 20 years ago, researcher Malcom Knowles identified characteristics of adults that affect the ways in which they learn, ways that are sometimes, but not always, different from the ways children learn. The following list identifies the characteristics of adult learners identified by Knowles (and reinforced by many other researchers since then) and shows how instructions could respond to them:[7]

If your audience is children rather than adults — for example, instructions for video games — which of these principles of adult learning still apply?

Self-concept. Adults like to be *self-directed.*	Instructions can provide tools for users to self-assess their prior knowledge and experience and then move to any appropriate place in the instructions.
Rationale. Adults want a *reason* for doing or learning something.	Instructions can provide reasons for using them as a whole as well as reasons for using subsections.
Experience. Adults have *prior knowledge* and *experiences* that help them complete tasks.	Instructions can use metaphors, analogies, and examples that draw on prior knowledge and experiences.
Readiness. Adult have *goals.*	Instructions can help users decide how the information fits with their goals.
Orientation. Adults focus on what is practical and useful	Instructions can focus on practical information — *what, why,* and *how.* They can also identify benefits.
Motivation. Adults are internally motivated by factors such as relationships and intellectual interests.	Instructions can provide users with options — how to approach a task, how to use the information, how to manage problems. Instructions can also address users in a way that encourages them to continue because they'll receive help.

What additional reasons might people have for ignoring instructions or a manual?

Even when certain tasks need to be completed, life sometimes intervenes. A common response is to set aside instructions and jump directly into the task. What causes this behavior? Lots of things: pressure to get the task done; negative prior experiences with instructions; overestimation of prior knowledge about how to complete the task without additional assistance; lack of time to read instructions; lack of ability to read; lack of interest in reading or willingness to read.

Addressing Aliteracy

Everyone reading this book has probably heard some version of the expression, "If all else fails, read the directions." While this expression is sometimes meant as a joke, it's a serious concern in the United States because of the growing rate of limited literacy and aliteracy.

You read in Chapter 4 about the widespread problem of limited literacy in the workplace, with approximately 50 percent of the adults in the United States being unable to complete very simple literacy tasks such as locating a single piece of information in a short article, entering a signature on a form, or locating information about eligibility for employee benefits in a table.

However, aliteracy is different. People who are aliterate *can* read, but they simply choose not to. Why? They don't want to, don't like to, don't want to take the time to, or don't think they need to.[8] What characterizes aliteracy?[9]

- Scanning text as a regular practice rather than reading text, whether print or electronic
- Depending on visuals rather than words for information
- Depending on icons, symbols, and logos rather than words for information
- Imbuing color, shape, position, and size with meaning to avoid written language
- Substituting various kinds of electronic communication (for example, Web sites, TV, radio, CDs, audio tapes, movies, videos, DVDs) for printed texts whenever possible

Do you know people who might be considered aliterate? What reasons do they give for not reading? How credible are these reasons?

People with limited literary use these strategies because for them reading is very difficult, and the strategies often simplify the task. Skilled readers also frequently use these strategies to reinforce, enhance, and sometimes speed their reading. However, aliterate readers consciously choose such strategies as an alternative to reading rather than using such strategies as part of their reading repertoire. Given this information, you have a critical challenge in designing instructions that respond to the large number of users who either have limited literacy or are aliterate.

You can learn more about aliteracy by going to www.english.wadsworth.com/burnett6e for links to sites with additional information.
CLICK ON WEBLINK
CLICK on Chapter 21/aliteracy

WEBLINK

Considering Task, Audience, and Genre

Analyzing task, audience, and genre is critical as you plan and design instructions. Skipping these analyses reduces the accessibility, comprehensibility, and usability of your instructions.

Analyzing the Task

> *In what ways might the accessibility, comprehensibility, and usability of instructions that come with a product influence the purchase of that product?*

Instructions — whether print or electronic, whether written, visual, or oral — help users complete tasks, which fall into four categories:

- Actions/behavior of personnel
- Assembly of objects or mechanisms
- Operation of equipment
- Implementation of a process

When planning instructions, you can analyze the task by asking a series of focused questions. Your responses to the questions in the first segment of Figure 21.1 should guide the way you represent the task, determining the chronological steps and influencing the kind and amount of detail you include. As with any written, oral, or visual text, you need to consider the context, specifically factors that affect accuracy or effectiveness, as listed in the second segment of Figure 21.1. You also need to ask yourself about constraints that affect completing the task, as listed in the third segment of Figure 21.1. Finally, part of your responsibility involves connecting these various task issues to audience needs, which you can do by responding to the questions in the final segment of Figure 21.1.

> *For what types of task would hard copy instructions be more desirable than electronic instructions?*
>
> *When are electronic instructions more valuable?*
>
> *When could either be used?*

Both the task and the context should influence your design. For example, will the instructions be used by a machinist on a manufacturing floor or by an office worker seated at a computer? Instructions on paper have unique concerns. For example, the quality of paper should suit the frequency of use. Instructions that will probably be used only once, such as for installing an air conditioner or assembling a rototiller, can be printed on inexpensive paper. However, instructions that will be used repeatedly should be on high-quality paper that can withstand frequent handling. Similarly, the task should influence the binding of multipage instructions. If users need to refer to the instructions while working on the task, the instructions should lie flat. Instructions that are likely to be updated from time to time can be placed in a loose-leaf binder so that revisions can be inserted. If revisions are relegated to a separate binder, they might be overlooked. For electronic instructions, accessibility, navigability, and usability are concerns. For example, will you provide software help files on the CD with the software or post the files on a Web site and assume all users will have Web access? The task analysis section in Figure 21.2 summarizes the critical task information that should be considered when preparing instructions.

FIGURE 21.1 **Questions for Task Analysis**

1. Representing the Task	■ ***History.*** What has come before? ■ ***Knowledge.*** What knowledge or experience is necessary to complete the task? ■ ***Steps.*** What steps are involved in the process or in completing the task? ■ ***Details.*** What details and explanations are necessary or helpful for users? ■ ***Outcome.*** What is the final outcome or product?
2. Understanding the Context for the Task	■ ***Situation.*** What specific situation necessitates the task? ■ ***Environment.*** What is the environment (physical, political) in which the task will be completed? ■ ***Interference.*** What will interfere with completion of the task? What will interfere with the accuracy or effectiveness of the task? ■ ***Help.*** What will aid completion of the task? What will increase the accuracy or effectiveness of the task?
3. Identifying Constraints on Task Completion	■ ***Safety.*** What safety precautions must be observed? ■ ***Complications.*** What complications might arise during performance of the task? ■ ***Time.*** How long should the task take? What might influence the time needed to complete the task? ■ ***Cost.*** What cost factors are involved? What might make completion of the task more or less expensive?
4. Relating Task and Audience	■ ***Attitude.*** What attitude are users likely to have toward completing the task? Toward using the instructions? ■ ***Understanding.*** What level of understanding do users need to complete the task? ■ ***Preparation.*** What preliminary or preparatory work must be completed by the users? ■ ***Resources.*** What resources (materials, tools, and equipment) do users need? ■ ***Skill.*** What level of skill do users need to operate the tools and equipment?

Analyzing the Audiences

Because instructions address users who actually do something, they need to address audience needs and experience. Writers of effective instructions pay attention to the tone appropriate for the audience and use consistent design and navigation features.

How many ways can you think of to test whether the instructions you write are appropriate for the intended audience? If you can't think of at least five, refer to Chapter 9 for additional ideas.

The term *user-friendly* implies that writers take a personal interest in the users. To a greater extent than any other type of technical communication (excluding correspondence), instructions attempt to establish a direct relationship with the audience through language, visuals, and design. Conversational tone, word choice, definitions, explanations, examples, visuals, and accessible design can all be used to establish this.

Most instructions and manuals are written in second person (using "you"), suggesting a link between the writer, speaker, or designer and the users. The imperative mood of verbs (e.g., "*Turn* the knob." "*Adjust* the fluid level.") is for the commands in individual steps. However, within these imperative steps, the vocabulary and level of detail can be adapted to the specific audience. For example, the instructions in the user reference manual for word-processing software approach the novice user in a direct, friendly manner:

> To enter text, just start typing. Each non-control character typed is entered into the text of your document. If you type beyond the right margin, notice that WordStar moves the word that wouldn't fit inside the margin to the next line, positioning the cursor after the word to allow you to continue typing. This is word wrap.[10]

The installation manual for the same software, although dealing with far more complex content, speaks directly to the technician:

> Patch "RUBFXF" below to non-zero. The contents of "RFIXER" will then be output immediately after a "DELETE" is input; this character, rather than the next cursor positioning string, should thus be replaced with backspace-space-backspace, reducing the consequences of your system's machinations. Try null (zero) in RFIXER first; if this doesn't work, try backspace (08) or space.[11]

Responses to the audience analysis questions in Figure 21.2 will help you construct and present comprehensible, usable instructions. More specifically, you will be able to respond to user attitude by justifying individual steps or entire instructions and using attention-getting devices; you'll be able to respond to user education by adjusting the level of vocabulary, defining terms, and providing appropriate visuals; you'll be able to respond to user experience by adjusting to prior knowledge as well as the amount and type of detail.

Adapting Task to Audience. When chunking and labeling are ignored in instructions, users have difficulty differentiating the background information from the task to be completed and may not be sure what they're actually supposed to do. Take, for example, the manager of the quality control department in a company that specializes in cleaning up hazardous waste sites. He was reviewing a set of instructions in preparation for an on-site inspection by a potential customer. During this preliminary review, he noticed a problem. In several sets of instructions used by technicians, such as the example in Figure 21.3a, the steps were preceded by a dense block of text that contained important safety

FIGURE 21.2 **Worksheet for Instructions**

Title of Instructions: ______

Task Analysis

Purpose of task (product created, behavior changed): ______

Environment where the task is completed: ______

Resources needed (materials, tools, equipment): ______

Steps to complete task: ______

Constraints to completing task (safety, time, resources, cost, etc.): ______

Other important task considerations: ______

Audience Analysis

User (organizational position/role): ______

User needs: ______

Attitude: *enthusiastic* ● - - - - - - - - - ● - - - - - - - - - ● - - - - - - - - - ● *resistant*

Sources of resistance: ______

Other user characteristics: ______

Education/experience needed to complete task: ______

User's education/experience with task: ______

Other important audience considerations: ______

Genre Analysis

Medium — print, electronic, aural, visual: ______

Context in which instructions will be used: ______

Format of instructions: ______

Testing (types and dates)

	Text-based testing	Expert-based testing	User-based testing
In-process dates			
Testing methods			
Final date			

information. Not only were general labels for task and equipment missing, but labels to signal warnings and safety precautions were also missing, which could be a serious liability problem.

The manager recognized that technicians were likely to resist reading the dense blocks of information, even though the warnings were important. He recommended that this preliminary information be redesigned so that employees would be likely to read and act on it. The revised version in Figure 21.3b shows how chunking and labeling the warnings and safety precautions increase the likelihood that users will better understand not only what they have to do but also how to do it safely.

Instructions accompanying products made outside the United States are sometimes written in stilted, nonidiomatic English. How do you react to this? Is your confidence in either the product or the accuracy of the instructions affected? If a product made by your company were intended for export, how would you avoid such problems in translation?

What size is the Kärcher information sheet in inches?

Considering Instructions for International Audiences. As the global economy makes products available to more people around the world, instructions need to be accessible to a broad range of users who come from very different cultures and have varying degrees of literacy. Two approaches can adapt instructions to international audiences:

- provide instructions translated into multiple languages
- use visuals that are likely to be understood in many cultures

Some companies have the resources to produce instructions in multiple languages; for example, Sony's Document Design Development Center in Tokyo produces complex manuals in more than 20 languages, ranging from Japanese, English, and most European languages to Arabic, Korean, and Russian.[12]

More modest is the instruction sheet provided by Kärchar to accompany its electric broom. The instruction sheet is printed in black ink on both sides of a single sheet of white paper (approximately 48cm × 62cm when open and approximately 12cm × 16cm when folded). The instructions are presented in 23 languages. In Figure 21.4, you see two of the panels from the instruction

FIGURE 21.3A **Original Version of Unchunked and Unlabeled Instructions**

Because this information is not chunked and labeled, users cannot easily differentiate what they need in order to perform the task safely and accurately.

The technicians who need to understand and act on this important information are likely to resist reading undifferentiated blocks of text.

The post assembly optical alignment of the SSL100 Nd:YAG laser head is performed by a laser technician after the laser head is assembled but prior to its installation in the system. Dangerous levels of laser radiation are emitted from the laser during the test. All windows must be covered so that the laser light will not exit. Doors must be lockable from the inside. The laser technician should not be disturbed or distracted while performing this procedure. The laser technician should be the only person in the room during the test. Laser safety glasses must be worn while the laser is on. The test station consists of an electrical power supply, a D.I. water recirculation system, a cooling system, and power and temperature controls. In addition to the test station, the following equipment is required: a HeNe laser and power supply, a laser power meter, an optics cleaning kit, and an I.R. source detector.

FIGURE 21.3B **Revised Version of Chunked and Labeled Instructions**

Task The post assembly optical alignment of the SSL100 Nd:YAG laser head is performed by a laser technician after the laser head is assembled but prior to its installation in the system.

⚠ DANGER Dangerous levels of laser radiation are emitted from the laser during the test. Take precautions:

- Cover all windows so that the laser light will not exit.
- Before the test, lock all doors from the inside.

Safety Precautions The safety of the laser technician is very important:

- The laser technician should not be disturbed or distracted while performing this procedure.
- The laser technician should be the only person in the room during the test.
- The technician must wear laser safety glasses while the laser is on.

Test Station The test station has four components:

- electrical power supply
- D.I. water recirculation system
- cooling system
- power and temperature controls

Required Equipment The post assembly optical alignment of the SSL100 Nd:YAG laser head requires four pieces of equipment:

- HeNe laser and power supply
- laser power meter
- optics cleaning kit
- I.R. source detector

Chunking and labeling increase the likelihood that the technicians will read this information.

The signal word DANGER with the international symbol that indicates risk to humans complies with ANSI guidelines (discussed later in this chapter) and increases the likelihood that the technicians will perform the task without injury to themselves or damage to equipment.

sheet: (1) panel #2 on the front side of the instruction sheet, which presents the first three steps and (2) panel #1 on the back side of the sheet, which presents a critical warning.

WEBLINK

The Kärcher Web site gives users the choice of selecting from nearly 40 countries (and languages) for viewing the site. Beyond the company's response to the international market, the site's use of Flash is subtle and sophisticated. For a link to the site, go to www.english.wadsworth.com/burnett6e.

CLICK ON WEBLINK

CLICK on Chapter 21/Kärcher

FIGURE 21.4A **Excerpt from Instruction Sheet in Multiple Languages[13]**

Each language is identified every time it is used.

The single line drawings make the parts easy to identify.

Numbers and arrows indicate the sequence of action and the physical direction for turning the handle to tighten it.

Lines and arrows indicate places for inserting one part into another.

Context (such as a tile walled in step #3) is only provided when necessary to show how the wall hanger is placed.

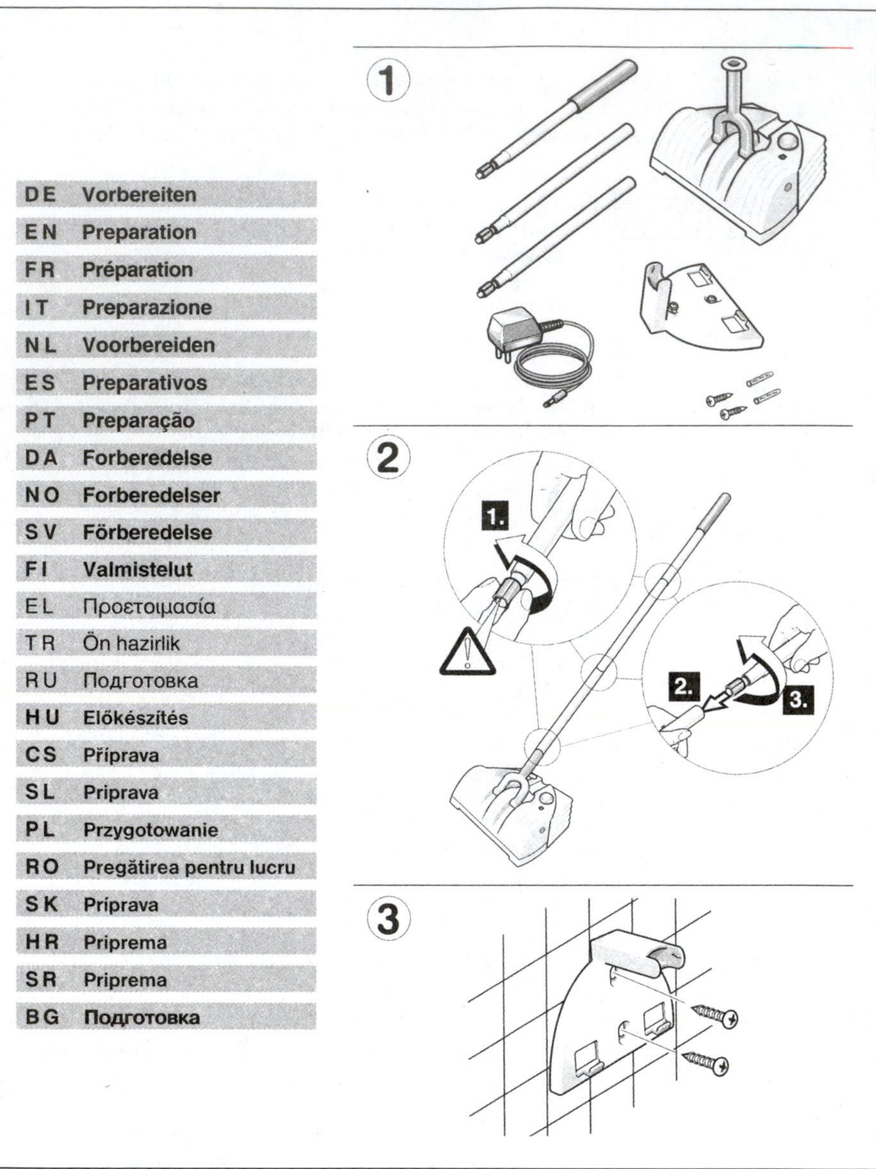

DE Vorbereiten
EN Preparation
FR Préparation
IT Preparazione
NL Voorbereiden
ES Preparativos
PT Preparação
DA Forberedelse
NO Forberedelser
SV Förberedelse
FI Valmistelut
EL Προετοιμασία
TR Ön hazirlik
RU Подготовка
HU Előkészítés
CS Příprava
SL Priprava
PL Przygotowanie
RO Pregătirea pentru lucru
SK Príprava
HR Priprema
SR Priprema
BG Подготовка
1
2
1.
2.
3.
3

Another way to approach instructions for an international market is to expect the visuals to carry much of the meaning. For example, Figure 21.5 is an excerpt taken from a Japanese-language manual produced by Fuji-Xerox, explaining how to replace a waste toner bottle in a photocopier. Although

FIGURE 21.4B **Excerpt from Instruction Sheet in Multiple Languages**[13]

DE Kurzschlussgefahr!
Keine leitenden Gegenstände (z.B. Schraubendreher oder Ähnliches) in die Ladebuchse stecken.

EN Risk of short-circuit!
Do not insert any conductive objects (e.g. screwdriver, etc.) into the charging socket.

FR Risque de court-circuit !
N'enfoncez aucun objet électroconducteur (par ex. un tournevis ou assimilé) dans la prise de recharge.

IT Pericolo di corto circuito!
Non inserire oggetti conduttori (per es. cacciaviti o simili) nella presa di carica.

NL Gevaar voor kortsluiting!
Steek geen geleidende voorwerpen (een schroevendraaier of iets dergelijks) in oplaadbus.

ES ¡Peligro de cortocircuito!
No introducir objetos electroconductores (por ejemplo destornilladores, etc.) en la conexión de carga de la batería.

PT Perigo de curto-circuito!
Não meter nenhum objecto condutor (por exemplo, chave de fenda ou coisa semelhante) no conector de carga.

DA Risiko for kortslutning!
Der må ikke stikkes genstande (f.eks. en skruetrækker eller lignende) ind i ladebøsningen.

NO Fare for kortslutning!
Det må ikke stikkes ledende gjenstander (f.eks. skrutrekker e.l.) inn i ladekontakten.

SV Risk för kortslutning!
Stick inga ledande föremål (t ex skruvmejsel eller liknande) i laddaruttaget.

FI Oikosulkuvaara!

EL Κίνδυνος βραχυκυκλώματος!
Μη βάζετε αγώγιμα αντικείμενα (π.χ. κατσαβίδια ή άλλα παρόμοια) στην υποδοχή φόρτισης.

TR Kısa devre tehlikesi!
Şarj soketine iletken cisimler (örn. tornavida veya benzeri cisimler) sokmayınız.

RU Существует опасность возникновения короткого замыкания! Не вставляйте в зарядное гнездо никаких предметов из проводящего материала (например, отвертку или нечто подобное).

HU Rövidzárlat veszélye!
Ne dugjon áramot vezető tárgyakat (pl. csavarhúzót vagy hasonlót) a töltőhüvelybe.

CS Nebezpečí zkratu!
Nestrkejte žádné vodivé předměty (např. šroubovák nebo pod.) do nabíjecí zdířky.

SL Nevarnost kratkega stika!
Ne vtikajte nobenih prevodnih predmetov (npr. izvijač ali kaj podobnega) v polnilno vtičnico.

PL Niebezpieczeństwo zwarcia elektrycznego!
Do gniazdka ładowania nie wolno wkładać żadnych przedmiotów przewodzących prąd (np. wkrętak lub temu podobne).

RO Pericol de scurtcircuitare!
Nu introduceți obiecte metalice (ca de exemplu șurubelnițe sau altele asemenea) în priza de încărcare.

SK Nebezpeèenstvo skratu!
Nestrkajte žiadne vodivé predmety (napr. skrutkovač alebo pod.) do nabíjacej zdierky.

HR Opasnost od kratkog spoja!
U utičnicu za punjenje ne umetati predmete koji su provodnici struje (primjerice odvijač ili slično).

SR Opasnost od kratkog spoja!
U utičnicu za punjenje ne umetati predmete koji su provodnici struje (na pr. odvijač ili slično)

The warning page is headed by the internationally recognized alert symbol ⚠, indicating that a hazard exists that might cause human injury.

A definition in the form of an example is provided for "conductive object."

Boldfacing distinguishes the risk statement from the imperative explaining safe behavior.

Examine the 23 languages. Try to identify each one.

the instructions include text (accessible to readers of Japanese), the conventions in the illustrations make the steps in the process easy to understand with international icons, numbered steps, human involvement, and close-ups.

FIGURE 21.5 Excerpt from Effective Japanese Manual[14]

Cautions are signaled with a ⚠, a widely recognized icon. The seven steps in the task are clearly numbered.

The arrows show the direction in which parts move.

The hand shows what the operator should do at each step.

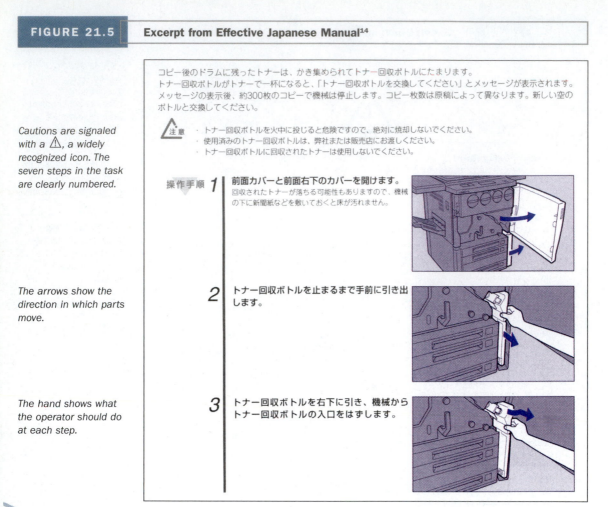

コピー後のドラムに残ったトナーは、かき集められてトナー回収ボトルにたまります。
トナー回収ボトルがトナーで一杯になると、「トナー回収ボトルを交換してください」とメッセージが表示されます。メッセージの表示後、約300枚のコピーで機械は停止します。コピー枚数は原稿によって異なります。新しい空のボトルと交換してください。

- トナー回収ボトルを火中に投じると危険ですので、絶対に焼却しないでください。
- 使用済みのトナー回収ボトルは、弊社または販売店にお渡しください。
- トナー回収ボトルに回収されたトナーは使用しないでください。

操作手順

1 **前面カバーと前面右下のカバーを開けます。**
回収されたトナーが落ちる可能性もありますので、機械の下に新聞紙などを敷いておくと床が汚れません。

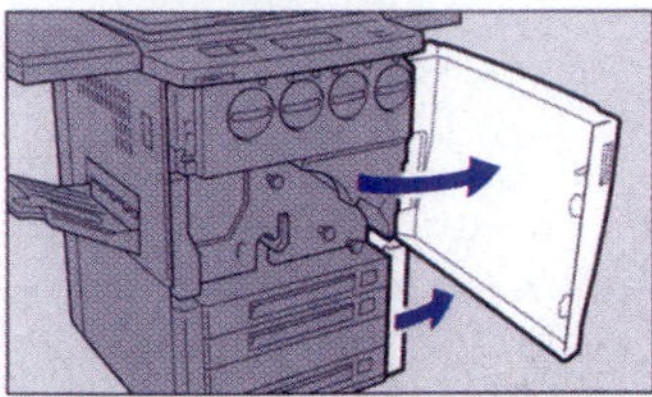

2 **トナー回収ボトルを止まるまで手前に引き出します。**

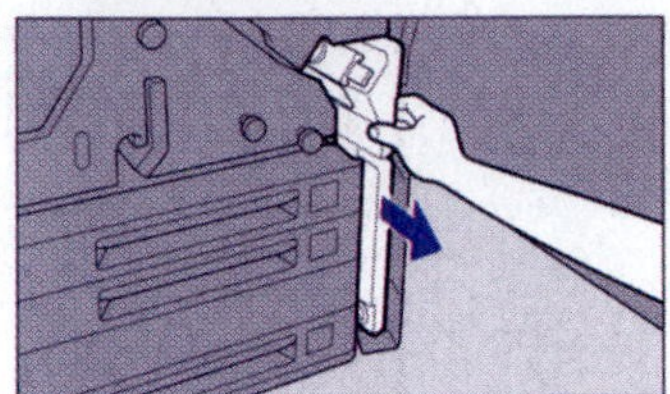

3 **トナー回収ボトルを右下に引き、機械からトナー回収ボトルの入口をはずします。**

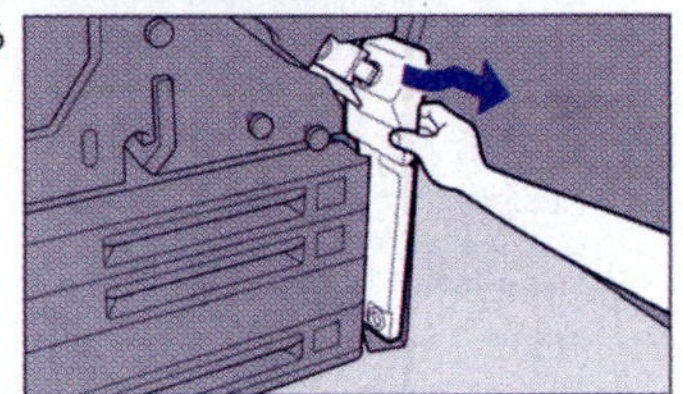

> Do you think instructions should be localized, or should they be the same regardless of country or culture?
>
> What aspects of instructions might be localized?
>
> How much should instructions be localized?
>
> What cultural and technical issues do you need to consider in localizing instructions?

Not all procedures are as easy to explain as the electric broom or the toner cartridge, so when designing instructions for an international audience, consider these suggestions:

- Provide an easy-to-use table of contents or common navigation structure as well as introductory visual maps that help users understand how to use the instructions.
- Be consistent in the way that words, links, and visuals are used.
- If words or visuals work equally well to convey information, choose the visuals. Whenever possible, illustrate text with visuals.
- Use visual cues to help users understand sequence and direction, for example, numbers to label the sequence of steps, arrows to indicate the direction knobs should be turned, and shading to highlight a key.

FIGURE 21.5 **Excerpt from Effective Japanese Manual**[14]

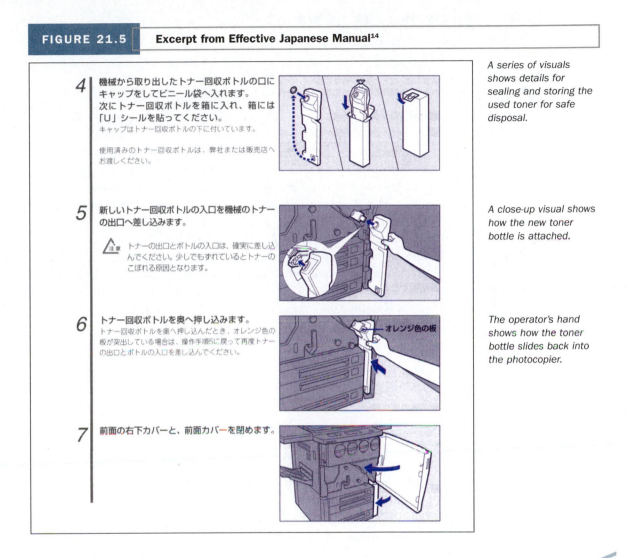

4 **機械から取り出したトナー回収ボトルの口にキャップをしてビニール袋へ入れます。次にトナー回収ボトルを箱に入れ、箱には「U」シールを貼ってください。**

キャップはトナー回収ボトルの下に付いています。

使用済みのトナー回収ボトルは、弊社または販売店へお渡しください。

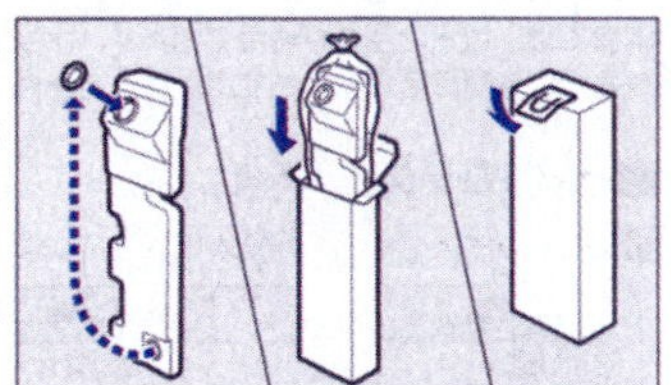

A series of visuals shows details for sealing and storing the used toner for safe disposal.

5 **新しいトナー回収ボトルの入口を機械のトナーの出口へ差し込みます。**

 トナーの出口とボトルの入口は、確実に差し込んでください。少しでもずれているとトナーのこぼれる原因となります。

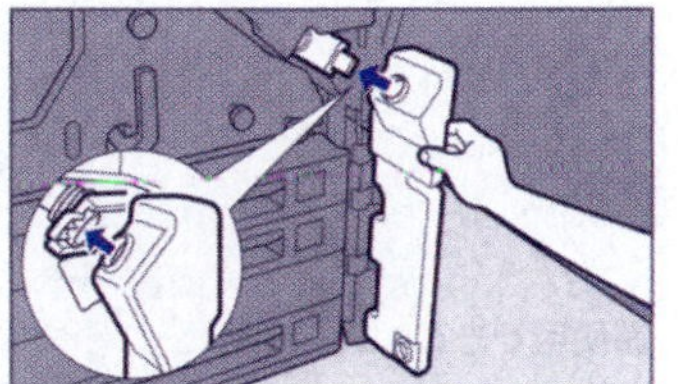

A close-up visual shows how the new toner bottle is attached.

6 **トナー回収ボトルを奥へ押し込みます。**

トナー回収ボトルを奥へ押し込んだとき、オレンジ色の板が突出している場合は、操作手順5に戻って再度トナーの出口とボトルの入口を差し込んでください。

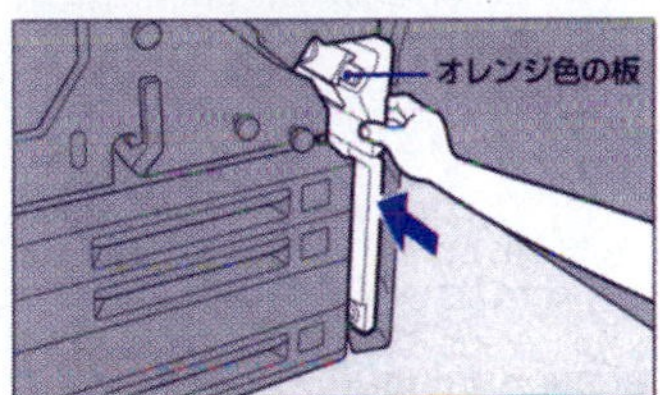

The operator's hand shows how the toner bottle slides back into the photocopier.

7 **前面の右下カバーと、前面カバーを閉めます。**

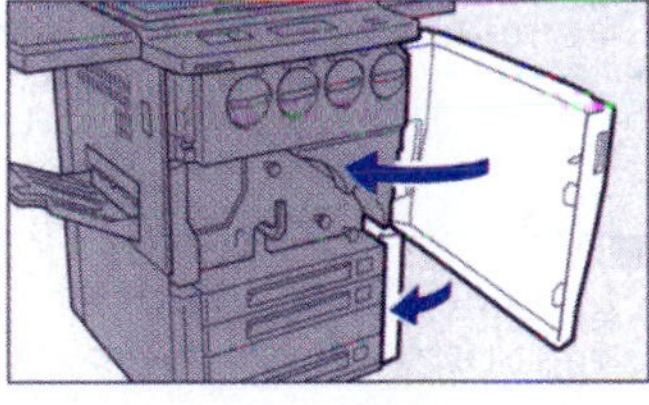

- Visually distinguish main steps from details and explanations by font size, type, style, or placement; for example, details can be in a smaller font, a sans serif font, italics, and/or indented, shaded, or boxed.
- Make sure that colors and icons conform to ISO standards and do not carry meanings that may be inappropriate or offensive to particular cultures.
- Place warnings, cautions, and dangers as close as possible to the related step. Visually distinguish them from steps in the instructions by icons; font size, type, or style; color; placement; and/or design cues such as shading or boxes.
- Have a native reader and writer carefully edit the document to make sure all usage is idiomatic.

Why do you think that icons can be as ambiguous as words? One participant in an online discussion group reported that "the broken-glass icon that marks fragile objects' packaging was misinterpreted by freight handlers as meaning 'damaged goods.' They treated the boxes accordingly."[15]

Analyzing the Genre

A number of aspects of genre — delivery options, context in which the instructions are used, and format — will affect the way you plan and design instructions. Your responses to the questions in the genre analysis section of Figure 21.1 will help you construct comprehensible, usable instructions.

Range of Options. Some companies provide users with both print and electronic documentation. More and more companies, however, provide only electronic support, which may include an electronic help system loaded on the hard drive or available on the Web. A help system typically contains hypertext links, more than one path to information, popup topics, and a keyword search capability. Some software programs like Framemaker automate the creation of help systems and can produce both electronic and print documentation.

When you are selecting the genre to use for any particular instructions, you need to consider both the instructions themselves (content, format, methods of delivery) as well as the situation in which they'll be used (purpose, task, audience). Want a surefire way to create unusable instructions? Ignore the situation in which they'll be used. Consider this remarkable range of instructions:

- Street signs and building signs directing people to go in a particular direction
- Package inserts that include directions for taking prescription and nonprescription medication, identification of people who shouldn't take the medication, and potential problems
- Quick reference guides that provide shortcuts, especially for using computer software
- Electronic help systems for software of all types
- Procedures for tasks, from building an aircraft to selecting the winning bid for a new highway project
- Human resources manuals, including policies ranging from family leave to procedures to report discrimination
- Installation instructions, from one-page sheets to detailed manuals
- Operation manuals for a range of tasks, from running a hydroponic greenhouse to managing an aluminum smelting operation
- Tutorials designed for individual, self-paced instruction
- Training sessions intended to instruct both employees and customers about policies, procedures, and tasks

> *What additional examples of instructions can you add to this list?*

Purpose and Delivery. Regardless of the genre, every instruction responds to factors that are synergistically integrated. For example, restaurant employees need to be reminded to wash their hands after using the restroom. Should the information be in text or visuals or both? Should the information be in English,

Spanish, Vietnamese, Chinese, or all four? Should the information be in the employee handbook, on a sign in each toilet stall and above each urinal, on a sign above the restroom sinks, on the wall in the kitchen, or some combination? Should the directions be presented as a stark order or should one or more reasons be included (for example, compliance with state law, reduction in the spread of hepatitis)?

Figure 21.6 summarizes several factors related to instructions. Writers and organizations may frequently assume instructions' purpose is simply getting tasks done, but they have far more complex purposes, as column one shows. Figure 21.6 also summarizes print, electronic, audio, and visual media for delivery. While these modes are presented as if they're independent, in fact they sometimes overlap. For example, a face-to-face seminar necessarily includes seeing and speaking and often incorporates print and electronic instructions as well.

Some instructions can appear in more than one mode of delivery. For example, instructions for filing tax returns are available in print and electronic form. What other instructions are typically delivered in more than one mode?

Figure 21.6 shows that not all instructions are imperative steps that tell users how to accomplish an immediate task. Sometimes information has an instructional function, but the information itself may be definitional and descriptive. Additionally, traditional imperative instructions may include definitions, descriptions, and sometimes more extended explanations.

WEBLINK

Providing training and development seminars, tutorials, and manuals is a major focus in many organizations. To learn more about ways to produce high-quality training and development manuals, go to www.english.wadsworth.com/burnett6e for a link.

CLICK ON WEBLINK

CLICK on Chapter 21/training manuals

Time is a critical and complex factor in instructions. For example, the time needed to complete a task can be very short (e.g., resetting a digital watch), or it can be several hours (e.g., installing a new window) or days (e.g., designing the electrical system for an office). Regardless of the time needed, however, some factors such as attention to detail and concern for accuracy remain the same. Additionally, how long a user needs to remember the instructions varies depending on the frequency of use and the kind of help available. For example, when you're installing a new software application on your computer, instructions that have high memorability aren't very important because you're likely to do the task only once, and even if you need to reinstall the software at some point, the instructions walk you through each step. However, for tasks such as using mammography equipment or operating a forklift, the memorability of the instructions is very important.

What additional examples can you identify?

Which instructions would be primarily written text?

Which would be primarily visuals?

Which would typically include an audio component?

FIGURE 21.6 **Examples of Purposes and Modes of Delivery**

Purposes		Delivery			
Writers/ organizations	**Users**	**Print Examples**	**Electronic Examples**	**Audio Examples**	**Visual Examples**
Ensure tasks are completed accurately by users Increase user ■ safety ■ understanding ■ satisfaction ■ productivity ■ compliance with standards	Respond to cues, prompts, reminders	■ Exit sign on highway ■ Reminder to employees to wash hands	■ Popup asking users if they want to delete a file ■ Prompt while using ATM	■ Voice cautions on airport people-movers and escalators ■ Audible pedestrian traffic signals at crosswalks	■ Icon signaling potential hazard ■ Icon to indicate crossing in safe
Provide users with ■ training ■ demonstrations of processes Provide legal protection for the organization	Complete immediate tasks	■ Instruction sheet to set up printer ■ Manual for replacing brake assembly	■ Electronic help for new software ■ Instructions directing the movement of the Mars Rover	■ Automated telephone banking transactions ■ GPS system in vehicles	■ Telemedicine diagnosis ■ Aircraft safety procedures card
Provide a positive image for the organization	Learn to do new tasks	■ Print tutorial to learn use new software ■ Print tutorial to learn CPR	■ Tutorial using Flash or PowerPoint ■ CD cookbook	■ Face-to-face seminar ■ Tele-seminar	
Provide an archival record for the organization	Understand practices, principles, and concepts	■ Maintenance manual for aircraft ■ Human resources policies about harassment	■ Background about quality control procedures ■ Background from CDC for international travelers about disease outbreaks, vaccination, food and water safety	■ Tutorial using Flash or PowerPoint with voice-over ■ Face-to-face seminar ■ Tele-seminar	

Distance, both geographic and cultural, is another factor in instructions. Instructions given and received in face-to-face situations can take advantage of paralinguistic cues such as facial expressions and pauses as well as the opportunity for direct questioning. With most instructions, though, some distance exists between the writer and the user, which increases the potential for misunderstanding due to various kinds of interference, whether cultural or technological.

Courtesy of NASA/JPL-Caltech

Mars Exploration Rover. Probably the longest distance for sending instructions has been with those conveyed to the Mars Exploration rovers. A team of eight rover drivers beamed instructions from a command post at Jet Propulsion Laboratory, more than 100 million miles removed from the rovers themselves.[16]

Effects of Using Genre Badly.

A number of problems can be related directly to instructions that are not accessible, understandable, and usable:

- ***Negative effects on marketing.*** Errors in instructions can translate into errors in marketing information, packing, ads, and choice of pre-release users (including reviewers as well as people involved in beta testing and at trade shows).
- ***Negative effects on training staff.*** Errors in instructions can translate into errors in training, leading to embarrassment during the sessions when the trainer and the manual disagree and to the need for additional training.
- ***Negative effects on support and field staff.*** Errors in instructions can translate into delays in preparing responses for marketing support, technical support, and Web site FAQs as well as many additional calls to marketing and technical support.[17]

Standard Components. No "correct" or "best" structure exists for lengthy instructions, whether they are print or electronic. Manuals are simply very long instructions with sections added to provide background and to answer anticipated questions from users. Whether print or electronic, whether user's manuals, repair manuals, operation manuals, or procedure manuals, most instruction manuals have a similar sequence of sections, shown in Figure 21.7. You can decide which introductory, instructional, and support sections to include by asking yourself what your readers would be willing to read and what they'll find most useful. You should modify and arrange these sections to meet your users' needs.

Locating Information in Manuals. A critical distinction between short directions and print manuals or electronic help is the amount of information for users and the need to provide several ways for users to access that information. For example, a person using a word-processing manual or electronic help system

FIGURE 21.7 Standard Components of Manuals

Introduction	
Background	Consider one or more of these subsections: ▪ Identify intended users (for example, "These instructions are designed for technicians who will be working with customers . . ."). ▪ Include information about how to use the manual. ▪ Provide overview, including a general definition, description, and functions of the equipment or process. ▪ Include the theory of operation for users who might want to know the *why* of things, not just the *what.* ▪ Include general background, such as project history.
Training section	Consider a training section that takes first-time users through a tutorial, including basic operations, terminology, and sample applications.
Instructions	
Step-by-step instructions	Include illustrated instructions. This critical section needs careful user testing.
Support	
Frequent users' guide Quick-reference cards	Abbreviated information is very useful: ▪ Provide a summary of steps, usually placed after the step-by-step instructions so that new users aren't tempted to follow these abbreviated instructions when they need the detailed ones. ▪ Provide a list of shortcuts for steps that will be repeated, like special keystrokes in a software program. ▪ Provide reminders about complex but frequent tasks such as a start-up procedure for a complicated machine. Some companies laminate the frequent users' guide and the quick-reference cards.
Troubleshooting and maintenance	Consider a troubleshooting section and a maintenance section. These sections save unnecessary service calls and give users more control. ▪ A troubleshooting section says, "If this happens, this is what might be wrong and what you should do." Troubleshooting sections often begin with a matrix that lists problems, probable causes, and possible solutions. ▪ A maintenance section recommends basic upkeep.

should be able to locate information about suppressing page numbers or creating headers in a number of ways: looking at the table of contents, the appropriate section preview, the glossary, the index, and maybe a quick-reference card. Figure 21.8 is a screen capture of an electronic menu, in this case showing the location of instructions for adding audio files to a project.

With few exceptions, print and electronic tools for locating information in manuals have similar purposes: Users need to have a sense of the overall document, know where they are in relation to the text as a whole, look up individual terms, see how many places a particular term appears, skim and scan the text, locate cross-referenced information, and check definitions of key terms. In some cases, the function and form in print and electronic documents are the same, for example, both having section headings that identify topically related information and indicate the hierarchy. In other cases, the function is the same, as with cross-referenced information, but the form is different. For instance, print cross-references usually indicate, "See additional discussion about food toxins on page 467" while electronic cross-references are hyperlinks like this — **food toxins** — that users simply click. Figure 21.9 summarizes important devices that help users locate information in print and electronic manuals.

FIGURE 21.8 **Electronic Menu for Locating Topics**[18]

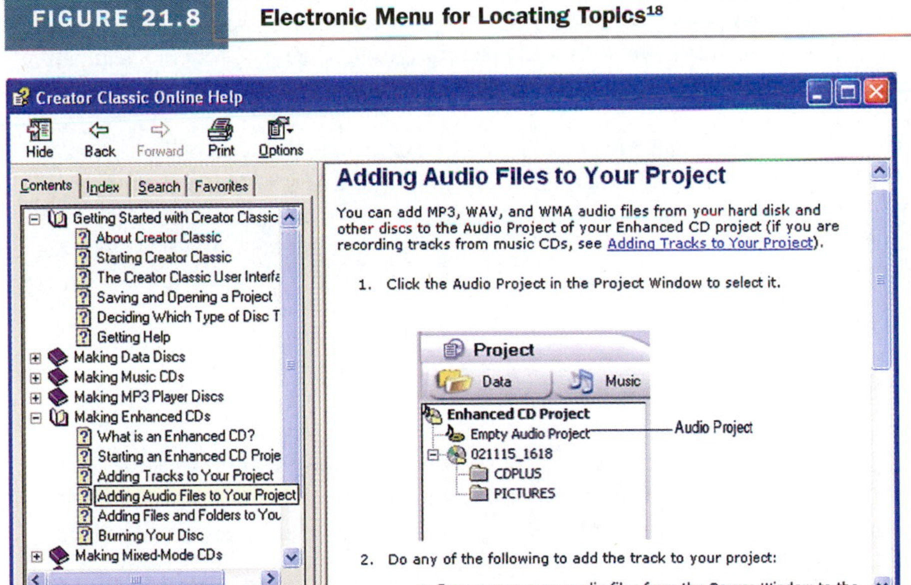

The menu uses indentation to show the hierarchy. Electronic menus serve the same function as tables of contents in print documents.

The content appears to the right of the menu, simplifing users' decisions about whether a particular topic is useful.

FIGURE 21.9 Devices for Locating Information in Print and Electronic Manuals

Section	Identify the Task and the Users
Table of contents Site map	Identify the main sections and subsections. Indicate the hierarchy with numbering, boldfacing, capitalizing, and indenting.
Section headings	Label all the sections and subsections with the same cues you used in the table of contents or site map to indicate the hierarchy.
Pagination	Number the pages in print manuals. In a dual-numbering system, page 6.7 refers to page 7 in section 6. (The specific form varies: 6.7 or 6/7 or 6·7 or 6-7.) When revising, only the page numbers in the affected section will require changing, which saves time and money.
Previews Reviews	Provide previews — brief paragraphs or numbered or bulleted lists — at the beginning of sections to let users know what's covered in that section. Provide reviews at the end of sections to summarize the key points that have been covered. Sometimes these reviews take the form of "Now you should be able to . . ." or "At this point you should have. . . ."
Cross-references Hyperlinks	Use cross-references and hyperlinks to direct users to other relevant information.
Glossary	Use a glossary as one of the early sections or as an appendix. Entries may contain definitions, examples, and explanations. Electronic help can include links to the glossary.
Index Search function	If possible, make electronic instructions searchable by using hyperlinks. For a print manual, use an index to alphabetically list and provide page numbers for all the important concepts, terms, and processes in the document.

Content Elements

The following content elements were compiled by business and industry professionals during technical communication seminars[19]; checking your instructions against this list may help you produce accessible, comprehensible, and usable instructions.

- Purpose with a title and goal statement or objective
- Necessary components: parts list, equipment list, materials list

- Accurate chronology, with time factors
- Clear, direct wording and consistent terminology
- Accurate, relevant details
- Rationale
- Stylistic and grammatical conventions
- Warnings and cautions (discussed in a separate section beginning on page 807)

Select some of the poorest instructions you have used. What in this list of essential content elements was missing from the poor instructions? How could the instructions be improved?

Purpose

Instructions need a precise title and a clear identification of purpose, goal, objective, or outcome. This identification can be achieved in a variety of ways, applied separately or in combination:

- Title may imply or state purpose:
  Operation Manual for Garden Tractor
- Title may be accompanied by visual that illustrates final objective
- Title may be supplemented by separately stated objective:
  Objective: To use mail-merge with word-processing software.

WEBLINK

Instruction sheets, manuals, and electronic instructions that accompany games are part of big business. Do you think these instructions differ in their content from more conventional workplace instructions? In content? In tone? To read guidelines for developing game manuals, go to www.english.wadsworth.com/burnett6e for a link.

CLICK ON WEBLINK

CLICK on Chapter 21/game manuals

Necessary Components

Technical instructions benefit by listing parts, materials, equipment, and definitions. For example, kits (for hobbyists who build model rockets or for industrial assemblers who build electronic monitors) should come with both a list of the parts and a list of tools or equipment needed to complete the assembly.

- A *parts list* identifies parts by name, part number, and quantity. Parts may be identified by a description or diagram if several are similar or if users are unfamiliar with part names. For example, a parts list that identifies "washers" and "lock washers" may need a diagram to differentiate the types; a list that identifies several capacitors may need labeled diagrams to distinguish among them.

What is the benefit of including drawings or diagrams of parts on a parts list?

- *Materials and equipment lists* specify what users require to complete a task. Such lists are helpful because users can order special tools or materials and organize their work areas.
- Most novice users appreciate *definitions* of unfamiliar parts and processes. Definitions can be presented as parenthetical definitions, as visuals, in a glossary, or as an appendix.

Chronological Order

Instructions should be presented in chronological order. Steps are easiest to follow if they are enumerated and separated. Not only should the overall sequence of steps in instructions be chronological, but each individual substep should also be in order, as the following examples show.

Inadequate — Not Chronological	***Improved — Chronological***
Anodize the aluminum housing after the surfaces have been deburred.	After the surfaces have been deburred, anodize the aluminum housing.

The version on the right is more logical because the steps are identified in the order in which they're done.

Effective instructions also specify details dealing with time. The following example illustrates the benefit of including chronological details:

Inadequate — Without Chronological Details	***Improved — With Chronological Details***
Pump the lever to prime the lantern.	Pump the lever for 15–20 seconds to prime the lantern.

Appropriate Diction

Instructions are useful only if users can read them. Select the simplest term that accurately conveys the information. One way to make instructions more direct is to use the verb form of a word, as the following list shows:

List 10 more nouns and their verb form.

Noun Form	***Verb Form***
Determine the calculation	Calculate
Begin the removal	Remove
Take a measurement	Measure

Another way to make diction clearer is to use a word consistently as the same part of speech, even though some words can easily be used as two or three parts of speech.

Filter the solvent. (*verb*)	Base the design on current cost. (*verb*)
Wash the filter. (*noun*)	Adjust the base. (*noun*)
Order a filter frame. (*adjective*)	Seal with a base coat. (*adjective*)

In such cases, appropriate usage is not a matter of being right or wrong but of being consistent. Conscientious writers try to employ the same word or phrase the same way within one set of instructions.

Writers also need to be consistent with terminology. Do not refer to a part as the "5 mm. tubing connector" in one place and the "Y connector" later in the same document. Once the term for a part or a step is established, use it throughout.

What are some predictable results of using directions with careless diction?

Appropriate Details

Generally, the instructions you construct should be as simple as possible without sacrificing accuracy. Instructions that are overly detailed violate Grice's maxims; they show little sense of audience needs and are as difficult to use as those that skimp on information and leave users unable to complete the task. Grice's maxims are a useful reminder: The details should be accurate and verifiable, sufficient, relevant, understandable, and well organized.

Sometimes appropriate details are in visuals, as in Figure 21.10, from a book for Guatemalan coffee farmers. The font choice, the simple line drawings, and the overall design of the document are intended to be accessible for an audience that may have limited literacy but needs technical information.

Explain whether equipment operators need to understand why each step in directions needs to be followed. Does understanding ensure that the instructions will be followed?

Can explanations take up too much space, making instructions too long and cumbersome?

Rationale for Steps

Should instructions specify only the required action, or should the action be explained or justified? The amount of detail you include depends on both the task and the audience. Explanations are essential in situations in which personal injury,

FIGURE 21.10 **Procedures for Guatemalan Farmers**[21]

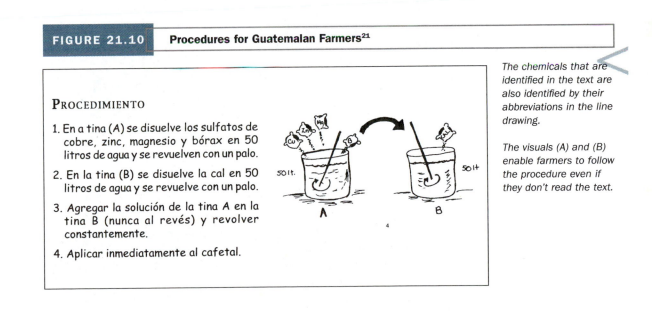

PROCEDIMIENTO

1. En a tina (A) se disuelve los sulfatos de cobre, zinc, magnesio y bórax en 50 litros de agua y se revuelven con un palo.
2. En la tina (B) se disuelve la cal en 50 litros de agua y se revuelve con un palo.
3. Agregar la solución de la tina A en la tina B (nunca al revés) y revolver constantemente.
4. Aplicar inmediatamente al cafetal.

The chemicals that are identified in the text are also identified by their abbreviations in the line drawing.

The visuals (A) and (B) enable farmers to follow the procedure even if they don't read the text.

Sign on St. Lucia Lake. The sign in this photo is posted at the end of a dock at a marina on St. Lucia Lake, the largest estuarine system in Africa, about 150 miles north of Durban.[20] What makes this sign effective (or ineffective)? Could it use more details or an explanation about why private boats cannot tie up to this dock?

equipment damage, or procedure malfunction might otherwise occur, but they are probably unnecessary if the audience is predisposed to comply with instructions. To determine whether justifications are necessary, ask yourself several questions:

1. Is the user/operator more likely to complete the step if a justification is included?
2. Does the inclusion of a justification influence the user's precision or accuracy?
3. Will the justification delay or interfere with the user's immediate understanding and implementation of the step?

The responses to these questions should guide your decision to include or omit justifications as part of the steps. If users are more likely to complete the step and with greater accuracy, then the justification will be beneficial. If the justification interferes with the users' understanding or implementation of the step, then it should be omitted.

Grammatical and Stylistic Conventions

The individual steps in instructions are written in parallel structure, with each statement using the same grammatical structure. The instructions in the following examples show how confusing nonparallel directions can be:

What misinterpretations might occur with the nonparallel example?

How could the nonparallel instructions be revised so they are parallel?

Nonparallel Structure

Always Observe Safety Rule

1. Wearing of safety glasses
2. Proper tool for the job
3. Use only tools that are functioning.
4. Maintain a clean, organized work area.
5. Questions relating to procedures or the proper handling of tools should be presented to your group leader.

Parallel Structure

Always Observe Safety Rules

1. Wear safety glasses.
2. Select proper tool for the job.
3. Use only properly tools.
4. Maintain a clean, organized work area.
5. Ask your group leader questions about procedures and tool handling.

Instructions use the imperative mood because individual steps are commands to the users, not statements about the process. The next example illustrates the difference between imperative mood and indicative mood.

> Clamp the specimen onto the flat plate. *(imperative mood)*
>
> The specimen is clamped onto the flat plate. *(indicative mood)*

Instructions that employ second person, referring to the user as *you,* are the most concise and effective. Sometimes the *you* is not stated, but the users, whether readers or listeners, understand that they are being directly addressed:

> [You] Use only properly functioning tools. (second person)
>
> [You] Make sure not to type anything after the target drive name, not even a space. (second person)
>
> [You] Wear safety glasses. (second person)

Instructions have the doer of the action as an implied subject of the sentence.

Visual Elements

Effective visuals are critical parts of instructions. The following visual elements were compiled by the same business and industry professionals who made the list of content elements earlier in the chapter.[22] Checking your instructions against this list may help you produce accessible, comprehensible, and usable instructions.

- Select appropriate visuals, especially for key parts and processes.
- Balance visual and verbal content.
- Select accurate visuals that are easily understood.
- Juxtapose labeled visuals with relevant text.
- Design an appealing, usable format.

Visuals in technical communications are usually placed as close as possible following or next to the text reference. Unless absolutely no possibility of confusion or ambiguity exists, visuals should always be referred to in the text by figure number or title, so users look at the correct visual for each step.

Appropriate Visuals

The combination of verbal and visual components is especially useful in instructions. Visuals can be used to illustrate a variety of elements:

- parts, tooling, equipment
- sequence of steps
- positioning of the operator and/or equipment

- development or change of object or equipment
- screens and pull-down menus in software development

Visuals are so important that one Web site advertises the quality of its visuals in promoting its manuals. Figure 21.11 presents a pair of pages from manuals, one with low-quality visuals and one with high-quality visuals. The high-resolution visuals are offered by a writer of manuals for the Mercedes-Benz all-terrain vehicle.

Why do you think that professional designers recommend keeping the visuals as simple as possible in instructions?

Why are drawings are often preferred over photographs?

No matter how clear and accurate the steps, the instructions will be virtually unusable if the accompanying visuals are unusable. Images need be high-resolution with distinguishable details, visible callouts, and clear figure-ground contrast.

The decision about whether to include visuals must be balanced by space requirements, download time, and production costs. The following checklist can help you determine whether visuals are appropriate for a particular set of

FIGURE 21.11 **Comparison of Low-quality and High-quality Visuals**[23]

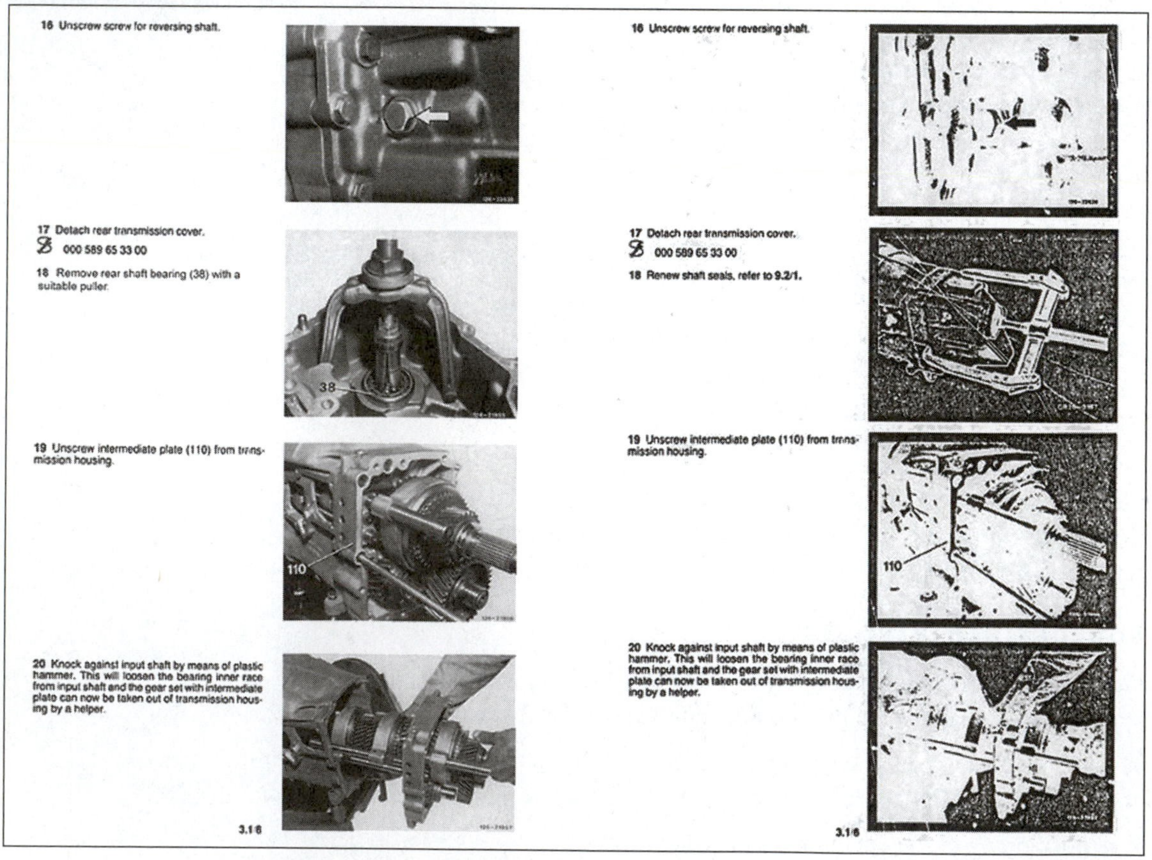

16 Unscrew screw for reversing shaft.

17 Detach rear transmission cover.

000 589 65 33 00

18 Remove rear shaft bearing (38) with a suitable puller.

38

19 Unscrew intermediate plate (110) from transmission housing.

110

20 Knock against input shaft by means of plastic hammer. This will loosen the bearing inner race from input shaft and the gear set with intermediate plate can now be taken out of transmission housing by a helper.

3.1 6

16 Unscrew screw for reversing shaft.

17 Detach rear transmission cover.

000 589 65 33 00

18 Renew shaft seals, refer to 9.2/1.

19 Unscrew intermediate plate (110) from transmission housing.

110

20 Knock against input shaft by means of plastic hammer. This will loosen the bearing inner race from input shaft and the gear set with intermediate plate can now be taken out of transmission housing by a helper.

3.1/6

instructions. A "yes" answer to any of these questions indicates that visuals should be integrated:

- Will a flowchart or series of visuals clarify the overall process for users?
- Will visuals enable users to clearly understand the end result?
- Will visuals help users correctly identify parts?
- Will visuals help users understand and implement individual verbal steps?
- Will including visuals emphasize safety and decrease risk?

Do you believe that the extra time, effort, and ink of creating a flowchart to accompany the steps is worthwhile?

What are some benefits?

When might a flowchart not be appropriate in instructions?

Visuals are often essential in illustrating and clarifying a sequence. Imagine trying to follow instructions for electronic troubleshooting, botanical identification, or celestial navigation without visual support. The visuals eliminate ambiguity by providing diagrams or maps where appropriate in the text. Figure 21.12 illustrates a series of steps and accompanying flowchart from a *Diagnostics and Troubleshooting Guide.* This process for troubleshooting a parallel printer is accompanied by a map that gives users a choice of step-by-step instructions or flowchart instructions, which some users will find faster and easier.

FIGURE 21.12 **Flowchart Accompanying Instructions**

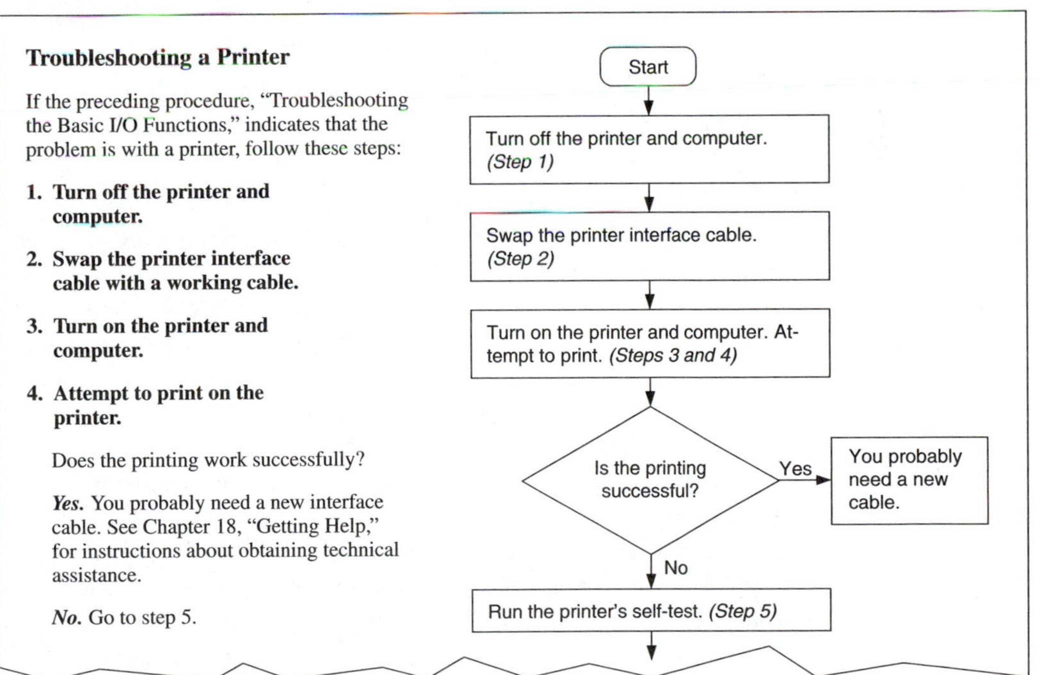
Troubleshooting a Printer

If the preceding procedure, "Troubleshooting the Basic I/O Functions," indicates that the problem is with a printer, follow these steps:

1. **Turn off the printer and computer.**
2. **Swap the printer interface cable with a working cable.**
3. **Turn on the printer and computer.**
4. **Attempt to print on the printer.**

   Does the printing work successfully?

   Yes. You probably need a new interface cable. See Chapter 18, "Getting Help," for instructions about obtaining technical assistance.

   No. Go to step 5.

Start
Turn off the printer and computer. (Step 1)
Swap the printer interface cable. (Step 2)
Turn on the printer and computer. Attempt to print. (Steps 3 and 4)
Is the printing successful?
Yes
You probably need a new cable.
No
Run the printer's self-test. (Step 5)

WEBLINK

Various kinds of visuals play a critical role in instructions. Medscape, an online continuing education resource for physicians and other health professionals, provides a bank of anatomical drawings to assist them in instructing their patients about a broad range of medical conditions and treatments. Go to **www.english.wadsworth.com/burnett6e** for a link that requires free site registration to view these simple line drawings.

CLICK ON WEBLINK

CLICK on Chapter 21/anatomical drawings

Visual and Verbal Balance

Some processes are more easily understood through a visual presentation than a verbal one. For example, suppose that you are eating in a restaurant when you choke on a piece of chicken. You give the international sign of choking — hands to your throat. Your friends have put off learning the Heimlich maneuver, but they see a "Choke Saver" poster on the restaurant wall. Which version of the poster would you want them to check?

1. *Entirely verbal* — well-organized chronological paragraphs including causal elements, clear topic sentences, and good chronological transitions.
2. *Verbal and visual* — sequence of captioned photographs showing a choking victim being saved by a trained person.
3. *Verbal and visual* — sequence of clear, captioned sketches showing a choking victim being saved by a trained person.
4. *Entirely visual* — sequence of clear sketches showing a choking victim being saved by a trained person, with arrows and inserted enlargements of critical positioning.

No doubt you would probably select version 4. Version 1 would take too long to read, and without the sketches the saver would not be sure of correctly positioning his or her hands on the victim's body. Version 2 would be an improvement, but photographs frequently show too much detail; also, people's clothing would obscure correct positioning. Version 3 would be a good choice, as long as the saver took the time to read necessary information in the captions. But, assuming the sketches and enlargements are accurate, version 4 would ensure the fastest and most accurate response.

Most instructions require a balance of verbal steps and visual support. Visuals should be included when they help the user complete the task more quickly or accurately and with less anxiety. *Access for Everyone: A Guide to Accessibility with References to ADAAG* [Americans with Disabilities Act Guidelines] balances visual and verbal information. Figure 21.13 is notable because the drawings are to scale, which is not always the case, and the information is easy to understand. This

FIGURE 21.13 **Excerpt from *Access for Everyone*[24]**

O Protruding Objects

Protruding objects should be located so that people who use canes can detect them. The standard cane sweep allows detection of the leading edges of objects up to 27in. (685mm) from the traveling surface. People using canes may not detect a protruding object in time where the bottom edge is higher than 27in. (685mm). Protruding objects must not reduce the required minimum clear width in a route.

O1 Check for objects that extend into the routes, including protrusions from the walls and ceilings and free-standing objects within the routes.

O2 Remove protruding or freestanding objects that reduce the minimum accessible route width and maneuvering space.

O3 Ensure that where the leading edge of a protruding object is between 27in. (685mm) and 80in. (2030mm) above the route surface, the object protrudes no more than 4in. (100mm) into the route and does not extend into the required minimum clear width of the route. **ADAAG: 1998 4.4; 2002 307**

Exception: Handrails are permitted to extend into the clear width 4.5in. (115mm) maximum.

Figure RT.12 shows a person using a cane to detect obstacles in a route passing an allowable protrusion.

O4 Where a sign or object is mounted on posts that are 12in. (305mm) or more apart, ensure that the lowest edge of the sign or object is

a. 27in. (685mm) or less above the traveling surface to ensure that a person using a cane will be able to detect the object **or**

b. 80in. (2030mm) or more above the traveling surface to provide sufficient headroom.

ADAAG: 1998 4.4; 2002 307

Figure RT.13 shows objects mounted between posts and the required clearances.

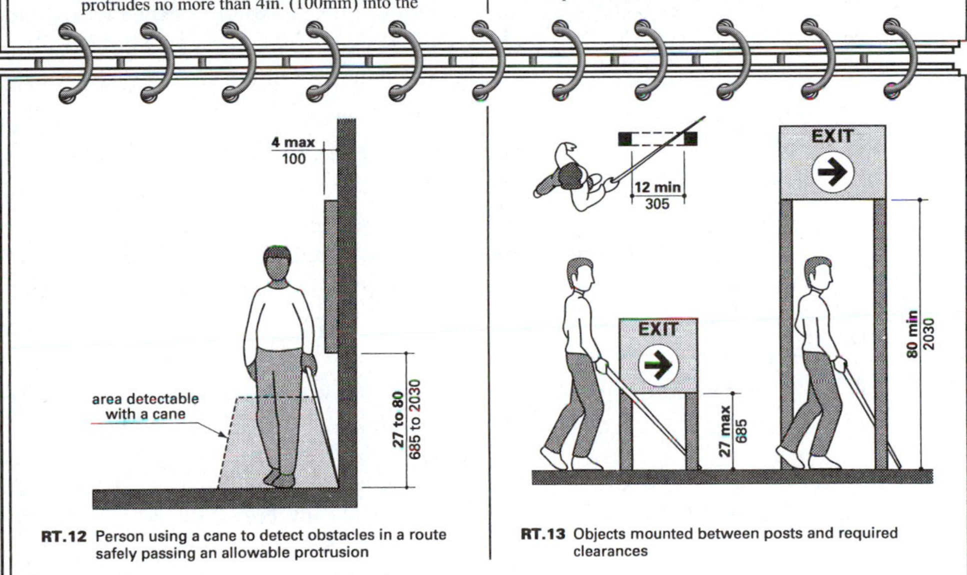

RT.12 Person using a cane to detect obstacles in a route safely passing an allowable protrusion

RT.13 Objects mounted between posts and required clearances

The headings and subheadings make the hierarchy clear.

Human figures are in proportion to the built environment.

The figure references are easy to use.

The language is clear and unambiguous.

The information in the text corresponds directly to the information in the accompanying visuals.

Americans with Disabilities Act Guidelines references, measurements, and exceptions are clearly identified.

careful attention to accuracy increases users' comprehension about the constraints in the built environment.

Accurate Visuals

Accuracy is critically important in any type of visuals in instructions. Visuals that cannot be easily understood are not much help to the user. Many problems can be eliminated if the writer and artist consider visuals as an integral part of the direction, not just a decorative addition.

The accuracy and appropriate use of each visual must be double-checked by both the artist and the writer. How helpful is a wiring diagram that omits some of the circuits? How can users confidently follow a step that reads, "Add two drops of oil in all set-screw holes, locations shown in the diagram," when the diagram does not label the set screws? How can users be expected to follow directions that say, "Calibrate by turning the middle knob to zero," with a diagram that shows only two knobs? Accuracy means completeness of the visual, inclusion of accompanying labels, and visuals and text that are related.

The size of visuals can affect users' ability to interpret these visuals accurately. If a drawing is too small, users won't be able to identify the important parts of the subject. Occasionally, a full-view drawing or photograph is accompanied by an enlargement of a crucial part. A drawing or photograph may have small parts identified by arrows, circles, boxes, or highlighted portions. Visuals should always be labeled with an indication of the scale (actual size, 1/2 scale, 1/5 scale, 1 inch = 1 foot, 1 inch = 10 miles, and so on).

Figure 21.14 illustrates one step in the process of installing a new hard drive in a desktop computer. The task is easier to understand and perform because the text is accompanied by a drawing that shows the connection points and labels each of the parts. Figure 21.14 contains two visual strategies to enhance interpretation. First, the graphic uses a cutaway to show the new hard drive's placement under an existing drive, with connections threaded under the existing

FIGURE 21.14 **Cutaway View and Enlargement of Image to Increase Technical Accuracy**[25]

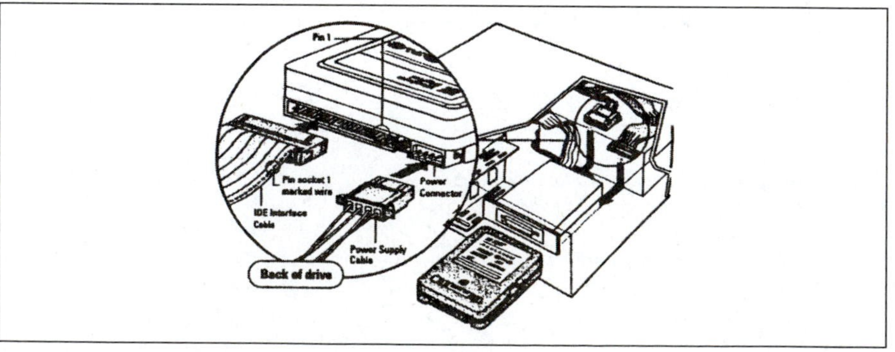

drive. Then an enlargement shows the installer the exact location on the back of the new drive for connecting the IDE interface cable and the power supply, a view not seen in the cutaway view.

Primarily Visual Instructions

In response to the issues raised by international users as well as by American users who prefer visual steps, many companies have experimented with primarily or totally visual instructions. For example, Figure 21.15 presents the first two steps of the primarily visual instructions for opening and replacing the ink cartridge for a printer.

Dependence on visuals challenges designers and writers because pictures, signs, and symbols do not have universal meanings. For example, could the up-arrow sign on shipping cartons be misinterpreted? Should the carton be placed with the arrow pointing up to indicate the top of the carton? Or should the carton be placed with the arrow pointing down, indicating the carton's most stable position? The answer depends on a person's perception.

Color coding is very important in primarily visual instructions, often replacing verbal emphasis and differentiating similar elements in a drawing. Problems arise, however, if instructions are to be reproduced (perhaps photocopied) in a manner that ignores the color. For this reason, color coding is often combined with a variety of design patterns.

FIGURE 21.15 **Visual Instructions**[26]

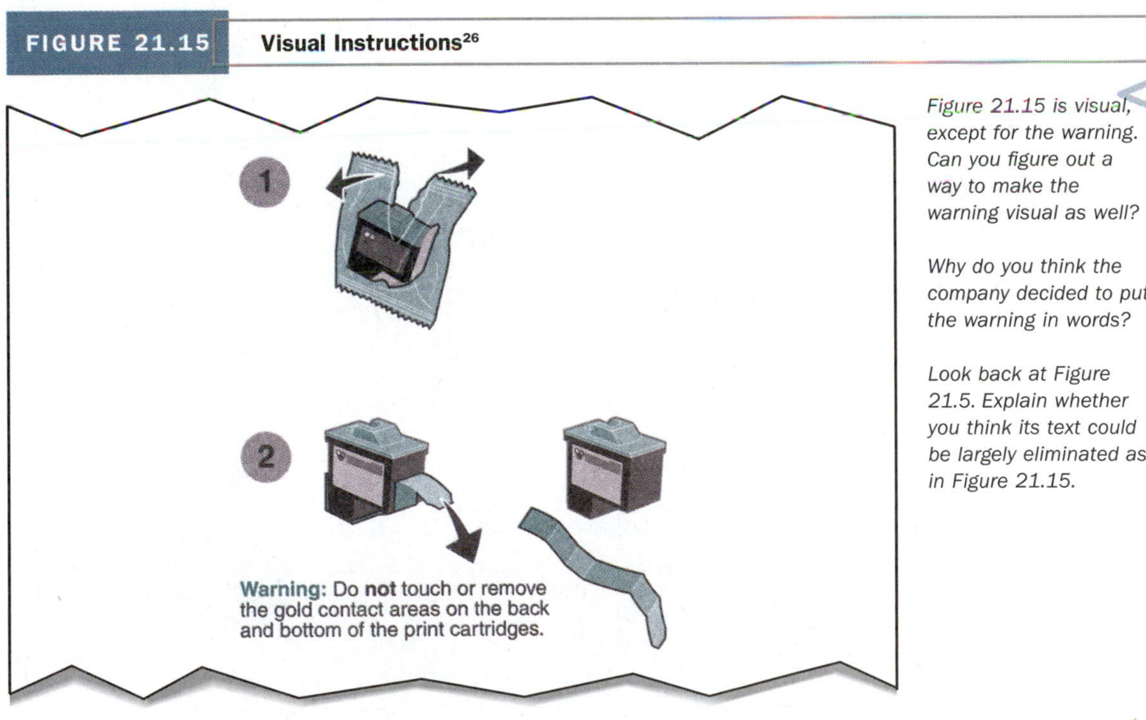
1
2
Warning: Do not touch or remove the gold contact areas on the back and bottom of the print cartridges.

Figure 21.15 is visual, except for the warning. Can you figure out a way to make the warning visual as well?

Why do you think the company decided to put the warning in words?

Look back at Figure 21.5. Explain whether you think its text could be largely eliminated as in Figure 21.15.

Do you agree that visuals carry more meaning than text for some audiences?

What problems might result if this premise is overgeneralized?

More varieties of primarily visual instructions exist than could ever be presented in this chapter; however, modular instructions developed by Deere & Company are particularly effective and can be adapted to many situations. The company's premise is that visuals often carry more meaning than text. Because many agricultural workers have limited literacy, the instructions make each step accessible and appealing, use simple language, and feature only easily understood visuals.

Because agricultural work is among the most dangerous in the country, Deere presents safety hazards in every owner/operator manual. Figure 21.16 illustrates two of the safety modules that are included in the owner/operator manual for a small tractor.

The basic grid for each page in a Deere manual has horizontal quarters; however, this layout is very flexible. Writers can use four modules on a page, or three, two, or one, depending on the amount of explanation and illustration that's necessary and the number of cautions and warnings that are needed.

FIGURE 21.16 **Safety Modules from the Owner/Operator Manual for a Deere & Company Tractor**[27]

Accessibility

- *The size and quality of the visuals are sufficient.*
- *The relevant visuals and text are placed together.*

Comprehensibility

- *In these figures, the especially important instructions use negative phrasing: "Do not let children operate the tractor." Explain whether this is more effective than using positive phrasing such as, "Only licensed adults drivers should operate the tractor."*
- *What terms presume that users have some prior knowledge about the equipment and its operation?*
- *Is the information in Figure 21.16 accessible to limited literacy workers?*

Usability

- *The steps are not numbered. Explain whether you think numbering is necessary here.*
- *The visuals are coordinated with the relevant text. Can either the visuals or text stand alone?*

PROTECT CHILDREN

Keep children and others away when you operate machine.

BEFORE YOU BACK UP:
—Stop the PTO.
—Look behind tractor for children.

Do not let children operate tractor.

Do not let children ride on tractor or any implement.

AVOID TIPPING

Do not drive where machine could slip or tip.

Stay alert for holes, rocks, and roots in the terrain, and other hidden hazards. Keep away from drop-offs.

Slow down before you make a sharp turn.

Driving forward out of a ditch or mired condition or up a steep slope could cause tractor to tip over backward. Back out of these situations if possible.

Use care when pulling loads or using heavy equipment. Use counterweights or wheel weights suggested in this operator's manual.

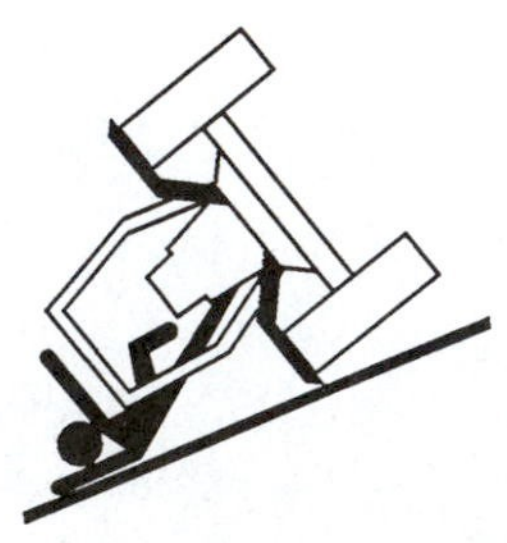

Product liability laws provide compensation for harm such as personal injury, death, property damage, or financial loss caused by a defective product. A manufacturer can be held responsible for any number of causes: faulty design, manufacture, installation, preparation, assembly, testing, or packaging. Or the defect may come from failure to provide either adequate directions or adequate cautions or warnings. Companies should protect their employees by providing accessible information about safety, protect their clients and customers by increasing the chances that work will be performed appropriately, and protect their own corporate interests by reducing risk of litigation resulting from inadequate instructional documents.

What is your ethical responsibility in preparing accurate, usable instructions? What is your legal responsibility?

WEBLINK

ANSI sets the standards for safety signs in the United States, but the International Organization for Standardization (ISO) is responsible for international standards. In 2002, ANSI and ISO started working toward standardizing their standards, a process that will take several years. Go to www.english.wadsworth.com/burnett6e for a link.

CLICK ON WEBLINK

CLICK on Chapter 21/ANSI and ISO formatting

Safety Standards

Because of the inherent risks associated with some instructions, writers have an obligation to provide users with sufficient notices about cautions, warnings, and dangers. As you read in Chapter 12, ANSI is responsible for publishing the safety standards used in the United States. The content for safety labels, "Product Safety Signs and Labels," usually referred to as ANSI Z535.4 standard, is explicit about what a safety label includes.[28]

1. The *alert symbol* ▲ indicates that a possible human injury hazard exists. It is omitted when the hazard or damage will affect only property.[29]
2. A *signal word* in combination with a background color indicates the degree or level of the hazard. How do you choose the appropriate signal word?
   - First, ask this question: If the safety label's message is ignored, how severe will the injury be? If the answer is death or serious injury, choose between DANGER (red) and WARNING (orange).
   - Then ask this question: If the message is ignored, how likely is it that an injury will occur? If an injury is highly likely, choose DANGER. If it is a possibility, choose WARNING.

⚠ DANGER

"DANGER indicates an imminently hazardous situation that, if not avoided, will result in death or serious injury. This signal word is to be limited to the most extreme situations."

⚠WARNING

"WARNING indicates a potentially hazardous situation that, if not avoided, could result in death or serious injury."

⚠CAUTION

"CAUTION indicates a potentially hazardous situation that, if not avoided, may result in minor or moderate injury. It may also be used to alert against unsafe practices."[32]

- If the answer to the question, "If the safety label's message is ignored, how severe will the injury be?" is not death or serious injury, choose CAUTION (yellow).[30]

3. The *graphic image* is recommended but optional. An image grabs attention and can often communicate across many (though certainly not all) cultures.
4. The *message* identifies three things:[31]
   - The specific hazard, including the degree or level of seriousness, is included.
   - The probable consequence of involvement with the hazard is included.
   - Ways to avoid the hazard are conveyed in words and, usually, images.

Figure 21.17 shows the basic structure of warning signs: alert symbol (if the risk is to humans), signal word, symbol, and message. The signs can be formatted horizontally or vertically.

Recent ANSI standards increase legibility. The message is black on a white background, rather than on a colored background, which reduced the figure-ground contrast. The new format, shown in Figure 21.18, also recommends an image.

WEBLINK

You may want to read more about how to construct warning signs that conform to ANSI standards. Go to www.english.wadsworth.com/burnett6e for a link.

CLICK ON WEBLINK

CLICK on Chapter 21/ANSI standards

FIGURE 21.17 **ANSI Z535.4 Formats for Warning Signs**[33]

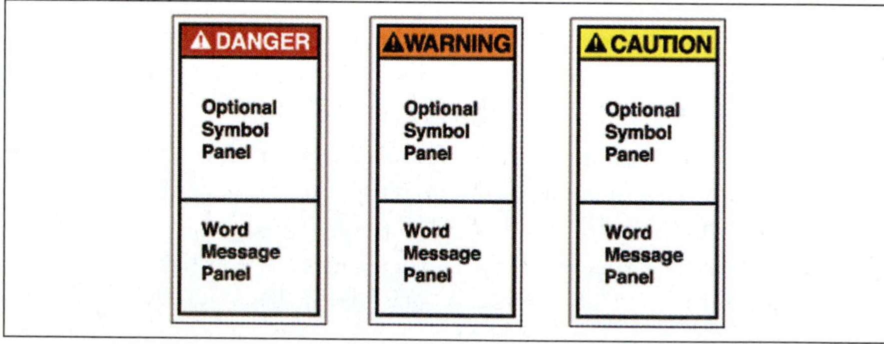

FIGURE 21.18 **Old Sign Format and New Sign Format**[34]

Old

New

Old

New

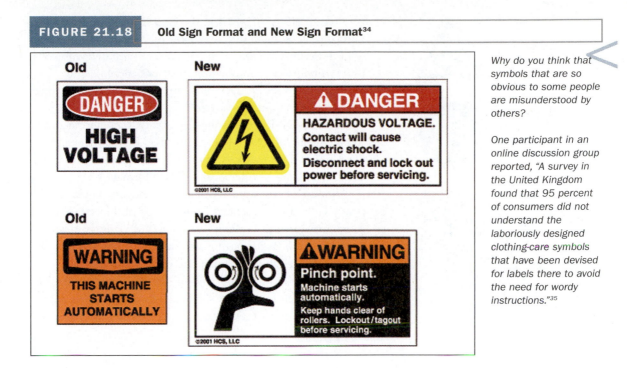

Why do you think that symbols that are so obvious to some people are misunderstood by others?

One participant in an online discussion group reported, "A survey in the United Kingdom found that 95 percent of consumers did not understand the laboriously designed clothing-care symbols that have been devised for labels there to avoid the need for wordy instructions."[35]

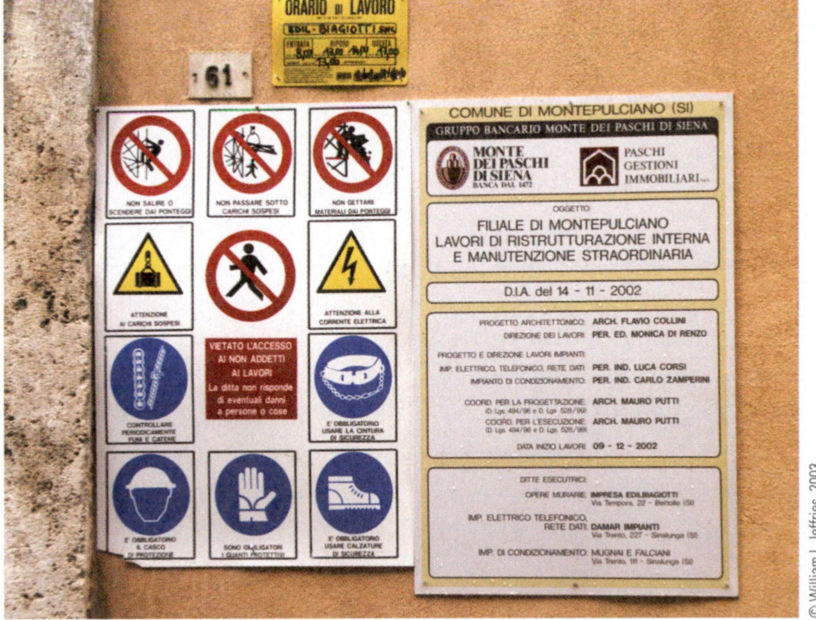

© William L Jeffries, 2003

This photo in Montepulciano, Italy, shows some of the ISO symbols that typically appear at construction sites throughout Europe.

What do you think each symbol means?

How much does knowing the location of the sign is a construction site influence your interpretation?

Which symbols might be open to multiple interpretations?

Warning Signs identify the hazard.

Mandatory Action Signs convey an action step that should be taken to avoid the hazard.

Probation Signs are used to convey a prohibited action.

Recent efforts of ANSI in the United States and ISO in Europe to coordinate their standards mean that people will need to become familiar with both sets of symbols symbols. ISO symbols use a combination of shapes and colors to communicate various kinds of warnings, which come in three categories, illustrated to the left.[36]

Unlike ANSI symbols and signs that always appear in a specific format, ISO symbols can be used by themselves. As the ANSI and ISO safety standards become similar, more and more ISO symbols will appear in ANSI formats.

Notices of cautions, warnings, and dangers should be clearly labeled and separated from the rest of the text. Some companies place all cautions, warnings, and dangers in a separate section at the beginning of the document. While this may satisfy a legal requirement and create an overall impression about the importance of careful operation, it is also very important to place cautions, warnings, and dangers in the text, in the actual place where the users will need this information.

WEBLINK

Designing symbols that are understood by everyone to mean the same thing is impossible. But designers of symbols need to try to create images that many people will interpret the same way. To read about a number of topics related to symbols design, go to www.english.wadsworth.com/burnett6e for a link.

CLICK ON WEBLINK

CLICK on Chapter 21/symbol design

Why don't people pay attention to cautions and warnings? Part of the reason could be that the wording of some cautions and warnings is so ridiculous that they cause humor rather than encourage safety. Organizations who don't conduct usability testing on the wording of their warnings risk having the serious intent of their warnings dismissed, as shown in these examples:

On the package for Top Cog fan belts (for automobiles) — Do not change the belt while the engine is running.

On Tesco's Tiramisu dessert — (printed on bottom of the box) Do not turn upside down.

On Nytol sleep aid — Warning: may cause drowsiness.

On Sainsbury's peanuts — Warning: contains nuts.

From a Pop-Tart box — Warning: Pastry Filling May Be Hot When Heated.[37]

Liability

Although workplace professionals have a number of choices about how they present cautions, warnings, and dangers, they need to know that providing inadequate safety information is a liability issue. The ethics sidebar below addresses questions about who is responsible and liable when something goes wrong.

ETHICS SIDEBAR

Warning Labels and Cartoons: Ethics in Instructional Texts

"If a technical writer's prose on a prescription drug data sheet is unclear, and a patient blows his brains out in a medicinally caused fit of depression, is the pharmaceutical company liable?"[38]

Researcher John Caher poses this question in his review of a court case relevant to technical professionals. The answer provided by the court is that technical professionals and their companies are responsible for the documents they produce. This case argued that a retired state trooper committed suicide after taking several different drugs because the warnings included with one of the drugs were insufficient. After a thorough review of the usage and warnings sections of the documents included with the drug, the court ruled that the instructions and warnings were legally sufficient to warn of potential problems. The court case, according to Caher, indicates both the ethical and legal requirements that technical professionals must recognize when including warning sections in instructional and informational texts.[39]

Warning labels are not the only element of instructional texts that technical professionals must write with caution. Cartoons or cartoon-like drawings are another feature found in many instructional texts. Such drawings have multiple benefits: They allow readers to visualize difficult steps; they emphasize dangerous procedures; and they can help overcome language and cultural differences for international audiences. These differences, however, can create ethical problems as well. These ethical problems, researcher Philip Rubens argues, arise because "certain cultural elements must be evoked to inform the viewer's interpretation of the [cartoon]."[40] If an audience does not have the necessary cultural knowledge to understand a cartoon, then confusion can occur. This confusion can be especially problematic when groups have similar images that have different meanings. For example, wavy lines can denote water as well as an electrical component. Confusing

Have you ever found a warning label in an instructional text unclear?
How do you make sure cartoons in an instructional text clearly communicate with your audience?

the difference in meaning of the wavy line would be potentially dangerous in an instructional text.

To create ethical instructional texts, technical professionals must recognize the various ways that an audience may interact with the document. Warning labels and cartoons provide avenues for communicating important information to an audience, but technical professionals must be aware of the varying interpretations. As Caher indicates in his review of warning labels, "When a technical writer's work is unclear, and an operator inadvertently reformats a hard drive, that's unfortunate. But if a technical writer's inaccuracy or imprecision claims a life, that's another matter altogether."[41]

How would you avoid sending mixed cultural messages in cartoons?

Adequacy

When you prepare instructions, you need to be sure that they satisfy the legal requirements for adequacy.[42] In general, if you ensure that your instructions and warnings are accurate, accessible, and appropriate, as described in Figure 21.19, you will be on your way to meeting the legal requirements for adequacy. You need to know that adequacy of instructions and adequacy of warnings are different — and you need to ensure that you provide both. You need to have instructions that provide accurate, accessible, and appropriate information, and you need to provide accurate, accessible, and appropriate warnings about any potential risks, hazards, or dangers. Who needs to be warned? Anyone who might reasonably be expected to use the product. How many clear and concise warnings do you need to include? Enough to make sure people are safe but not so many that they start to ignore the warnings. The less obvious the risk, hazard, or danger, the more important the warnings are.

WEBLINK

You need to consult a qualified attorney when making any legal decisions; however, you can educate yourself so that you can have more productive discussions with your attorney. One of the areas you need to be informed about is liability. To begin, go to **www.english.wadsworth.com/burnett6e** for useful links.

CLICK ON WEBLINK

CLICK on Chapter 21/warnings and liability

The user's manual that accompanies the AC-powered ionization smoke detector manufactured by BRK Electronics provides an example of a document

FIGURE 21.19 **Adequacy of Instructional Documents**

Accurate: Are the instructions accurate?

- ***Understand the product and its likely users.*** Know such things as the product's purpose and operation as well as users' probable experiences and capabilities.
- ***Describe the product's functions and limitations.*** Know what the product can be expected to do and not do.
- ***Fully instruct on all aspects of product ownership.*** Deal with factors such as assembly, installation, use and storage, testing, maintenance, and emergencies.
- ***Contain clear, correct, and tested instructions.*** Check the instructions using a variety of text-based, expert-based, and user-based testing.
- ***Identify risks, hazards, and dangers.*** Include information about risks, hazards, and dangers that are obvious and foreseeable (occurring when the product is used in the usual and expected manner) and those risks, hazards, and dangers that are not obvious.

Accessible: Are the instructions accessible?

- ***Present important directions or warnings so users spot and follow them.*** Make the warnings, cautions, and dangers conspicuous by using design factors such as placement, type size, type style, leading, line length, navigation bars, and cueing devices (such as indentation, icons, and boxes).
- ***Inform users in a timely manner of defects discovered after marketing.*** After the instructions are complete and the product is marketed, continue to determine the product's safety and effectiveness and have a plan for informing users if you discover problems.
- ***Reach product users.*** Consider the most effective way to get the instructions to the user: Package insert? Separate manual? Printed on the packaging? Web site? Electronic help?

Appropriate: Are the instructions appropriate?

- ***Use words and graphics that suit the intended audience.*** Make sure that all the textual and visual information can be understood by intended users.
- ***Appropriately warn of product hazards.*** Identify the nature of dangers, normal misuses that can cause danger, and the extent of harm that can result from misuse.
- ***Meet government, industry, and company standards.*** Cite specific state and/or federal standards that the product meets.
- ***Offset claims of product safety in advertising or other materials that don't include strong warnings.*** If the advertising downplays dangers, be especially careful to include warnings to counter these promotional claims.

that appears to meet these criteria for adequacy. The excerpts in Figure 21.20 from the user's manual are annotated to illustrate the way the BRK writers address product liability concerns in clear, easy-to-understand language with helpful visuals.

FIGURE 21.20 Instructions that Consider User's Needs as Legal Requirements[43]

AC POWERED IONIZATION SMOKE DETECTOR
WITH BATTERY BACK-UP
INPUT: 120 VAC, 60 Hz, .045A

USER'S MANUAL

BASIC INFORMATION ABOUT YOUR SMOKE DETECTOR

- **Put detectors inside and outside of every bedroom area and on every floor of your home.**
- **Put detectors close to the center of the ceiling when ceiling mounted.**
- The detector may beep when you put the battery in it.
- If the **indicator light** on the detector **is on, the detector is receiving AC power.** This does not ensure that the detector is working properly.
- **Test the detector weekly by holding the test switch button in for about 10 seconds until the alarm sounds. The alarm may not sound immediately** when you press the button. This checks all detector functions.
- If the detector **beeps once a minute, it needs a new battery.**

WARNING
GENERAL LIMITATIONS OF SMOKE DETECTORS:
WHAT SMOKE DETECTORS CANNOT DO

Smoke detectors have played a key role in reducing home fire deaths in the United States. However, according to the Federal Emergency Management Agency (an agency of the U.S. Government), they may not go off or give early enough warning in as many as 35% of all fires. What are some reasons smoke detectors may not work?

Smoke detectors will not work without power. Battery operated smoke detectors will not work without batteries, if the batteries are dead, if the wrong kind of batteries are used, or if the batteries are put in wrong. AC powered smoke detectors will not work if the power supply is cut off for any reason. Some examples are a power failure at the power station, a failure along a power line, a failure of electrical switching devices in the home, an open fuse or circuit breaker, an electrical fire, or any other kind of fire that reaches the electrical system and burns the wires. If you are concerned ab limitations of either batteries or power for your smoke detec-

WHAT THIS SMOKE DETECTOR CAN DO

This smoke detector is designed to sense smoke that comes into its sensing chamber. It does not sense gas, heat, or flame.

This smoke detector is designed to give early warning of developing fires at a reasonable cost. This detector monitors the air. When it senses smoke, it sounds its built-in alarm horn. It can provide precious time for you and your family to escape before a fire spreads. Such **early warning is only possible,** however, **if the detector is located, installed, and maintained as described in this User's Manual.**

⚠WARNING **This smoke detector is designed for use in a single residential living unit only.** In other words, it should be used **inside** a single-family home or apartment. It is not meant to be used in lobbies, hallways, basements, or another apartment in multi-family buildings, **unless there are already working detectors in each family unit. Smoke detectors placed in common areas outside of the individual living unit (such as on porches or in hallways) may not provide early warning to residents in multi-**

WHERE SMOKE DETECTORS SHOULD NOT BE PUT

Nuisance alarms occur when smoke detectors are put up where they will not work properly. **To avoid nuisance alarms, do not place detectors:**

- **In or near areas where combustion particles are present.** (Combustion particles are the by-products of something that is burning.) **Areas to avoid include kitchens with few windows or poor ventilation, garages** where there may be vehicle exhaust, **near furnaces, hot water heaters,** and **space heaters.**
- **Put up smoke detectors at least 20 feet** (6 meters) **away from places where combustion particles are normally present, like kitchens.** If a 20-foot distance is not possible, put the detector as far away from the combustion particles as possible, preferably on the wall. To prevent nuisance alarms, provide good ventilation in such places.
- **If smoke detectors are to be located in halls or rooms near or adjacent to kitchens where there is no wall above the doorway between rooms, mount detectors on an inside wall closest to the bedroom area and furthest from the kitchen.**

IMPORTANT: In mobile homes where a 20-foot distance is not possible, put smoke detectors as far away from combustion particles as possible. Provide good ventilation. **Do not, for any reason, disable the detector to avoid nuisance alarms.**

- **In air streams passing by kitchens.** Figure 6 shows how a detector can sense combustion products in normal air-flow paths. The picture shows how to correct this problem.

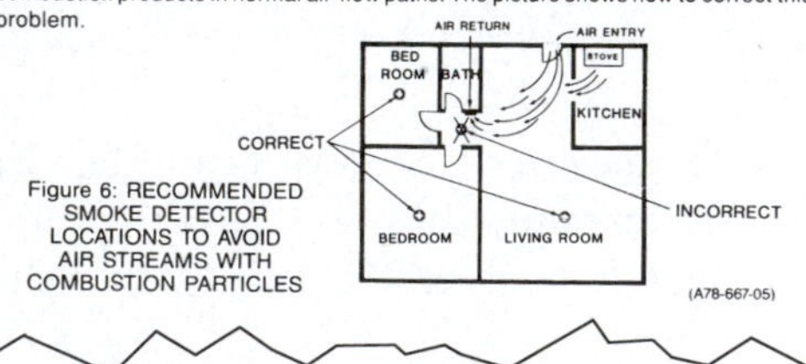

Figure 6: RECOMMENDED SMOKE DETECTOR LOCATIONS TO AVOID AIR STREAMS WITH COMBUSTION PARTICLES

Bulleted list indicates awareness of information needs of consumers using the product.

The manual uses clear, easy-to-understand vocabulary.

Clear, easy-to-read headings signal each section of the instructions.

Icons and boxes flag critical information.

The product's expected performance is clearly stated.

Typographic cues such as boldfacing signal especially important information.

Definitions or explanations of unfamiliar terms are provided.

Figures present information that is difficult to explain in words.

Incorporated figures illustrate information in the text. Limitations of performance are identified.

Figure labels give users critical information from the captions.

FIGURE 21.20 Instructions that Consider User's Needs as Legal Requirements (continued)

Directions for assembly and installation are provided.

Multiple cueing devices signal warnings, cautions, and dangers.

Terminology about wiring is specialized, since directions clearly state wiring should be performed only by a licensed electrician.

HOW THIS DETECTOR SHOULD BE PUT UP

This detector is made to be mounted on any standard 4-inch octagonal junction box.

Model 86RAC is made to be mounted on the ceiling, or on the wall if necessary. Model 86RAC can serve as a single-station stand-alone unit **or** be interconnected with other 86RAC units.

⚠WARNING **Detector installation must conform to the electrical codes in your area and to Article 760 of the U.S. National Electrical Code. Wiring should be performed only by a licensed electrician.**

⚠WARNING **The circuit used to power the detector must be a 24-hour 120 VAC 60Hz circuit. B[illegible] sure the circuit canno[illegible] turned off by a switc[illegible]round faul[illegible]**

Information is presented in a logical, step-by-step sequence.

7. Grasp the tab on the battery drawer and pull it straight out as shown in Figure 8.

⚠WARNING **The battery is positioned WRONG in the factory to keep it fresh until installation. It must be re-positioned to provide DC back-up power.**

8. Remove the battery and re-position it properly, as shown on the label in the drawer. Push the drawer straight in until it is flush with the housing.

⚠CAUTION This smoke detector comes with a "missing battery" indicator that will prevent the battery drawer from closing if a battery is not installed. This is to warn you that the smoke detector will not work under DC power until a new battery is installed.

Visuals show how the user interacts with the product.

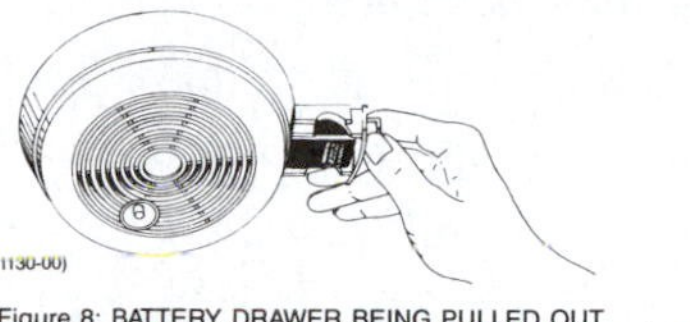

Figure 8: BATTERY DRAWER BEING PULLED OUT

Information about proper operation is easily accessible.

HOW TO TELL IF THE DETECTOR IS WORKING PROPERLY

When the indicator light (seen through the clear push button of the test switch) glows continuously, the detector is receiving AC power.

NOTE: **For interconnected Model 86RAC Detectors:**

When an interconnected system of Model 86RAC detectors goes into alarm under AC power, the indicator light will be OFF on the detector(s) sensing smoke, and will be ON all other detectors. When under DC power, no indication is provided.

Information about regular maintenance and testing is provided.

Test the detector weekly by pushing firmly on the test button until the horn sounds. This should take TEN seconds. If the alarm horn makes a continuous loud sound, the detector is working properly. **THIS IS THE ONLY WAY TO BE SURE THAT THE DETECTOR IS WORKING. TEST THE DETECTOR WEEKLY. IF THE DETECTOR FAILS TO TEST PROPERLY, HAVE IT REPAIRED OR REPLACED IMMEDIATELY.**

The manual warns against misuse.

⚠WARNING **Never use an open flame of any kind to test your detector.** You may set fire to and damage the detector, as well as your home. Also, do not use "aerosol" spray smoke detector testers. Build up of chemicals used in the spray can change detector sensitivity, or in some worst cases, impair detector functioning. The built-in test switch accurately tests all detector functions, as required by Underwriters' Laboratories.

Harm that could result from misuse is clearly specified.

Information about what to do in emergencies helps users.

⚠DANGER **If the alarm horn sounds a loud continuous sound and you have NOT pushed the test button, the detector has sensed smoke** or combustion particles in the air. **THE ALARM HORN IS A WARNING OF A POSSIBLY SERIOUS SITUATION. IT REQUIRES YOUR IMMEDIATE ATTENTION.**

The alarm could be caused by a nuisance situation. Cooking smoke or a dusty furnace, sometimes called "friendly fires," can cause the alarm to sound. If this happens, open a window or fa[illegible]the air to remove the smoke or dust. The alarm will turn itself off as soon as th[illegible]mpletely clea[illegible]T DISCONNE[illegible]HE POWER. TH[illegible] REMO[illegible]ECTIO[illegible]

Normal misuses that could cause danger are presented.

HOW TO TAKE CARE OF AND TEST THIS DETECTOR

Your smoke detector has been designed to be as maintenance-free as possible. To keep your detector in good working order, you must;

- **Test the detector weekly.** (See section "How to Tell if the Detector Is Working Properly."
- **Replace the battery once a year or immediately when the low battery "beep" signal sounds once a minute.** The low battery "beep" should last at least 30 days.

NOTE: For best performance, we recommend that you only use alkaline batteries (First Alert, Model FB2) as replacement batteries in this smoke detector. (Carbon zinc batteries are acceptable, but do not last as long.) First Alert batteries can be purchased at any retail store that sells batteries.

(If you cannot obtain a First Alert battery, the following batteries are also acceptable for proper smoke detector operation: Eveready #522, #1222, #216; Duracell #MN1604; or Gold Peak #1604P, #1604S.)

Individual and Collaborative Assignments

1. **Create an evaluation tool.**
   (a) Create a rubric to confirm that instructions fulfill audience needs, respond to the issues, and conform to conventions discussed in this chapter. Go to Chapter 1 to review Figure 1.2, Factors Affecting Accessibility, Comprehensibility, and Usability, to ensure that you include necessary factors. Account for the different kinds of instructions you write, such as instruction sheets, electronic instructions, manuals, training scripts, and tutorials.
   (b) Make sure your rubric is usable by other workplace professionals.
   (c) Test your rubric (which is itself a kind of instruction) and make revisions as needed.
   (d) Work with a small group to compare your rubrics and give each other ideas for improving each one.
   (e) Revise your rubric.
2. **Identify appropriate audiences.** Examine the variations in wording of a single step in the following three sets of instructions. Identify audiences appropriate for each version. Explain which version you think is the most effective and why.

Set I

Version A	Close the valve completely by turning the knob until the arrow points to the red dot.
Version B	Close the valve completely by turning the knob until the arrow points to the red dot. Failure to close the valve completely will result in pressure loss, decreasing the efficiency of the system.
Version C	So the system does not lose pressure, close the valve completely by turning the knob until the arrow points to the red dot.

Set II

Version A	Lower the safety bar before turning on the machine. If you operate the machine without the safety bar in place, you might severely injure your hand.
Version B	Lower the safety bar before turning on the machine to avoid personal injury.
Version C	Lower the safety bar before turning on the machine.

Set III

Version A	If the patient's pain persists, turn her on her side.
Version B	If the patient's pain persists, turn her on her side in order to alleviate the pressure.
Version C	If the patient's pain persists, turn her on her side. This change in position alleviates the pressure on the spine.

3. **Write the text for effective visuals.** In a small group, look at the instructions for changing the waste toner bottle in a photocopier (Figure 21.5). Write the steps (in English) for each of the seven illustrated steps.

4. **Revise the sign.** The sign in this photo is posted on the railing of a boat for ecological tours of St. Lucia Lake, Africa's oldest nature reserve. The area includes South Africa's largest populations of hippopotamus, crocodile, white-backed pelican, and pink-backed pelican. More than 530 species of birds frequent the lake region.[44] After you've read the sign in the photo, suggest an effective way to revise it to make the bulleted points parallel. Decide whether you should also revise it to eliminate the negatives or whether the negatives are effective. Do they increase or reduce compliance? If you think the negatives should be changed, make the revisions. If not, provide a rationale.

© William L Jeffries, 2003

5. **Revise an instructional document.** The following figure is taken from the draft of a project report submitted to the U.S. Department of Energy by an organization proposing a more efficient method of identifying the characteristics of substances at hazardous waste sites. The organization wanted both to convey its awareness of federal safety regulations for working at hazardous waste sites and to reinforce the idea that its employees had been given these guidelines. The information has not been formatted as instructions. Revise this excerpt so that the information is presented as

instructions, the sequence of information is logical and parallel, and the wording conveys the necessity of following the guidelines.

6.5.1 PPE Level D: Level D is the minimum PPE level. No respiratory protection is required; limited skin protection is required. Recommended equipment for Level D includes

- disposable coveralls over work clothes
- work gloves
- hard hat
- safety glasses or goggles
- chemical resistant boots or shoes

6.5.2 PPE Level C: Level C requires limited respiratory protection and skin protection from airborne hazards. Recommended equipment for Level C includes

- full-face purifying respirator
- chemical resistant clothing: overalls and long sleeve jacket; hooded, one- or two-piece chemical splash suit or limited use chemical resistant one-piece suit
- hard hat
- chemical resistant boots
- inner and outer chemical resistant gloves

6. **Revise instructions.** The following instructions, "Technical Instructions to Panel Physicians for Vaccination Requirements: Precautions in Administering Vaccines," are part of a 12-page set of instructions published by the Centers for Disease Control and Prevention. Read them and assess them using the rubric you've developed, taking special note of the audience (busy physicians, nurse practitioners, nurses, and public health officials). Then revise the instructions, considering both the content and the design.

Precautions in Administering Vaccines[45]

People administering vaccines should take the following necessary precautions to minimize the risk of spreading disease.

- They must wash their hands before and after seeing each applicant.
- They must wear gloves when administering vaccinations if they will have contact with potentially infectious body fluids or have open lesions on their hands.
- They must use sterile syringes and needles and preferably use disposable, autodestructible ones to minimize risk of contamination or needle stick.
- They must not mix different vaccines in the same syringe unless the vaccines are licensed for such use.
- They must discard disposable needles (not recap them) and syringes in labeled, puncture-proof containers for short-term disposal to prevent inadvertent needle stick injury or reuse.
- They must use an appropriate method, such as autoclaving or incineration, for long-term disposal of used needles and syringes.

7. **Attribute responsibility.** ConsumerAffairs.com reports a problem for some car owners:

   Toyota and Lexus owners' manuals specify that the oil should be changed every 7,500 miles or every six months, whichever comes first. Under severe driving conditions, oil should be changed every 5,000 miles or four months. [If this maintenance is not done,] a sludge problem occurs when oxidized oil builds up in an engine. It forms a mucky goo that can cause the engine to seize up.[46]

   The problem can be caused if car owners who drive a lot, especially under hard conditions, fail to change the oil much more often than the owner's manual specifies. Explain whether you believe that Toyota should cover the cover the cost of repairs under its five-year/60,000-mile power train warranty, whether or not owners have maintained their vehicles as specified in the owner's manual.

8. **Interpret a sign.** The sign in this photo is in Siena, Italy, to direct tourists who are interested in visiting the Duomo (the cathedral) and the Piazza del Campo (the main square in the center of the city, which some regard as the finest medieval square in Europe). What might be a tourist's initial reaction to the information in the sign? What information do you think the sign is conveying? Would you recommend any changes to the sign?

© William L. Jeffries, 2003

9. **Evaluate electronic help.** Most companies include electronic help files with their hardware and software products or have electronic help available on the Web for users.
   (a) Work with a small group to select an electronic help system for a product or process you regularly use or for one you are considering using or purchasing.
   (b) Use one of the rubrics created in Assignment 1 to evaluate a specific electronic help system.

(c) As a group, prepare and present an eight- to ten-minute oral presentation to summarize your assessment. Include screen shots of various aspects of the help system; annotate these screen shots with callouts (text labels and arrows) to illustrate its strengths and weaknesses. Make explicit recommendations for improving this specific system.

(d) At some place in the presentation, include these elements:

- Advantages and disadvantages of electronic help systems
- Criteria used for assessment and the assessment of this system
- Specific examples of ways in which this system fulfills the criteria
- Your recommendations for strengthening this system

10. Evaluate a manual.

(a) Locate a manual that is used by people in your field of study. In small groups (preferably with people who share your field of study), select criteria for evaluating manuals. Use the rubric you created in Assignment 1 (including accessibility, comprehensibility, and usability) to evaluate the manual you have selected.

(b) Prepare a short report that identifies the manual's strengths and recommends specific changes to eliminate weaknesses. Include discussion about what kinds of document testing could have eliminated many of the problems.

11. Rewrite poor or outdated instructions.

(a) Individually or in a small group, select a set of poor or outdated instructions (or, if you are particularly ambitious, select a procedure or activity that doesn't have instructions but needs them).

(b) Rewrite (or write) the instructions so they are accessible, comprehensible, and usable for the identified audience.

(c) Conduct text-based, expert-based, and user-based testing. Assess the testing results and make appropriate revisions. (Check Chapter 9, Usability Testing, to review options you have available.)

(d) Use the rubric you developed in Assignment 1 to check your draft as well as your final version.

12. Create visual instructions.

(a) Individually or in a small group, select a task or process that could effectively be presented entirely visually for a specified audience.

(b) Create a storyboard of your plan and then prepare the visual instructions.

(c) Test your instructions with representative users, and then analyze and interpret the test results.

(d) Revise your visual instructions so they are accessible, comprehensible, and usable for the identified audience.

Chapter 21 Endnotes

1 Adapted from (1) Kaner, C. (1995). Liability for defective documentation [Electronic version]. *Software QA, 2*(3), 8. Retrieved January 10, 2004, from http://www.badsoftware.com/baddocs.htm; and (2) Dick, D. (2000) Justification for documentation [Electronic version]. *Usability Interface, 6*(4). Retrieved January 11, 2004, from http://www.stcsig.org/usability/newsletter/0004-justification.html

2 SecureDoc Project. (2003, December 9). *Usable and safe operating manuals for consumer goods: A guideline.* Retrieved January 11, 2004, from http://216.239.53.104/search?q=cache:xiarLfYHdDoJ:www.tceurope.org/pdf/proof%2520Total%2520draft.pdf+manuals+causing+liability&hl=en&ie=UTF-8

3 Chilton, E. & Henley, P. (n.d.). *Jury instructions: Helping jurors understand the evidence and the law.* Retrieved January 10, 2004, from http://www.uchastings.edu/plri/spr96tex/juryinst.html

4 Adler, Stephen J. (1994). The jury: Disorder in the court 129. Quoted in Chilton, E. & Henley, P. (n.d.). *Jury instructions: Helping jurors understand the evidence and the law.* Retrieved January 10, 2004, from http://www.uchastings.edu/plri/spr96tex/juryinst.html

5 Chilton, E. & Henley, P. (n.d.). *Jury instructions: Helping jurors understand the evidence and the law.* Retrieved January 10, 2004, from http://www.uchastings.edu/plri/spr96tex/juryinst.html

6 Hanchak, N. A., Patel, M. B., Berlin, J. A., & Strom, B. L. (1996, June). Patient misunderstanding of dosing instructions. *Journal of General Internal Medicine, 11*(6), 325–28. Retrieved January 10, 2004, from http://www.ncbi.nlm.nih.gov:80/entrez/query.fcgi?CMD=Display&DB=PubMed

7 The brief discussion about principles of adult learning has drawn on these references: (1) Atherton, J. S. (2003). *Learning and teaching: Knowles' andragogy.* Retrieved January 8, 2004, from http://www.dmu.ac/uk/~jamesa/learning/knowlesa.htm; (2) Lieb, S. (1991, Fall). *Principles of adult learning.* Retrieved January 9, 2004, from http://www.honolulu.hawaii.edu/intranet/committees/FacDevCom/guidebk/teachtip/adults-2.htm; and (3) "Malcolm Knowles on andragogy." (n.d.). *Infed: The encyclopaedia of informal education.* Retrieved January 98, 2004, from http://www.infed.org/thinkers/et-knowl.htm

8 Kirsch, I. S., Jungeblut, A., Jenkins, L., & Kolstad, A. (1993). *Adult literacy in America, Educational Testing Services and the National Center for Educational Statistics.* Washington, DC: Government Printing Office.

9 Reading: It's incidental. (2001, May 29). *Metaforix Mail, 1*(42). Retrieved January 8, 2004, from http://www.metaforix.com/archives/mail_01_05-29.html

10 MicroPro International Corporation. (1981). *World star reference manual* (p. 2-2). San Rafael, CA: Author.

11 MicroPro International Corporation. (1981). *World star reference manual* (p. E-13.). San Rafael, CA: Author.

12 Author's professional visits to Japan, including interviews with officials at Sony's Document Design Development Center in Tokyo.

13 Kärcher. (2002). *Operating manual.*

14 Fumi Xerox. (1995). *A color 935/930* (pp. 286–87). No. TS-1543.

15 Rudy. (2001, May 21). Aliteracy and icons. Message posted to http://lists.evolt.org/archive/Week-of-Mon-20010521/032916.html

16 Adapted from (1) Chang, C. (2003, December 30). JPL's hopes riding on Mars rovers: First lander, Spirit, to plunge through Martian atmosphere. *Pasadena Star News.* Retrieved January 8, 2004, from http://www.pasadenastarnews.com/Stories/0,1413,206~22097~1862038,00.html; and (2) NASA-Jet Propulsion Laboratory. (n.d.). *Mars exploration rover mission.* Retrieved January 7, 2004, from http://marsrovers.jpl.nasa.gov/gallery/artwork/emerging_br.html

17 Adapted from (1) Kaner, C. (1995). Liability for defective documentation [Electronic version]. *Software QA, 2*(3), 8. Retrieved January 10, 2004, from http://www.badsoftware.com/baddocs.htm; and (2) Dick, D. (2000). Justification for documentation [Electronic version]. *Usability Interface, 6*(4). Retrieved January 11, 2004, from http://www.stcsig.org/usability/newsletter/0004-justification.html

18 Roxio (2003). Easy CD & DVD Creator (Version 6). [Computer software]. Santa Clara, CA: Roxio, Inc.

19 The results reported here are a compilation of data I obtained while conducting workplace seminars in organizations as varied as civil engineering firms, medical insurance companies, and government agencies. Seminar participants were invited to identify *content elements* they considered essential for effective instructions.

20 Living Lakes. (n.d.). *Saint Lucia.* Retrieved January 16, 2004, from http://www.livinglakes.org/stlucia/

21 Castañeda, P. y O. (2000). *El café ecológico: Algunas recomendaciones para su cultivo, procesamiento y comercializacion.* Guatemala: Vecinos Mundiales Gutemala.

22 The results reported here are a compilation of data I obtained while conducting workplace seminars in organizations as varied as civil engineering firms, medical insurance companies, and government agencies. Seminar participants were invited to identify *visual elements* they considered essential for effective instructions.

23 Pietschmann, H. (2003). *Example of good vs bad manuals.* Retrieved January 10, 2004, from http://www.4x4abc.com/G-Class/460prossex.html

24 Osterberg, A. & Kain, D. (2002). *Access for everyone: A guide to accessibility with references to ADAAG.* Ames, IA: Iowa State University, Facilities Planning and Management and the Department of Architecture.

25 Western Digital. (2001). *Western digital installation guide* (p. 14).

26 Lexmark. (2001). Installing the print cartridge. *From setup to printing* (p. 3).

27 Reproduced by permission of Deere & Company. Copyright © 1990 Deere & Company. All rights reserved.

28 American National Standard Institute. (1998). *Product safety signs and labels.*

29 Modified from (1) Safety Label Solutions. (n.d.). *ANSI Z535.4.* Retrieved January 11, 2004, from http://www.safety labelsolutions.com/Standards/SLSTHE_1/slsthe_1.HTM; and (2) HCS. (n.d.). *Standards FAQs: Signal words and colors.* Retrieved January 11, 2004, from http://www.safetylabel.com/safety labelstandards/ansi-signal-words.php#sas

30 Adapted from ANSI Z535.4. Safety Label Solutions. Retrieved January 11, 2004, from Safety label Solutions. (n.d.). *ANSI Z535.4.* Retrieved January 11, 2004, from http://www.safetylabelsolutions.com/Standards/SLSTHE_1/slsthe_1.HTM

31 Adapted from Safety label Solutions. (n.d.). *ANSI Z535.4.* Retrieved January 11, 2004, from http://www.safetylabelsolutions.com/Standards/SLSTHE_1/slsthe_1.HTM

32 Mavericklabel.com. (n.d.). *Signal words.* Retrieved January 11, 2004, http://pfl.labelserve.com/warning.html

33 Peckham, G. (n.d.). *Safety sign formats.* Retrieved January 11, 2004, from http://www.cemag.com/archive/02/05/peckham.html

34 Peckham, G. (n.d.). *Safety sign formats.* Retrieved January 11, 2004, from http://www.cemag.com/archive/02/05/peckham.html

35 Rudy. (2001, May 21). Aliteracy and icons. Message posted to http://lists.evolt.org/archive/Week-of-Mon-20010521/032916.html

36 Hazard Communication Systems. (n.d.). *ISO 3864 – The right standard for international labels.* Retrieved January 11, 2004, from http://www.weinigusa.com/safety/pg11.htm

37 Selected from (1) *Bad instructions.* (n.d.). Retrieved January 11, 2004, from http://www.thealchemist.info/bad_instructions.htm; and (2) *Humor house: Silly consumer warnings.* (2003, June 24). Retrieved January 11, 2004, from http://humor.smilezone.com/funnystuff/000675.htm

38 Caher, J. (1995). Technical documentation and legal liability. *Journal of Technical Writing and Communication, 25*(1): 5-10; quoted material from p. 5.

39 Caher, J. (1995). Technical documentation and legal liability. *Journal of Technical Writing and Communication, 25*(1): 5–10; quoted material from p. 7.

40 Rubens, P. (1987). The cartoon and ethics: Their role in technical information. *IEEE Transactions on Professional Communication, 30*(3): 196–201; quoted material from p. 197.

41 Caher, J. (1995). Technical documentation and legal liability. *Journal of Technical Writing and Communication, 25*(1): 5–10; quoted material from p. 10.

42 The information contained in this textbook should not be construed as legal advice, and readers should not act upon the legal information in this textbook without professional counsel. The discussion of legal adequacy in this chapter draws on the author's experience as an expert witness in such matters and on the following resources that are accessible to students.

(1) Bowman & Brooke. (n.d.). *Legally adequate warning labels: A conundrum for every manufacturer.* Retrieved January 16, 2004, from http://library.lp.findlaw.com/articles/file/00381/005582/title/sugject/topic/injury%20%20tort%20law_personal%20injury/filename/injurytortlaw_3_62

(2) Helyar, P. S. (1992). Products liability: Meeting legal standards for adequate instructions. *Journal of Technical Communication, 22:* 125–147.

(3) Karg, S. A. (1999). Products liability risk reduction 101 for clients [Electronic version]. *New Jersey Defense, 16*(4). Retrieved January 16, 2004, from http://www.nmmlaw.com/articles/products.html

(4) Lutz, W. (2002, April 1). Liability may arise despite warnings [Electronic version]. *National Law Journal.* Retrieved January 16, 2004, from www.jackscamp.com/publications/NLJ-WLutz-Products_Liab-04-01-02.pdf

(5) Eckert Seamans Cherin & Mellott, LLC, Product Liability Practice Group. (1997, June). *Product Liability Bulletin.* Retrieved January 16, 2004, from http://www.escm.com/new/pro/JUN97.HTM

43 First Alert. (n.d.). *User's manual.* Aurora, IL; Author. Reprinted by permission.

44 Living Lakes. (n.d.). *Saint Lucia.* Retrieved January 16, 2004, from http://www.livinglakes.org/stlucia/

45 Centers for Disease Control and Prevention. (n.d.). *Technical instructions to panel physicians for vaccination requirements.* Retrieved January 7, 2004, from http://www.cdc.gov/ncidod/dq/qdf/TI.pdf

46 ConsumerAffairs.com. (2002, February 11). *Toyota will cover sludge problems.* Retrieved January 11, 2004, from http://consumeraffairs.com/news02/toyota_sludge.html

PHOTO CREDITS

This page constitutes an extension of the copyright page. We have made every effort to trace the ownership of all copyrighted material and to secure permission from copyright holders. In the event of any question arising as to the use of any material, we will be pleased to make the necessary corrections in future printings. Thanks are due to the following photographers and agencies for permission to use the material indicated.

Part I
Pg. 1: © CORBIS

Chapter 1
Pg. 2: Detail showing a pile driver from *Codex Leicester* by Leonardo da Vinci © Seth Joel/CORBIS
Pg. 17: © AP Photo/John Moore
Pg. 20: © Paul A. Souders/CORBIS

Chapter 2
Pg. 36: Aerial helicopter: Leonardo da Vinci. © Réunion des Musées Nationaux / Art Resource, NY
Pg. 39: © Ryan McVay/Photodisc/Getty Images
Pg. 55: © William L. Jeffries, 2003
Pg. 57: (l) © Reza Estakhrian/Getty images; (r) © VCL/Chris Ryan/Getty images
Pgs. 60–61: Photos and excerpts from the Memorandum of Understanding and the Training Agreement appear courtesy of Iowa State University's Center for Transportation Research and Education.
Pg. 62: (l) © Alex Tossi/Alamy; (r) © Patrik Giardino/CORBIS
Pg. 65: © Walter Hodges/Getty Images
Pg. 70: © William L. Jeffries, 2003

Chapter 3
Pg. 76: Four winged flying machine: Leonardo da Vinci. Location: Bibliotheque de l'Institut de France, Paris, France. © Réunion des Musées Nationaux / Art Resource, NY

Chapter 4
Pg. 110: Wheel: Leonardo da Vinci. Sketch of machinery from *Codex Atlanticus.* Folio 387r. Location: Biblioteca Ambrosiana, Milan, Italy. © Art Resource, NY
Pg. 113: (t) © Omni Photo Communications Inc./Index Stock Imagery; (r) © Comstock Production Department/Alamy: (b) © R.W. Jones/CORBIS

Chapter 5
Pg. 142: Detail showing a pile driver from *Codex Leicester* by Leonardo da Vinci © Seth Joel/CORBIS
Pg. 145: © Desy Hamburg
Pg. 147: © Omnica Corporation — Irvine, CA
Pg. 155: © Lou Jones/Index Stock Imagery

Part II
Pg. 183: © Lester Lefkowitz/CORBIS

Chapter 6
Pg. 184: Aerial helicopter: Leonardo da Vinci. © Réunion des Musées Nationaux / Art Resource, NY
Pg. 187: Counter Clockwise from Top: © Double Exposure/Getty Images, © Spencer Grant/PhotoEdit, © David Young-Wolff/PhotoEdit, © Comstock Images/Alamy, © AP Photo/Michael Stravato
Pg. 192: © Frank Chmura/Index Stock Imagery
Pg. 197: Courtesy of NASA
Pg. 206: © Jose Luis Pelaez, Inc./CORBIS
Pg. 204: © The Wellcome Trust Medical Photographic Library

Chapter 7
Pg. 226: Four winged flying machine: Leonardo da Vinci. Location: Bibliotheque de l'Institut de France, Paris, France. © Réunion des Musées Nationaux / Art Resource, NY
Pg. 232: © Comstock Images/Getty Images
Pg. 235: © Christina Micek Photography

Chapter 8
Pg. 260: Wheel: Leonardo da Vinci. Sketch of machinery from *Codex Atlanticus.* Folio 387r. Location: Biblioteca Ambrosiana, Milan, Italy. © Art Resource, NY
Pg. 262: © Christina Micek Photography
Pg. 273: © Roger Ressmeyer/CORBIS
Pg. 276: © Helen King/CORBIS

Chapter 9
Pg. 304: Detail showing a pile driver from *Codex Leicester* by Leonardo da Vinci © Seth Joel/CORBIS
Pg. 307: from Top to Bottom: © Comstock Images/Alamy, © David Binder, © Stephen Simpson/Getty Images, © Tony Freeman/Photoedit, Courtesy of US Army Corps of Engineers, © Michael Rosenfeld/Getty Images, Courtesy of Carl Reese/CabinSafety.com
Pg. 312: Courtesy of Siemens
Pg. 326: © Christina Micek
Pg. 338: © Jean Dobbs
Pg. 339: (t) © Anton Vengo/Superstock
Pg. 339: (b) Courtesy of Dr. Arvid E. Osterberg

Part III
Pg. 343: © Morton Beebe/ CORBIS

Chapter 10
Pg. 344: Drawing of Light Reflected from a Mirror by Leonardo da Vinci © Alinari Archives/CORBIS
Pg. 346, 347: © William L. Jeffries 2002

Chapter 11
Pg. 376: Four winged flying machine: Leonardo da Vinci. Location: Bibliotheque de l'Institut de France, Paris, France. © Réunion des Musées Nationaux / Art Resource, NY
Pg. 382: © Tui De Roy/Minden Pictures
Pg. 404: © Courtesy of Conservation International

Chapter 12
Pg. 408: Wheel: Leonardo da Vinci. Sketch of machinery from *Codex Atlanticus.* Folio 387r. Location: Biblioteca Ambrosiana, Milan, Italy. © Art Resource, NY
Pg. 410: (l) © Chris Hallbeck; (r) © Alan Schein Photography/CORBIS
Pg. 433: Photos © Tom Myers
Pg. 444: (tl) © Dr. Richard Kessel & Dr. Gene Shih/Visuals Unlimited; (tm) © David Fleetham/Visuals Unlimited: (tr) © Dr. Stanley Flegler/Visuals Unlimited; (bl) © RMF/Visuals Unlimited; (bm) © Dr. Fred Hossler/Visuals Unlimited; (br) © Dr. Stanley Flegler/Visuals Unlimited
Pg. 445: (t) Courtesy of Iowa Department of Transportation; (b) © Courtesy Robert Simmon/NASA's Earth Observatory
Pg. 450: © Courtesy of Ron Doll/Falk Corporation
Pg. 454: Jeff Hester, Paul Scowen/NASA

Chapter 13
Pg. 466: Detail showing a pile driver from *Codex Leicester* by Leonardo da Vinci © Seth Joel/CORBIS
Pg. 469: clock-wise: © David Young-Wolff/PhotoEdit, © Michael Newman/PhotoEdit, and © Robin Nelson/PhotoEdit.
Pg. 509: (t) © Christina Micek; (b) Courtesy of Zachary Lavicky

Part IV
Pg. 515: © Robert Essel NYC/CORBIS

Chapter 14
Pg. 516: Four winged flying machine: Leonardo da Vinci. Location: Bibliotheque de l'Institut de France, Paris, France. © Réunion des Musées Nationaux / Art Resource, NY

Chapter 15
Pg. 546: Aerial helicopter: Leonardo da Vinci. © Réunion des Musées Nationaux / Art Resource, NY
Pg. 548: Courtesy NASA/JPL-Caltech
Pg. 551: Courtesy of National Aeronautics and Space Administration
Pg. 553: © Dennis MacDonald/PhotoEdit
Pg. 567: William L. Jeffries © 2002
Pg. 568: William L. Jeffries © 2002

Chapter 16
Pg. 578: Detail showing a pile driver from *Codex Leicester* by Leonardo da Vinci © Seth Joel/CORBIS
Pg. 581: © Royalty Free/CORBIS
Pg. 591: © Courtesy of Valencia Community College
Pg. 595: © Courtesy of Centro de Tecnologia Apropiada
Pg. 598: © Brandon Cole/Visuals Unlimited

Part V
Pg. 605: © Jon Feingersh/ CORBIS

Chapter 17
Pg. 606: Wheel: Leonardo da Vinci. Sketch of machinery from *Codex Atlanticus.* Folio 387r. Location: Biblioteca Ambrosiana, Milan, Italy. © Art Resource, NY
Pg. 608: © Yellow Dog Productions/Getty Images
Pg. 610: © David Young-Wolff/PhotoEdit
Pg. 612: (t) © Esbin-Anderson/The Image Works; (b) © Jeff Greenberg/PhotoEdit
Pg. 613: (t) © Michael Rosenfeld/Getty Images; (b) © Robert Essel NYC/CORBIS
Pg. 614: © Roger Ressmeyer/CORBIS
Pg. 631: Courtesy of Tara Barrett Tarnowski

Chapter 18
Pg. 644: Aerial helicopter: Leonardo da Vinci. © Réunion des Musées Nationaux / Art Resource, NY
Pg. 671: Courtesy of Lee Tesdell

Chapter 19
Pg. 676: Four winged flying machine: Leonardo da Vinci. Location: Bibliotheque de l'Institut de France, Paris, France. © Réunion des Musées Nationaux / Art Resource, NY
Pg. 681: (t) © Willie Hill Jr./The Image Works; (b) © Omni Photo Communications, Inc/Index Stock Imagery
Pg. 707: (t) © Fisher/Thatcher/Getty Images; (b) © Left Lane Productions/CORBIS

Chapter 20
Pg. 718: Wheel: Leonardo da Vinci. Sketch of machinery from *Codex Atlanticus.* Folio 387r. Location: Biblioteca Ambrosiana, Milan, Italy. © Art Resource, NY
Pg. 730: © Courtesy of USACE
Pg. 752: (t) © Jerry Kobalenko/Getty Images; (m) © Sean Sprague/The Image Works; (b) © Courtesy of Oskana Hylva

Chapter 21
Pg. 772: Detail showing a pile driver from *Codex Leicester* by Leonardo da Vinci © Seth Joel/CORBIS
Pg. 791: © Courtesy NASA/JPL-Caltech
Pg. 798, 809, 817, 819: © William L. Jeffries, 2003

INDEX